Encyclopedia of Environmental Science

by
John Mongillo
and
Linda Zierdt-Warshaw

ORYX PRESS
2000

The rare Arabian Oryx is believed to have inspired the myth of the unicorn. This desert antelope became virtually extinct in the early 1960s. At that time several groups of international conservationists arranged to have 9 animals sent to the Phoenix Zoo to be the nucleus of a captive breeding herd. Today the Oryx population is over 1,000, and over 500 have been returned to the Middle East.

4041 North Central at Indian School Road
Phoenix, Arizona 85012-3397
www.oryxpress.com

Published simultaneously in Canada
Printed and Bound in the United States of America

∞ The paper used in this publication meets the minimum requirements of American National Standard for Information Science—Permanence of Paper for Printed Library Materials, ANSI Z39.48, 1984.

Library of Congress Cataloging-in-Publication Data

Mongillo, John F.
Encyclopedia of environmental science / John Mongillo, Linda Zierdt-Warshaw.
p. cm.
Includes bibliographical references and index.
ISBN 1-57356-147-9 (alk. paper)
1. Environmenal sciences—Encyclopedias. I. Zierdt-Warshaw, Linda II. Title.

GE10.M66 2000
363.7'003—dc21 00-32657

Contents

Preface

The Encyclopedia of Environmental Science serves as a reference tool for high school students, college students, and others who wish to gain a better understanding of natural settings and the issues, problems, and concerns scientists and citizens have about the global environment. The environment is constantly changing and reports of events affecting the environment appear in magazines, newspapers, periodicals, newsletters, radio, television, and Web sites daily.

During the writing of this book, several areas of the world experienced catastrophic natural events or events caused by human activity that had the potential to adversely affect the environment. For example, India faced the worst cyclone in its history, and a technical problem at a nuclear fuel plant in Japan caused the evacuation of local residents who feared a possible leak of radiation into the environment. During the same period, residents of Turkey experienced two devastating earthquakes in a two-month period, while severe flooding associated with Hurricane Floyd damaged homes of numerous residents of North Carolina, displaced wildlife, completely destroyed agricultural crops, and killed thousands of livestock. In the near future, it is likely that other similar events will occur in various locations throughout the world as well.

It is impossible for any single text to comprehensively discuss all of the events and developments affecting the environment or to present the most current information about each topic; however, *The Encyclopedia of Environmental Science* is an excellent tool for developing a working knowledge of basic environmental topics, such as the components of an ecosystem, as well as topics and issues that have remained of concern to scientists and others over time, such as global warming, pollutants, pesticides, catastrophic natural events, key treaties and laws, factors affecting rainforests and other biomes, endangered and threatened species, hazardous wastes, alternative fuels, environmental activists, and technology.

The Encyclopedia of Environmental Science presents an introduction to environmental science by providing a framework of basic terminology and topics to assist the reader in developing an understanding of the global environment in general and how the human population and its technologies have affected Earth and its ecology. To develop this framework, the authors spent several years researching information about the environment from a variety of sources that included newspapers, high school and college textbooks, trade books, television reports, professional journals, national and international government organizations, nonprofit organizations, private companies and businesses, and encyclopedias. As a result of this research, the authors developed a comprehensive list of terms and topics representing what they found to be the most frequently recurring issues and themes of environmental science.

The encyclopedia presents coverage of key environmental terms and topics in an alphabetical listing. Coverage within individual entries is not limited to terms and their definitions but also includes biographies, histories, and numerous examples as well as cross-references (indicated through the use of SMALL CAPITAL LETTERS) that direct the reader to other articles within the encyclopedia that present additional information related to a particular topic. In addition, major entries are accompanied by a bibliographical listing called **EnviroSources**, which provides the reader with additional avenues for more up-to-date research and study about a particular topic. Unlike other reference books of this type, the **EnviroSources** extend beyond the standard bibliographic source materials available in print to provide the reader with the information needed to contact organizations and individuals via e-mail, Web sites, fax, telephone, or standard mail. Other source materials to which the reader may be directed include telephone hotlines, brochures, videotapes, audiotapes, and electronic bulletin boards.

The Major Environmental Topics

The Encyclopedia of Environmental Science provides terms, topics, and subjects covered in most high school and college environmental science courses. The major topics of environmental science include, but are not limited to,

- Agriculture, Crop Production, and Pest Control
- Atmosphere and Air Pollution
- Deserts, Forests, Grasslands, and other Biomes
- Ecology and Ecosystems
- Endangered and Threatened Plant and Wildlife Species
- Energy and Mineral Resources
- Environmental Laws, Regulations, and Ethics
- Oceans
- Nonhazardous and Hazardous Wastes
- Water Resources and Water Pollution
- Wetlands

Special Features

Many of the articles included in the encyclopedia are accompanied by one or more of the following features:

Time Capsule. The Time Capsule highlights one or more significant events related to the topic under discussion.

Tables, Charts, and Maps. Hundreds of photos, tables, maps, charts, and illustrations are used to enhance the text and provide additional information for the reader.

TechWatch. This section focuses on a discussion of a new technology that is being explored as a possible solution to an environmental problem such as acid mine drainage.

A Global View. This section presents a global perspective on a particular topic.

At a Glance. This feature provides additional information, in a concise manner, about a particular organism or environmental issue such as the size and scientific name of an endangered species or the geographic location of its native habitat.

EnviroTerms. This feature provides definitions for terms used within the article with which a reader may be unfamiliar.

EnviroLink. The EnviroLink feature is used to clarify terms mentioned in the article that may have a different meaning in another area of science.

Appendixes

Timeline. To understand the history of the environmental movement, the encyclopedia includes a comprehensive timeline that provides a general overview of people whose work has impacted how others view the environment, important laws and regulations, special events, and other environmental highlights over a period of more than 150 years.

Bibliography

Web Sites Sorted by Subject

Endangered Species by State

Environmental Organizations

Acknowledgments

The authors wish to acknowledge and express their appreciation for the contribution of the many nongovernment organizations, corporations, colleges, and government agencies that provided assistance with the research for this book. Many thanks to those who provided special assistance with particular topics and reviewed the topics and offered comments and suggestions. They include Bibi Booth, Bureau of Land Management; Professor Michael Kamrin, Michigan State University; Daniel Lanier, Environmental Professional; Dan Blaustein, writer and teacher, Chicago; Valerie Harms, environmental author, Montana; Tom Repine, West Virginia Geologic Survey; Peter Wright and Nancy Trautmann, Cornell University; Mary N. Harrison, University of Florida; Huanmin Lu, University of Texas, El Paso; Michael Mulligan, teacher, Sao Paulo, Brazil; and Eileen Reed. A special thanks to the following organizations that provided technical expertise and resources for photos and data: Government organizations and their representatives included the National Oceanic and Atmospheric Administration; Chuck Meyers, Office of Surface Mining; the U.S. Department of Agriculture; the U.S. Fish and Wildlife Service; the U.S. Department of Energy; the U.S. Environmental Protection Agency; the U.S. National Park Service; National Renewable Energy Laboratory; Tower Tech, Inc.; Earthday 2000; Marilyn Nemzer, Geothermal Education Office; the U.S. Agricultural Research Service; the U.S. Geological Survey; Glacier National Park; Monsanto; the CREST Organization; Shirley Briggs, Rachel Carson History Project; the General Motors Corporation; Yvonne Eglinton, Vortec Corporation; the National Interagency Fire Center/Bureau of Land Management; Susan Snyder, The Bancroft Library, University of California at Berkeley; Marine Spill Response Corporation; Lisa Bousquet, Roger Williams Park Zoo, Rhode Island; Netzin Gerald Steklis, Dian Fossey Gorilla Fund; International National Response Corporation; the U.S. Department of the Interior/Bureau of Reclamation, Bluestone Energy Services; OSG Ship Management Inc.; and Sweetwater Technology. In addition, the authors wish to thank Hollis Burkhart and Christine Beck for their typing and proofreading support, Muriel Cawthorn and Barbara Eisenberg for their assistance in photo research, and illustrators Christine Murphy, Susan Stone, and Carol Parker.

Please note the responsibility for the accuracy of the terms is solely that of the authors. If errors are noticed, please address them to the authors so corrections will be made in future revisions.

Abbey, Edward (1927–1989), An American novelist and nonfiction writer who wrote several books about the ENVIRONMENT that have been read by millions of people worldwide. Abbey's earliest two publications were the novels *Jonathan Troy* (1954) and *The Brave Cowboy* (1956); however, the book that first earned him an international reputation as a nature/environmental writer was *Desert Solitaire* (1968). In this book, Abbey details two six-month adventures he had while working as a ranger at what is today known as Arches National Park in Utah. (*See* National Park Service.) The book provides vivid details about the beauty of the DESERT, while also encouraging people to do what they can to preserve such natural ECOSYSTEMS.

Although Abbey produced many other books during his lifetime, his other major writing devoted to the environment was *The Monkey Wrench Gang* (1975), which introduced the term "monkeywrenching" to describe destructive actions used by "radical" environmentalists to end activities they see as damaging to the environment. In Abbey's novel, a group of environmentalists are determined to blow up a DAM in order to return a river to its natural state. Since the publication of Abbey's novel, the idea of monkeywrenching has been adopted by various individuals and organizations as a means of protecting the environment. Some extreme activities embraced by monkeywrenching supporters have included spiking trees, destroying bulldozers and other construction machinery by pouring sand or sugar in their FUEL tanks, and ramming and sinking boats and ships used in WHALING. (*See* ecotage.) Such tactics are often deemed ecoterrorism because they frequently result not only in loss of property, but in injury to people as well. One of the most well-known supporters of monkeywrenching is environmentalist DAVID FOREMAN, who founded the organization Earth First! in the 1980s. In 1985, Foreman wrote and published *Ecodefense: A Field Guide to Monkeywrenching*, in support of various monkeywrenching practices. *See also* ANIMAL RIGHTS; BROWER, DAVID ROSS; EHRLICH, PAUL RALPH; FRIENDS OF THE EARTH; GREENPEACE.

EnviroSources

Abbey, Edward. *Confessions of a Barbarian: Selections from the Journals of Edward Abbey, 1951–1989.* Boston, MA: Little Brown & Co., 1996.

———. *Desert Solitaire.* 1968. Reprint, New York: Ballantine Books, 1991.

———. *The Monkey Wrench Gang.* 1975. Reprint, New York: Avon Books, 1997.

Bishop, James. *Epitaph for a Desert Anarchist: The Life and Legacy of Edward Abbey.* New York: Touchstone Books, 1995.

Chang, Chris, et al. "Champions of Conservation: *Audubon* Recognizes 100 People Who Shaped the Environmental Movement and Made the Twentieth Century Particularly American." *Audubon* 100, no. 6 (November–December 1998): 81.

Ecotopia, Ecology Hall of Fame Website: www.ecotopia.org/ehof.

abiotic, A term used to describe the nonliving factors of the ENVIRONMENT, such as water, MINERALS, sunlight, temperature, and soil. ORGANISMS constantly interact with abiotic factors and sometimes, as with OXYGEN and water, depend upon them to sustain life. Other abiotic factors, such as some forms of ultraviolet radiation, other forms of ENERGY, and chemicals that are classified as POLLUTANTS, can make an area unfit for organisms.

Many abiotic factors are present in virtually all ECOSYSTEMS; however, the characteristics of these factors often differ depending upon an ecosystem's location in the world. For example, abiotic factors determined by CLIMATE, such as temperature, soil quality, and water availability, are very different in the TUNDRA of ARCTIC SIBERIA than in the RAINFOREST regions of the Amazon or in the DESERT regions of South Africa. (*See* biome.) Each SPECIES possesses ADAPTATIONS that allow it to live in a particular environment, resulting in the great BIODIVERSITY among Earth's organisms.

Sometimes, as few as one or two abiotic factors, called LIMITING FACTORS, can prevent certain organisms from surviving in an area or allow such survival. For example, an unusual group of BACTERIA, called thermoacidophilic bacteria (genus *Sulfolobus*), live in acidic hot springs where temperatures generally exceed 70°C (158°F). These bacteria, which synthesize their food from chemicals in the environment, cannot survive if temperatures drop below 55°C (131°F). Since few organisms can survive in environments with such high acidity and temperature, PH and temperature are limiting factors that prevent most species from surviving in a hot spring. *See also* ACID; BIOTIC.

acid, A substance that releases HYDROGEN ions (H^+) when in water. In an acidic solution the number of hydrogen ions is greater than the number of hydroxide ions (OH^-) and the PH is less than 7.0. Strong acids, such as hydrochloric acid (HCl), NITRIC ACID (HNO_3), and SULFURIC ACID (H_2SO_4), may have adverse effects on the ENVIRONMENT. At high enough concentrations, these acids can be corrosive to rocks and metals and harmful to the CELLS and tissues of ORGANISMS. By contrast, vinegar, which contains acetic acid (CH_3CO_2H), and most rainfall, which contains CARBONIC ACID (H_2CO_3), are weak acids. Acids also have a sour taste, turn blue litmus paper red, and are soluble in water.

Many acids form naturally when substances that dissolve in water, in the AIR, or at Earth's surface release hydrogen ions. For example, CARBON DIOXIDE (CO_2) gas combines with water vapor in the ATMOSPHERE to produce the carbonic acid that gives rainwater a slightly lower pH than pure water. Similarly, SULFUR DIOXIDE (SO_2) and NITROGEN DIOXIDE (NO_2) can combine with water vapor in the atmosphere to produce sulfuric acid and nitric acid, respectively. These substances can be deposited on Earth in ACID RAIN. *See also* ACID MINE DRAINAGE; ALKALI, WEATHERING.

> **EnviroTerm**
> **ion:** an electrically charged atom or molecule produced by a loss or gain of electrons.

Acid mine water is pumped up from underground coal mines to a watery pond-like area that contains anhydrous ammonia. The ammonia reduces the iron ions in the acid water to iron (Fe^{+3}) which will precipitate out of solution. The machine in the center churns up the water to expedite the oxidation of the ions in solution. The water is drained to the pond in the rear. *Credit:* Tom P. Cook, West Virginia Geological and Economic Survey

acid mine drainage, A water pollution problem resulting from the DISCHARGE of acidic water from COAL mines or mines containing MINERAL ores for iron, COPPER, LEAD, or zinc into streams and rivers. Acid mine drainage (AMD) also results when rainwater leaches through overburden or TAILINGS—the waste materials produced by MINING operations. (*See* leaching.)

Mine RUNOFF becomes acidic when water leaching through mine shafts and tailings causes chemical reactions to occur. The combination of AIR, DISSOLVED OXYGEN in the water, and the activities of CHEMOAUTOTROPHS (ORGANISMS that synthesize nutrients from inorganic chemicals), causes iron sulfide compounds in ores and waste rock to oxidize, producing a high concentration of SULFURIC ACID (H_2SO_4). (*See* oxidation.) When released into streams, the acidic solution is toxic to aquatic life. The ACID can also leach into and pollute GROUNDWATER.

AMD is a potential problem in any area where abandoned coal or metal mines or deposits of mine tailings are present. Some states with serious AMD problems are Pennsylvania, West Virginia, Colorado, Ohio, Wyoming, and Oklahoma. Outside the United States, AMD has been reported in Indonesia and in South Africa. Although not yet reported elsewhere, AMD is likely a problem in other countries that now have, or once had, heavy coal, zinc, iron, copper, or lead mining industries.

Several technologies are used to reduce AMD pollution. One method involves sealing abandoned mines to prevent water from flowing in or out, thus eliminating the discharge of acidic water into streams. Chemical treatment, in which limestone or lime is used to neutralize acids that form in mines, is most often used to eliminate AMD. Another successful method for reducing AMD involves using natural and human-made BOG-type WETLANDS to filter sulfuric acid from mine WASTEWATER before it enters streams and rivers. ORGANIC matter, BACTERIA, and ALGAE all work together to filter, adsorb, absorb, and PRECIPITATE out the HEAVY METAL ions and raise the PH level. More than 300 wetland water treatment systems have been built in the United States, many in coal mining regions.

The U.S. ENVIRONMENTAL PROTECTION AGENCY (EPA) has established regulations to limit acid levels of mine drainage. The regulations require the pH of discharge to be between 6.0 and 9.0. The average total iron content of the discharge must be less than 3 milligrams per liter and the average total MAN-

GANESE content less than 2 milligrams per liter. According to the EPA standards, new mines must be designed and operated to meet the standard of ZERO DISCHARGE. *See also* ACUTE TOXICITY; ADSORPTION.

EnviroLink

In geology, acid rock drainage is a term that encompasses both AMD and natural acid drainage which causes the same problems as AMD.

EnviroSources

Handbook for Constructed Wetlands Receiving Acid Mine Drainage. Environmental Protection Agency Pamphlet. (National Service Center for Environmental Publications: P.O. Box 42419, Cincinnati, OH 45242-2419; (800) 490-9198; Fax: (513) 489-8695.)

National Mine Reclamation Center. *Acid Mine Drainage: Control and Treatment.* Evansdale, WV: National Mine Reclamation Center, (304) 293-2867, ext. 444.

National Reclamation Center (West Virginia University, Evansdale) Web site: www.nrcce.wvu.edu/news/nsamd.html.

acid precipitation. *See* acid rain.

acid rain, Any precipitation or DEPOSITION having a PH lower than 5.6 as a result of contact with airborne particles. Acid rain is potentially damaging to ORGANISMS and the physical ENVIRONMENT when deposited on Earth. Because it includes snow and fog as well as rain, acid rain is sometimes called acid precipitation. Acid rain also exists in dry forms as acidic gases or particles. (*See* particulate matter.)

Acid rain resulting from industrial activities is a significant problem in much of the northeastern United States, eastern Canada, Germany, Scotland, and Scandinavia. Other countries suffering from some acid rain conditions include Brazil, Venezuela, India, China, and Japan. The prospect of increasing consumption of COAL in Asia makes the global acid rain threat even more real than ever. As an example, India's ENERGY requirements will result in a 300% increase in its usage of coal by the year 2020. The results will produce heavy acid rain conditions over many highly sensitive areas in the northeast region and coastal areas of India. Other industrialized areas of Indonesia, Malaysia, the Philippines, and Thailand will also experience heavy acid rain deposition by 2020. However, of all the countries in Asia, China will be the most likely to experience acid rain because it depends heavily on coal as an energy source. Also, Chinese coal has a very high of SULFUR content, a major contributor to acid rain. By 2020, most of China's eastern locations will be suffering from acid rain conditions.

Acid rain is caused by natural processes and human activities. Some acid rain results from particles released into the ATMOSPHERE through natural processes, such as FOREST FIRES and volcanic eruptions. (*See* volcano.) However, much acid rain derives from EMISSIONS resulting from the combustion of FOSSIL FUELS in AUTOMOBILES, homes, factories, and electric utility power plants. When fossil fuels burn, they DISCHARGE POLLUTANTS such as NITROGEN OXIDES (NO_x) and SULFUR DIOXIDE (SO_2). In the atmosphere, nitrogen oxides and sulfur dioxide combine with water vapor to produce NITRIC ACID (HNO_3) and SULFURIC ACID (H_2SO_4), respectively. Eventually, these ACIDS mix with rain, snow, or fog and are deposited on Earth's surface.

Even though automobiles and other GASOLINE-powered vehicles emit a great deal of nitrous oxide (N_2O) through their exhaust systems, electric utility power plants that burn PETROLEUM and coal are the major contributors to acid rain production. They account for about 70% of annual sulfur dioxide emissions and 30% of nitrogen oxides emissions in the United States.

Unpolluted rainfall generally has a pH of about 5.6 because it contains a weak acid—CARBONIC ACID (H_2CO_3). According to the U.S. ENVIRONMENTAL PROTECTION AGENCY (EPA), the eastern United States now has rainfall with an average pH of 4.5. However, because this is an average, the pH of rainfall in different locations within this region may vary. In 1978, rainfall having a pH of 2.0—5,000 times more acidic than normal—was measured in Wheeling, West Virginia.

Acid rain is corrosive. Long-term exposure to liquid and dry acid particles can damage or destroy FORESTS, streams, lakes, buildings, and metal and rock structures. Acid rain levels caused premature death of many tree SPECIES in the United States, Canada, Norway, Sweden, southwest Poland, the northwest Czech Republic, and southeast Germany. In sections of the Black Forest of Germany, conifers such as pine, fir, and spruce as well as some deciduous trees such as beech, maple and oak are leafless. (*See* coniferous forest; deciduous forest.) The trees are dead, damaged, or deformed from acidic conditions.

WILDLIFE is also harmed by acid rain. When acid rain falls on soils containing alkaline substances, such as particles of limestone and sandstone, the acid is neutralized. (*See* alkali.) In contrast, when acid rain falls on soils composed of igneous or metamorphic rocks, which do not contain the MINERALS or "buffers" that can neutralize the acids, deposited acids accumulate in and overload the soil. Acids in soil can undergo chemical changes and release toxic metals in forms that can be absorbed by PLANTS. The plants may be damaged or destroyed; animals feeding on these plants are then harmed as well. Acid rain also leaches vital nutrients, such as potassium and calcium, from soil making it less fertile. (*See* leaching.)

Acid in streams and lakes can dissolve metals and form toxic chemicals. Acidic buildup in lakes disrupts the reproductive processes of fishes and AMPHIBIANS. Acids can also disrupt FOOD CHAINS. Studies have shown that lakes having a pH between 6.0 and 6.5 can support a variety of organisms; however, when the pH drops to 4.7 or less, MICROORGANISMS begin to die, with large-scale fish kills soon following. Hundreds of lakes in Canada and the northeastern United States (especially upstate New York) are acidified to the point where they contain little or no life. Many lakes in Scandinavia and central Europe are in the same condition.

In addition to destroying wildlife HABITAT, acid rain also damages human-made structures. Acid causes particles in statues, monuments, and buildings to chemically weather, par-

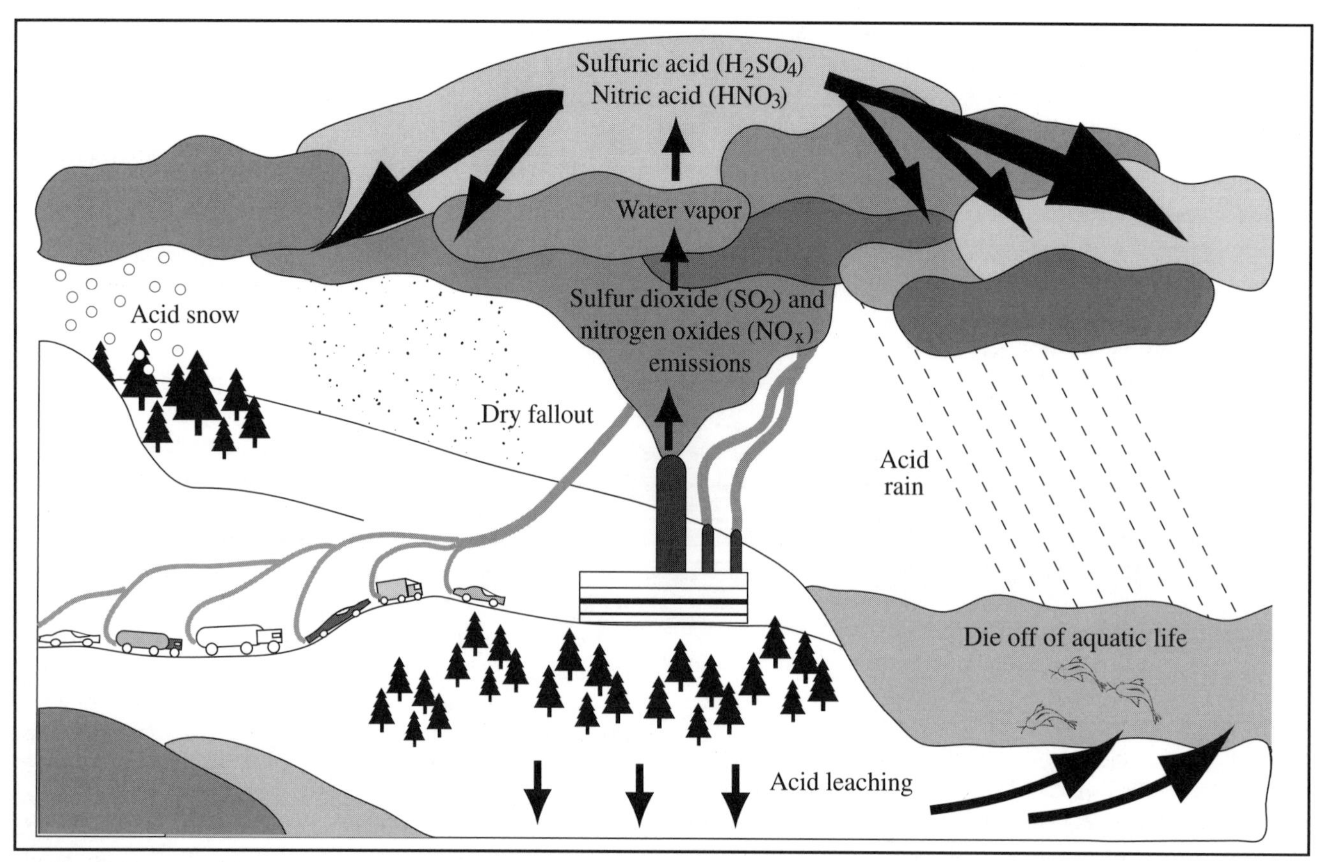

Acid Rain. Acid rain is caused by natural processes and human activities. Acid rain derives from emissions resulting from the combustion of fossil fuels in automobiles, homes, industries, and electric utility power plants. When fossil fuels burn, they discharge pollutants such as nitrogen oxides (NO_x) and sulfur dioxide (SO_2).

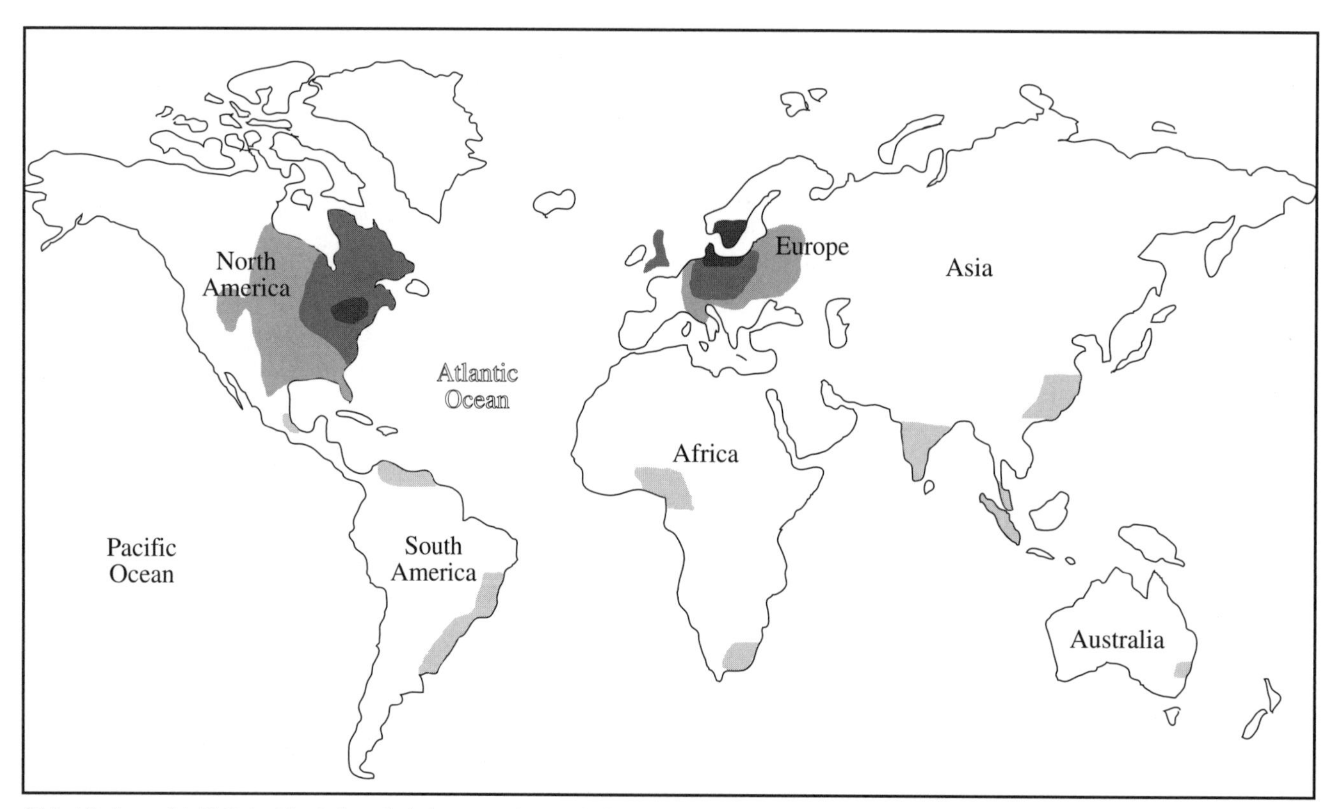

Global Pattern of Acid Rain. The darkest shaded areas on the map indicate the most severe acid rain conditions. The lighter shaded areas range from light to moderate acid rain conditions.

ticularly if the structures contain limestone or marble—rocks easily dissolved by acids. Such activity is illustrated by the Parthenon in Greece, which has been severely damaged as a result of the corrosive nature of acid rain.

Several methods are used to reduce acid rain damage caused by sulfur dioxide emissions. SCRUBBERS placed in industrial smokestacks can remove many harmful emissions, preventing their release into the atmosphere. Using low-sulfur fuels and adding limestone to fuels also reduces sulfur dioxide emissions. The spreading of crushed limestone or lime over lakes and forests suffering the effects of acid rain has been used to neutralize acidic water and soil, but this method is very expensive and requires many repeated treatments.

Because airborne pollutants can travel hundreds or thousands of kilometers from their sources, controlling acid rain is complicated. Pollutants released in one state or country are often deposited as acid rain in another. Acid rain problems in Canada and the northeastern United States, for example, have been traced to emissions from industries in the midwestern United States. In 1981, the Quebec Ministry of Environment informed U.S. officials that about 60% of the sulfur dioxide polluting Canada's air and water derived from the United States. In Europe, emissions from industries in England and Germany fall as acid rain in Scandinavia.

Controlling acid rain requires local and international cooperation. In 1989, the United States and 27 other nations signed the United Nations Sofia Agreement, which requires a reduction in nitrogen oxide emissions. In 1990, amendments to the CLEAN AIR ACT of the United States included provisions for significantly reducing industrial emissions of sulfur dioxide and nitrogen oxides by the year 2000. Nations are also working together to develop new technologies to reduce acid rain. As an example, the United States and the Ukraine are developing a reburning technology for power plant boilers that use fossil fuels to generate energy. The noxious gases produced in one section of the boiler will be piped into a "reburn" zone where they will be exposed to extremely high temperatures that will convert the gases into harmless materials. Research on reburning technology continues with hopes that the method may lead to a further reduction of acid rain. *See also* ACUTE TOXICITY; CHRONIC TOXICITY; CRITERIA POLLUTANT; ROCK CYCLE.

TechWatch

Engineers at Northeastern University in Boston, Massachusetts, may have developed a unique way to reduce acid rain emissions from coal-burning power plants. Presently, chemicals that form acid rain are passed through vats of limestone SLUDGE before they reach the smokestack. This method is effective in removing sulfur dioxide, but the vats do not catch much nitrogen oxide. The engineers at Northeastern are experimenting with a patented system in which calcium and magnesium carbonates are sprayed into the combustion chamber of the power plant. Once heated to a specific temperature, these particles expand. Their increased surface area maximizes their ability to absorb sulfur dioxide.

EnviroSources

Forster, Bruce. *The Acid Rain Debate: Science and Special Interest in Policy Formation*. Natural Resources and Environmental Policy Series. Iowa State University Press, 1993.

Hocking, Colin, Jan Coonrod, and Jacqueline Barber. *Acid Rain Guide*. Berkeley, CA: Gems, 1990.

Huttermann, Aloys, and Douglas Godbold, eds. *Effects of Acid Rain on Forest Processes*. Wiley Series in Ecological and Applied Microbiology. New York: John Wiley & Sons, 1994.

Rose, John, ed. *Acid Rain: Current Situation and Remedies*. Environmental Topics, vol. 4, Gordon & Breach Science Publications, 1994.

Somerville, Richard C.J. *The Forgiving Air: Understanding Environmental Change*. Los Angeles: University of California Press, 1998.

Tyson, Peter. *Acid Rain*. New York: Chelsea House, 1992.

U.S. Environmental Protection Agency, Acid Rain Hotline (202) 233-9620.

U.S. Environmental Protection Agency, Acid Rain Program Web site: www.epa.gov/docs/acidrain/effects/enveffct.html.

U.S. Geological Survey Web site: www.pubs.usgs.gov/gip/acidraid/2.html

activated charcoal. *See* adsorption.

active solar heating system. *See* solar heating.

acute toxicity, A life threatening reaction that occurs within 96 hours or less from the ingestion or contact with a toxic substance or radiation. Adverse effects may appear after only one exposure or dose but it may also occur after a few short exposures or doses. Acute toxicity may occur from misuse of cleaning products, SOLVENTS, or PESTICIDES, as well as ACIDS, ALKALIS, and some forms of RADIATION. For example, ingestion of ethylene glycol, a component of antifreeze, can be acutely toxic in humans and other MAMMALS. *See also* AGENCY FOR TOXIC SUBSTANCES AND DISEASE REGISTRY; CHRONIC TOXICITY.

EnviroSource

Agency for Toxic Substances and Disease Registry Web site: www.atsdr.cdc.gov/cxcx3.html

Adamson, Joy Gessner (1910–1980), Austrian-born WILDLIFE painter, photographer, and NATURALIST who wrote several books and established a foundation to focus international attention on the need for wildlife preservation. In the mid-1950s, Joy Adamson and her husband George, a game warden in Kenya, adopted an orphaned lion cub. The Adamsons raised the lioness, named Elsa, to adulthood within their camp. Once the lioness was grown, Kenyan officials, fearing that the lion might attack a human, encouraged the Adamsons to place Elsa in a ZOO. However, Joy Adamson did not want Elsa confined, and thus gained government approval to help Elsa develop the skills needed to survive in the wild.

Adamson wrote about her work in the books, *Born Free* (1960) and *Living Free* (1961). Both books received international attention and were made into films. Using the money made from the books and films, Adamson established the Elsa Wild Animal Appeal in 1969, an organization dedicated to the preservation and humane treatment of wild and captive animals.

After Elsa's release into the wild, she bore three cubs before dying from an infection. Adamson adopted the cubs and trained them for eventual release into the wild. She reported on this work in *Forever Free* (1963). Through her writings and the Elsa Wild Animal Appeal, Adamson received worldwide support that allowed her to continue raising orphaned cubs for later release back into the wild. Similar programs have since been established throughout the world to assist abandoned or injured animals of other SPECIES. In addition, Adamson's methods have been used in captive breeding programs in which ENDANGERED SPECIES are bred and born in captivity, with the goal of establishing a hardy POPULATION for later release into a natural ENVIRONMENT.

In 1980, Adamson lost the support of the Kenyan government when one of her released cubs attacked a human. Later the same year, Joy Adamson was killed during a dispute with an employee. However, the work of the Elsa Wild Animal Appeal continues today. *See also* CALIFORNIA CONDORS; CAPTIVE PROPAGATION; FISH AND WILDLIFE SERVICE; NATIONAL AUDUBON SOCIETY; WORLD WILDLIFE FUND.

EnviroSources

Adamson, Joy. *Born Free.* New York: Harcourt, Brace, & World, 1960.

———. *Forever Free.* New York: Harcourt, Brace, & World, 1963.

———. *Living Free.* New York: Harcourt, Brace, & World, 1961.

Elsa Wild Animal Appeal: P.O. Box 4572, North Hollywood, CA 91617.

adaptation, Any trait of an ORGANISM that makes it suited to life in its ENVIRONMENT. Adaptations include physical features, physiological processes, and behavioral traits. Physical features include the type of body covering an organism has, its color, and the structures it possesses for movement and food acquisition. Adaptations also may include physiological processes such as PHOTOSYNTHESIS, or whether the organism is ectothermic or endothermic. (*See* EnviroTerms.) Adaptive behavioral traits may include nest-building, migration, and courting rituals.

Adaptations result from EVOLUTION as organisms compete with one another in response to changing environmental conditions. (*See* competition.) In this process, those individuals who are most likely to survive possess traits that make them most suited to their environment. These organisms, in turn, are most likely to reproduce and pass these beneficial traits to their offspring. Scientists use the term "natural selection," a process first described by CHARLES DARWIN in 1859, to describe the process by which certain traits in a SPECIES appear or disappear as the environment favors individuals who produce the greatest numbers of surviving offspring.

Adaptations allow each species to occupy a slightly different environmental NICHE within an ECOSYSTEM. When species compete to fill the same niche, nature generally selects one species over the other. The outcompeted species will then either die, move to an environment where it can meet its needs, or make use of traits it possesses to survive in a slightly different niche. Through natural selection, the species that has adapted will pass the traits that permitted its survival to its offspring, which also will fill the new niche. Over time, the traits of the organism that adapted to the new niche may diverge enough from its original ancestral species as to allow a new species to emerge. This process of filling environmental niches through the development of new species is called ADAPTIVE RADIATION and is responsible for Earth's great BIODIVERSITY. *See also* COMPETITIVE EXCLUSION; CONVERGENT EVOLUTION; EXTINCTION; GENE POOL; GENETIC DIVERSITY; WILSON, EDWARD OSBORNE.

EnviroTerms

ectothermic: Term used to describe organisms whose metabolic processes are not sufficient to regulate body temperature and thus have body temperatures that are largely controlled by the surrounding environment. Ectotherms include fishes, REPTILES, and AMPHIBIANS.

endothermic: Term used to describe organisms whose body temperatures are regulated and maintained at a fairly constant level by their metabolic processes. Endotherms include birds and MAMMALS.

adaptive radiation, An evolutionary process that results in the emergence of new forms of ORGANISMS (at the SPECIES, genus, family, or order level) from a common ancestral POPULATION, thus increasing Earth's BIODIVERSITY. (*See* evolution.) Adaptive radiation, also called divergent evolution, occurs when organisms develop ADAPTATIONS that enable them to successfully exploit a HABITAT or NICHE that other members of their population cannot. The genes responsible for the new adaptations are then passed to offspring, who may inherit the beneficial trait. Over time, through the process of natural selection, organisms may develop enough traits in response to increased environmental pressures that a new form of organism emerges, distinct from the ancestral population.

The process of adaptive radiation is evident in the 14 distinct finch species observed on the GALÁPAGOS ISLANDS by CHARLES DARWIN. Each species had differences in its beak and foot structure that enabled it to eat foods and live in habitats distinct from those of other finch species. Despite these differences, Darwin hypothesized that all the finches shared a common ancestor. These observations helped Darwin develop his theory of evolution based on natural selection. Much of the diversity observed among BATS, frogs, and mice also results from adaptive radiation. *See also* CONVERGENT EVOLUTION; GENE POOL; GENETIC DIVERSITY; WILSON, EDWARD OSBORNE.

Addo National Elephant Park, An 11,718-hectare (28,955-acre) WILDLIFE preserve established in the Eastern Cape Province of South Africa in 1931 to provide a fenced and protected HABITAT for African ELEPHANTS—a SPECIES at risk of EXTINCTION. The park was set up to save the last survivors of the once numerous Eastern Cape elephants. Prior to the park's establishment, the elephant POPULATION had declined primarily because so many of the animals had been killed to obtain their IVORY. Since 1931, 11 protected elephants have successfully bred, increasing their numbers to about 240. As a result of the increased population size, the Addo National Elephant Park has achieved international recognition for its CONSERVATION success.

As a national park, it receives funding from the government of South Africa. Much of this funding results from money generated through a successful ECOTOURISM industry that has developed in and around the park. Visitors to Addo can observe an ECOSYSTEM that supports not only African elephants, but also antelope, rare black RHINOCEROSES, Cape buffalo, and many REPTILE, AMPHIBIAN, bird, and INSECT species. Some other animals include the Cape grysbok, duiker, ZEBRA, black-backed jackal, bat-eared fox, warthog, ostrich, small-spotted genet, yellow mongoose, tortoises, and the vervet monkey. *See also* AFRICAN WILDLIFE FOUNDATION; ENDANGERED SPECIES; SERENGETI NATIONAL PARK; TSAVO NATIONAL PARK.

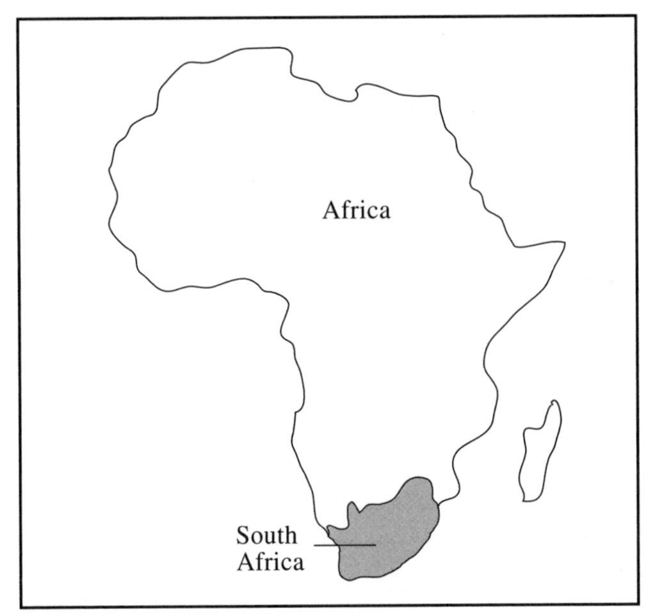

Addo Elephant National Park was established as a wildlife preserve in the Eastern Cape region of South Africa in 1931.

EnviroSources

Addo National Elephant Park Web site: www.jan.ne.jp/~kawabe/addo/addopark_e.html.

South African National Parks Board Web site: www.africa.com/~venture/saparks/index.html.

U.S. Fish and Wildlife Service, Species List of Endangered and Threathened Wildlife Web site: www.fws.gov/r9endspp/lsppinfo.html.

Adriatic Sea, The northern section of the MEDITERRANEAN SEA which is subject to periodic water pollution from ALGAL BLOOMS. The sea is bordered by Italy, Slovenia, Croatia, Bosnia and Herzegovina, Yugoslavia, and Albania. The Adriatic Sea is about 800 kilometers (500 miles) long and 160 kilometers (100 miles) wide and flows into the Mediterranean Sea at the Strait of Otranto, which is about 70 kilometers (42 miles) in width.

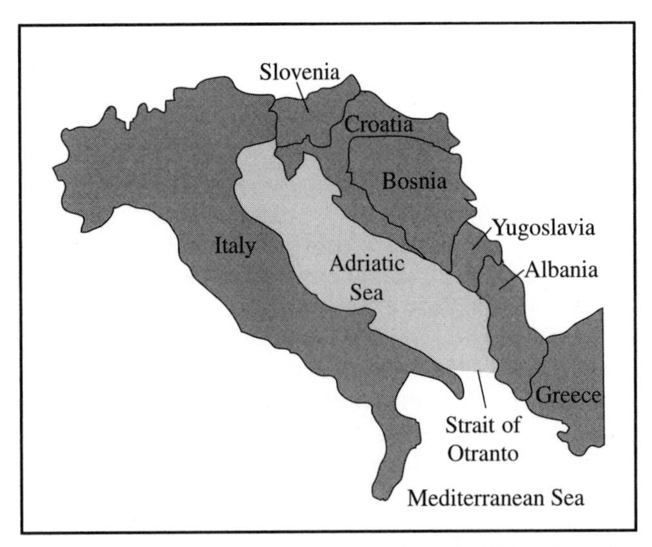

The Adriatic Sea is located in the northern section of the Mediterranean Sea which is subject to periodic water pollution from algal blooms.

The Adriatic Sea is very narrow and shallow along the Italian coast. Most of the sea bed is less than 200 meters (600 feet) below the surface of the sea. Since it is shallow and not an open sea, TOXIC CHEMICALS and HAZARDOUS SUBSTANCES, such as ARSENIC, PHOSPHATES, and NITRATES, can build up causing EUTROPHICATION. The Adriatic Sea has a very low DISSOLVED OXYGEN content which threatens fish and other marine ORGANISMS. Pollution also causes regular outbreaks of algal blooms. In the late 1980s and early 1990s algal blooms caused severe and persistent coastal destruction that ruined COMMERCIAL FISHING and tourism activities which are major industries in the region. Today, major outbreaks of algal blooms are being avoided, but more needs to be done to prevent toxic chemicals from entering the sea via rivers and streams from the bordering countries. *See also* ALGAE; OCEAN.

adsorption, A process in which gaseous materials adhere to the surfaces of solid materials. Adsorption is used to remove and collect wastes or POLLUTANTS from AIR and water and thus make these substances fit for use by people and other living things. Usually, adsorption involves the use of activated carbon (also called activated charcoal). As air or WASTEWATER containing dissolved liquid, solid, or gaseous matter passes over the activated carbon, the matter collects on the carbon's surface. The process of adsorption is used to remove VOLATILE ORGANIC COMPOUNDS, odors, and toxic substances from air or water and in emission control systems of motor vehicles. *See also* FILTRATION.

aeration, The application of OXYGEN to WASTEWATER or soil to assist in the decay of ORGANIC matter or to assist in PLANT growth, respectively. Modern SEWAGE TREATMENT PLANTS use aeration tanks to inject AIR into wastewater. In the oxygen-rich water, AEROBIC MICOORGANISMS consume organic materials. Without aeration, the DISSOLVED OXYGEN content would be too low to allow DECOMPOSITION of organic waste material by such microorganisms. The lack of oxygen in the SEWAGE would kill the aerobic BACTERIA. Soil aeration increases the porosity of the soil to allow the diffusion of oxygen into the soil, thus allowing plant roots to "breathe." *See also* DECOMPOSER; PRIMARY, SECONDARY, AND TERTIARY TREATMENTS.

aerobic, A term used to describe biological processes that require the presence of OXYGEN. Most living ORGANISMS are able to survive only in the presence of available oxygen. These organisms use oxygen to carry out metabolism—the process of taking in materials, using them to produce ENERGY, and then removing wastes. Aerobic respiration takes place in the CELLS when oxygen is used to break down food and release its chemical energy. *See also* AERATION; ANAEROBIC; DISSOLVED OXYGEN.

EnviroTerm

aerobic decomposition: The breaking down of materials by aerobic organisms. It is an important process in treating WASTEWATER in SEWAGE TREATMENT PLANTS.

aerogenerator, A device, sometimes called a wind TURBINE, that transforms the renewable ENERGY in wind into ELECTRICITY, in a manner that does not produce or release any POLLUTANTS into the ENVIRONMENT. An aerogenerator consists of blades on a shaft that is connected to an electric generator. As wind strikes the turbine blades, they turn. The mechanical energy of the turning blades is then transferred to the shaft, which causes coils of wire surrounded by magnets inside the generator to rotate, resulting in electricity.

Aerogenerators have two basic designs: those making use of blades resembling the propellers of airplanes, which turn vertically, and those making use of blades that resemble egg beaters, which turn horizontally. Currently, most aerogenerators in use worldwide use the horizontal design. Each of these can produce between 250,000 and 500,000 kilowatts of electricity annually—enough to power 50 homes for 1 year.

Depending upon their intended use, aerogenerators vary in size. For example, large boats or ships may use one or more small aerogenerators to recharge the BATTERIES on the vessel. Larger aerogenerators may be used singly or in small numbers to generate electricity for individual homes; wind farms use many large aerogenerators to produce electricity for entire communities. Currently, the state of California is the world's greatest producer of electricity generated by aerogenerators. Other countries using this ALTERNATIVE ENERGY RESOURCE include Japan, Denmark, the Netherlands, Sweden, Germany, Spain, Italy, and the United Kingdom. *See also* WIND POWER.

aerosol, Any liquid droplets or solid particles suspended in a gas. The diameter of an aerosol particle is less than 10 microns (10 millionths of a meter). Aerosols may incorporate POLLUTANTS that can adversely affect the health of ORGANISMS if taken into respiratory tissues. Aerosols in the ATMOSPHERE can also alter WEATHER and CLIMATE because of their ability to reflect or absorb solar radiation.

Many aerosols enter the atmosphere through natural processes such as volcanic eruptions. Such aerosols can reflect or absorb enough solar radiation to cool the AIR at Earth's surface. The use of products packaged in aerosol spray cans also releases aerosols into the atmosphere. Such products include lubricants, cosmetics, insecticides, paints, and DISINFECTANTS.

CHLOROFLUOROCARBONS (CFCs) were once used as propellants in aerosol cans and as coolants in refrigeration and air conditioning units. However, these gases were discovered to cause OZONE DEPLETION. Scientists are concerned about this depletion because thinning of the OZONE layer is believed to contribute to GLOBAL WARMING. As a result, CFCs have been banned and eliminated from use in many parts of the world, including the United States. The MONTREAL PROTOCOL, which was signed in 1987 and amended in 1990, called for a ban on aerosol spray cans by the year 2000. The international agreement was signed by more than 90 countries. In addition to banning aerosols and CFCs, the amended Montreal Protocol also bans several other substances that cause ozone layer depletion. *See also* HALONS; MOUNT PINATUBO; OZONE HOLE; VOLCANO.

African Wildlife Foundation, An international organization established in 1961 to protect African WILDLIFE. The African Wildlife Foundation (AWF) devotes its time to educating people worldwide about threatened African wildlife and measures that can prevent the EXTINCTION of certain SPECIES. Much of the AWF's efforts are focused on the endangered African ELEPHANT, mountain GORILLA, and black RHINOCEROS.

One of the greatest problems facing many ENDANGERED SPECIES in Africa is POACHING. Often animals are killed to obtain only parts of their bodies, such as the IVORY tusks of elephants or the hands or heads of gorillas. (*See* Convention on International Trade in Endangered Species of Wild Flora and Fauna.) To educate people about such species, the AWF publishes *Wildlife News* magazine. The group also administers education centers and youth programs and is involved in ECOTOURISM—specifically safaris that provide a closeup view of Africa and its wildlife.

The AWF operates two colleges that train individuals who wish to work in wildlife CONSERVATION as wardens or rangers in Africa's national parks or WILDLIFE REFUGES. The AWF also equips people working in these areas, especially those working to bring an end to the poaching of elephants, gorillas, rhinoceroses, and other endangered species. *See also* ADAMSON, JOY GESSNER; ADDO NATIONAL ELEPHANT PARK; FOSSEY, DIAN; GREENPEACE; NATIONAL AUDUBON SOCIETY; RED LIST OF ENDANGERED SPECIES; WORLD CONSERVATION MONITORING CENTRE; WORLD WILDLIFE FUND.

EnviroSource

African Wildlife Foundation: 1717 Massachusetts Ave., NW, Washington, DC 20036; (800) 344-TUSK.

afterburner, An AIR POLLUTION instrument used in industrial incinerators and some jet engines. The afterburner injects FUEL into incinerator or engine exhausts to increase the combustion efficiency and decrease the amount of smoke, POLLUTANTS, and odors emitted into the ENVIRONMENT. An afterburner may be attached to or separate from the incinerator or engine on which it works. *See also* CATALYTIC CONVERTER; EMISSION STANDARDS; INCINERATION.

age structure, A POPULATION study of an area in which the age of its people is the key factor being examined, usually to predict how the population's needs—consumption of NATURAL RESOURCES, means of waste disposal and treatment, and land use needs related to housing, educational and health facilities, and recreational space—are likely to change. Age structure studies usually are done for large areas, such as cities or entire countries. For example, a city may request an age structure study to determine the population size of school-age children for a given year in the future. Results of the study can help community officials make decisions about land use issues and the budgeting of tax dollars. If, for example, the study predicts an increasing population, it may indicate that open spaces need to be reserved for use as parks or other recreational areas. It may also indicate that land needs to be reserved for construction of new schools, medical facilities, or shopping centers. Such decisions have a variety of impacts on the ENVIRONMENT. For example, new construction may require clearing of land and simultaneous destruction of WILDLIFE HABITAT to make available space for buildings. In addition, increased populations place a greater demand on water and ENERGY resources, the need for improved SEWAGE TREATMENT PLANTS, and the allocation of space to accommodate an increase in SOLID WASTES. By contrast, stable or decreasing population may indicate that money can be used to preserve open areas or that funds should be allocated to a different societal need.

Results of age structure studies are often shown in a histogram—a bar graph that shows what percentages of individuals in a population fall within various age ranges. A histogram may also include a vertical axis within each bar, which divides the age data by gender. *See also* BIRTH RATE; DEATH RATE; EMIGRATION; IMMIGRATION; POPULATION DENSITY.

Agency for Toxic Substances and Disease Registry, An agency of the U.S. Department of Health and Human Services (DHHS) whose primary objective is to prevent adverse human health effects associated with exposure to HAZARDOUS SUBSTANCES from waste sites and other sources of pollution present in the ENVIRONMENT. The agency also identifies communities where people might be exposed to hazardous substances in the environment. They produce toxicological profile information fact sheets for hazardous substances found at NATIONAL PRIORITIES LIST (NPL) sites. These are hazardous United States waste sites, now abandoned, that are scheduled for cleanup under the SUPERFUND Act. These hazardous substances are ranked based on frequency of occurrence at NPL sites, TOXICITY, and potential for human exposure. The agency maintains Web pages on topics such as the toxicologic hazard of Superfund hazardous waste sites, industrial chemicals and terrorism, chemical hazards that occurred during the war in Croatia. They also post their own top 20 hazardous substances list. *See also* HAZARDOUS WASTE; POLLUTANT; TOXIC CHEMICAL; TOXIC WASTE; TOXICOLOGY.

Top 20 Hazardous Substances

1. Arsenic
2. Lead
3. Mercury, Metallic
4. Vinyl Chloride
5. Benzene
6. Polychlorinated Biphenyls (PCBs)
7. Cadmium
8. Benzo(a)pyrene
9. Benzo(b)fluoranthene
10. Polycyclic Aromatic Hydrocarbons
11. Chloroform
12. Aroclor 1254 (PCB)
13. Dichlorodiphenyltrichoroethane
14. Aroclor 1260 (PCB)
15. Trichloroethylene
16. Chromium
17. Dibenzofurans
18. Dieldrin
19. Hexachlorobutadiene
20. Chlordane

EnviroSource

Agency for Toxic Substances and Disease Registry Web site: www.atsdr.cdc.gov/cxcx3.html.

Agenda 21, The major document that resulted from the UNITED NATIONS EARTH SUMMIT held in Rio de Janeiro, Brazil, in 1992. Agenda 21 outlines principles of SUSTAINABLE DEVELOPMENT and provides recommendations that countries can follow to achieve this goal. Agenda 21 seeks to encourage developed nations to assist developing nations in industrialization using means that will not harm the ENVIRONMENT or require a nation to give up its sovereignty. In addition, Agenda 21 established broad principles to help nations preserve their natural ECOSYSTEMS. Agenda 21 encourages decision-making at the community level for issues related to economic and social development to help individuals develop an awareness of the interrelationships between people, the economy, and the environment. *See also* BIODIVERSITY TREATY.

EnviroSources

Sitarz, Daniel. *AGENDA 21: The Earth Summit Strategy to Save Our Planet.* Boulder, CO: Earth Press, 1993.

United Nations Earth Summit Web site: www.un.org/dpcsd/earthsummit.

Agent Orange, A powerful DEFOLIANT that was widely used from the 1940s through the 1960s to remove unwanted PLANTS from many ENVIRONMENTS. Agent Orange was a mixture of two herbicides: 2,4-dichlorophenoxyacetic ACID (2,4-D) and 2,4,5-trichlorophenoxyacetic acid (2,4,5-T)—compounds composed of CARBON, HYDROGEN, OXYGEN, and CHLORINE. In plants, the mixture caused a specific auxin (hormone) to collect in leaf nodes and to close leaf veins, thus eliminating the leaf's ability to obtain needed nutrients. Agent Orange was very effective, but became suspected of causing many human illnesses.

Agent Orange was named for the orange band painted around the containers in which it was stored. This mixture, and its relatives Agent Blue and Agent White, was first used in the United States in the mid-1940s to remove unwanted vegetation from FORESTS, rangelands, and water pipes. Its use peaked during the Vietnam War when it was sprayed from military aircraft onto farms, jungles, MANGROVE SWAMPS, and other heavily foliated areas throughout Vietnam and Laos to destroy vegetation that could provide cover for enemy soldiers.

Agent Orange was the most effective defoliant used in Vietnam; it also became one of the most controversial when it was shown that 2,4,5-T caused adverse effects in humans. Additional concern developed when it was claimed that Agent Orange often was contaminated with DIOXIN, a potentially deadly, carcinogenic chemical formed as a byproduct during 2,4,5-T production. (*See* cancer; cancinogen.)

In the late 1960s, hints of the dangers posed to humans by Agent Orange surfaced when Vietnamese newspapers reported increased numbers of miscarriages and birth defects in human POPULATIONS living near areas where Agent Orange had been used. From the mid- to late 1970s, U.S. and Australian soldiers who had served in areas of Vietnam where Agent Orange had been used began reporting a variety of ailments, including headaches, nausea, muscle weakness, depression, liver damage, skin lesions, cancers, and birth defects among their children. However, no convincing scientific evidence of a connection between Agent Orange and these illnesses, other than skin lesions, was shown. Protracted court cases were filed by veterans and their families claiming injury from Agent Orange exposure. While no uniform legal resolution was reached, many out-of-court settlements were made. In 1979, the U.S. ENVIRONMENTAL PROTECTION AGENCY suspended domestic use of 2,4,5-T. The suspension was later changed to a permanent ban of the substance. *See also* ACUTE TOXICITY; CHRONIC TOXICITY; DISEASE; MUTAGEN; PESTICIDE; TOXIC CHEMICAL.

EnviroSources

Saign, Geoffrey C. *Green Essentials: What You Need to Know about the Environment.* San Francisco, CA: Mercury House, 1994.
U.S. Environmental Protection Agency Web site: www.epa.gov.

agricultural pollution, The release of POLLUTANTS deriving from farming or ranching practices into the ENVIRONMENT. (*See* agriculture.) Agricultural pollution may result from the use of AGROCHEMICALS such as FERTILIZERS and PESTICIDES, from RUNOFF of wastes and nutrients from feedlots or ranches, and from agricultural practices that allow topsoil EROSION.

Fertilizers. Chemical preparations called fertilizers are added to soil to replace nutrients removed by PLANTS. Fertilizer use promotes plant growth; however, when carried in runoff into aquatic ECOSYSTEMS, fertilizers result in EUTROPHICATION. Eutrophication is the accumulation of nutrients in a body of water. Too many nutrients in an aquatic ecosystem can promote a POPULATION explosion of ALGAE called an ALGAL BLOOM.

During an algal bloom, the algae population can exceed the CARRYING CAPACITY of the ecosystem. When this happens, large numbers of algae suddenly die, stimulating the DECOMPOSITION process. (*See* decomposer.) As large numbers of BACTERIA and FUNGI carry out decomposition, they may use so much of the water's DISSOLVED OXYGEN that the ecosystem becomes unable to support other aquatic ORGANISMS.

Feedlot and Ranch Runoff. Animal and food wastes are rich in the nutrients that promote plant and algae growth. Like fertilizers, these wastes and their nutrients can be carried in runoff from feedlots and ranches into aquatic ecosystems, causing the same problems as fertilizers.

Pesticides. Chemical substances collectively called pesticides are used to kill unwanted organisms such as INSECTS, fungi, weeds, and rodents. Pesticides are effectively used to protect crops and livestock from being eaten, sickened, or damaged by pest organisms; however, pesticides can harm humans, other organisms, and the environment. For example, many pesticides are toxic to people if ingested, inhaled, or absorbed through the skin. Such pesticides can cause illness, DISEASE, or even death.

Many pesticides are not organism specific. Thus, a pesticide used to kill one type of organism may also sicken or kill other organisms. Sometimes, a pesticide used to kill a pest insect, such as a beetle, may kill other insects that benefit the environment because of their roles in pollination. In addition, many pesticides are persistent (do not degrade quickly) in the environment. Over time, these pesticides can build up in soil, water, or in the CELLS and tissues of organisms (a process called BIOACCUMULATION). As the concentration of a persistent pesticide increases, the potential danger it poses to other organisms or their offspring may also increase.

Topsoil Erosion. Loose and exposed topsoil is easily carried away by wind, running water, and other agents of erosion. When small soil particles are carried in the air, they become particulate pollutants. (*See* particulate matter.) These small pieces of matter can irritate the eyes and the respiratory system.

Soil carried by wind and water may be deposited into bodies of water, where the soil settles at the bottom, in a process called SEDIMENTATION. Sedimentation can benefit an environment by building soil and allowing for the growth of plants. However, if too much sediment is deposited and too many plants grow, the ecosystem may undergo ECOLOGICAL SUCCESSION, a process that leads to the development of a different type of ecosystem as one community of organisms replaces an existing community. (*See* biological community.)

Some particles that enter water remain suspended in the water. Too many suspended particles prevent sunlight from penetrating the water. Without sunlight, PHOTOSYNTHESIS cannot occur, and the algae, plants, and other PRODUCERS in the water cannot survive. The suspended particles may also be taken into the respiratory or digestive systems of organisms living in the water, disrupting the function of these important body systems. (*See* siltation.)

To help reduce some of the pollution problems associated with agriculture, many farmers and ranchers have begun using farming and ranching methods that are more environmentally friendly. For example, many farmers reduce their use of fertilizers through crop rotation, a practice in which the types of crop plants grown on a particular parcel of land are periodically alternated with another type of crop plant that has different nutrient needs or is left unplanted. This farming practice helps to lower the use of chemical fertilizers by preventing the depletion of any given nutrient from the soil. In addition, alternating the growth of crop plants with plants known as LEGUMES may also help to replenish soil as bacteria that live on the roots of these plants convert nitrogen gas from the air into nitrogen compounds that are needed for plant growth. (*See* nitrogen cycle.) INTEGRATED PEST MANAGEMENT (IPM) is a farming practice that is being used increasingly throughout the world to reduce pesticide use. In IPM, farmers use BIOLOGICAL CONTROLS such as natural predators to help reduce the numbers of pest organisms that could potentially damage crops. (*See* predation.) In addition, farmers often make use of farming methods such as CONTOUR FARMING, TERRACING, STRIP CROPPING, and no-till or low-till agriculture to reduce the effects of erosion. All of these practices involve plowing the land in a different way to reduce the ability of wind or running water to carry away soil. Ranchers also help reduce erosion by periodically moving grazing animals to different pastures to prevent overgrazing, the removal of the ground cover in an area to the point where soil becomes exposed and vulnerable to the agents of erosion. *See also* AGROFORESTRY; BORLAUG, NORMAN ERNEST; GREENBELT MOVEMENT; POLYCULTURE; SOIL CONSERVATION; SUSTAINABLE AGRICULTURE.

EnviroSources

Alternative Farming Systems Information Center U.S. Department of Agriculture: National Agriculture Library, 10301 Baltimore Boulevard, Room 304, Beltsville, MD 20705; e-mail: afsic@nalusda.gov.

Montgomery, John H. *Agrochemicals Desk Reference: Environmental Data.* Albany, GA: Lewis, 1993.

Olson, Richard K., ed. *Integrating Sustainable Agriculture, Ecology, and Environmental Policy.* Binghamton, NY: Food Products Press, 1992.

Soule, Judith D., and Jon K. Piper. *Farming in Nature's Image: An Ecological Approach to Agriculture.* Washington, DC: Island Press, 1992.

agriculture, The practice of growing PLANTS and raising animals for human use. People began practicing agricultural about 10,000 years ago, primarily on small parcels of land with the goal of meeting the food needs of their immediate families. This practice, called SUBSISTENCE AGRICULTURE, still occurs on small farms throughout the world; however, much of today's agriculture is practiced by fewer individuals on larger parcels of land, and the products derived therein are sold for the economic gain of both individuals and entire nations. Several environmental concerns have arisen with regard to the practice of agriculture. For most among these concerns are questions regarding how to ensure that agricultural practices can continue to meet the food and food product demands of a growing global human POPULATION, how much land is needed to meet such needs now and in the future, how CLIMATE impacts the productivity of agriculture, and how potentially harmful practices, such as SLASH AND BURN and the use of various AGROCHEMICALS, can be minimized or eliminated through the use of more sustainable agricultural practices that pose less harm to the ENVIRONMENT and its ORGANISMS. (*See* sustainable agriculture.)

Land Use and Food Needs. Much of the agriculture practiced throughout the world today still has the goal of producing food and food products for a growing world population; however, agriculture also may be carried out to obtain useful products such as fibers, chemical substances, or medicines that are derived from plants or animals. Activities involved in agriculture include cultivating the soil, growing and harvesting crops, and breeding and raising livestock. All of these activities require land, which is in limited supply. At the same time, the global human population is growing at a rapid rate. Much of this growth is occurring in developing regions, where people do not have access to the economic resources or technologies needed to provide abundant crops and livestock food or enough arable land to make such practices feasible.

In some parts of the world, poor agricultural practices are decreasing the amount of arable land available for agriculture. For example, as much as 6 million hectares (14.8 million acres) of soil is transformed into DESERT each year through the activities of humans. Much of this transformed land was once occupied by dense FORESTS. The process of DESERTIFICATION started when these forest were cleared to create more land for crops or livestock grazing. DEFORESTATION combined with the overgrazing of livestock and poor IRRIGATION practices have been shown to have disastrous effects on the ability of land to support the growth of crops. To help reduce the amount of arable land that is lost throughout the world each year as a result of such practices, governments worldwide working through the United Nations are helping their people develop a knowledge of more sustainable agricultural practices such as CONTOUR FARMING, AGROFORESTRY, TERRACING, no-till and low till agriculture, and the use of shelterbelts. (*See* EnviroTerms.)

Climate. A lack of arable land is not the only problem that threatens the ability of farmers to produce sufficient crops toFeed the global population. Each year, billions of dollars worth of agricultural products are lost due to WEATHER disasters such as severe flooding and DROUGHT. Many of these problems are associated with the periodic climate changes associated with EL NIÑO-LA NIÑA; however, many scientists believe that increased incidents of severe flooding and drought also are associated with GLOBAL WARMING—a progressive increase in the temperature of the ATMOSPHERE that is associated

Two farm owners check organically grown peanuts. *Credit:* United States Department of Agriculture. Photo by Bill Tarpenning.

with the increased release of POLLUTANTS such as CARBON DIOXIDE and other GREENHOUSE GASES into the atmosphere.

Several examples of the potential devastation of crops and livestock in response to drought and flooding were observed in the United States and Europe during the last decade of the twentieth century. In the early 1990s, heavier-than-normal snowfalls in the northern central United States led to significant flooding of the Mississippi River and its tributaries. The flooding damaged billions of dollars worth of homes from the northern to the southern United States. In addition, the flooding also entirely destroyed crops growing on farms located near the FLOODPLAINS of the Mississippi River and its tributaries. In the mid-1990s severe FLOODS in France destroyed much of that country's grape crop, a crop that drives the winemaking industry of that nation. During the summers of 1998 and 1999, severe drought associated with El Niño and La Niña led to crop failures throughout the central and northeastern United States. In 1999, several northeastern states lost most of their crops to severe drought. This problem was compounded in September 1999 when several coastal mid-Atlantic states that had received sufficient rainfall to support the growth of crops during the summer lost their crops to flooding associated with Hurricane Floyd. In less than 24 hours, the torrential rains that accompanied this hurricane flooded farmlands to depths of more than 3 meters (10 feet) in some areas. In addition, huge numbers of livestock were lost. In North Carolina alone, crop and livestock losses from this single storm were greater than $2 billion. Surrounding states also suffered severe crop losses due to the heavy rains associated with the storm.

Agrochemicals. Another environmental concern related to agriculture stems from the use of agrochemicals, which often produce immediate benefits to crops that result in higher yields; however, these same substances often pose problems to other organisms and the environment. FERTILIZERS, for example, provide crop PLANTS with nutrients that stimulate their growth. These same chemicals also stimulate the growth of unwanted plants, or weeds, that have the potential to outcompete crop plants for HABITAT, sunlight, water, and nutrients. In addition, nutrients in fertilizers that are carried off farmland in RUNOFF and enter aquatic ECOSYSTEMS can stimulate the growth of plants and ALGAE living in these ecosystems. When such growth is excessive, a population explosion of algae called an ALGAL BLOOM may result. The immediate problems associated with an algal bloom occur as the algae outcompete plants and other organisms for resources such as sunlight and nutrients; however, the major problem related to algal blooms occurs when the algae die and DECOMPOSITION begins. During this process, the BACTERIA and FUNGI that degrade the dead algae can use up so much of the DISSOLVED OXYGEN in the water, that other organisms living in the water cannot meet their oxygen needs to carry out respiration. Unable to carry out this vital life process, the organisms suffocate and die.

PESTICIDES used on croplands also often pose threats to the environment. Many pesticides are not organism specific and thus can kill organisms other than those for which they were intended. For example, a pesticide may kill the aphids that feed on the leaves of crop plants, but the same pesticide also may kill the butterflies and bees that pollinate these same crops, thus reducing the number of seeds that will result from these plants. In addition, the pesticides may be toxic to other animals that feed at higher levels of the FOOD CHAIN.

Use of pesticides also may harm organisms without killing them directly. For example, the pesticide DICHLORODIPHENYLTRICHOROETHANE (DDT) was widely used throughout the United States and Europe during the middle part of the twentieth century. DDT was extremely effective at killing the INSECTS for which it was used; however, in time, DDT was shown to have some extremely damaging long-term effects on other animal populations, specifically birds of prey. (*See* predation.) During the 1960s, scientists began to observe a decline in populations of BALD EAGLES and PEREGRINE FALCONS. While some of the decline resulted from loss of habitat as a result of increased human population sizes, the major cause for the decline was traced to the use of DDT. It was determined that DDT accumulated in the CELLS and tissues of organisms as it moved through the food chain. (*See* bioaccumulation.) The effect of the chemical in bald eagles and peregrine falcons was not the deaths of the animals that ingested the substance, but a disruption in the calcium formation of their eggs. Eggs laid by these birds had thin, brittle shells that rarely survived the incubation process to produce healthy hatchlings. As a result, both SPECIES of birds were threatened with EXTINCTION. To resolve the problem, several developments occurred. The first was a ban on the use of DDT in the United States in 1972 (followed soon after by similar bans in other nations). In addition, healthy wild birds from populations of both species were taken into captivity, where they were bred in large numbers for later release back into the wild. (*See* captive propagation.) These efforts proved successful enough that in the 1990s both species were removed from the ENDANGERED SPECIES LIST maintained by the FISH AND WILDLIFE SERVICE (FWS).

No-till soybeans (dark areas) are growing among wheat residue. *Credit:* United States Department of Agriculture. Photo by Gene Alexander.

Many pesticides have adverse health effects to humans. Humans may experience these effects from contact with the pesticides as a result of their application or from ingesting pesticide residues left on foods following harvest. Health problems associated with pesticides range from skin rashes, headaches, nausea, and diarrhea, to more severe problems such as CANCER, seizure disorders, and death. To reduce the problems associated with pesticide use, many farmers are beginning to use BIOLOGICAL CONTROLS to control pest populations. One part of this practice involves using natural predators to help keep the pest populations in check. For example, aphid populations that feed on leaves may be kept in check by releasing ladybugs into the environment. The ladybugs feed on the aphids without causing harm to the environment.

Many farmers throughout Europe and the United States are beginning to practice ORGANIC farming—the growth of plants without the use of synthetic chemicals such as pesticides and fertilizers. In place of synthetic substances, farmers promote plant growth by returning nutrients to soil through natural decomposition processes, use natural substances such as organic and inorganic MULCHES to control weeds, and control pest populations through the use of biological controls or other methods that do not require chemicals. Organic farming is believed to benefit the environment by making use of only natural materials. In addition, many people who fear that their food may be contaminated with pesticide residues are willing to pay a higher price for crops that are grown without the use of chemical pesticides. *See also* AGRICULTURAL POLLUTION; AGROCLIMATOLOGY; AGROECOLOGY; AGROFORESTRY; BORLAUG, NORMAN ERNEST; CLEAR-CUTTING; COVER CROP; DUST BOWL; FOOD AND AGRICULTURAL ORGANIZATION; GREEN MANURE; GREENBELT MOVEMENT; NATURAL DISASTERS; SOIL CONSERVATION; SUSTAINED YIELD.

EnviroTerms

low-till agriculture: A farming practice in which only those portions of the cropland in which crops are actually planted are plowed or turned over. Low-till agriculture reduces soil erosion from wind and running water by minimizing the amount of soil that is directly exposed to the elements.

no-till agriculture: A farming practice in which crops are planted without plowing or turning over soil. No-till agriculture reduces soil erosion by not removing ground cover plants and thus eliminating exposure of the soil to the elements.

shelterbelts: Row of trees planted around the perimeter of a farm or a section of cropland for the purpose of slowing wind and thus reducing soil erosion.

EnviroSources

Abromovitz, Janet, et al. *State of the World: 1998.* New York, NY: Norton, 1998.

Bernstein, Leonard, Alan Winkler, and Linda Zierdt-Warshaw. *Environmental Science: Ecology and Human Impact.* Menlo Park, CA: Addison-Wesley, 1995.

Montgomery, John H. *Agrochemicals Desk Reference: Environmental Data.* Albany, GA: Lewis, 1993.

agroclimatology, A branch of METEOROLOGY in which WEATHER and CLIMATE are studied to determine their impact on AGRICULTURE. By studying climate factors such as average temperature, daily hours of sunlight, and precipitation, agroclimatologists can identify which crops or crop varieties are best suited to the natural conditions of a region. These data can be used to develop SUSTAINABLE AGRICULTURE practices (farming methods that promote long-term crop yields without decreasing soil fertility) that aid in SOIL CONSERVATION and can lessen the need for excessive water use through IRRIGATION. *See also* AGROECOLOGY; HYBRIDIZATION; HYDROPONICS; SUSTAINED YIELD.

agroecology, A science that aims to apply principles of ECOLOGY—interactions between ORGANISMS and their ENVIRONMENTS—to AGRICULTURE. The goal of agroecology is to develop sustainable, or long-term, agriculture practices that produce quality agricultural products without harming soil, organisms, or other aspects of the environment. Such practices make use of BIOLOGICAL CONTROLS (living organisms) and other INTEGRATED PEST MANAGEMENT (IPM) techniques in place of chemical PESTICIDES that can decimate organisms other than their targets. Other goals of agroecology include making use of nature's BIOGEOCHEMICAL CYCLES (the biological and chemical processes that naturally cycle substances through the environment) and practicing ENERGY and SOIL CONSERVATION techniques. Agroecology also strives to improve IRRIGATION practices to conserve water resources and reduce the SALINIZATION of soil, which often results from irrigation. Using such knowledge, farmers and ranchers can reduce their use of AGROCHEMICALS and keep soil healthy while reducing AGRICULTURAL POLLUTION and conserving NATURAL RESOURCES.

Techniques of agroecology have been implemented with positive results in many parts of the world. For example, Chinese cotton farmers have reduced pesticide use by as much as 50% using biological controls. Many rice farmers of Indonesia turned to IPM after many of the pesticides used on this crop were banned from use by the government. In both countries, crop yields increased or remained the same, while costs involved in growing the crops decreased. Similar results have been achieved by agricultural workers using similar practices in other parts of the world. *See also* AGROCLIMATOLOGY; AGROECOSYSTEM; AGROFORESTRY; BORLAUG, NORMAN ERNEST; GREEN REVOLUTION; GREENBELT MOVEMENT; HYBRIDIZATION; HYDROPONICS; SUSTAINABLE AGRICULTURE.

agroecosystem, All the ORGANISMS, ABIOTIC factors, and interactions among them that occur on land used for AGRICULTURE and adjacent areas. Abiotic factors of the agroecosystem include AIR, SURFACE WATER and GROUNDWATER supplies, and the cultivated and uncultivated soil of the region. An agroecosystem includes lands used for crops, pasture, and livestock, as well as adjacent lands that provide HABITAT to native WILDLIFE. Thus, an agroecosystem is unique because it includes POPULATIONS of both NATIVE SPECIES and introduced SPECIES. *See also* AGROFORESTRY; COMPETITION; ECOSYSTEM; ECOTONE; EXOTIC SPECIES.

agroforestry, A SUSTAINABLE AGRICULTURE practice in which crops and trees are cultivated in the same ENVIRONMENT. The growth of crops and trees in the same field benefits both types of PLANTS. Crops benefit from the moisture given off by the trees through TRANSPIRATION; trees benefit from the return of nutrients to the soil when the wastes of harvested crops are permitted to decompose through natural processes. (*See* decomposition.) Agroforestry also benefits the environment by preserving natural FORESTS and the HABITAT they provide to ORGANISMS and by reducing the soil EROSION that often occurs when forests are cleared to open land for AGRICULTURE. (*See* deforestation.)

Agroforestry was widely practiced throughout Africa, Asia, and Latin America for centuries. However, farmers in these regions were encouraged to give up the practice for what were considered the more efficient farming methods associated with the GREEN REVOLUTION, an agricultural movement that encouraged the use of BIOTECHNOLOGY to develop new varieties of crop plants while also encouraging the use of chemicals that stimulate increased crop yields and reduce losses caused by pest organisms. As a result of these pressures, large sections of forest area were cleared to create land for crop cultivation, while the use of agrochemicals, such as FERTILIZERS and PESTICIDES, and use of IRRIGATION increased. In the short term, areas that practiced these methods experienced increased crop yields; however, the long-term effects of these practices devastated many areas, especially tropical forest areas that were cleared for farming. These areas experienced extreme soil quality degradation from the combined effects of increased exposure to the baking heat of the sun and heavy rainfalls. In many of these areas, topsoil has been greatly eroded by wind and running water. In others, years of irrigation, followed by evaporation of irrigation water, left the soil infertile as a result of SALINIZATION.

To avoid further soil degradation, many African, Asian, and Latin American peoples have returned to agroforestry as an alternative to these "bare earth" farming techniques. This return began in Kenya with the GREENBELT MOVEMENT, when volunteers led by environmentalist Wangari Muta Matthai began planting trees in areas that had been previously deforested and on farms (as shelterbelts) to help reverse the effects of deforestation and erosion. Efforts were later expanded as greenbelt practitioners set up full-scale agroforestry projects in which they planted both trees and crops, while also decreasing or eliminating the use of synthetic agrochemicals. Similar projects have begun throughout many other regions of the world that once made use of agroforestry methods. *See also* AGRICULTURAL POLLUTION; BORLAUG, NORMAN ERNEST; CLEAR-CUTTING; DEBT-FOR-NATURE SWAP; DESERTIFICATION; SLASH AND BURN; SUSTAINABLE DEVELOPMENT.

EnviroSources

Breton, Mary Joy. *Women Pioneers for the Environment.* Boston, MA: Northeastern University Press, 1998.

Seymour, John, and Herbert Girardet. *Blueprint for a Green Planet.* New York: Prentice Hall, 1987.

air, The mixture of gases that forms Earth's ATMOSPHERE, supplies the gases needed for respiration and PHOTOSYNTHESIS, and contributes to WEATHER and CLIMATE. Air contains hundreds of different gases. The approximate composition of dry air, by volume at sea level, contains: NITROGEN (N_2), which makes up about 78.08%; OXYGEN (O_2), which makes up about 20.95%; and argon (Ar), which makes up about 0.90%. The remaining percent of gases in air are called "trace gases" because they exist only in tiny amounts. These trace gases include OZONE (O_3), CARBON DIOXIDE (CO_2), water vapor (H_2O), METHANE (CH_4), nitrous oxide (N_2O), CHLOROFLUOROCARBONS (CFCs), and HALONS (brominated CFCs). The temperature and chemistry of the atmosphere are largely controlled by its trace gases.

Composition of Air

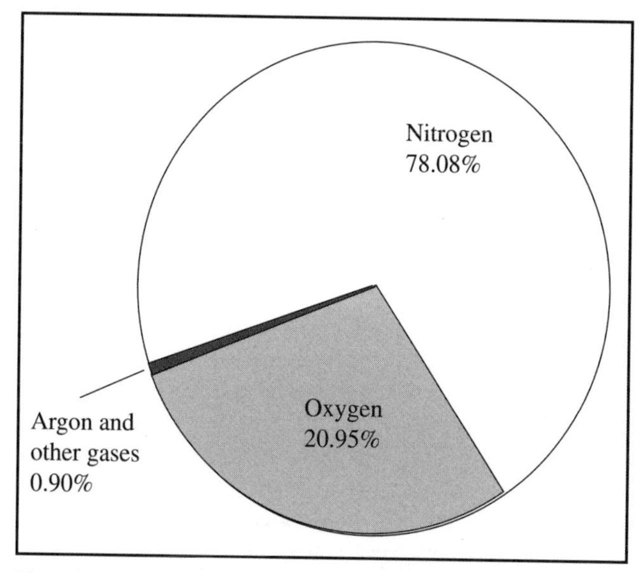

Most of the air surrounding Earth is composed of nitrogen and oxygen.

The gases in air are continually cycled through the ENVIRONMENT by natural processes such as photosynthesis, respiration, and DECOMPOSITION. Natural processes such as FOREST FIRES and volcanic eruptions release gases into the air. (*See* volcano.) Gases are also released into the air through human activities such as the burning of FOSSIL FUELS to generate ELECTRICITY or the burning of GASOLINE to power AUTOMOBILES. Both of these activities release large amounts of carbon dioxide, CARBON MONOXIDE (CO), water vapor, and NITROGEN OXIDES (NO_X) into the air. Scientists are concerned that these practices, in combination with DEFORESTATION and other human activities, are altering the makeup of the atmosphere and contributing to such environmental problems as the formation of ACID RAIN and increased GLOBAL WARMING. *See also* AEROSOL; AIR POLLUTION.

air pollution, CONTAMINANTS in AIR that have been emitted from mobile or stationary sources, that degrade natural air quality, and often present a health hazard. Outdoor air quality is affected by many human and natural activities. Manufacturing companies, power plants, small businesses, AUTOMOBILES, FOREST FIRES, and VOLCANOES all are potential sources of air pollution, as is any activity that releases potentially harmful materials into the air. POLLUTANTS in the air can create SMOG and ACID RAIN, cause respiratory DISEASE and other serious illnesses, damage the protective OZONE layer in the upper ATMOSPHERE, and contribute to CLIMATE change and GLOBAL WARMING. Air pollution costs citizens billions of dollars every year in health care costs and lost productivity.

An early warning about the limits of the ENVIRONMENT'S capacity to absorb pollutants was the severe air pollution generated during the Industrial Revolution, when the burning of COAL to run mills and machinery released huge quantities of contaminants into the air. It was not until after World War II, however, that pollution came to be viewed by many as a threat to the health of the planet. By the 1960s, the expanding human POPULATION, the growing heavy industry, and the increasing truck and automobile use were producing wastes in such quantities that natural dispersion and RECYCLING processes could not always keep pace. Airborne industrial wastes created acid rain, and automobile EMISSIONS produced severe air pollution problems, including smog, in many urban and suburban communities. The contribution of air pollutants to global environmental problems, such as climate change, global warming, and OZONE DEPLETION, has since prompted international meetings such as the UNITED NATIONS EARTH SUMMIT, and international agreements such as the MONTREAL PROTOCOL and the Kyoto Protocol.

Air pollution is different from other forms of pollution in that once the contaminants are in the air, exposure cannot be easily avoided. If high levels of outdoor air pollution are occurring in a city, for example, a large portion of the population will be exposed. Air pollution hurts the human body both by directly inflaming and destroying lung tissue and by weakening the lungs' natural defenses. Exposure to air pollution can cause irritation of the eyes, nose, mouth, and throat, and can cause coughing and sneezing. More serious, it can also worsen, and may cause, lung diseases such as asthma, bronchitis, and emphysema. In some cases, it can even contribute to the premature death of people with heart and lung disease. For people who are already sick or especially sensitive, air pollution may cause discomfort, limited activities, increased use of medications, more frequent visits to doctors and hospitals, and shortened lifespan. There is growing scientific evidence that suggests that air pollution has long-term effects on the lungs' ability to function and encourages the development of lung disease.

Over the past decade, there has been increasing recognition that minority and economically disadvantaged populations are disproportionately subjected to a variety of environmental health hazards, including air pollution. The nation's most severe air pollution problems, including those caused by ozone, are typically found in urban areas, which house many minority populations: 86.1% of African Americans and 91.2% of Hispanics live in urban settings, as com-

Air Pollution

Pollutant	Source
Carbon monoxide	Combustion of automobile engines
Fluorides	Manufacturing of fertilizer
Hydrogen sulfide	Oil wells, refineries
Lead	Leaded gasoline, lead smelting
Mercury	Pesticides, fungicides, chemicals
Nitrogen dioxide	Produced by combinations of nitrogen and oxygen during combustion of automobile engines, fertilizers
Particulate matter	Forest fires, fuel combustion, incineration, vehicles
Photochemicals	Photochemical reactions in the atmosphere
Sulfur dioxide	Combustion of fossil fuels, petroleum refining, smelting
Volatile organic compounds	Evaporation from solvents and liquid fuel
Greenhouse gases	Combustion of automobile engines
Ozone	Vehicles, factories, landfills

pared with 70.3% of whites. Researchers have found that higher percentages of African Americans and Hispanics than whites also live in areas that do not comply with NATIONAL AMBIENT AIR QUALITY STANDARDS for PARTICULATE MATTER, CARBON MONOXIDE, ozone, SULFUR DIOXIDE, and LEAD.

Once in the environment, pollutants may be dispersed via air, water, soil, and living ORGANISMS. Dispersion pathways vary greatly, depending upon both the emission source and the pollutant. Rates and patterns of dispersion also depend on environmental factors, such as wind conditions, altitude of the emission, topographical features, and whether the pollutant source is fixed or mobile. Many pollutants therefore show extremely complex dispersion patterns, especially in urban environments, where there are a large number of emission sources and wide variations in environmental conditions. This complexity often makes it very difficult to model or measure air pollution patterns and trends, or to predict levels of human exposure.

Indoor air is often even more polluted than the air outside homes and workplaces, sometimes resulting in SICK BUILDING SYNDROME or MULTIPLE CHEMICAL SENSITIVITY. This trend has been shown to be true across the United States, even in neighborhoods without heavy industrial pollution. Indoor air pollution may result from contaminants in faulty heating units and gas stoves, cleaners and SOLVENTS, cigarette smoke, wall coverings and paints, and improperly stored CHEMICAL products. Another significant indoor health hazard results from RADON gas, which in many areas of the United States rises into buildings from underlying rock.

The U.S. ENVIRONMENTAL PROTECTION AGENCY (EPA) has developed health-based national air quality standards for six CRITERIA POLLUTANTS: carbon monoxide, ozone, NITROGEN DIOXIDE, sulfur dioxide, particulate matter, and lead. Toxic or hazardous air pollution consists of substances in the air that are known or suspected to cause CANCER, genetic mutation, birth defects, or other serious illnesses in people, even at relatively low exposure levels. Toxic air pollutants may exist as particulate matter or as gases. Exposure to these toxic contaminants is nationally regulated through the use of pollution controls such as SCRUBBERS and ELECTROSTATIC PRECIPITATORS rather than air quality standards.

One of the most encouraging environmental developments of recent years is the trend toward preventing, rather than just treating, air pollution. Although scientists have found various ways to treat wastes in order to protect the environment, there is growing realization that whenever possible, avoiding wastes altogether is far more beneficial. The passage of the CLEAN AIR ACT (CAA) in 1970 clearly signaled our nation's intention to address air pollution. Between 1987 and 1996, lead emissions decreased by 50%, primarily as a result of the widespread switch to unleaded GASOLINE as a motor vehicle FUEL. During the same period, decreases were also seen in atmospheric concentrations of other pollutants, including sulfur dioxide (14%), particulate matter (12%), and ozone (18%). Nevertheless, as of 1996, transportation sources were still responsible for 47% of pollutant emissions. When examined by individual pollutant, 79% of carbon monoxide pollution was attributable to transportation, as was 31% of nitrogen oxides, 41.5% of VOLATILE ORGANIC COMPOUNDS, 2.8% of particulate matter smaller than 10 microns, and 14.6% of lead.

Many activities and programs have succeeded the CAA to further limit allowable DISCHARGES into the air. Preventing pollution can actually save money in a variety of ways, so the EPA has designed several nonregulatory, innovative pollution prevention programs. For example, its Energy STAR programs seek to prevent emissions of air pollutants associated with climate change and acid rain, while also promoting profitable investments in energy-efficient technologies. The Ag STAR program focuses on METHANE, emitted into the air when animal manure ferments. Such emissions waste a usable energy supply, produce odors, and contribute to climate change. The EPA program recovers methane gas from livestock manure for reuse by farmers. The Climate-Wise program challenges organizations from all sectors of the economy to find creative ways to limit or reduce GREENHOUSE GAS emissions. Such actions may include raw-material substitution, process improvements, and switching to fuels with lower CARBON contents. *See also* AIR POLLUTION CONTROL ACT; CATALYTIC CONVERTER; ENERGY; ENVIRONMENTAL JUSTICE; ENVIRONMENTAL RACISM; FOSSIL FUELS; GLOBAL ENVIRONMENT MONITORING SYSTEM; GLOBAL WARMING POTENTIAL; GREENHOUSE EFFECT; HYDROCARBON; INCINERATION; NATIONAL EMISSIONS STANDARDS FOR HAZARDOUS AIR POLLUTANTS; NATIONAL POLLUTANT DISCHARGE ELIMINATION SYSTEM; SOURCE SEPARATION; TOXIC SUBSTANCE CONTROL ACT; UNITED NATIONS ENVIRONMENT PROGRAMME; WORLD HEALTH ORGANIZATION.

EnviroSources

State and Territorial Air Pollution Program Administrators, Association of Local Air Pollution Control Officials Web site: www.4cleanair.org.

U.S. Environmental Protection Agency, AIRSData Web site: www.epa.gov/airsweb.
U.S. Environmental Protection Agency, Office of Air and Radiation Web site: www.epa.gov/oar.

Air Pollution Control Act, The first national law of the United States devoted to reducing and controlling AIR POLLUTION. The Air Pollution Control Act (APCA) was passed in 1955. Its main function was to provide funding and technical assistance to states to help them regulate air pollution locally.

Oregon was the first state to establish a statewide AIR quality control program in 1952. Before 1955, laws regarding air pollution control were enacted locally and generally dealt with regulating the releases of factory smoke and soot. Although the APCA was a federal act, it left the authority for regulating air quality with individual states, thus air pollution remained a local concern. The APCA remained the main piece of federal air pollution legislation until the passage of the CLEAN AIR ACT in 1970. *See also* NATIONAL AMBIENT AIR QUALITY STANDARDS.

air quality standards. *See* National Ambient Air Quality Standards.

Alar, The trade name for daminozide, a chemical PESTICIDE that provides more color and firmer texture to fruits and makes them less likely to fall off PLANTS prior to harvest. Alar has been used mostly on apples, but also on other fruits, such as cherries and Concord grapes. In 1989, the manufacturer of Alar voluntarily withdrew the product from use by fruit growers because of public concerns about reports that suggested the substance might be carcinogenic, particularly to children. *See also* AGROCHEMICAL; CANCER; CARCINOGEN; CARSON, RACHEL LOUISE.

EnviroSources

National Resources Defense Council: 40 West 20th St., New York, NY 10011; (212) 727-2700.
Rosen, J.D. "Much Ado about Alar." *Issues in Science and Techology* 7, no. 1 (1990): 85–90.

Alaska National Interests Lands Conservation Act, A federal law enacted in 1980 to set aside and protect WILDERNESS and other land and water areas in Alaska deemed to have significant natural, scenic, historic, archeological, geological, scientific, cultural, recreational, and WILDLIFE value to the United States as a whole. The Alaska National Interests Land Conservation Act (ANILCA) also is known as Public Law 96-487.

The ANILCA increased the size of the former Arctic National Wildlife Range by more than 100%; its current size is almost 8 million hectares (19 million acres). This range was created as a PROTECTED AREA in 1960, largely to protect the calving grounds of the Porcupine Caribou herd. The wildlife range also provides nesting grounds for a variety of migratory birds. The ANILCA also changed the name of the range to its current name—the ARCTIC NATIONAL WILDLIFE REFUGE.

The ANILCA contains provisions not only for wilderness CONSERVATION. The act also permits native and nonnative Alaskans to use some park resources as they did before the park's creation. In addition, certain Alaskan residents are permitted to use NATURAL RESOURCES on federal lands in Alaska, including resources located within national parks and WILDLIFE REFUGES. These provisions were included to ensure that the economic and social needs of Alaska's people could be met, while also protecting the region's ENVIRONMENT. *See also* ALASKA PIPELINE; PUBLIC LAND.

Alaska Pipeline, A 1,300-kilometer (808-mile) pipeline that transports PETROLEUM across the state of Alaska, providing about 10% of the oil used in the United States, as well as oil for export to several countries in Asia. The Alaska Pipeline, also known as the Trans-Alaska Pipeline, extends from Prudhoe Bay, near the ARCTIC Circle, to the Port of Valdez in southern Alaska. The diameter of the pipeline is 1.5 meters (5 feet) and it carries approximately 2 million barrels of oil each day. Oil flow is maintained by 11 pump stations located along the pipeline.

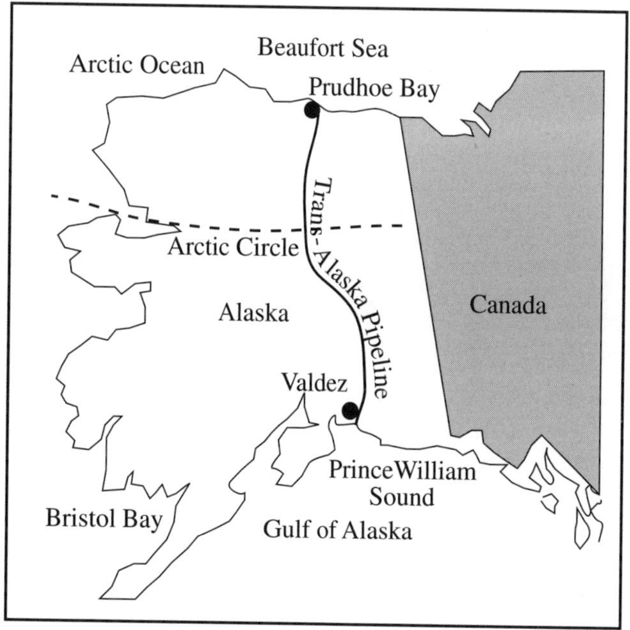

The Alaska Pipeline extends from northern to southern Alaska.

Construction of the Alaska Pipeline was approved by the U.S. government in 1973; however, both before and after its construction, the pipeline has been a subject of concern to citizens and environmentalists. One concern was the cost of construction—the project, originally budgeted at $900 million, was completed in 1977 at a cost of more than $8 billion. Many people, concerned about U.S. dependence on FOSSIL FUELS believe this money would have been better spent on developing nonpolluting, sustainable, ALTERNATIVE ENERGY RESOURCES. Another concern was how pipeline construction and operation might impact Alaska's fragile TUNDRA ECOSYSTEM, specifically its PERMAFROST layer. To help prevent permafrost damage, more than half (655 kilometers, or 407 miles) of the pipeline is located above ground. Where the pipeline is below ground, it is buried in thaw-stable, nonpermafrost areas. In total, the Alaska Pipeline is buried under or crosses over more than 800 rivers and streams. Workers constructed 13 bridges along the route, including a $30-million, 700-meter-long (2,297-foot) bridge that passes over the Yukon River.

A major concern about the Alaska Pipeline is the threat of an OIL SPILL and the impact it would have on the ENVIRONMENT. This concern was intensified following the EXXON VALDEZ OIL SPILL in Prince William Sound in 1989. This spill immediately devastated beaches and OCEAN water quality and killed enormous numbers of WILDLIFE. Effects of the spill on fishing in the area and beach cleanliness are still observed today. Since 1991, there have been reports skeptical of the safe delivery of the oil in the pipeline. One report stated that certain sections of the pipeline were aging and could break, causing oil to spill onto the delicate tundra environment and local WATERSHEDS. Faulty pump stations and electrical failures have also been reported. Environmentalists have asked the president of the United States to appoint an independent task force to inspect the pipeline and evaluate its ability to deliver oil across Alaska without undue risk to the environment. Although the recommendations have been reviewed by the government, no such plan has yet been initiated. *See also* ARCTIC NATIONAL WILDLIFE REFUGE.

EnviroSources

Alaska Forum for Environmental Responsibility: e-mail afervdz@alaska.net.

Kosova, Weston. "Alaska: The Oil Pressure Rises." *Audubon* 99, no. 6, (November–December 1997): 67–74.

alcohol. *See* ethanol; methanol.

aldrin, A chemical PESTICIDE belonging to the family of CHLORINATED HYDROCARBONS that includes the closely related compound dieldrin. As pesticides and under some conditions, aldrin and dieldrin may be harmful to the ENVIRONMENT and to human health. The main effects from short-term exposure to high levels or DOSES of aldrin and dieldrin are headache, dizziness, irritability, loss of appetite, nausea, muscle twitching, convulsions, and loss of consciousness. Death may occur at extremely high exposures or doses.

From the 1950s to the early 1970s, aldrin and dieldrin were widely used as soil insecticides to control root worms, beetles, and termites. Dieldrin has been used in AGRICULTURE for soil and seed treatment and in public health to control DISEASE spread by INSECTS such as mosquitoes and tsetse flies. Dieldrin has also had veterinary use as a sheep dip and has been used in the treatment of wood and the moth-proofing of woolen products.

The ENVIRONMENTAL PROTECTION AGENCY (EPA) considers aldrin and dieldrin probable CARCINOGENS. Most uses for aldrin and dieldrin were banned in 1974 and these chemicals are no longer produced in or imported into the United States. Because aldrin and dieldrin are very persistent (degrade slowly in the environment), they are of a special concern.

At the 1997 UNITED NATIONS EARTH SUMMIT, several governments agreed to negotiate a legally binding global convention to reduce and/or eliminate persistent ORGANIC POLLUTANTS (POPs), which include pesticides such as aldrin, HEPTACHLOR, and DICHLORODIPHENYLTRICHOROETHANE (DDT). Though several governmental agreements to eliminate POPs have occurred in some regions of the world, little implementation had occurred by the late 1990s.

Chlorinated pesticides such as aldrin contribute to global environmental problems. This type of organic chemical can disrupt ECOSYSTEMS and injure human health through its toxic effects. Through a process called BIOACCUMULATION, aldrin becomes more concentrated in organisms at the higher levels of the FOOD CHAIN, with the result that higher concentrations of the chemical occur in predator animal species. Once in the atmosphere or the food chain, such POPs can travel great distances before any significant degradation has occurred. Eventually, such chemicals will concentrate in human blood, milk, fat, and semen. ARCTIC people are particularly susceptible to chlorinated pesticide contamination.

The UNITED NATIONS ENVIRONMENT PROGRAMME, the WORLD HEALTH ORGANIZATION, and the Intergovernmental Forum on Chemical Safety are among the international organizations that seek reform in the worldwide use of POPs. The list of POPs considered most damaging to the environment includes several organochlorine pesticides: DDT, toxaphene, chlordane, heptachlor, endrin, mirex, aldrin, and dieldrin. *See also* AGROCHEMICAL; TOXIC CHEMICAL.

EnviroSources

Agency for Toxic Substances and Disease Registry, Division of Toxicology: 1600 Clifton Road, E-29, Atlanta, GA 30333; Web site: atsdr1.atsdr.cdc.gov:8080/atsdrhome.html.

United States Environmental Protection Agency Web site: www.epa.gov.

algae, A diverse group of PROTISTS consisting of more than 20,000 SPECIES worldwide. Algae live in moist HABITATS, such as lakes and ponds, OCEANS, damp soil, or within the tissues of ORGANISMS. Like PLANTS, algae are PRODUCERS—they synthesize their nutrients and energy through PHOTOSYNTHESIS. As producers, algae form the base of a variety of FOOD CHAINS and FOOD WEBS. Algae also play a role in cycling CARBON DIOXIDE and OXYGEN through the ENVIRONMENT.

The CELLS of all algae contain the green pigment chlorophyll. (*See* EnviroTerms.) Algae also may contain other pigments that mask the color of chlorophyll, giving the algae a red, brown, yellow, blue, or purple appearance. Color, along with cell organization, is used to classify algae into six groups within two broad categories.

Unicellular Algae. Three groups of algae—diatoms, dinoflagellates, and euglenoids—are unicellular. Diatoms are gold-colored microscopic algae that are a common component of PHYTOPLANKTON, a major food source of aquatic organisms. Diatoms generally are enclosed within shells composed of silica and sometimes are called golden algae.

Dinoflagellates are solely aquatic, and unlike diatoms, exist in a variety of colors. Many dinoflagellates are an important food source for marine CONSUMERS; however, several species produce TOXINS that harm marine organisms directly or build up in the tissues of organisms to become increasingly more toxic as they move through the food chain. (*See* bioaccumulation.) Some dinoflagellates reproduce rapidly during warm periods, producing a dangerous ALGAL BLOOM known as a RED TIDE. When a red tide occurs, the toxins produced by dinoflagellates are present in large enough amounts to be toxic to many fishes, shellfishes, and to people who eat organisms exposed to the toxins.

Euglenoids are a unique type of algae that have plantlike and animal-like traits. Euglenoids contain chlorophyll and can carry out photosynthesis; however, euglenoids also have locomotive and digestive structures that permit them to obtain nutrients by feeding on other organisms. Thus, the euglenoid NICHE can be that of either producer or consumer.

Multicellular Algae. The multicellular algae, or seaweeds, consist of three groups of organisms that are identified primarily by their colors—green, brown, and red. Of these, the green algae, which include more than 7,000 species, are the most diverse. Unlike brown algae and red algae, green algae include a small number of unicellular species. Most green algae live in freshwater habitats; others live in the ocean, or in moist land habitats such as in soil, on tree trunks, or on the surfaces of other organisms. (*See* symbiosis.)

Some green algae live in symbiotic association with other organisms. For example, ZOOXANTHELLAE are marine algae that live in a mutualistic association with coral polyps. (*See* mutualism; coral reef.) In this relationship, the polyps benefit by receiving a constant supply of nutrients; the algae receive shelter and the raw materials needed for photosynthesis. LICHENS are a symbiotic association between green algae and FUNGI. The organisms involved in this association are so dependent upon each other that many scientists consider a lichen to be a single organism. Unlike most green algae, lichens are terrestrial and often live on the surfaces of rocks or trees. These unusual organisms are widely dispersed and often inhabit harsh ENVIRONMENTS, such as the ARCTIC TUNDRA.

Brown algae are almost exclusively ocean dwelling. The largest brown algae are the kelps, which have structures resembling the leaves and stems of plants. Large kelps may grow to be more than 60 meters (197 feet) in length and may form vast underwater FORESTS that provide habitat to a variety of marine species. Many brown algae have structures called air bladders that help them float near the water's surface, where they are exposed to the sunlight needed for photosynthesis. Brown algae generally live in cold ocean waters along rocky coasts.

The red algae are a group of diverse species of a variety of colors. Like brown algae, red algae are primarily marine organisms and live along rocky coasts. However, red algae live in warm, tropical waters. The red and blue pigmentations of these algae can absorb light at great depths. Thus, red algae are adapted to living deeper in the ocean than are other algae species.

Importance of Algae. Algae have great importance to the environment. Because they manufacture their nutrients through photosynthesis, algae are important producers for their ECOSYSTEMS. Algae also are responsible for producing much of the OXYGEN in Earth's ATMOSPHERE. In fact, as much as 80% of Earth's atmospheric oxygen may result from the photosynthetic activities of algae.

Algae are commercially important in many parts of the world. For example, algae are a major food source for people living in coastal areas worldwide. In Japan, algae are such a popular food that tons of algae are now grown and harvested on AQUACULTURE farms. Some algae is harvested for use in making food products. For example, kelps growing along the coasts of the United States are harvested to obtain a substance called alginic ACID that is used in the production of ice cream and puddings.

Many algae have commercial value in the making of nonfood products. For example, alginic acid is used to make pharmaceutical products and toothpaste. Diatoms also contain a substance that is used to make toothpaste. This substance, called diatomite, is an abrasive formed from diatom shells. In many areas, large deposits of diatomite that formed when the shells of dead diatoms accumulated and became buried by sediment during the Cenozoic era are quarried to make abrasives, paints, and insulation materials. Diatomite is also widely dispersed in some soils. Soils rich in diatomite are called diatomaceous earth. *See also* PLANKTON.

EnviroTerm

chlorophyll: The green pigment in the cells of algae, plants, and some other organisms that captures light energy to drive the process of photosynthesis.

EnviroSource

Margulis, Lynn, and Karlene V. Schwartz. *Five Kingdoms: An Illustrated Guide to the Phyla of Life on Earth.* 2d ed., New York: W. H. Freemand and Co., 1996.

Rees, Robin, ed. *The Way Nature Works.* New York: Macmillan, 1992.

algal bloom, A POPULATION explosion of ALGAE in a water ENVIRONMENT. Algal blooms threaten the health of the ECOSYSTEM in which they occur and may adversely affect surrounding environments.

Growth of algae is encouraged by PHOSPHORUS, a nutrient needed by photosynthetic ORGANISMS. When the phosphorus level in water increases, often due to the DISCHARGE of untreated WASTEWATER or the RUNOFF from lands treated with FERTILIZERS, a rapid and uncontrolled growth of algae, called an algal bloom, can occur. The initial effects of an algal bloom are observed as large masses of algae covering the water's surface. These masses prevent sunlight from penetrating the water, making it difficult for PLANTS living in deeper water to carry out PHOTOSYNTHESIS. Unable to carry out photosynthesis, many plants, as well as the organisms that feed on the plants, die.

When algal growth is extreme, algae out compete other organisms for resources, such as OXYGEN and CARBON DIOXIDE. In addition, masses of algae can clog pipes or channels that carry water away from the affected body, making the water unfit or unavailable for use by people and other organisms. (*See* potable water.) Algal blooms become most harmful when the algae die, because the MICROORGANISMS that carry out DECOMPOSITION use up much of the water's DISSOLVED OXYGEN. This can kill other aquatic organisms by robbing them of the oxygen they need for their life processes. (*See* biochemical oxygen demand.)

Algal blooms often result from the activities of people. Many products used by people, such as DETERGENTS, fertilizers, and PESTICIDES are rich in phosphorus. Raw or untreated SEWAGE, including that collected in storm drains, also can have a high phosphorus content. The discharge of these phosphorus-rich substances into freshwater ecosystems, through dumping or in runoff from farms, yards, and roadways, can cause an algal bloom. Many algal blooms can be prevented by reducing such runoff and eliminating wastewater discharge into aquatic ecosystems. *See also* ADRIATIC SEA; EUTROPHICATION; RED TIDE.

alien species. *See* exotic species.

alkali, A basic chemical compound, such as sodium carbonate, potassium hydroxide, or sodium hydroxide that can neutralize ACIDS and is used in LIMING. Alkaline substances have a pH of more than 7.0 and are soluble in water. Alkaline products are used to treat soil and water that are too acidic. Baking soda, used in toothpaste and other household products, and caustic soda, used to clean clogged drain pipes, are examples of common alkalis. The term alkalinity refers to the degree to which a substance exhibits the properties of an alkali. *See also* ACID MINE DRAINAGE; ACID RAIN.

alkaline. *See* alkali.

alkalinity. *See* alkali.

allergy, An abnormal sensitivity in some individuals to particular substances, especially proteins, in the ENVIRONMENT. Substances that initiate allergic responses are called "allergens." Common environmental allergens include PLANT pollen, mold spores, feathers, animal fur, and dust mites. INSECT stings; and contact with plant oils from poison ivy, poison sumac, and poison oak; chemicals that are present in some foods; as well as natural and synthetic chemicals present in soaps, DETERGENTS, cosmetics, pharmaceutical products, and fragrances also may initiate allergic responses in susceptible individuals.

The symptoms and severity of allergic reactions vary among individuals and according to the allergen. Minor to moderate reactions include skin rashes, hives, nasal congestion, itchy and watery eyes, and a runny nose. In more severe reactions, individuals may suffer an asthma attack, anaphylactic shock, or even death. *See also* ACUTE TOXICITY; AIR POLLUTION; CHRONIC TOXICITY; DISEASE; ENVIRONMENTAL MEDICINE; MULTIPLE CHEMICAL SENSITIVITY; SICK BUILDING SYNDROME.

EnviroTerm

pollen: The male reproductive CELLS of angiosperms (flowering plants) and gymnosperms (plants that form naked seeds) which are commonly carried in moving air.

EnviroSources

Moeller, Dade W. *Environmental Health.* Cambridge, MA: Harvard University Press, 1997.

Vos, J.G., et al. *Allergic Hypersensitivities Induced by Chemicals: Recommendations for Prevention.* Cleveland, OH: CRC Press, 1996.

alligators, Freshwater REPTILES belonging to the order Crocodilia. Worldwide, there are only two alligator SPECIES, both of which have been classified as threatened or ENDANGERED SPECIES at different times. They are the American alligator (*Alligator mississipiensis*) and the Chinese alligator (*Alligator sirensis*). The American alligator is native to the United States, from North Carolina through southern Florida and as far west as the Rio Grande. The Chinese alligator is native to the Yangtze River valley of China.

Alligators are fierce predators that feed mostly on fishes and MAMMALS, occasionally including humans. (*See* predation.) Their predatory habits have led to many alligators' being killed

Populations of American alligators are again thriving following years of protection by the United States Fish and Wildlife Service.

to protect livestock and people. Alligators once were commonly killed for sport, for food, or to obtain their skin, which is used to make products such as footwear, wallets, and belts. As a result of overhunting, the Chinese alligator has become rare and is nearing EXTINCTION. During the 1960s, overhunting also led to the near extinction of the American alligator. However, POPULATIONS of American alligators have steadily increased since 1973, when they were replaced on the ENDANGERED SPECIES LIST of the U.S. FISH AND WILDLIFE SERVICE.

Today, the American alligator population is thriving in the wild and its recovery is considered a success story of the ENDANGERED SPECIES ACT. To help maintain wild U.S. alligator populations, while also providing people with products from alligators, many of these animals are now raised on farms in a manner similar to livestock. *See also* CROCODILES; POACHING; SEA TURTLES; THREATENED SPECIES.

EnviroSource

Chadwick, Douglas H., and Joel Sartore. *The Company We Keep: America's Endangered Species.* Washington, DC: National Geographic Society, 1995.

alpha particle. *See* radiation.

alpine tundra, One of two types of northern BIOME. The ARCTIC TUNDRA circles the North Pole, while the alpine tundra is located south of the pole on mountain slopes in high ALTITUDES. Because tundra hosts fragile ECOSYSTEMS, environmentalists believe that any heavy construction, MINING, or use of vehicles in the tundra can permanently damage this HABITAT, leaving barren areas that will take a long time to recover.

The alpine tundra includes windswept, treeless areas that extend from the treeline to the highest mountain peaks. The CLIMATE is exceedingly harsh with wind speeds in excess of 160 kilometers (100 miles) per hour. The mean annual temperature is below freezing, with the frost-free season approximately 1.5 months long.

The alpine tundra often appears as barren rock or a cover of thin soils, though places have deep soils and abundant PLANT cover. A variety of plant communities exist, including: low-growing shrubs, cushion plants, small forbs, and meadows of sedges and grasses. Most are slow-growing perennials, with 90% of the total plant structure located in the roots for storing nutrients and energy for use during periods of scarce resources. Alpine WILDLIFE includes big-horned sheep, mountain goats, pikas, marmots, and ptarmigans. Butterflies, beetles, and grasshoppers are abundant.

EnviroSources

Above the Timberline: The Alpine Tundra Zone. Canadian Film Distribution Center: (800) 388-6784.

Ann H. Zwinger, and Beatrice E. Willard. *Land above the Trees: A Guide to American Alpine Tundra.* New York: Harper and Row, 1972.

alternative energy resource, An ENERGY source, such as WIND POWER, GEOTHERMAL ENERGY, BIOMASS, SOLAR ENERGY, nuclear power, or HYDROELECTRIC POWER (water), which is used as an alternative to FOSSIL FUELS to generate electrical power. Alternative energy resources can be renewable or nonrenewable. The principal RENEWABLE RESOURCES are solar energy, biomass, wind, and hydroelectric power. The NONRENEWABLE RESOURCES include geothermal energy and NUCLEAR ENERGY. (Geothermal resources are considered to be nonrenewable or unsustainable if underground hot water and steam are used at rates exceeding their rates of replenishment.)

Although the technology for expanding the use of alternative energy sources is growing, these resources still do not supply very much of the world's energy. In the late 1990s, fossil fuels still supplied about 60% of the world's ELECTRICITY and about 80% of its nonelectrical energy needs such as heating buildings and FUELS for vehicles. Of the renewable sources, biomass supplied about 20% of the nonelectrical energy and hydropower supplied about 20% of electricity. However, economic analysts and environmentalists predict that by the mid-twenty first century, alternative energy sources such as biomass and direct sunlight could supply about 60% of the world's electricity and about 40% of nonelectrical energy. The pace at which use of alternative energy grows will depend on the support for research and development offered by industries and governments.

EnviroSources

National Alternative Fuels Hotline: U.S. Department of Energy, P.O. Box 12316, Arlington, VA 22209; (800) 423-1363.

U.S. Department of Energy Web site: www.doe.gov.

U.S. Department of Energy Alternative Fuels Data Center Web site: www.afdc.nrel.gov.

alternative fuel, FUEL that the DEPARTMENT OF ENERGY (DOE) classifies as an alternative to GASOLINE. Alternative fuels cause less environmental damage than traditional fuels do. All alternative fuels reduce OZONE-forming tailpipe EMISSIONS in motor vehicles. The DOE currently recognizes several substances as alternative fuels. METHANOL (M85) and ETHANOL (E85) are fuel mixtures that contain no less than 70% alcohol. Methanol (M85) is produced from NATURAL GAS, COAL, and BIOMASS, and denatured ethanol (E85) is produced from sources such as corn and other grains. Other alternative fuels include compressed natural gas and liquefied natural gas, both produced from underground reserves, and liquefied PETROLEUM gas, which is produced as a byproduct of petroleum refining and natural gas processing. Other alternative fuels include HYDROGEN, coal-derived liquid fuels, fuels derived from biological materials such as ethanol, and SOLAR ENERGY. The DOE plans to expand this list as new fuels are developed and approved for use. *See also* ALTERNATIVE ENERGY RESOURCES; BIOFUEL; GASOHOL; RENEWABLE RESOURCES; SYNTHETIC FUEL.

EnviroSources

U.S. Department of Energy, Alternative Fuel Vehicle Fleet Buyers' Guide: www.afdc.doe.gov/ or www.fleets.doe.gov.

U.S. Department of Energy, National Alternative Fuels Data Center Hotline: (800) 423-1363.

altitude, Elevation or height above sea level. As altitude increases or decreases there are changes in ECOSYSTEMS and HABITATS. Environmental changes that occur when altitude increases or decreases are similiar to those changes that occur when latitude increases or decreases. A TUNDRA BIOME, for example, is found in the high latitudes and also at high altitudes. ABIOTIC factors such as soil, precipitation, and temperature change with increased latitude or altitude. These factors determine the vegetation and the ORGANISMS found in the given ecosystem. *See also* ALPINE TUNDRA.

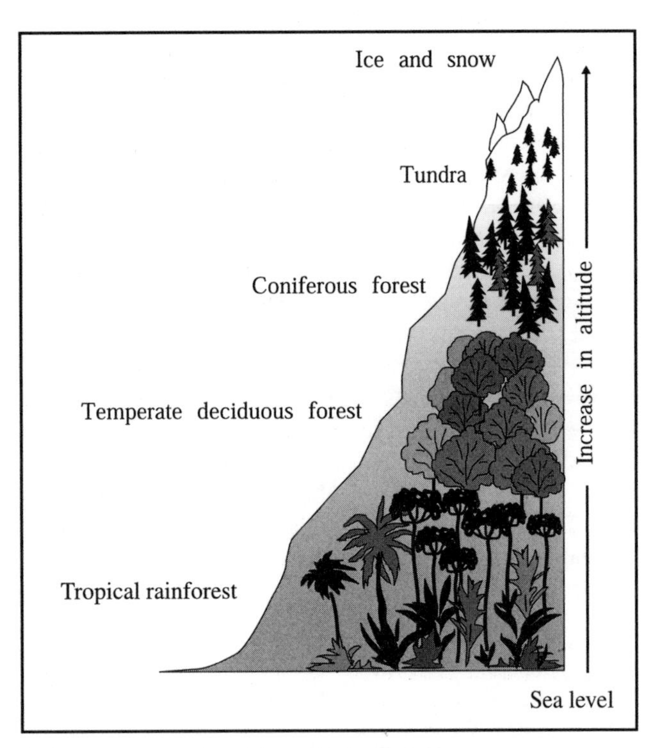

A pictorial representation of how altitude affects biomes.

aluminosis. *See* aluminum.

aluminum, A light metallic element that can damage the ENVIRONMENT and cause human health problems at high levels. Elemental aluminum does not exist in nature; it is always found in compounds. In the environment, ACID RAIN can dissolve and leach aluminum compounds into the soil as TOXINS. These POLLUTANTS can then enter aquatic HABITATS such as lakes, streams, and rivers through RUNOFF and potentially cause adverse effects on PLANTS, and fish.

People who breathe in high levels of aluminum particles in the AIR may have respiratory problems including coughing and asthma. Excessive levels of ingesting aluminum particles in the air can cause aluminosis, a lung disorder. Although claims have been made linking aluminum to nervous disorders, kidney damage, and Alzheimer's DISEASE, they have not been supported scientifically. Low-level exposure to aluminum from food, air, water, or contact with skin is not thought to be harmful to humans.

Aluminum is the most abundant metal in Earth's crust. It is produced by electrolysis of bauxite ore, much of which is found in tropical FORESTS. This lightweight corrosion-resistant, malleable metal is used in cooking utensils, foil wrapping, beverage and food containers, electrical appliances, and building materials. It is used in paints and fireworks and to produce glass, rubber, and ceramics. It is used in several chemical forms including aluminum nitrate, aluminum oxide, aluminum hydroxide (used in antacids), and aluminum chlorohydrate (used in deodorants). Aluminum sulfate, or alum, is used in wastewater treatment plants as a coagulant to remove particles.

RECYCLING of aluminum cans has reduced the number of cans deposited in landfills. About 50% of all aluminum cans are melted down and recycled.

The ENVIRONMENTAL PROTECTION AGENCY (EPA) recommends that the concentration of aluminum in the drinking water supply not exceed 0.2 parts per million because of taste and odor problems.The FOOD AND DRUG ADMINISTRATION (FDA) has determined that aluminum cooking utensils, aluminum foil, antiperspirants, antacids, and other household aluminum products are generally safe. The OCCUPATIONAL SAFETY AND HEALTH ADMINISTRATION (OSHA) has set a maximum allowable concentration for aluminum dust in the workplace. *See also* HEAVY METALS.

EnviroSources

Agency for Toxic Substances and Disease Registry, Division of Toxicology: 1600 Clifton Road NE, Mailstop E-29, Atlanta, GA 30333; (800) 447-1544; fax: (404) 639-6315; Web site: www.atsdr1.atsdr.cdc.gov:8080/atsdrhome.html.

U.S. Environmental Protection Agency Web site: www.epa.gov.

Amazon rainforest. *See* rainforest.

American Zoo and Aquarium Association, An organization that provides accreditation to ZOOS, aquariums, WILDLIFE parks, and oceanariums in North America. The mission of the American Zoo and Aquarium Association (AZA) is to advance zoological parks and aquariums as centers for CONSERVATION, education, recreation, and scientific study. To reach these goals, the AZA works with the U.S. FISH AND WILDLIFE SERVICE (FWS), the Animal and Plant Health Inspection Service, the NATIONAL MARINE FISHERIES SERVICE (NMFS), the INTERNATIONAL WHALING COMMISSION (IWC), and numerous international wildlife and conservation organizations.

The AZA was organized in 1924 as the American Association of Zoological Parks and Aquariums (AAZPA). The AAZPA's goal was to help zoo and aquarium professionals exchange information. In 1966, the AAZPA merged with the National Recreation and Park Association, but it separated from this group in the 1970s.

As an independent organization, the AAZPA stressed wildlife conservation and protection of wildlife HABITAT. The group also developed an accreditation process for its new and existing members. In 1980, the AAZPA became involved in the

Species Survival Plan (SSP), which requested member institutions to conduct research and field work and to educate the public about THREATENED SPECIES and ENDANGERED SPECIES.

The AAZPA changed its name to the American Zoo and Aquarium Association in 1994. As the AZA, the group continues its work in accreditation and wildlife preservation through its representation of most professionally operated zoos and aquariums in North America. The AZA also works worldwide with wildlife and conservation centers and is involved in more than 1,200 conservation projects in more than 60 countries. Through the work of its members, the AZA continues to educate the public about the need for protecting Earth's BIODIVERSITY. *See also* CAPTIVE PROPAGATION; RED LIST OF ENDANGERED SPECIES; WORLD CONSERVATION MONITORING CENTRE; WORLD CONSERVATION UNION.

EnviroSource

American Zoo and Aquarium Association Web site: www.aza.org.

ammonification. *See* nitrogen cycle.

***Amoco Cadiz* oil spill,** The DISCHARGE of more than 204,000 metric tons (225,000 tons) of PETROLEUM into the Atlantic Ocean off the coast of Portsall, Brittany (France), in 1978 that resulted from the grounding and breakup of the *Amoco Cadiz* supertanker. The *Amoco Cadiz* disaster occurred after the OIL TANKER lost its steering ability during a severe storm on March 17, 1978. While being assisted by a tugboat, the tanker became grounded on rocks, broke apart, and spilled its entire cargo of crude oil into the OCEAN.

The environmental impact of the *Amoco Cadiz* OIL SPILL was severe. Oil was deposited along about 325 kilometers (200 miles) of coastline, often to depths as great as 30 centimeters (12 inches). WILDLIFE suffered: As many as 30,000 sea birds died. The mass of fishes and CRUSTACEANS killed by the spill exceeded 225,000 metric tons (248,000 tons), which in turn, disrupted FOOD CHAINS and FOOD WEBS. The spill also had long-term effects on the fishing industry. Years after its occurrence, some petroleum remained dispersed in the ocean. Many fishes and crustaceans absorbed enough of this petroleum into their tissues to make them unfit for human consumption.

The responsibility for the *Amoco Cadiz* oil spill was difficult to determine. The tanker was owned by a Liberian shipping company; the cargo belonged to Amoco, a U.S. company. After years of investigation and litigation, a federal judge in Chicago attributed primary responsibility for the disaster to the Standard Oil Company of the United States, of which Amoco is a part. As a result of this verdict, French claimants, who had filed suit against the responsible party, were awarded more than $85 million in damages from Amoco.

To date, the *Amoco Cadiz* spill remains one of the largest oil spills involving an oil tanker. Its discharge was eight times that of the EXXON VALDEZ OIL SPILL in Prince William Sound. The largest oil spill involving a tanker—the 1983 discharge from the *Castillo de Beliver*—exceeded the *Amoco Cadiz* spill by 4,545 metric tons (5,000 tons) of oil. *See also* OCEAN DRILLING PROGRAM; OIL SPILL EQUIPMENT.

amphibians, Cold-blooded, egg-laying VERTEBRATES such as frogs, toads, and salamanders that are declining in POPULATION in parts of Australia, the western United States and Canada, Central America, and South America. Amphibians are a class of vertebrates (Amphibia) whose characteristics include a three-chambered heart, cold bloodedness, and a lack of hair,

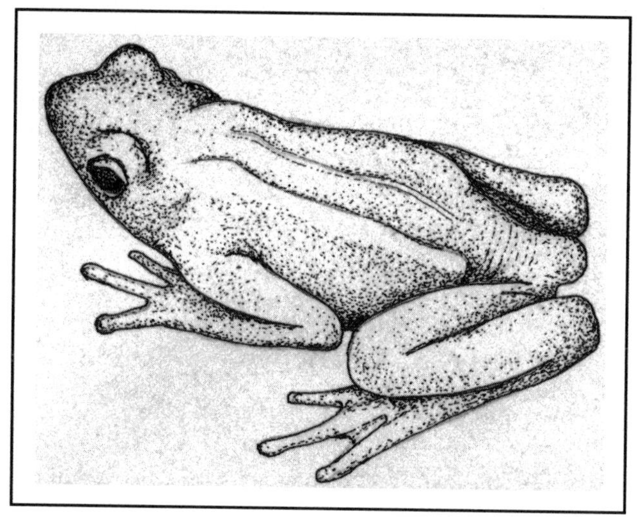

The golden frog is an amphibian.

scales, or feathers. Many SPECIES lay their eggs in water. After hatching, the young are aquatic HERBIVORES that draw oxygen from the water through gills. Later, the amphibians develop lungs, become INSECTIVORES, and live on land as adults. There are about 4,000 species of amphibians found in many ECOSYSTEMS, including DESERTS, FORESTS, and GRASSLANDS, and at various ALTITUDES. They are abundant in the tropics, but can also be found in the temperate zones and in the higher latitudes of northern North America.

Although many populations of amphibians are thriving, scientists are concerned that in some places populations are declining. Researchers believe that these declines are caused by HABITAT loss, PESTICIDES, DISEASES, and excessive ultraviolet RADIATION resulting from OZONE DEPLETION.

In one study, researchers found a correlation between ultraviolet-B RADIATION and the destruction of amphibian eggs and larva. Amphibians lay their unshelled eggs in open, shallow water, leaving them vulnerable to increases in solar ultraviolet-B radiation. Over generations, the resulting destruction of eggs can contribute to the gradual decline of adult populations in a given area. Ultraviolet radiation may also be contributing to the destruction of aquatic INSECTS and other ORGANISMS that are prey for amphibians.

The destruction of habitats, such as ponds and WETLANDS, is probably the main reason for the decline in certain species, particularly those living in tropical areas of Africa, Asia, and the Americas. In one study in the United States, CLEAR-CUTTING of FORESTS was shown to lead to the deaths of a large number of salamanders.

A parasitic fungus called chytrid may also be associated with the decline of amphibians in Central America and Australia. The fungus infects the skin of amphibians, which causes respiration problems in those amphibians that breathe through their skin. Another fungus, saprolegnia, has destroyed tadpole populations in the western United States.

POLLUTANTS such as ACID RAIN, industrial chemicals, and pesticides also are a likely cause of population declines in amphibians such as frogs and toads. Airborne pesticides blown in from agricultural areas impact the reproduction and development of amphibians. Although more long-term studies are needed, there is growing evidence that some amphibian species are declining because of the environmental changes detailed above. *See also* ENDANGERED SPECIES; ENDANGERED SPECIES ACT; ENDANGERED SPECIES LIST; FUNGI; PARASITISM; THREATENED SPECIES.

Some Endangered Amphibians

Species	*Location*
Japanese giant salamander	Japan
Shenandoah salamander	Virginia
Stephen Island frog	New Zealand
Panamanian golden frog	Panama
Monte Verde toad	Costa Rica
Wyoming toad	Wyoming

Source: Division of Endangered Species, U.S. Listed Vertebrate Animal Species Index.

EnviroSources

Amphibians Web site: www.frogweb.gov/.

North American Reporting Center for Amphibian Malformations Web site: www.npwrc.usgs.gov/narcam/.

U.S. Fish and Wildlife Service, Division of Endangered Species, U.S. Listed Vertebrate Animal Species Index Web site: www.fws.gov/vertdata.html

anaerobic, A term that refers to the absence of OXYGEN in a process or ENVIRONMENT. Most PLANTS and animals, including humans, obtain their ENERGY using both AEROBIC and anaerobic processes; however, other ORGANISMS, such as yeast (a type of fungus) and some kinds of BACTERIA, obtain all of their energy by means of anaerobic respiration. They are able to live without oxygen. Habitats, independent of oxygen, such as the muddy bottom of a marsh are referred to as anoxic. Water without DISSOLVED OXYGEN is also *anoxic. See also* ANAEROBIC DECOMPOSITION.

anaerobic decomposition, The breaking down of ORGANIC matter by MICROORGANISMS that live in an OXYGEN-free, or anoxic, ENVIRONMENT. Anaerobic decomposition occurs in natural environments such as MARSHES, WETLANDS, and rice paddies.

Anaerobic decomposition is put to use in anaerobic digesters—containers in which complex organic molecules are broken down to gaseous byproducts in oxygen-free conditions, with very little SOLID WASTE produced in the process. Anaerobic digesters are used in some municipal SEWAGE and WASTEWATER treatment plants because they are more compact, are more ENERGY efficient, and produce much less solid waste than traditional trickling filter and activated SLUDGE treatment plants. Furthermore, anaerobic SEWAGE TREATMENT PLANTS generate significant quantities of biogases, such as METHANE, along with liquid and solid ORGANIC FERTILIZERS, which can be sold. Biogas and organic fertilizer production from human, animal, or vegetable wastes is therefore another important application of anaerobic decomposition. Some countries such as India and China use anaerobic digesters to convert animal wastes into methane for use as a FUEL. *See also* BIOFUEL; PRIMARY, SECONDARY, AND TERTIARY TREATMENTS.

anaerobic digester. *See* anaerobic decomposition.

animal rights, A term to describe the belief that all animals are entitled to consideration of their best interests, regardless of whether they are perceived as attractive or useful to humans. According to the animal rights organization People for the Ethical Treatment of Animals (PETA), a belief in animal rights also includes the acknowledgment that animals are not free for use by humans as a source of food, clothing, entertainment, or experimentation.

Many animal rights proponents distinguish between animal welfare and animal rights. Animal welfare theories accept the proposition that animals have interests, but allow for those interests to be compromised for human benefit as long as animals are treated humanely in the process. Animal rights proponents, by contrast, believe that animals have inviolable interests that cannot be sacrificed simply for the benefit of humans. This philosophy does not hold that rights are absolute, however; an animal's rights, like those of a human, must be limited, since the rights of one group frequently conflict with those of others.

According to many animal rights organizations, animals do not always have the *same* rights as humans. Because the interests of animals are not always the same as ours, some human rights would be irrelevant to animals' lives. For instance, a dog does not have the right to vote, since that right would be meaningless to a dog. However, most animal rights activists believe that a dog has a right to live without pain inflicted on it, and that humans must respect and consider that interest.

The philosophy of animal rights activists often conflicts significantly with that of people whose livelihoods, lifestyles, and/or diets use the products of livestock farming (meat production, leather, and fur industries, among others). Another point of general conflict with respect to animal rights involves the use of animals in medical, consumer-product, and military experimentation, which is sometimes also supported by government funding. *See also* ANTHROPOCENTRIC; BIOCENTRIC; DEEP ECOLOGY; DOLPHINS AND PORPOISES; SPECIESISM.

EnviroSources

Animals' Agenda: (410) 675-4566; Web site: www.envirolink.org/arrs/aa/index.html.

People for the Ethical Treatment of Animals: (757) 622-PETA; Web site: www.peta.com.

Antarctic Treaty, An international agreement, initially signed in 1959 and amended in 1991, that binds its signatories to work together to preserve the NATURAL RESOURCES of ANTARCTICA until the year 2041. The treaty states that Antarctica should be used exclusively for peaceful purposes and to promote the exchange of scientific data. It prohibits any military activity, the explosion of NUCLEAR WEAPONS, and the disposal of radioactive material. On-site inspections ensure that the treaty is observed. Antarctica is the only continent completely governed by international agreement. Before the Antarctic Treaty was signed, Australia and several other nations laid claims to parts of Antarctica; however, disagreements among these nations arose, posing serious legal and political problems. To address these issues, the Antarctic Treaty was prepared during the International Geophysical Year (IGY) in 1957–58. The 12 signatory nations (Argentina, Australia, Japan, Belgium, Chile, France, New Zealand, Norway, South Africa, the United Kingdom, the United States, and the former Soviet Union) present in Antarctica at that time set aside their differences and adopted the treaty in 1959. An additional 31 nations signed the treaty in 1961 when it took effect. The signatory nations making territorial claims are Argentina, Chile, France, Australia, United Kingdom, New Zealand, and Norway. *See also* BIODIVERSITY TREATY; CONVENTION ON INTERNATIONAL TRADE IN ENDANGERED SPECIES OF WILD FLORA AND FAUNA; UNITED NATIONS EARTH SUMMIT.

Scientists conduct studies of Weddell's seals in Antarctica. The hood over the head of the seal prevents the animal from moving. *Credit:* NOAA

EnviroSource

Environmental Treaties and Resource Indicator Web site: www.sedac.ciesin.org/entri/

Antarctica, The ice-covered continent surrounded by the Southern Ocean, with the South Pole near its center. Antarctica is threatened with CLIMATE change and OZONE DEPLETION. The area of Antarctica is about 13,824,000 square kilometers (5,330,000 square miles), making it the world's fifth-largest land mass. It is 1.5 times the size of the United States.

Antarctica is the highest, coldest, driest, windiest, least-populated place on Earth. Its elevation averages 2.4 kilometers (1.5 miles) above sea level, 1.5 kilometers (about 1 mile) higher than the global average; however, only about 2% of Antarctica's land is visible above the thick ice sheets that blanket the continent. During the coldest months, low temperatures have reached –88°C (–125°F). Along the coast, the annual mean temperature is about –15°C (5°F). Each year, Antarctica receives less than 3 centimeters (about 1 inch) of water in the form of snow, a precipitation rate similar to that of the Sahara Desert. (*See* desert.) Winds reaching 300 kilometers (180 miles) per hour blow downhill out of the continental interior.

The OXYGEN-rich Antarctic bottom water spreads out over the world's ocean floors, causing deep-ocean temperatures to be cooled to less than 2°C (35.6°F), and helping to drive ocean circulation. Antarctic waters provide essential nutrients to the rest of the world's oceans, supporting life systems thousands of kilometers away.

Antarctica can be divided into East Antarctica, West Antarctica, and the Antarctic Peninsula. East Antartica, the largest region, is a stable shield of very ancient rocks; the discontinuous west would resemble rock "islands" if its overlying ice were removed. The 4,800-kilometer (2,900-mile) Transantarctic Mountains divide East and West Antarctica. They rival the Rocky Mountains in height, yet ice covers all but their tips.

The base of the thicker East Antarctic ice sheet lies above sea level; by contrast, the western ice sheet's base is below sea level. They react very differently to changes in the ENVIRONMENT: The East Antarctic ice sheet is stable and responds slowly, but the western ice sheet may alter very rapidly.

The Antarctic ice cap, averaging 2.3 kilometers (1.4 miles) in thickness, covers 98% of the continent and contains about eight times more ice than the ARCTIC region. It is the largest body of fresh water in the world, holding nearly 70% of Earth's total supply, enough to cause a global SEA LEVEL RISE of almost

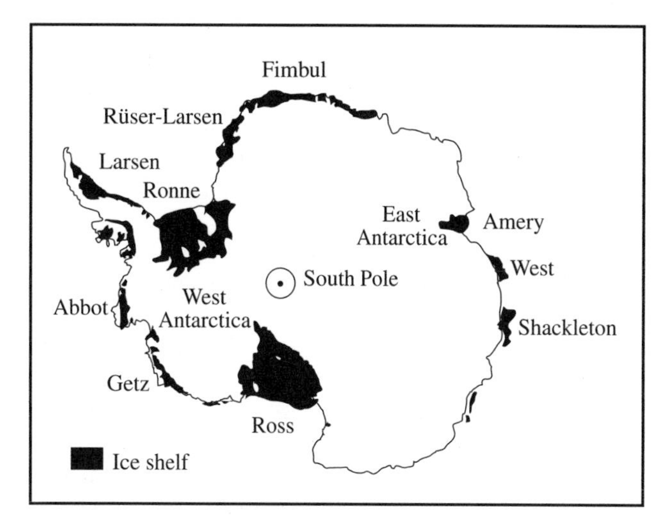

There are several ice shelves in Antarctica.

70 meters (215 feet). The ice sheets' weight is so great that in many areas ice has actually pushed the land below sea level; without its ice cover, Antarctica would eventually rebound another 450 meters (1,476 feet) above sea level.

During the Antarctic summer, sunlight lasts six months; during winter, the region receives little or no sunlight. Solar RADIATION is weak because of the oblique angle at which it strikes Earth's poles and because sunlight must travel through a thick layer of ATMOSPHERE, which reflects and absorbs the radiation. Of the solar radiation that does reach Antarctica, ice reflects 80–90%.

Annually, Antarctica actually loses more SOLAR ENERGY than it retains; only for a short time at summer's height is there a net ENERGY gain. The continent would be even colder if it did not receive heat from warm, moist equatorial AIR that travels toward the poles. This migration is an essential part of Earth's heat balance; by acting as a heat SINK, a reservoir, Antarctica helps control global climate and WEATHER.

Antarctica has only a limited number of finely balanced ECOSYSTEMS, most of which are marine based. Its extreme climatic conditions prevent the establishment of higher PLANT and animal SPECIES on land; as a result terrestrial ecosystems are very simple. On land, portions that are not permanently covered by ice are home to BACTERIA, LICHENS, mosses, and small INVERTEBRATES, the largest of which is the midge, a species of tiny INSECT. Many survive the freezing temperatures because their body fluids contain glycerol, an antifreeze that allows survival at temperatures way below freezing.

No indigenous people have ever occupied Antarctica. The major groups of Antarctic FAUNA are sea birds, penguins, SEALS, and WHALES. All Antarctic MAMMALS are marine, including seals and whales; the latter are protected in the Southern Ocean's International Whale Sanctuary. Most Antarctic mammals are part-time residents that migrate northward in the winter.

In the entire Antarctic FOOD WEB, there are only three or four levels of species. The top CARNIVORES are killer whales and leopard seals, which feed on penguins, squid, and other seals. Antarctic squid and fish make up the intermediate level. PHYTOPLANKTON, such as diatoms, and ZOOPLANKTON, mainly KRILL, form the lowest TROPHIC LEVELS of the food web, appearing in profusion during the Antarctic summer.

The Southern Ocean contains 400–650 million metric tons (440–715 million tons) of krill, which are critically important in the Antarctic FOOD CHAIN; many fish, sea birds, squid, seals, and whales depend on them as their sole food source. Krill swarms are seasonal; if they fail to appear, dependent species may experience drastic breeding failures.

The ANTARCTIC TREATY was signed by 12 nations in 1959 and ratified in 1961 to allow peaceful and scientific studies of this unique region. Today, 42 signatories represent over 80% of the world's population, including superpowers, developed nations, and developing nations.

Despite protective accords, Antarctica is nevertheless threatened by OVERFISHING, pollution and other damage from scientific bases, and waterborne POLLUTANTS transported by ocean currents. Increasing levels of tourism also pose a threat to WILDLIFE, particularly on the Antarctic Peninsula.

In 1985, scientists discovered that each year more than half the natural OZONE was being lost from the stratosphere over Antarctica following the seasonal return of sunlight. Research showed that the culprit was human-created air pollutants, particularly CHLOROFLUOROCARBONS (CFCs), which were releasing more than 455,000 metric tons (0.5 million tons) of CHLORINE into the atmosphere every year, destroying thousands of ozone molecules.

In 1998, National Aeronautics and Space Administration (NASA) satellites revealed that the area of OZONE DEPLETION over Antarctica was larger than ever, encompassing 26.9 million square kilometers (10 million square miles), 5% more than the previous record set in 1996. Increased ultraviolet radiation due to ozone layer depletion may result in elevated skin CANCER and cataract rates; suppression of the human immune system; disruption of plant life, including increased susceptibility to DISEASE; reduction in phytoplankton growth; and the eventual reduction of numbers of aquatic species, including krill.

Today, climate change is the greatest environmental threat to Antarctica and its fragile ecosystems. GLOBAL WARMING visibly affects polar regions of the planet by reducing the amount of water held in ice. Records indicate that the average temperature on the Antarctic Peninsula has increased by 2.5°C (4.5°F) in the past 40 years; the area of the Wordlie Ice Shelf on the peninsula's western coast has already been reduced by two-thirds. Significant rises in temperature could promote further widespread break-up of the ice sheets, which would have enormous consequences both for the Antarctic environment and global climate. The resultant sea level rise could also severely affect coastal areas around the world. *See also* AIR POLLUTION; FLUOROCARBONS; GLACIER; INTERNATIONAL WHALING COMMISSION; PLANKTON.

EnviroSources

Antarctica Project: P.O. Box 76920, Washington, DC 20013.

Crossley, Louise. *Explore Antarctica.* Cambridge, UK: Cambridge University Press, 1995.

Greenpeace International, Antarctic Web site: www.greenpeace.org/~comms/98/antarctic.

Heacox, Kim. *Antarctica: The Last Continent.* (National Geographic Destinations). Washington, DC: National Geographic Society, 1999.

International Centre for Antarctic Information and Research Web site: www.icair.iac.org.nz.

Lucas, Mike, *Antarctica.* New York: Abbeville Press, 1996.

Monteath, Colin. *Antarctica: Beyond the Southern Ocean.* Hauppauge, NY: Barrons Educational Series, 1997.

Virtual Antarctica Web site: www.terraquest.com/va.

Wheeler, Sara. *Terra Incognita: Travels in Antarctica.* New York: Modern Library, 1999.

anthracite. *See* coal.

anthropocentric, A term, meaning "human-centered," which describes the belief that humans are superior to, and thus more important than, all other ORGANISMS. People having an anthropocentric view evaluate nonhuman SPECIES according to how useful or desirable they are to humans and make decisions about how to manage environmental resources based on how humans are benefited, rather than how all species are impacted. Anthropocentrism contrasts with biocentrism, a philosophy in which all species are deemed to have intrinsic natural value and therefore a right to exist. *See also* ANIMAL RIGHTS; BIOCENTRIC; DEEP ECOLOGY.

aphotic zone. *See* ocean.

aquaculture, The cultivation of fish, shellfish, or aquatic PLANTS in natural or controlled marine or freshwater ENVIRONMENTS. About 20% of all commerical fish are raised in an aquaculture environment, and the industry will continue to grow throughout the twenty-first century. Today, aquaculture is a multi-million dollar business. Much of the trout, catfish, and shellfish consumed in the United States are products of aquaculture.

Practiced since ancient times, aquaculture today is used to cultivate a wide variety of aquatic products, including fish, such as catfish and trout, for food; ornamental fish, such as carp and koi, for aquariums; bait fish for the fishing industry; sporting fish for restocking lakes and ponds; oysters for obtaining pearls; and mussels and seaweed for food. The science of aquaculture is far behind that of its terrestrial counterpart—AGRICULTURE; however, worldwide, aquaculture has grown dramatically in the past 20 years.

Workers harvest catfish from an aquaculture farm in Mississippi.
Credit: United States Department of Agriculture. Photo by Ken Hammond.

Animal Aquaculture. The practice of cultivating and raising fish for food probably began as early as 4,000 years ago in China. Today, fish farming, an important industry in the United States, the Philippines, Japan, China, India, Israel, and Europe, is by far the most common form of aquaculture. Fish farming is the practice of raising fish in captivity in order to improve their growth and reproduction, similar to the way livestock is raised on land. Most fish farms consist of many enclosures, ponds, lakes, tanks, pens, and long, narrow channels, each containing fish at varying stages of development. Fish culturists manage the aquatic environments by circulating clean water through the enclosures and protecting the fish from predators, DISEASE, and parasites. (*See* predation; parasitism.) In these types of farm operations, fish are grown to maturity and then harvested for food.

Share of Global Aquaculture Productions. 1994

Country	Percent Share of Global Production
China	57
India	9
Japan	4
Indonesia	4
Thailand	3
United States	2
Philippines	2
Korea, Republic of	2
Other Countries	17

Source: Food and Agriculture Organization of the United Nations, *The State of World Fisheries and Aquaculture*, 1996, Rome: FAO 1997 p. 12.

Another type of fish farm common in the United States is known as a fish ranch. In a fish ranch, many fishes, particularly sport fish SPECIES such as salmon, are hatched in small ponds and then released into rivers. The fish then migrate downstream to the OCEAN where they will reach adulthood. Once these fishes mature, they instinctively return to the river from which they were released in order to reproduce. When they do so, they are captured and harvested for food.

Plant Aquaculture. The aquatic plants raised in aquaculture include ornamental plants, such as pond lilies, and NATIVE SPECIES of plants used for HABITAT restoration; however, the vast majority of the plants systematically grown in aquaculture operations are seaweed, a type of ALGAE. The cultivation of seaweed, is particularly popular in China and Japan, where it is an important food source. Since the seventeenth century, for example, Japanese aquaculturalists have grown their own seaweeds. Traditionally, farmers cultivated the algae by placing long bamboo sticks into rivers. When small seaweed plants began to grow, the sticks were removed and brought to the sea, where the plants thrived in a mixture of fresh and salt water. Today, the cultivation of algae in Japan and elsewhere is highly mechanized. Small seaweed plants are first grown in special hatching tanks. The plants are then trucked to the coasts, where they are mechanically tended to until harvest.

There are several economic and social forces behind the tremendous growth and interest in aquaculture. The main reason, perhaps, is the recognition that the world's oceans, lakes, and rivers cannot produce enough food to satisfy the world's appetite for fish and other types of seafood. Worldwide, about 80 million metric tons (88 million tons) of fish and shellfish are consumed per year. By the year 2010, world demand for edible seafood is projected to be 110 million to 120 million metric tons (121 million to 132 million tons). The oceans, lakes, and rivers of the world will not be able to meet this demand, so any shortages in seafood can only be made up through aquaculture.

Another force behind the growth in aquaculture is the increased interest, particularly within the United States, in eating a healthful diet. Numerous studies have recognized that fish and seafood are low in sodium, fat, and cholesterol. Additional studies have found that certain fish contain fatty substances that have the effect of reducing cholesterol in the body.

Finally, aquaculture may be one possible solution to the world's food supply problems. The amount of farmable land is limited and continues to shrink everywhere; however, because two-thirds of Earth is covered with water, many people believe that the numbers of plants and animals that can be grown in aquatic environments is almost limitless. In addition, some of the problems faced by aquaculturalists are not the same as those faced by farmers on land. Aquaculture, for instance, is less affected by the impact of CLIMATE change such as DROUGHTS, FLOODS, and extreme temperature changes.

There are problems, however, with some of the fish farms, particularly shrimp-raising farms. Disease outbreaks, chemical pollution, and the environmental destruction of MARSHES and MANGROVES have been linked with fish farm activities. Environmentalists believe more sustainable practices are needed to control the potential for pollution and damage of NATURAL RESOURCES. *See also* COMMERCIAL FISHING; OVERFISHING; SUSTAINABLE DEVELOPMENT.

EnviroSources

National Oceanic and Atmospyeric Administration, NOAA Fisheries Contact Web site: www.nmfs.gov/.
U.S. Department of Agriculture Web site: www.usda.gov.
U.S. Department of Agriculture, National Agricultural Library: 10301 Baltimore Ave., Beltsville, MD 20705.

aquatic biome. *See* biome; ocean.

aquifer, A natural underground water resource that is an important source for drinking water and IRRIGATION in many parts of the world. Aquifers store large quantities of water below ground where the water is better protected from the effects of CONTAMINANTS than exist at the surface; however, aquifers require protection and careful planning to ensure that they do not become polluted by GROUNDWATER contamination as a result of inappropriate land uses or accidental chemical leaks or spills.

Aquifers may consist of soil, which is made up of separate grains, or rock containing fractures or other channels. In an aquifer, the groundwater flows through the spaces between

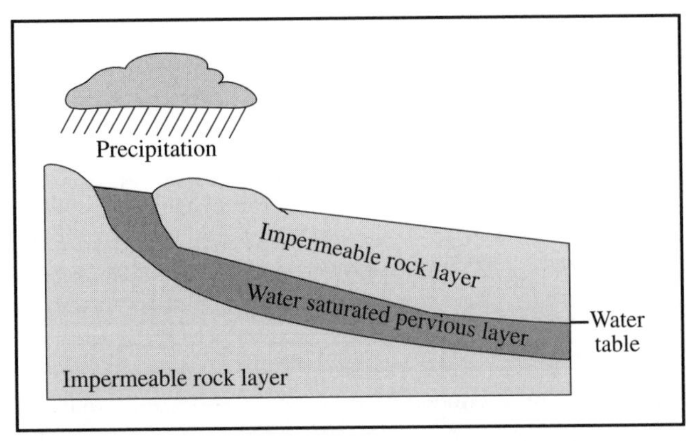

An aquifer is a natural underground water resource for drinking water and irrigation.

the soil particles or within the rock openings. Within many aquifers, groundwater flows horizontally through the natural spaces in the soil or rock until it reaches a DISCHARGE area, such as an OCEAN, lake, or stream, where the water reaches the surface of the land. Aquifers also include recharge areas where water enters the ground and travels downward into the aquifer, replenishing the supply of water. Recharge areas may include SURFACE WATER bodies or WETLANDS. (*See* water cycle.)

Generally, water is obtained from aquifers through WELLS that are drilled or dug into the ground. Groundwater is usually pumped out of an aquifer from a well, although in some areas groundwater may naturally flow to the surface through a well. The process of removing groundwater from an aquifer is sometimes called withdrawal, and is often measured in cubic meters (or gallons) per day.

Two principal types of aquifers are confined aquifers and unconfined aquifers. A confined aquifer has a layer of soil or rock above it (a "confining layer") that blocks the upward or downward flow of water. An unconfined aquifer has no upper confining layer and permits the level of the water to rise and fall without restriction.

As a groundwater resource, generally a desirable aquifer must be able to provide enough water to support the planned uses. Aquifers may range in extent from less than a square kilometer to hundreds of square kilometers. Aquifers used to supply towns or cities must be quite large and capable of providing many millions of liters of water per day. When the amount of groundwater withdrawn from an aquifer exceeds the ability of the aquifer to recharge, the available supply of water begins to diminish. *See also* OGALLALA AQUIFER; SUBSIDENCE; ZONE OF AERATION; ZONE OF DISCHARGE; ZONE OF SATURATION.

EnviroSources:

National Groundwater Association Web site: www.h2o-ngwa.org. www.h2o-ngwa.org.

U.S. Geological Survey, Water Resources of the United States Web site: www.water.usgs.gov.

Aral Sea, A salt-water lake in Central Asia that has been greatly altered as a result of diversion of the water from its feeder streams. The Aral Sea lies between the countries of Kazakstan and Uzbekistan (areas that are largely DESERT) and is fed by the Syr Darya and Amu Darya Rivers. Until the mid-1980s, the Aral Sea was the fourth-largest lake in the world and a commercially important fishing region; however, since the 1960s, massive amounts of water have been diverted from the lake's feeder streams to irrigate cotton fields in the region. This water diversion has reduced the volume of the Aral Sea by 50–60%. This dramatic change became clear in 1987 when a strip of land once covered by water emerged, dividing the Aral Sea in two.

The Aral Sea lies between Kazakhstan and Uzbekistan (countries that are largely desert) and is fed by two major rivers.

The reduction in the volume of the Aral Sea has devastated the region. Diversion of fresh water away from the Aral Sea, combined with the natural process of evaporation, has caused the lake's water to become more saline. (*See* salinization.) The fishing industry has collapsed as the carp, bass, sturgeon, and perch that once flourished in the water have disappeared along with most other lake ORGANISMS. Human respiratory DISEASES and mortality rates in nearby towns have increased as dust mixed with particles of salt and PESTICIDES that accumulated as the exposed lake bed dried out became airborne. (*See* erosion.) In addition, the accumulation of salts in soil once covered by lake water has led to increased DESERTIFICATION in the area. *See also* AIR POLLUTION; COMMERCIAL FISHING; IRRIGATION; MONO LAKE; PARTICULATE MATTER.

EnviroSource

Gore, Al. *Earth in the Balance: Ecology and the Human Spirit.* New York: Houghton Mifflin, 1992.

Arctic, A remote region around the North Pole consisting of OCEAN, ice, and TUNDRA, which is faced with a unique set of environmental issues. The Arctic comprises the northernmost portions of the Northern Hemisphere, including parts of Canada, Russia, the Scandinavian countries, and Greenland. The Arctic Circle, situated at about 62° north latitude, represents the approximate northern limit of FORESTS, where the tundra begins. The polar ice cap that surrounds the North Pole may vary in size over time and constitutes only a small portion of the entire Arctic region.

A popular image of the Arctic depicts the region as remote, barren, and environmentally clean; however, this view of the Arctic does not represent modern conditions accurately. Despite the relatively small human POPULATIONS of the Arctic (approximately 10 million inhabitants), it has experienced pollution and other impacts from human activities since the Industrial Revolution.

The Arctic has experienced FALLOUT of PARTICULATE MATTER such as dust and dirt from the ATMOSPHERE for centuries, leaving combustion products and chemical residues in the ice. The DEPOSITION of atmospheric particulates affects water, sediments, and soil, from which PLANTS and animals take up these POLLUTANTS. Researchers first observed an atmospheric feature known as "Arctic haze" in the 1970s. A haze that covers virtually the entire Arctic, this AIR POLLUTION appears to originate largely from industrial areas of Europe and Asia, where COAL- and PETROLEUM-burning plants emit large quantities of particulates and haze-producing gases.

In modern times, naval activities have resulted in the dumping of RADIOACTIVE WASTES in the Barents and Kara Seas of the Arctic. POLYCHLORINATED BIPHENYLS (PCBs), several metals, and

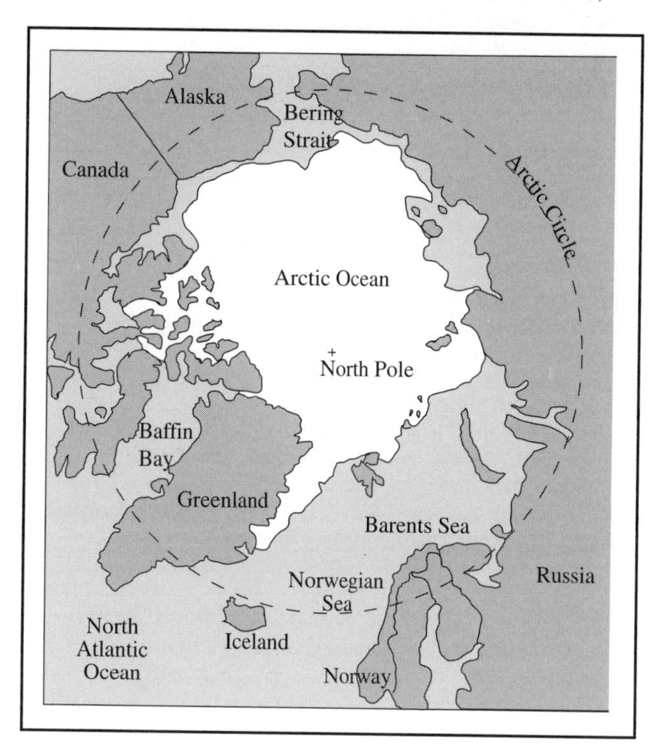

The Arctic comprises the most northernmost portions of the Northern Hemisphere, including parts of Canada, Russia, the Scandinavian countries, and Greenland.

some PESTICIDES that bioaccumulate are now widespread in Arctic WILDLIFE. The BIOACCUMULATION phenomenon has particularly affected the native human populations of the Arctic, because humans occupy a high position in the FOOD CHAIN as CONSUMERS of the fish and wildlife in which toxins accumulate. For instance, MERCURY concentrations are abnormally high in the bodies of many native Arctic people whose diets are rich in seafood. Similarly, among Arctic natives, PCBs have become common in mothers' milk and are passed on to infants through nursing.

The Arctic serves an important function as a record and as an indicator of global CLIMATE changes. Scientists have obtained ice cores from thick accumulations of Arctic ice. The layers of deposited material within the ice cores reveal a history of climate conditions. During the 1990s, climatologists observed indications in the Arctic of GLOBAL WARMING. Ice samples obtained from GLACIERS and sediment samples collected from Arctic lake bottoms have yielded evidence of a recent marked rising temperature trend in the Arctic. Although some of the temperature increase is attributable to nonhuman causes, such as SOLAR ENERGY bursts, much of the Arctic warming may be a result of the GREENHOUSE EFFECT, a process by which gases in the atmosphere trap heat.

Many organized efforts have begun to investigate and seek solutions to Arctic environmental problems. The Arctic Council formed in the late 1990s to increase awareness and understanding of Arctic issues among the affected circumpolar nations. Government representatives from Canada, Denmark, Finland, Iceland, Norway, Sweden, the former Soviet Union, and the United States held a conference on the protection of the Arctic environment. The Arctic Monitoring and Assessment Programme studies pollutants and their effects on the Arctic ENVIRONMENT. Conservationists have expressed concern over the development of petroleum fields in the ARCTIC NATIONAL WILDLIFE REFUGE, located in Alaska, because of the environmental damage that could result from oil drilling and OIL SPILLS. *See also* BERING SEA; CONSERVATION; *EXXON VALDEZ* OIL SPILL.

EnviroSources

Arctic Council Web site: www.nrc.ca/arctic/index.html.

Arctic Monitoring and Assessment Programme (Norway) Web site: www.gsf.de/UNEP/amap1.html.

Arctic National Wildlife Refuge Web site: energy.usgs.gov/factsheets/ANWR/ANWR.html.

Conefrey, Mick and Tim Jordan. Icemen: Mick Conefrey and Tim Jordan TV Books, (Companion Volume to the Documentary Series). New York: HarperCollins, 1999.

Institute of Arctic and Alpine Research Web site: instaar.colorado.edu

Institute of the North (Alaska Pacific University) Web site: www.institutenorth.org/.

Inuit Circumpolar Conference Web site: www.inac.gc.ca/decade/circum.html.

Nunavut Web site: www.acs.ucalgary.ca/~dgwhite/nu.html.

Pielou, E.C. *A Naturalist's Guide to the Arctic*. Chicago: University of Chicago Press, 1994.

Smithsonian Institution Arctic Studies Center Web site: www.mnh.si.edu/arctic.

World Conservation Monitoring Centre Arctic Programme Web site: www.wcmc.org.uk/arctic.

Arctic National Wildlife Refuge, An 8-million-hectare (19-million-acre) area in northeastern Alaska that was set aside in 1980 to protect the HABITAT of native and migratory SPECIES. The Arctic National Wildlife Refuge (ANWR) is the northernmost site administered by the U.S. NATIONAL PARK SERVICE. The area was designated a WILDLIFE REFUGE by the ALASKA NATIONAL INTERESTS LANDS CONSERVATION ACT of 1980.

Prior to 1980, less than half the area now occupied by the ANWR was set aside as a WILDLIFE range. This range, the Arctic National Wildlife Range, was established in 1960, primarily to protect the calving grounds of porcupine caribou. Today, the expanded refuge also protects the habitat of the Dall's sheep, Arctic lynx, moose, Arctic fox, and polar bears as well as many migratory birds that flock to the region during the summer months.

Despite the ANWR's protected status, attempts have been made to open some parts of the refuge to oil exploration and development. In 1991, the U.S. Senate Energy Committee proposed that the area be opened to PETROLEUM development and NATURAL GAS drilling. This proposal was defeated in the Senate. A second attempt to open the area to oil development was made several years later, but was abandoned when President Clinton declared the ANWR off-limits to oil drilling. *See also* ALASKA PIPELINE; BIOSPHERE RESERVE; OIL SPILL; TUNDRA; YELLOWSTONE NATIONAL PARK; YOSEMITE NATIONAL PARK.

EnviroSources

Arctic National Wildlife Refuge: U.S. Fish and Wildlife Service Web site: www.r7.fws.gov/nwr/arctic/twildlf.html.

Kosova, Weston. "Alaska: The Oil Pressure Rises." *Audubon* 99, no. 6 (November–December 1997): 66–74.

Army Corps of Engineers, The division of the U.S. Army responsible for providing engineering, management, and technical support to the U.S. Department of Defense, federal agencies such as the U.S. ENVIRONMENTAL PROTECTION AGENCY (EPA), and state and local governments. Although they are called engineers, members of the corps have training in many areas. Among the science and technology fields in which corps members have training are BOTANY, chemistry, computer science, GEOLOGY, ZOOLOGY, and various engineering fields.

As a division of the U.S. Army, the Army Corps of Engineers has a military function; however, the corps also has many noncombat responsibilities, such as managing the design and construction of training and housing facilities for army and air force personnel. Environmental responsibilities of the corps include the planning, design, and construction of projects affecting FLOOD control, water supply, HYDROELECTRIC POWER, WILDLIFE protection, recreation, environmental restoration, and river and harbor navigation. The corps also protects the nation's waterways and WETLANDS.

The Army Corps of Engineers often directly serves U.S. citizens. An example is their role in providing relief and recovery work to victims of NATURAL DISASTERS such as TORNADOES, HURRICANES, or floods. The corps also assists the EPA and other federal agencies in TOXIC WASTE cleanup projects, such as those addressed by the SUPERFUND. *See also* COMPRE-

HENSIVE ENVIRONMENTAL RESPONSE, COMPENSATION, AND LIABILITY ACT.

EnviroSource

U.S. Army Corps of Engineers Web site: www.usace.army.mil/whatwedo/.

arsenic, A gray, metallic element that naturally occurs in compounds with other elements. Some of these compounds are highly toxic. Because arsenic is a natural part of our ENVIRONMENT, living ORGANISMS are often exposed to some amount of it. Very low levels of it are always present in soil, water, food and AIR. Some marine species have high contents of arsenic but in forms that are not toxic. Most arsenic compounds have no smell or special taste, even when present in drinking water.

There are both inorganic and ORGANIC compounds of arsenic. Organic arsenic compounds contain CARBON; inorganic arsenic compounds are formed when arsenic combines with OXYGEN, SULFUR, or CHLORINE. Inorganic arsenic compounds are used to preserve wood and as ingredients in insecticides and herbicides.

Humans can be exposed to arsenic when it is released into the environment as a byproduct of SMELTING COPPER and LEAD ores and through the application of insecticides and herbicides. Other common sources of exposure include sawdust or smoke from wood containing arsenic preservatives and from contaminated drinking water.

High levels of inorganic arsenic can cause thickening and discoloration of the skin. Other health problems may include stomach pain, nausea, vomiting and diarrhea, and numbness in the hands and feet. Arsenic is a CARCINOGEN and can cause death in cases of extreme DOSES. As an example, in Bangladesh, arsenic levels in the drinking water were high enough to cause cancer in 1 out of 10 people ingesting the drinking water. Some people may be affected by small levels of arsenic, others may not. Young children, the elderly, people with long-term illnesses, and unborn babies may be at greatest risk, as they are often most sensitive to chemical exposures.

The ENVIRONMENTAL PROTECTION AGENCY (EPA) sets limits on the amount of arsenic that industrial sources can release. The EPA has restricted or banned many uses of arsenic in PESTICIDES and has set a limit of 0.05 parts per million (ppm) of arsenic in the drinking water supply; this limit is presently under review and may be lowered. The OCCUPATIONAL SAFETY AND HEALTH ADMINISTRATION (OSHA) established a maximum permissible exposure limit for workplace airborne inorganic arsenic of 10 micrograms per cubic meter averaged over an eight-hour day. *See also* HAZARDOUS SUSTANCES; HEAVY METAL; PARTS PER BILLION AND PARTS PER MILLION; TOXIC SUBSTANCE CONTROL ACT.

EnviroSources

Agency for Toxic Substances and Diseases, Registry Division of Toxicology: 1600 Clifton Road NE Mailstop E-29, Atlanta, GA 30333; Web site: www.atsdr1.atsdr.cdc.gov:8080/atsdrhome.html.

Michael Kamrin (Institute for Environmental Toxicology): C-231 Holden Hall, Michigan State University, East Lansing, MI 48824-1206; e-mail: Kamrin@pilot.msu.edu.

artesian well, A type of water WELL that permits GROUNDWATER to rise to the surface without mechanical pumping, as a result of the naturally occurring hydraulic pressure and force of gravity (forces called "head"), similar to a spring. Sometimes referred to as "flowing wells," artesian wells will result only in areas where the proper conditions exist for such a rise of groundwater to occur, such as in relatively low areas with areas of higher elevation nearby. Often artesian wells are found in the area of a confined AQUIFER—an aquifer bounded above and below by impermeable (resisting the passage of water) layers such as clay or semi-impermeable layers of rocks such as shale.

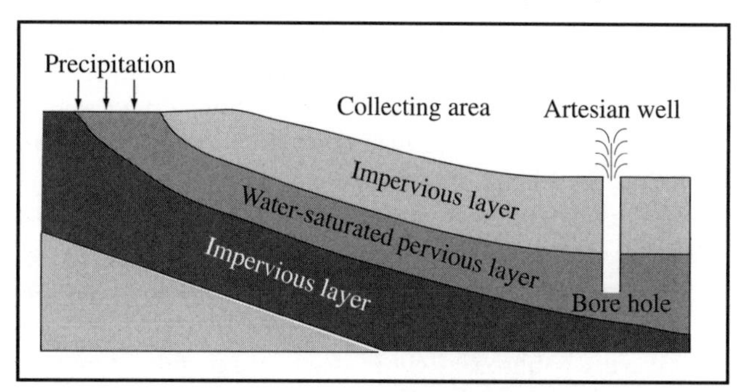

The artesian well allows groundwater to rise to the surface without mechanical pumping.

The term "artesian well" is commonly used in a nontechnical way to refer to any moderately deep water well that is drilled into soil or bedrock, rather than dug. In general, dug wells are more susceptible to pollution originating near the surface than are drilled wells, which are better sealed from infiltration of CONTAMINANTS. Also, where a confined aquifer is the water source, the confining layers of clay or rock may partially protect the groundwater from contaminants at the surface.

asbestos, A family of fibrous, silicate MINERALS that was widely used in the manufacture of a variety of products worldwide until it was discovered to be a health hazard to humans. Because of its unhealthful effects, asbestos use is now banned in the United States and many European countries. However, its use continues in many developing nations throughout the world.

Most asbestos derives from the minerals amosite, crocidolite, and chrysotile. More than 90% of the world production of asbestos has been of the chrysotile variety, or "white asbestos," the fibrous form of the mineral serpentine. Deposits of chrysotile exist in the United States in several western states, as well as the Italian Alps and Canada, but chrysotile is mined principally east of the Ural Mountains in Russia. Canada is the second leading producer of chrysotile asbestos.

Crocidolite, which is mined in South Africa, has accounted for less than 5% of world asbestos production. MINING of crocidolite in western Australia ceased in 1966. Research has indicated that the type known as "blue asbestos," which includes crocidolite, presents a much greater health hazard than does chrysotile. Amosite, also known as "brown asbestos," is a minor variety that has been mined in South Africa.

The softness, resistance to heat and flame conduction, and resistance to reaction with ACIDS led to its use in fireproof cloth for oven mitts and protective clothing for firefighters and other workers exposed to excessive heat. Asbestos was also used to make brake linings for AUTOMOBILES, roofing and siding shingles for homes, and insulating material for water pipes in homes and buildings. In the United States today, asbestos is used primarily by the National Aeronautics and Space Administration (NASA) in the manufacture of rubber liners for the solid-fuel boosters of the space shuttle, where it protects the boosters from the extreme heat of liftoff.

Asbestos is fibrous, light in mass, and readily carried in the AIR. These properties allow asbestos fibers to be easily inhaled. When inhaled, asbestos can become lodged in the tissues lining the lungs, where it causes irritation and scarring, a condition known as asbestosis. The scarring makes breathing difficult and can eventually be fatal. Inhalation of asbestos is also clearly associated with development of CANCERS of the lungs, and there is limited evidence linking it with cancers of the esophagus, and gastrointestinal system. Asbestos is also easily carried in water. This has led to restrictions on disposal of asbestos and asbestos-containing products. Several European governments have recognized the potential dangers of asbestos-containing materials (ACM) and have taken steps to protect workers from the health effects of ACM.

Based on studies linking asbestos to cancer and other DISEASES, the ENVIRONMENTAL PROTECTION AGENCY (EPA) banned the use of asbestos in insulation, fireproofing, or decorative materials in 1978. The next year, the EPA began a program to assist states in removing asbestos insulation that was flaking off pipes and ceilings in school buildings throughout the United States. This process was expanded in 1985, when the federal government passed the Asbestos Health Emergency Response Act, which required all elementary and secondary school buildings to be inspected to determine if ACM had been used in their construction. If ACM was found, the act required the materials to be contained or removed to prevent the asbestos fibers from breaking free and becoming airborne.

The main asbestos type used in U.S. building materials is chrysotile. The EPA's assessment of asbestos insulation as a serious health hazard generated controversy. Critics of the EPA maintained that chrysotile asbestos, unlike other forms of the substance, is not a health hazard in the buildings or schools.

A low-level concentration (background level) of asbestos fibers exists in the air and water virtually throughout the world, the result of natural WEATHERING of common abestos-containing rocks. This natural release of asbestos has occurred throughout Earth's history, such that a person breathes about one million asbestos fibers each year.

Asbestos was a common insulation material used in ship construction. The European Union does not condone sending vessels contaminated with asbestos to developing countries for disposal; however, some companies have sent older asbestos-contaminated ships to countries that do not have asbestos regulations—for example, some Asian countries such as India—for dismantling as scrap (shipbreaking), disregarding the potential danger to scrapyard workers. The Basel Convention on the Control of the Transboundary Movement of Hazardous Wastes and Their Disposal, whereby member nations agreed not to transport HAZARDOUS WASTES to other countries for dumping, is considered by some applicable to asbestos-containing ships. International organizations concerned with such potential exposure of workers to ACM and other hazards during shipbreaking include the International Maritime Organization, the Basel Convention, and the International Labor Organization.

EnviroSources

Benarde, Melvin A. *Asbestos: The Hazardous Fiber.* Clevelend, OH: CRC Press, 1990.

Castleman, Barry I., and Stephen L. Berger. *Asbestos: Medical and Legal Aspects.* 4th ed. Wilmington, NC: Aspen, 1996.

Interagency Panel on Ship Scrapping (U.S. Department of Defense) Web site: www.denix.osd.mil.

Saign, Geoffrey C. *Green Essentials: What You Need to Know about the Environment.* San Francisco, CA: Mercury House, 1994.

U.S. Environmental Protection Agency, Small Business Ombudsman Hotline: 401 M St., SW, Mail Code 1230-C, Washington, DC 20460; (800) 368-5888.

U.S. Environmental Protection Agency, Toxic Substances Control Act Assistance, Information Service TSCA Hotline: 401 M St., SW, Mail Code 1230-C, Washington, DC 20460. (202) 554-1404.

asbestosis. *See* asbestos.

ash, A noncombustible PARTICULATE MATTER formed during high-temperature combustion, which becomes a POLLUTANT when expelled into the AIR as fly ash in FLUE GAS or when it drops to the bottom of the combustion unit as bottom ash. Depending upon its source, fly ash may consist of nontoxic or TOXIC CHEMICAL particles, as well as dust and soot. Some electrical power plants that burn COAL produce a nontoxic fly ash that is used to make a building material similar to cement. Many municipal waste and industrial incinerators produce toxic ash that must be disposed of as HAZARDOUS WASTE in appropriate waste disposal facilities. *See also* AIR POLLUTION; INCINERATION; RESOURCE CONSERVATION AND RECOVERY ACT.

EnviroSources

U.S. Environmental Protection Agency Web site: www.epa.gov.

assimilative capacity, The ability of a natural body of water to receive and DISCHARGE WASTEWATERS or TOXIC CHEMICALS without adverse effects and without damage to ORGANISMS or humans who consume the water.

EnviroTerm

assimilation: The ability of a body of water to purify itself of POLLUTANTS.

Aswan High Dam, A huge rockfill DAM on the Nile River in southern Egypt that is undergoing ecological consequences that include EROSION along the river bed. Built in 1970, the Aswan High Dam is located about 7 kilometers (4 miles) south of a smaller Dam which was built in 1902, but did not have an adequate RESERVOIR volume.

The reservoir formed by the Aswan High Dam is called Lake Nasser after the former president of Egypt, Gamal Abdel Nasser, who died in 1970. The lake is over 480 kilometers (330 miles) long and 16 kilometers (10 miles) wide and stores sufficient water to irrigate 2.8 million hectares (7 million acres). The High Dam, which hosts 12 TURBINES, provides about half of Egypt's ENERGY supply. The dam was built to regulate the yearly FLOODS of the Nile and to create a reservoir to prevent FAMINE during severe DROUGHTS. During the late 1980s, Egypt was unaffected by the drought that hit much of the rest of Africa. The High Dam also prevented several floods in the 1990s.

The cost to build the dam was $1 billion. One-third of the money was given by the former Soviet Union as a gift, a few years after a war between Egypt and the countries of Israel, France, and Great Britain. The United States initially offered to provide funds, but later withdrew the offer.

The creation of Lake Nasser submerged a large part of the Lower Nubia destroying monuments and archaeological sites along the Nile River. Over 90,000 people had to be relocated. The impact on the region's ECOLOGY is still being evaluated. Egyptian farmers now use about 0.9 million metric tons (1 million tons) of artificial FERTILIZER as a substitute for the nutrient-filled sediment that was previously used to enrich their land, creating the potential for RUNOFF pollution by AGROCHEMICALS. There has already been erosion of the Nile Delta and the shrimp catch has decreased in the MEDITERRANEAN SEA as a result of the restriction in the flow of the river. *See also* IRRIGATION; SEDIMENTATION.

EnviroSource

North Atlantic Treaty Organization Country Database Web site: www.nato.int/ccms/general/countrydb/egypt.html.

atmosphere, The protective layer of gases that surrounds Earth and plays a key role in Earth's ECOSYSTEM and BIOGEOCHEMICAL CYCLES. The atmosphere contains the gases NITROGEN (78.1%), OXYGEN (20.1%), and other trace gases, including water vapor, argon, CARBON DIOXIDE, neon, helium, METHANE, HYDROGEN, ammonia, CARBON MONOXIDE, and OZONE. The atmosphere also contains PARTICULATE MATTER and AEROSOLS.

Earth's atmosphere is divided into several layers; however, most environmentalists are concerned mainly with two layers—the troposphere and the stratosphere. The troposphere is the region of the atmosphere closest to Earth. It includes

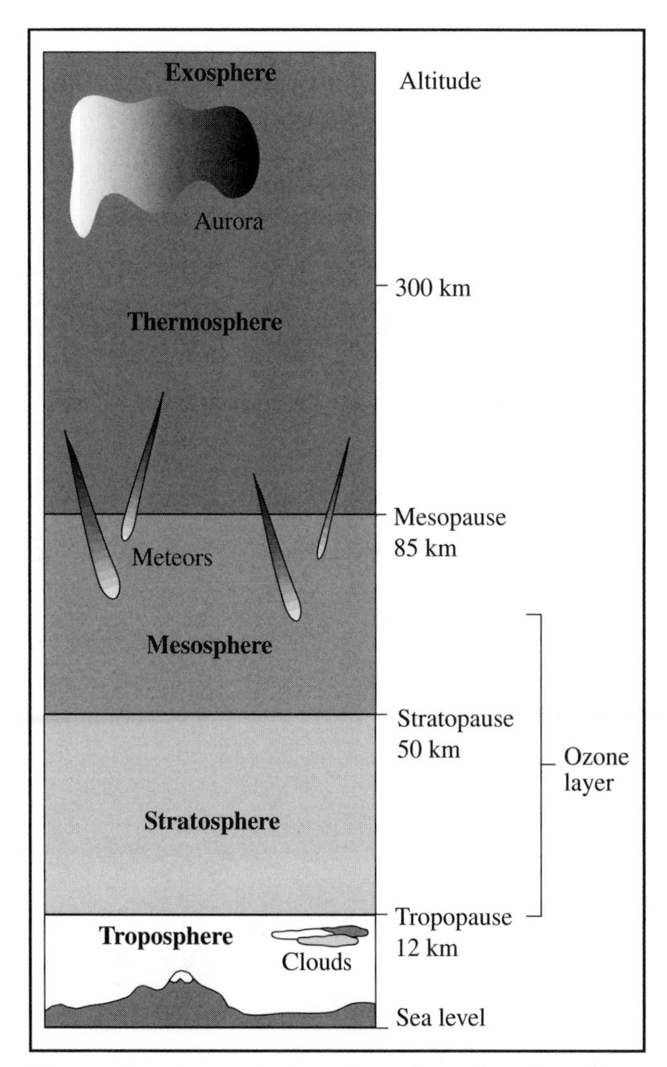

The atmosphere, the protective layer of gases that surrounds Earth, plays a key role in Earth's ecosystem and biogeochemical cycles.

water vapor and clouds and is a source of Earth's WEATHER and CLIMATE. It is also the place where most of Earth's AIR POLLUTION is found.

The troposphere extends from Earth's surface to an ALTITUDE of about 12 kilometers (7 miles), although this height varies with latitude. Temperatures decrease with altitude in the troposphere. As warm AIR rises, it cools, falling back to Earth, a process known as convection. Huge air movements mix the gases in the troposphere very efficiently. Chemicals, particulate matter, and other substances carried in the gases of the troposphere can be washed back to Earth by precipitation. (*See* water cycle.) An area called the tropopause separates the troposphere and the stratosphere.

Above the tropopause is the stratosphere. The stratosphere is a region that extends from 12 kilometers (7 miles) to about 50 kilometers (30 miles) above Earth. Commercial airlines fly in the lower stratosphere where there are strong steady winds and little water vapor. Warm air remains in the upper stratosphere, and cool air remains lower. There is much less mixing of gases in this region than in the troposphere and convection does not occur.

Nearly 90% of the Earth's ozone is in the stratosphere. CHLOROFLUROCARBONS (CFCs) and other compounds, however, rise into the stratosphere where they can deplete the ozone layer. The ozone forms a shield that absorbs ultraviolet radiation from the sun. As a result, the temperatures in the stratosphere increase with altitude. Unlike the tropsophere, most substances, such as CFCs and other particulates, that enter the stratosphere remain in this layer.

The outer layers of the atmosphere include the mesosphere, thermosphere, and the exosphere. The stratopause separates the stratosphere and the mesosphere. The mesosphere begins at about 50 kilometers (30 miles) above Earth and extends to 85 kilometers (51 miles). In this layer there is a drop in temperature, little ozone, and very little oxygen and nitrogen. Most meteors are incinerated in this layer. Above the mesophere is the thermosphere that begins at 85 kilometers (51 miles) to an altitude of about 700 kilometers (435 miles) above Earth. Temperatures become warmer in this layer. Auroras occur in the thermosphere. Beyond the thermosphere is the exosphere about 700 kilometers (435 miles) above sea level. *See also* ACID RAIN; CARBON CYCLE; GREENHOUSE GASES; NITROGEN CYCLE; OZONE DEPLETION.

EnviroTerms

air pressure: The pressure exerted by the combined forces of the ATMOSPHERE and gravity. Air pressure equals 1013.2 millibars (14.7 pounds per square) inch at sea level on Earth. Air pressure decreases with a rise in ALTITUDE. As an example, AIR pressure in the stratosphere can range between about 250 millibars at its lowest layer to about 2 millibars near the beginning of the mesosphere.

air mass: A large volume of moving AIR that is either a high- or low-pressure system.

atoll. *See* coral reef.

atomic energy. *See* nuclear energy.

Atomic Energy Act, Legislation passed in 1946 that placed the responsibility for producing atomic or NUCLEAR ENERGY within one government agency in the United States, the Atomic Energy Commission (AEC). The act was revised in 1954 to allow the AEC to license private companies to use radioactive materials to build NUCLEAR REACTORS. The AEC was abolished in 1974, and a new agency, the NUCLEAR REGULATORY COMMISSION (NRC), was created.

Industrialized nations that have nuclear power plants have developed environmental and health laws to govern the management of nuclear FUEL and byproducts. Other governmental nuclear energy regulatory authorities include the United Kingdom Atomic Energy Authority, the Russian Ministry of Atomic Energy, the Swedish Nuclear Power Inspectorate and the Radiation Protection Institute, and the Australian Nuclear Science and Technology Organization.

Despite the existence of such laws and authorities, the standards and practices observed in nuclear materials management are not consistently protective of the ENVIRONMENT throughout the world. Grave concerns exist about the questionable methods of nuclear waste disposal used in the former Soviet Union. Through the International Atomic Energy Agency, member nations have established agreements on the use, transport, and management of radioactive materials used in the nuclear power industry. *See also* NUCLEAR WASTE POLICY ACT; RADIATION; RADIOACTIVE WASTE; URANIUM.

Atomic Energy Commission. *See* Atomic Energy Act; Nuclear Regulatory Commission.

attainment area, Any large area identified by the U.S. ENVIRONMENTAL PROTECTION AGENCY (EPA) that has an ambient AIR quality that is as good or better than the NATIONAL AMBIENT AIR QUALITY STANDARDS prescribed in the CLEAN AIR ACT (CAA). The area can be in attainment for one or more of six POLLUTANTS that include PARTICULATE MATTER, SULFUR DIOXIDE, CARBON MONOXIDE, photochemical oxidants, NITROGEN DIOXIDE, and HYDROCARBONS. An area may be identified as an attainment area for one pollutant and a NONATTAINMENT AREA for others. A nonattainment designation is assigned with respect to one or more particular pollutants. *See also* AIR POLLUTION; CRITERIA POLLUTANTS.

EnviroSource

U.S. Environmental Protection Agency Web site: www.epa.gov.

Audubon, John James (1785–1851), A Haitian-born, French-American ornithologist who was noted for his sketches and paintings of birds and was one of the United States' first NATURALISTS. Audubon's detailed compositions depicted scenes he had witnessed in nature, although he has been criticized during the twentieth century for having killed animals to use as models.

After being educated in France, Audubon moved to his family's plantation outside Philadelphia. Here, he experimented with bird banding and studied avian migration. In search of new birds to study, Audubon explored the Ohio and Mississippi Rivers and the Great Lakes in the 1820s. The resultant paintings—more than 400—were compiled and published in Edinburgh, Scotland, as *Birds of America* between 1827 and 1838. Scottish naturalist William MacGillivray collaborated with Audubon on the project, and supplied most of the scientific data for the text, *Ornithological Biography*, which appeared in five volumes between 1831 and 1839. Audubon never accepted offers for the individual prints that comprise *Birds of America*; he insisted that the book be sold intact, which is one reason that his prints are so rare today. Audubon also interspersed his bird studies with writings on American life during the nation's turbulent beginnings.

In 1843, Audubon traveled from St. Louis up the Missouri River to Fort Union, and then overland to the Yellowstone River. The purpose of the trip, his last great adventure before his death in 1851, was to gather specimens for a new series of paintings of American MAMMALS. Although he never achieved his dream of reaching the West Coast, he returned to St. Louis attired in Native American hunting dress and accompanied by live animals to serve as models for his paintings. With the help of his sons, he completed *Quadrupeds of North America* in 1845. Throughout his life, Audubon remained extremely interested in observing animals in their natural ENVIRONMENT. The NATIONAL AUDUBON SOCIETY has been created in his name. The organization's main goal is the CONSERVATION of WILDLIFE. *See also* LEOPOLD, ALDO; MUIR, JOHN; SIERRA CLUB.

EnviroSource

National Audubon Society Web site: www.audubon.org.

automobile, The most popular motor vehicle used for transportation in many parts of the world and which is a major contributor to AIR POLLUTION such as photochemical SMOG.

Researchers predict that worldwide motor vehicle ownership will increase from 600 million cars in 1999 to about 800 million by the year 2010. The annual production of automobiles in 1998 reached about 39 million. Europe manufactures about 35% of all automobiles in the world, followed by Asian countries (30%) and the United States (22%). The rest are produced mostly in Latin American countries, particularly Brazil. The United States and other developed countries purchase about 75% of all new cars. Industry experts believe that the biggest growth of automobile sales in the next decade will be in China and India.

Automobiles offer consumers many advantages such as fast transportation and convenience. In addition, the production, marketing, and sales of automobiles play an important role in the global economy; however, the disadvantages include adverse effects on the ENVIRONMENT and human health due to air pollution from automobile EMISSIONS. The primary POLLUTANTS include CARBON DIOXIDE (CO_2) CARBON MONOXIDE (CO), NITROGEN OXIDES (NO_x), HYDROCARBONS, VOLATILE ORGANIC COMPOUNDS (VOCs), SULFUR DIOXIDE (SO_2), and PARTICULATE MATTER. Carbon monoxide affects the human body's ability to absorb OXYGEN. Emissions of VOCs and nitrogen oxides into the ATMOSPHERE are the chief source of smog conditions in many major cities.

According to the INTERNATIONAL COUNCIL FOR LOCAL ENVIRONMENTAL INITIATIVES (ICLEI), automobiles emit far more carbon dioxide and other pollutants per kilometer traveled than other alternative forms of transportation. In most cities, the single largest source of carbon dioxide emissions is the transportation sector. Carbon dioxide is considered the most significant GREENHOUSE GAS and is the leading contributor to GLOBAL WARMING.

With so many automobiles and other motor vehicles on the road there are other problems as well. Many countries are experiencing congested roads and highways, high NOISE POLLUTION levels, and frequent vehicle accidents. And governments must expend funds to keep cars on the road: major government expenditures are needed to build roads, bridges, and parking areas. Large quantities of PETROLEUM resources are also used to manufacture automobiles and automobile FUEL. In all, automobiles and other motor vehicles consume about 50% of the world's production of petroleum, a NONRENEWABLE RESOURCE.

Many countries are trying to find ways to reduce automobile use. The ICLEI states that public transit in the form of buses, trains, streetcars, or subways is a far more efficient means of transportation than the single occupant vehicle. Public transit uses less ENERGY and emits less greenhouse gases and other pollutants per kilometer traveled. Other ideas in reducing car use include mandatory car pooling, developing emission-free vehicles, and restricting the use of cars on certain days of the week. *See also* AIR POLLUTION CONTROL ACT; ALTERNATIVE FUEL; CATALYTIC CONVERTER; CLEAN AIR ACT; ELECTRIC VEHICLES; FOSSIL FUEL; FUEL CELL; GASOHOL; GREENHOUSE EFFECT.

EnviroSources

Cars and Their Enviromental Impact Volvo Web site: www.environment.volvocars.com/ch1-1.htm.

DeCicco, John, and Martin Thomas. *The Green Guide to Cars and Trucks.* Washington, DC: American Council for an Energy-Efficient Economy, 1999.

Kennedy, Donald, and Richard Bates, eds. *Air Pollution, the Automobile, and Public Health.* Washington, DC: National Academy Press, 1989.

National Center for Vehicle Emissions Control and Safety Web site: www.colostate.edu/Depts/NCVECS/ncvecs1.html.

U.S. Environmental Protection Agency, Air Toxics from Motor Vehicles (fact sheet) Web site: www.epa.gov/oms/02-toxic.htm.

U.S. Environmental Protection Agency, National Vehicle and Fuel Emissions Laboratory: 2000 Traverwood Drive, Ann Arbor, MI 48105; (734) 214-4333.

U.S. Enviromental Protection Agency, Office of Mobile Sources Web site: www.epa.gov/oms.

autotroph, An ORGANISM that is able to synthesize nutrients from inorganic substances such as CARBON DIOXIDE and water, or METHANE, ammonia, and HYDROGEN sulfide, obtained from the ENVIRONMENT. There are two types of autotrophs: the photoautotrophs, which synthesize nutrients through PHOTOSYNTHESIS, and the CHEMOAUTOTROPHS, which synthesize nutrients through CHEMOSYNTHESIS.

Photoautotrophs include all PLANTS, blue-green BACTERIA (CYANOBACTERIA), and PROTISTS classified as euglenoids and ALGAE. All photoautotrophs use ENERGY from sunlight to drive their food-making process, which is called photosynthesis. In this process, the light energy is used to combine water and carbon dioxide to make glucose. OXYGEN and water vapor are released to the environment as waste products. In contrast, chemoautotrophs are SPECIES of bacteria that use chemical energy obtained from substances in the environment to synthesize their nutrients. For example, chemosynthetic bacteria that live near HYDROTHERMAL VENTS on the OCEAN floor synthesize their nutrients from hydrogen sulfide that is present in the heated water released from the vent. These organisms form the base of the FOOD CHAIN for their ECOSYSTEM which includes a variety of unusual organisms such as tube worms and white crabs.

Autotrophs serve a vital role in ecosystems because they provide food either directly or indirectly for all organisms that share their environments. For example, many nonautotrophic organisms, called HETEROTROPHS, obtain their nutrients by feeding directly on autotrophs. In turn, these heterotrophs may serve as food and provide nutrients to other organisms. (*See* predation.) Because they produce nutrients to support themselves and other POPULATIONS in the environment, autotrophs are PRODUCERS and always form the base of the food chain. *See also* CONSUMER; FOOD WEB; TROPHIC LEVEL.

B

bacteria, Single-celled, microscopic ORGANISMS that fill a variety of environmental NICHES and sometimes serve as agents of DISEASE. Bacteria are among the most abundant organisms on Earth and are classified in a variety of ways. One way to name and classify bacteria is according to their shapes—round, rod, or spiral. Round bacteria are cocci; rod-shaped bacteria are bacilli; bacteria with a spiral shape are spirilla. Bacteria also may be classified according to likely evolutionary relationships or according to their roles in the ENVIRONMENT—the aspect of bacteria that will be examined here.

A group of bacteria called CYANOBACTERIA, or blue-green bacteria, are environmentally important as PRODUCERS. The bacteria in this group are generally round or spiral in shape. Like PLANTS, cyanobacteria contain chlorophyll and are able to synthesize nutrients through PHOTOSYNTHESIS. In this process, cyanobacteria take in CARBON DIOXIDE (CO_2) from the environment and return OXYGEN (O_2), a waste product of photosynthesis. As producers, cyanobacteria are at the base of the FOOD CHAIN and provide food to many CONSUMERS. Cyanobacteria also are environmentally important for their role in changing NITROGEN gas (present in AIR) into nitrogen compounds that can be used by other organisms. *(See* nitrogen cycle.) The cyanobacteria that carry out this role live inside small nodules on the roots of plants called LEGUMES.

Another group of bacteria synthesizes nutrients via CHEMOSYNTHESIS. Chemosynthetic bacteria lack chlorophyll and do not require sunlight to drive their food-making process. Instead, these bacteria make food by oxidizing inorganic chemicals obtained from the environment. Usually, the chemicals used are SULFUR, ammonia, iron, or nitrites. *(See* nitrate.) Chemosynthetic bacteria have been discovered living in the deep OCEAN, near thermal vents in the ocean floor. *(See* hydrothermal ocean vent.) The vents release the chemicals these bacteria use for chemosynthesis. In turn, the bacteria serve as the producers for the ocean vent ECOSYSTEM.

Another environmentally important bacterial group are the DECOMPOSERS—organisms that obtain nutrients by feeding on the wastes or remains of other organisms. As they feed, decomposers break down the complex ORGANIC matter contained in wastes and remains into simpler substances. Some of these substances are used as food by the bacteria; the rest are returned to the environment, usually the soil, where they can be used by plants and soil organisms.

Bacteria that serve as decomposers are sometimes called "bacteria of decay." The feeding process of these organisms serves two important functions: First, by breaking down organic matter, these bacteria prevent wastes from building up on Earth's surface. Second, bacteria of decay act as natural recyclers by returning important nutrients to the environment. *(See* recycling.)

Many bacteria are important because of the symbiotic relationships they have with other organisms. *(See* symbiosis.) *ESCHERICHIA COLI (E. coli)*, for example, normally live in the intestinal tracts of humans. The intestinal tract provides *E. coli* with a warm, safe environment and a constant food source. In return, the bacteria aid in the digestive process and help make vitamins needed by the human body. Bacteria that live in the intestines of cows, sheep, and horses have a similar relationship with their hosts. In these animals, the bacteria help the animals break down the cellulose contained in the plant matter on which the animals feed. Symbiotic relationships between bacteria and other organisms are often site specific. For example, *E. coli* outside of the human digestive tract causes disease in humans. Symbiotic associations between bacteria and other organisms may also be concentration specific. In such cases, overgrowth of usually benign bacteria can cause disease.

Not all relationships between bacteria and other organisms are beneficial. Many bacteria are PATHOGENS to other organisms. As pathogens, the bacteria cause disease in the organisms they infect. Human diseases caused by bacteria include strep throat, Lyme disease, diphtheria, *SALMONELLA* food poisoning, Legionnaire's disease, and tuberculosis. Bacteria also cause many diseases in plants and in animals other than humans. For example, anthrax and BRUCELLOSIS are diseases that affect cattle, sheep, and sometimes humans and are caused by bacteria. *See also* BIOLOGICAL WEAPON; BIOTECHNOLOGY; DECOMPOSITION; NITROGEN FIXATION; SICK BUILDING SYNDROME.

EnviroSource

Margulis, Lynn, and Karlene V. Schwartz. *The Five Kingdoms: An Illustrated Guide to the Phyla of Life on Earth.* 2d ed. New York: W. H. Freeman and Co., 1996.

Baikal, Lake, The world's oldest and deepest lake that is threatened by a variety of pollution problems. Lake Baikal contains about 20% of Earth's liquid fresh water and almost 80% of the fresh water within Russia. Located in south-central SIBERIA near the Mongolian border, Lake Baikal measures almost 640 kilometers (400 miles) long, averages 85 kilometers (50 miles) wide, and reaches depths as great as 1,740 meters (5,712 feet). It has been estimated that almost 2,000 PLANT and animal SPECIES live in or around the lake, many of which are unusual ORGANISMS supported by nutrients derived from a hydrothermal vent at the lake's bottom. *(See* chemoautotroph; chemosynthesis.)

Lake Baikal is supplied with fresh water by hundreds of rivers and streams, such as the Serenga River—one of the major tributaries; however, the streams and rivers contain discharged POLLUTANTS from paper mills, MINING sites, ship-building sites, and SEWER systems. *(See* discharge.) The lake also suffers from RUNOFF of PESTICIDES and FERTILIZERS from farms. The pollutants have damaged many HABITATS in and around the lake, resulting in a decrease in many WILDLIFE POPULATIONS. COMMERCIAL FISHING has also declined. In addition, changes in water level have led to shore EROSION and degradation in nearby WETLANDS.

In 1988 and again from 1990 through 1992, environmentalist DAVID ROSS BROWER led delegations to Lake Baikal at the request of the Soviet government. The purpose of the trips was to develop plans to aid in the protection and restoration of the area. In 1994, the Russian government called for preventive measures to control the destruction of the lake. How and when the measures will be implemented remains unclear. *See also* AGRICULTURAL POLLUTION; ARAL SEA; RHINE RIVER.

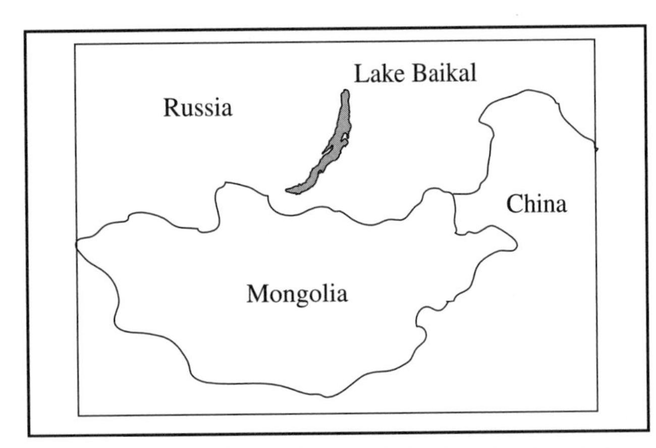

Lake Baikal contains about 20% of Earth's liquid fresh water and almost 80% of the fresh water within Russia.

EnviroSources

Baikal Watch Web site: www.earthisland.org/baikal/bcci.html

Earth Island Institute: 300 Broadway, Suite 28, San Francisco, CA 94133; (415) 788-3666; e-mail baikalwatch@earthisland.org.; Web site: www.earthisland.org/baikal/baikal.html

bald eagles, A SPECIES of fish eagle (*Haliaeetus leucocephalus*) restricted to North America that is native to every U.S. state except Hawaii. Once endangered in all of the lower 48 states, the bald eagle's status was upgraded to threatened in 1994, two decades after the United States banned the PESTICIDE DICHLORODIPHENYLTRICHOROETHANE (DDT) and enacted laws to protect both eagles and their HABITAT. About half the current U.S. POPULATION of 70,000 bald eagles live in Alaska. British Columbia hosts an additional 20,000, making northwestern North America the species' greatest stronghold. They thrive there in part because of the region's large SALMON population.

The bald eagle was removed from the endangered species list of the U.S. Fish and Wildlife Service in 1999.

Immature bald eagles are dark brown; the head and tail feathers turn predominantly white when the birds reach four to five years of age. Bald eagles live up to 40 years in the wild and are monogamous (remaining faithful to their mates until death). Each spring, females lay one to three eggs, incubating them for approximately 35 days. The birds nest in large trees near rivers or coasts, and often use the same nest year after year. Some nests eventually become quite large, reaching more than 3 meters (9 feet) in diameter with a mass of 1,814 kilograms (4,000 pounds); a typical nest is about 1.5 meters (5 feet) in diameter. Some bald eagles migrate to follow seasonal food supplies, exhibiting some of the most complex migration patterns among North American birds.

When European settlers first arrived in North America, bald eagles nested along the Atlantic Coast from Labrador to the tip of Florida and along the Pacific Coast from Baja California to Alaska. They inhabited the lands bordering large rivers, major lakes, and lake clusters in the continental interior. Some scientists estimate their population numbered around half a million individuals. As recently as the late nineteenth century, for example, it was estimated that there was a bald eagle nest for every 1.6 kilometers (1 mile) of CHESAPEAKE BAY shoreline. In the mid-twentieth century, bald eagle populations began to decline, largely as a result of the widespread use of DDT, which the eagles ingested in their prey, causing them to produce eggs with very thin shells. *(See* predation.) In 1972, the United States banned DDT. Although many other countries followed suit, but DDT is still used in some devel-

Bald Eagles At a Glance	
Scientific Name	*Haliaeetus leucocephalus*
Status	Threatened 1994 to 1999; endangered prior to 1994
Population Size	94,500 in the wild (most in northwestern United States and Canada)
Wingspan	2.4 m (8 ft.)
Mass	6.8 kg (15 lbs.)
Lifespan	40 years in wild; more in captivity.

oping nations to control DISEASE-bearing INSECTS. Additional chemical threats also have been identified: POLYCHLORINATED BIPHENYLS (PCBs), HEPTACHLOR, and dieldrin, have all been found in bald eagle eggshells in the Great Lakes region. Other environmental poisons, such as MERCURY, also contributed to bald eagle mortality, as has LEAD, which eagles ingested in the form of lead shot in prey birds. The United States banned the use of lead shot for waterfowl hunting in 1991. In 1995, Canada announced a similar ban for national WILDLIFE areas; this ban was extended to cover the entire nation in 1997.

As part of an extensive bald eagle recovery and reintroduction program, the U.S. FISH AND WILDLIFE SERVICE (FWS) assembled a large colony of bald eagles at its Patuxent Wildlife Research Center in Maryland in an effort to return healthy eagles to the wild. They removed each bald eagle pair's first clutch of eggs and incubated them artificially. The bald eagles would then usually lay a second clutch, which the birds were allowed to incubate themselves. In all, 124 bald eagles were hatched at Patuxent. The program came to an end in 1988, after bald eagles had again begun to reproduce successfully in the wild. With these and other recovery methods, as well as habitat improvement and the banning of DDT, bald eagle populations have steadily increased. The number of nesting pairs in the lower 48 states has increased from less than 450 in the early 1960s to more than 4,500 in the 1990s. *See also* ALDRIN; BIOMAGNIFICATION; CAPTIVE PROPAGATION; ENDANGERED SPECIES; ENDANGERED SPECIES ACT; ENDANGERED SPECIES LIST; FOOD CHAIN; MIGRATORY BIRD TREATY ACT; THREATENED SPECIES.

EnviroSources

American Bald Eagle Information Web site: www.baldeagleinfo.com.
Chadwick, Douglas H., and Joel Sartore. *The Company We Keep: America's Endangered Species.* Washington, DC: National Geographic Society, 1995.
U.S. Fish and Wildlife Service Web site: www.fws.gov.

Baltic Sea, A sea located in northern Europe that is suffering from a wide range of environmental problems. The Baltic Sea is surrounded by several countries that include Denmark, Sweden, Finland, Germany, and Poland. Untreated commercial and municipal wastes as well as agricultural RUNOFF washed into numerous rivers have been discharged into the Baltic Sea.

A group called the Baltic Marine Environment Protection Commission was formed in 1974 and is governed by the Helsinki Commission, which was established to address environmental issues in the Baltic. This international effort seeks to protect the marine ENVIRONMENT from pollution derived from land, vessels, and airborne sources. Water quality issues in the Baltic include EUTROPHICATION, agricultural runoff, OCEAN DUMPING, and industrial and municipal discharges. High levels of CHLORINATED HYDROCARBONS in the water have resulted in declines in marine MAMMALS such as SEALS and fish such as SALMON, cod, and herring. Some fish are so contaminated with MERCURY that they are no longer sold in markets. Pulp and paper mills DISCHARGE thousands of metric tons of chlorinated compounds, used in the process of making paper, into the Baltic Sea. Other problems include high levels of HEAVY METALS and OVERFISHING.The Helsinki Commission is preparing the fishing industry for the possibility of encountering chemical WEAPONS dumped into the Baltic. The commission has explained to the fishing industry how to deal with chemical containers that may be brought to the surface accidentally during fishing.

Some progress has been made in cleaning POLLUTANTS from the Baltic Sea. New SEWAGE TREATMENT PLANTS have been built, and restrictions have been placed on paper mills that use CHLORINE products in manufacturing paper. However, environmentalists believe that more needs to be done. *See also* ADRIATIC SEA; AGRICULTURAL POLLUTION; NORTH SEA; RHINE RIVER.

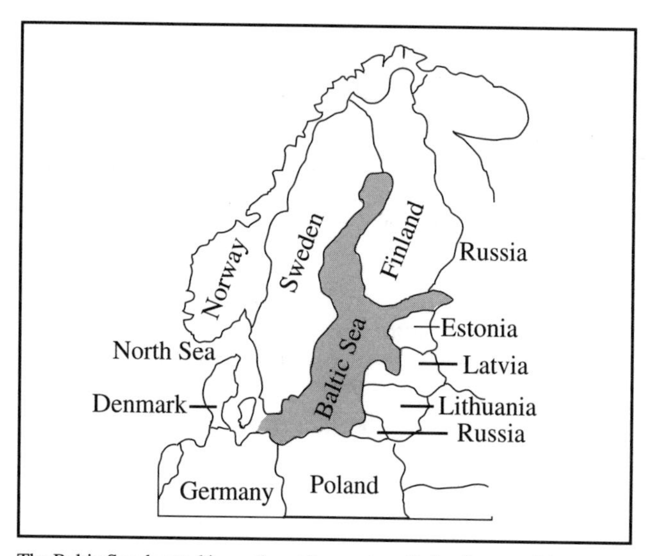

The Baltic Sea, located in northern Europe, is suffering from a wide range of environmental problems.

barrel. *See* petroleum.

basalt. *See* rock cycle.

base. *See* alkali.

bats, Animals, many of which are threatened or endangered, that make up the second largest order of MAMMALS. All bats are classified in the order Chiroptera, from the Greek words meaning "hand" and "flight," which refers to the ability of bats to fly using fleshy wing-like forelimbs. Bats are the only mammals capable of sustained flight.

Bats live in almost all regions of the world, except the true arctic. The illustration shows a fishing bat.

There are two major suborders of bats: megabats and microbats. Of the 850 to 900 bat SPECIES known, about 150 are megabats. Megabats are generally, though not always, larger in size than microbats. The two bat groups also differ in skeletal structure and tooth form.

Bats live in most regions of the world, except the ARCTIC. Most species live in areas with tropical or subtropical CLIMATES. Some microbats make their homes in temperate regions, and in the summer, a few species of microbats migrate to sub-Arctic regions.

Most bats are nocturnal, or active at night. Nocturnal bats navigate using echolocation, a process in which a bat gives off a high-pitched sound through its mouth or nostrils and then uses the resulting echo to sense the position, distance, and other traits of objects.

Bats have a variety of roles in the ENVIRONMENT. For example, many microbats feed on INSECTS, helping to keep insect POPULATIONS in check. Many tropical megabats are fruit eaters. Some of these bats decimate large quantities of fruits, making the fruit unfit for harvest and use by humans; however, some fruit-eating bats are vital to ECOSYSTEM health because of their roles in pollination or seed dispersal.

Endangered Bats of the World At a Glance

Name	Where Found
Bumblebee bat	Thailand
Sanborn's lesser long-nosed bat	Mexico/Central America/ United States and possessions
Mexican long-nosed bat	Mexico/Central America/ United States and possessions
Rodriguez flying fox fruit bat	Rodriguez Island
Singapore roundleaf horseshoe bat	Malaysia
Gray bat; Hawaiian hoary bat; Indiana bat; little Mariana fruit bat; Mariana fruit bat; Virginia big-eared bat; Ozark big-eared bat	United States and possessions

Members of three bat groups are CARNIVORES and are often called "vampire bats" because they feed on the blood of endothermic animals such as livestock, birds, and occasionally humans. Bats that feed in this way do not generally kill their food source; however, they can transmit DISEASES, such as rabies, which can be fatal to the host animal or to species to which the host transmits the disease. A few bat species are OMNIVORES that feed on small insects, birds, mice, and AMPHIBIANS, in addition to fruit.

Of the animals included on the ENDANGERED SPECIES LIST maintained by the U.S. FISH AND WILDLIFE SERVICE bats are among Earth's most threatened mammals. As of April 1997, 13 bat species were designated ENDANGERED SPECIES. Nine of these species are native to the United States or its possessions. *(See* **At a Glance** for global list.)

The main threats to bat populations are exposure to chemicals in the environment and HABITAT loss. Like the BALD EAGLE and the PEREGRINE FALCON, many bats have been threatened by DICHLORODIPHENYLTRICHOETHANE (DDT), an insecticide that was widely used in the United States throughout the 1950s and 1960s. The adverse effects of DDT on bald eagles and peregrine falcons resulted mostly in problems with eggshell development following BIOACCUMULATION of the PESTICIDE in the FOOD CHAIN; however, many bats have a CHRONIC TOXICITY to DDT, and were directly poisoned by the pesticide when it was ingested through a food source. The ban on DDT use in the United States in 1972 has eliminated the threat of DDT poisoning for most bat species; however, these same species often have low tolerances to other AGROCHEMICALS that are still used and are persistent in the environment.

For many endangered bat species, loss of habitat is the main threat to their existence. To help two of these species, three national WILDLIFE REFUGES have been established in the United States solely for the purpose of preserving bat habitat. Two of these refuges, the Blowing Wind Cave and the Fern Cave, are located in Alabama. These refuges protect the Indiana bat and Gray bat. The Indiana bat receives additional protection in Missouri through the efforts of the Pilot Knob National Wildlife Reserve. *See also* ENDANGERED SPECIES ACT; HERBIVORE; INSECTIVORE.

EnviroSources

Bat Conservation International: P.O. Box 162603, Austin, TX 78716-2603.

Tuttle, Merlin D. *America's Neighborhood Bats.* Austin, TX: University of Texas Press, 1988.

battery, A portable unit that contains two or more chemical cells that store and produce direct electric current. The most familiar are the small household batteries, used in a variety of tools, games, and appliances are alkaline batteries. *(See* alkali.)

There is now much interest in developing other types of batteries for ELECTRIC VEHICLES (EVs). The two kinds of batteries now used in EVs are lead-acid batteries and nickel-metal hydride batteries. (*See* acid.) Lead-acid batteries are similar to batteries used in GASOLINE-powered motor vehicles for ignition and lighting. Since they were invented in the mid-1800s, lead-acid batteries have been significantly improved and now have greater power and longer life. For electric vehicles, lead-acid batteries have the advantage of proven performance and relatively low cost because they are produced in such high volume already. Nickel-metal hydride batteries are commonly used in laptop computers and for other small applications. Both lead-acid batteries and the nickel-metal hydride batteries are rechargeable.

The downside of batteries is that they are heavy, contain TOXIC CHEMICALS, and can produce toxic EMISSIONS if damaged. They are a source of HEAVY METAL TOXICITY in landfills. And disposal of the batteries can present a SOLID WASTE issue if they cannot be recycled. *See also* FUEL CELLS; LEAD; RECYCLING.

Bay of Fundy, A large inlet of the Atlantic OCEAN that is a noted marine ECOSYSTEM for more than 800 species of animals, including 100 species of fish. The funnel-shaped Bay of Fundy is 270 kilometers (168 miles) long and 80 kilometers (50 miles) wide at its mouth. At its head, it forms two basins, Chignecto Basin and Minas Basin. Located in southeastern Canada, between New Brunswick and Nova Scotia, the Bay of Fundy is famous for its high TIDES that reach 12–15 meters (40–50 feet)—the world's highest tides. It is known as one of the marine wonders of the world.

The area is critically important as a migratory staging area for millions of birds; it is also a significant summering and wintering area. In any given year, 40–70% of the world POPULATION of semipalmated sandpipers stage there. The endan-

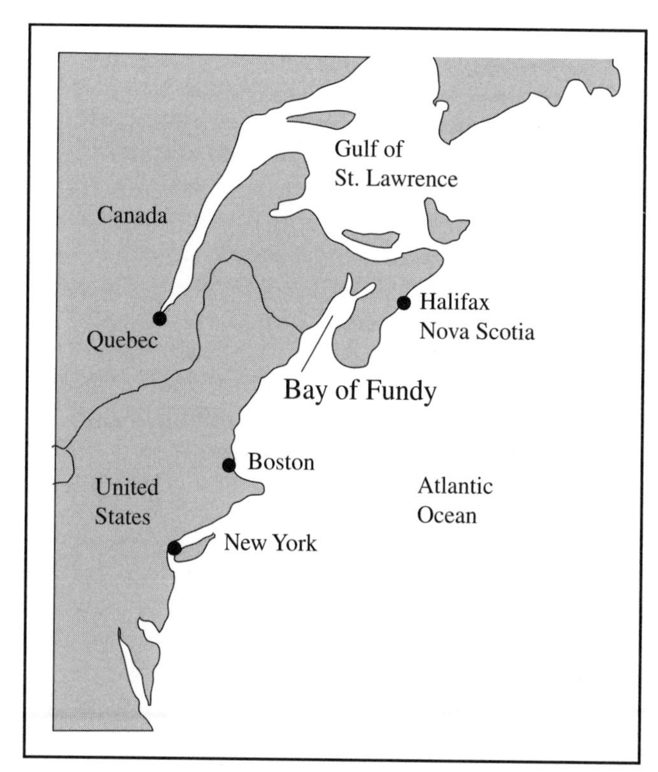

The Bay of Fundy is famous for its tidal bore and for high tides that reach 12 to 15 meters.

gered right WHALE uses the mouth of the bay as a nursery area for mother-calf pairs and juveniles. Sei, fin, minke, and humpback whales and harbor porpoises are found there from June to October. The Canadian government is trying to balance the preservation of the ENVIRONMENT with the needs of humans. Tourism and COMMERCIAL FISHING contribute to the economy of the area. *See also* DOLPHINS AND PORPOISES; ECOTOURISM; ENDANGERED SPECIES.

EnviroSources

Conservation Council of New Brunswick Web site: www.web.net/~ccnb.

Tourism New Brunswick: Dept. 262, P.O. Box 12345, Woodstock, New Brunswick, E7M 5C3; Web site: www.bayoffundy.com.

Bennett, Hugh Hammond (1881–1960), An American soil scientist and advocate for SOIL CONSERVATION measures in AGRICULTURE. Bennett was the first chief of the Soil Conservation Service in the U. S. DEPARTMENT OF AGRICULTURE. He observed and reported the effects of soil EROSION in the United States and other foreign countries. Bennett gave many speeches to different groups throughout the country. In one of his speeches at the Ohio State University in 1933, he explained how the erosion of topsoil was becoming a major problem in the United States. He stated that although erosion was an old process, early people had discovered ways to control erosion. To emphasize his point, he told the following story:

> In 1813, Thomas Jefferson, writing about his farm in Albemarle County, Virginia, said: 'Our country is hilly and we have been in the habit of plowing in straight rows, whether up or down hill, in oblique lines, or however

they lead, and our soil was all rapidly running into the rivers. We now plow horizontally following the curvature of the hills and hollows on dead level, however crooked the lines may be. Every furrow thus acts as a reservoir to receive and retain the waters, all of which go to the benefit of the growing plant instead of running off into the streams.'

Bennett went on to say that many of Jefferson's fields were still in good shape, but those of some of the farms adjacent to his estate were terribly gullied.

Bennett's presentations and writings always incorporated the importance of topsoil. He described soil as the "feeding zone of PLANTS which provided food for humans and livestock as well as timber and wood products for human shelter." During the 1920s and 1930s, Bennett provided detailed studies on how soil erosion was ruining millions of hectares of land in the United States causing hardships for many farmers and ranchers. Hillsides were stripped of trees; gullies were formed; and streams were filled in with sediments from the RUNOFF. *(See* sedimentation.) Good soil was being washed away. Farms and ranchlands were being abandoned. The yearly cost of soil erosion in the United States during that time was reaching $400 million a year. Bennett believed that unless soil conservation measures were implemented soon, the good soil would continue to diminish. In time, Bennett's advocacy of soil conservation had an impact on state governments, farmers, and ranchers who began to reexamine the importance of soil as a major NATURAL RESOURCE.

Hugh Hammond Bennett, referred to by some as the "Father of Soil Conservation," and others in the soil conservation movement founded the Soil Conservation Society of America, in 1941, now called the Soil and Water Conservation Society. *See also* AGROFORESTRY; DESERTIFICATION; DUST BOWL.

EnviroSources

Brink, Wellington. *Big Hugh: The Father of Soil Conservation*. New York: Macmillan, 1951.

Soil and Water Conservation Society: 7515 NE Ankeny Road, Ankeny, Iowa 50021; Phone (515) 289-2331; Fax (515) 289-1227; Web site: www.swcs.org

benthos, The assemblage of ORGANISMS that live on the floors of marine ECOSYSTEMS, where they play a vital role in FOOD CHAINS and FOOD WEBS and sometimes provide HABITAT for other organisms. Benthos include PLANTS and animals. Plants, which have no means of locomotion, root in the floor. Benthic animals may crawl, as do crabs and lobsters, or burrow into the floor, as do clams and aquatic worms. Some fishes, such as rays, skates, and flounders, are benthic organisms that partially bury themselves on the floor—an ADAPTATION that provides camouflage as the fish await prey. *(See* predation.) Other benthic animals, such as sponges, sea anemones, and barnacles, are sessile and live attached to surfaces, including those provided by other animals. *See also* COMMENSALISM; CORAL REEF; CRUSTACEANS; OCEAN; SYMBIOSIS.

benzene, A liquid HYDROCARBON that is flammable and toxic to humans and other ORGANISMS under some conditions. Benzene (C_6H_6) exists naturally in PETROLEUM, GASOLINE, and tobacco. It is released into the ENVIRONMENT through natural processes such as volcanic eruptions and FOREST FIRES as well as through industrial processes. In industry, benzene (sometimes called benzol) is widely used as a SOLVENT and in the making of PETROCHEMICAL products such as resins, nylon, synthetic fibers and rubber, lubricants, dyes, DETERGENTS, pharmaceuticals, and PESTICIDES.

Benzene in AIR, water, or soil may threaten the health of humans and other organisms. Benzene is released into the air in EMISSIONS from COAL and oil-burning power plants, in motor vehicle exhaust, and in tobacco smoke. Benzene also enters the air through evaporation from gasoline, benzene waste and storage sites, and industrial solvents. People living near HAZARDOUS WASTE sites, petroleum refining plants, petrochemical manufacturing sites, or gasoline service stations may be exposed to benzene in the air. Benzene is released into water and soil through industrial DISCHARGES, improper disposal of products containing benzene, and from leaks in underground gasoline storage tanks.

The U.S. Department of Health and Human Services lists benzene as a CARCINOGEN. Long-term exposure to high levels of airborne benzene can cause leukemia. Automobile mechanics exposed to benzene, a gasoline additive, can suffer high rates of anemia. Other health problems associated with short-term exposure to high levels of benzene include drowsiness, dizziness, headaches, tremors, confusion, and unconsciousness. Long-term exposure to such levels may result in stomach irritation, dizziness, rapid heart rate, convulsions, coma, or death. *See also* ACUTE TOXICITY; CANCER; CHRONIC TOXICITY; HAZARDOUS SUBSTANCE; VOLCANO.

benzol. *See* benzene.

Bering Sea, A northern section of the Pacific Ocean that lies between the United States and Russia and provides 56% of the production of U.S. fisheries. Besides Russia and the United States, the Bering Sea provides fishing grounds for Japan, Korea, Norway, and Poland. In 1999, the total commercial value of the catch exceeded $1 billion. More than 455 million kilograms (1 billion pounds) of pollack were caught, the largest number of caught fish in COMMERCIAL FISHING.

Environmentalists are concerned that commerical fishing in the Bering Sea by too many vessels will cause depleting stocks of fish, a similar fate that has occured in sections of the North Atlantic where cod and haddock fish HABITATS have been overfished. *(See* overfishing.) Both the Russian and international fish areas of the Bering are now becoming depleted. And even though the U.S. fishing area is sustainable for now, it may be threatened in the future. Another concern is that overfishing threatens the survival of other marine animals, such as sea lions, who depend on fish for their diet. As an example, the population of Bering sea lions has plummeted

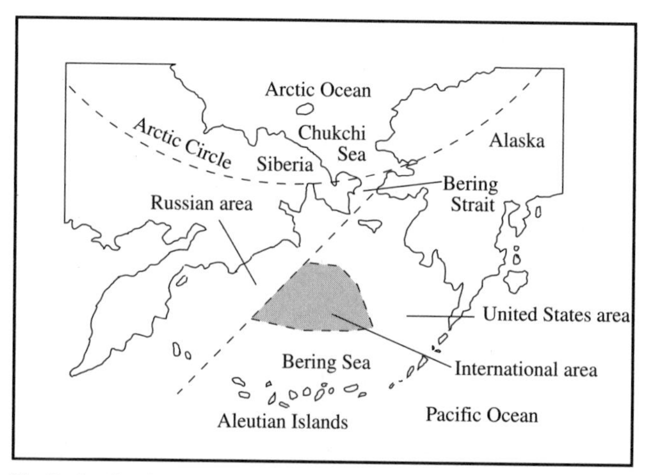

The Bering Sea, located in the northern section of the Pacific Ocean that lies between the United States and Russia, provides 56% of the United States' fisheries production.

within the last 30 years. *(See* seals and sea lions.) Another issue is that too many vessels dump and kill nontargeted fish in their catches and that large drag nets are destroying bottom-dwelling habitats of marine ORGANISMS.

The Bering Sea is a rich marine ECOSYSTEM that includes herring, mackerel, KRILL, WHALES, halibut, SALMON, Steller's sea lions, and northern fur seals. It has the largest international aggregation of sea birds in the world, representing about 40% of all breeding sea birds in the United States. In 1997, the WORLD WILDLIFE FUND launched a campaign to protect regions where the birds breed.

The Bering Strait, is an area that separates the American and Russian peninsulas. It has been an area of controversy between those groups that want to drill for oil and those groups that are concerned about the fragile ENVIRONMENT. During the ICE AGE, about 20,000 years ago, the Bering Strait was a "land bridge" that stretched between Asia and North America.

EnviroSource

National Oceanic and Atmospheric Administration e-mail: bering@pmel.noaa.gov.

beryllium, A hard, gray-white, metallic element present in certain rocks, COAL, soil, and volcanic dust. Beryllium is toxic to humans when inhaled in large amounts. Beryllium (Be) is mined primarily in the United States and Brazil, where it is obtained from the MINERALS bertrandite and beryl. Beryllium compounds are made into alloys for use in electrical and electronics parts, machinery, and molds for plastics. Pure beryllium is used to make NUCLEAR WEAPONS, NUCLEAR REACTORS, instruments for aircraft and aerospace vehicles, mirrors, and x-ray machine windows.

Beryllium-containing dust is often released to the ENVIRONMENT during the MINING process. Inhalation of dust containing high levels of beryllium or its compounds can cause lung CANCER, pneumonitis (an acute inflammation of the lungs), and berylliosis (a chronic DISEASE characterized by noncancerous growths within the lungs). Beryllium is released into the AIR in EMISSIONS from burning coal and PETROLEUM and from foundries and ceramics plants that use beryllium in their products. People living near landfills containing HAZARDOUS WASTES also may risk exposure to airborne beryllium. *See also* CARCINOGEN; CHRONIC TOXICITY; HAZARDOUS SUBSTANCES; VOLCANO.

EnviroSources

Agency for Toxic Substances and Disease Registry Web site: atsdr1.atsdr.cdc.g..rofiles/phs8807.html.

Environmental Health Center (National Safety Council): 1025 Connecticut Ave., NW, Suite 1200, Washington, DC 20036; Web site: www.nsc.org/ehc/ew/shems/berylliu.html.

best available control technology, EMISSIONS standards established as part of the CLEAN AIR ACT (CAA) of 1970 that require stationary (nonmoving) sources of AIR POLLUTION in regions meeting the NATIONAL AMBIENT AIR QUALITY STANDARDS (NAAQS) to make use of equipment and technologies that will reduce the release of POLLUTANTS into the AIR. Stationary sources of air pollution include such things as ELECTRICITY generating power plants and factories. Best available control technology (BACT) methods include such things as the design and layout of a facility, use of pollution control equipment such as SCRUBBERS, and general operational procedures of the facility.

Before such stationary facilities can release any pollutants into the air, they are required by the Clean Air Act to obtain permits from the U.S. ENVIRONMENTAL PROTECTION AGENCY (EPA) that allow for the release of specific pollutants in designated amounts. Permits are issued for a specific time period that is not to exceed five years. If new sources of pollution, such as new equipment or a change in methods used to produce a product, are planned for use at a facility, the facility in its permit application must demonstrate that emissions from this source will not violate the NAAQS of the CAA. If a new facility is unable to meet the BACT standards because doing so would be cost prohibitive, the facility must then meet the New Source Performance Standards (NSPS) that have been established by the EPA. Unlike the BACT standards, the NSPS standards consider the financial burden (i.e., cost increase) that pollution control methods and devices may place on a business.

Since their creation, the BACT standards have decreased, often significantly, the amount of pollutants released by stationary sources. Overall, the release of SULFUR DIOXIDE (SO_2) into the ATMOSPHERE from stationary sources has decreased by 25% since 1970. During the same period, PARTICULATE MATTER releases have been lowered by as much as 50%. Decreases in LEAD, sulfur dioxide, OZONE, CARBON MONOXIDE, and NITROGEN OXIDES—the five other pollutants classified as CRITERIA POLLUTANTS—also have been observed. *See also* ACID RAIN; AIR POLLUTION CONTROL ACT; ELECTROSTATIC PRECIPITATOR; NITROGEN DIOXIDE.

beta particle. *See* radiation.

Bhopal incident, A horrific accident that occurred in Bhopal, India, in 1984 when toxic gases leaked from a Union Carbide chemical manufacturing plant, killing an estimated 3,000 people and injuring many thousands more. Unofficial estimates suggest even higher casualties.

Bhopal is located in central India and has a POPULATION of about 850,000. The highly toxic cloud from the plant covered a 60-square-kilometer area (23 square miles) for less than one hour, killing more than 2,000 people instantly. The released chemical, methyl isocyanate, is used to make common PESTICIDES. During the accident, more than 27 metric tons (30 tons) of methyl isocyanate were released, causing irritation to the eyes, throat, and lungs from inhalation of the gas. Extreme exposure caused coughing, swelling of the lungs, bleeding, visual impairment, respiratory problems, and death.

Reports showed that the plant had a good safety record before the disaster. The tragedy was due to a number of human mistakes and mechanical problems: There was no emergency plan for evacuating the local people to safer areas; safety equipment failed to operate and control the gas from escaping; and there were weak pollution control and early warning systems. The Bhopal catastrophe set in motion the Responsible Care Program, established in Canada in 1985 by the chemical industry to implement safety standards, pollution prevention techniques, and improvement of community relations in areas where chemical plants are located. In the United States, Congress passed legislation, such as the EMERGENCY PLANNING AND COMMUNITY RIGHT-TO-KNOW ACT in 1986. Part of that law states that manufacturers must list the types and quantities of chemicals released into the AIR, land, and water by their operations. That information is included in an annual report, the TOXICS RELEASE INVENTORY published by the U.S. ENVIRONMENTAL PROTECTION AGENCY. *See also* AGENCY FOR TOXIC SUBSTANCES AND DISEASE REGISTRY; HAZARDOUS MATERIALS TRANSPORT ACT; HAZARD RANKING SYSTEM; HAZARDOUS SUBSTANCE; SANDOZ CHEMICAL SPILL; TOXIC CHEMICAL.

EnviroSources

TOXNET (National Library of Medicine, Specialized Information Services Division): 8600 Rockville Pike, Bethesda, MD 20894; Web site: www.toxnet.nlm.nih.gov.

bioaccumulation, The gradual increase of a chemical in the tissues of a living ORGANISM. Chemicals accumulate in living things when they are taken into the body and stored faster than they are broken down or excreted. Bioaccumulation is a normal and essential process for the growth of organisms. All animals, including humans, bioaccumulate many vital nutrients on a daily basis. These nutrients include Vitamins A, D, and K; trace minerals; and essential fats and amino acids. Bioaccumulation can be a vital process by which the body accumulates needed chemicals. What concerns toxicologists is the bioaccumulation of substances such as HEAVY METALS to levels in the body that can cause harm. (*See* toxicology.) Understanding the process of bioaccumulation is very important in protecting human beings and other organisms from the adverse effects of chemical exposure, and it has become a critical consideration in the regulation of chemicals.

Bioaccumulation results from an equilibrium between chemical concentrations in the outside ENVIRONMENT and those within an organism. The extent of bioaccumulation depends on the concentration of a chemical in the environment; the amount of the chemical entering an organism from the food, AIR, or water; and the time it takes for the organism to absorb the chemical, and then excrete, store, and/or break it down or to excrete or store it.

The bioaccumulation process begins when a chemical enters the CELLS of an organism. The chemical will at first move into an organism more rapidly than it is stored, degraded, or excreted. With constant exposure, the concentration inside the organism gradually increases. Eventually, the concentration of the chemical inside the organism will reach an equilibrium—the amount of chemical entering the organism will be the same as the amount leaving. And, although the amount inside the organism remains constant, the chemical continues to be taken in, stored, broken down, and excreted. If the environmental concentration of the chemical increases, the amount inside the organism will increase. However, if the concentration in the environment decreases, the amount inside the organism will also decline. If the organism moves to a clean environment, in which there is no exposure to CONTAMINANTS, then the chemical eventually will be eliminated from the body. Exposure to large amounts of a chemical for a long period of time, however, may overwhelm the equilibrium and potentially cause harmful effects.

Bioaccumulation varies among individual organisms as well as among SPECIES. Large, fat, long-lived individuals or species with low rates of metabolism or excretion of a chemical will tend to bioaccumulate chemicals that are stored in fat more than small, thin, short-lived organisms. Thus, an old lake trout may bioaccumulate much more POLYCHOLORINATED BIPHENYLS (PCBs) than a young bluegill in the same lake. *See also* BALD EAGLE; BIOCONCENTRATION; BIOMAGNIFICATION; FOOD WEB; PEREGRINE FALCON; PESTICIDES.

EnviroTerms

storage: The temporary deposit of a chemical in body tissue or an organ. Storage is just one facet of chemical BIOACCUMULATION. The term also applies to other natural processes, such as the storage of fat in hibernating animals or the storage of starch in seeds.

uptake: The entrance of a chemical into an ORGANISM, such as by breathing, swallowing, or absorbing it through the skin.

EnviroSource

Michael Kamrin (Institute for Environmental Toxicology): C-231 Holden Hall, Michigan State University, East Lansing, MI 48824-1206; e-mail Kamrin@pilot.msu.edu.

bioassay, The process of using living ORGANISMS to test the effects of a substance or condition by comparing the organism's status or conditions before and after exposure. Bioassays can be used for short-term or long-term tests. Short-term tests of living organisms or their CELLS can yield results within a few days. Long-term assays of living organisms such as fish, rats, and mice last for several weeks to a lifetime.

Bioassays are used to test for a variety of effects. In the early 1990s, several large fish kills occurred in the ESTUARY waters off the coasts of North Carolina and Maryland. At first scientists were puzzled as to what was causing the death of millions of fish. The puzzle was solved when scientists at North Carolina State University carried out a bioassay. They collected water samples from the estuary and deposited them into laboratory aquariums. Healthy fish were added to the tanks and in a short time the fish were dead verifying that the water samples from the estuary were toxic. Other parts of their bioassay allowed them to rule out a number of causative agents such as FUNGI and PESTICIDES as being responsible for the fish kills. What they did discover was that a MICROORGANISM, *Pfiesteria piscicida*, a dinoflagellate, was attacking and killing the schools of fish. *(See* algae.) Additional research is now being done to find out more about this microorganism and how it affects fish and human activity in the estuaries. *See also* ACUTE TOXICITY; ALLERGY; ANIMAL RIGHTS; CHRONIC TOXICITY; TOXIC CHEMICAL.

biocentric, A term, literally meaning "life-centered," that is used by supporters of DEEP ECOLOGY to describe their belief that all ORGANISMS have natural value and therefore a right to exist. According to biocentric principles, all SPECIES, including those that people may find unattractive, nonuseful, or harmful, should have the same opportunities for protection and survival as species that are considered desirable. Biocentrism contrasts sharply with anthropocentrism, which measures the value of a species by its usefulness to humans. *See also* ANIMAL RIGHTS; ANTHROPOCENTRIC; FOREMAN, DAVID.

biochemical, A naturally occuring ORGANIC chemical derived from living things or synthetic variety that is identical to such naturally occuring chemicals. Biochemicals are important to INTEGRATED PEST MANAGEMENT (IPM) programs. For example, some biochemicals, including enzymes, pheromones, and hormones, can function as nontoxic PESTICIDES. Synthetic or natural biochemicals can also be used as INSECT repellants, to lure and trap insect pests, to prevent mating, and to disrupt the life cycles of insects. *See also* AGRICULTURAL POLLUTION; AGROCHEMICAL; BIOLOGICAL CONTROL.

biochemical oxygen demand, A measure of water quality based on the amount of OXYGEN consumed in the biological processes that break down ORGANIC matter in water. The biochemical oxygen demand (BOD) is an expression of how much DISSOLVED OXYGEN is required by MICROORGANISMS to carry out the AEROBIC DECOMPOSITION of natural or POLLUTANT-based organic matter. The higher the BOD, the greater the amount of oxygen that is depleted. The amount of oxygen used in the breakdown of organic material is measured in milligrams per liter of oxygen consumed for a period of five days at a temperature of 20°C (68°F). In clean water, the BOD is low and the dissolved oxygen level is high; however, in polluted water the BOD will be high and the dissolved oxygen will be low. Low oxygen levels can lead to adverse effects on aquatic life. *See also* BACTERIA.

bioconcentration, The process by which the concentration of a chemical in an ORGANISM becomes higher than its concentration in the AIR or water around the organism. The increase exceeds the rate of excretion or metabolism. Bioconcentration usually refers to chemicals foreign to the organism. As an example, FISH, oysters, and clams can concentrate HEAVY METALS through the gills or sometimes the skin in fresh or salt water. *See also* BIOMAGNIFICATION.

bioconversion, The process of converting the stored chemical ENERGY in BIOMASS by changing the form or concentration of the energy. The most widely used method of bioconversion involves the direct burning of biomass, such as wood, PEAT, leaf litter, or animal dung, to release its stored chemical energy as heat for dwellings, cooking, or to produce steam that drives machinery. Many people rely on the direct burning of biomass for energy. For example, more than one billion people living in developing nations, primarily in Africa and South America, obtain most of their heat by burning wood. Cattle dung is a primary FUEL source in much of India. Scotland and several other northern European nations rely largely on peat for heat energy.

In industrialized nations, bioconversion often involves converting solid biomass into BIOFUELS such as METHANE gas, which forms through the DECOMPOSITION of PLANT matter, and liquid ETHANOL, an alcohol formed by the distillation and processing of carbohydrates. The advantage of biofuels over the direct burning of biomass is that biofuels yield more energy, are usually less polluting, and are more convenient to use than biomass. *See also* ALTERNATIVE ENERGY RESOURCE; ALTERNATIVE FUEL; COAL; COGENERATION; FOSSIL FUELS; FUEL WOOD; GASOHOL; METHANOL; SYNTHETIC FUEL.

biodegradable, A term used to describe matter that can be broken down in the ENVIRONMENT through biological processes. Biodegradable materials generally are composed of natural compounds, such as PLANT and animal products, that are broken down through DECOMPOSITION by MICROORGANISMS (BACTERIA and FUNGI). In this process, compounds contained in the matter are released back into the environment, where they can be reused by ORGANISMS to carry out their life processes.

Many products used by people are composed of synthetic compounds that do not readily break down in the environment. *(See* EnviroTerms.) Examples include most plastics, synthetic rubber, PESTICIDES, and some PETROCHEMICAL products. Among the largest environmental problem associated with use of these materials is how to dispose of them. Because these products persist in the environment, they quickly fill the limited space of open dumps and landfills. When improperly disposed of, many NONBIODEGRADABLE materials become hazards to WILDLIFE that ingest or become entangled in them. Compounds in these products also may be toxic to organisms.

To reduce the problems associated with nonbiodegradable synthetic materials, concerned citizens and environmentalists have lobbied large companies to use biodegradable substances in the making and packaging of their products whenever possible. This movement has met with some success, and companies that have adopted such practices often advertise their products as "biodegradable" or "environmentally friendly." *(See* ecolabeling.) Environmental groups also have worked to increase public awareness of problems associated with nonbiodegradable materials, while also encouraging such practices as reducing, reusing, and RECYCLING such materials when possible. *See also* BIOGEOCHEMICAL CYCLES; DEEP-WELL INJECTION; INCINERATION; NITROGEN CYCLE; OCEAN DUMPING; PHOTODEGRADABLE PLASTIC; SANITARY LANDFILL; SOLID WASTE.

EnviroTerm
synthetic: Not made in nature; human made.

EnviroSources

Green Seal Web site: www.greenseal.org.
MacEachern, Diane. *Save Our Planet: 750 Everyday Ways You Can Help Clean Up the Earth.* New York: Dell, 1995.

biodiversity, Term used to refer to the variety of PLANTS, animals, FUNGI, PROTISTS, and BACTERIA living on Earth or in a specific area. Some people use the term "biodiversity" to refer to the variations existing within SPECIES; however, this is more appropriately termed GENETIC DIVERSITY. The distribution of ORGANISMS within ECOSYSTEMS is generally described as ecosystem diversity.

Most often, biodiversity refers to the number of different species inhabiting Earth. Currently, more than 1.1 million species of existing organisms have been identified by scientists; however, many scientists believe that the number of species that actually inhabit Earth may be as much as ten times greater than the number of known species. These unknown species are believed to exist mostly in the regions of Earth that have not yet been thoroughly explored, such as remote areas in tropical RAINFORESTS, among CORAL REEFS, and in the deep OCEAN.

Earth's Biodiversity. The number of known species breaks down as follows: animals 1.03 million species, three-quarters of which are INSECTS; plants, almost 250,000 species; fungi, 69,000 species; protists, almost 58,000 species, of which almost 27,000 species are ALGAE; and monerans or bacteria, 4,800 species. In addition, scientists have identified about 1,000 different kinds of VIRUSES, which are considered organisms by some scientists, but considered nonliving things by others.

The rapid rate at which EXTINCTION of species is occurring in parts of the world drives scientists to identify as many of Earth's existing species as possible, while also learning as much as they can about these organisms. In addition, scientists also are interested in how the biodiversity of different types of ecosystems compare and why differences exist. They do know that the greatest biodiversity of land ecosystems is

Biodiversity of Known Species

Group	Number of species	Percentage
Animals	1,030,000	72.9%
Plants	250,000	17.7%
Fungi	69,000	4.9%
Protists	58,000	4.1%
Monerans	4,800	0.3%

found in tropical rainforests, while the most diverse aquatic ecosystems are coral reefs. By contrast, DESERTS and TUNDRA ecosystems have fairly low biodiversity when compared to other terrestrial ecosystems, and the deep ocean has less diversity than more shallow ocean waters.

Maintaining Biodiversity. Maintaining Earth's biodiversity is a major goal of ENVIRONMENTAL SCIENCE. Biodiversity is essential to ecosystem health. Because of the complex interactions that occur between Earth's organisms and the physical environment, the health of ecosystems, in turn, is essential to maintaining the health of the BIOSPHERE. For example, organisms and the processes that keep them alive are essential to the maintenance of a proper balance of atmospheric gases such as OXYGEN, CARBON DIOXIDE, and NITROGEN. *(See* atmosphere.) Earth's WATER CYCLE also is largely dependent upon organisms.

In addition, many of the products humans rely on to maintain life or to make their lives more comfortable also are derived directly or indirectly from Earth's living things. For example, many life-saving medications have been derived from plants. The drugs colchinine, taxol, and vinbalstine, which are used to prevent or treat different types of CANCER are derived from the autumn crocus, Pacific yew, and rosy periwinkle, respectively. The antibiotic penicillin originated from the penicillin mold—a fungus—although it can today be made synthetically. Quinine, an antimalarial drug, is derived from the yellow cinchona, while the drug known as L-dopa, which is used in the treatment of Parkinson's disease, comes from the velvet bean. Scientists point to the use of plants in making medicines as one of the reasons that preserving biodiversity is so important. They also point out that many species that have not yet been identified may contain similar substances. Other products derived from organisms include a variety of foods, food products such as spices, wood, paper, dyes, and fibers used in the manufacture of clothing.

To maintain Earth's biodiversity, scientists recommend that people learn as much as they can about the ENVIRONMENT and use its resources wisely. This may help to prevent the extinction of many species in the future. In addition, many programs have been established to protect those species that have been identified as being on the verge of extinction. Some of these programs set aside areas as protected HABITATS and CAPTIVE PROPAGATION programs. Such programs can be expensive to implement and are not always successful. The failure to take steps to preserve biodiversity, however, may be even more costly to the future of humans and the environment. *See also* ADAPTATION; BIODIVERSITY TREATY; BIOGEOCHEMICAL CYCLE; CONSERVATION; ENDANGERED SPECIES; ENDANGERED SPECIES ACT; ENDANGERED SPECIES LIST; NICHE; THREATENED SPECIES; WILSON, EDWARD OSBORNE; WORLD CONSERVATION MONITORING CENTRE; WORLD CONSERVATION UNION.

EnviroSources

Ehrlich, Paul, and Anne Ehrlich. *Extinction: The Causes and Consequences of the Disappearance of Species.* New York: Random House, 1981.

Margulis, Lynn, and Karlene V. Schwartz. *Five Kingdoms: An Illustrated Guide to Life on Earth.* 2d ed. New York: W. H. Freeman and Co., 1988.

Wilson, Edward O. *The Diversity of Life.* Cambridge, MA: Belknap Press, 1992.

Biodiversity Treaty, An international agreement developed to protect Earth's BIODIVERSITY by decreasing the number of SPECIES that become extinct each year. *(See* extinction.) This treaty resulted from the UNITED NATIONS EARTH SUMMIT held in Rio de Janeiro, Brazil, in 1992, where it was signed by more than 150 nations. Its major provisions call for a worldwide inventory of threatened and ENDANGERED SPECIES, cooperative efforts among nations to protect such species, and financial assistance from wealthy developed nations to less wealthy developing nations to help protect potentially valuable species. To achieve this last goal, the developed nations that signed the treaty agree to compensate developing nations for PLANTS and animals used to make products.

The Biodiversity Treaty has received mixed reactions from citizens and governments. While many people applaud the treaty's goals, critics emphasize that the treaty contains no provisions for how its goals are to be met. This criticism was emphasized during the Earth Summit, when U.S. President George Bush refused to sign the treaty, claiming it was too vague. Despite this criticism, the United States endorsed the treaty in 1993, when it was signed by newly elected President Bill Clinton. *See also* BACTERIA; ENDANGERED SPECIES ACT; ENDANGERED SPECIES LIST; FUNGI; MAN AND THE BIOSPHERE PROGRAMME; PROTISTS; RED LIST OF ENDANGERED SPECIES; THREATENED SPECIES; WORLD CONSERVATION MONITORING CENTRE; WORLD CONSERVATION UNION.

biofuel, A solid, liquid, or gaseous FUEL derived from BIOMASS. Biofuels are used as an alternative to FOSSIL FUELS and include biogas, biodiesel, and METHANE. About 5% of the ENERGY consumed in the United States is provided by biofuels. Most of the biofuels are produced from wood wastes from logging operations but they can also be produced from corn and sugar crops. In France, Italy, and Germany, biodiesel fuels are produced from domestic oilseeds and cottonseeds. Biofuels are cleaner than fossil fuels, releasing fewer GREENHOUSE GASES, such as CARBON DIOXIDE and SULFUR, and PARTICULATE MATTER into the ATMOSPHERE. *See also* ALTERNATIVE ENERGY RESOURCE; ALTERNATIVE FUEL; FUEL WOOD; GREENHOUSE EFFECT.

biogeochemical cycle, The movement of large amounts of chemicals such as NITROGEN, potassium, calcium, CARBON, HYDROGEN, sodium, SULFUR, and PHOSPHORUS from living ORGANISMS through the ATMOSPHERE, HYDROSPHERE, and LITHOSPHERE, and back to the organisms. The cycle includes the processes of production and DECOMPOSITION and is essential for the continuation of life.

Interference with the biogeochemical cycle caused by ACID RAIN, water pollution, AIR POLLUTION, PESTICIDES, and HAZARDOUS WASTES may harm the ENVIRONMENT and living organisms. As an example, excessive MINING of MINERALS for metals that are not BIODEGRADABLE, can lead to contamination of ECOSYSTEMS and FOOD CHAINS. *See also* CARBON CYCLE; CONTAMINANT; FOOD WEB; NITROGEN CYCLE; PHOSPHORUS; SINK; WATER CYCLE.

biological community, All the POPULATIONS of ORGANISMS that naturally live within a specific, well-defined area or HABITAT. A biological community may be identified with a major physical characteristic of an area. For example, the organisms of a pond may be referred to as a pond community. Such organisms include all the animals and PLANTS living within or at the surface of the pond, including any MICROORGANISMS (BACTERIA, PROTISTS, and FUNGI) that are present. Communities also may be named for their dominant SPECIES, as in a pine FOREST community, in which the dominant species are pine trees. Other organisms of this community would likely include animals such as deer, toads, salamanders, earthworms, and rabbits; plants such as mosses and ferns; fungi such as bracket fungi, puffballs, molds, and mushrooms; and bacteria and protists that live in the soil, AIR, water, and on or in other forest organisms.

Organisms in a biological community are linked to each other through a variety of interactions, including the feeding relationships that make up a simple FOOD CHAIN or a more complex FOOD WEB. The various organisms in a community also are in COMPETITION with one another for resources such as living space, water, sunlight, mates, and food. Other interactions that occur within a biological community include PREDATION, PARASITISM, COMMENSALISM, and MUTUALISM. *See also* ABIOTIC; BIOREGION; BIOTIC; ECOSYSTEM; NATIVE SPECIES; NICHE; SYMBIOSIS.

biological control, Methods used to reduce damage caused by pest ORGANISMS to crops, livestock, and timber, by using the pest's natural vulnerabilities to ecological interactions such as PREDATION, PARASITISM, COMPETITION, and pathogenic ORGANISMS. Biological control, or biocontrol, often involves the release of PATHOGENS, predators, or parasites of a pest SPECIES into the ENVIRONMENT where the pest is causing a problem. The goal of this activity is to reduce both the pest POPULATION size and the use of chemical PESTICIDES.

In California, wasps that prey on fruit-boring INSECT larvae are being used as a biological control. The wasps are commercially bred and purchased by fruit farmers for release on their orchards. Once released, the wasps feed on the larvae, thus eliminating their potential for damaging fruit. The ladybug beetle, a predator of many PLANT-eating pests, also is being commercially bred for use by farmers. The ladybugs are released into crop fields, where they consume pest organisms, thus reducing their populations. Recently, scientists imported black ladybugs from Japan for release into several hemlock FORESTS of the northeastern United States. The black ladybugs are a natural predator of a parasitic insect known as the woolly adelgid that feeds on the fluids of the trees, causing their deaths. The adelgids are an EXOTIC SPECIES that was first introduced to the United States from Japan many years ago.

Other methods of biological control involve disrupting pest reproduction. The use of sex pheromones, substances that organisms produce to attract mates, has proved an effective means for luring and trapping many insects. Scientists have successfully synthesized pheromones for several species of fruit flies including the Mediterranean fruit fly (Medfly) and the Oriental fruit fly. Once captured, the flies are killed or sterilized using chemicals or RADIATION. If sterilized, the organisms are released back into the environment, where they are unable to reproduce, thus limiting population growth of the pest species. *(See* sterilization, reproductive.) Controlled breeding techniques such as HYBRIDIZATION and GENETIC ENGINEERING techniques such as transgenesis also are being used as biological controls to produce plant varieties that are naturally pest or disease resistant.

INTEGRATED PEST MANAGEMENT (IPM) is a technology for controlling pest populations that incorporates use of biological controls. IPM also makes use of other methods that serve not only to reduce pest populations, but also to reduce EROSION and nutrient deficiencies in soil. These methods include such activities as POLYCULTURE farming, which uses increased crop diversity to reduce the likelihood that pest organisms will find a host; the use of low-till or no-till AGRICULTURE to help maintain populations of natural enemies of pest organisms in soil; and the controlled use of chemical pesticides, when necessary, to target specific pest populations. Biological controls and IPM methods are now practiced in many parts of the world because of bans on the use of specific pesticides that pose harm to people and the environment. *See also* AGRICULTURAL POLLUTION; AGROECOLOGY; BACTERIA; BIOTECHNOLOGY; BLIGHT; DICHLORODIPHENYLTRICHOROETHANE; DISEASE; FEDERAL INSECTICIDE, FUNGICIDE, AND RODENTICIDE ACT; FUNGI; HYBRID; VIRUS.

EnviroSources

Saign, Geoffrey C. *Green Essentials: What You Need to Know about the Environment.* San Francisco, CA: Mercury House 1994.

Seymour, John, and Herbert Girardet. *Blueprint for a Green Planet.* New York: Prentice Hall, 1987.

biological diversity. *See* biodiversity.

biological magnification. *See* biomagnification.

biological reserve. *See* Carara Biological Reserve.

biological weapon, Any type of armament that employs tasteless, odorless biological agents, which are invisible to the human eye. Biological weapons are much stronger than the most lethal CHEMICAL WEAPONS, which use toxic substances such as chlorine gas to kill or disable. Biological weapons can be made from widely available PATHOGENS, such as anthrax, pneumonic plague, and botulism, which can be obtained from legitimate biomedical research or from infected humans or animals. Biological weapons can contaminate an area for between several hours and several weeks. An attack or the threat of an attack can cause a widespread shutdown of civil and economic activity in the targeted area. These agents can cause long-term pandemic DISEASE and widespread death. Biological weapons are not as controllable or predictable as chemical weapons. They are more dependent on temperature, topography, and WEATHER. Most must be inhaled or ingested to be effective.

The history of the use of biological weapons is difficult to assess because of problems with verification, propaganda, and the incidence of naturally occurring endemic or epidemic diseases during hostilities. The Geneva Protocol of 1925 and the Biological and Toxin Weapons Convention of 1972 attempted to stop the use of biological weapons but have not been entirely effective even though biological warfare has been renounced by 140 countries.

In 1990, U.S. President George Bush issued Executive Order 12735 to declare a national emergency to stop the proliferation of chemical and biological weapons and to outline sanctions against countries violating international law regarding their use. Today, Iraq still has an arsenal of biological weapons even after attempts were made during Operation Desert Storm and the Persian Gulf War to persuade the government to disarm the weapons.

In 1996, a Biological Weapons Convention Review Conference was held. The United States, the United Kingdom, and the Russian Federation are the depositary governments for the convention, which more than 120 countries have signed.

EnviroSources

An Act to Implement the Convention on the Prohibition of the Development, Production, and Stockpiling of Bacteriological (Biological) and Toxin Weapons and Their Destruction, by Prohibiting Certain Conduct Relating to Biological Weapons, and for Other Purposes. Washington, DC: Government Printing Office, 1990.

Chemical and Biological Defense Information Analysis Center Web site: www.cbiac.apgea.army.mil.

Federation of American Scientists Biological Weapons Control Web site: www.fas.org/bwc.

biomagnification, The process through which a chemical becomes more and more concentrated in ORGANISMS as it moves up through a FOOD CHAIN—the food linkage from PLANTS or other PRODUCERS to increasingly larger animal SPECIES. In a typical aquatic food chain, for example, ALGAE is eaten by a water flea, which is eaten by a minnow, eaten by a trout, and finally consumed by an osprey or human being. If each level results

This illustration is an example of biomagnification. Notice the changes in the DDT (parts per million) concentration as it moves through the food chain. *Credit:* Michael Kamrin, Michigan State University, *Ecotoxicology for the Citizen.*

in increased biomagnification of a substance, then an animal at the top of the food chain, through its regular diet, may have accumulated a much greater concentration of a chemical than was present in organisms lower in the food chain. *(See* bioaccumulation.) A study of DICHLORODIPHENYLTRICHOROETHANE (DDT), for example, showed that where levels of DDT in soil were 10 parts per million (ppm), DDT reached a BIOCONCENTRATION of 141 ppm in earthworms and 444 ppm in robins. Through biomagnification, the concentration of a chemical in the animal at the top of the food chain may be high enough to cause death or adverse effects on behavior, reproduction, or DISEASE resistance and thus endanger that species, even though contamination levels in the AIR, water, or soil are low. *See also* PARTS PER BILLION AND PARTS PER MILLION.

biomass, Living matter or the remains of living matter, which can be used as FUEL; also the total amount of living matter in a particular area at any given time. The amount of biomass present in an ECOSYSTEM is of importance to scientists because biomass contains stored ENERGY that provides a source of fuel, known as BIOFUEL. The most suitable biomass for use as fuel derives from PRODUCERS (PLANTS), because they comprise the TROPHIC LEVEL with the greatest amount of stored energy. This biomass includes such matter as wood, herbaceous plants, and excess food crops that can be burned as a direct source of energy. The biomass in municipal SOLID WASTES also are often burned directly as fuel. These sources along with aquatic crops, timber, and agricultural waste products may also undergo the thermochemical or biochemical processes of gasification or liquefaction to render fuels in a gaseous or liquid form. *(See* EnviroTerm.)

Because they are derived from RENEWABLE RESOURCES, biomass fuels have the advantage over FOSSIL FUELS of being a sustainable energy resource if harvested carefully. By contrast, overharvesting of such fuels can lead to environmental problems such as DEFORESTATION, as has occurred in many parts of India, China, and Kenya, and some South American countries that use FUEL WOOD as their primary fuel source. In addition, like fossil fuels, burning biomass fuels (either directly or in a gaseous or liquefied form) releases great amounts of POLLUTANTS to the ENVIRONMENT. These pollutants include CARBON DIOXIDE, CARBON MONOXIDE, NITROGEN OXIDES, HYDROCARBONS, METHANE, and some HEAVY METALS or radioactive substances that may have been taken into the biomass source from soil. *See also* AIR POLLUTION; ALTERNATIVE ENERGY RESOURCE; ALTERNATIVE FUEL; ECOLOGICAL PYRAMID; ETHANOL; METHANOL; SYNTHETIC FUEL.

EnviroTerm

thermochemical: Term used to describe chemical reactions that either require the addition of heat to take place or give off heat as they occur.

EnviroSources

American Bioenergy Association Website: www.biomass.org.
Saign, Geoffrey C. *Green Essentials: What You Need to Know about the Environment.* San Francisco, CA: Mercury House, 1994.

biomass fuel. *See* biofuel; biomass.

biome, A large geographic area with a distinctive CLIMATE and ECOSYSTEM. Earth contains hundreds of different kinds of ecosystems. For the sake of convenience, ecologists divide these ecosystems into separate biomes, which are named according to a particular form of vegetation and associated animals that are adapted to the local conditions of the community. Most ecologists agree that Earth's land ecosystems can be grouped into six biomes: DESERT, GRASSLAND, TAIGA, temperate forests, RAINFORESTS, and TUNDRA. The OCEAN, lakes, and other waters would be aquatic biomes.

Climate is the main determining factor of a biome. A region's climate influences the kinds of PLANTS that can grow there, which, in turn, determine the types of animals that can live there. Climate refers to the WEATHER conditions in an area over a long period of time. Many factors make up a region's climate; the two main factors are temperature and precipitation, such as rainfall or snow. *(See* water cycle.)

The temperature of a region is determined by the amount of sunlight that reaches it and the angle at which the sunlight strikes Earth. Not all parts of the world receive the same amount of ENERGY from the sun. The amount of sunlight is primarily related to the latitude of a region. The higher its latitude—that is, the farther away from the equator and the closer to the North or South Poles—the colder the climate.

Temperature has an important effect on the amount of precipitation in an area. For instance, areas surrounding the equator (latitudes from 0° to 23.5° North or South) generally have large amounts of rain. This is because sunlight is most intense near the equator; when the sun's rays warm the AIR, the warm air rises and then cools. Eventually, the air cools enough for water vapor to condense and fall as precipitation.

Types of Biomes. Typical deserts are very hot and dry terrestrial environments that occur between 15° and 35° N or S. Deserts are found in the southwestern United States, the western coast of South America, northern and southern Africa, central Asia, and Australia. Precipitation in the world's deserts is very low, averaging less than 25 centimeters (10 inches) per year. This is why deserts are inhospitable to most forms of life. Although precipitation is low in all deserts, temperature tends to vary. In tropical deserts, such as the Sahara Desert in northern Africa, temperatures are extremely hot all year round. By contrast, in temperate deserts, such as the Mojave Desert in the southwestern United States, temperatures are hot during the summer, but relatively cold in the winter, with temperatures reaching as low as 0°C (32°F). Plants and animals that live in deserts are adapted for surviving in these extremely dry and sometimes very hot or very cold conditions. Typical desert plants, such as CACTI, for example, have small or needlelike leaves that help the plant conserve water. Animals usually adapt behaviorally in desert conditions. For instance, many animals, such as spadefoot toads, burrow into mud banks and only emerge from their holes during periods of rainfall.

Grasslands are ecosystems dominated by grasses and some shrubs. Few trees can exist in grassland climates because severe DROUGHTS are common. In general, all grasslands are rather warm and dry. Average precipitation in temperate and tropical grasslands (SAVANNAS) ranges from 25–75 centimeters (10–30 inches) annually. In tropical grasslands, such as those in Africa and South America, temperatures are fairly consistent throughout the year, remaining around 30°C (86°F). In temperate grasslands, such as those in the central United States and central Europe, temperatures show great fluctuation, from about 0°C (32°F) in the winter to about 25°C (77°F) in the summer. Grasses have adapted to survive during the long droughts. Their root systems, for instance, form thick underground mats, which enable them to survive with little precipitation and brush fires. Many grassland animals, such as horses, giraffes, and buffalo, have adapted to feed on grasses. Some animals such as prairie dogs and badgers are able to dig extensive underground burrows to avoid detection by predators.

The taiga is the world's largest land biome. It stretches in a broad band across the Northern Hemisphere just below the ARCTIC Circle between 50° and 60° N. The taiga is a very cold and rather dry biome. Average temperatures range from about 10°C (14°F) in the winter to about 10°C (50°F) during the summer months. There is little precipitation in the taiga. Average precipitation ranges from 30–40 centimeters (14–16 inches) per year, but it is mostly snow. The dominant vegetation is conifers—evergreen trees such as pines, firs, cedars, and spruces. The trees are adapted with needle-shaped leaves that help the plants to survive the cold temperatures and low precipitation. Many animals adapt to the harsh winters in the taiga by burrowing underground. Other animals migrate south to warmer temperatures where more food is available. Common SPECIES include WOLVES, black bear, moose, lynx, and other MAMMALS with thick fur.

There are generally two kinds of temperate forests on Earth, TEMPERATE RAINFORESTS and TEMPERATE DECIDUOUS FORESTS. Temperate deciduous forests are characterized by seasonal changes in temperature; trees drop their leaves in the fall in advance of winter drought. Temperate deciduous forests primarily occur between 30° and 50° N and are found in parts of North America, Europe, and northeast Asia. Much of the eastern half of theUnited States, for instance, is considered to be temperate forest. Average precipitation ranges from about 75–150 centimeters (30–60 inches) per year. Temperatures can reach as high as 35°C (95°F) during the summer months and dip well below freezing in the winter. Common plants include maple, birch, and oak trees, and a variety of other plants that have adapted to survive seasonal changes. Animals show adaptations to the seasons as well. Common animals include migratory birds, deer, foxes, squirrels, and rabbits.

North America's only temperate rainforest is located in the Pacific Northwest. These unique ecosystems also can be found in parts of South America, New Zealand, and Australia. The Pacific Northwest's temperate rainforest occurs between 50° to 60° N; however, it never freezes because the Pacific Ocean moderates the coastal temperature and provides ample moisture. This mixture of moderate temperature and high rainfall makes the area perfectly suited to the growth of evergreen trees, such as spruce and fir.

Tropical rainforests are restricted to the equatorial parts of the world, such as in South America, Africa, and southeast Asia. This biome is characterized by heavy rainfall, about 250 centimeters (100 inches) per year, and high, stable temperatures year round of about 27°C (about 81°F). Tropical rainforests are crowded with many kinds of plants and have the greatest BIODIVERSITY of any biome. Most plants have large, broad leaves in order to maximize capture of the ample sunlight. Animals in tropical rainforests tend to specialize on one food item in order to avoid COMPETITION between SPECIES. Common plants include palm trees, orchids, and bromeliads. Animals common to tropical rainforests include monkeys, parrots, sloths, tree frogs, and a host of INSECTS.

The tundra is a treeless biome characterized by a very cold and dry climate. It is found across the world within the Arctic Circle. Temperatures range from about –25°C (–13°F) in the winter to about 6°C (43°F) in the summer. Precipitation averages less than 25 centimeters (10 inches) per year. The summer is very short in the TUNDRA and ice dominates the landscape for much of the year. Below the surface of the ground is a layer of permanently frozen soil called PERMAFROST. Although it is a treeless environment, it is not entirely without plants. During the summer, tough shrubs, mosses, and grasses grow. Plants in the tundra are very well adapted to the cold winters; they are short and grow very close to the ground. Flowering plants grow and flower very quickly to make use of the short growing season. Common animals include migratory birds, reindeer, Arctic fox, wolves, caribou, Arctic hare, and many types of insects. Many of the year-round animal residents have thick skin and fur to protect them from the cold. Other species build underground burrows to survive the harsh conditions.

Today, many of the world's ecosystems are being poisoned by pollution or cleared for AGRICULTURE, ranching, and other forms of development. Tropical rainforests are the most severely threatened ecosystems. The lush tropical rainforests of South America, Africa, and Asia previously covered about 20% of Earth's surface. Today, they cover only about 7%. With the disappearance of these HABITATS, plants and animals become extinct. *(See* extinction.) *See also* ALPINE TUNDRA; ARCTIC; BIOSPHERE; FORESTS; OLD-GROWTH FORESTS; PRAIRIE.

EnviroSources

Brown, Lauren. *Grasslands.* Audubon Society Nature Guides. New York: Knopf, 1985.

Committee for the National Institute for the Environment Web site: www.cnie.org/nle/biodv-6.html.

Langewiesche, William. *Sahara Unveiled: A Journey across the Desert.* Vintage Departures. New York: Vintage Books, 1997.

McMahon, James. *Deserts.* Audubon Society Nature Guides. New York: Knopf, 1985.

Sayre, April Pulley. *Temperate Deciduous Forest.* Exploring Earth's Biomes. Breckenridge, CO: Twenty-First Century Books, 1995.

Terborgh, John. *Diversity and the Tropical Rainforest.* Scientific American Library, no. 38. New York: W.H. Freeman & Co, 1992.

Yahner, Richard H. *Eastern Deciduous Forest: Ecology and Wildlife Conservation.* Wildlife Habitats, vol 4. Minneapolis: University of Minnesota Press, 1996.

Zwinger, Ann H., and Beatrice E. Willard. *Land above the Trees: A Guide to American Alpine Tundra.* Boulder, CO: Johnson Books, 1996.

biomonitoring, The use of ORGANISMS to measure the suitability or safety of an EFFLUENT that will be discharged into water. *(See* discharge). Biomonitoring is used to assess the effects of pollution and other factors on POPULATIONS in a particular HABITAT or ECOSYSTEM. Biomonitoring is also performed in the workplace to assess workers' exposure to chemical substances. In the work ENVIRONMENT, urine and blood samples are taken to determine exposure to TOXIC CHEMICALS. *See also* ACUTE TOXICITY; BIOASSAY; CHRONIC TOXICITY; OCCUPATIONAL SAFETY AND HEALTH ADMINISTRATION.

bioregion, An area that has boundaries consisting of natural rather than artificial borders and has a natural ECOSYSTEM with its characteristic FLORA, FAUNA, and environmental conditions. Many conservationists believe that for WILDLIFE CONSERVATION to be successful, it must be centered around identifying and preserving the natural features of the bioregion in which ORGANISMS live. Persons who believe that environmental policies should be centered around bioregions rather than regions determined by political or economic boundaries are called "bioregionalists." *See also* BIOCENTRIC; DEEP ECOLOGY.

bioremediation, The degradation, DECOMPOSITION, or stabilization of POLLUTANTS by MICROORGANISMS, such as BACTERIA, FUNGI, and CYANOBACTERIA. Each year, enormous quantities of ORGANIC and inorganic chemicals are released into the ENVIRONMENT. In some instances, these releases are deliberate and regulated, such as industrial EMISSIONS. In other cases, such as chemical or OIL SPILLS, they are accidental and sometimes unavoidable. Either way, many of the substances released into the environment are toxic and can present dangers to human health, as well as to aquatic and terrestrial ECOSYSTEMS. Government policies and environmental groups continue to apply pressure on industry to reduce TOXIC WASTE production and pollution, but today BIOTECHNOLOGY does have a way to detoxify many types of industrial EFFLUENTS: bioremediation. In the process of bioremediation, various toxic substances, such as PETROLEUM, POLYCHLORINATED BIPHENYLS (PCBs), GASOLINE and other FUELS, and PESTICIDES, are transformed into harmless substances by the action of microorganisms. Various PLANT SPECIES are also used to detoxify HEAVY METALS, a process known as PHYTOREMEDIATION.

Bioremediation research in the laboratory includes isolating and identifying microorganisms with bioremediation potential, developing genetically engineered microbes with CONTAMINANT-degrading abilities, and performing bioremediation experiments to determine the most effective ways to detoxify particular chemical compounds. The first oil spill to be treated by bioremediation occurred in 1990 when the OIL TANKER, *Mega Borg*, had an accident off the coast of Galveston, Texas. Approximately 95,000 barrels of oil spilled into the Gulf of Mexico. The U.S. Coast Guard mixed oil-eating microbes with sea water nutrients and spread the solution on one section of the oil spill. Bioremediation also successfully helped to clean up oil from one of the largest oil spills in history, the 1989 *EXXON VALDEZ* OIL SPILL disaster which occurred in Alaska's Prince William Sound.

One of the earliest and best-documented examples of bioremediation occurred in South Carolina. In 1975, an accident at a military fuel storage facility released thousands of liters of jet fuel. Scientists and engineers were able to contain the spill, but they could not prevent some of the fuel from soaking into the permeable soil and entering the underlying WATER TABLE.

The science and practice of bioremediation is advancing quickly. As bioremediation research and technology advances, industry and environmental agencies may increase use of microbes to clean up numerous leaking storage tanks and other hazardous land and water waste sites. *See also* HAZARDOUS SUBSTANCE; HAZARDOUS WASTE; HAZARDOUS WASTE TREATMENT.

EnviroSources

Bioremediation Consortium Web site: www.rtdf.org/public/biorem.
U.S. Environmental Protection Agency Web site: www.epa.gov.

biosphere, The portion of Earth that contains life including portions of Earth's ATMOSPHERE, LITHOSPHERE, and HYDROSPHERE. The word "biosphere" comes from the Greek word "bios," meaning "life," and the Latin word "sphere," which means "total range." Thus, the biosphere contains every living SPECIES on Earth.

Humans are living ORGANISMS and are also part of the biosphere. Human CELLS carry out cellular respiration and other metabolic processes and help cycle matter and ENERGY through the biosphere; however, humans, unlike other organisms, possess culture and technology. Some of their activities—such as dumping wastes into OCEANS, burning FOSSIL FUELS, and harvesting essential NATURAL RESOURCES such as timber—have an adverse impact on ECOSYSTEMS and on themselves. Today, many people feel that we must conserve the biosphere's resources and coexist peacefully with other organisms to ensure the health and success of future generations. (*See* conservation.)

Compared to the total size of Earth, the biosphere is quite small. Living things can be found in the AIR as far as 8 kilometers (5 miles) from Earth's surface, on and just below the surface, and within bodies of water. Spores and microscopic organisms are sometimes carried high into the atmosphere by air currents, and evidence exists that BACTERIA are present in PETROLEUM deposits at depths of about 2,000 meters (7,000 feet) below the surface, but many living things live within the upper 100 meters (300 feet) of Earth's crust.

The biosphere is characterized by the interactions between BIOLOGICAL COMMUNITIES and their ENVIRONMENTS. Terrestrial PLANTS obtain water from the soil; lizards, turtles, frogs, and other ectothermic animals warm their bodies in the sun; HERBIVORES such as deer, rabbits, and birds eat the leaves, seeds, and berries of plants; and predatory species such as WOLVES and great white sharks feed on other animals. (*See* predation.) Such interactions between organisms and between organisms and the nonliving components of the environment define an ecosystem. (*See* abiotic.)

The major metabolic processes that occur in the biosphere are PHOTOSYNTHESIS and respiration. Plants, ALGAE, and other PRODUCERS, through the process of photosynthesis, use the sun's energy to manufacture food. CONSUMERS, primarily animals and FUNGI, obtain this stored energy by feeding on producers or other animals that have consumed producers. Respiration, a metabolic process carried out by the cells of most living things, releases the stored energy, which is then used to FUEL reproduction, digestion, excretion, photosynthesis, and other biological processes.

Organisms, through their metabolic activities, are important components of several BIOGEOCHEMICAL CYCLES that help circulate water and other essential nutrients through the biosphere. CARBON, for example, is a naturally occurring element and the primary component of all living matter. Carbon in the environment exists primarily as CARBON DIOXIDE (CO_2) gas in the atmosphere and oceans. During photosynthesis, plants and other producers use carbon dioxide, water, and sunlight to make energy-rich sugars. When this food is eaten by consumers, carbon, as well as other substances and energy, is transferred from one organism to the next. Carbon returns to the air, water, and soil during the process of respiration, which releases carbon dioxide as a waste product. This continuous biogeochemical process, known as the CARBON CYCLE, transfers carbon throughout the biosphere. Water, NITROGEN, OXYGEN, PHOSPHORUS, and calcium are circulated in similar ways. *See also* BIOSPHERE RESERVES; CARNIVORE; MAN AND THE BIOSPHERE PROGRAMME; NITROGEN CYCLE; SYMBIOSIS; WATER CYCLE.

biosphere reserve, A multifunctional area encompassing terrestrial or coastal/marine ECOSYSTEMS, or a combination thereof, which is internationally recognized within the framework of the MAN AND THE BIOSPHERE PROGRAMME, begun in 1971 by United Nations Educational, Scientific, and Cultural Organization (UNESCO). Biosphere reserves are designed to meet one of the most challenging issues facing the world today: how to conserve the diversity of PLANTS, animals, and MICROORGANISMS that make up the living BIOSPHERE while at the same time meeting the needs of increasing human POPULATIONS. Collectively, biosphere reserves constitute an international network. Nominated by national governments, they must meet a minimum set of criteria and adhere to certain conditions before being admitted into the network.

Each biosphere reserve is intended to fulfill three basic functions, which are complementary and mutually reinforcing: the CONSERVATION of landscapes, ecosystems, SPECIES, and BIODIVERSITY; the development of socioculturally and ecologically sustainable economic and human infrastructure; and the provision of logistical support for research, education, and information exchange related to both conservation and development. The SUSTAINABLE DEVELOPMENT function, in particular, is a distinguishing feature of biosphere reserves.

To fulfill their conservation, development, and logistical roles, biosphere reserves typically comprise three distinct zones: a core area devoted to strict legal protection of landscapes, ecosystems, HABITATS, and species; a surrounding buffer zone, which permits activities that are compatible with conservation objectives; and an outer transition area, where sustainable resource management practices are developed in cooperation with local populations. Where they exist, transition zones tend to make up the largest percentage of the reserve's total area.

Biosphere reserves span a great variety of natural ENVIRONMENTS, from pristine mountains to human-impacted plains, coastal regions to vast inland FOREST, and tropical DESERTS to polar TUNDRA. To qualify for designation as a reserve, an area should be representative of a major biogeographic region and include a gradation of human effect on the region's natural systems. It should also contain landscapes, ecosystems, and/or species that require particular conservation and should provide opportunities to explore region-appropriate approaches to sustainable development. The biosphere reserve core area must already be a legally protected area under the laws of the host country.

Planning and management of each reserve involves the participation of public authorities, local communities, and private interests. National committees are responsible for preparing reserve nominations for consideration by the program's International Coordinating Council. The director general of UNESCO notifies the nominating nation of the council's decision.

As of 1996, there were 337 biosphere reserves located in 85 countries, with a total area of 219,891,487 hectares (547 million acres), of which 47 reserves were located in the United States. Most U.S. biosphere reserves are located within national parks, such as Denali National Park in Alaska, or national forests, such as San Bernardino National Forest in California.

Individual biosphere reserves remain under the jurisdiction of the countries in which they are located; however, exchanges between biosphere reserves, such as of research results or experiences in resolving specific issues, and cooperative activities, such as environmental education and specialist training, are encouraged. The international network of biosphere reserves is expanding every year, both in terms of the number of sites and the number of countries hosting reserves, and the network is constantly evolving to meet the dual needs of conservation and sustainable development for local communities. *See also* BIOREGION; FOREST SERVICE; NATIONAL PARK SERVICE.

EnviroSources

Man and the Biosphere Programme Web site: www.mabnet.org.
United Nations Educational, Scientific, and Cultural Organization (UNESCO) Web site: www.unesco.org.

biotechnology, The use of ORGANISMS or biological principles to solve practical problems. Biotechnology often involves manipulating organisms into making products or carrying out activities for humans. In AGRICULTURE, biotechnology has been used to develop PLANTS and animals that yield more food and to produce foods with certain qualities. Biotechnology also has been used to produce new medicines, to aid in waste disposal, to clean up toxic materials from the ENVIRONMENT, and in MINING.

Biotechnology that involves selective breeding—the deliberate breeding of organisms that have specific traits to produce these traits in the offspring—has been in use since ancient times. This practice began in agriculture, where it was used to develop diverse varieties of plants and animals used for food. (*See* hybrid; hybridization.) Selective breeding has been used to produce hardier and stronger animals for use in farm work or to produce animals, such as cattle, that have more meat or produce more milk than earlier varieties. Many countries of the world, including the United States, Japan, Mexico, and many nations of Europe and Africa continue to use selective breeding for these purposes. For example, scientists have developed new varieties of crop plants, such as wheat and corn, that are more resistant to DISEASE, frost, DROUGHT, and some TOXIC CHEMICALS than traditional varieties. Selective breeding also is used to produce new varieties of flowering plants that have certain coloration or other traits and to develop animals with traits, such as larger or smaller size, longer or shorter hair, or certain temperament characteristics to be kept as pets.

More recently, biotechnology has involved GENETIC ENGINEERING, the manipulation of the genes of organisms to produce offspring with certain traits or to derive certain products from the organisms. One type of genetic engineering used to derive products from organisms is gene splicing, or recombinant deoxyribonucleic acid (DNA) technology. DNA is a complex chemical compound (nucleic acid) that is responsible for transfering the chemical codes that determine an organism's traits from one generation to the next. Recombinant DNA technology involves the use of certain enzymes, called restriction enzymes, to cut open a strand of an organism's DNA. Once opened, a gene (a piece of DNA that carries instructions for certain traits) from another organism is inserted into the first strand in place of the organism's own DNA. This new gene drives the organism to produce the trait directed by the new gene as it reproduces. Scientists have used this technology to cause BACTERIA to make medicines such as human insulin, human interferon, human growth hormone, as well as vaccines used in veterinary medicine.

Transgenic organisms are plants, animals, bacteria, or FUNGI that have been genetically altered to contain a gene from a different SPECIES. Trangenesis is achieved by injecting a gene from one organism into a fertilized egg or into embryonic CELLS of another organism. Once injected, the gene becomes part of the host cell's DNA and is inherited by all cells formed during embryonic development. This new gene can then be inherited by offspring of the organism in the same way as the organism's own genes, carrying the desired trait from parent to offspring. Transgenesis has been used to develop new varieties of food, such as fruits and vegetables.

Recently, biotechnology has been used in waste management. For example, scientists have created bacteria that consume oil or toxic chemicals resulting from spills. Some plastics have been made more BIODEGRADABLE—that is, they more easily break down in the environment through natural processes—by adding cornstarch to them during their manufacture process. When disposed in soil, MICROORGANISMS feed on the cornstarch, helping to break the plastic apart. Polymers (very large molecules) of lactic ACID that have properties similar to plastics also have been made. Unlike traditional plastics, products made from the lactic acid plastic (which is derived from bacterial fermentation of corn stalks) is biodegradable. Inks derived from soybeans are now being widely used in the publishing industry. Unlike traditional inks, which are PETROCHEMI-

CAL products, these SOY INKS do not persist in the environment and are not toxic to organisms.

The mining industry has made use of biotechnology in the mining of zinc, LEAD, COPPER, and other metals. Most metals in the environment exist in combination with other elements as MINERALS. Copper, for example, generally is obtained from the mineral chalcopyrite. A bacteria can use copper molecules present in this mineral to form copper sulfate ($CuSO_4$). Once formed, the copper sulfate is treated with chemicals to retrieve the copper. Microbiological mining processes are currently used only with low-grade ores that cannot be profitably mined using more traditional methods and do not account for a large percentage of metal production; however, these practices are expected to become more widely used as high-grade deposits of metals are depleted. *(See* nonrenewable resources.)

Some uses of biotechnology are not without controversy. Many people oppose procedures that alter the genetic composition of organisms. *(See* biocentric.) One concern is that in the environment, genetically altered organisms will outcompete and replace naturally occurring species, thus disrupting the natural balance of ECOSYSTEMS. Another concern is that pathogenic organisms may be produced and be accidentally released into the environment, causing potentially life-threatening epidemics. *(See* pathogen.) Still other people object to manipulation of the genetic traits of organisms on religious grounds. To address some of these concerns, the National Institutes of Health developed rules in 1976 to regulate how organisms are used in recombinant DNA experiments. In addition, the U.S. FOOD AND DRUG ADMINISTRATION must approve the use of foods developed through transgenesis and requires that approved products be clearly identified, prior to their sale. *(See* ecolabeling.) In addition, the United States and many other countries recently developed laws prohibiting the use of some genetic engineering techniques, such as cloning, on humans. *See also* BIOLOGICAL CONTROL; BIOREMEDIATION; IRRADIATION; PRIMARY, SECONDARY, AND TERTIARY TREATMENTS; STERILIZATION, REPRODUCTIVE.

EnviroSources

Arms, Karen. *Environmental Science.* Austin, TX: Holt, Rinehart and Winston, 1996.

Brown, Lester R., et. al. *State of the World: 1998.* New York: W.W. Norton, 1998.

Chiras, Daniel. *Environmental Science: A Framework for Decision Making.* Menlo Park, CA: Addison-Wesley, 1989.

Seymour, John, and Herbert Giardet. *Blueprint for a Green Planet: Your Practical Guide to Restoring the World's Environment.* New York: Prentice Hall, 1987.

biotic, A term used to describe the living components of the ENVIRONMENT—BACTERIA, FUNGI, PROTISTS, PLANTS, and animals. The biotic factors, or ORGANISMS, in an environment constantly interact with one another and sometimes, as occurs in feeding relationships, depend upon one another for their survival. In addition, the living components of an environment continuously interact with ABIOTIC factors such as sunlight, soil, water, MINERALS, and AIR. *See also* BIODIVERSITY; BIOLOGICAL COMMUNITY; BIOSPHERE; ECOSYSTEM; SPECIES.

biotic community. *See* biological community.

birth rate, A measure of the number of births per 1,000 members of a POPULATION over a one-year period of time that is used to calculate population growth. Changes in birth rates are of concern to environmental scientists because they may indicate changes in environmental conditions. For example, increased birth rate may indicate plentiful resources, low COMPETITION, and favorable CLIMATE conditions. A decline in birth rate may be an indicator of a scarcity of resources, DISEASE, or other factors that indicate a population or SPECIES is in decline.

Birth rate is calculated by dividing the annual number of births by the total population at midyear and multiplying the resulting product by 1,000. An example is shown. The example assumes that in a one-year period a total of 25 individuals were born in a population that had 1,500 individuals at midyear.

$$\text{Birth Rate} = \frac{\text{Number of Births} \times 1{,}000}{\text{Population Size}} \qquad \text{Birth Rate} = \frac{25 \times 1{,}000}{1{,}500}$$

$$\text{Birth Rate} = 0.0166 \times 1{,}000 = 16.6$$

Birth rate is one of the factors used to determine population growth or decline. Other factors considered include immigration, which increases population size and DEATH RATE and EMIGRATION, which decrease population size. *See also* AGE STRUCTURE; COMPETITION; NATURAL RESOURCE; POPULATION DENSITY; ZERO POPULATION GROWTH.

bison, A North American herd animal, also known as the American buffalo *(Bison bison)*, which is closely related to domestic cattle and sheep. Bison once roamed North America in the tens of millions, but were close to EXTINCTION by the end of the nineteenth century, primarily due to intensive hunting.

The size and color of bison vary with POPULATION locations, but most experts believe all populations belong to the same SPECIES, with differences in appearance resulting from the diversity of the ENVIRONMENTS in which bison live. Both male and female bison have a set of hollow, curved horns. Males often weigh more than 0.9 metric tons (1 ton) and stand 2 meters (6 feet) tall at the shoulder. They have a large head, a woolly hump, and narrow hips. Despite their size and bulkiness, bison are agile and can sprint at speeds of up to 38 kilometers (30 miles) per hour. Bison are hardy animals that can live up to 20 years; they are able to endure long periods of poor grazing. They usually travel in small family groups composed of a bull and a few cows. WOLVES are their main nonhuman predator. *(See* predation.)

It is believed that bison crossed over the land bridge that once connected Asia with North America. They slowly moved southward, eventually traveling as far south as Mexico and as far east as the Atlantic Coast, extending south to Florida. The largest herds lived on the plains and PRAIRIES from the Rocky

The bison is a native species to North American prairies.

Mountains east to the Mississippi River, and from the Great Slave Lake in Canada to Texas. *(See* grassland.)

In the early 1800s, an estimated 65 million bison lived in North America. With westward expansion of the American frontier, systematic reduction of the bison herds began around 1830, when buffalo hunting became a significant industry in the area. Hunting for meat, hides, and sport had a devastating effect on the bison population, and by 1890, fewer than 1,000 remained. Even with the establishment of YELLOWSTONE NATIONAL PARK, protection of the bison did not begin until the U.S. Army arrived in 1886 to guard the park's resources.

Today, Yellowstone National Park is home to the largest wild, free-roaming bison herd in the United States, descendants of those that once roamed the continent. As a result of the protection of bison within Yellowstone, approximately 1,500 animals, many of which had been transplanted from other areas, remained in the park by 1954. In 1968, manipulation of Yellowstone bison herds was discontinued, and the population was allowed to fluctuate with environmental conditions such as WEATHER and food availability. In May 1984, Congress enacted a law making bison hunting in Yellowstone National Park illegal. Eight years later, money was appropriated to purchase 21 bison from private herds to further enhance the Yellowstone herd, which has since grown steadily.

During harsh winters, Yellowstone's bison migrate to lower elevations outside the park, where they can forage. Pushed by harsh winters, more than 1,900 Yellowstone bison tried to leave the park over the last four years. These animals were killed by state and federal officials to prevent the spread of BRUCELLOSIS, a bacterial DISEASE that can cause domestic cows to abort their calves. Incidents of brucellosis were first detected in Yellowstone bison in 1917; the disease has since been detected in the region's elk. In the winter of 1996–1997, NATIONAL PARK SERVICE officials killed 1,100 animals; many others died from natural causes. As a consequence, the Yellowstone bison population declined to just 2,200 animals of the 3,500 estimated to inhabit the park in 1995. Federal and state agencies claim that the killing is necessary and unavoidable; opponents assert that there has never been a confirmed case of brucellosis transmission from wild bison to cattle. Environmental groups believe that the U.S. FOREST SERVICE should modify grazing allotments north and west of Yellowstone National Park and use land exchanges and CONSERVATION EASEMENTS to set aside more public land to which bison and other WILDLIFE can migrate without contacting cattle.

Additional bison populations inhabit the National Bison Range in the Flathead Valley of Montana, the Wichita Mountains National Wildlife Refuge in southwest Oklahoma, the Fort Niobrara National Wildlife Refuge in northern Nebraska, and the Sullys Hill National Wildlife Refuge in northwestern North Dakota. Many private herds, maintained primarily for meat production, also have boosted the bison's overall population. Currently the 200,000 individuals of the overall population remains smaller than the great herds that once ranged North America, but this is enough to ensure the continued survival of the species. *See also* BALD EAGLES; CALIFORNIA CONDORS; ENDANGERED SPECIES; ENDANGERED SPECIES ACT; ENDANGERED SPECIES LIST; HERBIVORE; NATIONAL WILDLIFE REFUGE SYSTEM.

EnviroSources

National Park Service Web site: www.nps.gov.
Peacock, Doug. "The Yellowstone Massacre." *Audubon* 99, no. 3 (May–June 1997): 40–49.

bituminous coal. *See* coal.

Black Sea, An inland sea located in southeastern Europe that is suffering from OVERFISHING and increasing in levels of TOXIC WASTE. About 90% of the water in the deep bottom of the Black Sea is anoxic—totally devoid of DISSOLVED OXYGEN.

The Black Sea is surrounded by six countries (Ukraine, Russia, Georgia, Turkey, Bulgaria, and Romania), has an area of about 460,000 square kilometers (178,000 square miles) and has a maximum depth of 2,200 meters (7,250 feet). Almost one-third of the entire land area of continental Europe, which includes major parts of 17 countries, drains into the Black Sea. This area around the sea is home to about 160 million people. Three major European rivers, the Danube, Dnieper, and Don, all flow into the Black Sea. Waters from the Black Sea eventually empty into the Atlantic Ocean after flowing through the narrow Bosporus Channel into the Sea of Marmara, then to the Dardanelles, the Aegean Sea, and the MEDITERRANEAN SEA.

Within the last 30 years, the Black Sea has suffered catastrophic degradation of its NATURAL RESOURCES, which has affected COMMERCIAL FISHING, tourism, and swimming. Nutrients discharged into the Black Sea from agricultural activities have caused an overproduction of ALGAE, which has lead to EUTROPHICATION. As a result, BACTERIA in the bottom waters consume all available OXYGEN and the sea is virtually dead below a depth of about 180 meters (550 feet). The loss of marine life and the overexploitation of remaining fish stocks have caused a sharp decline in commercial fisheries. Uncontrolled SEWAGE pollution has led to frequent beach closures and considerable losses in the tourist industry. In some places untreated municipal SOLID WASTE and industrial WASTEWATER is being dumped directly into the sea or onto valuable WETLANDS. OIL TANKER accidents and OIL SPILLS from PETROLEUM drilling activities have added to the problems.

In 1993, the environmental problems of the Black Sea were addressed at the first meeting of the Black Sea Environmental Programme Steering Committee in Bulgaria. The overall objectives were to improve the capacity of Black Sea countries to support the development and implementation of new environmental policies and laws, to control land-based sources of pollution and the dumping of waste, and to provide a plan in the case of accidents such as oil spills.

Most experts believe that the protection of the Black Sea cannot be achieved on a unilateral basis, but rather through joint management. Besides the six countries that border the sea, the responsibility of controlling aquatic and airborne pollution has to be shared among the other 11 countries that have a major part of their land in the Black Sea DRAINAGE BASIN. Only concerted international action can do anything to protect the BIODIVERSITY of the Black Sea. If polluted conditions do not improve, some scientists predict that the Black Sea will be lifeless by the year 2050. *See also* ADRIATIC SEA; AGRICULTURAL POLLUTION; NORTH SEA; OCEAN DUMPING.

blight, The common name for any fungal DISEASE that affects PLANTS. Blights spread rapidly, infecting and killing numerous plants, including those used for food or other products. The Irish potato FAMINE of the mid-1840s resulted from a water mold that decimated the potato crop; as a result, more than one million people died in only one year. In 1904, a FUNGUS introduced to the United States from Asia destroyed most of the American chestnut POPULATION—trees that once produced vast supplies of timber and edible nuts. Fire blight is a fungal disease that infects apples and pears. To prevent blight, many crops and decorative plants are treated with fungicides—PESTICIDES designed to kill fungi. *See also* DUTCH ELM DISEASE; PARASITISM.

blue-green bacteria/algae. *See* cyanobacteria.

bog, A type of WETLAND that accumulates appreciable PEAT deposits and often serves as an intermediate stage of ECOLOGICAL SUCCESSION. Bogs depend primarily on precipitation as their water source and are usually acidic and rich in PLANT material. Grasses, sedges, and mosses are typical bog vegetation. Animal life supported by bogs include a variety of INSECTS, waterfowl, AMPHIBIANS, and REPTILES. These animals, in turn, may support several species of predatory birds and MAMMALS. A bog can exist for a long period of time; however, some may become filled with sediment, and provide suitable HABITAT for shrubs and trees and eventually a FOREST ECOSYSTEM in the final stage of the ecological succession. *See also* ACID; MARSH; PIONEER SPECIES; SEDIMENTATION; SWAMP.

The pitcherplant grows in bog-like soils.

boreal forest. *See* taiga.

Borlaug, Norman Ernest (1914–), An agricultural scientist, forester, and PLANT pathologist who developed several HYBRID strains of wheat and other grains that were used to increase food production throughout the world. Many of the plants developed by Borlaug had traits that allowed them to be grown in CLIMATES that did not support other, related food plants. Borlaug also introduced new agricultural technologies, such as the use of machinery and PESTICIDES, and encouraged wider use of IRRIGATION to help increase crop production of people living in many developing nations. This work served as the basis of the green revolution. *(See* EnviroTerm.)

Borlaug began his professional career working as a forester for the U.S. FOREST SERVICE in the 1930s. *(See* forestry.) He later spent 15 years working as a plant pathologist, first with the Rockefeller Foundation and later for the government of Mexico. It was during this period that Borlaug developed several of his hardy, hybrid grains, which helped improve food yields in a number of developing countries throughout the world. For his work in this area, Borlaug was awarded the Nobel Peace Prize in 1970.

The work done by Borlaug improved the lives of many people worldwide by providing them better access to food; however, it also introduced new environmental problems to areas that implemented the technologies. For example, increased use of pesticides in some areas have led to pollution problems such as RUNOFF from croplands carried chemicals into nearby waterways. In addition, Borlaug stressed that his methods could not be sustainable nor completely successful if the growth of the global human POPULATION was not slowed. *See also* AGRICULTURE; AGROFORESTRY; EHRLICH, PAUL RALPH; GREENBELT MOVEMENT; ZERO POPULATION GROWTH.

EnviroTerm

green revolution: Term used to describe the attempt to significantly increase crop production and diversity in certain developing nations (particularly Mexico, the Philippines, India, and Pakistan) from the 1960s–1980s. The green revolution stressed the planting of high-yielding HYBRID grains, along with increased use of FERTILIZERS, PESTICIDES, and IRRIGATION.

botany, The branch of biology that deals with the study of PLANTS, including their roles in the ENVIRONMENT. Scientists who work in this field are called "botanists." One of the main concerns of botany is the classification of plants into taxonomic groups based on their physical traits and evolutionary relationships. Botanists also study the structure and physiology of plants, factors that affect plant growth, and plant DISEASES. *See also* BIODIVERSITY; FLORA; GENETIC ENGINEERING; TAXONOMY.

bottle bill, Legislation that helps reduce LITTER and encourages RECYCLING by implementing a refundable container deposit. Such laws require that consumers pay a small deposit when they purchase a beverage container and receive a refund when the container is returned. Store owners or special outlets are responsible for collecting the containers and storing them for collection or recycling.

Oregon passed the first bottle recycling bill in 1972. Massachusetts, Maine, Delaware, Michigan, Connecticut, Vermont, Iowa, New York, and California are other states that have bottle laws. Those in favor of such laws claim that bottle bills reduce litter: most containers can be refilled several more times, and ENERGY is conserved in producing new containers.

Those opposed to such laws claim that bottle bills increase costs for labor to handle the returned containers, create sanitary problems in handling the returnables, and raise beverage costs for the consumer. Opponents also contend that many kinds of containers cannot be recycled or refilled. Despite this opposition, public support for bottle deposit laws is high, and none of the bottle bill laws in the United States have been repealed.

Outside of the United States, there are container deposit programs in Sweden, Denmark, Norway, and the Netherlands. Several provinces in Canada also have bottle-deposit laws.

Internationally, many of the packaging reform movements that emerged in the 1980s and 1990s began in the form of voluntary campaigns. In Japan, for instance, the PET Bottle Recycling Promotion Association recommended the voluntary boycott of colored bottles, laminated aluminum foil labels, and the use of low-environmental-impact (environmentally friendly) inks. Several Japanese firms formed a panel to promote recycling plastic containers and packaging and to prepare a report on recycling systems to the Japanese government and the plastic packaging industry.

Australians also have sought to implement voluntary recycling programs supported by government. The National Producer Responsibility Bill was suggested as a means of accelerating a movement toward recyclable packaging in the private sector. The Australian and New Zealand Environmental Council has worked with the Packaging Council of Australia to establish a covenant that encourages manufacturers to reduce, voluntarily, the amount of packaging and to use recycled material. *See also* SOY INK.

bovine spongiform encephalopathy, A transmissible, fatal, neurodegenerative illness in cattle, also known as mad cow disease, that first came to the attention of the scientific community in 1986 when it appeared in British cattle. Over the next 10 years, approximately 160,000 cases of bovine spongiform encephalopathy (BSE) were confirmed in Great Britain. Epidemiological studies indicated that the source of the DISEASE was cattle feed made from the carcasses of livestock, such as sheep, that were infected with a similar disease. The use of carcasses in feed preparation had been introduced in the early 1980s. Others speculate that the initial appearance of the disease was a spontaneous occurrence in cattle.

Researchers do not completely understand the infectious agent that causes BSE, which affects the brain and spinal cord of cattle, causing the formation of lesions characterized by abnormal, spongy tissue. Many believe it is caused by a prion, an infectious agent similar to a VIRUS, but lacking some viral components. The agent is very stable, resisting even the high heat of pasteurization and sterilization, and is equally unaffected by freezing and desiccation. *(See* EnviroTerms; sterilization, pathogenic.) The disease is fatal for cattle within weeks or months of its onset.

By October 1996, BSE had been reported in 10 areas outside Great Britain. In Switzerland, France, Portugal, and the Republic of Ireland, the disease occurred in native cattle, probably because their feed had been imported from Great Britain. In Germany, Italy, Denmark, Canada, the Falkland Islands, and Oman, the only cases appeared in cattle that had been imported from Great Britain. To date, no cases have been reported in the United States.

A newly recognized form of Creutzfeldt-Jakob disease (CJD), the human version of mad cow disease, was identified in 10 human patients and reported by Great Britain to the WORLD HEALTH ORGANIZATION in March 1996. In contrast to typical occurrences of CJD, this variant form affected mainly young patients and caused a relatively long illness as compared to classic CJD. Public health experts and disease specialists concluded that while no definite link between BSE and the variant of CJD could be established on the basis of current knowledge, circumstantial evidence suggested that exposure to BSE through ingestion or handling of affected cattle tissue was the most likely cause.

In July 1988, Great Britain banned the use of ruminant proteins in the preparation of cattle feed, and in November 1989 it banned the use of cattle brains, spinal cords, tonsils, thymuses, spleens, and intestines in foods for human consumption. British cattle herds are now continuously monitored for BSE. Affected individuals are destroyed, and the incidence of the disease is decreasing. *See also* AGRICULTURE; DEPARTMENT OF AGRICULTURE; *ESCHERICHIA COLI*; PATHOGEN; *SALMONELLA*.

EnviroTerms

desiccation: A method of food preservation that involves drying or removing all moisture from the food.

pasteurization: A method of food preservation that involves heating the food to a specific temperature for a specific period of time as a means of killing MICROORGANISMS that could cause disease, spoilage, or other undesired effects such as fermentation.

EnviroSources

Centers for Disease Control Web site: www.cdc.gov/ncidod/diseases/cjd/bsenet.htm.

CJD Information Webring: www.webring.org/cgi-bin/webring?ring=cjdring;list.

U.S. Department of Agriculture, Food Safety and Inspection Service Web site: www.fsis.usda.gov.

World Health Organization: Avenue Appia 20, CH 1211, Geneva 27, Switzerland; Web site: www.who.int.

brackish, A term used to describe water that is a combination of fresh water and salt water. While the salinity of salt water is 35,000 parts per million, the average salt content in brackish water ranges from 1,500 to 6,000 parts per million, still saltier than POTABLE WATER which is about 300 to 500 parts per million. Brackish water ENVIRONMENTS are fluctuating environments. The salinity is variable depending on the TIDES, the amount of fresh water entering from rivers and streams or as rain, and the rate of evaporation. As a result, ORGANISMS that inhabit brackish waters are very tolerant of changes in salinity. ADAPTATIONS to assist survival under such conditions vary. Some SPECIES, for instance, avoid periods of high salinity by moving upstream or burrow under the sediment.

ESTUARIES are the best known brackish water environments. An estuary is formed where a river meets the OCEAN. Typically, estuarine waters are tremendously productive, but contain few species. This apparent contradiction occurs because relatively few fishes and INVERTEBRATES can tolerate the fluctuations in salinity. On the other hand, those animals that can live there do so in enormous numbers.

Another important brackish water HABITAT is the MANGROVE SWAMP. Mangrove swamps, or mangals, are characteristic of the tropics. Mangrove PLANTS are some of the most remarkable plants in the world. They form dense FORESTS that provide habitat for a tremendous variety of organisms both above and below the water's surface. Monkeys, snakes, and a huge variety of birds are to be found in the canopy, while the aquatic roots of mangroves support oysters and barnacles, as well as many fishes, snails, and CRUSTACEANS. The mud that collects within mangrove roots is also a prime habitat for fiddler crabs and various clams.

The temperate equivalent of the mangrove swamp is the SALT MARSH. Salt marshes are tidal habitats periodically covered by the sea. Salt marshes are important habitats, especially as feeding grounds for various sea birds, fishes, and crustaceans. Instead of trees, the major plants in a salt marsh are grasses, such as the hardy cordgrass. Unfortunately, salt marshes, mangrove swamps, and other estuaries are threatened by pollution, agricultural and commercial development, and OVERFISHING. *See also* CHESAPEAKE BAY; OCEAN DUMPING.

breeder reactor, A NUCLEAR REACTOR that generates at least as much fissionable FUEL as it consumes, while also generating steam that can drive a TURBINE to produce ELECTRICITY. Because they produce fuel, breeder reactors could greatly extend the useful life of URANIUM reserves; however, a disadvantage of breeder reactors is that the electricity they generate is generally more costly than that produced by other nuclear reactors. Breeder reactors are also expensive to construct and require more complex engineering tasks to make them operational. Another disadvantage is that PLUTONIUM–which is produced by breeder reactors—is much more radioactive than uranium, making disposal of breeder reactor wastes difficult and the threat of a nuclear disaster more imposing. Plutonium can also be used to build NUCLEAR WEAPONS.

A breeder reactor is designed to produce fissionable plutonium-239 (Pu-239) from nonfissionable uranium-238 (U-238). The core of the reactor consists of bundles of plutonium-filled FUEL RODS surrounded by an outer layer of U-238 fuel rods. The U-238 fuel rods are bombarded by high-speed neutrons, which split apart the U-238 and cause a CHAIN REACTION in which some U-238 is transformed into Pu-239. Only high-speed neutrons can cause U-238 to split in a way that releases enough neutrons to continue the chain reaction and produce plutonium in sufficient amounts for extraction and processing for later use as a fuel.

The use of high-speed neutrons in breeder reactors requires the use of special coolants other than water. The use of water would slow down the neutrons, preventing them from converting U-238 into Pu-239. Instead of using water to cool its core, a breeder reactor uses liquid sodium or helium gas, substances that do not affect neutron speed.

Issues over cost and safety have caused many countries to abandon plans to construct breeder reactors. In 1983, the United States ceased funding for the nation's first breeder reactor in Clinch River, Tennessee, because of cost concerns and growing opposition to the use of NUCLEAR ENERGY by U.S. citizens. In September 1995, a nuclear accident occurred during the startup testing of the Monju breeder reactor in Japan. The accident led to decreased support for nuclear energy and caused the Japanese government to abandon subsidies for development of future reactors. *See also* CHERNOBYL; THREE MILE ISLAND.

EnviroSources

Nuclear Information and Resource Service Web site: www.nirs.org

U.S. Department of Energy, Office of Nuclear Energy, Science and Technology Web site: www.ne.doe.gov.

U.S. Nuclear Regulatory Commission Web site: www.nrc.gov.

Wolfson, Richard. *Nuclear Choices: A Citizen's Guide to Nuclear Technology.* Cambridge, MA: MIT Press, 1993.

British thermal unit. *See* energy.

Brower, David Ross (1912–), A leading spokesperson for and writer on ECOLOGY and CONSERVATION issues. Born in Berkeley, California, David Brower is the founder of the Earth Island Institute (EII), an organization started in 1982 to build awareness of how people can conserve, preserve, and restore the ENVIRONMENT, activities EII members refer to as Earth CPR. Brower has also written several books, including *Only a Little Planet* (1975), *For Earth's Sake: The Life and Times of David Brower* (1990), and *Let the Mountains Talk, Let the Rivers Run* (1995).

Brower's conservationist activities began in 1933 when he joined the SIERRA CLUB. He later became a member of the Sierra Club's board of directors and served as its first executive director (1952 through 1969). In 1970, after leaving the Sierra Club, Brower founded the conservation group FRIENDS OF THE EARTH (FOE). The goals of the FOE were to raise global awareness of such environmental problems as ACID RAIN, GLOBAL WARMING, and pollution, and to advocate environmentally friendly activities such as resource conservation and SUSTAINABLE AGRICULTURE. To help reach these goals, Brower facilitated the founding of independent FOE branches in other countries. Today, the FOE operates in 53 countries worldwide. The same year that he began the FOE, Brower also founded the political group, the League of Conservation Voters (LCV). The LCV is made up of representatives from environmental and conservation organizations who work to back political candidates who promote and support activities and legislation that benefit the environment.

For his work on conservation issues, David Brower has gained worldwide fame. In 1988 and again from 1990 through 1992, Brower was invited by the Soviet government to lead delegations to LAKE BAIKAL in SIBERIA. The purpose of the trip was to aid in the protection and restoration of the area. In 1993, Brower conducted a speaking tour called "Vision for Action," which traveled throughout the United States and Europe. Today, Brower's work continues through the Ecological Council of Americas (ECA), a group he cofounded in 1994. The ECA is a network of American organizations that seek ways to integrate economic concerns into solutions to environmental problems. For his work on conservation issues, David Brower was nominated for the Nobel Peace Prize in 1978, 1979, and again in 1998, at which time he was nominated with PAUL EHRLICH.

EnviroSources

Brower, David. *For Earth's Sake: The Life and Times of David Brower.* Layton, Utah: Gibbs Smith, 1990.

———. *Let the Mountains Talk, Let the Rivers Run: A Call to Those Who Would Save the Earth.* New York: Harper Collins, 1995.

———. *Only a Little Planet.* New York: McGraw-Hill. 1972.

Brower Web site of the Earth Island Institute: earthisland.org/brower/about.htm

Earth Island Institute Web site: www.earthisland.org/

League of Conservation Voters Web site: www.lcv.org/

Sierra Club Web site: www.sierraclub.org

brownfield, A term that usually refers to industrial property that is perceived as undesirable to own or use because of the possibility that environmental contamination exists within the boundaries of the property. Old industrial properties, such as factories that may have been abandoned after going out of business, may be termed "brownfields" because the soil or GROUNDWATER on the property may contain oil or hazardous materials that would be costly to clean up and could harm the health of ORGANISMS or the ENVIRONMENT. Often such properties are not utilized to their best potential. When industrial properties are abandoned (i.e. left unused), often the local government cannot obtain taxes from the unused properties, which creates a burden for the community. Even though there may be interest in purchasing and redeveloping such properties, the prospective purchasers and bankers often fear the consequences of the COMPREHENSIVE ENVIRONMENTAL RESPONSE, COMPENSATION, AND LIABILITY ACT (CERCLA), which could make them legally responsible for cleanup costs. In the 1990s, state and local governments implemented new brownfields laws that helped businesses to investigate, clean up, and purchase brownfields by removing some of the CERCLA difficulties and making it easier to obtain funding for brownfields investigations and cleanup activities. *See also* HAZARDOUS SUBSTANCE; HAZARDOUS WASTE; LOVE CANAL; RESTORATION ECOLOGY; SUPERFUND; TIMES BEACH SUPERFUND SITE.

EnviroSources

U.S. Environmental Protection Agency, Brownfields Web site: www.epa.gov/swerosps/br/

browser. *See* herbivore.

brucellosis, A DISEASE caused by BACTERIA of the genus *Brucella* that affects cattle, pigs, sheep, goats, BISON, dogs, and sometimes humans. Brucellosis is a worldwide problem in livestock. Transmission occurs through contaminated feed, animal wastes, or direct contact. In affected livestock, the disease causes spontaneous abortion. *Brucella* bacteria can be transmitted to humans in milk or through direct contact with infected animals. Ingesting only pasteurized milk and aged cheeses lowers the risk of contamination; however, people such as meat packers, ranchers, and veterinarians, who work closely with animals prone to brucellosis, are at higher risk of infection. People infected with brucellosis generally suffer from fever and chills, headache, malaise, and sometimes diarrhea. Initial symptoms may last from one week to several months. Recurrences of symptoms generally occur after a remission of 2 to 14 days, with the cycle repeating for months or even years. The disease is rarely fatal and once diagnosed can be treated with a combination of antibiotics and steroids. Even with treatment, relapses are common. *See also* AGRICULTURE; BOVINE SPONGIFORM ENCEPHELOPATHY; CONTAMINANT; ENVIRONMENTAL MEDICINE; *ESCHERICHIA COLI*; PATHOGEN; VIRUS; WORLD HEALTH ORGANIZATION.

Global View

Brucellosis infection in humans is known by different names throughout the world. Other names for this disease include undulant fever, Malta fever, Mediterranean fever, and Gibraltar fever.

BTU. *See* energy.

buffalo. *See* bison.

Bureau of Land Management, An agency of the U.S. DEPARTMENT OF THE INTERIOR that is responsible for managing 107 million hectares (264 million acres) of PUBLIC LAND—about one-eighth of the land in the United States—and about 122 million additional hectares (300 million acres) of lands with MINERAL reserves. The Bureau of Land Management (BLM) also is responsible for wildfire management on a total of 157 million hectares (388 million acres). The BLM's mission is to "sustain the health, diversity, and productivity of the public lands for the use and enjoyment of present and future generations."

Most of the lands managed by the BLM are located in the western United States, including Alaska, and represent many ECOSYSTEMS, including GRASSLANDS, FORESTS, mountains, TUNDRA, and DESERT. The BLM manages a wide variety of NATURAL RESOURCES, including ENERGY and minerals, timber, fish and WILDLIFE HABITAT, and WILDERNESS areas. It also manages cultural heritage resources.

The BLM's origins date to the Land Ordinance of 1785 and the Northwest Ordinance of 1787, which provided for the survey and settlement of lands ceded to the federal government by the 13 original colonies following the Revolutionary War. As the United States acquired additional lands from other countries, the Congress directed that they be explored, surveyed, and made available for settlement. In 1812, Congress established the General Land Office within the Department of the Treasury to oversee the disposition of these lands and to encourage settlement of the land by enacting laws such as the Homestead Act of 1862 and the GENERAL MINING LAW OF 1872. With the exception of the General Mining Law of 1872 and the Desert Land Act of 1877 (which has since been amended), all of these early laws have been repealed or superseded by other statutes.

In the late nineteenth century, new federal land management priorities created the first national parks, national forests, and WILDLIFE REFUGES. By withdrawing these lands from settlement, Congress formalized a shift in the purposes that public lands were expected to serve. There was a new recognition that such lands should be held in public ownership because of their natural resource values. In the early twentieth century, Congress took additional steps toward recognizing these values. For example, the Mineral Leasing Act of 1920 permitted leasing, exploration, and production of certain commodities, such as COAL, PETROLEUM, NATURAL GAS, and sodium. The Taylor Grazing Act of 1934 established the U.S. Grazing Service to manage the public rangelands; the Oregon and California Act of 1937 required SUSTAINED-YIELD management of the forests of western Oregon.

In 1946, the U.S. Grazing Service merged with the General Land Office to form the Bureau of Land Management. At the time, there were more than 2,000 unrelated and often conflicting laws governing management of the public lands. In 1976, Congress passed the Federal Land Policy and Management Act (FLPMA) to provide the BLM a unified legislative mandate, which recognized the value of the remaining public lands by declaring that they would stay in public ownership. Through the FLPMA, Congress also developed "MULTIPLE USE management" as a land management term and philosophy.

Public lands are increasingly viewed from the perspective of the recreational opportunities they offer, their cultural resources, and—in an increasingly urban world—their vast open spaces; however, against this backdrop, the more traditional land uses of grazing, timber production, and MINING are still in high demand. The public, constituent groups, and other government agencies have proven eager to participate in collaborative decision-making. *See also* BIOREGION; BIOSPHERE; BIOSPHERE RESERVE; CONSERVATION; ECOSPHERE; FOREST FIRE; LAND TRUST; SUSTAINABLE DEVELOPMENT.

EnviroSources

Bureau of Land Management Web site: www.blm.gov.
Public Lands Foundation Web site: www.publicland.org.

Bureau of Reclamation, An agency of the U.S. DEPARTMENT OF THE INTERIOR whose mission is to manage, develop, and protect water and related resources in an environmentally and economically sound manner. The Bureau of Reclamation functions as the fifth-largest electric utility in the 17 western states and is the nation's second-largest wholesale water supplier, administering 348 RESERVOIRS with a total storage capacity of 245 million acre-feet. *(See* EnviroTerms.) The agency operates 58 HYDROELECTRIC POWER plants that generate about 42 billion kilowatt hours of ELECTRICITY annually, and it delivers 38 trillion liters (10 trillion gallons) of water to more than 31 million people each year. It provides 1 out of 5 western farmers with IRRIGATION water for 4 million hectares (10 million acres) of farmland that produce 60% of the nation's vegetables and 25% of its fruits and nuts. In partnership, it also manages 308 recreation sites that are visited by 90 million people each year.

Inadequate precipitation in the West forced early settlers to use irrigation for AGRICULTURE. At first, they diverted water from rivers and streams for this purpose, but in many arid areas, demand outstripped supply. As the need for water increased, settlers sought ways to store storm RUNOFF for later use in drier seasons. Supporters of irrigation believed that water management programs would also encourage western settlement. Pressure mounted for the federal government to invest in water storage and irrigation projects.

The Reclamation Act of 1902 created the U.S. Reclamation Service within the U.S. GEOLOGICAL SURVEY (USGS) to help alleviate water problems. (At the time, irrigation projects were known as "reclamation" projects because they "reclaimed" arid lands for human use.) Revenue from the sale of federal lands was the initial source of the program's funding. In the early years, many projects encountered problems, such as unsuitable soils poor settlement patterns, high land preparation and construction costs, settlers who were inexperienced in irrigation farming, and expensive drainage projects. In 1907, the secretary of the interior removed the Reclamation Service from the USGS and created an independent bureau within the department.

The bureau's construction of water facilities peaked during the Great Depression and after World War II. The last major authorization for construction projects occurred in the late 1960s, after which the growing American environmental movement began to voice strong opposition to water development projects. After constructing more than 600 DAMS and reservoirs, including the well-known Hoover and Grand Coulee Dams, The Bureau of Reclamation's initial mission was accomplished. Few, if any, new federally funded dams were likely to be built in the United States because of high financial costs and adverse environmental impacts. Between 1988 and 1994, the agency underwent a major reorganization as construction on projects that had been authorized in the 1960s and earlier was completed. The bureau's programs shifted emphasis from construction to operation and maintenance of existing facilities. Its current mission also places greater emphasis on water CONSERVATION, RECYCLING, and reuse; developing partnerships with its customers, states, and Native American tribes; and transferring title and operation of some facilities to local beneficiaries who can operate them more efficiently. *See also* GROUNDWATER; MONO LAKE; OGALLALA AQUIFER; POTABLE WATER; SURFACE WATER.

EnviroTerm

acre-foot: A unit of measure equal to 325,851 gallons of water, enough water to supply a family of four for one year.

EnviroSource

Bureau of Reclamation Web site: www.bor.gov.

cacti, A PLANT family (*Cactaceae*) whose members are generally adapted to life in DESERT or near-desert ENVIRONMENTS. There are about 2,000 SPECIES of cacti, most of which are native to the southwestern United States, Mexico, Central America, and several countries of South America. Of the 2,000 known cacti species, approximately 20 are at risk of EXTINCTION, primarily because of overcollecting by humans and damage caused by grazing animals.

Cacti have several ADAPTATIONS to help them conserve water and thrive in arid regions. Most cacti are succulents—plants with swollen, fleshy stems that store large amounts of water. The stems also serve as the sites where PHOTOSYNTHESIS occurs. In most plants, photosynthesis and TRANSPIRATION occur in leaves—structures that are absent in cacti. Instead of leaves, many cacti have spines that help to conserve water and protect cacti from predators. *(See* predation.) The root structure of cacti also differs from that of many other plants. Instead of penetrating deep into soil, as do the roots of many trees, the roots of cacti spread out laterally, near the soil's surface. By remaining near the surface, the roots can quickly absorb water made available during infrequent rains.

Cacti vary greatly in size and shape. The largest, the giant saguaro *(Cereus giganteus)*, is native to the United States and receives protection within the boundaries of the Saguaro National Monument in Arizona. Some cacti, such as the night-blooming cereus and the endangered Chisos Mountain hedgehog cactus, also produce beautiful flowers. Many cacti, such as the night-blooming cereus and the peyote, contain chemicals that are used to make medications or beverages having alcohol-like qualities. These traits have made the preservation of cacti important to scientists and pharmaceutical companies. Other cacti, such as the prickly pear, produce edible fruits. These diverse characteristics have made many species of cacti attractive to collectors of rare and exotic plants, thus contributing to the threatened and endangered status of many cacti. *See also* ENDANGERED SPECIES; THREATENED SPECIES.

Endangered Cacti At a Glance

Common Name	Where Found
Tobusch fishhook cactus	Texas
Star cactus	Texas, Mexico
Fragrant prickly-apple	Florida
Nellie cory cactus	Texas
Key tree-cactus	Florida Keys
Arizona agave	Arizona

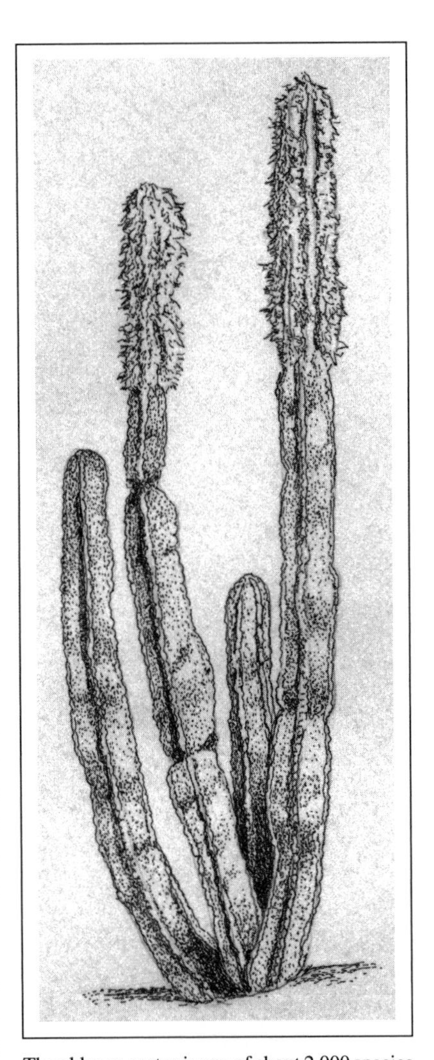

The old man cactus is one of about 2,000 species of cacti, most of which are native to the southwestern United States, Mexico, Central America, and the southernmost countries of South America.

EnviroSources

Middleton, Susan, David Liittschwager, and the California Academy of Sciences. "The Endangered 100." *Life* (September 1994): 50–63.

U.S. Fish and Wildlife Service Web site: www.fws.gov/ur9endspp/plants3.html.

cadmium, A highly toxic HEAVY METAL present in paints, electroplating materials, nickel-cadmium storage BATTERIES, PESTICIDES, alloys, and the CONTROL RODS of NUCLEAR REACTORS. Cadmium (Cd) compounds are found in nature commonly as components of zinc ores. Cadmium may enter soil or water during zinc MINING or be present in ASH formed through the INCINERATION of waste products that have been transported to landfills. Cadmium enters the AIR when coal and household waste are burnt, during the metal mining and refining processes, when wastes from households or industries are disposed, when fertilizer is applied to the soil, and when hazardous waste sites leak. ORGANISMS may be exposed to cadmium in contaminated food or water. *(See* contaminant.) Because of its high TOXICITY, cadmium is listed as a HAZARDOUS SUBSTANCE by the U.S. ENVIRONMENTAL PROTECTION AGENCY (EPA).

Cadmium requires careful disposal to avoid exposing organisms to the metal. High levels of cadmium exposure may cause death in MAMMALS, birds, and fishes. In PLANTS, cadmium may slow growth. Chronic effects of cadmium exposure in humans may include kidney or heart damage, decreased fertility, and changes in physical appearance or behavior. *See also* ACUTE TOXICITY; CHRONIC TOXICITY; STERILIZATION, REPRODUCTIVE.

EnviroSources

National Institute of Environmental Health Services Research on Environmental Related Diseases Web site: www.nieh.nih.gov

National Institute of Health Web site: www.nih.gov

calcium cycle, The natural circulation of calcium through its BIOGEOCHEMICAL CYCLE. The chemical element calcium is indispensable to life on Earth. Calcium is perhaps best known for its role in the formation and structure of bones and teeth. In addition to this structural role, it also plays a role in several other vital functions in the body, including nerve transmission, muscle function, and ENERGY production. In PLANTS, calcium is used in the formation of CELL walls and other cell structures.

Much of the calcium on Earth is tied up in sedimentary deposits of calcium carbonate, or limestone, as well as various other calcium-rich rocks. *(See* rock cycle.) In the calcium cycle, the element cycles continuously back and forth between the living (BIOTIC) and the nonliving (ABIOTIC) components of the ENVIRONMENT. The calcium cycle, along with the iron cycle and the PHOSPHORUS cycle, is an example of a sedimentary cycle. Sedimentary cycles vary somewhat, but each cycle consists fundamentally of a solution phase and a rock or sediment phase. The calcium cycle begins when WEATHERING and EROSION break down limestone and other calcium-containing rocks. Limestone often dissolves very easily in water. When this happens, a solution is formed, making calcium available for ALGAE and terrestrial plants. Calcium is then passed on through the FOOD CHAIN as animals eat plants or other animals. When ORGANISMS die and decompose, calcium is once again released into the environment for RECYCLING. *See also* CARBON CYCLE: DECOMPOSITION; NITROGEN CYCLE; SINKHOLE; SUBSIDENCE; WATER CYCLE.

California condors, The largest flying bird in North America, whose wild POPULATION had fallen to only nine individuals by 1985. The California condor (*Gymnogyps californianus)* is the subject of an intense and sometimes controversial effort to save the SPECIES from EXTINCTION. Faced with rapidly declining numbers, scientists collected eggs laid in the wild and removed wild birds for CAPTIVE PROPAGATION with hopes of ultimately restoring condor populations to their native ENVIRONMENT.

The California Condor usually nests in caves or on cliffs that have nearby trees for roosting and clear approaches for take-offs and landings.

A type of vulture, California condors have bare heads and necks, dull gray-black feathers, and blunt claws. The birds can soar on thermal AIR currents for hours, reaching speeds of more than 88 kilometers (55 miles) per hour and ALTITUDES of 4,575 meters (15,000 feet). They usually nest in caves or on cliffs that have nearby trees for roosting and clear approaches for take-offs and landings. Typically, an adult pair lays one egg every other year, with the fledgling dependent upon its parents through the next breeding season. Like all vultures, condors are carrion eaters. *(See* scavenger.) They prefer large, dead animals, such as deer and cattle, but also eat the remains of rodents and fish.

Condors were probably never numerous in North America. The species once ranged along the entire Pacific Coast from British Columbia to Baja California; fossil condors have been found as far east as Texas, Florida, and New York. More recently, condors were confined to a horseshoe-shaped area north of Los Angeles. They have been recognized as a declining species since the 1890s. One estimate put their number at 100 in the early 1940s; another indicated there were 50 to 60 in the early 1960s. By the late 1970s, the estimate had dropped to 25 to 30 birds.

Factors in the condor's decline include collection of condors and their eggs, poisoning by bait intended to kill livestock predators, poisoning due to ingestion of LEAD bullet fragments in animal carcasses, and collisions with power lines. In addition, roads, cities, and houses have replaced much of the open country condors need to locate food. Their low rate of reproduction and slow maturation have increased the species vulnerability to these and other threats.

The U.S. FISH AND WILDLIFE SERVICE (FWS) listed the bird as an ENDANGERED SPECIES in 1967 under a law that preceded the ENDANGERED SPECIES ACT of 1973. The FWS, the California Department of Fish and Game, and the NATIONAL AUDUBON SOCIETY, among other government and private groups, began a joint effort in 1979 to study and preserve the species. As part of this program, biologists captured birds, weighed and measured them, and fitted some with tags and radio transmitters so they could be monitored after release. Biologists learned about the condor's feeding, mating, and chick-rearing habits, as well as their HABITAT needs. They also determined that if a California condor pair lost an egg, they would quickly produce a second or even a third one.

To accelerate the wild condors' egg production, biologists began removing eggs from wild nests in 1983. The eggs were taken to the San Diego Wild Animal Park and the Los Angeles Zoo for hatching. *(See* zoo.) The first California condor was hatched in captivity in 1983. Chicks were raised in boxes that simulated the caves used by condor parents in the wild, and zookeepers minimized human contact by feeding the chicks with hand puppets resembling adult condors. Meanwhile, researchers began capturing young wild condors to start breeding them as quickly as possible.

Until 1985, biologists planned to leave some condors in the wild to serve as role models for captive-hatched birds slated for reintroduction to the wild; however, when several members of the remaining wild breeding pairs unexpectedly disappeared during the winter of 1984–85, the wild population plummeted from 15 to only 9 birds. Biologists therefore decided to capture all remaining wild California condors and bring them into the captive-breeding program. The FWS and its partner organizations trapped the last of the nine remaining wild condors in 1987. Two of these birds quickly mated, producing the first captive-bred condor chick the following year.

In the fall of 1988, the FWS began a three-year reintroduction experiment using Andean condors (*Vultur gryphus*), which also are endangered, as surrogates for their related North American species. Between December 1988 and January 1989, 13 female Andean condors were released, equipped with radio transmitters. (Only females were released to prevent the South American condors from reproducing in a habitat to which they were not native.) The Andean birds helped scientists to perfect release techniques and identify potential environmental threats before the California condors were reintroduced.

California Condors At a Glance	
Scientific Name	*Gymnogyps californianus*
Status	Endangered
Population Size	134 in the wild; 107 in captivity
Wingspan	2.5 – 3 m (8.5 – 9.5 ft)
Mass	7.25 – 10.8 kg (16 – 24 lbs)
Life Span	40 – 50 years in wild; more in captivity

In 1991, two California condors were released into the Los Padres National Forest's Sespe Condor Sanctuary in California, along with two Andean condors meant to assist in forming a condor social group. Six more California condors became ready for release later that year and the two Andean condors were recaptured for eventual return to South America.

Today, there are 134 California condors living in the wild (in protected areas of California and Arizona) and in breeding facilities. The FWS plans to continue its release program, particularly since the condors originally released are not expected to nest until sometime between 2000 and 2010. The search continues for appropriate habitat in remote areas for such releases, and additional breeding facilities have been established as well. In addition to the San Diego Wild Animal Park and the Los Angeles Zoo, California condors also are kept at the World Center for Birds of Prey in Boise, Idaho. *See also* ADAMSON, JOY GESSNER; FOREST SERVICE; HABITAT; MIGRATORY BIRD TREATY ACT; NATIVE SPECIES; POACHING; WOLVES.

EnviroSources

Big Sur Chamber of Commerce, California Condor Reintroduction Program in Big Sur California Web site: www.bigsurcalifornia.org/condors.html.

Graham, Frank, Jr. "The Day of the Condor." *Audubon* 102, no.1 (January–February 2000): 46–53.

Los Angeles Zoo, Programs for Endangered Species Web site: www.lazoo.org/conserve.html.

Los Angeles Zoo, Timeline for California Condor Recovery Program Web site: www.lazoo.org/ctime.htm.

San Diego Zoo, Center for Reproductive and Endangered Species Web site: www.sandiegozoo.org/cres.

U.S. Fish and Wildlife Service Web site: species.fws.gov.

cancer, A group of DISEASES characterized by the formation of abnormal CELLS that spread throughout the body, where they destroy or disrupt the functions of other body tissues. Cancers are classified according to the types of tissues they invade (breast, colon, skin) and according to the type of cell from which they originate. The development of cancer occurs when the genetic material (deoxyribonucleic acid or RIBONUCLEIC ACID) in a cell is altered in such a way as to promote the formation of abnormal cells. The mechanisms that trigger this alteration include inherited genetic abnormalities and environmental factors such as exposure to certain types of chemicals, RADIATION, or VIRUSES.

Cancer forms from a single cell that reproduces uncontrollably in the body. In time, the reproduced cells form a mass called a tumor. Once formed, a malignant, or cancerous, tumor can spread to invade nearby tissues or distant sites in the body. To spread, secondary growths called metastases form on the tumor and then break free, enabling them to enter a body cavity such as the abdomen or be carried by blood or lymph to other organs and tissues of the body.

Many chemicals have been determined to be carcinogenic, or cancer-causing, in humans or other animals. *(See* carcinogen.) Others are suspected of causing cancer, although no direct cause and effect relationship has been established. Use of such chemicals is monitored by the U.S. ENVIRONMENTAL PROTECTION AGENCY (EPA) and accidental releases of these chemicals into the ENVIRONMENT must be reported to the EPA for documentation in its annual TOXIC RELEASE INVENTORY (TRI). Included among these chemicals are: ARSENIC, friable ASBESTOS, BENZENE, BERYLLIUM, CADMIUM, CARBON TETRACHLORIDE (CCl_4), CHLORDANE, DIOXIN, mustard gas, POLYCHLORINATED BIPHENYLS (PCBs), saccharine, VINYL CHLORIDE, and the PESTICIDE DICHOLORODIPHENYLTRICHOROETHANE (DDT).

Other environmental causes of cancer include exposure to some types of ionizing radiation, ultraviolet RADIATION, and VIRUSES. Ionizing radiation includes the X-RAYS and gamma rays that are given off by radioactive elements. Ultraviolet (UV) radiation is the radiation given off by the sun and includes UV-A and UV-B rays. The OZONE layer that surrounds Earth protects living things from much of the damaging UV radiation given off by the sun; however, this layer is known to be thinning and incidents of skin cancer in humans are occurring with greater frequency. Burkitt's lymphoma is a form of cancer that affects children living in South Africa and has been linked to a virus. In this case, the viral agent is the Epstein-Barr virus, which is believed to be the causative agent responsible for infectious mononucleosis. This same virus also has been identified as the cause of nasopharyngeal cancer in Asian populations. *See also* AIR POLLUTION; ENVIRONMENTAL MEDICINE; FOOD AND DRUG ADMINISTRATION; MUTATION; SICK BUILDING SYNDROME.

EnviroSources

Bennett Information Group. *The Green Pages: Your Everyday Shopping Guide to Environmentally Safe Products.* New York: Random House, 1990.

Saign, Geoffrey C. *Green Essentials: What You Need to Know about the Environment.* San Francisco, CA: Mercury House, 1994.

canned hunt, A "sporting" practice in which individuals pay a fee to hunt and kill a selected type of animal that is contained within an enclosure from which it cannot escape. Animals used in canned hunts are often EXOTIC SPECIES native to countries of Asia or Africa. Such SPECIES may include gazelles, Cape buffalo, ZEBRAS, wild boars, and various types of deer, elk, goats, and large cats. The animals, some of which are included on the ENDANGERED SPECIES LIST, are acquired from animal dealers and brokers, wild-animal ranches, ZOOS, or from private breeders who breed, raise, and train animals for use in canned hunts using methods that remove the animals' natural fear of humans.

Canned hunts have been denounced by many ANIMAL RIGHTS groups, CONSERVATION organizations, individual citizens, hunting associations, and government officials. In 1995, the Captive Exotic Animal Protection Act, which would make it illegal to use exotic animals for canned hunts in areas consisting of 405 hectares (1,000 acres) of land or less, was reintroduced for consideration by the U.S. Congress. While the act has much support in Congress, it is strongly opposed by the National Rifle Association, which claims that the act is an attempt by animal rights extremists to abolish sport hunting. Despite this claim, the act is supported by some sport hunters' groups who believe that canned hunts, because they do not allow an animal a means of escape, are not true sport.

EnviroSource

Director of Captive Wildlife of the Humane Society of the United States: 2100 L St., NW, Washington, DC 20037; (202) 452-1100; Web site: www.hsus.org.

canopy. *See* forest.

captive breeding. *See* captive propagation.

captive propagation, The controlled mating and breeding of captive animals and PLANTS in ZOOS, aquariums, botanical gardens, and private research institutions. Captive propagation, or captive breeding, programs are especially useful for increasing the POPULATIONS of rare and ENDANGERED SPECIES.

Throughout the world, animal and plant SPECIES are disappearing, largely because their HABITATS are being destroyed by human activities, such as CLEAR-CUTTING FORESTS and filling in WETLANDS, or pollution. Animals are also hunted for food, medicines, and the use of their skins. This is troublesome to ecologists because when a population becomes too small, the genetic variation in the SPECIES is reduced. When this occurs, a species can have difficulty adapting to changes in its ENVIRONMENT and may eventually become extinct. *(See* adaptation; extinction.)

Various environmental laws have been enacted to help protect endangered species; however, the only hope for saving some species from extinction is captive propagation. For example, the CALIFORNIA CONDOR, North America's largest bird, once ranged from California to Florida; however, human encroachment on the bird's habitat has led to a severe decrease in the condor population. In 1980, scientists discovered that only 27 condors remained in the wild. Five years later, the population plummeted to about nine individuals. In 1987, the last nine wild condors were captured, leaving the animal effectively extinct in the wild. Today, through captive propagation at the Los Angeles Zoo and the San Diego Zoo scientists have increased the population of this rare bird. About 100 condors are still being held in captivity, but scientists have reintroduced about 13 condors into the wild.

Other animal species that have benefited from being placed in captive propagation programs include the red wolf, the golden lion tamarin, the Arabian ORYX, the Guam kingfisher, the Mongolian wild horse, and the Hawaiian goose (also called the nene). Overall, captive propagation programs have been quite successful; however, they are both time consuming and expensive. While some critically endangered species are rescued by such programs, hundreds of others move closer to extinction. *See also* BIODIVERSITY; ENDANGERED SPECIES ACT; THREATENED SPECIES; WORLD CONSERVATION UNION.

EnviroSources

Los Angeles Zoo, Programs for Endangered Species Web site: www.lazoo.org/conserve.html.

San Diego Zoo, Center for Reproductive and Endangered Species Web site: www.sandiegozoo.org/cres.

Carara Biological Reserve, An area in Costa Rica that has been established as part of the country's national park system to help preserve BIODIVERSITY. The Carara Biological Reserve is located in the Puntarenas Province near the Pacific Coast and contains 4,700 hectares (11,614 acres) of land. The reserve is primarily an ECOTONE, or transition area, between the humid RAINFOREST to its south and the dry FOREST region to its north.

Reserves are regions rich in NATURAL RESOURCES. In the Carara Biological Reserve, the primary resources are its FLORA and FAUNA, many SPECIES of which are almost exclusive to this ECOSYSTEM. For example, the area provides a HABITAT for the scarlet macaw, a member of the PARROT family found in only one other region of Costa Rica. The reserve also provides a habitat for various monkeys, sloths, armadillos, and iguanas. The Tarcoles River which borders the reserve, also provides a home for CROCODILES.

The Carara Biological Reserve has become a popular area for ECOTOURISM. Through the 1990s, the park boasted more than 26,000 visitors annually. Despite its protected status, human encroachment along the reserve's borders has threatened the habitat of some of the species the park aims to protect. Illegal MINING activities and damage caused by FOREST FIRES also threatens the survival of some park species. *See also* ADDO NATIONAL ELEPHANT PARK; AFRICAN WILDLIFE FOUNDATION; BIOSPHERE RESERVE; GALÁPAGOS ISLANDS; MADAGASCAR; SERENGETI NATIONAL PARK; WILDLIFE REFUGE.

carbohydrate. *See* photosynthesis.

carbon, A naturally occurring nonmetallic element that is present in the CELLS of all ORGANISMS. Carbon contains usually six neutrons and six protons and can combine with other elements to form a variety of compounds such as CARBON DIOXIDE and CARBON MONOXIDE. Carbon makes up about 19% of the mass of the human body and is an essential component of proteins, carbohydrates, fats, and nucleic acids. All FOSSIL FUELS, such as COAL, PETROLEUM, and NATURAL GAS, contain carbon as a principal element. Graphite and diamond are composed almost exclusively of carbon. *See also* CARBON CYCLE; CARBON TETRACHLORIDE; CHLOROFLUOROCARBONS; HYDROCARBONS; RIBONUCLEIC ACID.

EnviroTerm

Carbon-14. A radioactive ISOTOPE of CARBON (carbon-14) contains eight neutrons rather than six neutrons. It emits beta particles when it underdoes RADIOACTIVE DECAY.

carbon cycle, A natural process in which CARBON circulates and is recycled within the ENVIRONMENT. Carbon is the primary component of all living matter. It is a naturally occurring element that, like water, NITROGEN, OXYGEN, and PHOSPHORUS, is an essential nutrient for living ORGANISMS. Like these other substances, carbon is continuously cycled back and forth between organisms and the environment. This never-ending, cyclical process is known as the carbon cycle and is part of Earth's BIOGEOCHEMICAL CYCLE.

Carbon in the environment exists primarily as CARBON DIOXIDE (CO_2) gas in the ATMOSPHERE and OCEANS. PLANTS, ALGAE, CYANOBACTERIA, and other PRODUCERS absorb carbon dioxide from the AIR and water for the process of PHOTOSYNTHESIS. During photosynthesis, carbon dioxide, water, and sunlight react to form ENERGY-rich sugars. Producers store some of the sugars; the rest provide energy, which the organisms use to carry out their biological activities. Producers also make carbon available to other organisms in the BIOLOGICAL COMMUNITY. Animals and other CONSUMERS (BACTERIA, PROTISTS, AND FUNGI) obtain carbon when they feed upon producers or other consumers. In this way, carbon is passed along the FOOD CHAIN.

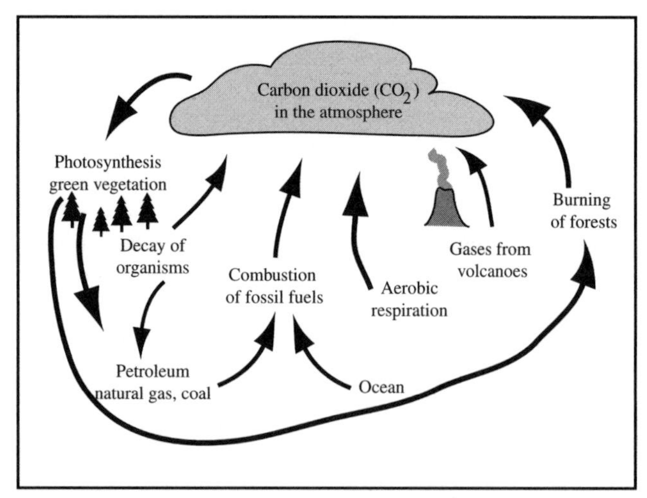

The carbon cycle is a natural process in which carbon is cycled within the environment.

Carbon returns to the environment during respiration, the energy-making process in CELLS. During cellular respiration, carbon-containing food molecules are broken down inside the cells to make the energy that organisms need to carry out their life processes. One waste product of this chemical reaction is carbon dioxide gas, which is released by the organisms and returned to the AIR, water, and soil. It is estimated that about 10% of the total amount of carbon dioxide in the air cycles back and forth between the atmosphere and organisms each year through photosynthesis and cellular respiration. Large quantities of carbon are also found in the solid earth. Long-term WEATHERING and EROSION of rocks and sediments, as well as volcanic eruptions, also return small amounts of carbon to the atmosphere.

Human Activity and the Carbon Cycle. Humans are living organisms and contribute to the carbon cycle by eating plants and animals and carrying out cellular respiration. People contribute to the carbon cycle in other ways, too. Scientists are very concerned that many human activities are altering the natural cycling of carbon and damaging the environment. The combustion of FOSSIL FUELS is one important way that humans interfere with the carbon cycle. Fossil fuels—COAL, PETROLEUM, and NATURAL GAS—form over millions of years when plants become buried in sediments. *(See* sedimentation.) Gradually, pressure and heat transform the plants into carbon-rich fossil fuels. Whenever factories, AUTOMOBILES, and power plants burn fossil fuels for energy, carbon dioxide is released into the atmosphere. Approximately two-thirds of all global EMISSIONS of carbon dioxide are produced from the burning of fossil fuels.

Another significant source of carbon dioxide in the atmosphere results from DEFORESTATION—the partial or total CLEAR-CUTTING of FORESTS and vegetation from an area of land. Deforestation is especially severe in the tropical RAINFORESTS of South America, Asia, and Africa, where SLASH-AND-BURN AGRICULTURE is practiced. In this method, trees and vegetation are cut down and burned in order to open the land for the cultivation of crops. The problem is that when plants are burned, large amounts of carbon dioxide are released. Deforestation also indirectly increases carbon dioxide levels in the atmosphere because when plants are cut down, they immediately stop absorbing it from the atmosphere. The result is an overall net increase in atmospheric carbon dioxide.

Level of Carbon Emissions, 1950-1998

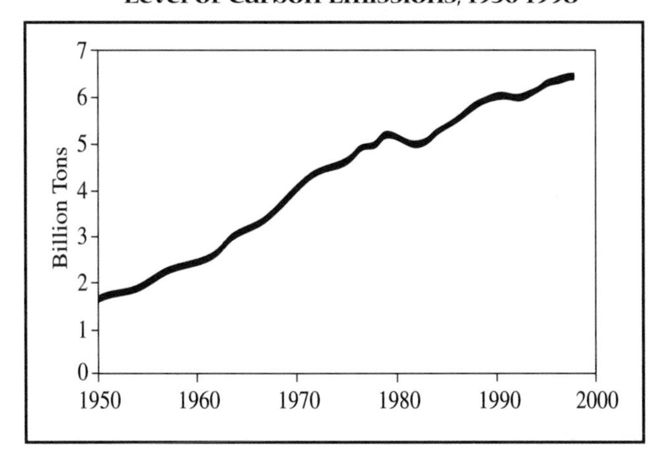

Source: Worldwatch Institute

Excess carbon dioxide in the atmosphere can have negative effects on the environment. Carbon dioxide emissions are of concern to scientists because carbon dioxide is the GREENHOUSE GAS that contributes most to the GREENHOUSE EFFECT. The greenhouse effect is a natural phenomenon that maintains the global CLIMATE by trapping heat near the surface of Earth. Like the glass panes of a greenhouse, greenhouse gases, including carbon dioxide, METHANE, and NITROGEN OXIDES (NO_x)—keep heat in and make Earth hospitable to organisms. Scientists fear that if greenhouse gas levels in the atmosphere continue to increase, it may result in increased temperatures on Earth, a phenomenon known as GLOBAL WARMING. Scientists are investigating the affects of even minor increases in global temperature on all ECOSYSTEMS. *See also* FOOD WEB; ORGANIC.

EnviroSources

U.S. Geological Survey, Global Change and Climate History Web site: climchange.cr.usgs.gov/info/carbon/

U.S. Geological Survey, U.S. Global Change Research Program, Carbon Cycle Science Program Web site: geochange.er.usgs.gov/usgcrp/ccsp.

carbon dioxide, A colorless, odorless GREENHOUSE GAS that is vital to life on Earth. Each molecule of carbon dioxide contains one atom of CARBON and two atoms of OXYGEN. Carbon dioxide exists in very small amounts in the ATMOSPHERE, making up only 0.04% of all atmospheric gases. PLANTS, ALGAE, and other PRODUCERS absorb carbon dioxide from the AIR or water for use in PHOTOSYNTHESIS, and both plants and animals release it as a waste product.

A Carbon Dioxide Molecule

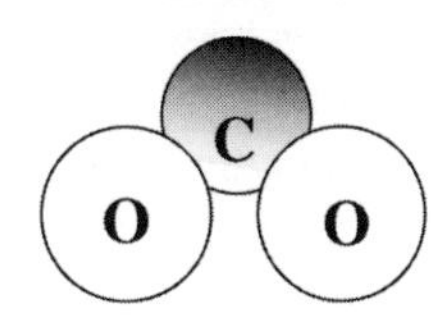

The carbon dioxide cycle is a basic part of Earth's BIOGEOCHEMICAL CYCLE. It moves between ECOSYSTEMS and the atmosphere in the CARBON CYCLE. Carbon enters an ecosystem when plants and other producers absorb carbon dioxide. Some of this carbon dioxide is used for photosynthesis; the rest is used in the cell structure of the organism. Animals and other CONSUMERS obtain carbon when they eat plants or other consumers that have eaten plants. Carbon dioxide cycles naturally back into the atmosphere when living things decompose and as a byproduct of cellular respiration, the process in which food is burned for ENERGY in most ORGANISMS.

Carbon dioxide also plays a major role in controlling temperatures at Earth's surface. As a greenhouse gas, carbon dioxide in the atmosphere contributes to the GREENHOUSE EFFECT by trapping heat near Earth's surface, much as the glass walls of a greenhouse trap heat. Without the greenhouse effect, heat from the sun would reflect back into space and Earth would become too cold to sustain life.

Environmental scientists are concerned because the amount of carbon dioxide in the atmosphere is increasing.

Sources of Carbon Dioxide

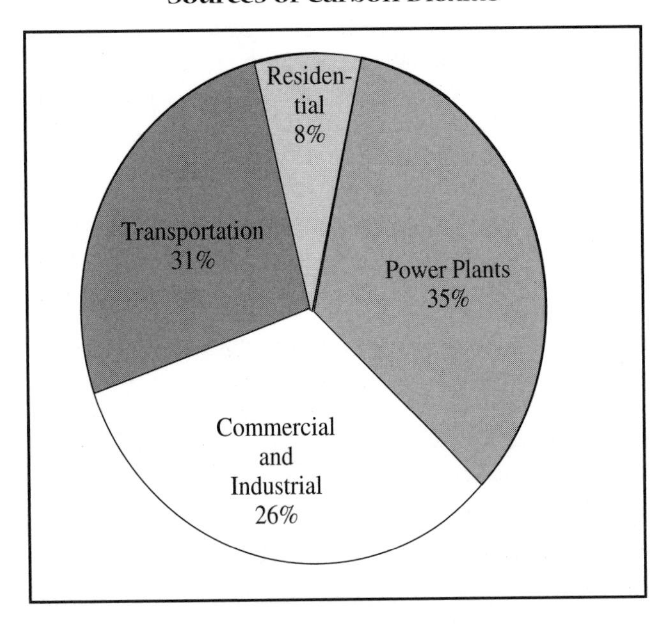

More carbon dioxide is being released into the atmosphere than ever before. According to the WORLD RESOURCES INSTITUTE, carbon dioxide EMISSIONS have increased from about 0.90 billion metric tons (1 billion tons) to more than 5.4 billion metric tons (6 billion tons) over the past 47 years, primarily due to the combustion of FOSSIL FUELS—PETROLEUM, COAL, and NATURAL GAS—by AUTOMOBILES, factories, and electrical power plants. Additional increases in carbon dioxide levels are caused by DEFORESTATION. When trees are cut down and burned to clear land for AGRICULTURE, carbon dioxide is released. In addition, as FORESTS are cut down, there are fewer trees and other plants to absorb carbon dioxide from the atmosphere.

Many scientists are worried that the increased levels of carbon dioxide in the atmosphere are contributing to GLOBAL WARMING. In 1997, according to one research study, carbon dioxide in Earth's atmosphere reached 336.6 parts per million (ppm), the largest amount in 160,000 years. *(See* parts per million and parts per billion.) In the future, if carbon dioxide levels reach 550 ppm the world's average surface temperature will increase by 1–3.5°C (1.8–6.3°F) during the twenty-first century. The increase will have an impact on human POPULATIONS and ECOSYSTEMS. The countries with the highest industrial emissions of carbon dioxide, in order, are the United States, China, Russia, Japan, Germany, India, Ukraine, and the United Kingdom. Total U.S. carbon dioxide emissions are four times higher than Japan. Some countries with the lowest industrial emissions of carbon dioxide include Norway, Sweden, Denmark, and Singapore. *See also* AIR POLLUTION; CARBON CYCLE; GLOBAL WARMING POTENTIAL; RAINFOREST; SLASH AND BURN.

EnviroSources

U.S. Department of Energy Carbon Dioxide Information Analysis Center: Oak Ridge National Laboratory, P.O. Box 2008, Mail Code 6335, Building 1000, Oak Ridge, TN 37831; e-mail: cdp@orn/.gov.

U.S. Environmental Protection Agency (global warming) Web site: www.epa.gov/global warming.

U.S. Geological Survey, U.S. Global Change Research Program Carbon Cycle Science Program Web site: www.geochange.er.usgs.gov/usgcrp/ccsp/.

carbon dioxide pellets, Small, pill-shaped particles composed of CARBON DIOXIDE (CO_2) that are used to remove contaminated radioactive materials from metal equipment, making the equipment safe for reuse. (*See* radioactive waste). Stripper SOLVENTS were once used for metal surface cleaning; however, the solvents created additional wastes as they worked.

Carbon dioxide pellets were successfully used at one site to clean up 6,800 kilograms (14,000 pounds) of URANIUM-contaminated scrap metal. To work, the pellets are blasted at metal surfaces at speeds of about 330 meters (1,000 feet) per second. The pellets penetrate the metal's surface and form a gas that causes contaminated particles to fall off the metal; these particles are then collected for disposal. All work is done in an enclosed area to protect workers and the ENVIRONMENT. *See also* CONTAMINANT; RADIATION; RADIOISOTOPES OR RADIOACTIVE ISOTOPE.

carbon monoxide, A colorless, odorless, tasteless, and toxic gas composed of one atom of OXYGEN and one atom of CARBON. Carbon monoxide gas, a major AIR POLLUTANT, is formed from the incomplete combustion of FOSSIL FUELS in AUTOMOBILES, industry, and homes. The main source of carbon monoxide in the ATMOSPHERE is GASOLINE-powered motor vehicles.

Carbon monoxide forms when carbon-containing matter is burned in an ENVIRONMENT lacking a sufficient oxygen supply. When present in the air, especially in an enclosed space, carbon monoxide is easily taken into the body where it reacts with hemoglobin in the blood. In the blood, it forms a compound (carboxyl-hemoglobin) that disrupts oxygen transport to the body's CELLS, robbing them of oxygen. Symptoms of carbon monoxide poisoning include drowsiness, nausea, headaches, dizziness, and redness of the skin. High concentrations of carbon monoxide are deadly, particularly in children and people with heart DISEASE or respiratory problems.

A Carbon Monoxide Molecule

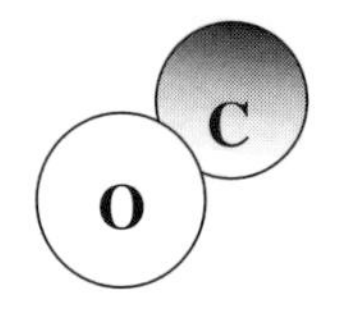

Sources of carbon monoxide in homes include unclean furnaces, fireplaces, and burning cigarettes. Use of carbon monoxide detectors can reduce the risk of carbon monoxide poisoning from these sources. The use of CATALYTIC CONVERTERS in automobiles has reduced carbon monoxide EMISSIONS into the air; however, carbon monoxide remains a major pollution problem in urban areas. The United States produces more than 60 million metric tons (66 million tons) of carbon monoxide each year. *See also* AIR POLLUTION; CARBON DIOXIDE; FUEL WOOD; HYDROCARBONS; SICK BUILDING SYNDROME.

carbon tetrachloride, A VOLATILE ORGANIC COMPOUND (VOC) derived from METHANE (CH_4) that is harmful to the ENVIRON-

MENT because it contributes to the breakdown of Earth's OZONE layer, which allows more of the harmful ultraviolet RADIATION from the sun to reach Earth. Carbon tetrachloride is also a probable human CARCINOGEN. Carbon tetrachloride is used in some fire extinguishers. The chemical was once widely used as a degreasing agent in industry, as a dry cleaning solution, and in homes as a spot remover for clothing, furniture, and carpeting. Carbon tetrachloride was also once used as a fumigant PESTICIDE to kill INSECTS in grain. In the 1960s, the U.S. ENVIRONMENTAL PROTECTION AGENCY (EPA) identified carbon tetrachloride as a probable human carcinogen and also linked high levels of exposure to the chemical to kidney problems in humans. Because of its potential dangers to the environment and human health, the EPA encouraged discontinuance of most uses of carbon tetrachloride in the mid-1960s. *See also* CANCER; CHRONIC TOXICITY; GLOBAL WARMING POTENTIAL; OZONE; OZONE DEPLETION.

EnviroSource

U.S. Environmental Protection Agency, Pollution Prevention Information Clearinghouse: 401 M St., SW, Mail Code 3404, Washington, DC 20460; e-mail: ppic@epamail.epa.gov.

carbon-14 dating, A method of estimating the age of ORGANIC matter such as PLANTS or animals by computing the rate of their nuclear disintegration based on the proportion of carbon-14 they contain. Carbon dating is an important tool for studying prehistoric life, past climatic changes, and global changes.

All living things contain carbon-14; however, once an ORGANISM is dead, it is no longer in the CARBON CYCLE and it stops absorbing carbon-14. As a result, the radioactive carbon in the organism begins to decay at a certain known rate. The known rate of disintegration, or the HALF-LIFE of carbon-14 is about 5,760 years. Utilizing the carbon-14 method, scientists can date organisms that were alive up to 50,000 years before the present time. In materials older than 50,000 years, there would be little carbon-14 left to measure. Scientists use carbon-14 dating to study organic deposits in lakes and ponds and FLOODPLAINS. Analyzing data from these deposits can show what kinds of organisms lived in the area and in what CLIMATES. For example, much of what is known of the people, age, and climate of the ancient city of Machu Picchu in the mountains of Peru was determined by carbon-14 dating. *See also* GLOBAL WARMING; RADIATION.

carbonic acid, A weak ACID that forms in the ATMOSPHERE when CARBON DIOXIDE (CO_2) reacts with water vapor (H_2O). Pure water is neutral, meaning it has a PH of 7.0. Many naturally occurring substances, chiefly carbon dioxide, combine with water vapor in the atmosphere to produce weak acids such as carbonic acid. These acids cause rainwater to be slightly acidic. Thus, natural, unpolluted rainwater has a pH of 5.0–5.6. By contrast, ACID RAIN has a pH below 5.0. *See also* NITRIC ACID; SULFURIC ACID.

carcinogen, Any agent (matter or ENERGY) that has the capability of causing or promoting CANCER—a condition characterized by abnormal, uncontrolled CELL growth—in humans or other animals. Many carcinogens are naturally present in the ENVIRONMENT. For example, light energy given off by the sun in the form of ultraviolet-B rays has been identified as a carcinogen leading to skin cancer in humans. Sustained or repeated exposure to these rays has been found to cause MUTATIONS by altering the deoxyribonucleic acid (DNA) of skin cells, resulting in as many as three different types of cancer. In addition, all ionizing RADIATION, including X-RAYS and the gamma rays given off by radioactive elements, have been found to be carcinogenic in humans and other animals. Other natural carcinogens include some VIRUSES—particles composed of a nucleic acid surrounded by a protein coat that are capable of replicating only when inside living cells—and some MICROORGANISMS. Examples of such viruses and microorganisms include the hepatitis-B virus, which can cause liver cancer, and the mold known as *Aspergillus flavus,* which often contaminates stored nuts and grains. This mold gives off a chemical substance called aflatoxin, which has been linked to liver cancer.

Many carcinogenic chemicals are released into the environment through the activities of humans. For example, tobacco smoke contains a variety of chemicals that have been linked to cancers of the mouth and lungs. Some insecticides and herbicides have also been found to contain cancer-causing chemicals. Many occupations place people at increased risk of exposure to chemical carcinogens. Such occurrences were documented as early as 1775, when British scientist Sir Percivall Pott observed that people who made their living as chimney sweeps often developed cancer of the scrotum. The cancer was attributed to carcinogens present in chimney soot and dust. *(See* particulate matter.) People working in the PETROLEUM and PETROCHEMICAL industries often are exposed to many types of HYDROCARBONS, including BENZENE, that have been discovered to be carcinogenic.

Many substances that have been determined to be carcinogenic have been banned from use in the United States and in other countries. An example is ASBESTOS, which was once widely used as an insulating material in buildings as well as in the linings of automobile brakes. After discovering that inhalation of asbestos particles could result in the development of lung cancer, the U.S. ENVIRONMENTAL PROTECTION AGENCY (EPA) banned further use of the substance in 1989. In addition, strict guidelines regarding the handling of asbestos during the renovation of buildings that contained the substance have been established. In addition to banning some carcinogenic chemicals, the EPA strictly regulates the use of others and also maintains a list of carcinogenic chemicals as part of its Superfund Amendments and Reauthorization Act (SARA). *(See* Times Beach Superfund.) Any release of the chemicals included on this list to the environment must be reported to the EPA for inclusion in its annual TOXIC RELEASE INVENTORY (TRI), which is available to the public upon request. *See also* AIR POLLUTION; ALDRIN; ARSENIC; BERYLLIUM; CADMIUM; CARBON TETRACHLORIDE; CHEMICAL WEAPONS; CHLORDANE; CHROMIUM; DICHOLORODIPHENYLTRI-

CHOROETHANE (DDT); HEPTACHLOR; NUCLEAR ENERGY; OCCUPATIONAL SAFETY AND HEALTH ADMINISTRATION; PESTICIDE; POLYCHLORINATED BIPHENYL; SICK BUILDING SYNDROME; VINYL CHLORIDE.

EnviroSources

Benarde, Melvin A. *Asbestos: The Hazardous Fiber.* Cleveland, OH: CRC Press, 1990.

Saign, Geoffrey C. *Green Essentials: What You Need to Know about the Environment.* San Francisco, CA: Mercury House 1994.

Caribbean Sea, A sea located in the western Atlantic Ocean east of Central America, which encompasses several groups of islands, including the Bahamas, Greater Antilles, and Lesser Antilles. The Caribbean area has experienced many marine OIL SPILLS that cumulatively amounted to hundreds of thousands of barrels. Other water pollution concerns in the Caribbean include SOLID WASTE dumping, municipal SEWAGE DISCHARGES, and storm water RUNOFF.

In 1994, an international organization, the Regional Marine Pollution Emergency Information and Training Centre, established an office in Curaçao, Netherlands Antilles, to help Caribbean countries prevent, prepare for, and respond to marine pollution incidents. For Caribbean nations, many of which are not wealthy, the financing of effective pollution planning and response programs represents a difficult task in view of other national priorities. Important environmental activities of the 1990s have included the assessment of pollution levels in the Caribbean region, the organization of Caribbean cleanup organizations to develop strategies to prevent pollution caused by plastics and GARBAGE, and use of satellite surveillance to monitor the sea for oil pollution indicators. *See also* CARTAGENA CONVENTION; OCEAN DUMPING

carnivore, An animal such as a bird, fish, MAMMAL, or REPTILE that feeds on other animals. Carnivores often live and hunt in groups, and are secondary and high-order CONSUMERS in the FOOD CHAIN or FOOD WEB. Carnivores such as WOLVES, lions, LEOPARDS, CHEETAHS, foxes, jackals, and TIGERS play an important part in predator-prey relationships as they ensure a balance between predator and prey POPULATIONS. Carnivores tend to accumulate TOXIC CHEMICALS in their body tissues as a result of the BIOACCUMULATION process, because they are at the upper levels of food chains and food webs. *See also* HERBIVORE; OMNIVORE; PREDATION; SCAVENGER.

EnviroSource

U.S. Fish and Wildlife Service, Species List of Endangered and Threathened Wildlife Web site: www.fws.gov/r9endspp/lsppinfo.html.

carrying capacity, The maximum POPULATION size that a specific HABITAT or ECOSYSTEM can support over time without degrading the ENVIRONMENT. The carrying capacity for a SPECIES is determined by the availability of resources in an environment over time and the amount of COMPETITION that exists among individuals and species for those resources. In a balanced ecosystem, resources such as food, water, and living space are present in amounts that can sustain a species almost indefinitely; however, if a population exceeds the carrying capacity of its environment, the environment often is degraded (its resources exhausted) to a point where the environment is no longer able to support that population. In response to this problem, many members of the population will either move to a different environment where they can meet their needs or they will die. *(See* emigration.) In time, the population will usually once again stabilize and reach a size that can be supported by available resources.

Natural populations rarely exceed the carrying capacity of their environments; however, the growing global human population is adversely affecting the environment's ability to sustain some natural populations. As humans outcompete other species for resources such as living space and introduce substances into the environment that make it less able to support native WILDLIFE, scientists have observed an increase in the rate of EXTINCTIONS. In addition, some scientists are concerned that if the human population continues to grow at its current rate, humans may one day reach their carrying capacity. As this occurs, humans themselves will begin to die as resources such as food, clean water, and clean AIR become unable to support a growing population. To help prevent this from occurring, many individuals and societies are promoting ZERO POPULATION GROWTH as a goal for the human population. *See also* BIRTH RATE; DEATH RATE; ECOLOGICAL SUCCESSION; EHRLICH, PAUL RALPH; FAMINE; HABITAT; IMMIGRATION; POLLUTANT; POPULATION DENSITY; POTABLE WATER; STERILIZATION, REPRODUCTIVE; SUSTAINED YIELD.

EnviroSources

Ehrlich, Paul. *The Population Bomb.* San Francisco, CA: Sierra Club, 1968.

———. *Population Explosion.* New York: Touchstone, 1991.

Ehrlich, Paul, and Anne Ehrlich. Extinction: *The Causes and Consequences of the Disappearance of Species.* New York: Random House, 1981.

Carson, Rachel Louise (1907–1964), A marine biologist, educator, and writer whose book *Silent Spring,* published in 1962, helped initiate the period sometimes referred to as the "age of ecology." In this book, Carson wrote about the potential dangers to the ENVIRONMENT and its ORGANISMS posed by the increased use of chemical substances, especially PESTICIDES and FERTILIZERS. Prior to this work, Carson's writings had dealt largely with the characteristics, and importance of OCEANS and the ORGANISMS that lived there. Her early books, *Under the Sea Wind* (1941), *The Sea Around Us* (1951), and *The Edge of the Sea* (1955), had enjoyed great success and won Carson a National Book Award as well as international acclaim as a writer.

From an environmental perspective, Carson's most significant writing was her final book, *Silent Spring*. In this work, Carson detailed the ways in which many chemicals being manufactured by industry and used in AGRICULTURE could be harmful to organisms. Among the most notable of these was the pesticide DICHOLORODIPHENYLTRICHOROETHANE (DDT), which at the time was widely used throughout the world. Carson warned that continued use of this pesticide would have many undesirable effects on organisms other than the INSECTS it was intended to destroy, because the chemical did not readily break down in the environment. As a result of her research, Carson

Rachel Carson was a marine biologist, educator, and writer whose book *Silent Spring,* published in 1962, helped begin a period sometimes referred to as the "Age of Ecology." *Credit:* Photo used by permission of the Rachel Carson History Project. Bob Hines, photographer.

championed a campaign to protect WILDLIFE by banning the use of DDT and other widely used chemical CONTAMINANTS.

When initially published, both *Silent Spring* and Carson were widely criticized by many working in the food and chemical industries; however, in time, it was shown that Carson's concerns about the use of DDT in particular were warranted. It became clear that the use of DDT was responsible for the declining POPULATIONS of several SPECIES of predatory birds, including the BALD EAGLE and PEREGRINE FALCON, which were nearing EXTINCTION. Studies determined that through the process of BIOACCUMULATION, concentrations of DDT accumulated in the tissues of predatory birds, which were at the highest levels of the FOOD CHAIN. *(See* predation.) While the pesticide did not appear to adversely affect the birds in which it accumulated, it did interfere with reproduction by causing the shells of the eggs they laid to be thin and brittle, preventing the hatching of new offspring. When scientists established the role DDT played in the decline of several bird populations, they began echoing Carson's call for a ban on the substance. In 1972, six years after Carson's death, Congress enacted legislation banning further use of DDT in this country and also began regulating use of other chemicals that could potentially harm the environment. Although DDT could still be manufactured and sold outside the United States, many other countries issued similar bans.

Carson began her professional career as a ZOOLOGY instructor at the University of Maryland, where she worked from 1931–36. After leaving her teaching career, she accepted a position with the U.S. Bureau of Fisheries (now part of the U.S. FISH AND WILDLIFE SERVICE), where she worked primarily as a writer and editor of bureau publications from 1936–52. It was while working there that Carson began learning of significant declines in certain bird and MAMMAL species, inspiring her to begin researching causes for these declines. Four years later, this research culminated in *Silent Spring*. *See also* ACUTE TOXICITY; BORLAUG, NORMAN ERNEST; CHRONIC TOXICITY; DOUGLAS, MARJORY STONEMAN; GREENBELT MOVEMENT; HAZARDOUS SUBSTANCE; TOXIC CHEMICAL.

EnviroSources

Breton, Mary Joy, *Women Pioneers for the Environment.* Boston, MA: Northeastern University Press, 1998.

———. *Lost Woods: The Discovered Writing of Rachel Carson.* Ed. Linda Lear. Boston, MA: Beacon Press, 1998.

Carson, Rachel L. *The Edge of the Sea.* Mariner Books, 1998.

———. *The Sea Around Us.* Oxford University Press, 1989.

———. *Silent Spring.* Houghton Mifflin Company, 1987.

———. *Under the Sea Wind.* Penguin, 1996.

Graham, Frank, Jr. "Sounding the Alarm: Rachel Carson." *Audubon* 100, no. 6 (November–December 1998): 83.

Cartagena Convention, An international agreement adopted in Cartagena, Columbia, in 1983 for the protection and development of the marine ENVIRONMENT of the Caribbean region. The Cartagena Convention was ratified by 19 countries to protect the marine environment of the Gulf of Mexico, the CARIBBEAN SEA, and areas of the Atlantic Ocean below 30° north latitude and within 200 nautical miles of the Atlantic Coast of the United States. The convention addresses various aspects of marine pollution. Its members must adopt measures aimed at preventing, reducing, and controlling OCEAN pollution from ships, OCEAN DUMPING, and seabed activities; airborne pollution; and pollution from land-based sources and activities. In addition, members of the Cartagena Convention must take appropriate measures to protect and preserve rare or fragile ECOSYSTEMS, as well as the HABITAT of depleted, threatened, or ENDANGERED SPECIES. *See also* CARIBBEAN SEA; CLEAN WATER ACT; COASTAL ZONE MANAGEMENT ACT; MARINE PROTECTION, RESEARCH, AND SANCTUARIES ACT; THREATENED SPECIES.

catalyst. *See* catalytic converter.

catalytic converter, Equipment placed in the exhaust systems of some internal combustion engines to reduce the levels of AIR POLLUTION from exhaust gases such as CARBON MONOXIDE. Catalytic converters are standard equipment in AUTOMOBILES that use unleaded FUEL in the United States and many countries in Europe. As exhaust passes through the equipment, the metal catalysts in the converter oxidize carbon monoxide into CARBON DIOXIDE and water vapor. *(See* EnviroTerm.)

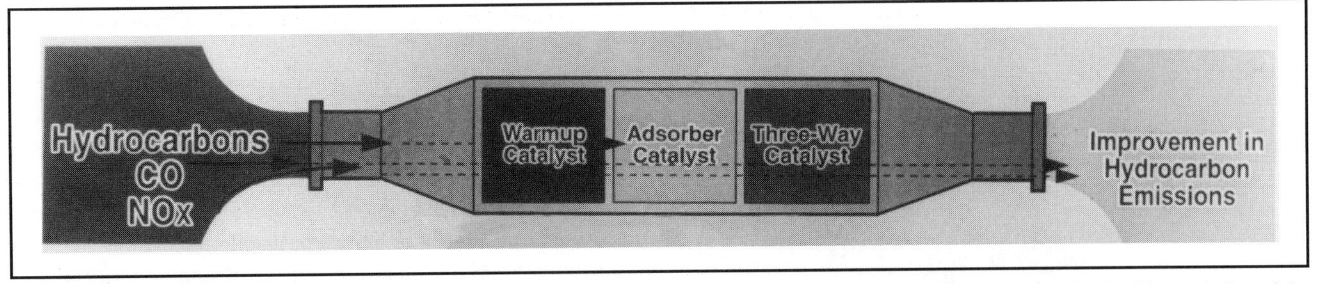

The illustration shows how gases are converted through the catalysts in a catalytic converter to improve on emissions. *Credit:* Used with permission of the General Motors Media Archives.

Catalytic converters do not work well in the presence of LEAD additives. In many developing countries, where leaded GASOLINE is sold, cars are not equipped with catalytic converters. As a result, car exhausts produce much of the air pollution in those countries. *See also* OXIDATION; OXIDIZING AGENT; OXYGEN.

EnviroTerm

catalyst: A substance such as vanadium that is used to speed up a chemical reaction but does not change its own form or composition in the reaction.

cell, The structural and functional unit of which all living things are comprised, the presence of which is an indicator that something is an ORGANISM. Many cells, such as those of BACTERIA, some PROTISTS, and some FUNGI, exist as independent units capable of carrying out all the functions needed to sustain life. Other cells, such as those of multicellular PLANTS, protists, fungi, and animals, may be specialized to carry out specific functions for the entire organism. Such cells work with other types of cells to carry out all the functions needed to sustain life.

There are two main types of cells: prokaryotic cells and eukaryotic cells. Prokaryotic cells are believed to be the first to have evolved and are characteristic of monerans, or bacteria. The cells of prokaryotes tend to be smaller than eukaryotic cells. In addition, prokaryotic cells lack a membrane-bound nucleus and membrane-bound organelles. In such cells, the genetic material is suspended within the cell's cytoplasm.

Eukaryotic cells are present in every type of organism, except for monerans. Eukaryotic cells generally share three features: a cell membrane; cytoplasm, which makes up most of the cell's internal ENVIRONMENT; and a nucleus bound by a membrane. The genetic material of eukaryotic cells is contained within the nucleus. Eukaryotic cells may also have several types of organelles, which carry out specific functions for the cell. Like the nucleus, the cell's organelles also are enclosed within membranes. The types of organelles contained within the cell and the presence or absence of a rigid, outer cell wall can determine the type of organism to which a cell belongs. For example, a cell wall is present in plant cells but absent in animal cells. This cell wall provides the cell with rigidity and helps support the plant as it grows. Other structures of plant and animal cells are shown in the diagrams. *See also* CANCER; MUTATION; RIBONUCLEIC ACID.

An Animal Cell

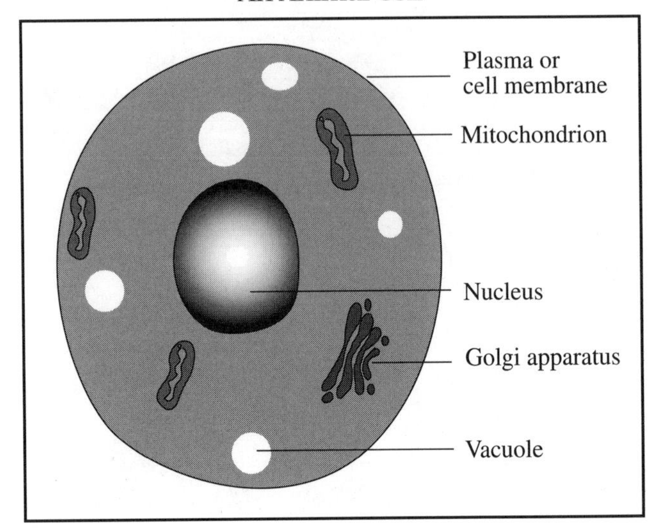

A Plant Cell

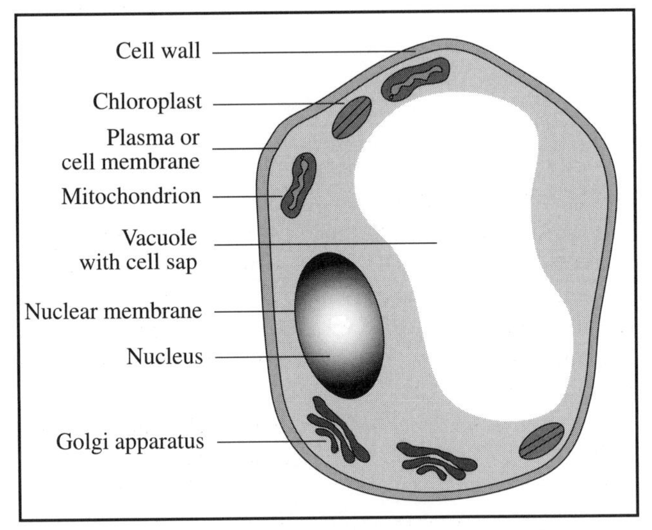

Cells have membrane-bound spaces called vacuoles for food, enzymes, and other material. Plant cells usually have one large vacuole. Animal cells contain many vacuoles.

EnviroTerms

cytoplasm: The gel-like material located between the cell membrane and nucleus that makes up most of a cell's volume. The organelles of a cell are suspended within the cytoplasm.

genetic material: Nucleic acid, usually DNA (but sometimes RNA), that carries the code which determines the traits of an ORGANISM.

During reproduction, the genetic material carries the traits of an organism to its offspring.
membrane: Thin, semi-permeable layer that surrounds and encloses a cell or various cell structures. Organelles, such as those of eukatyotic cells that are enclosed within a membrane are described as being membrane-bound.
nucleus: Large, membrane-bound organelle that contains the deoxyribonucleic acid (DNA) of a cell and regulates all of the cell's activities.
organelle: Any of several small structures located within eukaryotic cells that carry out specialized functions within the cell, similar to the functions of organs in more complex ORGANISMS. Examples of organelles include vacuoles, which store food or waste materials and mitochondria, which carry out cellular respiration.

EnviroSource

Margulis, Lynn, and Karlene U. Schwartz. *Five Kingdoms: An Illustrated Guide to the Phyla of Life on Earth.* 2nd ed. New York: W.H. Freeman and Co., 1996.

cesium-137, A hazardous RADIOISOTOPE of the metal cesium found in the FALLOUT that occurs after a nuclear explosion and the RADIOACTIVE WASTE from a NUCLEAR REACTOR. The HALF-LIFE of cesium-137 is about 30 years—the time required for the radioactivity to be reduced to one-half its original level. Cesium-137 and its decay byproducts—beta and gamma particles—may be health hazards because the RADIATION can enter the body causing damage and death at high enough levels. Radioactive cesium-137 waste was found at the HANFORD NUCLEAR WASTE SITE in the state of Washington and was also released during the CHERNOBYL accident.

CFCs. *See* chlorofluorocarbons.

chain reaction, A sequence wherein one reaction leads to another in atomic FISSION. A neutron collides with an unstable nucleus causing it to undergo fission, which in turn releases one or more additional neutrons which cause still other neurons to split, and so on.

Chain reactions release great amounts of heat ENERGY. The amount of energy given off is dependent upon the rate of the reaction. To regulate the reaction rate, CONTROL RODS containing CADMIUM (Cd) absorb neutrons without undergoing fission themselves.

chaparral, A BIOME sometimes referred to as a scrubland, which is dominated by scrub oaks, evergreen bushes, and thorny shrubs. The CLIMATE consists of long, hot, dry summers with low precipitation and cool, rainy winters. *(See* water cycle.) Chaparrals are found in central and southern California and throughout countries in southern Europe that are located along the coast of the MEDITERRANEAN SEA. Other chaparrals are found in Africa, Chile, Mexico, and parts of Australia. The chaparral is home to small animals such as lizards, snakes, rabbits, and birds, and some browsing animals such as sheep and goats. *(See* herbivore.)

Fast-burning fires can occur during the dry season and can sweep through large areas of the chaparral; the fires are necessary for some plants to regenerate from roots, bulbs, and seeds lying deep beneath the soil's surface. Destruction of chaparral has resulted from clearing the land for AGRICULTURE and the overuse of PESTICIDES and FERTILIZERS, which can damage sensitive ECOSYSTEMS. Clearing chaparral vegetation on slopes allows RUNOFF to result in hillside EROSION and FLOODING. *See also* GRASSLAND.

cheetahs, (Acinonyx jubatus) An ENDANGERED SPECIES of the cat family that is protected by the CONVENTION ON INTERNATIONAL TRADE IN ENDANGERED SPECIES OF WILD FLORA AND FAUNA (CITES) and the ENDANGERED SPECIES ACT of the United States. The cheetah is the world's fastest land animal with the ability to run at speeds up to 96 kilometers (60 miles) per hour over short distances. At one time cheetahs roamed throughout Africa and Asia, but its overall range has shrunk and continues

Namibia, Africa, has the largest population of cheetahs.

to do so. Today, Namibia, Africa, has the largest POPULATION of cheetahs. Smaller populations exist in Botswana, Zimbabwe, South Africa, Kenya, and Tanzania. Some are protected in reserves such as the SERENGETI NATIONAL PARK in Tanzania.

In 1900, there were about 100,000 cheetahs worldwide. Now, some researchers estimate that the wild cheetah population is between 10,000 and 15,000 animals; others estimate that the number is between 3,000 and 10,000. Several hundred cheetahs live in captivity in ZOOS.

The main threat to the cheetah's survival is the encroachment on its HABITAT by AGRICULTURE, but hunting them for skins has always been a threat. In the mid-1980s, an estimated 5,000 cheetah skins were traded annually. Although the animals are protected today, widespread POACHING of cheetahs and illegal trade still continue. In addition to habitat loss and poaching, the cheetah's survival is also threatened by COMPETITION with the larger predators such as lions, hyenas, and LEOPARDS. *(See* predation.)

The cheetah's habitat is the open GRASSLAND or SAVANNA of Africa. It once ranged across Asia to India. Cheetahs range from 112 to 130 centimeters (44 to 50 inches) in length. They weigh from 40 to 65 kilograms (88 to 145 pounds). The common cheetah has spots. The king cheetah, a variety of cheetah, has black striped fur and lives in South Africa. All cheetahs have short fur except around the neck. The cheetah is a solitary animal that hunts day or night. Its usual prey is large antelope, wildebeest calves, gazelles, young GORILLAS and other apes, birds, and hares.

In 1991, Laurie Marker formed the Cheetah Conservation Fund and moved to Namibia to observe these animals. The Cheetah Conservation Fund is the only international organization dedicated solely to the survival of the cheetah in the wild. *See also* RED LIST OF ENDANGERED SPECIES; WORLD WILDLIFE FUND.

EnviroSources

Caro, Timothy M., and George Schaller, eds. *Cheetahs of the Serengeti Plains*. Chicago: University of Chicago Press, 1994.
Cheetah Conservation Fund: 4649 Sunnyside Ave N, Suite 325/ Seattle, WA 98103; Web site: www.cheetah.org.
U.S. Fish and Wildlife Service, Species List of Endangered and Threathened Wildlife Web site: www.fws.gov/r9endspp/ lsppinfo.html.
World Wildlife Fund: 1250 24th St, NW/Washington, DC 20037; (800) 225-5993; Web site: www.worldwildlife.org/.

chemical oxygen demand (COD), A measurement of water quality, expressed in parts per million, required to oxidize all CARBON compounds, in a sample. It is used as a measurement to determine the amount of ORGANIC matter in WASTEWATER or SEWAGE. The chemical oxygen demand measurement records the amount of OXIDIZING AGENT that is reduced during the process. *See also* OXIDATION; PARTS PER BILLION AND PARTS PER MILLION; SEWAGE TREATMENT PLANT.

chemical weapon, A special weapon, that consists of a device that releases poisonous or lethal gases, intended to debilitate or kill enemy troops. Chemical weapons were used by militaries throughout the twentieth century. Some chemical weapons use munitions, including rockets, artillery, and shells, that contain nerve gas and mustard gas. The chemical agents—poisonous gases—used in such weapons generally cause asphyxiation (OXYGEN depravation) or act on the central nervous system. Modern weapons designed to deliver chemical agents include specially equipped missiles and bombs.

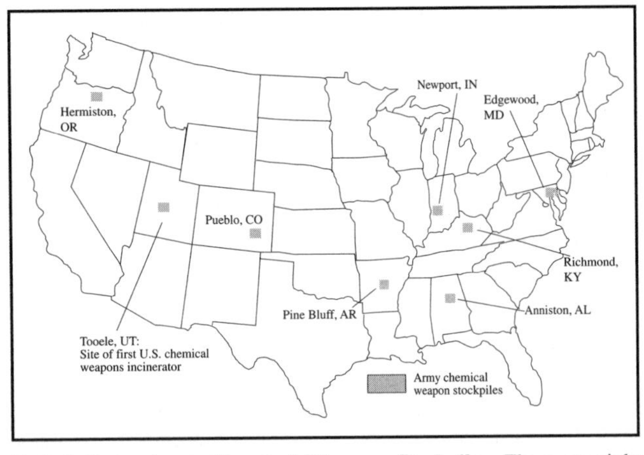

United States Army Chemical Weapons Stockpiles. There are eight chemical weapons storage sites in continental United States and another one on the Johnston Atoll, an island southwest of Hawaii. The other storage sites are located in Alabama, Kentucky, Utah, Maryland, Indiana, Arkansas, Colorado, and Oregon.

Over the years, several governments accumulated stockpiles of unused chemical weapons. As the chemical arsenals aged, the possibility that a spill would occur or that a chemical agent might leak from the deteriorating containers grew more likely. During the last quarter of the twentieth century, many governments took steps to destroy or dispose of chemical weapons.

There are eight chemical weapons storage sites in the continental United States and another one on the Johnston Atoll, an island southwest of Hawaii. The other storage sites are located in Alabama, Kentucky, Utah, Maryland, Indiana, Arkansas, Colorado, and Oregon.

About 45 percent of the nation's original stockpile of chemical weapons is stored in Tooele, Utah. A huge incinerator was built to work around the clock to eliminate the stockpile. Toxic liquids are burned at temperatures reaching 1480°C (2,700°F). The smoke is reburned and filtered, with a 99.99% incinerator rate for toxics. According to the Army, as of June 1998, the Tooele Chemical Agent Disposal Facility has safely destroyed over 1,350,000 kilograms (3,000,000 pounds) of chemical materials, or approximately 13% of the nation's stockpile of chemical weapons since the beginning of operations in August 1996. Nevertheless, some environmental groups are concerned about the operation.

EnviroSources

Chemical Stockpile Disposal Project (CSDP) Web site: www-pmcd.apgea.army.mil/graphical/CSDP/index.html
Tooele Chemical Agent Disposal Site Facility Web site: www.deq.state.ut.us/eqshw/cds/tocdfhp1.htm.

chemoautotroph, An ORGANISM that synthesizes its ENERGY and nutrients by oxidizing inorganic compounds such as METHANE (CH_4), ammonia (NH_3), and HYDROGEN SULFIDE (H_2S), through a process known as CHEMOSYNTHESIS. *(See* oxidation). The BACTERIA classified as chemoautotrophs are PRODUCERS, organisms capable of making their own food; however, they differ from other producers (PLANTS, ALGAE, and CYANOBACTERIA) in that they derive their energy from chemical reactions, rather than from sunlight. *(See* photosynthesis.) As producers, chemoautotrophs are environmentally important because they form the base of the FOOD CHAIN in their ECOSYSTEMS. These ecosystems include harsh environments that generally are not exposed to sunlight such thermal vent communities of the deep ocean and some deep lakes, in caves, or in soil. (*See* hydrothermal ocean vents.)

Among the organisms that carry out chemosynthesis are the methane-oxidizing bacteria classified as *Methylomonas* (commonly called METHANOGENS), the SULFUR-oxidizing bacteria of the genus *Thiobacillus*, and the NITROGEN-oxidizing bacteria known as *Nitrosomas* and *Nitrobacter*. The nitrogen-oxidizing bacteria also are responsible for nitrification, the conversion of ammonia into nitrogen compounds that can be used by plants and other organisms. *See also* AUTOTROPH; BAIKAL LAKE; BIOGEOCHEMICAL CYCLE; CONSUMER; FOOD WEB; HETEROTROPH; NITROGEN CYCLE.

chemosynthesis, A process in which ORGANISMS generate ENERGY from inorganic chemicals, such as salts, HYDROGEN SULFIDE (H_2S), and CARBON DIOXIDE (CO_2). Many types of chemosynthetic organisms exist on Earth. All are BACTERIA—that is, they are composed of a single CELL that lacks a central nucleus. SPECIES that carry out chemosynthesis live in very hostile and difficult ENVIRONMENTS, such as hot SULFUR springs, saline lakes, and OXYGEN-depleted soil. Often, these species serve as the primary PRODUCERS for their ECOSYSTEM.

Chemosynthetic organisms are important to life on Earth. Near HYDROTHERMAL OCEAN VENTS on the OCEAN floor, chemosynthetic bacteria obtain energy through the synthesis of hydrogen sulfide. They form the base of the FOOD CHAIN for a rich ecosystem that supports unusual organisms such as tube worms and white crabs. Another group of chemosynthetic bacteria help convert atmospheric NITROGEN into a form that can be used by growing plants. Without these nitrogen fixers, life on Earth could not exist. The METHANOGENS are also important chemosynthetic species. Methanogens produce METHANE (CH_4) from carbon dioxide and are responsible for much of the methane that exists in the ATMOSPHERE. *See also* AUTOTROPH; CHEMOAUTOTROPH; NITROGEN CYCLE; NITROGEN FIXATION.

Chernobyl, A town in the Ukraine whose nuclear power plant suffered a catastrophic accident in 1986. The fire and explosion of the power plant created nuclear FALLOUT and RADIATION that contaminated large areas of northern Europe, particularly the United Kingdom, Finland, and Sweden. Some reports indicated that as many as 20 countries received high levels of fallout from the accident. The initial fire, which lasted for several days, killed 31 people, but a larger number of people are still expected to die from prolonged exposure to the resultant radiation. Health experts expect to see a rise of CANCER deaths over the next 50 years, which may total as many as 40,000 cases. Most of these deaths will occur in the immediate vicinity of the NUCLEAR REACTOR; however, no one can predict how many will perish as a result of RADIATION SICKNESS. Another health concern is the increased number of people that have been diagnosed with tuberculosis since the accident.

The former Soviet government charged several officials at the plant with gross negligence because the proper safety procedures had not been implemented. The explosion occurred during a test, when FUEL RODS became so hot that they caused a steam explosion that blew off the top of the reactor, ejecting radioactive FUEL and burning CONTROL RODS into the ATMOSPHERE. From the air, helicopters dropped tons of limestone and sand to control the fire. Risking their own lives, lab technicians, firefighters, and others worked hard to control the fire. Several gave their lives and many were commended for their bravery. More than 600,000 workers were involved in the cleanup.

Today, the nuclear power reactor has been sealed with concrete and is visited by researchers to learn more about the effects of nuclear fallout and its impact on people and the ENVIRONMENT. Since 1986, more than 300,000 people have been evacuated from towns and villages near Chernobyl. The land is still contaminated with radiation, which has devastated farms and waterways. The cost of the cleanup so far has been estimated at over $350 billion. The reactor will be radioactive for 100,000 years. As a result of this accident, public support for NUCLEAR ENERGY decreased substantially in several countries. The accident has also promoted more work in safety technology on existing nuclear reactors. Chernobyl will be closed permanently in December 2000. *See also* THREE MILE ISLAND.

EnviroSources

Chernousenko, V.M.M. *Chernobyl: Insight from the Inside*. New York: Springer-Verlag, 1992.

Condon, Judith. *Chernobyl and Other Nuclear Accidents*. Austin, TX: Raintree/Steck Vaughn, 1998.

Chernobyl Radiation Disaster Information Web site: www.chernobyl.com/.

International Atomic Energy Agency International Chernobyl Assessment Project Web site: www.iaea.org/worldatom/inforesource/other/chernobook/index.html.

Marples, David R. *Chernobyl and Nuclear Power in the U.S.S.R.* New York: St. Martin's Press, 1986.

Nuclear Energy Agency, Chernobyl Executive Summary Web site: www.oecdnea.org/html/rp/chernobyl/c0e.html.

Chesapeake Bay, A waterway located on the Atlantic Coast of the United States, which contains some of the most important ESTUARIES in North America. This great bay and its WATERSHED encompass more than 165,000 square kilometers (64,000 square miles) along the coasts of New York, New Jersey, Pennsylvania, Delaware, Maryland, Virginia, and the District of Columbia. About 40 rivers and many streams contribute to the Chesapeake Bay, including the Susquehanna, Patuxent, and Potomac.

The Susquehanna River formed the Chesapeake Bay beginning about 18,000 years ago, as the great continental GLACIERS over North America began to melt, causing great rivers to flow toward the Atlantic Ocean. The prehistoric Susquehanna, a bigger and more powerful river than its modern descendant, eroded the land as it flowed seaward. As the glaciers melted, the sea level rose, flooding the coast and forming the great bay.

The estuaries along the Chesapeake's shores contain many salt-water and fresh-water WETLANDS of great economic and ecological importance. Significant fish SPECIES and CRUSTACEANS associated with the Chesapeake Bay include oysters, blue crabs, crayfish, striped bass, speckled trout, shad, and white perch. Annual seafood harvests from Chesapeake Bay may total more than 45 million kilograms (100 million pounds). These species, and many others, breed or feed in the bay. Some types of development, such as the construction of DAMS and roads, have hampered the ability of fish that breed in fresh water (anadravious fish such as shad and herring) to travel upstream to their established spawning grounds far inland.

At the end of the twentieth century, about 15 million people lived in the Chesapeake Bay region. The Chesapeake Bay has suffered from the impact of human activities within the watershed, many of which originate upstream along the tributary waterways within urban areas of Pennsylvania and New

York. For example urban and agricultural RUNOFF, which carries excess nutrients, sediment, and pollutants into the estuaries, affects the water quality of the bay. OVERFISHING has depleted important commercial fish POPULATIONS, thus impacting the ECOLOGY of the bay. Development of the shoreline has reduced coastal wetlands that serve important functions for water quality and WILDLIFE HABITAT. Some of the ecological impacts of water pollution and other activities are evident in the marked decline in fish and shellfish harvests during the latter part of the twentieth century.

Improvements in the water quality of the Chesapeake Bay are expected to result from improved agricultural practices, runoff control, and SEWAGE treatment within the bay's watershed during the 1980s and 1990s. In addition to water pollution control, government regulations that encourage fisheries management, such as placing limits on fish harvests, will help depleted fish populations to recover. To protect the Chesapeake Bay regions of Virginia and Maryland, toxic DISCHARGES from federal facilities will be reduced by 75% by the year 2000. Some of the discharges of chemicals include LEAD, COPPER, CADMIUM, POLYCHLORINATED BIPHENYLS (PCBS), and CHLORDANE.

Efforts have also been made to preserve the bay's wetlands, which serve critical functions in filtering pollution and providing habitat for FLORA and FAUNA. Environmental engineers, commercial fishers, fisheries, biologists, and wetland ecologists have noted a curtailment of the decline in aquatic environmental quality since the implementation of better environmental practices in the Chesapeake Bay region.

The Chesapeake Bay Commission is a tristate organization that cooperates with the legislatures of Pennsylvania, Maryland, and Virginia to restore and protect the Chesapeake Bay. The commission was formed between 1980 and 1985 to support the development and implementation of technologies, programs, and legislation that contribute to the health of the bay. The commission has supported environmental legislation and practices such as bans on the use of PHOSPHATES in DETERGENT, the implementation of fish passages for shad migration, and the establishment of riparian buffer areas to control nutrient-laden RUNOFF from agricultural areas.

EnviroSource

U.S. Geological Survey, Coastal and Marine Geology Web site: marine.usgs.gov/.

chlordane, A PESTICIDE belonging to the CHLORINATED HYDROCARBON group that may be harmful to the ENVIRONMENT and human health. Chlordane is a complex mixture of more than 50 chemicals and was used mainly to prevent termite infestations in homes and other buildings. Chlordane was also used on corn and other crops.

Use of chlordane was discontinued because of concerns that it might cause CANCER. Evidence showed that continued human exposure can lead to BIOACCUMULATION of the TOXIN in body fat. A persistent pesticide, chlordane stays in the environment for many years, posing a danger to WILDLIFE. It is still found in food and AIR. Those at highest risk of exposure are people who live in houses that were treated with chlordane for termites. Many of the homes are in the South and Southwest.

Chlordane can enter the body through the skin, through the lungs if inhaled, and through the digestive tract if swallowed. People at or near waste sites may be exposed by touching chlordane in the soil or by breathing chlordane that evaporates into the air. Chlordane is very insoluble in water. High exposures of chlordane affect the nervous system, the digestive system, and the liver. Swallowing the pure compound or breathing air heavily contaminated with chlordane vapors can cause headaches, irritation, confusion, weakness, vision problems, and other nervous effects as well as upset stomach, vomiting, stomach cramps, and diarrhea.

The U.S. ENVIRONMENTAL PROTECTION AGENCY (EPA) considers chlordane to be a human CARCINOGEN. Chlordane levels in the workplace are regulated by the OCCUPATIONAL SAFETY AND HEALTH ADMINISTRATION. *See also* ALDRIN; AGENT ORANGE.

EnviroSource

Agency for Toxic Substances and Disease Registry, Division of Toxicology, 1600 Clifton Road, E-29, Atlanta, GA 30333

chlorinated hydrocarbons, Any of a group of insecticides that are toxic to a wide variety of living ORGANISMS, linger in the ENVIRONMENT, and can bioaccumulate in the FOOD CHAIN. Chlorinated hydrocarbons are among the most widely used groups of insecticides; ORGANOPHOSPHATES and carbamates are also widely used. Chlorinated hydrocarbon compounds contain CHLORINE, HYDROGEN, and CARBON. Insecticides in this group include DICHOLORODIPHENYLTRICHOROETHANE (DDT), ALDRIN, dieldrin, HEPTACHLOR, kepone, CHLORDANE, lindane, and endrin. This group of insecticides is the most persistent—they break down very slowly. As an example, DDT may remain in the environment for more than 10 years after an application. These insecticides accumulate in the fatty tissues of animals and some can break down into other substances that also may cause adverse effects such as neurotoxicity, CANCER, and reproductive illnesses at high exposure levels. *See also* BIOACCUMULATION BIOMAGNIFICATION; HAZARDOUS WASTE; NEUROTOXIN; PESTICIDE; TOXIC CHEMICAL.

chlorination, The process of using CHLORINE gas or other chlorine-containing substances as a disinfectant to treat drinking water and WASTEWATER by killing MICROORGANISMS. It is also used in private and public swimming pools.

In the early 1900s, the public drinking water supply was unsafe in many cities in the United States. Untreated drinking water was the cause of illnesses and deaths due to DISEASES such as cholera and typhoid fever. These diseases were caused by PATHOGENS in water.

Chlorination was first used in the 1900s in water treatment plants in the United States as a result of the efforts of ABEL WOLMAN. Wolman was a research scientist with the Maryland State Department of Health who tested chlorination in a small water treatment plant in Maryland. His studies revealed that chlorinating drinking water decreased the number of deaths due to infectious diseases. Today, chlorination is widely used by many countries in the treatment of drinking water, but its use is being reevaluated. If scientists can find a substitute that can prevent waterborne disease, it will be replaced.

EnviroTerm

dechlorination: The removal of CHLORINE from a substance by chemically replacing it with HYDROGEN or hydroxide ions.

chlorine, A naturally occurring element in the halogen group of elements. It has a light green-yellow color and an obnoxious odor. At high levels, chlorine is harmful to living ORGANISMS. Chlorine is prepared by the electrolysis of melted sodium chloride or brine.

Chlorine gas is used to control waterborne DISEASES such as typhoid fever, cholera, and dysentery. In addition to its use in water and waste treatment plants, chlorine gas is used to make ethylene dichloride and other chlorinated SOLVENTS, polyvinyl chloride (PVC) resins, CHLOROFLUOROCARBONS (CFCs), and propylene oxide. Chlorine has been used as a bleaching chemical to whiten cotton and paper, but this use is being phased out. Chlorine is used in nonflammable solvents for degreasing and dry cleaning. It is also used in swimming pools.

In 1997 the UNITED NATIONS ENVIRONMENT PROGRAMME identified persistent ORGANIC POLLUTANTS as a major threat to human health and the ENVIRONMENT. Many of these pollutants contain chlorine-based chemicals. Chlorine is used in CHLORINATED HYDROCARBONS, one of the subgroups of insecticides that may be toxic to a wide variety of living organisms. These chlorine-based insecticides are among the three most widely used groups of insecticides, which also include ORGANOPHOSPHATES and carbamates. Chlorinated hydrocarbon compounds contain chlorine, HYDROGEN, and CARBON. Insecticides in this group include DICHLORODIPHENYLTRICHOROETHANE (DDT), ALDRIN, dieldrin, HEPTACHLOR, kepone, CHLORDANE, lindane, endrin, mirex, hexachloride, and toxaphene. This group of insecticides is the most persistent of the three groups—that is they break down very slowly in the environment. They also accumulate in the fatty tissues of animals and can break down into other substances. At high levels, these chemicals and their products may cause adverse health effects including neurotoxicity, CANCER, and reproductive illnesses.

Chlorine is also found in CFCs. These are synthetic chemical compounds composed of chlorine, carbon, and FLUORINE that have been found to be destructive to Earth's OZONE layer. To prevent further damage to the ozone layer, in 1987 more than 90 nations signed the MONTREAL PROTOCOL, a treaty setting limits on CFC production. OZONE DEPLETION begins when CFC molecules enter the stratosphere and are acted upon by ultraviolet (UV) RADIATION. UV radiation causes CFC molecules to break apart and release chlorine atoms into the ATMOSPHERE. The U.S. demand for chlorine is expected to decline over the next several years because of environmental concerns for chlorinated ORGANIC chemicals.

In some cases, the process of CHLORINATION and the use of chlorinated products can be replaced by chlorine-free processes and chlorine-free products. Sometimes this can be beneficial for the environment, but in many cases, especially that of PVC, in which a lot of investigations have been done, known and controlled dangers are exchanged for other, sometimes unknown, dangers, which can have a much greater impact on humans and ECOSYSTEMS.

Exposure to chlorine can occur in the workplace or in the environment following releases into the AIR, water, or land. The effects of chlorine on human health and the environment depend on how much chlorine is present and the length and frequency of exposure as well as the health of a person or condition of the environment when exposure occurs. Breathing small amounts of pure chlorine gas for short periods of time adversely affects the human respiratory system. Effects range from coughing and chest pain to water retention in the lungs. Chlorine irritates the skin, the eyes, and the respiratory system. Some studies show that workers develop adverse effects from repeated inhalation of chlorine, but others do not. Laboratory studies show that repeated exposure to high levels of chlorine in the air can adversely affect the immune system, the blood, the heart, and the respiratory system of living organisms. *See also* PRIMARY, SECONDARY, AND TERTIARY, TREATMENTS.

chlorofluorocarbon, Any of a number of synthetic chemical compounds composed of CHLORINE, CARBON, and FLUORINE, found to be destructive to Earth's OZONE layer. Chlorofluorocarbons (CFCs) were first synthesized by Thomas Midgley in 1928. Liquid CFCs are nontoxic, nonflammable, and nonreactive with other chemicals. These properties led to widespread use of CFCs as coolants for air conditioning and refrigeration units, propellants for products packaged in AEROSOL spray cans, electronics cleaning SOLVENTS, and foaming agents used in insulation and packaging materials.

A Molecule of Chlorofluorcarbon (CFC)

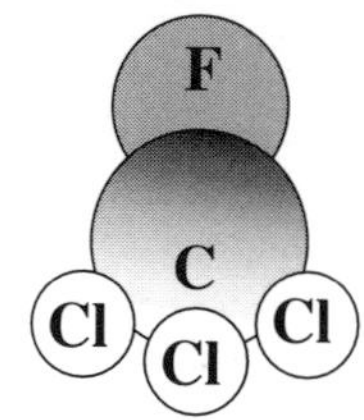

Chlorofluorocarbons (CFCs) were first synthesized by Thomas Midgley in 1928.

In 1976, the National Academy of Science reported that CFCs released from aerosol spray cans were damaging the ozone layer in the stratosphere. (*See* atmosphere.) The ozone layer shields Earth and its ORGANISMS from damaging ultraviolet (UV) RADIATION given off by the sun, particularly UV-B radiation. UV-B damages the CELLS of organisms and can cause skin CANCER and cataracts in humans. Additional evidence of stratospheric OZONE DEPLETION was reported in 1984.

Ozone layer depletion begins when CFC molecules enter the stratosphere and are acted upon by UV radiation. UV radiation causes CFC molecules to break apart and release chlorine atoms into the atmosphere. When a chlorine atom (Cl) collides with an ozone molecule (O_3), a chemical reaction occurs, resulting in the formation of chlorine monoxide (ClO) and OXYGEN (O_2) gas. Thus, the original molecule of ozone is broken apart. In addition to releasing chlorine from CFCs, UV radiation breaks apart oxygen molecules (O_2) to form unstable oxygen atoms (O). When an oxygen atom collides with a chlorine monoxide molecule, a reaction occurs in which the oxygen atoms unite, again forming oxygen molecules, and an atom of chlorine is released, and is thus able to break apart additional ozone molecules. These reactions occur repeatedly and simultaneously, allowing a single chlorine atom to destroy many ozone molecules leaving a layer of mostly oxygen atoms, which do not protect Earth from damaging radiation.

Individual chlorine atoms do not remain in the stratosphere forever. Eventually, they make their way to a lower atmospheric layer, where they react with gases such as METHANE (CH_4) to form hydrogen chloride (HCl) molecules. These hydrogen chloride molecules combine with water vapor to form hydrochloric ACID, which falls to Earth as precipitation. *(See* water cycle.) The complete cycle of chlorine atom from the release of CFC into the atmosphere to the chlorine atom's return to Earth in the form of rain takes one to two years.

Nations throughout the world have been monitoring global ozone levels. The National Aeronautics and Space Administration (NASA) monitors ozone levels from space using the Total Ozone Mapping Spectrometer (TOMS). *(See* spectrometry.) TOMS instruments have shown a decrease in stratospheric ozone at all latitudes outside the tropics.

To prevent further damage to the ozone layer, in 1987 more than 90 nations signed the MONTREAL PROTOCOL, a treaty setting limits on CFC production. Three years later, the protocol was amended to completely phase out CFC production as well as the production of other ozone-depleting chemicals by the year 2000. Many companies are researching compounds that might replace CFCs. Current replacements include HYDROCHLOROFLUOROCARBONS and hydrofluorocarbons. These alternatives also destroy ozone, but not as readily as do CFCs. *See also* GLOBAL WARMING POTENTIAL.

EnviroSources

U.S. Environmental Protection Agency, Office of Air and Radiation Web site: www.epa.gov/oar.

U.S. Environmental Protection Agency, Stratospheric Protection Division Web site: www.epa.gov/ozone.

Worldwatch Institute. *State of the World 1997.* Report on Progress Toward a Sustainable Society. New York: W.W. Norton and Co., 1997.

chromium, A steel-gray element that is naturally found in rocks, soil, animals, and in volcanic dust and gases. Large amounts of chromium are released into the ENVIRONMENT as a result of steel production, chemical manufacturing, municipal INCINERATION; and in the SLUDGE that is a byproduct of SEWAGE treatment. Some forms of chromium are used for making dyes and pigments, chrome plating, leather tanning, and wood preserving. Exposure to chromium happens mostly from breathing contaminated AIR in the workplace or ingesting water or food from soil near waste sites. Chromium can cause allergic responses in the skin, and ingesting high DOSES can cause stomach upset, ulcers, convulsions, and lung and kidney damage, or can lead to death.

The U.S. ENVIRONMENTAL PROTECTION AGENCY (EPA) has set a maximum level for chromium in drinking water at 100 micrograms of chromium per liter of water. The OCCUPATIONAL SAFETY AND HEALTH ADMINISTRATION recommends that levels of chromic ACID and chromium compounds in workplace air should not exceed 100 micrograms per cubic meter of air for any period of time. Small amounts of chromium are an essential nutrient in the human diet. The National Research Council recommends a dietary intake of chromium of not more than 50–200 micrograms per day, since cases of chromium deficiency are rare in the United States.

EnviroSources

Mike Kamrin, Institute for Environmental Toxicology, C231 Holden Hall, MSU, East Lansing, MI 48824-1206; (517) 353-6469. Web site: www.kamrin@pilot.msu.edu.

Agency for Toxic Substances and Diseases Registry, Division of Toxicology, 1600 Clifton Road NE, Mailstop E-29, Atlanta, GA 30333; (404) 639-6000.

chronic toxicity, The ability of a chemical substance or form of ENERGY to have long-term adverse health effects in humans or other ORGANISMS. As an example, smoking over a period of years can cause chronic lung CANCER. Chronic toxicity can result from repeated exposure to energy or chemicals in the ENVIRONMENT. For example, repeated exposure to some forms of RADIATION, such as the ultraviolet-B rays given off by the sun, may damage skin CELLS and result in skin cancer. Similarly, repeated exposures to chemicals or HEAVY METALS such as MERCURY and CADMIUM, which can accumulate in the CELLS and tissues of organisms, can also adversely effect health over time. Chronic toxicity may also result from only one exposure to a single harmful substance that has permanent, long-lasting effects on the body. Examples of such substances are PARTICULATE MATTER and ASBESTOS fibers present in AIR, which can cause scarring of lung tissue when inhaled. *See also* ACUTE TOXICITY; AIR POLLUTION; BIOACCUMULATION; RADIATION SICKNESS; TOXIC CHEMICAL.

CITES. *See* Convention on International Trade in Endangered Species of Wild Flora and Fauna.

Clean Air Act, A comprehensive federal environmental law that regulates EMISSIONS into the AIR from sources that are stationary, such as power plants and factories, as well as those that are mobile, such as AUTOMOBILES. Under this law, the U.S. ENVIRONMENTAL PROTECTION AGENCY (EPA) sets limits on how much of a POLLUTANT can be contained in the air anywhere in the United States, to ensure that all Americans have the same basic health and environmental protections. The Clean Air Act (CAA) was originally passed in 1970; amendments were enacted in 1977 and 1990.

The CAA authorizes the EPA to establish NATIONAL AMBIENT AIR QUALITY STANDARDS to protect public health and the ENVIRONMENT. The goal of the 1970 act was to set and achieve such standards in every state by 1975. The setting of maximum pollutant standards was coupled with the requirement that each state develop a state implementation plan (SIP) applicable to industrial pollutant sources operating in that state.

The CAA was amended in 1977 primarily to set new target dates for achieving attainment of the air quality standards, since many areas failed to meet the original 1975 deadline. The 1990 amendments to the CAA were intended to meet unaddressed or insufficiently addressed AIR POLLUTION problems, such as ACID RAIN, ground-level OZONE, SMOG, stratospheric OZONE DEPLETION, and air TOXINS. The 1990 amendments also mandated the application of BEST AVAILABLE CONTROL TECHNOLOGY (BACT) to reduce air toxins.

A major breakthrough in the 1990 CAA amendments was the establishment of a permit program for larger sources that release pollutants into the air. The 1990 act includes a list of 189 hazardous air pollutants selected by Congress on the basis of potential health and/or environmental hazards; the EPA must regulate these air toxins. The 1990 act also allows the EPA to add new chemicals to the list as necessary.

A few common air pollutants, such as smog and PARTICULATE MATTER, are present throughout the United States and are known as CRITERIA POLLUTANTS because the EPA has set permissible standards for these substances, first on the basis of health-based criteria and secondarily to prevent environmental and property damage. Geographic areas that meet or exceed the health-based (primary) standard are known as ATTAINMENT AREAS; those that do not meet the standard are considered NONATTAINMENT AREAS, and are classified on the basis of severity—from marginal to extreme. Although the EPA has been regulating criteria air pollutants since passage of the 1970 CAA, many urban areas are still in nonattainment for at least one criteria pollutant. As of 1993, it was estimated that about 90 million Americans were living in nonattainment areas.

The CAA authorizes state governments to take the lead in carrying out its requirements, since effective pollution control often requires an in-depth understanding of local industries, geography, housing patterns, and other factors. It tailors cleanup requirements to the severity of the pollution in nonattainment areas and sets realistic deadlines for reaching cleanup goals.

Under the current CAA, air pollution is managed by a national permit system; permits are issued to large pollution sources by state governments or, if a state fails to perform satisfactorily, by the EPA. The permit includes information on which pollutants are being released, how much may be released, and what steps the source's owner or operator is taking to reduce pollution, including plans for pollution monitoring. The permit system simplifies and clarifies the obligations of businesses to clean up air pollution and ultimately also reduces paperwork. States use the permit system to ensure that power plants, factories, and other pollution sources meet their cleanup goals with respect to criteria pollutants. The CAA allows individual states to have stronger pollution controls than those prescribed in the law, but states may not have weaker controls. The EPA assists the states by providing scientific research, expert studies, engineering designs, and funding to support clean air programs.

The 1990 CAA has many other features designed to clean up air pollution as efficiently and inexpensively as possible, letting businesses make choices about the best way to reach pollution cleanup goals. These flexible programs are called market-based approaches. For example, the CAA's acid rain cleanup program lets businesses choose how they will reach their pollution reduction goals, and includes pollution allowances that can be traded, bought, and sold.

The BHOPAL INCIDENT inspired the 1990 CAA requirement that factories and other businesses also develop plans to prevent accidental releases of highly TOXIC CHEMICALS. The act established the Chemical Safety Board to investigate and report on accidental releases of hazardous air pollutants from industrial plants. The 1990 law also enables the EPA to fine violators and brings CAA enforcement powers into line with other U.S. environmental laws.

Public participation is a significant component of the 1990 CAA requirements. At every step, the public is given opportunities to take part in determining how the law will be carried out. For example, members of the public may take part in hearings on state and local plans for cleaning up air pollution in particular areas.

In the mid-1990s, the EPA reviewed the current air quality standards for ground-level ozone and particulate matter. On the basis of new scientific evidence, revisions were made to both standards in 1997. The EPA is also developing a new program to control regional haze, which is largely caused by particulate matter. *See also* AEROSOL; AFTERBURNER; AIR POLLUTION CONTROL ACT; ALLERGY; ALTERNATIVE ENERGY RESOURCE; ATMOSPHERE; CATALYTIC CONVERTER; FLUOROCARBON; GREENHOUSE GAS; HYDROCARBON; INCINERATION; MONTREAL PROTOCOL; OZONE HOLE; SOURCE SEPARATION; VOLATILE ORGANIC COMPOUND; VOLCANO.

EnviroSources

American Meterological Society, History of the Clean Air Act Web site: www.ametsoc.org/AMS/sloan/cleanair/index.html.

U.S. Environmental Protection Agency, Office of Air and Radiation Web site: www.epa.gov/oar/sect812/index.html.

U.S. Environmental Protection Agency, Plain English Guide to the Clean Air Act Web site: http://www.epa.gov/oar/oaqps/peg_caa/pegcaain.html.

clean coal technology. *See* coal.

Clean Water Act, Legislation first created as the WATER POLLUTION CONTROL ACT in 1948. This law was renamed the Clean Water Act (CWA) in 1972 and has been amended since with changes that further protected SURFACE WATER resources. The 1977 amendment set the basic structure for regulating DISCHARGES of POLLUTANTS into waters of the United States. The law gave the U.S. ENVIRONMENTAL PROTECTION AGENCY (EPA) the authority to set EFFLUENT standards on an industry basis and to set water quality standards for all CONTAMINANTS in surface waters. The NATIONAL POLLUTANT DISCHARGE ELIMINATION SYSTEM, a provision of the CWA, makes it unlawful for any person to discharge any pollutant from a POINT SOURCE into navigable waters unless a permit is obtained from the EPA.

As a result of the CWA and its amendments, U.S. water quality conditions have improved. About 60% of U.S. waters met the act's designated-use goals in 1992, compared with 36% in 1972, according to the EPA. In the same period, the quality of 98% of river miles and 96% of lake acreage remained the same or improved, according to the Association of State and Interstate Water Pollution Control Administrators.

Other goals of the CWA include the elimination of discharges of chemicals, and achievement of water quality that is suitable for fishing and swimming. The CWA also established water quality standards, restrictions on WASTEWATER discharges, and requirements for prevention and cleanup of OIL SPILLS.

Industrial or municipal wastewater requires special treatment to eliminate releases of chemicals, and discharges of wastewater into waterways. Other results of the CWA include upgrades of municipal SEWAGE systems, the protection of WETLANDS, and improvements in stormwater discharges. The ARMY CORPS OF ENGINEERS and the EPA enforce the CWA.

EnviroSources

Sierra Club, "Happy 25th Birthday, Clean Water Act" Web site: sierraclub.org/wetlands/cwabday.html.

U.S. Environmental Protection Agency regional Web site: www.epa.gov/region5/defs/index.html.

clear-cutting, A tree-harvesting method that involves cutting down all trees and vegetation in an area. Clear-cutting is a controversial subject. Some foresters like the process because it is easy to cut an entire FOREST and then replant trees. They believe that the cleared land is ideal for the regeneration of the shade-intolerant tree SPECIES because direct sunlight is needed for seed germination. They claim that in small areas of 10 hectares (25 acres) or less, where the ground is level and precipitation is light to moderate, clear-cutting is a feasible method for harvesting shade-intolerant trees.

Clear-cutting over vast areas, however, causes destruction of HABITATS, heavy soil EROSION, and soil depletion. In tropical forests, clear-cutting is especially a problem in areas where tree species are poorly suited to regeneration. The poor soil of the tropical forest is exposed to hot temperatures that accelerate the decay of ORGANIC matter in the soil and reduce the moisture and nutrients important for tree growth. Important nutrients in the soil are also lost in RUNOFF during rains. Landslides can occur when clear-cutting is practiced along unstable hilly slopes in locations where there is heavy rainfall. Land-

A clear-cutting operation in Oregon. Clear-cutting is a harvesting method of cutting down all trees and vegetation in an area. *Credit:* United States Department of Agriculture. Photo by Clarence Knezevich.

slides develop because there are no trees and other vegetation to hold the soil in place. Once the vegetation is removed, most of the roots of the cut trees do not regenerate, but decay, losing the ability to hold the soil in place. Sometimes forests that are clear-cut do not regenerate. Clear-cutting may also prevent the original ECOSYSTEM from being restored in that area.

Alternatives to clear-cutting include strip cutting and SELECTION CUTTING. In selection cutting, only the mature trees are marked and cut down, while the immature ones are left to grow. In this method, trees of different sizes and types are left to maintain the ecosystem. In strip cutting, trees are cut in narrow rows or strips. The uncut areas remain for use in recreation and as WILDLIFE habitats. The uncut trees provide seeds and afford young trees protection from sun and wind.

Today, clear-cutting is being evaluated more closely as it relates to the surface and soil of the land, the number of trees to be cut, and the kinds of trees on the land. *See also* FORESTS; FOREST MANAGEMENT; FORESTRY; RAINFORESTS; SLASH AND BURN.

climate, The WEATHER conditions—temperature, precipitation, humidity, AIR pressure, winds, sunlight—in an area over a long period of time. Weather refers to what is happening in the ATMOSPHERE at a particular time. Climate, on the other hand, is the average weather in an area over a sustained period of time. Climate is important to all life on Earth because it determines the types of ORGANISMS that can survive in a particular area.

A variety of factors, including latitude, air circulation, OCEAN CURRENTS, and the local geography, help regulate and determine a region's climate. The latitude of a place—its distance from the equator, either north or south—is perhaps the most important factor because it has the most direct influence on average yearly temperatures. Latitude helps determine temperature because it influences the amount of sunlight an area receives. Scientists recognize three general climate zones based on how much sunlight is received: polar, temperate, and tropical. At the equator (0° latitude), for example, the sun is directly overhead, and its rays strike Earth's surface directly. This is why tropical areas, between 23.5° north latitude and 23.5° south latitude, consistently have the hottest year-round temperatures. On the other hand, in polar zones (between 66.5° N and the North Pole or between 66.5° S and the South Pole) the sun is lower in the sky, and sunlight strikes Earth at a more oblique angle, causing it to spread out over a much larger area. Some sunlight is also lost because it is reflected by polar ice. These factors explain why polar regions have the lowest annual temperatures. Between the tropical and polar zones are the temperate zones. The continental United States, for example, is in a temperate zone. In temperate zones, weather changes with the seasons: winters are cold and summers are hot. The spring and fall seasons are characterized by mild temperatures.

The world's OCEANS also have a significant effect on climate, particularly in coastal regions. This is mainly because water heats up and cools down more slowly than land. Because of this, many coastal areas are warmer in the winter and cooler in the summer than inland areas of similar latitude.

Ocean currents, moving streams of ocean water, also have a great effect on coastal climate. Warm currents, such as the Gulf Stream, which passes up the Atlantic Coast of the United States, originate near the equator and flow toward higher latitudes, warming the regions they pass. When the currents cool and flow back toward the equator, they cool the air and land nearby. Winds blowing from the sea contain more moisture than those blowing from land. Thus, coastal regions tend to have wetter climates than places inland, particularly where the prevailing winds blow onto the coast.

Air circulation patterns also factor into an area's climate, primarily with respect to the amount of precipitation—rain, snow, or fog—the area receives. For instance, in tropical regions, air at the surface is warmed rapidly, causing it to expand and rise. When warm air rises, it picks up moisture that has evaporated from the oceans and land. As the air continues to rise, it slowly cools, losing some of its ability to hold water, which then falls as heavy rain. *(See* water cycle.)

The warming and cooling of air in the tropics also helps determine the amount of rain that falls in other parts of the world because when warm air rises from the tropics it forces cooler air in the atmosphere to move toward the poles. Eventually, the cooler air falls back to Earth at latitudes of about 30° N or S, warming as it falls. As this warm, dry air moves across the surface of the Earth, it creates extremely dry conditions. This is one reason why many of the world's DESERTS, including those in the southwestern United States, occur near 30° latitude.

ALTITUDE is also a factor in a region's climate. Mountainous regions are cooler than lowlands or flatter areas of land. At the same latitude, the climate is much cooler in the mountains than at sea level. Mountains also affect the climate of nearby areas. For example, on the side of a mountain facing the wind, air rises up the side and cools, dropping its moisture. On the other side, the air descends and heats up, drying the land, often forming desert-like conditions.

Large cities can also affect local climate. When sunlight strikes areas of vegetation, much of the ENERGY is used to evaporate moisture. In cities, on the other hand, sunlight is absorbed by streets, buildings, parking lots, cars, trucks, and buses. These objects then radiate heat back into the atmosphere. Exhaust from AUTOMOBILES and other motor vehicles and EMISSIONS from factories and power plants trap this heat, creating a HEAT ISLAND effect. Summer temperatures in a city, for instance, can sometimes be 10°C (18°F) higher than in surrounding rural areas.

Over a human's lifetime, climate generally remains fairly stable, except for the changing of the seasons from year to year. Sometimes there can be noticeable differences in weather in a given year, but the differences are not large enough to constitute climate change.

Over a longer time scale of thousands to millions of years, however, climate fluctuates constantly. Scientists, for instance, have identified several ICE AGES—extended periods of extremely cold temperatures—during Earth's geologic history. The last ice age began about two million years ago and ended about 11,000–12,000 years ago. Similarly, fossils of tropical

PLANTS and animals found in today's polar and temperate regions, reveal much warmer worldwide climates in the past. Research into the causes of climate change suggests a variety of possible explanations including catastrophic events such as volcanic eruptions and large meteorite impacts, CONTINENTAL DRIFT, and changes in Earth's tilt upon its axis. *(See* volcano.)

Scientists are now studying how human activities may be affecting climate. In particular, scientists are very concerned about how various human activities are contributing to a gradual increase in worldwide temperatures, a phenomenon known as GLOBAL WARMING.

Temperatures on Earth have been relatively stable for several thousand years. Most experts believe that this stability is due to the GREENHOUSE EFFECT, a natural process in which atmospheric GREENHOUSE GASES trap sunlight and warm the planet. Without the greenhouse effect, Earth would be much too cold to sustain life. Scientists, however, have calculated that Earth's temperature has risen about 0.5°C (1°F) in the past century. In addition, 7 of the 10 hottest years of recorded average surface temperatures have occurred in the last 20 years. Research indicates that by the year 2100 the temperature on Earth will increase between 1°C and 3°C (1.8°F to 6.3°F). In the last 100 years, the global sea level has risen 10–25 centimeters (4–10 inches). According to the U.S. ENVIRONMENTAL PROTECTION AGENCY (EPA), there will be a SEA LEVEL RISE of about 12 centimeters (20 inches) by the year 2100. Climate changes in the twenty-first century will have impacts on many ECOSYSTEMS, particularly WETLANDS and coastal areas. As a result of these changes, fish and WILDLIFE HABITAT will be at risk. Climate change will remain an important environmental issue in the twenty-first century. *See also* BIOME; CARBON CYCLE; GLOBAL WARMING POTENTIAL; METEOROLOGY.

EnviroTerms

climatic cycle: Periodic changes in CLIMATE, such as a series of dry years following a series of years with heavy rainfall.

climatic year: A period used in meteorological measurements. The climatic year in the United States begins on October 1 each year.

EnviroSources

Climate Institute, 324 4th St., Washington, DC 20002.

Global Climate Coalition, 1275 K St., NW, Washington, DC 20005; (202) 682-9161.

Graedel, Thomas E., and Paul Crutzen. *Atmosphere, Climate and Change.* Scientific American Library Paperback, no. 55. New York: W. H. Freeman and Co., 1997.

Greenpeace International (climate) Web site: http://www.greenpeace.org/~climate.

Houghton, John Theodore. *Global Warming: The Complete Briefing.* Cambridge, UK: Cambridge University Press, 1997.

Houghton, John Theodore, et al., eds. *Climate Change 1995: The Science of Climate Change,* Cambridge, UK: Cambridge University Press, 1996.

United Nations Intergovernmental Panel on Climate Change Web site: www.ipcc.ch.

U.S. Environmental Protection Agency (Global Warming) Web site: www.epa.gov/globalwarming.

U.S. Geological Survey, Climate Change and History Web site: geology.usgs.gov/index.shtml.

climate change. *See* climate.

climax community. *See* ecological succession.

coagulation. *See* primary, secondary, and tertiary treatments.

coal, A blackish ORGANIC substance and the most abundant FOSSIL FUEL, which is used primarily to produce ELECTRICITY and to a lesser degree heat for buildings. Environmental issues associated with coal include AIR POLLUTION from coal-fired power plants and the impact of coal MINING on NATURAL RESOURCES.

According to the World Coal Institute, about 36% of the world's electricity is produced by burning coal. Coal is a major FUEL for generating electricity in Poland (97%), South Africa (93%), Australia (85%), China (80%), India (75%), and the United States (57%). In the year 2000, coal use is expected to rise in southeast Asia, where coal will be the major fuel for producing electricity.

Most of the coal reserves are found in large deposits in the mid-latitudes of the Northern Hemisphere, but there are a few coal deposits in the Southern Hemisphere. About 100 countries have coal reserves. Most of the world's largest deposits are in North America, eastern Europe, Russia, China, and Africa. The largest producers and users of coal are China, the United States, India, and South Africa. Recent estimates indicate that the world's supply of coal should last for another 250 to 400 years, at current production levels.

Coal is a mixture of CARBON and various other materials that was formed from the accumulation of partially decayed PLANTS in large, shallow SWAMPS, lakes, and MARSHES millions of years ago. It is found in beds and seams both near the surface and underground. The transformation of organic deposits to coal involved compaction and compression by burial under hundreds of meters of sediments. The formation of PEAT was the first step in the coal-making process. Over time, the peat was compacted beneath other deposits. As a result, water was squeezed out of the peat and gases such as METHANE were expelled into the ATMOSPHERE. Over thousands of years, the continued burial and compression caused the peat to alter into different grades of coal: lignite, bituminous, and anthracite. Lignite is the lowest grade and has low heat content. Bituminous is a higher grade than lignite and is the most abundant of the three types of coal. Anthracite, or hard coal, is the highest-grade coal, with a high carbon content and high heat content.

Coal is the dirtiest of all fossil fuels and emits several air POLLUTANTS when it burns. Coal-fired plants emit pollutants such as CARBON DIOXIDE, SULFUR DIOXIDE (SO_2), and NITROGEN OXIDES (NO_x) into the atmosphere. Studies indicate that about 70% of all sulfur dioxide EMISSIONS and 35% of carbon dioxide pumped into the atmosphere comes from coal-burning power plants. Other air pollutants released from coal-burning plants include VOLATILE ORGANIC COMPOUNDS (VOCs), soot, ASH, and other PARTICULATE MATTER. HEAVY METALS such as CADMIUM

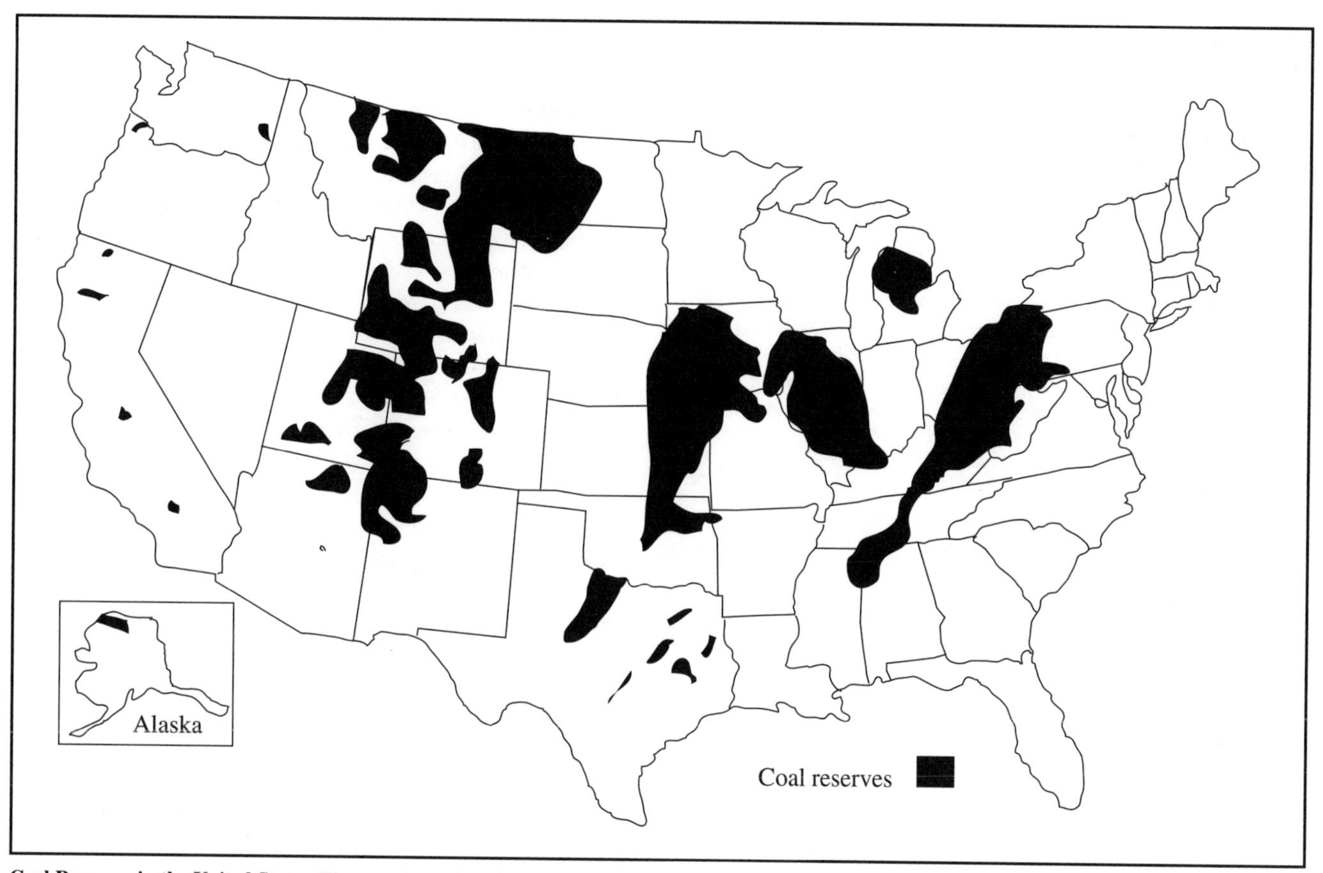

Coal Reserves in the United States. There are large deposits of coal in the United States. Most of the coal is used to generate electricity.

(Cd) and MERCURY are also released. These plants also produce bottom ash that needs to be collected and disposed of in landfills.

Coal mining activities can also cause serious environmental problems. Common problems include the destruction of ECOSYSTEMS and HABITATS. In addition, the removal of vegetation involved in SURFACE MINING makes an area more prone to soil EROSION and landslides. Water pollution can also occur when RUNOFF of wastes enters nearby streams. Water LEACHING through mine TAILINGS picks up MINERALS and carries them into GROUNDWATER reserves or into lakes and streams. To correct such problems in the United States, the SURFACE MINING CONTROL AND RECLAMATION ACT, federal legislation enacted in 1977, requires companies involved in surface mining operations to restore mined lands back to their natural conditions after the mining operations cease. It also prohibits surface mining on certain lands, such as national forests.

Underground mines and open-pit mines can be a serious environmental problem if left abandoned. ACID MINE DRAINAGE (AMD) is a water pollution problem resulting from the DISCHARGE of acidic water from coal mines or mines containing mineral ores for iron, COPPER, LEAD, or zinc into streams and rivers. Acid mine drainage also results when rainwater leaches through overburden or tailings—the waste materials produced by mining operations.

As a result of concerns about harmful emissions of coal burning, particularly the burning of high-SULFUR and low-quality coal, the United States, Europe, and Japan initiated research and development programs in the 1980s to generate technologies, projects, and devices for controlling harmful emissions. These programs also sought ways to increase the efficiency of coal combustion.

In the United States, Clean Coal Technology Program Legislation, passed in 1986 and supervised by the U.S. DEPARTMENT OF ENERGY, directed industry to burn coal more efficiently and reduce emissions from domestic coal-fired plants. The program also sought ways to reduce the release of pollutants that contribute to ACID RAIN. One of the options in clean coal technology programs is coal gasification, the conversion of coal to a gaseous product by one of several available technologies; this gas is considered an ALTERNATIVE FUEL. Other options include FLUIDIZED BED COMBUSTION systems and the installation of emission control systems such as SCRUBBERS in smokestacks to remove fly ash and other PARTICULATE MATTER. Additional options include improved boilers to burn coal more efficiently and more widespread use of coke in iron- and steel-making processes. (Coke is a coal product that has been heated in an oven to remove trapped gases. The coke burns with intense heat and is cleaner burning than coal.)

Countries such as China and many countries in southeast Asia are looking at clean coal technologies to help reduce air pollution and acid rain problems; however, environmentalists believe the best way to reduce air pollutants is to replace coal-fired plants with those that employ cleaner-burning fossil fuels or to eliminate the plants entirely and use renewable ENERGY sources such as WIND POWER and SOLAR ENERGY. The problem is that those developing nations that have large coal reserves may find it cheaper to continue to use coal rather than alternatives which might cost too much to use or take too long to develop. *See also* ACID; GREENHOUSE EFFECT; GREENHOUSE GASES; RECLAMATION.

TechWatch

As part of the DEPARTMENT OF ENERGY's Clean Coal Technology Program, Custom Coals International of Pittsburgh, Pennsylvania, is testing COAL-cleaning processes that will produce low-cost coals with low SULFUR content. One such process produces a coal product by crushing and screening mined coal and then applying a separation technique to remove about 90% of the pyritic sulfur. This special coal product can reduce SULFUR DIOXIDE EMISSIONS to levels that meet the compliance standards set by the new CLEAN AIR ACT.

EnviroSources

Coal Age Magazine Web site: http://www.coalage.com.

Gray, Robert G. et al, eds. *Coal Conversion and the Environment: Chemical, Biomedical, and Ecological Considerations*. Washington, DC: U.S. Department of Energy, 1981.

van Krevelen, Dirk W. *Coal: Typology - Physics - Chemistry - Constitution.* 3d ed. New York: Elsevier Science Publishers, 1993.

U.S. Department of Energy, Office of Fossil Energy Web site: www.doe.gov.

U.S. Geological Survey, National Coal Resources Data System Web site: www.energy.er.usgs.gov/coalqual.htm.

coal gasification. *See* coal; synthetic fuel.

Coalition for Environmentally Responsible Economies, A nonprofit organization comprised of investors, environmental groups, religious organizations, public pension trustees, labor unions, and public interest groups. The coalition for Environmentally Responsible Economies (CERES) promotes responsible corporate activity that ensures a safe and sustainable future for our planet. In 1989, responding to the disastrous *EXXON VALDEZ* OIL SPILL, CERES developed a comprehensive 10-point environmental guideline for corporations, known as the Valdez Principles. The 10 principles were intended to encourage a corporate commitment to a healthy ENVIRONMENT. CERES urges all corporations to sign the principles, which include protection of the BIOSPHERE, sustainable use of NATURAL RESOURCES, reduction and disposal of wastes, wise use of ENERGY, risk reduction, marketing of safe products and services, damage compensation, disclosure, employment of environmental directors and managers, and annual environmental audits.

By signing the principles, a company makes a pledge to uphold the principles by using only environmentally friendly policies and methods and to submit an annual progress report to CERES and to the public. Although CERES has no enforcement power, it identifies companies that are not following the principles and urges them to do better. *See also* OIL SPILL.

EnviroSource

Coalition for Environmentally Responsible Economies CERES, 711 Atlantic Ave., Boston, MA 02111; (617) 451-0927. Web site: http://www.ceres.org; e-mail ceres@igc.apc.org.

coast. *See* oceans.

Coastal Zone Management Act, A 1972 federal law enacted in response to the increasing pressure of development upon the nation's coastal resources. *(See* ocean.) The law established a voluntary national program within the Department of Commerce to encourage coastal states (i.e., those bordering the ocean or a Great Lake) to develop and implement coastal zone management plans. Such plans were intended to ensure CONSERVATION, protection, and, where possible, restoration and enhancement of coastal NATURAL RESOURCES such as WETLANDS, ESTUARIES, dunes, and CORAL REEFS, as well as the fishes and other WILDLIFE that populate coastal HABITATS. To encourage states to participate, the law made federal financial assistance available to any coastal state or territory that was willing to develop and implement a comprehensive coastal zone management program. To be eligible for federal approval, each state's plan was required to define coastal zone boundaries, identify uses of the area proposed for state regulation, and establish guidelines to prioritize uses within the coastal zone. The act also established the National Estuarine Research Reserve System, a PROTECTED AREA network that today encompasses 171,997 hectares (425,000 acres).

Amendments enacted to the Coastal Zone Management Act (CZMA) in 1975 made minor technical revisions in the administration of the financial assistance program, and 1976 amendments established the Coastal Energy Impact Program. The CZMA was reauthorized and again amended as part of the Omnibus Budget Reconciliation Act of 1990; the amendment created the Coastal Zone Enhancement Program. The current statute requires that to the maximum extent possible, federal activities that directly affect any land or water use or natural resource within a state's coastal zone be consistent with the policies of that state's coastal zone management plan. Application of the federal consistency requirement has been complicated by the difficulty of defining which actions "directly affect" the coastal zone. More recently, the administration of President Bill Clinton has proposed new legislation, the Coastal Management Enhancement Act of 1999.

The Secretary of Commerce delegated the administration of the CZMA and individual state programs to the NATIONAL OCEANIC AND ATMOSPHERIC ADMINISTRATION'S (NOAA) Office of Ocean and Coastal Resource Management (OCRM). The OCRM presently oversees programs in all coastal states except for Illinois, Indiana, Georgia, Minnesota, Texas, and Ohio. Coastal programs for the latter four nonparticipating states are in development. In addition to resource protection, the CZMA specifies that coastal states with OCRM-approved programs may manage coastal zone development, denying or restricting any development that is inconsistent with the state's coastal zone management programs. The provisions of the act do not apply to states that are not participants, nor to those whose plans have not yet received OCRM approval. *See also* MARINE MAMMAL PROTECTION ACT; MARINE PROTECTION, RESEARCH, AND SANCTUARIES ACT.

EnviroSources

National Oceanic and Atmospheric Administration, Coastal Zone Management Program Web site: www.nos.noaa.gov/ocrm/czm.
U.S. Fish and Wildlife Service Web site: www.fws.gov.

coastline. *See* oceans.

cogeneration, The MULTIPLE USE of ENERGY from a single source. In producing ELECTRICITY, many power plants rely on the simple conversion of thermal (heat) energy to mechanical energy. PETROLEUM, COAL, or other FOSSIL FUELS are first burned to heat and boil water, generating steam. The high-energy steam is then directed to a fan-like TURBINE, which spins and turns an electric generator. Substantial electricity can be produced in this way, but it is not the only form of energy produced; huge amounts of heat energy are also produced and usually wasted during the process. Cogeneration systems make use of energy that would otherwise be wasted. A simple example of cogeneration would be a car heater, which works by using the heat generated by the running engine to warm the inside of the car.

According to the second law of thermodynamics, there is always a loss of useful heat energy during an energy conversion. This means that most machines are inefficient, with more energy being used to run the machine than is produced as useful energy. For example, a great deal of energy is wasted as heat that is generated by the friction between moving parts. A GASOLINE internal-combustion engine, for instance, is about 20% efficient, meaning that only one-fifth of the energy input is changed to energy output. The remaining energy input is transformed into heat loss. A useful cogeneration system is one that can utilize otherwise wasted energy and so operate at a higher efficiency.

Cogeneration systems are most commonly used by power plants. In cities such as New York and Detroit, for instance, electric companies are now using heat from their powerful generators to heat their own buildings. In Europe, heat from power plants is also used to supply heat to many industrial and commercial buildings. Large industries are also now using cogeneration. During the production of paper, for instance, a large amount of steam is given off as waste; cogeneration systems in these factories can now utilize the wasted steam to generate electricity to operate other parts of the paper plant. *See also* ALTERNATIVE ENERGY RESOURCES; ALTERNATIVE FUEL; LAWS OF THERMODYNAMICS.

coliform bacteria, MICROORGANISMS present in the intestinal tracts of humans and other VERTEBRATES that become an environmental problem when discharged into bodies of water, usually in SEWAGE, or when they contaminate food. Coliform BACTERIA, such as *ESCHERICHIA COLI*, live within the intestinal tracts of humans and animals, such as cattle and horses. In the digestive tract, these bacteria aid in the digestive process of the host by breaking down cellulose or other matter the host cannot digest (*see* mutualism); however, when ingested in contaminated water or food, these bacteria become PATHOGENS that can cause acute DISEASE and possibly death, even for the ORGANISMS in which they normally live.

The presence of coliform bacteria in water is evidence of pollution. Such pollution is common in some developing nations, where POTABLE WATER is in short supply and people obtain drinking water from sources that have been exposed to farm RUNOFF or are used by animals as drinking and cooling ponds or by humans for bathing. To help prevent the spread of fecal coliform in the United States, drinking water supplies are routinely tested and assigned a coliform index, a rating of water purity based on a count of fecal bacteria.

Illness caused by coliform bacteria can result from ingesting food contaminated with these bacteria. Food becomes contaminated through poor hygiene practices of humans and through poor handling practices at meat slaughtering and processing plants. In the United States, inspectors working for the U.S. FOOD AND DRUG ADMINISTRATION (FDA) and state and city health departments are responsible for examining meat products, meat processing facilities, and restaurants to prevent the spread of bacteria in foods; however, this effort is limited and does not protect the entire food supply from contamination. To help offset this problem, restaurants serving meats that could be contaminated with coliform bacteria are encouraged to cook such foods until their temperatures reach 68.3°C (155°F) to kill bacteria that may be present in these foods. Individuals are encouraged to do the same in their homes or to cook meats until the juices run clear. *See also* CHLORINATION; CONTAMINANT; DISCHARGE; IRRADIATION; PRIMARY, SECONDARY, AND TERTIARY TREATMENTS; *SALMONELLA*; WASTEWATER TREATMENT.

commensalism, A type of symbiotic association between two SPECIES of ORGANISMS in which one species derives benefit, while the other neither benefits nor is harmed. (*See* symbiosis.) Many commensal relationships exist in nature. Often, these relationships provide the benefiting species with increased access to resources such as food, water, and sunlight. The commensal relationship between barnacles and some WHALES, for example, provides the barnacles with greater access to food. In this relationship, the barnacle, which is a sessile, filter-feeding animal, attaches to a whale. As the whale swims, water containing the small MICROORGANISMS the barnacle consumes moves over the barnacle. The relationship between barnacles and whales also aids in the colonization of barnacles throughout the ocean as the barnacles are carried to new locations by the whales. Many species of orchid are EPIPHYTES—plants that grow in the canopy of tropical RAINFOREST trees. The trees provide the orchids with a HABITAT that allows them to obtain adequate sunlight and maintain exposure to the AIR from which they obtain the water and MINERALS they need for PHOTOSYNTHESIS and growth. The trees do not appear to be affected by the relationship. *See also* COMPETITION; MUTUALISM; PARASITISM; PREDATION.

commercial fishing, Industry in which fish are harvested, either in whole or in part, for sale or trade. Fish and shellfish

Region	Gross Tonnage (1,000 metric tons) 1970	1992	Percent Growth, 1970–92
Asia	4,802.3	11,012.5	129
Former Soviet Union	3,996.7	7,765.5	94
Europe	3,097.4	3,018.3	-3
North America	1,076.9	2,560.0	138
South America	361.5	816.5	126
Africa	244.0	699.1	187
Oceania	37.1	122.3	230
World	**13,615.9**	**25,994.2**	**91**

The gross tonnage of fish caught between 1970 and 1992. *Source:* Food and Agriculture Organization of the United Nations.

are among the economically important features of the world's OCEANS and are vital sources of protein for the world's people.

Since ancient times, fishing has provided employment and economic benefits to many people. For most of our history, the wealth of aquatic resources has been assumed to be unlimited. By the late 1960s, however, it became clear that fishery resources could no longer sustain rapid and often uncontrolled exploitation and development and that new approaches to fisheries management embracing CONSERVATION and environmental considerations were urgently needed.

Worldwide, commercial landings of fish nearly quintupled between 1950 and 1989, from 20 million metric tons (22 million tons) to nearly 100 million (110 million tons). During this period, commercial fishing fleets had become so large and efficient that fish populations declined to a level that threatened stock reproduction, and fishing became unprofitable without subsidies. Although overall catch had remained constant in the late 1990s, the increased landings of low-value SPECIES masked the decline of more commercially valuable species.

In some regions of the world, declines of commercially important fish stocks have become so severe that the welfare of coastal communities and regions has become threatened. Perhaps the most dramatic depletions of fish stocks have been in the western Atlantic Ocean, where commercially viable quantities of cod have all but vanished from the Grand Banks.

As in many nations, managers of U.S. marine fisheries are struggling to reverse the OVERFISHING trends, improve economic performance, strengthen the conservation of protected species, and maximize harvests while maintaining productive stocks. Early attempts at management were compromised by largely unregulated foreign and domestic fleets. By 1976, the overexploitation of several stocks in offshore U.S. waters led to the passage of the Magnuson-Stevenson Fishery Conservation and Management Act; the prevention of overfishing was the act's standard for new fishery management plans. This law established a U.S. Exclusive Economic Zone (EEZ), in which foreign fishing could be controlled, and set up a conservation and management structure for U.S. fisheries

In 1982, the UNITED NATIONS CONVENTION ON THE LAW OF THE SEA was adopted to provide a new framework for the better management of global marine resources. The new legal regime gave coastal nations rights and responsibilities for the management and use of fishery resources within their EEZs, which embrace about 90% of the world's marine fisheries. Such extended national jurisdiction was a limited step toward the efficient management and SUSTAINABLE DEVELOPMENT of fisheries

In 1995, the governing bodies of the FOOD AND AGRICULTURE ORGANIZATION of the United Nations recommended a global Code of Conduct for Responsible Fisheries, which established principles and standards applicable to the conservation, development, and management of all fisheries. The code provides a necessary framework for national and international efforts to ensure sustainable exploitation of aquatic living resources in harmony with the ENVIRONMENT.

In 1996, the Sustainable Fisheries Act was enacted as an amendment to the Magnuson-Stevenson Act. Among its provisions were mandatory overfishing elimination and stock rebuilding, the establishment of a program to protect essential fish HABITAT, and the establishment of a new national standard for bycatch reduction.

Increases in the amount and/or intensity of fishing have not been solely responsible for stock declines. In evaluating fish populations, scientists examine human-induced effects as well as the effects of a host of natural occurrences. Natural environmental changes—in salinity or temperature—for example, may affect biological productivity of a fishery resource in a largely unpredictable fashion, yielding wide fluctuation in annual production.

Equally important, and at times surpassing the effects of natural variation, are the effects of human activities. Such factors include the destruction of coastal spawning habitats, fishing practices that kill immature fish and nontarget species, impoundments along migratory routes, harmful land use practices, pollution and SEDIMENTATION, and continually increasing fishing pressure. *See also* AQUACULTURE; OIL POLLUTION; TUNA; WHALING.

EnviroSources

National Oceanographic and Atmospheric Administration Fisheries Web site: www.nmfs.gov/.

United Nations Food and Agriculture Organization Fisheries Web site: www.fao.org/waicent/faoinfo/fishery/fishery.htm.

Commoner, Barry (1917–), One of the founders and leaders of the modern environmental movement; Commoner was called the "Paul Revere of ecology" by *Time* magazine in 1970 and "the dean of the environmental movement, who has influenced two generations" by *Earth Times*.

Born in Brooklyn, New York, Commoner became a specialist in cellular physiology, and served as a professor of PLANT physiology at Washington University in St. Louis, where he pioneered the use of a technique called "electron-spin resonance spectroscopy" in biology. Commoner is best known as an advocate of stricter control of environmental pollution. He ran for U.S. president on the Citizens Party ticket in 1980.

Commoner, an educator and biologist, is known for his leadership in the movement to oppose NUCLEAR WEAPONS testing in the 1950s, the science information movement of the 1960s, the ENERGY debates of the 1970s, and the organic farming/PESTICIDES, waste management/RECYCLING, and TOXIC CHEMICAL issues of the 1980s and 1990s. The research team at his Center for the Biology of Natural Systems made major advances in environmental science: developing alternatives to INCINERATION, promoting recycling, studying DIOXIN, and analyzing the economics of RENEWABLE RESOURCES.

Commoner is the author of many books, including *The Closing Circle* (1971), *The Politics of Energy* (1979), *Making Peace with the Planet* (1993), and *Science and Survival.*

EnviroSource

Commoner, Barry. *The Closing Circle.* New York: Knopf, 1971. *Earth Times*: www.earthtimes.org/book_info.htm.

community. *See* biological community.

competition, An interactive relationship present in all ECOSYSTEMS that occurs when two or more ORGANISMS, POPULATIONS, or SPECIES try to use the same resources. Examples of resources for which animals compete include food, water, living space, social status, and mates. PLANTS compete for water, soil nutrients, living space, and sunlight.

Competition is intraspecific when it involves members of the same species. Such competition exists in the feeding patterns of WOLVES, which pursue their prey as a team, but may compete so fiercely for the kill that a pack member may be injured or killed. Competition involving different species is interspecific, as occurs when wolves and coyotes both try to consume the remains of the same animal or when small groundlevel plants compete with taller plants, such as trees, for sunlight, water, and MINERALS in soil. *See also* ADAPTATION; CARRYING CAPACITY; COMMENSALISM; COMPETITIVE EXCLUSION; MUTUALISM; NATURAL RESOURCE; NICHE; PARASITISM; PREDATION; SYMBIOSIS.

competitive exclusion, The tendency for one POPULATION of ORGANISMS in an ECOSYSTEM to outcompete another population that has the same requirements for survival. The principle of competitive exclusion holds that two or more different SPECIES that have overlapping NICHES cannot be supported indefinitely in the same ecosystem. Instead, the best adapted, or most "fit," of the competing species will always outcompete the other, requiring the less fit organisms to change their lifestyle in a way that reduces COMPETITION, move to a new area, or die. In extreme cases, populations of an outcompeted species may decrease to the point of EXTINCTION if they cannot adapt to a slightly different niche or move to an area where less competition exists. *See also* ADAPTATION; ADAPTIVE RADIATION; CARRYING CAPACITY; COMMENSALISM; EVOLUTION; NATURAL RESOURCE; SYMBIOSIS.

composting, The process of RECYCLING various ORGANIC residues using AEROBIC DECOMPOSITION of organic matter by MICROORGANISMS such as BACTERIA, FUNGI, and INVERTEBRATES. The process is completed when the recycled organic materials are transformed into a dark soil that can be used as a natural FERTILIZER. Composting can take place in backyards and in large municipal facilities. Scientists believe that composting is an effective and inexpensive option in reducing the volume of organic wastes sent to landfills for disposal. (*See* sanitary landfill.)

For many years, SOLID WASTE composting plants have been operating in western Europe, Israel, and Japan. In the United States, yard waste composting is a key tool in addressing the municipal solid waste stream because yard waste accounts for nearly one-fifth or 27 million metric tons (29.8 million tons) of the municipal solid waste generated. Some communities have begun to conduct large-scale centralized composting of yard waste in an effort to save landfill capacity. Individuals are also helping to reduce waste by composting yard waste in their backyards, rather than bagging grass clippings or other yard wastes for municipal sanitation pickup. Composting yard waste, which is classified as source reduction, has seen tremendous growth in the past eight years. In 1985 in the United States, the amount of yard waste recovered was negligible (less than 45,450 metric tons, 50,000 tons, or 0.05%). By 1995, the amount of yard waste recovered had grown to 8.2 million metric tons (9 million tons, or 30.3%).

The process of composting can take place quickly or slowly depending on the methods used. A pile of organic matter will decompose slowly if not maintained. To get active composition completed in two to six weeks requires three key ingredients: AERATION, adequate moisture, and proper temperature.

OXYGEN, through the process of aeration, is needed by the microorganisms or invertebrates to successfully decompose the materials in the compost pile. Aeration provides the oxygen that would otherwise be lacking in the pile. Turning the compost pile is an effective means of adding sufficient oxygen for efficient decomposition.

The moisture content of the material in the pile should be between 40 and 60%. Compost experts use the "squeeze test." They squeeze a handful of material and compare the moisture content to the ideal level—that of a well-wrung sponge. If the moisture content is below 40%, the activity of the microorganisms will slow down. If the pile is too wet, above 60%, aeration is hindered, decomposition slows down, LEACHING occurs, and there is a definite odor due to ANAEROBIC DECOMPOSITION. A wet pile can be corrected by adding more dry material.

Microorganisms generate heat as they decompose organic materials. An efficient compost pile will generate temperatures between 32°C and 60°C (90°F to 140°F). If the temperature is too high, it will inhibit the activity of the organisms. Cool temperatures will advance the decomposition process, but at a slower rate.

Another factor in the composting process is the ratio of CARBON to NITROGEN, two important elements in composting. The bulk of the organic material should be mostly carbon, with just enough nitrogen to help in the decomposition. The

High school teacher records daily measurements of compost temperatures. *Credit:* Nancy Trautmann.

ratio should be 30 parts carbon to 1 part nitrogen (30:1). Carbon is considered the "food" and nitrogen the "digestive enzyme." Leaves are a good source of carbon while grass and manures are sources of nitrogen.

In large-scale composting, municipalities collect yard waste and take it to a central location. There the material is piled into rows, between 1 meter (3 feet) and 3 meters (9 feet) high. About once a month, heavy equipment is used to mix the piles for aeration. The temperature and moisture of the piles are checked twice a week. The finished compost is then sold, given away, or used in public works projects. *See also* DECOMPOSERS; HUMUS.

EnviroSources

BioCycle, Journal of Composting & Recycling, The JG Press, Inc., 419 State Avenue, Emmaus, PA 18049, (610) 967-4135; or e-mail: biocycle@jgpress.com. http://www.jgpress.com.

The Composting Council, 114 South Pitt Street, Alexandria, Virginia 22314; (703) 739-2402; fax: (703) 739-2407; e-mail: ComCouncil@aol.com.

Composting News, Ken McEntee, Editor, 13727 Holland Road, Cleveland, OH 44142, (216) 362-7979, fax: 216-362-6553; or e-mail: mcenteemedia@compuserve.com; Web site: www.recycle.cc.

Cornell Composting Web site: http://www.cfe.cornell.edu/compost/Composting_Homepage.html.

U.S. Environmental Protection Agency Office of Solid Waste and Emergency Response (composting) Web site: www.epa.gov/epaoswer/nonhw/compost/index.htm.

Comprehensive Environmental Response, Compensation, and Liability Act, A federal law of the United States enacted in 1980 to address problems associated with the release of HAZARDOUS SUBSTANCES into the ENVIRONMENT such as those derived from chemical spills or those present at inactive or abandoned waste disposal sites. The Comprehensive Environmental Response, Compensation, and Liability Act (CERCLA) often is referred to simply as the SUPERFUND because enactment of the law established a large monetary fund to pay for the cleanup of hazardous sites. The responsibility for identifying sites qualifying for cleanup falls to the U.S. ENVIRONMENTAL PROTECTION AGENCY (EPA) as does the designation of what substances pose a substantial threat to humans, other ORGANISMS, or the environment because of their TOXICITY or other hazardous properties. CERCLA also provides the EPA the authority to recover costs incurred during cleanup of a particular site from the party it deems responsible for the damage.

Sites identified by the EPA as needing cleanup under provisions of CERCLA are identified on the NATIONAL PRIORITIES LIST (NPL). The first site deemed by the EPA to be a federal environmental disaster area requiring cleanup under the Superfund provisions was the LOVE CANAL community neighboring Niagara Falls, New York, in 1980. Cleanup and RECLAMATION of this site, which was rendered noninhabitable by humans as a result of years of inadequate disposal of TOXIC

CHEMICALS by a company that had left the area almost 30 years earlier, took 10 years to complete.

Since the passage of CERCLA, thousands of other sites have been listed on the NPL. Many of these sites have undergone cleanup and been removed from the list; however, more than 1,000 remain on the list awaiting cleanup action. Included among these are many abandoned federal facilities such as munitions plants and former military installations. To help speed the cleanup process at such sites and increase the amount of money available for this purpose, Congress in 1986 passed the Superfund Amendments and Reauthorization Act. This act increased the money available for cleanup to more than $8 billion and also authorized use of funds collected from potentially responsible parties to be used in the cleanup process. *See also* ARMY CORPS OF ENGINEERS; DEEP-WELL INJECTION; EMERGENCY PLANNING AND COMMUNITY RIGHT-TO-KNOW ACT; HANFORD NUCLEAR WASTE SITE; HAZARD RANKING SYSTEM; HAZARDOUS WASTE; OIL SPILL; POLLUTANT; RESOURCE CONSERVATION AND RECOVERY ACT; TIMES BEACH SUPERFUND SITE.

EnviroSource

U.S. Environmental Protection Agency, Superfund Web site: www.epa.gov/superfund.

conifer. *See* coniferous forest.

coniferous forest, A FOREST type characterized by a range of conifers (cone-bearing trees), including varieties of spruce, cedar, pine, fir, sequoia, and hemlock, with some hardwood oak and maple SPECIES typically intermixed. The soil in a coniferous forest is acidic, and much of the ground is covered with litter from fallen tree needles. The coniferous forest ENVIRONMENT exists in the alpine, temperate, and subtropical regions of North America, Europe, and Asia. The northernmost coniferous forest is the TAIGA. Coniferous trees are also found at high ALTITUDES. The animals that live in a coniferous forest include HERBIVORES such as carbiou, moose, elk, squirrels, mice, and hares and predators including wolverines, foxes, hawks, owls, and eagles.

Douglas firs grow in coniferous forests.

The rainy climate of the Pacific Northwest of North America contributes to the growth of one of the most productive softwood forests in the United States, which extends in a narrow band from southern Alaska to central California. Most American OLD-GROWTH FORESTS are in this region. The tallest stands of coniferous forest are the giant redwood forests of the Pacific Coast; many of the trees that grow there exceed 100 meters (300 feet) in height, with diameters of 1.5–3 meters (4–9 feet). Other important coniferous forests are located in the Alaskan interior, the southeastern United States, the Rocky Mountains, Canada, northern Asia, Russia, and northern Europe. ACID RAIN has had a severe impact on coniferous forests, particularly in Canada and northern Asia. Other environmental problems include overharvesting of trees and the use of PESTICIDES to reduce INSECT infestations. *See also* ACID; DECIDUOUS FOREST.

Animals of the Coniferous Forest

American kestrel (*Falco sparvenus)*
Andean condor (*Vulture gryphus)*
Collared peccary (*Pecari tajacu)*
Eastern milk snake (*Lampropeltis triangulum)*
Fisher (*Martes pennanti*)
Florida king snake (*lampropeltis*)
Great horned owl (*Bubo virginianus)*
Kodiak brown bear *(ursus arctos middendorffi)*
Red tailed hawk (*Buteo jamaicencis)*
Reeves muntjac (*Muntiacus reevesi)*
Rocky Mt. bighorn (*Ovis canaclensis)*
Siberian tiger (*Panthera tigris altaica)*
Wild boar (*Sus scrofa)*

consent decree, A legal document, approved and issued by a judge, formalizing an agreement between the U.S. ENVIRONMENTAL PROTECTION AGENCY (EPA) and the parties identified as being responsible for site contamination. The decree describes cleanup actions that the responsible parties are required to perform and/or the costs incurred by the government that the parties will reimburse. If a settlement between the EPA and a responsible party includes cleanup actions, it must be in the form of a consent decree, a binding document. *See also* BROWNFIELDS; HARZARDOUS WASTE; SUPERFUND.

EnviroSource

U.S. Environmental Protection Agency, Office of Enforcement and Compliance Assurance Web site: www.es.epa.gov/oeca/osre/decree.html

conservation, The wise use and management of NATURAL RESOURCES in a manner that provides the maximum benefit of these resources to humans over a period of time. Natural resources are classified as renewable—those which are regularly replaced through natural processes, such as PLANTS and WILDLIFE—or nonrenewable—those that cannot be replaced, such as FOSSIL FUELS (PETROLEUM, COAL, and NATURAL GAS) and MINERALS. The principle of conservation includes the sustainable use of natural resources for economic purposes as well as the preservation and maintenance of natural resources deemed to have aesthetic or cultural value, such as wildlife, WILDERNESS areas, national parks, and historic sites.

The idea of conserving natural resources is not new; however, emphasis on conserving natural resources and preserving nature has increased greatly over the last century. This emphasis arose in part because people began to become more aware of how many of their activities have adversely impacted the environment, particularly land, water, and AIR. The significance of these adverse effects has become more apparent as the global human POPULATION has continued to increase, placing a higher demand on many of Earth's finite resources.

Throughout the world, governments and citizens are working to increase awareness of the importance of the conservation of natural resources. Such awareness has focused on such areas as reducing, reusing, and RECYCLING natural resources; decreasing pollution of air, water, and soil by making use of more ENERGY-efficient appliances and vehicles and disposing of waste materials properly; and preserving BIODIVERSITY by decreasing practices such as draining WETLANDS, clearing forests, and constructing DAMS along rivers that may potentially destroy the critical HABITAT of various SPECIES.

The United States has created several agencies and departments to oversee the development and use of the nation's natural resources. Included among these are the NATIONAL PARK SERVICE (NPS), the agency charged with managing the federal lands of the United States; the ENVIRONMENTAL PROTECTION AGENCY (EPA), which works with state and local governments to ensure that environmental legislation are enforced; and the FISH AND WILDLIFE SERVICE (FWS), which identifies ORGANISMS at risk of EXTINCTION and works with other agencies to develop plans to protect such species and increase their numbers.

The United States is not alone in its conservation efforts. Other nations throughout the world have begun to emphasize the need for conservation as a means of ensuring the availability of resources for their peoples and the economic growth of their nation. Often, these nations develop treaties and international laws to achieve such goals. The MONTREAL PROTOCOL, for example, is an agreement signed by more than 90 nations in which the nations agree to phase out the production and use of chemical substances such as CHLOROFLUOROCARBONS (CFCs) that have been linked to the breakdown of Earth's OZONE layer. The CONVENTION ON INTERNATIONAL TRADE IN ENDANGERED SPECIES OF WILD FLORA AND FAUNA (CITES) is an international agreement that is intended to help preserve BIODIVERSITY and allow populations of threatened and ENDANGERED SPECIES to recover by making trade in those species or parts of and substances derived from those species illegal in all participating nations. *(See* threatened species.)

Conservation is not a concern only of governments. Many organizations and citizens throughout the world also have become involved in conservation efforts. The goal of many of these groups is to increase awareness of how human activities can adversely affect the environment and provide individuals with information about how even small changes in their lifestyles can help to ensure that natural resources and natural settings will be available for the use and enjoyment of future generations. *See also* AGENDA 21; ALASKA NATIONAL INTERESTS LANDS CONSERVATION ACT; ALTERNATIVE ENERGY RESOURCE; ALTERNATIVE FUEL; ANTARCTIC TREATY; BIODIVERSITY TREATY; BIOREGION; BIOSPHERE RESERVE; BOTTLE BILL; CARARA BIOLOGICAL RESERVE; CARTAGENA CONVENTION; COASTAL ZONE MANAGEMENT ACT; COUNCIL ON ENVIRONMENTAL QUALITY; DEBT-FOR-NATURE SWAP; EARTH DAY; EARTHWATCH; ECOLABELING; EMISSIONS STANDARDS; ENDANGERED SPECIES ACT; ENDANGERED SPECIES LIST; ENVIRONMENTAL IMPACT STATEMENT; ENVIRONMENTALLY SENSITIVE AREA; FOREST SERVICE, FRIENDS OF THE EARTH; GREENBELT MOVEMENT; GREENPEACE; INTERNATIONAL BIOLOGICAL PROGRAMME; INTERNATIONAL WHALING COMMISSION; MARINE MAMMAL PROTECTION ACT; MARINE PROTECTION, RESEARCH, AND SANCTUARIES ACT (MPRSA); NATIONAL AUDUBON SOCIETY; NATIONAL WILDLIFE REFUGE SYSTEM; NATURAL RESOURCES CONSERVATION SERVICE; RED LIST OF ENDANGERED SPECIES; SAFE DRINKING WATER ACT; SOIL CONSERVATION AND DOMESTIC ALLOTMENT ACT; SUSTAINABLE DEVELOPMENT; TRADE RECORDS ANALYSIS OF FLORA AND FAUNA IN COMMERCE; TSAVO NATIONAL PARK; UNITED NATIONS DEVELOPMENT PROGRAMME; UNITED NATIONS EARTH SUMMIT; UNITED NATIONS ENVIRONMENT PROGRAMME; WATER POLLUTION CONTROL ACT; WORLD CONSERVATION MONITORING CENTRE; WORLD CONSERVATION UNION; WORLD RESOURCES INSTITUTE; WORLD WILDLIFE FUND.

conservation easement, A voluntary agreement between a landowner and a "conservancy"—usually a private, nonprofit organization—that allows a landowner to limit or regulate the amount of development on a U.S. property while retaining private ownership of the land. CONSERVATION easements enable landowners to protect their property from unwanted development while retaining ownership of the land. By granting a conservation easement a landowner can assure that the property will be protected forever, regardless of who owns the land in the future. Many landowners receive a federal income tax deduction for the gift of a conservation easement if the easement is perpetual and donated "exclusively for conservation purposes."

The activities allowed by a conservation easement depend on the landowner's wishes; for example, the owner may stipulate that no further development be allowed on the land. In other circumstances some additional development may be allowed, but the specified amount and type of development are less intense than would otherwise be allowed.

The conservancy monitors the property, generally once a year, to assure that the easement is not being violated by landowners. If the easement has been breached, the conservancy will take whatever steps are necessary to uphold the terms of the easement, including pursuing legal remedies.

U.S. Internal Revenue Service regulations require that conservation easements have "significant" conservation values. Eligible areas include FORESTS, WETLANDS, ENDANGERED SPECIES HABITAT, beaches, scenic areas, and more. The conservancy also has its own criteria for accepting easements. At the invitation of the landowner, the conservancy staff will evaluate the property to determine whether it meets these criteria.

EnviroSources

Landtrust Alliance *Conservation Options: A Landowners Guide.* Pamphlet.

Little Traverse Conservancy Web site: www.landtrust.org; or email: ltc@landtrust.org.

consumer, An ORGANISM that cannot synthesize its own nutrients and must therefore obtain them by consuming other organisms. Consumers include all animals and FUNGI and those SPECIES of PROTISTS and BACTERIA that do not carry out PHOTOSYNTHESIS or CHEMOSYNTHESIS. Depending upon the TROPHIC LEVEL at which they feed, consumers, which are also called HETEROTROPHS, may feed on PRODUCERS, other consumers, or both.

Most consumers are classified as HERBIVORES, CARNIVORES, or OMNIVORES, on the basis of the types of organisms they eat. Herbivores feed only on producers and their products. Examples include grazing animal, such as horses, cattle, BISON, and the ZEBRAS of the African plains, as well as browsers, such as giraffes. Seed-eating birds, ALGAE-eating fishes and shellfishes, and fruit-eating monkeys and BATS also are herbivores.

Carnivores feed on other consumers or their products. Many carnivores are meat eaters; however, carnivores may feed on nonmeat products such as eggs and milk as well. Many, though not all, meat-eating carnivores are predators that hunt and kill their prey. For example, lions, polar bears, PEREGRINE FALCONS, and the toothed WHALES are predators. *(See* predation.)

Omnivores are consumers that feed on both plant and animal products. Examples include humans, some bears, coyotes, raccoons, and some PARROTS. SCAVENGERS and DECOMPOSERS are other types of consumers. These organisms feed on DETRITUS—the wastes or remains of other organisms. Examples of scavengers include vultures and hyenas. Because they feed on meat products, scavengers are carnivores; however, they differ from other carnivores in that they do not kill the organisms they eat. Instead, they feed on the remains of animals that have died in the wild or the carrion, or animal parts, left behind by predators.

Fungi and many species of bacteria are DECOMPOSERS. Decomposers obtain their nutrients by breaking down the complex ORGANIC compounds present in the wastes or remains of other organisms into simpler organic and inorganic compounds they can use as food. Although scavengers and decomposers use similar sources for their food, the two types of organisms differ in how they obtain their nutrients. Decomposers release chemicals that break down or digest matter before it is absorbed or taken into the "body." Scavengers ingest the matter before it is broken down through digestion.

Both decomposers and scavengers play an important ecological role by clearing the wastes and remains of other organisms from the ENVIRONMENT. Decomposers also help to RECYCLE nutrients back into the environment, especially the soil, where the nutrients can be used by PLANTS and other organisms. In this way, decomposers are involved in Earth's nutrient cycles. *See also* AUTOTROPH; BIOGEOCHEMICAL CYCLE; BIOLOGICAL COMMUNITY; CHEMOAUTOTROPH; DECOMPOSITION; DETRIVORE; ECOLOGICAL PYRAMID; FOOD CHAIN; FOOD WEB; NICHE.

contaminant, Any HAZARDOUS SUBSTANCE, MICROORGANISM, radioactive material, or chemical that has an adverse affect on AIR, water, or soil that can result in injury, DISEASE, or death in an ORGANISM. Some sources of contaminants include *GIARDIA LAMBLIA*, VIRUSES, BENZENE, ASBESTOS, LEAD, CHLORDANE, MERCURY, DIOXINS, and alpha and beta particles. *See also* POLLUTANT; RADIATION.

continental drift, A hypothesis proposed in 1912 by Alfred Wegener (1880–1930), a German meteorologist *(see* meteorology), which postulated that all of Earth's masses had once been joined together as one large land mass called Pangaea. The hypothesis further states that Pangaea broke apart about 200 million years ago and that the continents have gradually

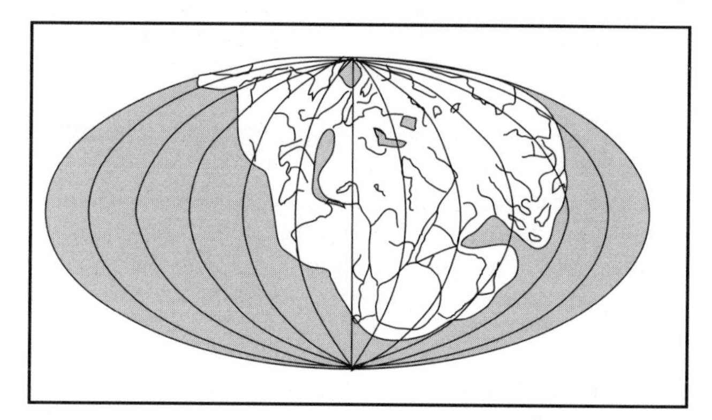

Pangaea. Pangaea as it may have looked about 200 million years ago.

drifted into the positions they occupy today. Wegener's was the first detailed study of continental drift. He incorporated evidence from rocks and fossils from different parts of the world. Wegener's hypothesis was rejected by many scientists of his generation because he was not able to explain how or why the drifting occurred.

During the 1960s through the 1980s, the basic concepts and hypotheses of continental drift became generally accepted by earth scientists. Theories of continental drift, PLATE TECTONICS, and sea-floor spreading developed greatly during this period, marking a revolution in earth science. According to the theory of plate tectonics, Earth's crust consists of at least 15 sections, or plates, that "float" on Earth's mantle, the semisolid layer just below the crust. As the plates move slowly over time, the positions of the continents and the OCEAN basins change, and this movement continues today. *See also* GEOLOGY.

EnviroSources

Moores, Eldridge, ed. *Shaping the Earth: Tectonics of Continents and Oceans: Readings from Scientific American Magazine.* New York: W.H. Freeeman & Co., 1990.

Stewart, John A. *Drifting Continents and Colliding Paradigms.* Indiana Bloomington, IN: University Press, 1990.

Windley, Brian F. *The Evolving Continents.* 3d ed. New York: John Wiley & Sons, 1995.

continental shelf, The portion of the OCEAN floor, between the shoreline and the continental slope, where the sea floor angles away from the edge of a continent. The continental shelf extends about 160 kilometers (100 miles) off shore and reaches a depth of about 200 meters (600 feet). The continen-

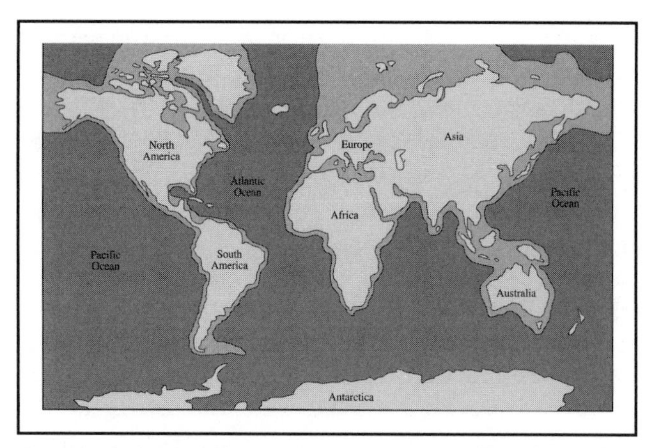

The Continental Shelves of the World. The continental shelf is the most productive area of the ocean for commercial fishing and offshore oil drilling.

tal shelf is the most productive area of the ocean with respect to COMMERICAL FISHING and offshore oil drilling. Environmentalists are concerned that oil and gas exploration on the continental shelf, along with leaks and OIL SPILLS associated with offshore oil drilling, will be hazardous to marine HABITATS and WILDLIFE. *See also* PETROLEUM.

contour farming, A SOIL CONSERVATION practice in which soil is plowed according to the natural shape of the land to prevent topsoil EROSION due to running water and wind. In contour farming, or contour plowing, land is plowed across or perpendicular to slopes to create a furrow in which water that runs downhill will come to rest. This helps prevent the water from carrying away topsoil and provides needed water to crops. As crops begin growing, erosion from wind is also reduced because PLANT roots hold soil in place and foliage serves as a protective ground cover that prevents wind from contacting the soil. Contour farming is used as a method of soil conservation worldwide.

Contour farming is sometimes carried a step further through the planting of rows of crops in strips alternating with low-growing, erosion-resistant plants such as clover or grasses that help hold soil in place with their roots and foliage. This practice is generally referred to as contour strip farming. *See also* AGRICULTURE; DUST BOWL; NATURAL RESOURCES CONSERVATION SERVICE.

control rod, A cylinder made of CADMIUM or boron that is part of the NUCLEAR REACTOR core. By absorbing neutrons, control rods moderate the rate of nuclear FISSION in a nuclear reactor. The control rod is raised or lowered in the reactor core to regulate the number of neutrons available for additional reactions. The rods absorb neutrons effectively without much re-radiation. *See also* NUCLEAR ENERGY.

Convention on International Trade in Endangered Species of Wild Flora and Fauna, An agreement signed by 60 nations in 1973, which prohibits the international trade

Contour farming is a soil conservation practice in which soil is plowed according to the natural shape of the land to prevent topsoil erosion. *Credit:* United States Department of Agriculture. Photo by E.W. Cole.

of any endangered PLANT or animal SPECIES and any products derived from them such as medicines, furs, and tusks. The convention also regulates the trade of species it deems in need of protection. The agreement was amended in 1979 to include the Convention on Conservation of Migratory Species of Wild Animals, which protects animals that migrate across borders of different nations. Illegal trade is monitored by TRADE RECORDS ANALYSIS OF FLORA AND FAUNA IN COMMERCE. Today, the convention has more than 100 member-nations. *See also* ENDANGERED SPECIES; ENDANGERED SPECIES ACT; ENDANGERED SPECIES LIST; MIGRATORY BIRD TREATY ACT; RED LIST OF ENDANGERED SPECIES; THREATENED SPECIES.

EnviroSources

Convention on International Trade in Endangered Species of Wild Flora and Fauna Web site: www.cites.org.

convergent evolution, The emergence of structures that have similar functions (and possibly similar appearances) in unrelated ORGANISMS. Biologists call structures resulting from convergent evolution analogous structures. Analogous structures evolve in organisms that live in similar ENVIRONMENTS and thus require similar ADAPTATIONS. Examples of analogous structures include the wings of birds and INSECTS and the winglike structures of BATS. All these structures evolved as adaptations for flight even though birds, insects, and bats are not closely related. The streamlined bodies of fishes and WHALES are adaptations that evolved to allow these unrelated organisms to move easily through water. Convergent evolution contrasts with divergent evolution, or ADAPTIVE RADIATION, in which related organisms become less similar as they adapt to differing environments. *See also* EVOLUTION; GENETIC DIVERSITY; HYBRID; HYBRIDIZATION.

cooling pond, A pond designed to hold heated water discharged from a nuclear or industrial electric power plant. *(See* discharge.) The water that acts as a coolant comes from lakes, rivers, or OCEANS, is used by the power plant and then can be treated and reused or discharged directly into the ocean, or a lake, river, canal, cooling pond, or COOLING TOWER.

A cooling pond is a shallow RESERVOIR with a large surface area for removing heat from water. The purpose of the cooling pond is to transfer heat in the cooling water to the AIR. Cooling ponds are used where land is relatively inexpensive, when water needed for cooling is scarce or expensive, or where there are strict THERMAL POLLUTION laws. In some cases, water in the pond can be reused in the power plant, thus reducing the overall water-withdrawal requirement.

EnviroTerm

coolant: AIR, water, or other materials used to conduct heat away from internal-combustion engines or NUCLEAR REACTORS.

cooling tower, A tall, cylindrical- or rectangular-shaped structure that dissipates waste heat into the ATMOSPHERE from a COAL-fired or nuclear electric power plant. Cooling towers, some reaching a height of more than 165 meters (500 feet) are also used in industrial plants and for large air-conditioning systems in buildings. There are two common types of cooling towers—the wet cooling tower and the dry cooling tower. The dry cooling tower works in a manner similar to the way a car radiator cools the engine. Hot water from the power plant flows through an arrangement of coils inside the tower. A fan blows AIR over the coils cooling the water in the pipes. The recycled water is then returned to the power plant.

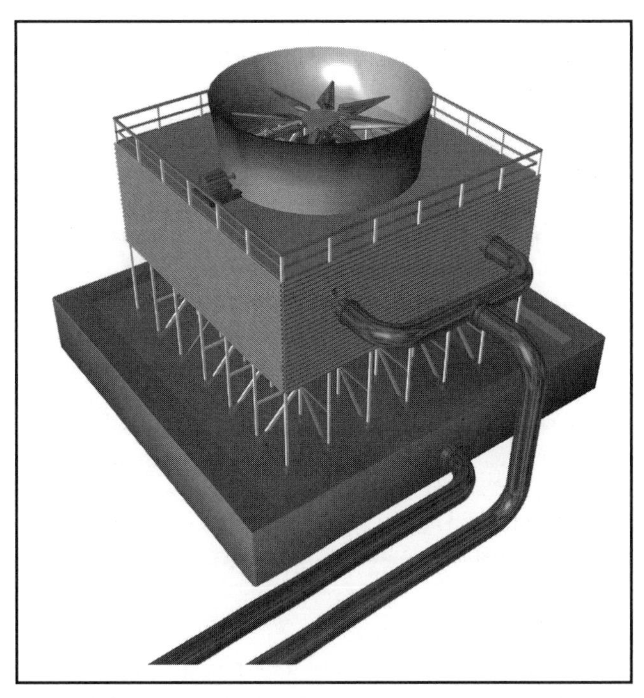

Cooling towers dissipate waste heat from a coal-fired or nuclear electric power plant and other installations. *Credit:* Tower Tech Inc.

In a wet cooling tower, hot water from the power plant is piped to the top of the tower, where the water is sprayed. The water droplets fall onto a series of louvers or baffles inside the tower. These louvers increase the surface area of the droplets, allowing for greater evaporation into the atmosphere. The tower exhausts hot moist air through its top and upward into the atmosphere. Any left-over water is collected into a catch basin, or pond, or is discharged into a COOLING POND, lake, or river. *(See* discharge.)

Cooling towers can be a public health hazard because they often discharge chemicals such as ACIDS, copper chloride, CHROMIUM, and ARSENIC, and biological CONTAMINANTS into the public WASTEWATER stream and into the surrounding air. If a tower contains galvanized metal, then particles of LEAD and zinc are released as well.

Some scientists now believe that the deadly Legionnaire's disease caused by the legionella BACTERIA develops in conventional cooling towers, in which water is exposed to direct sunlight. In the 1970s, scientific researchers traced a deadly outbreak of Legionnaire's disease to a Pennsylvania cooling tower with an open basin and louver design. Since then, there have been additional incidents in which dozens of people have become ill or died of Legionnaire's disease, such as occured in an outbreak in Mankato, Minnesota, in July 1995. The U.S. ENVIRONMENTAL PROTECTION AGENCY has stepped up investigative efforts after recognizing that conventional cooling towers often discharge pollutants. *See also* AIR POLLUTION; NUCLEAR ENERGY; NUCLEAR REACTOR.

copper, A metal element that is an important MINERAL in the human diet. Copper is soluble in water and can be absorbed into the body. Very high levels are required to cause adverse effects such as damage to the liver.

Copper is used principally to make electric wire and other electrical products because it is an excellent conductor of ELECTRICITY, is easy to bend, and is very resistant to corrosion. Much SMELTING is necessary to separate the copper from other materials. The smelter stacks DISCHARGE SULFUR DIOXIDE and other particles. Smelting operations can contaminate water resources and adversely affect HABITATS and ORGANISMS. Copper from mine TAILINGS can leach into the soil and cause contamination of SURFACE WATER and GROUNDWATER resources. *See also* LEACHING; TOXIC CHEMICALS.

coppicing, In FOREST MANAGEMENT, the process of cutting back trees to near ground level to allow the growth of thin shoots that will be harvested for FUEL WOOD and fencing. Coppicing is conducted at regular intervals to allow the shoots to grow to maturity. This can take 2 to 15 years for each harvest. *See also* CLEAR CUTTING; SELECTION CUTTING; SHELTERWOOD HARVESTING; SILVICULTURE.

coral reef, A vital marine ECOSYSTEM composed of living coral polyps that are attached to a structure formed from the skeletal remains of previous coral generations. Coral reefs are being threatened by POLLUTION from coastal SEWAGE and industrial chemicals at an accelerating rate. A study released in 1999, by the WORLD RESOURCES INSTITUTE, stated that nearly 60% of Earth's living coral reefs are threatened by human activity, including coastal development, OVERFISHING, and inland pollution. By 1999, scientists believe that worldwide, about 10% of the coral will have died and disappeared.

Corals and dinoflagellates (golden brown ALGAE) live together in a symbiotic partnership. *(See* symbiosis.) The algae grow inside the CELLS of the corals. Wastes produced by the corals are used as nutrients by the dinoflagellates. In turn, through PHOTOSYNTHESIS the dinoflagellates provide OXYGEN and food for the coral and enhance the rate at which corals build their skeletons. The coral and algae grow upward and over each other during the building process. The coral secrete a limy skeleton that is the basic structure of the reef.

New colonies and new coral structures are constantly being built on top of the dead skeletons of older colonies, but the rate of growth is slow. Some coral may grow only about 7.5 centimeters (3 inches) a year. Coral reef construction requires warm, well-lit tropical water free from pollution and turbidity. Extreme water temperatures, dim light, and salinity can cause the coral to die. *(See* salinization.)

The loss of coral HABITATS has alarmed environmentalists because it is one of the world's critical ecosystems. Coral reefs

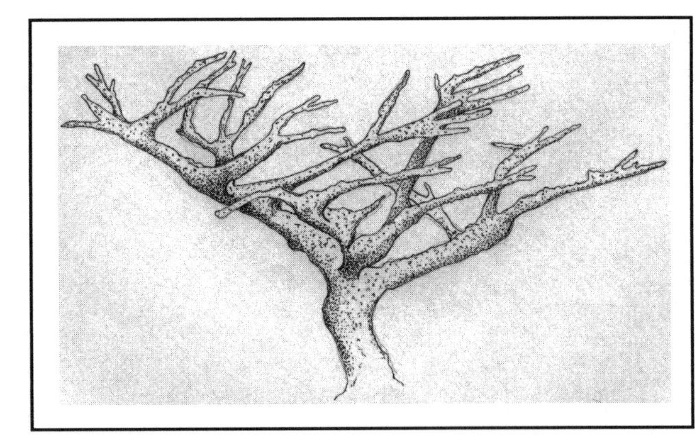

The loss of coral habitats has alarmed environmentalists because it is one of the world's most diverse ecosystems. The Elkhorn coral is one type of coral.

may host more numerous and diverse PLANT and animal SPECIES than any other ecosystem. About 25% of all marine species live in the coral reef ecosystem. The reefs are important food sources. Studies indicate that as many as 25% of the fish harvested in developing countries are taken from coral reefs. Coral reefs also protect coastlines from OCEAN storms. The porous structure is ideal for absorbing and dissipating the ENERGY of strong storm waves approaching coastal land areas.

Overfishing threatens many of the coral reef areas relied upon for food and livelihoods in developing countries. OIL SPILLS, MINING coral for building materials, and NATURAL DISASTERS such as HURRICANES can damage reefs. Another type of reef damage occurs when people destroy the coral reefs when harvesting fish for food or collecting them for aquariums. Fishers sometimes use explosives to dislodge the fish living in the reef. Coral reefs are also threatened and damaged by tourism, such as when boaters drop anchors on reefs and when people dive to collect specimens, which is illegal. In 1999, a Florida importer was found guilty of smuggling Philippine corals into the United States, in what prosecutors called the first federal conviction for trafficking in internationally protected coral.

Coral bleaching is a phenomenon that causes coral to lose color and die. Some scientists believe that this problem occurs when the dinoflagellates die off as a result of abnormally high water temperatures due to GLOBAL WARMING. Highly sensitive corals can live only in water between 18°C and 30°C (64°F and 86°F). Bleaching has occurred when water temperature has increased by 1°C (1.8°F) or more. When water temperature increases the oxygen also decreases. The OXYGEN-deprived coral becomes brittle and stunted. Massive bleaching of coral reefs occurred during 1983, 1987, and 1991 in the Pacific, Caribbean, and Indian Oceans. Devastating coral bleaching was recorded in 1998 in regions including Australia, India's Bay of Bengal, the Gulf of Thailand, Florida, and the Seychelles Islands off East Africa. However, more conclusive research of warming cycles will need to be done to find out if bleaching is actually due to high ocean temperatures.

Coral reefs are found worldwide in tropical and subtropical areas that are approximately 25° north and 25° south of the equator. Coral reefs grow in shallow, tropical water along the coasts of 110 countries. Over one-half of them are located

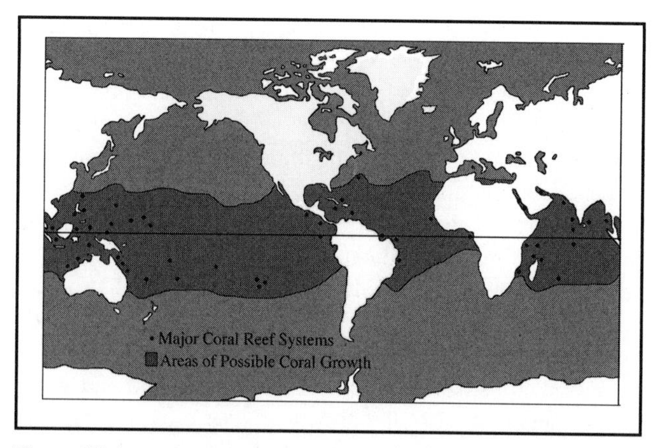

The world's largest coral site is Australia's Great Barrier Reef, which extends more than 1,600 kilometers along the northeastern coast of Australia.

in the Indian and western Pacific area. Another large area of coral reefs is located in the western Atlantic and Caribbean region. The coral growing near the coast are known as fringe reefs, while those farther out in the ocean are barrier reefs. Atolls—ring-shaped coral islands that enclose a shallow body of water, or lagoon—are found in the South Pacific, the Indian Ocean, and the CARIBBEAN SEA.

The world's largest coral site is Australia's Great Barrier Reef, which extends more than 1,600 kilometers (1,000 miles) along the northeast coast of Australia. This huge reef is home to different species of coral, birds, and fish. To protect the ecosystem, the Australian government made it a marine reserve in 1975, encompassing more than 2,500 coral reefs.

In the continental United States, the Florida reefs are where the most extensive living coral are located. The largest are off Key Largo, and have been designated a marine sanctuary by the NATIONAL OCEANIC AND ATMOSPHERIC ADMINISTRATION. The Key Largo reefs exist at the northernmost fringe of coral development in the Caribbean. The coral sites in Florida have been infected by human VIRUSES from SEPTIC TANKS. The viruses have affected swimmers and boaters who have contracted illnesses ranging from ear and eye infections to sore throats and respiratory DISEASES. *See also* BIODIVERSITY; OIL POLLUTION; SYMBIOSIS; ZOOXANTHELLAE.

EnviroSources

Coral Forest (A quarterly newsletter about coral protection): (415) 788-7333.

The Coral Reef Alliance Web site: http://www.coral.org. 2014 Shattuck Avenue, Berkeley, CA 94704; (510) 848-0110.

Davidson, Osha Gray. *The Enchanted Braid: Coming to Terms with Nature on the Coral Reef.* New York: John Wiley & Sons, 1998.

Greenpeace, *Coral Reef Network Directory*, 1995, 1436 U St., NW, Washington, DC 20009; Web site: http://www.greenpeace.org.

Massa, Renato. *The Deep Blue Planet: The Coral Reef.* Austin: Raintree Steck Vaughn, 1998.

National Oceanic and Atmospheric Administration. *25 Things You Can Do to Save Coral Reefs.* Pamphlet. Office of Public Affairs, 14th and Constitution Ave. NW, Room 6013, Washington, DC 20230; Web site: http://www.noaa.gov.

Reef Relief: (305) 294-3100, P.O. Box 430, Key West, FL 33041; Web site: http://www.reefrelief.org/.

Sale, Peter F., ed. *The Ecology of Fishes on Coral Reefs.* San Diego: Academic Press, 1994.

Steene, Roger. *Coral Seas.* Firefly Books, 1998.

core sample, A sample collected with a special tube-like instrument so material can be analyzed. The removed sample is broken down in vertical layers. In the case of soils, the sample can be used to study chemical composition, radioactivity, color, structure, and texture. Core samples can reveal EROSION patterns and evidence of CLIMATE changes, and can also be used to analyze packed ice in GLACIERS and ice sheets. *See also* ICE AGE; RADIATION.

Council on Environmental Quality, A board within the executive branch of the U.S. government that is charged with the responsibility of formulating and recommending national policies that will promote improvement in environmental quality. The Council of Environmental Quality (CEQ) was established in 1969 as part of the NATIONAL ENVIRONMENTAL POLICY ACT (NEPA). The chair of the CEQ is appointed by the president of the United States with approval and confirmation of the U.S. Senate.

Since its inception, the main responsibilities of the CEQ have included administering the provisions of the NEPA, setting regulations for the NEPA such as guidelines regarding what data must be included as part of an ENVIRONMENTAL IMPACT STATEMENT, and advising the president on environmental matters of national and international concern. The CEQ also aids the president in the preparation of an annual report on environmental quality. *See also* ENVIRONMENTAL JUSTICE; ENVIRONMENTAL PROTECTION AGENCY.

Cousteau, Jacques Yves (1910–1997), A French scientist, environmentalist, and inventor, whose filmed OCEAN adventures brought the ocean world into millions of households worldwide. In 1933, Cousteau joined the French Navy as a gunnery officer. It was during this time that he began his marine explorations and commenced work on development of a breathing machine for divers. In 1943, he and French engineer Emile Gagnan perfected the aqualung, an underwater breathing apparatus (also known as scuba gear) that allowed a diver to stay submerged for several hours. Divers first used the aqualung to locate and remove mines after World War II.

In 1950, Cousteau became president of the French Oceanographic Campaigns. That same year, he purchased the 364-metric ton (400-ton), former minesweeper, called the *Calypso*, to facilitate his investigations. He transformed it into a floating laboratory, equipping it with up-to-date devices, such as underwater television cameras. To finance his expeditions and increase public awareness of his undersea investigations, Cousteau produced many films and books. Among his films are *The Silent World* (1956) and *World Without Sun* (1966), each of which won an Academy Award for Best Documentary. His books include *The Living Sea* (1963), *Dolphins* (1975), and *Jacques Cousteau: The Ocean World* (1985).

In 1952 and 1953, Cousteau took the *Calypso* to the Red Sea and dived at 50 meters (165 feet) below the surface, to create the first color film of the ocean ENVIRONMENT at that depth. One of his most renowned discoveries was the hull of an ancient Greek wine freighter, buried in deep mud 45 meters

(148 feet) below the surface off the coast of Marseilles, France. The *Calypso* also conducted the first offshore oil survey performed by divers.

In 1957, Cousteau was appointed director of the Oceanographic Museum of Monaco, founded the Underseas Research Group at Toulon, and headed the Conshelf Saturation Dive Program, an experiment in which people lived and worked underwater for extended periods of time. In 1968, he was invited to create a television series. For the next eight years, the award-winning *The Undersea World of Jacques Cousteau* introduced the public to the realm of the world's oceans. In 1974, Cousteau founded the Cousteau Society, a nonprofit, membership-supported organization whose mission was to protect ocean life. Membership in this group has since grown to more than 300,000 worldwide. Cousteau was awarded the Medal of Freedom by President Ronald Reagan in 1985; in 1989, he was honored by France with membership in the French Academy.

It was primarily in later life that Cousteau tried to teach the world's people the importance of safeguarding natural treasures. In 1992, he was an official guest at the United Nations Conference on Environment and Development. The following year, he was appointed to the United Nations High-Level Advisory Board on Sustainable Development and agreed to serve as a WORLD BANK advisor on environmentally SUSTAINABLE DEVELOPMENT. That same year, the president of France named him chairman of the newly created Council on the Rights of Future Generations; Cousteau resigned this post in 1995 in protest of France's resumption of NUCLEAR WEAPONS testing in the Pacific. "Future generations would not forgive us for having deliberately spoiled their last opportunity and the last opportunity is today," he said at a 1992 environmental gathering. On January 11, 1996, the *Calypso* sank in Singapore harbor after being struck by a barge; the Cousteau Society is currently raising funds to build *Calypso II*. Jacques Cousteau died in 1997. *See also* DOLPHINS AND PORPOISES; EARLE, SYLVIA; GREENPEACE; OCEANOGRAPHY; UNITED NATIONS EARTH SUMMIT; WHALES.

EnviroSources

Chang, Chris, "Champions of Conservation: *Audubon* Recognizes 100 People Who Shaped the Environmental Movement and Made the Twentieth Century Particularly American," *Audubon* 100, no. 6 (November-December 1998): 88.

Cousteau, Jacques Yves. *Exploring the Wonders of the Deep.* Chatham, NJ: Raintree/Steck-Vaughn, 1997.

Cousteau Society Web sites: www.scubaworld.com/cousteau, www.cousteau.org, and www.dolphinlog.org/index.html.

Earle, Sylvia. "Cousteau Remembered." *Popular Science* 251, no. 4 (October 1997): 81.

cover crop, PLANTS used primarily to improve soil quality and soil texture by supplying needed nutrients to the soil. Cover crops are not raised for harvest or resale or cash crop but can be used as a forage crop and feed source. Cover crops can have stabilizing effects on the AGROECOSYSTEM by holding soil and nutrients in place, conserving soil moisture, and by increasing water PERCOLATION and retention.

Cover crops can play an important role in the biological control of INSECTS. Cover crops in orchards and vineyards buffer the system against pest infestations by increasing beneficial insect POPULATIONS, thereby reducing the need for chemical spraying. Cover crops can attract and sustain a wide variety of beneficial insects.

Using cover crops in AGRICULTURE is not a new farming practice. Before the production of manufactured FERTILIZERS, cover crops were commonly used to improve soil structure and productivity. LEGUME cover crops such as clover made NITROGEN available for future use by conventional crops.

Cover crops can also be used to reduce wind and water EROSION and to control the growth of weeds by shading and interfering with the germination and growth of these pesty plants. As an example, a special rye plant produces chemicals that suppress EXOTIC SPECIES and undesirable plants.

Unfortunately, cover crops can also become weeds if not controlled. Cover crops can be killed by plowing and mowing. It is critical that cover crops be controlled in the spring to prevent them from interfering with the cash crops.

EnviroSource

Kellogg Biological Station Extension Office (Michigan State University): 3700 E. Gull Lake Dr. Hickory Corners, MI 49060-9516; (616) 671-2412; (616) 671-4485.

cradle-to-grave, A term used in the RESOURCE CONSERVATION AND RECOVERY ACT (RCRA) to describe the way that HAZARDOUS SUBSTANCES must be identified, documented, and tracked during the lifetime of a product. The system keeps track of how the product is manufactured, treated, transported, and finally disposed of in a special document called a waste manifest.

criteria pollutant, Any of the six primary AIR POLLUTANTS, such as LEAD, SULFUR DIOXIDE, CARBON MONOXIDE, PARTICULATE MATTER, OZONE, and NITROGEN OXIDES (NO_x), for which NATIONAL AMBIENT AIR QUALITY STANDARDS (NAAQS) have been established by the U.S. ENVIRONMENTAL PROTECTION AGENCY (EPA). The standard for each pollutant is based on estimates of the highest level that can be tolerated by humans without any ill effects. Short-term standards are set for each pollutant to protect against ACUTE TOXICITY and long-term standards are set to protect against CHRONIC TOXICITY. *See also* AIR POLLUTION; CLEAN AIR ACT; ELECTROSTATIC PRECIPITATOR; SCRUBBER.

critical mass, The smallest amount of radioactive material that can undergo nuclear FISSION without an external source of neutrons. As little as 11 kilograms (24 pounds) of PLUTONIUM-239 can cause a fission reaction releasing neutrons and gamma RADIATION. When the critical mass of fissionable FUEL is assembled in a NUCLEAR REACTOR, CONTROL RODS absorb some of the neutrons, thus preventing the reaction from producing enough ENERGY to cause an explosion. *See also* CADMIUM; NUCLEAR WEAPONS; RADIOISOTOPE OR RADIOACTIVE ISOTOPE.

crocodiles, The largest living REPTILES of the order Crocodilia, two SPECIES of which are endangered. Crocodiles are narrowly distributed in tropical and subtropical regions. The largest member of the order, which may grow to lengths of 9 meters (30 feet), is the salt-water crocodile, which inhabits coastal

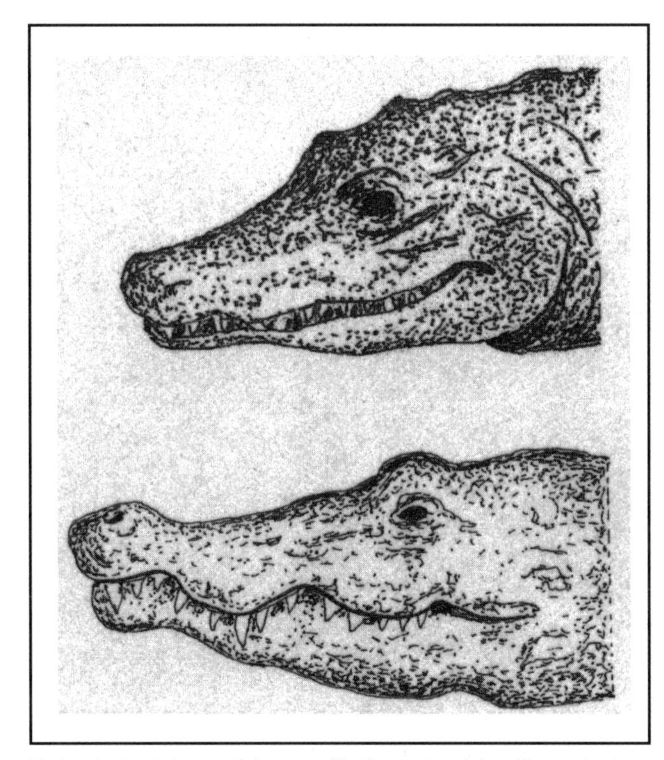

Notice the head shapes of the crocodile (bottom) and the alligator (top) are different.

waters of Australia, Malaysia, southern China, and India. The mugger, or SWAMP crocodile, inhabits inland swamps in the same regions. The Nile crocodile, which is native to Africa, has been hunted almost to EXTINCTION in the lower Nile River, but exists in greater numbers in the upper Nile. The American crocodile (*Crocodylus acutis*) is the only crocodile native to the United States and is found in the EVERGLADES NATIONAL PARK and the Florida Keys. This SPECIES also is found in some parts of Brazil. Three smaller crocodile species are narrowly distributed in parts of Cuba, Venezuela, and the countries of Central America.

Crocodiles are fierce predators. Unlike ALLIGATORS, which generally live in fresh-water HABITATS, crocodiles generally inhabit marine or BRACKISH waters. Crocodiles also are distinguished from alligators by the shapes of their snouts, which are narrower, longer, and more triangular than those of alligators, and by their teeth, several of which protrude from the closed mouth of the crocodile, but not of the alligator.

Worldwide, the major threat to crocodiles is overhunting for its skin, eggs, or a substance from their musk glands that is used to make perfumes. Both the Nile crocodile and American crocodile are ENDANGERED SPECIES, primarily due to overhunting. Loss of habitat from human encroachment into coastal areas also threatens both crocodile species. An additional problem facing the American crocodile has resulted from the diversion of the Everglades' waters for IRRIGATION. This practice has reduced the amount of fresh water flowing into the tidal MARSHES and ESTUARIES in which the crocodile lives, making the water too saline for both the crocodile and the animals on which it feeds. Efforts are under way to reverse some of the damage caused by this water diversion; however, progress has been slow. Nevertheless, American crocodile POPULATIONS have shown a slow increase, largely due to controls on hunting, since their placement on the ENDANGERED SPECIES LIST in 1975. *See also* CARNIVORE; PREDATION; SALINIZATION.

EnviroSource

Crocodilia Natural History and Conservation, sponsored by the Crocodile Specialist Group Web site: www.flmnh.ufl/edu/natsci/herpetology/brittoncrocs/cnhc.html.

crop rotation. *See* soil conservation.

crude oil. *See* petroleum.

crustaceans, A class of arthropods (INVERTEBRATE animals with jointed appendages) containing about 35,000 SPECIES. Crustaceans are an important component of FOOD CHAINS and FOOD WEBS. As arthropods, crustaceans are invertebrates with jointed appendages and an exoskeleton composed largely of chitin. Members of this class, which include shrimps, crabs, water fleas, and wood lice, are distributed throughout the world, primarily in fresh-water and marine HABITATS. Wood lice are an example of terrestrial crustaceans.

Many crustaceans, especially shrimps, lobsters, and crabs, are an important food source for people. Others, such as the KRILL which are one of the many organisms that make up PLANKTON, serve as a major food source for marine ORGANISMS, including many species of WHALES. *See also* ENDANGERED SPECIES; INDICATOR SPECIES; OCEAN; ZOOPLANKTON.

Lobsters are crustaceans that hunt for food on the ocean bottom. Some can grow up to 2 meters long and live more than 60 years.

cryptosporidium, A protozoan PATHOGEN that is associated with the DISEASE cryptosporidiosis and can be transmitted in drinking water. Cryptosporidiosis can cause diarrhea, stomach cramps, vomiting, and fever. The disease can be fatal in humans who have weak immune systems.

According to the U.S. ENVIRONMENTAL PROTECTION AGENCY (EPA), the largest outbreak of cryptosporidiosis in the United

States took place in Milwaukee, Wisconsin, in 1993. Milwaukee's water supply comes from Lake Michigan and is treated and filtered by a treatment plant before being distributed to homes and businesses. During 1993 there was a period of heavy rains and RUNOFFS making the water treatment plant ineffective. As a result more than 400,000 people were affected by the disease and 4,000 needed to be hospitalized. The disease caused over 50 deaths. The original source of the contamination is still uncertain. *See also* CONTAMINANT; *ESCHERICHIA COLI; GIARDIA LAMBLIA;* PRIMARY, SECONDARY, AND TERTIARY TREATMENTS.

EnviroSource

Solo-Gabriele, H., and S. Neumeister. "U.S. Outbreaks of Cryptosporidiosis." *Journal of the American Water Works Association* (September 1996).

cyanobacteria, BACTERIA that serve as PRODUCERS because of their ability to synthesize nutrients through PHOTOSYNTHESIS. Cyanobacteria, or blue-green bacteria, were once classified as blue-green ALGAE; however, since these ORGANISMS lack a membrane-bound nucleus and membrane-bound organelles, they are not true algae, but members of the Kingdom Monera.

Cyanobacteria may be Earth's oldest organisms and have been discovered as fossils dating back 3.5 billion years. Today, cyanobacteria are distributed worldwide in aquatic HABITATS, in soil, and living on rocks and trees, either in colonies or united as filaments.

Cyanobacteria serve several roles in the ENVIRONMENT. As producers, they form the base of a variety of aquatic and terrestrial FOOD CHAINS and FOOD WEBS. As they carry out photosynthesis, cyanobacteria remove CARBON DIOXIDE from the AIR and release OXYGEN and water vapor as wastes. The oxygen is used by AEROBIC organisms for respiration. Thus, cyanobacteria are involved in Earth's carbon-oxygen cycle.

Some cyanobacteria are responsible for NITROGEN FIXATION, a process that changes the free NITROGEN in the ATMOSPHERE to nitrogen compounds, such as NITRATES, that can be used by most organisms. In this way, cyanobacteria make up an important part of Earth's NITROGEN CYCLE.

Another important role of cyanobacteria lies in their association with other organisms. Some cyanobacteria live in SYMBIOSIS with FUNGI, forming an organism known as a LICHEN. In this relationship, the cyanobacteria provide the fungi with nutrients and oxygen. The fungi provide the cyanobacteria with carbon dioxide, water, and a HABITAT. The association permits both the cyanobacteria and fungi to survive in environments where neither organism could survive individually. Often, this environment is on tree trunks or bare rocks.

Like cyanobacteria, lichens serve as a food source for many types of organisms, including the large moose of the TUNDRA and TAIGA. In addition, as they carry out their life processes, lichens secrete digestive ACIDS that serve to break apart the surfaces on which they live. When this surface is ORGANIC, such as a tree trunk, nutrients derived from the breakdown of the surface are returned to the environment, where they may be used by other organisms. When these surfaces are inorganic, such as a rock, the acids cause WEATHERING that contributes to soil formation. *See also* BIOGEOCHEMICAL CYCLE; ECOLOGICAL SUCCESSION; NICHE; PIONEER SPECIES.

EnviroTerms

nucleus: The central part of a CELL. The nucleus contains the genetic material of a cell and controls cell processes.

organelle: Any membrane-bound cell structure that carries out specific functions in the cell.

EnviroSources

Margulis, Lynn, and Karlene V. Schwartz. *Five Kingdoms: An Illustrated Guide to the Phyla of Life on Earth.* 2d ed. New York: W. H. Freeman and Co., 1988.

Wilson, Edward O. *The Diversity of Life.* Cambridge, MA: Belknap Press, 1992.

cyclone, An area of low AIR pressure in the ATMOSPHERE. The air turns around a cyclone in a counterclockwise direction in the Northern Hemisphere. In the Southern Hemisphere the winds blow in a clockwise direction around the cyclone. Violent tropical cyclones are also called HURRICANES in North America. Small but violent twisters or cyclones are called TORNADOES in many parts of the world.

In 1999, Cyclone Vance had moved along Australia's northwest coast with wind speeds of 230 kilometers (142 miles) an hour, ravaging towns, closing ports, and halting MINING operations. Cyclone Vance caused extensive flooding, damaged hundreds of houses, and toppled power lines with its heavy rains and wind. The damages were estimated in the tens of millions. Fortunately no one was killed.

However, in December 1999, the most powerful cyclone yet recorded in Australia sideswiped a thinly populated coastal region north of Perth, causing relatively light damage before veering back out to sea. Again as in Cyclone Vance, no injuries were reported in the aftermath of 288 kilometers-per-hour (180-miles per hour) winds.

Cyclones can be horrific. In October 1999, a cyclone that struck the Indian coastal state of Orissa caused the deaths of 9,885 people. The cost of repairing and replacing the property damaged by the cyclone was set at $1.5 billion by government officials. *See also* FLOOD; NATURAL DISASTER.

EnviroTerm

anticyclone: An area of high-pressure air that descends in the center of the cyclone. The air moves in a clockwise direction in the Northern Hemisphere and in a counterclockwise direction in the Southern Hemisphere.

D

dam, A structure, often human built, that restricts the flow of water. Dams are constructed on a waterway, such as a stream or river, usually to alter the flow of water, by slowing it, rerouting it, or creating a new surface water body, such as a RESERVOIR. Dams are used to create sources of HYDROELECTRIC POWER or to store water for human needs. Human-built dams represent remarkable works of engineering, but also have the potential to create new problems, sometimes in unforeseen ways.

In spite of their many advantages, dams sometimes have created or contributed to environmental and social problems. In the 1990s, the construction of the Gabcikovo Dam on the DANUBE RIVER in Slovakia produced severe impacts on the ECOSYSTEM of the Danube valley. By retaining and diverting water, this dam caused the WATER TABLE to lower and dried WETLANDS downstream of the dam. The depressed water table adversely affected farms that depended on GROUNDWATER for IRRIGATION; diversion of the river reduced fish harvests; and the loss of wetlands resulted in WILDLIFE endangerment. In the James Bay region of Quebec, Canada, controversy surrounds a series of existing and proposed hydroelectric dams, the JAMES BAY HYDROPOWER PLANT. Although these dams provide a source of clean electrical power (no FUEL or radioactive material is involved), they alter the environment substantially. The lakes created by the dams have flooded some of the traditional hunting grounds of the native James Bay Cree people, and they have also altered local weather conditions. In addition, the lakes appear to contribute to the release of METHYL MERCURY, a toxic substance derived from industrial pollution, into the FOOD CHAIN.

Various types of dam structures exist. A gravity dam consists of a wall constructed across the waterway in such a way that water is permitted to flow only under, not over, the wall. This construction directs much of the force of the water toward the base of the dam. A special type of dam called an arch dam features a wall that curves upstream, against the flow of water. Typically built in small canyons, arch dams direct the force of the water toward the canyon walls. Buttress dams have a system of beams, or buttresses that support the downstream side of the dam. A buttress dam, which can be made relatively large, is often used in wide valleys. When great quantities of soil or rock are available for construction, engineers may choose to build an embankment dam, which typically consists of an earthen wall.

People have built dams for a variety of purposes. Early dams helped to corral fish, making them easier to catch. Agricultural societies use dams to improve irrigation, by storing water or by redirecting water into irrigation ditches. For centuries, mill dams have provided the power to turn grindstones used to mill grain. Dams constituted a very important power source during the Industrial Revolution, when water propelled by the force of gravity was used to operate machinery. Hydroelectric dams harness the power of flowing water to generate ELECTRICITY. In addition, dams can provide FLOOD control by storing excess water during times of heavy RUNOFF and later regulating the release of that water.

The planning, design, and construction of a dam require a great deal of care and consideration for the environmental and social effects of the project; the builders and the community must therefore weigh the positive and potential negative impacts of dam construction. *See also* AQUIFER; ASWAN HIGH DAM; THREE GORGES DAM.

EnviroSources

American Rivers: www.amrivers.org
U.S. Department of the Interior, Bureau of Reclamation Hoover Dam Web site: www.hooverdam.com
U.S. Bureau of Reclamation Web site: www.usbr.gov

Danube River, One of the longest waterways in Europe that is threatened by major water pollution problems. The Danube River, which begins in Germany, flows 2,850 kilometers (1,770 miles) through several European countries including Austria, Bulgaria, Croatia, the Czech Republic, Hungary, and the Ukraine and then empties into the BLACK SEA off the coast of Romania. Some of the major cities along the way include Vienna, Budapest, and Belgrade. The river is an important source of drinking water for millions of people.

The Hoover dam is one of the largest dams in the world. It spans the Colorado River and provides water and electricity for Arizona, Colorado, and Nevada. *Credit:* U.S. Department of Interior, Bureau of Reclamation, Andrew Pernick, Photographer.

The Danube River basin or WATERSHED covers an area of 817,000 square kilometers (315,000 square miles) and includes a rich variety of WETLANDS and FLOODPLAINS that are HABITATS for many PLANT and animal SPECIES.

Most of the widespread pollution in the river originates from agricultural and industrial sources. Industrial DISCHARGES from smelters, paper mills, and chemical plants have polluted many streams and tributaries and contaminated drinking water supplies. *(See* contaminant.) The RUNOFF of FERTILIZERS, PESTICIDES, and animal wastes from farmlands have caused EUTROPHICATION in many sections of the river.

The Danube is also the recipient of untreated WASTEWATER and fecal PATHOGENS. *(See* coliform bacteria.) Much of the SEWAGE from several major cities and communities along the Danube is discharged directly into the river without treatment. As a result of the pollution, most of the species of fish are listed as threatened, including all species of sturgeon. Some of the bird species that are threatened include the red-breasted goose, the Dalmatian pelican, and the pygmy cormorant.

In 1994, several Danube countries and the European Union met and signed the Danube River Protection Convention. The major goal of the convention was to find ways to reduce the pollution loads in the Danube and Black Sea, to develop cooperation between the Danube countries on pollution prevention measures by exchanging data, and to ensure the sustainable use of water resources.

In 1997, the Danube Pollution Reduction Programme was initiated by various communities in the Danube River basin in support of the Danube River Protection Convention. The program's primary objective was to establish programs and projects that would reduce pollution in the basin and Black Sea. *See also* MEDITERRANEAN SEA; NORTH SEA; RHINE RIVER.

Darwin, Charles Robert (1809–1882), A British NATURALIST who developed the theory of EVOLUTION based on natural selection. This theory states that new SPECIES emerge over time as a result of new traits developed by individual ORGANISMS that provide them with survival advantages in a particular ENVIRONMENT. Darwin was not the first scientist to propose a theory of evolution; however, he was the first to relate this process of change to the ADAPTATIONS organisms develop to help them meet the challenges of their environments.

The ideas that led Darwin to his theory of evolution resulted largely from observations he made during his five-year tenure as a naturalist aboard the H.M.S. *Beagle*. The *Beagle* was a scientific research ship commissioned to explore the Southern Hemisphere. One of the crew's research sites was the GALÁPAGOS ISLANDS located off the western coast of Ecuador. While studying the FLORA and FAUNA of the islands, Darwin observed that PLANTS and animals of the Galápagos were similar in type to those found in other parts of the world (such as South America), but differed enough from their mainland counterparts to be recognized as different species. He also observed that similar kinds of plants and animals living among the islands of the Galápagos archipelago had traits that were distinct enough from one another as to be considered different species. As an example, Darwin pointed to the variations he observed among finches living on the islands.

Darwin observed 14 species of finches. Differences among them were most observable in the structures of their feet and beaks; Darwin observed that the latter features were uniquely adapted to the type of food each finch species ate. Thus, the variations in each species provided each a survival advantage in its environment. Darwin hypothesized that all of the finches he observed were related and had evolved from a common ancestral species. He suggested that over time, nature selected for some variations, making individuals with those variations more suited to life in a slightly changed environment than were members of the original finch species. These variations were then passed to the offspring of those individuals through the reproductive process, and in time, the traits and preferred HABITATS of the finches were dissimilar enough that two distinct species emerged. With more time and successive generations, this process was repeated enough times that 14 distinct species of Galápagos finches had evolved.

In 1859, Darwin published his theory in the book, *The Origin of the Species by Means of Natural Selection*. At the time, scientists did not have a clear understanding of how traits could be passed from one generation to the next; however, as they learned about the role of genes in this process and how MUTATIONS and other factors could alter the characteristics of an organism, support for Darwin's theory increased. Today, most scientists accept Darwin's theory of evolution and use the theory to explain the great BIODIVERSITY that exists among Earth's organisms. *See also* ADAPTIVE RADIATION; COMPETITIVE EXCLUSION; CONVERGENT EVOLUTION; EXTINCTION; NICHE; TAXONOMY; WILSON, EDWARD OSBORNE.

EnviroSources

Darwin, Charles. *The Autobiography of Charles Darwin, 1809–1882.* Reprint 1992. Ed. by Nora Barlow. New York: W.W. Norton and Co., 1993.

———. *The Origin of the Species by Means of Natural Selection.* Reprint 1859. Oxford, UK: Oxford University Press, 1998.

———. *The Voyage of the Beagle: Charles Darwin's Journal of Researches.* New York: Penguin, 1989.

DDT. *See* dichlorodiphenytrichlorethane

death rate, A measure of the number of deaths per 1,000 members of a POPULATION over a one-year period of time, which is used to calculate overall population growth or decline. A change in the death rate, or mortality rate, is of interest to environmental scientists, because it may be linked to an improvement or decline in environmental conditions. For example, scarcity of food, increased incidence of DISEASE, and drastic CLIMATE changes may all increase the death rate of a population. By contrast, a decline in predators or favorable growth conditions may decrease death rate. *(See* predation.)

Death rate is determined by dividing the annual number of deaths by the total population size at mid-year, and multiplying the resulting product by 1,000. An example of how death rate is calculated is shown. The example assumes that in a one-year period 15 individuals from a total population of 1,200 individuals died.

$$\text{Death Rate} = \frac{\text{Number of Deaths} \times 1{,}000}{\text{Population Size}} \qquad \text{Death Rate} = \frac{15 \times 1{,}000}{1{,}200}$$

$$\text{Death Rate} = 0.0125 \times 1{,}000 = 12.5$$

Death rate is one of the factors used to determine changes in population size. Other factors considered include BIRTH RATE and IMMIGRATION, which increase population size, and EMIGRATION, which decreases population size. *See also* AGE STRUCTURE; CARRYING CAPACITY; LIMITING FACTOR; POPULATION DENSITY; ZERO POPULATION GROWTH.

debt-for-nature swap, A practice in which an organization or government assumes the debt of a region in exchange for a promise from the debtor that proceeds from the exchange will be dedicated to projects that benefit the ENVIRONMENT. Most environmental projects addressed through this mechanism involve the acquisition of land, often for use as nature preserves or national parks. Other exchanges have been made to help free up funds to clean polluted areas, train personnel such as park rangers, or for land RECLAMATION and REFORESTATION projects.

Several environmental organizations and government agencies have been involved in debt-for-nature swaps worldwide. Among them is Conservation International, which purchased $5 million in debt from the African island of MADAGASCAR in 1992. Funds from the swap were to be used to continue the reforestation projects and park personnel training begun in 1991 through a similar arrangement between Madagascar and the WORLD WILDLIFE FUND. Other countries that have used debt-for-nature swaps to address environmental problems include Zambia, Poland, Costa Rica, Brazil, Ecuador, Bolivia, the Dominican Republic, and the Philippines. *See also* CONSERVATION; DEEP ECOLOGY; LAND TRUST.

EnviroSource

Conservation International Web site: www.conservation.org.

decibel. *See* noise pollution.

deciduous forest, A FOREST major life zone or BIOME found in the middle latitudes that is characterized by hardwood, broadleaf tree SPECIES that lose their leaves each year. The deciduous forest biome is one of the most altered biomes on Earth. Woods from deciduous forests have been used for building materials and firewood, and sections of forests have been cleared for farming where land is scarce and because the soil is rich. (*See* clear-cutting.) Many communities throughout Europe and the United States have been built on land that was once deciduous forests. These human activities have led to the decline and loss of these forests in many parts of the world.

Oak (left) and maple leaves are found in deciduous forests.

The major dominant PLANT species include oak, hickory, maple, basswood, and beech, with some pine trees. The animals that live in the deciduous forest include foxes, coyotes, squirrels, chipmunks, woodchucks, opossums, raccoons, toads, turtles, snakes, mice, black bears, deer, and a variety of birds including hawks and owls. Precipitation in a decidous forest averages between 75 centimeters to 200 centimeters (30–80 inches) a year.

The two expressions of this forest are the TEMPERATE DECIDUOUS FOREST and the tropical deciduous forest, of which the temperate variety has the much greater range and worldwide distribution. In both types, leaf fall is caused by seasonal DROUGHT. In temperate regions, winter drought is due to the increase of frozen water; in tropical regions, the availability of water is restricted because of dry-season reductions in precipitation.

Tropical deciduous forests, also known as dry tropical forests, occur in areas of 15–20 degrees north or south latitude, on the fringe of the tropical RAINFOREST biome. Dry-season rainfall in these forests is usually less than 3 centimeters (1.2 inches). The forest has a canopy of trees that are 20–25 meters (60–75 feet) tall; small-leafed, deciduous understory trees are typically 3–10 meters (9–30 feet) high. Bamboo thickets and palms are common, and the shrub layer is dominated by members of the pea family. Elevations are usually moderate. *See also* CONIFEROUS FOREST; DEFORESTATION; OLD-GROWTH FOREST; SHELTERWOOD HARVESTING.

decomposer, An ORGANISM that obtains its nutrients and ENERGY by breaking down large ORGANIC compounds derived from the wastes or remains of other organisms into simpler organic and inorganic molecules. Decomposers help to naturally recycle nutrients through the ENVIRONMENT. In this way, decomposers play an important part in Earth's BIOGEOCHEMICAL CYCLES.

Organisms that carry out DECOMPOSITION are often called saprobes or saprophytes. These organisms include BACTERIA of decay and FUNGI, such as molds, mildews, smuts, rusts, mushrooms, and puff balls. Some animals, such as earthworms and the larvae or adult forms of some INSECTS, also obtain nutrition and energy through decomposition.

Decomposers obtain nutrition mostly from the remains of dead organisms. By feeding on and breaking down these remains, decomposers help to cleanse the environment of accumulated wastes. Because decomposers often feed on the remains of predatory organisms, decomposers are generally placed at the highest TROPHIC LEVEL. (*See* predation.) Thus, decomposers form the last link in FOOD CHAINS and FOOD WEBS. *See also* DETRITUS; DETRIVORE; RECYCLING.

decomposition, The mechanical, thermal, or chemical process by which substances are broken down into simpler mixtures, compounds, molecules, or elements. Decomposition occurs naturally in the ENVIRONMENT through the actions of nonliving, or ABIOTIC, agents and the chemical activities of DECOMPOSERS—ORGANISMS, such as BACTERIA and FUNGI, that obtain their nutrients and ENERGY by breaking down complex ORGANIC compounds derived from the wastes or remains of other organisms into simpler organic and inorganic components. In some environments, decomposers obtain nutrition by breaking down inorganic compounds.

As they carry out their digestive processes, most decomposers break down complex organic substances into simpler organic and inorganic substances. Some of this matter is used by the decomposers to meet their nutritional and energy needs; matter not needed by the organisms is released into the environment, where it may be taken in and used by other organisms. Thus, decomposition is a vital part of Earth's BIOGEOCHEMICAL CYCLES. In addition to RECYCLING matter for use by organisms, decomposition aids in soil formation. It also contributes to the formation of rocks and other matter as substances deposited in soil are acted upon by mechanical and chemical processes, heat, and pressure. (*See* rock cycle, weathering.)

Some human activities can alter the rates at which decomposition occurs. For example, many food preservatives are intended to slow decomposition and keep foods safe to eat and available for sale for longer periods of time. IRRADIATION of food products also helps prevent spoilage resulting from the activities of MICROORGANISMS that serve as PATHOGENS or decomposers. *See also* COMPOSTING; DETRITUS; DETRIVORE; DISEASE.

decontamination, The removal of harmful matter or ENERGY from exposed ORGANISMS or the ENVIRONMENT. Examples of harmful matter include toxic substances, noxious fumes, and PATHOGENS, such as BACTERIA and VIRUSES. Forms of energy that often are harmful include RADIATION and excessive heat.

Decontamination can be relatively simple. For example, some pathogens are removed from surfaces by cleaning the surfaces with DISINFECTANTS. Noxious fumes may be eliminated from an enclosed area by simply venting the area with fresh AIR and allowing the fumes to dissipate. Other decontamination processes are complex. For example, the decontamination of OCEAN water and beaches following an OIL SPILL may involve the work of many people and much equipment. Such cleanups often begin with containment of the PETROLEUM using oil spill containment booms and then skimming some oil from the surface. In addition, the affected area may require treatment with special chemicals that break apart the oil or bacteria that "digest" the oil. *(See* biotechnology.) Similarly, the cleanup of a TOXIC WASTE site may involve many different decontamination methods. Such complex cleanups may take months or years to complete. *See also AMOCO CADIZ* OIL SPILL; ARMY CORPS OF ENGINEERS; BHOPAL INCIDENT; BUREAU OF RECLAMATION; COMPREHENSIVE ENVIRONMENTAL RESPONSE, COMPENSATION, AND LIABILITY ACT; *EXXON VALDEZ* OIL SPILL; FEDERAL EMERGENCY MANAGEMENT AGENCY; HANFORD NUCLEAR WASTE SITE; HAZARDOUS WASTE; LOVE CANAL; OIL SPILL EQUIPMENT; PLUME; PRIMARY, SECONDARY, AND TERTIARY TREATMENTS; RECLAMATION; SANDOZ CHEMICAL SPILL; SUPERFUND; TIMES BEACH SUPERFUND SITE.

deep ecology, A branch of the environmental movement that seeks to protect nature not because of its value to humans but because of its intrinsic value. The term "deep ecology" was coined in 1973 by Norwegian philosopher Arne Naess. Naess stressed the need to understand the perspectives of self-realization and BIOCENTRIC equality. Self-realization requires people to recognize that they are connected with something greater than themselves and that the greatest potential of life results from its diversity. *(See* biodiversity.) The idea of biocentric equality, or biocentrism, holds that all natural things have a right to exist. This is seen as being an intrinsic value, rather than one based on a natural being's or thing's perceived value to humans.

Although the term "deep ecology" was not introduced until 1973, NATURALIST ALDO LEOPOLD is generally called "the father of deep ecology." This designation is a result of Leopold's essay, "A Land Ethic," which appeared in his book, *A Sand County Almanac*, published in 1949. In this essay, Leopold stressed the need for people to recognize the intrinsic value of nature and to establish a system of values or moral principles about how land is used.

To deep ecology proponents, the efforts of most environmental groups are too limited because they strive to protect only those segments of nature that have been deemed valuable to humans. Deep ecologists refer to such work as "shallow ecology." The environmental group most linked with the deep ecology philosophy or movement is Earth First! which was cofounded by DAVID FOREMAN in 1980. Deep ecology opponents often criticize Earth First! as being too radical in its efforts to protect the ENVIRONMENT. *See also* ANIMAL RIGHTS; ANTHROPOCENTRIC; ECOLOGY; GREENPEACE.

EnviroSources

Devall, Bill, and George Sessions. *Deep Ecology: Living as if Nature Mattered.* Layton, UT: Peregrine Smith Books, 1985.

Leopold, Aldo. *A Sand County Almanac*. New York: Oxford University Press, 1949.

Leopold Education Project Web site: www.lep.org

deep-well injection, A method of HAZARDOUS WASTE disposal in which wastes are pumped into concrete or steel-encased shafts deep beneath Earth's surface. Currently, deep-well injection is the most widely used method for disposing of hazardous wastes on land. Of the 21 million metric tons (23 million tons) of hazardous wastes disposed of on land in 1995, 19 million metric tons (21 million tons) were disposed of via deep-well injection.

Deep-well injection is generally considered to be harmless to humans and the ENVIRONMENT. Nevertheless, some problems have been associated with the technique. For example, in 1961 the U.S. Army created an injection site near Denver, Colorado, to dispose of nerve gas originating from the Rocky Mountain Arsenal. *(See* chemical weapon.) The well was dug to a depth of more than 3,658 meters (12,000 feet) and was injected with more than 2 million liters (528,000 gallons) of water. About 636 liters (168 gallons) of nerve gas were pumped into the well over a four-year period, and the opening to the well was sealed in 1966. Although no materials leaked from the well, the injection process is believed to have triggered a series of EARTHQUAKES and tremors in the Denver region. Prior to construction and use of the well, the region had not experienced an earthquake in more than 80 years. Because of the potential earthquake problem, the site was permanently sealed, closed, and abandoned in 1985.

Despite occasional problems, deep-well injection has been safely used for waste disposal for more than 30 years. The process is usually used for liquid wastes that have been treated to immobilize or destroy their hazardous contents. Before an injection site is selected, geological studies are conducted in the proposed area to ensure that deposited wastes cannot leach from the site and migrate to the surface or contaminate GROUNDWATER. *(See* leaching.) An ideal site is one that will trap wastes in the pores of a permeable rock layer situated between impermeable rock layers, with all layers located well below the WATER TABLE. *(See* permeability.) The disposal site must be located in a geologically stable area, away from faults or regions of volcanic activity. *See also* CONTAMINANT; RESOURCE CONSERVATION AND RECOVERY ACT.

EnviroSources

Operations Management International, The Key West Project Web site: www.keywestwastewater.com/default.htm.
Rocky Mountain Arsenal, Remediation Venture Office Web site: www.army.mil/rma.html.
University of Colorado, Colorado Internet Center for Environmental Problem Solving, Superfund and the Rocky Mountain Arsenal Web site: www.colorado.udu/conflict/CASES/rymnars.html.

defoliant, A specialized herbicide that removes leaves from trees or other PLANTS. Once the leaves are destroyed, the trees or plants will die because they can no longer carry out PHOTOSYNTHESIS. The United States used defoliants during the Korean conflict of the 1950s and the Vietnam War of the 1960s. The chemicals were dropped from airplanes onto heavily foliated areas to eliminate places where enemy soldiers could hide and ambush American troops. The main defoliants used were AGENT ORANGE, Agent White, and Agent Blue.

While the use of defoliants by the military served its purpose during war time, the chemicals' effects are still seen today over millions of hectares of SWAMP and FOREST lands in Korea and Vietnam. For example, more than 50% of the MANGROVE forests of South Vietnam treated with defoliants in the 1960s had not recovered 30 years later. An additional 5% of hardwood forests were also destroyed. The plants that have reclaimed the affected areas are mostly weeds. In addition to devastating the plants, the defoliants also destroyed the HABITATS of many fishes and other animals.

People who lived in the regions of Asia where defoliants were used suffered as well. Destruction of the forests resulted in economic losses because wood from the area could not be harvested, sold, and used for FUEL. In addition, many of the people in these regions, particularly those living in the Saigon area of South Vietnam, suffered health problems from direct exposure to the chemicals, from drinking or bathing in contaminated water or eating contaminated food. *(See* contaminant.) The health problems have ranged from headaches, nausea, and diarrhea to severe skin rashes, miscarriages in pregnant women, birth defects in children, and various types of CANCER. Similar health problems have been observed in U.S. and Australian soldiers who were directly or indirectly exposed to defoliants used during the Vietnam War. Despite these reports, however, there is little direct scientific evidence linking defoliant use to many of these health problems. *See also* PESTICIDE.

EnviroSources

Saign, Geoffrey C. *Green Essentials: What You Need to Know about the Environment.* San Francisco, CA: Mercury House, 1994.
U.S. EPA Web site: www.epa.gov

deforestation, The complete or partial removal of trees and vegetation from an area of land. Deforestation is a consequence of POPULATION growth. As the human population grows, more land and resources are needed to support it.

FORESTS of the world, covering one-quarter to one-third of Earth's land surface, are important NATURAL RESOURCES for people. Each year, millions of hectares of forest are cleared for the purposes of AGRICULTURE, timber production, FUEL-WOOD collection, livestock grazing, and MINING. Although clearing forests serves many purposes for humans, scientists are concerned about the short- and long-term effects of deforestation on the ENVIRONMENT. As of this date, studies show that many forests have remained stable in Europe and the United States, but forest cover is declining greatly in Asia, Africa, and South America.

Agriculture accounts for most of the world's deforestation. In many countries, large-scale farms, ranches, and plantations are rapidly replacing tropical RAINFORESTS. The deforestation of the rainforests has already caused significant changes in soil quality. Despite the abundance of vegetation, tropical rainforests have surprisingly poor soil. Most of the nutrients available in tropical rainforests are already locked up in the ORGANISMS that live there. When trees are cleared, the already nutrient-poor soils of the rainforest become even more infertile. The hot, tropical sun also bakes exposed soils into a hard, brittle surface, which can more easily be eroded by rainfall.

Cattle ranching and other types of livestock production account for a significant amount of deforestation. In South and Central America alone, some 52,000 square kilometers (20,000 square miles) of forest—an area the size of New Jersey—are cleared each year to create grazing pastures for cattle. After a few years, the soils become destroyed and eroded by the movements of vehicles and cattle. *(See* erosion.) Eventually, the land is abandoned and a new patch of forest is cleared.

Forests are also cut and cleared for timber, paper pulp, and other wood products, or to establish TREE FARMS and plantations. Trees also provide fuel wood for many people. About 1.5 billion people in developing countries depend on fuel wood as their major source of FUEL.

Deforestation is most severe in the tropical rainforests of Africa, Asia, Central America, and South America. About 18 million hectares (45 million acres) of tropical rainforest are cleared annually. Over half of the world's original tropical rainforests have already been destroyed by deforestation. The remaining rainforests, almost 1.8 billion hectares (4.5 billion acres), cover only 5% to 7% of the world's land. At the current rate of deforestation, scientists predict that no rainforests will remain by the middle of the twenty-first century. When a tropical rainforest is destroyed, much more than trees are lost. Deforestation destroys the HABITATS of countless animal and PLANT SPECIES. Many species have now become extinct or are being pushed to the brink of EXTINCTION because their habitats are being destroyed and cut down. *(See* endangered species; threatened species.)

Deforestation can also lead to changes in local and global CLIMATE. Tropical rainforests help control local climates by maintaining humidity and offering protection from the wind. Water evaporates from the leaves of trees and other vegetation, enters the ATMOSPHERE, and eventually falls back to land during heavy rainfalls. When these forests are cleared, this important source of water vapor is also lost. Eventually, this can lead to severe DROUGHT and DESERTIFICATION of an area. Continued deforestation of the world's tropical rainforests may also alter global climate by contributing to the GREENHOUSE EFFECT. The greenhouse effect is a natural phenomenon that maintains global temperatures. It works because several gases, primarily CARBON DIOXIDE (CO_2), form a blanket in the atmosphere and trap heat in, much like the glass walls of a greenhouse. The loss of trees contributes to the greenhouse effect in two ways. First, when trees are burned, large amounts of carbon dioxide are released into the air. What's more, the loss of trees and other plants means that less carbon dioxide is being absorbed from the atmosphere for PHOTOSYNTHESIS. Overall, deforestation contributes to 25% of global carbon dioxide EMISSIONS. Scientists fear that more GREENHOUSE GASES in the atmosphere will trap more heat in the AIR around the planet, thus causing GLOBAL WARMING.

Currently, governments, environmental organizations, and scientists are working to slow the rates of deforestation. Efforts include establishing tropical forest reserves to protect forested land, improving FOREST MANAGEMENT of unprotected forests, improving agricultural techniques, and curbing the demand for tropical forest products that are not easily replenished. *See also* OLD-GROWTH FOREST; SELECTION CUTTING; SLASH AND BURN.

EnviroSources

Brown, Katrina, et al. *The Causes of Tropical Deforestation: The Economic and Statistical Analysis of Factors Giving Rise to the Loss of the Tropical Forests*. Vancouver: University of British Columbia, 1994.
The Conservation Atlas of Rainforests. 2 vols. 1991.
Gash, J. H. C., et al., eds., *Amazonian Deforestation and Climate*. New York: John Wiley and Sons, 1996.
Gentry, Alwyn, ed. *Four Neotropical Rainforests*. New Haven, CT: Yale University Press, 1991.
Graewohl, Judy, and Russell Greenberg. *Saving the Tropical Forests*. London, UK: Earthscan Publication, 1988.
Greenpeace International Web site: www.greenpeace.org/~forests
Hutchinson, B. A., and B. B. Hicks, eds. *The Forest-Atmosphere Interaction*. D. Reidel, 1985.
Myers, N., ed. *Rainforests*. Emmaus, PA: Rodale Press, 1993.
O'Brien, Karen L. *Sacrificing the Forest: Environmental and Social Struggles in Chiapas*. Boulder, CO: Westview Press, 1998.
Steen, H. K., and R. P. Tucker, eds. *Changing Tropical Forests*. Durham, NC: Forest History Society, 1992.
Whitmore, T. C. *Tropical Deforestation and Species Extinction*. Boca Raton, FL: Chapman & Hall, 1992.

denitrification, A vital part of the NITROGEN CYCLE that occurs in the soil or water. The process of denitrification occurs when the nutrient forms of NITROGEN—ammonia and NITRATES—in the soil are removed and converted by ANAEROBIC BACTERIA to nitrogen gas (N_2), which is released into the ATMOSPHERE. Denitrification is used in the secondary treatment of removing POLLUTANTS from SEWAGE. *See also* PRIMARY, SECONDARY, AND TERTIARY TREATMENTS.

denitrifying bacteria. *See* denitrification.

Department of Agriculture, An executive department of the U.S. government and the leading agency supporting research and technology development in the areas of PLANT and animal GENOMES, plant DISEASE management and control, human nutrition and food safety, and animal reproduction. Key agencies of the department that have environmental and NATURAL RESOURCES responsibilities include the U.S. FOREST SERVICE, the NATURAL RESOURCES CONSERVATION SERVICE, and the Animal and Plant Health Inspection Service.

President Abraham Lincoln established the Department of Agriculture (USDA) as part of the cabinet in 1862. President Lincoln called it the "people's department," since at the time 90% of Americans were farmers who needed good seed and reliable information to grow their crops. Reserves of PUBLIC LANDS available for homesteading and AGRICULTURE were becoming depleted, but instead of moving from worn-out sites to other croplands, farmers sought the aid of scientific technologies to increase crop production and income. As a result, the USDA's responsibilities grew rapidly. The growth and expansion of the USDA continued through the Great Depression of the 1930s, when the problem of surplus farm products became acute.

Today, only 2% of Americans are farmers, but USDA programs serve all Americans, in both rural areas and cities. The USDA's multifaceted mission is to ensure a safe food supply; to care for agricultural lands, FORESTS, and rangelands; to support sound development of rural communities; to provide economic opportunities for farm and rural residents; to expand global markets for agricultural and forest products and services; and to work to reduce hunger in America and the world.

The department's 30 agencies manage more than 200 programs, including the food stamp program, which benefits millions of families; the AmeriCorps national service program; and programs to address environmental and CONSERVATION issues on both federal and private lands. The USDA also hosts the National Agricultural Library, the nation's primary source for agricultural information and one of the world's largest and most accessible agricultural research libraries. It is one of only four national libraries of the United States; the other three are the Library of Congress, the National Library of Medicine, and the National Library of Education.

Farmers and ranchers work with the USDA to produce healthy crops while caring for the quality of soil and water. Consumers benefit from USDA research, which helps ensure that the United States has the most available, highest-quality, and least expensive food supply of any country in the world. Future generations are assured that they will inherit a national forest system where they will be able to hike, camp, and enjoy the splendor of America's outdoors. Thanks to the work of the USDA, people throughout the world look to the United States as the world's largest agricultural exporter and as the world's largest donor of foreign food aid. *See also* DUST BOWL; EXOTIC SPECIES; FOOD AND DRUG ADMINISTRATION; HYBRIDIZATION.

EnviroSource

U.S. Department of Agriculture Web site: www.usda.gov

Department of Energy, A department of the U.S. government established in 1977 to regulate and manage the nation's

ENERGY policy. The Department of Energy (DOE) consolidated many of the federal government's responsibilities for energy and national defense materials into one agency by replacing earlier energy-related agencies of the federal government.

The main responsibilities of the DOE include providing technologies and developing policies that achieve efficiency in energy use while maintaining environmental quality and a secure national defense. The DOE is a world leader in the research and development of programs and technologies that generate energy from FOSSIL FUELS, nuclear FUELS, and ALTERNATIVE ENERGY RESOURCES such as SOLAR ENERGY, WIND POWER, and BIOFUELS. In addition, the DOE administers comprehensive environmental management programs involving the cleanup of sites contaminated with HIGH-LEVEL RADIOACTIVE WASTES and other CONTAMINANTS resulting from energy generation. The DOE also manages the nation's HYDROELECTRIC POWER plants, such as the Bonneville Power Administration of the northwestern United States. *See also* INTERNATIONAL ATOMIC ENERGY COMMISSION; NUCLEAR ENERGY; NUCLEAR REGULATORY COMMISSION; TENNESSEE VALLEY AUTHORITY.

EnviroSource

U.S. Department of Energy, Center for Environmental Management Information, P.O. Box 23769, Washington, DC 20026-3769, (800) 736–3282; http://www.doe.gov (Most DOE sites have a public information office that generates notices of public meetings. For the DOE site nearest your home, consult the blue pages of your telephone directory.)

Department of the Interior, A department of the U.S. government that was established by Congress in 1849 to administer the nation's internal affairs. At its creation, the Department of the Interior (DOI) was charged with a wide range of responsibilities, including the construction of Washington, D.C., and the water supply system; exploration of the western WILDERNESS; the administration of some hospitals and universities, and the management of Native American affairs, PUBLIC LANDS, patents, and pensions. The unifying theme among the DOI's seemingly disparate responsibilities was the internal development of the nation and the welfare of its people. In 1873, Congress transferred territorial oversight responsibilities from the secretary of state to the secretary of the interior, and in 1879, the U.S. GEOLOGICAL SURVEY was established as a DOI agency. Its responsibilities included providing the federal government with information about locations of MINERAL and water sources.

With the American CONSERVATION movement's inception in the early twentieth century, the focus of the DOI began to shift toward the preservation and wise use of the nation's fragile cultural and NATURAL RESOURCES, including its ECOSYSTEMS. In 1916, President Woodrow Wilson signed legislation creating the NATIONAL PARK SERVICE. In 1940, the Bureau of Fisheries, formerly located within the U.S. Department of Commerce, and the Bureau of Biological Survey, formerly within the U.S. DEPARTMENT OF AGRICULTURE, were transferred to the DOI and consolidated as the U.S. FISH AND WILDLIFE SERVICE. Today, the DOI is composed of eight agencies, including the BUREAU OF LAND MANAGEMENT, the Bureau of Indian Affairs, and the BUREAU OF RECLAMATION.

As the nation's principal conservation and land management agency, the DOI is now responsible for most of our nation's public lands and natural resources. Today, the DOI's mission is to provide for the appropriate management, preservation, and use of the nation's public lands and natural resources; to carry out related scientific research and investigations in support of these objectives; to develop and use resources in an environmentally sound manner and obtain an equitable monetary return on those resources; and to carry out federal trust responsibilities with respect to American Indians and Alaska natives. The DOI generates revenue from many sources on the public lands, including the development of mineral resources (such as NATURAL GAS, PETROLEUM, COAL, and metals), the sale of timber, the issuance of grazing permits, and the collection of recreation fees. *See also* ARCTIC NATIONAL WILDLIFE REFUGE; WILDLIFE REFUGE; YELLOWSTONE NATIONAL PARK; YOSEMITE NATIONAL PARK.

EnviroSource

U.S. Department of the Interior Web site: www.doi.gov

deposition, The laying down of sediments that have been carried away from another location. Deposition is a natural process that often results when agents of EROSION such as wind, water, and GLACIERS stop moving, and thus drop any sediment, including soil, they are carrying. As a result of deposition, Earth's surface is built up in a new area, often aiding in the development of new HABITAT. Deposition often occurs in glacial moraines, the FLOODPLAINS of rivers, river deltas, and alluvial fans. Deposition also may occur very quickly at the bottoms of mountain slopes as a result of such events as mudslides and avalanches. The materials carried during these mass movements of land can be extremely destructive to habitats and result in loss of both human property and life. Poor land management practices by humans often increase the likelihood of such disasters by loosening soil and rock materials along slopes. *See also* FLOOD; NATURAL DISASTER; SEDIMENTATION; SILTATION.

EnviroTerm

moraine: A mass of rock or other sediment deposited at the sides or end of a GLACIER.

desalination, The process of removing dissolved salts from seawater and BRACKISH water. Desalination, also known as desalinization, is very important in many DROUGHT-prone areas of the world because it can transform unusable water from the OCEAN into valuable fresh water.

The average salt content of ocean water has about 35,000 parts per million (ppm) of dissolved materials of which most is sodium chloride (NaCl), common salt. Brackish water has about 1,500 to 6,000 ppm. POTABLE WATER usually ranges from 300 to 500 ppm and is drinkable. Desalinization reduces the salt content of these waters to about 500 ppm. There are two general methods for removing these salts: distillation and RE-

A model of the Taunton River desalination plant to be built in Massachusetts in 2001. The reverse osmosis treatment plant will provide 20 million liters (5.3 million gallons) of water a day using the brackish water of the lower Taunton River. *Credit:* Bluestone Energy Services, Inc.

VERSE OSMOSIS. Distillation is the simpler and more inexpensive method. In distillation, fresh water is simply evaporated from salt water, leaving the salts behind. In reverse osmosis, seawater is forced under high pressure through a filter, which traps the salt crystals. This method purifies about 45% of the salt water passed through the filter. The remaining, more concentrated, salt water is pumped back into the sea. Reverse osmosis is commonly used in very dry areas, such as the Persian Gulf, where fresh water supplies are limited.

Desalination became important during the nineteenth century when steam-powered ships were in use. Scientists were searching for dependable supplies of fresh water that would not damage the steam engines. In 1869, the first patent for a desalination process was granted. That same year, the first desalination plant was built near the Red Sea by the British government to supply fresh water to ships that came to port. Today, there are approximately 7,500 desalination plants worldwide. Most of these are located in countries around the Persian Gulf, such as Saudi Arabia and Kuwait, but a few smaller plants are now operating in southern California. Generally, desalination is not used to produce a region's total water supply but rather to supplement existing fresh water supplies or in emergency situations, such as after long droughts.

Although there are many desalination plants scattered throughout the world, together they produce less than 1% of the world's fresh-water supply. Desalination is a very expensive process that requires a lot of ENERGY. For example, in the United States, desalinized water produced by the reverse osmosis method costs about $3 per 3,800 liters (1,000 gallons). This is about four to five times what the average U.S. citizen pays for fresh drinking water, and over 100 times the price paid by farmers for IRRIGATION water.

Because of the high costs, desalinized water is mainly used to supply drinking water in the home. Small, reverse osmosis desalination units capable of producing a few liters per day can be purchased for home use. In areas such as the Persian Gulf, where energy is inexpensive, using desalinized water for these purposes is quite practical. *See also* SALINIZATION; WATER CYCLE.

desert, A major life zone or BIOME where the extreme climatic conditions prevent water from being readily available to living things. It is an arid land where the rate of water evaporation is greater than the rate of precipitation, which is very little. Deserts occur in very hot, dry (arid) regions, and in very cold regions, where precipitation is less than approximately 25 centimeters (10 inches) per year. Extremes of temperature often occur in deserts, ranging from very hot daytime temperatures to near-freezing temperatures at night. Deserts constitute the least productive type of ECOSYSTEM because the lack of water limits productivity.

Examples of deserts include the Gobi Desert of central Asia, the Sahara Desert of northern Africa, the Gibson Desert of western Australia, Patagonia, and the Sonoran and Mojave Deserts of the southwestern United States.

Sometimes characterized by sandy soils, deserts may feature sand dunes created by the action of wind on loose sand. Many deserts feature a surface, called "desert pavement," composed of coarse pebbles and stones that have been left behind from the gradual blowing away of the smaller grains of soil. Often deserts are susceptible to EROSION, which can be extreme in the absence of PLANT cover and in the event of sudden thunderstorms accompanied by heavy rain.

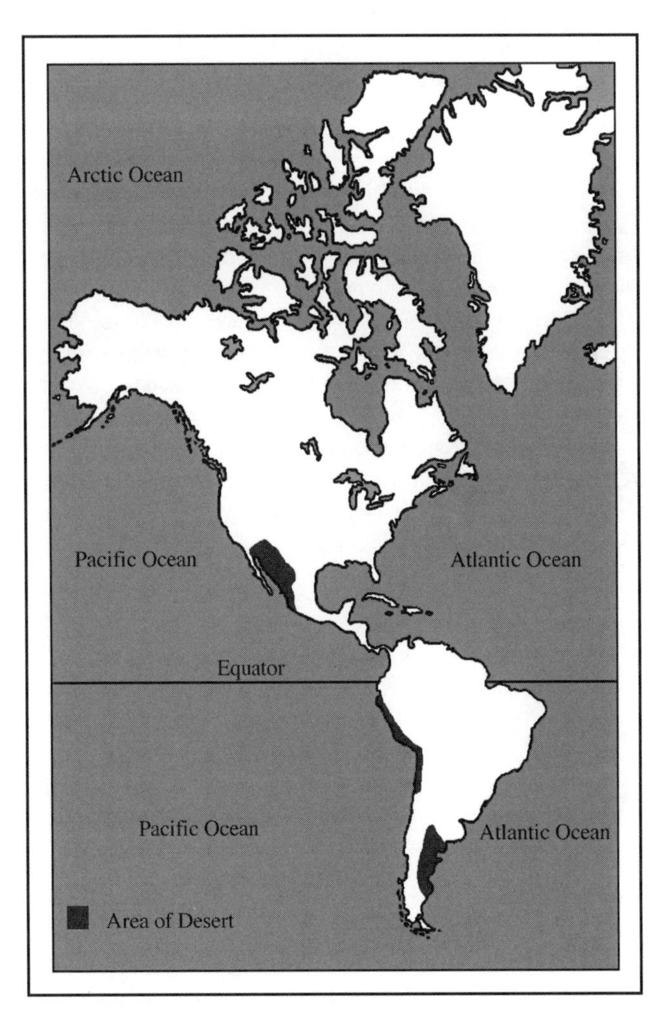

Areas of desert in the Western Hemisphere.

Desert plants and animals are specially adapted to the harsh conditions and typically are scarce, the limited water supply supporting only small POPULATIONS. Biological ADAPTATIONS enable desert ORGANISMS to endure long periods without water or to obtain water that is not readily accessible. Many desert plants feature exterior surfaces (such as a cuticle) that reduce moisture loss to the dry desert AIR and have the ability to remain dormant for long dry spells. CACTI are examples of well-adapted desert plants. Some varieties of desert plants have developed the ability to rapidly flower and reproduce during brief periods of moisture. Animals found in various deserts include lizards, snakes, tortoises, gerbils, camels, and a variety of INSECTS, such as tarantulas and scorpions. Some of these desert animals are principally nocturnal—that is, they are primarily active during the night when it is cooler.

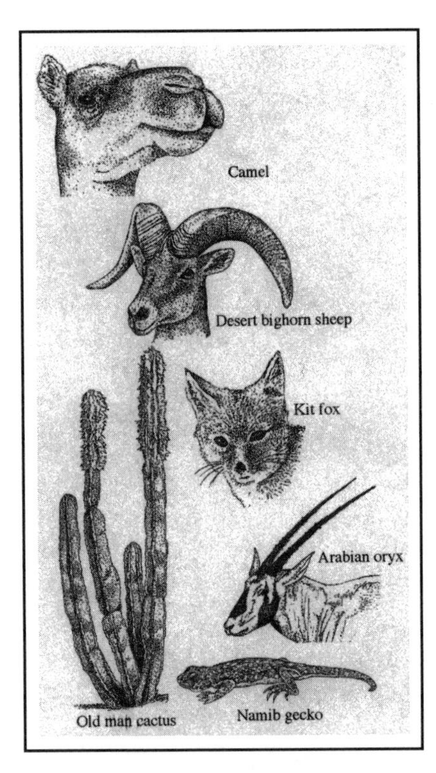

Historically, many human cultures have existed in deserts throughout the world. Aboriginal groups occupied vast arid portions of the Australian continent for thousands of years. In the southwestern United States, many generations of the Native American Pueblo peoples successfully inhabited desert regions. The Near East was well known for the nomadic tribes that roamed the Arabian and Persian Deserts. The Kalahari Desert of South Africa is the home of the San (Bushmen) cultures.

EnviroSources

Foster, Lynne. *Adventuring in the California Desert*. Sierra Club Adventure Travel Guide. San Francisco, CA: Sierra Club Books, 1997.

Langewiesche, William. *Sahara Unveiled: A Journey across the Desert*. Vintage Departures. New York: Vintage Books, 1997.

McMahon, James. *Deserts*. Audubon Society Nature Guides. New York: Knopf, 1985.

desertification, A process by which a nondesert area becomes progressively more like a DESERT and which is now a major problem in parts of North and South America, Mediterranean Europe, southern and western Asia including China and Mongolia, Australia, and many countries in Africa. In these areas, much of the rangelands and croplands are affected by EROSION and desertification. Desertification represents problems that are not easy to fix once the land becomes sandy, coarse, stony, and loses its capactity to hold water. Other problems of desertification include the lowering of the WATER TABLE, which makes it difficult to pump water for crops, and the drying up of lakes and ponds used for IRRIGATION. There is also an increase of salinity in the soil and the loss of native PLANTS that can help cut down on wind and soil erosion.

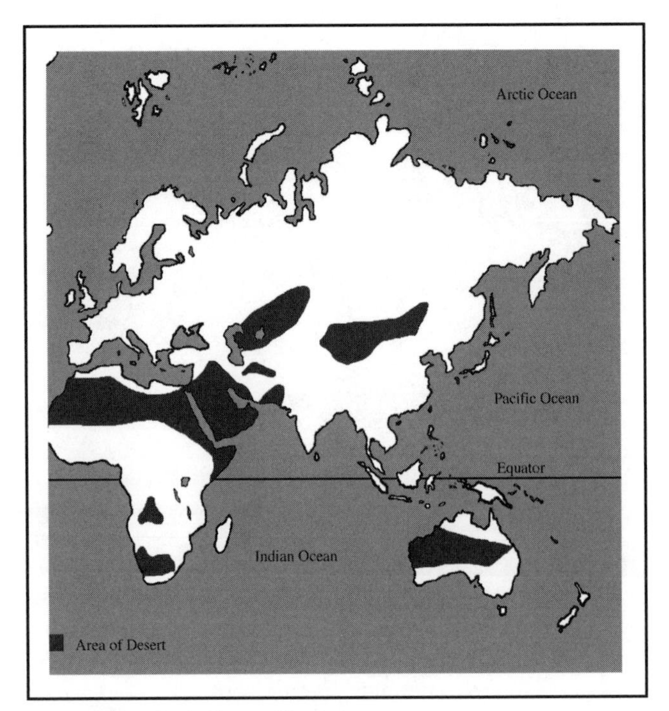

Areas of desert in the Eastern Hemisphere.

In arid or semiarid regions, desertification may occur as a result of human activities or climatic change. The great deserts of the world have developed in response to long-term climatic trends; however, the extent of deserts has increased as a result of human activities in many areas, and at a more rapid pace than would occur without human influence.

Commonly, desertification occurs on the edges of existing deserts, dry areas, or at the edges of GRASSLANDS and proceeds outward, creating a larger area of desert conditions. This progressive growth of a desert does not take place as a uniform expansion, but occurs in increments: one patch followed by another. Naturally occurring DROUGHTS hasten the transformation.

Some ancient agricultural practices have contributed to the growth of deserts. Artificial irrigation by diversion of water from streams or rivers, while an effective agricultural technique, can cause the accumulation of salts in poorly drained soil. (*See* salinization.) Over time, as this process makes the irrigated soil increasingly unsuitable for farming, erosion of the topsoil can occur. This sequence of changes may result ultimately in desertification. Overgrazing of livestock also contributes to desert conditions by eliminating plant cover and exposing soil to erosion.

SLASH-AND-BURN AGRICULTURE, which occurs in some tropical regions, is an example of a human practice that may lead to desertification. In this type of farming, areas of FOREST may

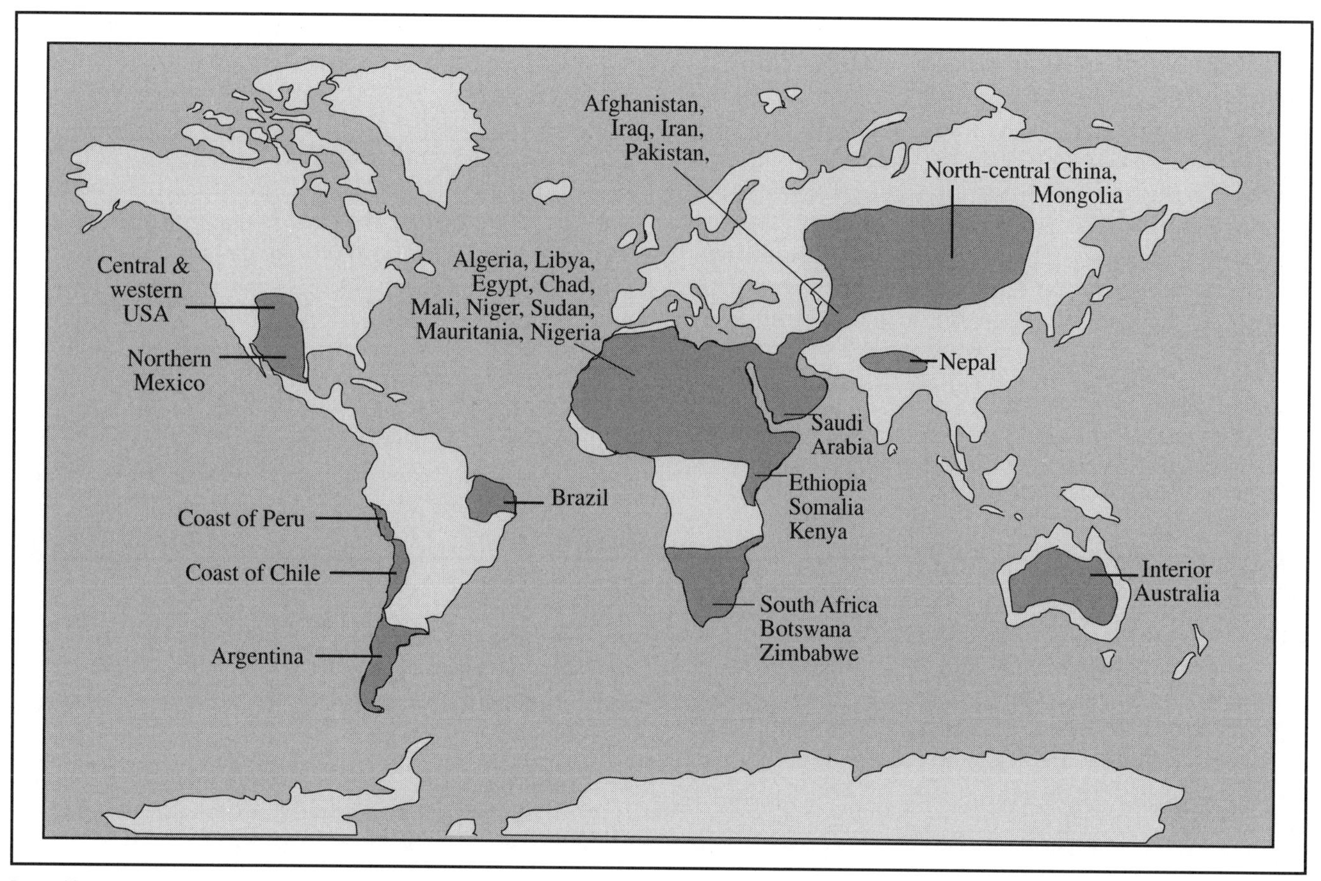

Desertification, shown in the dark tint, is a major problem in parts of North and South America, Mediterranean Europe, southern and western Asia including China and Mongolia, Australia, and many countries in Africa.

be cleared and then burned to permit planting of crops. The burning produces a highly fertile but short-lived soil that fails to support AGRICULTURE after a few harvests because the nutrients provided through burning are quickly used by crop plants. When large areas are cleared in this fashion, rapid soil erosion and the failure of the original forest to reclaim the land may result in eventual desertification.

During the twentieth century, scientists recognized the impact of human activities in the enlargement of desert areas. During the 1960s, a major drought that occurred in the vast sub-Saharan region of north-central Africa known as the Sahel increased world awareness of desertification. In the Sahel, the drought caused starvation, livestock death, agricultural land loss, and relocation of large human POPULATIONS. This episode, and similar events elsewhere in the world, brought about the realization that arid land that is near the limit of its ability to support a growing human population cannot easily recover from the stress of periodic droughts and will become susceptible to desertification.

Another factor identified as contributing to modern desertification is the decreasing OZONE layer in Earth's ATMOSPHERE. The size of the ozone layer, because it regulates the amount of ultraviolet RADIATION reaching Earth, appears to have an important relationship to changes in global temperature. Damage to the ozone layer occurs from the release into the atmosphere of CHLOROFLUOROCARBONS, chemicals commonly used in air conditioning and refrigeration systems. The GLOBAL WARMING trend observed during the late twentieth century may be correlated with OZONE DEPLETION and increasing desertification. *See also* CLIMATE; DEFORESTATION; DUST BOWL; SUSTAINABLE AGRICULTURE.

EnviroSource

U.S. Department of Agriculture Web site: www.usda.gov

detergent, A compound used as a synthetic cleansing agent in household cleaners to remove oil, soil, and dirt. Some contain compounds, such as PHOSPHATE, that kill useful BACTERIA and encourage ALGAE growth when they are discharged into rivers and streams and other waters from SEWAGE TREATMENT PLANTS. *(See* discharge.) Heavy discharges of phosphate-laden detergents can cause EUTROPHICATION— a major problem in waterways near highly populated areas. More than 20 states have banned or restricted the level of phosphate-containing laundry detergents; however, some of the bans do not include dishwasher products containing phosphates. Today the detergent manufacturers are switching to more nonphosphate ingredients in their detergents. Another step to control phosphates and other nutrients such as NITROGEN is to upgrade equipment at sewage treatment plants to control these EFFLUENTS from being discharged into waterways. *See also* POLLUTANT; SEWAGE.

detritus, ORGANIC matter, such as leaf LITTER, that collects on FOREST and OCEAN floors and on the bottoms of lakes and ponds, and which derives from the wastes or remains of ORGANISMS. Detritus provides nutrients and ENERGY to organisms called detritus feeders and contributes to the formation of nutrient-rich soil. Detritus feeders fall into two groups: the DETRIVORES, which are animals, and the DECOMPOSERS, which are primarily BACTERIA and FUNGI. As detritus undergoes DECOMPOSITION, matter that is not used by detrivores or decomposers is returned to the environment, where it can be used by other organisms. *See also* BIOGEOCHEMICAL CYCLE; HUMUS.

EnviroLink

In GEOLOGY, detritus is the small particles, such as grains of sand and silt, that are worn away from rock as a result of WEATHERING. This detritus is primarily inorganic in composition.

detrivore, An animal that feeds on ORGANIC matter (DETRITUS) derived from the wastes or remains of ORGANISMS. Detrivores obtain their ENERGY and nutrients from detritus and help cleanse the ENVIRONMENT by preventing a buildup of wastes in natural and artificial ECOSYSTEMS.

Many INSECTS and insect larvae are detrivores. For example, the housefly and its larvae (maggots) are detrivores that feed on discarded foodstuffs in GARBAGE or on the remains of organisms that have died in the wild. Larger detrivores of the African GRASSLANDS include hyenas and vultures, which feed on the carrion, or uneaten remains, of animals killed by predators. *(See* predation.) Many SPECIES of crabs, snails, and catfishes are detrivores in aquatic ecosystems and often are included in home aquariums because of the role they play in keeping the environment clean. *See also* BIOGEOCHEMICAL CYCLE; DECOMPOSER; DECOMPOSITION; EUTROPHICATION; SCAVENGER.

dichlorodiphenyltrichoroethane (DDT), An organochlorine PESTICIDE that was banned in the United States in 1972, but is still used in many developing nations for nonagricultural pest control. Dichlorodiphenyltrichoroethane (DDT) is used to kill INSECTS by disrupting their nervous systems.

DDT is a broad-spectrum (designed to kill a variety of pests), fat-soluble, environmentally persistent chemical with a HALF-LIFE of 15 years. Because it is fat soluble, BIOMAGNIFICATION occurs as DDT moves along the FOOD CHAIN. Thus, animals at the top of the food chain acquire a much higher concentration of the compound than did the insects that originally consumed it. This creates a particular problem for birds of prey, such as BALD EAGLES and PEREGRINE FALCONS; dichlorodiphenylethylene (DDE), produced when DDT degrades, causes affected birds to lay eggs with shells that are too thin to incubate successfully. DDT also is highly toxic to fishes, can cause sex changes in bird SPECIES, and can damage the heart, liver, and nervous systems of MAMMALS. In humans, DDT has been associated with reduced lactation in mothers and is believed to cause CANCER. *(See* carcinogen.) RACHEL CARSON's 1961 book, *Silent Spring*, first sounded the alarm about DDT's hazards to a worldwide audience.

Not all effects of DDT use have been negative. DDT also has played an important role in improving world health. In developing nations, millions more people would have died of malaria if DDT had not been used to control parasite-bearing mosquitoes. *(See* parasitism.) During the 1950s, the WORLD HEALTH ORGANIZATION encouraged use of DDT as part of an ambitious campaign aimed at total eradication of the DISEASE. The program was subsequently eliminated or drastically reduced in 37 countries. In the tropics, where human lives are at stake because of malaria and other diseases, the benefits of DDT are perceived to outweigh the environmental costs, and its use has therefore been continued.

Use of DDT often causes environmental effects outside the nation that used it. For example, high levels of DDT residues still appear in birds that migrate between North America and South America. About 30,000 metric tons (33,000 tons) of DDT are produced each year in countries such as Russia, India, Mexico, and China. The WORLD WILDLIFE FUND has set a goal to create worldwide conditions in which DDT production can be phased out by 2007, if not earlier. *See also* ACUTE TOXICITY; CHRONIC TOXICITY; FEDERAL INSECTICIDE, FUNGICIDE, AND RODENTICIDE ACT; HYDROCARBON; PESTICIDE REGULATION; PREDATION; TOXICITY.

EnviroSources

Carson, Rachel L. *Silent Spring*. Boston, MA: Houghton Mifflin, 1987.

Food and Agricultural Organization of the United Nations Web site: www.fao.org.

U.S. Environmental Protection Agency ("Pesticides and Public Health") Web site: www.epa.gov/history/publications/formative6.html.

U.S. Fish and Wildlife Service Web site: www.fws.gov

World Wildlife Fund Web site: www.panda.org.

dieldrin. *See* aldrin.

digester. *See* anaerobic decomposition.

digester gas. *See* anaerobic decomposition.

dioxin, A general term for a group of chemical compounds composed of CARBON, HYDROGEN, CHLORINE, and OXYGEN that were described as a serious health threat by the U.S. ENVIRONMENTAL PROTECTION AGENCY (EPA) in 1984. Health problems associated with dioxin exposure include severe skin lesions (chloracne), liver damage, appetite and weight loss, and digestive disorders. The most well known dioxin is 2,3,7,8-TCDD, or tetrachlorodibenzo-para-dioxin. Tests on laboratory animals suggest that 2,3,7,8-TCDD may be a CARCINOGEN.

Dioxin does not occur naturally; it is a byproduct formed during the manufacturing of pulp, paper, plastics, and certain PESTICIDES and insecticides. Dioxin may also form and be released into the ENVIRONMENT when industrial or municipal wastes are incinerated, from fires involving chlorinated BENZENES and POLYCHLORINATED BIPHENYLS (PCBs), or from improper disposal of chemical wastes.

Dioxin received international attention in 1978 when the U.S. government evacuated hundreds of people living in the LOVE CANAL community of New York from their homes. Studies showed that dioxin from buried chemical wastes was being released into the environment and causing a variety of illnesses in residents of the area. In 1983, a similar evacuation occurred in Times Beach, Missouri, when soil tests showed a dioxin level more than one million times higher than those in soils from other urban areas. *(See* Times Beach Superfund Site.) This dioxin originated from a dust-controlling oil that was placed on the town's unpaved roads. Other dioxin contaminations have been discovered in fish obtained from Lake Ontario, Saginaw Bay, Michigan rivers, and WATERSHEDS in Maine, Wisconsin, and Minnesota. In addition, 28 HAZARDOUS WASTE sites included on the NATIONAL PRIORITIES LIST are known to be contaminated with 2,3,7,8-TCDD.

Dioxin contamination is not unique to the United States. Many U.S. and Australian soldiers who fought in the Vietnam War, as well as residents of Saigon and other parts of Vietnam, have suffered a variety of illnesses, including chloracne and various types of CANCER, since the war ended. Many people believe these illnesses resulted from exposure to AGENT ORANGE, a dioxin-containing herbicide that was used as a DEFOLIANT during the war. In 1976, an explosion of a chemical plant in Sevoso, Italy, spread a dioxin-containing cloud over the countryside. Immediate effects of the contamination were observed in numerous animals. In addition, more than 30,000 people living in the region have shown higher-than-normal incidences of cancer, including liver cancer, leukemia, and lymphoma. Minute amounts of 2,3,7,8-TCDD have also been detected in the breast milk of women from several European countries and regions of the United States. *See also* COMPREHENSIVE ENVIRONMENTAL RESPONSE, COMPENSATION, AND LIABILITY ACT.

EnviroSource

Saign, Geoffrey C. *Green Essentials: What You Need to Know about the Environment.* San Francisco, CA: Mercury House, 1994.

discharge, An outflow of water. Discharge refers to the amount of water that flows past a particular point in a river or stream over a period of time. Use of the term is not restricted with respect to the course or location of water or its point of origin; it can describe the flow of water from a waste pipe or from a DRAINAGE BASIN. *See also* EFFLUENT; RUNOFF; SEWER.

disease, A breakdown of the body's natural defenses or regulatory systems that results in an impairment of health. Disease results from a variety of factors, including genetic abnormalities, PATHOGENS (disease-causing ORGANISMS), improper nutrition, exposure to chemicals or other substances in the ENVIRONMENT, and from exposure to some forms of ENERGY.

Some organisms from all kingdoms have the potential to cause disease in humans. For example, *ESCHERICHIA COLI* and *SALMONELLA* are BACTERIA that cause illness in humans when ingested. Giardiasis is one disease caused by PROTISTS. Other diseases caused by protists include malaria, African sleeping sickness, and toxoplasmosis. Contact with the leaves of PLANTS such as poison ivy, poison oak, and poison sumac result in allergic reactions in some people *(see* allergy), while other plants contain TOXINS that may produce symptoms ranging from minor stomach irritation to death if eaten. Fungal diseases in humans include thrush, some forms of pneumonia, ringworm, and athlete's foot. Other FUNGI are highly toxic and can kill humans when eaten in only small amounts. Animals that pose risk to humans include poisonous snakes and fish. In addition, many VIRUSES (although not living things) present in the environment cause diseases in humans ranging from the common cold or flu, to some forms of CANCER, and Acquired Immune Deficiency Syndrome (AIDS).

Improper nutrition, sometimes resulting from FAMINE, can lead to a variety of health problems in humans. Often, an illness can be traced to a lack of a specific nutrient in the diet. For example, KWASHIORKOR results from insufficient intake of proteins, goiter from insufficiency of salt, and rickets from an insufficiency of vitamin D.

Forms of energy linked to disease include exposure to ionizing RADIATION, such as X-RAYS or gamma rays, and exposure to too much ultraviolet radiation, the form of radiation given off by the sun. Extreme exposure to ionizing radiation, such as that which may result following the detonation of a NUCLEAR WEAPON or through the release of radiation from a malfunctioning nuclear power plant, can quickly devastate human and nonhuman POPULATIONS, resulting in many deaths. Such exposures may also bring about RADIATION SICKNESS. In addition, exposure to radiation can alter the genetic makeup of CELLS leading to abnormalities (MUTATIONS) that appear in offspring or various forms of cancer. Exposure to too much ultraviolet radiation can cause problems ranging from sunburn, to glaucoma, to skin cancer. *See also* ACUTE TOXICITY; AGENT ORANGE; AGRICULTURAL POLLUTION; AIR POLLUTION; ASBESTOS; BIOACCUMULATION; BIOLOGICAL WEAPON; BLIGHT; BOVINE SPONGIFORM ENCEPHALOPATHY; BRUCELLOSIS; CARCINOGEN; CHEMICAL WEAPON; CHRONIC TOXICITY; COLIFORM BACTERIA; CRYPTOSPORIDIUM; DOSE/DOSAGE; FOOD AND DRUG ADMINISTRATION; *GIARDIA LAMBLIA*; HAZARDOUS SUBSTANCE; LETHAL DOSE–50%; MICROORGANISMS; MINAMATA DISEASE; MULTIPLE CHEMICAL SENSITIVITY; MUTAGEN; NICOTINE; OCCUPATIONAL SAFETY AND HEALTH ADMINISTRATION; OIL POLLUTION; OPPORTUNISTIC ORGANISM; PARASITISM; POLLUTANT; POTABLE WATER; TOXIC CHEMICAL; WARNING LABEL; WORLD HEALTH ORGANIZATION.

disinfectant, Any physical or chemical substance that kills MICROORGANISMS or prevents their growth. Disinfectants, also known as antiseptics, are primarily used in homes as cleaning agents, in medical facilities to take care of cuts and injuries that occur on the human body, and by industry for the preservation of food or medicine. Some common disinfectants used in the care and treatment of cuts and bruises and for sterilization of medical tools include heat, iodine, boric acid, alcohol, and mercurochrome. *(See* sterilization, pathogenic.) CHLORINE is used in the sterilization of water, particularly in public water systems. Smoking and salting are common methods used to reduce the growth of microorganisms in foods. IRRADIATION, or the application of RADIATION on food, is also being used to prevent the growth of some microorganisms. *See also* DISEASE; PATHOGEN; PRIMARY, SECONDARY, AND TERTIARY TREATMENTS.

disintegration. *See* radioactive decay.

dissolved oxygen, OXYGEN atoms that are freely available in water. The amount of dissolved oxygen in water is expressed in milligrams of oxygen per liter of water (mg/L). The amount of oxygen that can dissolve in water is related to temperature; the lower the temperature, the more oxygen that can be dissolved. Dissolved oxygen is vital to fishes and other aquatic life. Dissolved oxygen levels are considered an important indicator of a body of water's ability to support aquatic life. Water with dissolved oxygen of less than 5 milligrams per liter generally cannot support living ORGANISMS and is considered polluted. Secondary and advanced waste treatment of water is generally designed to ensure adequate dissolved oxygen in waste-receiving waters. *See also* BIOCHEMICAL OXYGEN DEMAND; PRIMARY, SECONDARY, AND TERTIARY TREATMENT; SEWAGE; WASTEWATER.

divergent evolution. *See* adaptive evolution.

diversity. *See* biodiversity.

DNA. *See* biotechnology; ribonucleic acid.

Dobson unit, A unit used to measure the concentration of OZONE levels in the ATMOSPHERE and which is a good indicator of OZONE DEPLETION. The measurement has been used worldwide for more than 30 years.

In the late 1950s, scientists measured the concentration of ozone in the stratosphere in ANTARCTICA. At that time the concentration of ozone was approximately 300 Dobson units. By the 1980s, however, the concentration of ozone had dropped to about 200 Dobson units. Since then, the concentration of ozone in Antarctica has reached a high of 250 Dobson units and a low of about 90 Dobson units. Seasonal variations and other natural conditions can produce large swings in ozone levels and changes in the concentration of ozone. For example, measurements in Leningrad, Russia, have indicated ozone levels as high as 475 Dobson units and as low as 300. *See also* OZONE HOLE.

dodo birds, A native bird of the island of Mauritius in the Indian Ocean that was driven to EXTINCTION around the 1680s. Of the 45 bird SPECIES originally found on the island, only 21 have managed to survive.

In the 1500s, trading ships landed on the island and sailors hunted the dodos (*Raphus cucullatus*) for food. Later nonnative animals such as rats, pigs, and monkeys were introduced to the island; the animals preyed on the dodos and ate their eggs. *(See* predation.) The bird's HABITAT was also being destroyed as the island's FORESTS were cut down and the land was converted to farming plantations. The combination of overhunting, destroyed habitats, and the introduction of nonnative animals took its toll on the dodo POPULATION. Within 200 years of the arrival of humans on Mauritius, the dodo became extinct.

The dodo bird was a flightless bird that became extinct. It lived on many islands in the Indian Ocean.

Scientists researching the effects of the dodo's extinction on the island's ECOSYSTEM have discovered that a certain species of tree was becoming quite rare on Mauritius. The remaining trees were about 300 years old, and there was a marked absence of new growth or younger trees. Since the average life span of this tree was about 300 years, it was clear that the species would soon die and become extinct. The scientists discovered that the dodo ate the fruit of this tree and that its seeds passed through the dodo's digestive system. Only through this process would the seed germinate and grow. In their efforts to save the remaining trees, the scientists discovered that domestic turkey gullets could imitate the action of the dodo's digestive system. The turkeys are now being used to process the seeds to begin a new generation of the tree, which has been named the dodo tree. If these seedlings survive to produce their own seeds, there is a good chance that the dodo tree species will be saved. *See also* DEFORESTATION; EXOTIC SPECIES; NATIVE SPECIES.

EnviroSource

U.S. Fish and Wildlife Service, Species List of Endangered and Threathened Wildlife Web site: www.fws.gov/r9endspp/lsppinfo.html.

dolphins and porpoises, Aquatic MAMMALS comprising the smallest members of the order Cetacea—the toothed WHALES. As toothed whales, dolphins and porpoises are CARNIVORES, mostly predators of fishes and squids. *(See* predation.) Dolphins and porpoises represent three whale families: the true dolphins (more than 30 SPECIES), the porpoises (6 species), and the river dolphins (5 species).

True Dolphins and Porpoises. True dolphins and porpoises are marine. These animals look very similar, but are distinguished by their sizes and the shapes of their snouts and teeth. True dolphins, which average 2–3 meters (7–10 feet) in length, are generally larger than porpoises. True dolphins also have longer, more pointed snouts than do porpoises, which have short, blunt snouts. The other significant difference between the true dolphins and porpoises is in their teeth: Dolphins have cone-shaped teeth, and porpoises have teeth with flat surfaces like the molars in humans.

The white-sided dolphin, like all dolphins, is highly intelligent and can communicate with others of its own species.

True dolphins and porpoises often travel in large groups called pods or herds. Studies of these animals have shown that they are highly intelligent and able to communicate with others of their species using high-pitched sounds. Because of their intelligence, these animals have often been captured and raised in captivity at aquariums, amusement parks, and ZOOS, where they are frequently trained to perform tricks. In many areas (such as Florida, Hawaii, and Bermuda), the animals are placed in enclosures where tourists can swim with them. Trained dolphins and porpoises also have been used by the U.S. Navy in search-and-rescue and surveillance missions. Use of dolphins and porpoises for all of these purposes is often protested by ANIMAL RIGHTS and DEEP ECOLOGY supporters, who believe the animals should not be removed from their native HABITAT for the benefit of people.

True dolphins and porpoises range throughout the world's OCEANS. Unlike many other whale species, these animals exist in fairly large numbers. Currently, the major threats to their survival are pollution, DISEASE, capture for display (which is regulated by the MARINE MAMMAL PROTECTION ACT), and periodic capture in the nets of commercial fishers, especially those seeking TUNA. *(See* commercial fishing.) This last threat decreased as worldwide awareness of the problem led to changes in fishing techniques and the development of nets that resulted in the capture of fewer dolphins and porpoises.

River Dolphins. River dolphins have smaller brains and tend to be less social than true dolphins. Unlike their marine relatives, river dolphins are freshwater animals that make their habitat in the rivers of South America and southeastern Asia. The limited habitat of river dolphins has placed these animals under greater environmental pressure than their relatives. Currently, two species of river dolphins—the Chinese river dolphin, which lives throughout China, and the Indus River dolphin, which lives only in the Indus River and its tributaries in Pakistan—are included on the ENDANGERED SPECIES LIST. The primary threat to these animals is human encroachment on their habitat. *See also* AMERICAN ZOO AND AQUARIUM ASSOCIATION; ENDANGERED SPECIES; MANATEES; RED LIST OF ENDANGERED SPECIES; SEA TURTLES; SEALS AND SEA LIONS.

EnviroSources

Atwood, Nicolas. "Feeding Flipper." *Audubon* 99, no. 6 (November–December 1997): 12.

Hoyt, Erich. *Meeting the Whales: The Equinox Guide to Giants of the Deep.* Manitoba: Camden House, 1991.

Seidman, David. "Swimming with Trouble." *Audubon* 99, no. 5 (September–October 1997): 78–82.

dose/dosage, The quantity of a chemical or radiation an ORGANISM is given or exposed to and is generally expressed as the amount of chemical for a given weight of organism.

The most important factor that influences TOXICITY is the dose, or amount of chemical that enters the body. Every chemical is toxic at a high enough dose. The dose of a chemical plays a major role in determining toxicity. Generally, there is no effect at low doses, but as the dose is increased, a toxic response may occur. The higher the dose, the more severe the toxic response that occurs. At a high enough dose, a chemical can be fatal. The dose-response relationship is different for every chemical and may vary for each individual. *See also* LETHAL DOSE–50%.

EnviroSource

Professor Michael Kamrin, Institute for Environmental Toxicology, C-231 Holden Hall, Michigan State University, East Lansing, MI 48824-1206; e-mail kamrin@pilot.msu.edu.

Douglas, Marjory Stoneman (1890–1998), A writer and conservationist who led the campaign for the preservation of the Florida Everglades. Born in Minnesota, Douglas moved to Florida in 1915, where she worked as a journalist for the *Miami Herald,* a newspaper founded by her father. During her journalism career, Douglas wrote several CONSERVATION articles; however, it was the publication of her book, *The Everglades: River of Grass* (1947), that earned her national recognition as an environmentalist. The book remains as the definitive work on the fragile Everglades ECOSYSTEM.

The same year this book was published, Douglas served as a member of the committee working to promote the designation of the Everglades as a national park. The group achieved its goal on December 6, 1947, when President Harry S. Truman designated 600,000 hectares (1.5 million acres) of the 1.6-

million-hectare (4-million-acre) region as the EVERGLADES NATIONAL PARK. Following this success, Douglas continued to be the leading advocate for the preservation of this WETLAND ecosystem. In 1969, she founded Friends of the Everglades, an activist group dedicated to protecting the Everglades WILDERNESS from damage resulting from water diversion, pollution, and consequent HABITAT destruction. The Friends of the Everglades successfully petitioned the state and federal governments to establish a multimillion-dollar fund to help restore much of the Everglades to its predevelopment condition. Despite this initial success, the group was unsuccessful in its 1996 lobby for a penny-a-pound sugar tax proposed to help fund cleanup efforts related to polluted RUNOFF emanating from sugar plantations located to the north of the park.

Douglas has also authored or coauthored several other books about conserving ecosystems. They include *Florida: The Long Frontier* (1967), *Coral Reefs of Florida* (1988), and *Freedom River* (1994). These books earned her national recognition. In 1993, she was awarded the National Medal of Freedom by President Bill Clinton in recognition of her work. In addition, the building that houses the state Department of Environmental Protection in Tallahassee, Florida, has been named for Douglas as have several public schools. Following her death in 1998, Douglas's ashes were scattered over the park she had devoted her life to protecting. *See also* ALLIGATORS; CARSON, RACHEL LOUISE; IRRIGATION; SWAMP.

EnviroSources

Breton, Mary Joy. *Women Pioneers for the Environment.* Boston, MA: Northeastern University Press, 1998.

Douglas, Marjory Stoneman. *The Everglades: River of Grass.* 50th anniversary ed. 1947. Reprint, Sarasota, FL: Pineapple Press, 1997.

Douglas, Marjory Stoneman, and John Rothchild. *Voice of the River.* Sarasota, FL: Pineapple Press, 1988.

Levin, Ted. "Defending the 'Glades: Marjory Stoneman Douglas." *Audubon* 100, no. 6 (November–December 1998): 84.

Voss, Gilbert L., and Marjory Stoneman Douglas. *Coral Reefs of Florida.* Sarasota, FL: Pineapple Press, 1998.

drainage basin, An area of land, also known as a WATERSHED, that provides fresh water to a river, lake, or pond. Drainage basins are environmentally important because they regulate the quality and amount of water that enters lakes and rivers. The size, shape, and vegetation of a drainage basin influence how it functions. For example, lakes and rivers located in drainage basins that have little vegetation tend to have poor water quality and are often prone to FLOODS. In drainage basins with lots of vegetation, the waters of rivers, streams, and lakes are usually clear. In addition, heavily forested drainage basins tend not to have flooding problems because water flow is slower and because they contain enriched soils that can absorb large amounts of water. *See also* AQUIFER; EROSION; PERMEABILITY; RUNOFF; WATER CYCLE; WATER TABLE.

EnviroSource

Committee for the National Institute for the Environment Web site: www.cnie.org/biodv-6.html

dredging, The underwater excavation of sediments such as sand, mud, silt, and other solid matter from canals, rivers, harbors, and other waterways. Dredging is usually carried out to make waterways deeper and to facilitate the construction of DAMS, dikes, and levees. Once dredged, the material is often deposited on land and used to raise the level of lowlands or to create new land areas.

While dredging is certainly a useful process, it can pose several environmental problems. For example, in highly industrialized areas, material at the bottom of a waterway often contains harmful CONTAMINANTS and wastes. When these areas are dredged, the sediments become suspended in the water, contaminating the aquatic ECOSYSTEM. The dredged material is then deposited on land or buried, where it can cause additional environmental hazards.

Dredging is accomplished by means of dredges, large floating devices containing equipment necessary for the removal of sediments. Many types of dredges have been developed over the years. Bucket-ladder dredges consist of a continuous chain of scoop-like buckets that rotate around a rigid ladder. To remove sediments, the ladder is first lowered to the bottom at a slant. As the chain moves along, the empty buckets on the underside of the ladder scoop up sediments and other material. The loaded buckets then return along the ladder's upper side and drop the dredged material into large containers. Another type of dredge is the dipper dredge. A dipper dredge is essentially a power shovel that can push and remove large amounts of material from the bottom. Suction dredges remove sediments that are mixed with water. They are most often used in areas where sediments are highly mobile and not entirely settled on the bottom. One type is the trailer suction dredge, which can be dragged along behind a ship as it cruises up and down a waterway or other area, sucking up material as it moves along. *See also* HEAVY METALS; SEDIMENTATION.

driftnet, A fine, nylon mesh net that is used to catch fish, mostly TUNA, SALMON, and squid. Driftnets, which are set below the surface at depths of 13 meters (39 feet) can extend in length from 2.5 to 50 kilometers (1.5–30 miles). These oversized nets allow a small fishing crew to haul in tons of fish; however, the nets catch many other SPECIES, known as bycatch, that are not wanted by the fishers. Government agencies and scientists believe that too many other species are being unnecessarily killed using these nets. Thousands of animals such as dolphins, SEALS, WHALES, SEA TURTLES, sea birds, and nontargeted fish are killed each year in these nets. Some of these species, including the leatherback turtles and MANATEES, are in a decline. Sometimes driftnets also break away and are lost in the OCEAN; however, even these "lost" nets continue to entangle and kill fish and other marine life.

In April 1990, major tuna companies refused to accept and purchase any tuna caught using dolphins as "spotters" or to import tuna caught with driftnets. In the latter part of that year, the U.S. Senate adopted amendments to the Magnuson-Stevenson Fishery Conservation Act of 1976 that mandated a ban on importing any fish or fish products caught with driftnets in the South Pacific and made it illegal for any U.S. citizen to use driftnets. The amendments became effective in 1991.

At the end of 1991, the General Assembly of the United Nations adopted a resolution that called for a 50% reduction of driftnet operations by mid-1992. Although driftnets of more than 2.5 kilometers (1.5 miles) in length are now banned by the United Nations, some fishers still use such nets. Much of this practice takes place in the MEDITERRANEAN SEA and Atlantic Ocean. One of the reasons that driftnet enforcement is ineffective in those areas is because of poor monitoring of fishing fleets by governments.

Many European governments want the use of all driftnets, even short driftnets, banned. For example, Spain has banned domestic use of short drift nets. Other countries, however, do not want a total ban of the nets. In some countries, more than 60% of the fishing fleets use short drift nets, and fleet owners believe the ban would cause hardship and that fishers would lose their jobs. *See also* COMMERCIAL FISHING; DOLPHINS AND PORPOISES; MARINE MAMMAL PROTECTION ACT; OVERFISHING.

EnviroSources

Earthtrust: e-mail earthtrust@aloha.net.
National Marine Fisheries Service Web site: www.nmfs.gov.

drought, A period of abnormally dry WEATHER resulting from a lack of rainfall. *(See* water cycle.) The severity of a drought is determined by such factors as the decrease in moisture in an area, the duration of the dry period, and the size of the area affected. Droughts of brief duration generally do not have significant long-lasting effects on ORGANISMS and the ENVIRONMMENT; however, lengthy droughts can have a severe impact.

People are largely affected by drought as water usage from RESERVOIRS, AQUIFERS, and WELLS exceeds its replacement through natural processes. When this happens, restrictions on how water may be used or water rationing may be implemented as means of conserving water. If the drought continues, especially in developing regions where water storage facilities such as artificial reservoirs may be scarce, people may be forced to move, or migrate, to areas where water supplies are more plentiful. In addition, severe drought may make food scarce as crops fail to grow from the lack of water.

People are not the only organisms adversely affected by drought. As small lakes and ponds dry up, aquatic organisms lose their HABITAT, while terrestrial organisms may lose their water source. As the water flow through larger streams and rivers diminishes, organisms that live in and around those bodies of water also are affected. In addition, as PLANTS continue to remove available water from soil to meet their growth needs, the depletion of soil moisture makes soil more vulnerable to EROSION. Areas that experience repeated drought conditions and extremely dry soil conditions are at increased risk of long-term soil damage through DESERTIFICATION.

Most severe droughts occur in regions adjacent to areas with a DESERT or near-desert CLIMATE that are located at about 10° to 20° of latitude. An example is the Sahel region of Africa, which suffered severe drought conditions from 1968 to 1974. This drought led to major crop failures, resulting in a FAMINE that caused the deaths of more than one-half million people and millions of animals raised as livestock. During the last year of this drought, the African nation of Ethiopia also suffered a severe drought that was responsible for the deaths of more than 100,000 people in that country. In the United States, the most severe drought recorded affected the GRASSLANDS, or Great Plains, region of the Midwest between 1933 and 1935. This dry period, combined with poor farming methods that increased the soil's vulnerability to erosion, devastated the area and resulted in frequent dust storms that led much of this region to become known as the DUST BOWL. In 1988, the Great Plains suffered its second worst drought. Crop failures from this drought were so severe that the United States had to import grain from other countries for the first time in its history. *See also* AQUIFER; EL NIÑO/LA NIÑA; GLOBAL WARMING; GROUNDWATER; IRRIGATION; NATURAL RESOURCES CONSERVATION SERVICE; OVERDRAFT; SOIL CONSERVATION; SOIL CONSERVATION AND DOMESTIC ALLOTMENT ACT.

EnviroSources

National Center for Atmospheric Research Web site: www.ncar.ucar.edu/.
National Oceanographic and Atmospheric Administration Web site: www.noaa.gov/.
National Weather Service Web site: www.nws.gov.

Dubos, Rene (1901–1982), A French-American microbiologist who was a leading ecologist, microbiologist, educator, and Pulitzer Prize–winning author. He was the first scientist to study BACTERIA in the soil where they lived and not in an indoor laboratory. Dubos's work included isolating antibacterial substances in MICROORGANISMS. In 1939, Dubos isolated antibacterial substances from the microorganisms *Bacillus brevis*. One of the substances, called Gramicidin was used effectively in treating infectious wounds in humans. Dubos was also known for his research and writings on antibiotics, acquired immunity, and tuberculosis.

In 1975, Dubos founded the Dubos Center for Human Environments in collaboration with a nonprofit organization, Total Education in the Total Environment. The center is a research organization that is involved with environmental education. Its motto is "Think locally; act globally." Dubos believed that the center can help citizens and decision-makers develop plans to solve environmental problems in a cooperative spirit and at the same time help create new environmental values.

EnviroSource

Dubos, Rene Jules. *Man Adapting* (Silliman Milestones in Science.) New Haven, CT: Yale University Press, 1982.

Dust Bowl, The disastrous result of severe wind EROSION in the southern Great Plains region of the United States during the 1930s that resulted from a combination of poor agricultural practices and years of sustained DROUGHT. Beginning in

1931, powerful dust storms carrying millions of tons of black soil repeatedly swept across western Kansas, the panhandles of Texas and Oklahoma, and the eastern portions of Colorado and New Mexico, rendering millions of hectares of farmland useless. The agricultural devastation helped prolong the Great Depression, whose effects were felt worldwide. John Steinbeck's novel, *The Grapes of Wrath* (1939), chronicled both the Dust Bowl and the resultant migrations of many affected farmers.

The first major influx of farmers into the southern plains took place in the late 1800s and early 1900s. Attracted to the fertile land, the farmers were unaware that the plains were subject to regular cycles of rain and drought. They rushed to convert all of the southern plains into profitable farmland, intensively plowing and planting the land in wheat.

During years of adequate rainfall, the land produced bountiful crops. In the summer of 1931, however, the rains ceased; crops withered and died. The ground cover that once held soil in place was gone, cleared for crop plantings. As the land dried up, huge amounts of wind-blown sand and dust covered the landscape; the term "Dust Bowl" was coined by a journalist to describe the devastated region. Violent storms—"black blizzards"—rolled in without warning, blocking the sun's light and thrusting entire communities into darkness. Afterward, dust lay everywhere—in food, water, homes, and in the respiratory and digestive systems of people and animals.

WEATHER records indicate that there were 14 dust storms in 1932; the next year there were 38. By 1934, the storms were arriving with alarming frequency, and the dust was beginning to cause illness. Animals were found dead in the fields, the insides of their stomachs coated with thick layers of dirt. An epidemic of human respiratory DISEASE, dubbed "dust pneumonia," raged throughout the plains.

President Franklin D. Roosevelt created New Deal programs and organizations to assist Dust Bowl residents, including the Federal Emergency Relief Administration, the Federal Surplus Relief Corporation, the Works Progress Administration, the Civilian Conservation Corps, and the Drought Relief Service. In 1936, an innovative plan was proposed by HUGH BENNETT, a leading agricultural expert, to conserve valuable topsoil. Now known as "the father of soil conservation," he persuaded Congress to approve a federal program that would pay farmers to use new, soil-conserving farming techniques. Congress responded by passing the SOIL CONSERVATION AND DOMESTIC ALLOTMENT ACT of 1935, and by 1937, SOIL CONSERVATION efforts were in full force. By 1938, the rate of soil loss in the southern plains had been reduced by 65%.

Even after the rains finally returned in late 1939 and the southern plains again yielded great harvests, the relationship between American farmers and the federal government endured. A complex, and frequently controversial system of commodity price supports and subsidies emerged from the Dust Bowl period to form the basis for twentieth-century federal farm policy.

Sixty years after the Dust Bowl, wind EROSION continues to threaten the sustainability of U.S. NATURAL RESOURCES by stripping away lighter soil constituents, such as ORGANIC matter, clays, and silts—the most fertile parts of the soil. On U.S. croplands, about 70 million hectares (173 million acres) are eroded each year by wind and water at rates that are more than twice the estimated limits for sustainable agricultural production. *(See* sustainable agriculture.) On average, wind erosion is responsible for about 40% of this loss, a proportion that increases substantially in times of drought. As recently as 1996, for example, wind erosion again severely damaged agricultural lands throughout the Great Plains. *See also* AGRICULTURE; AGROCLIMATOLOGY; CLIMATE; CONTOUR FARMING; EL NIÑO/LA NIÑA; FAMINE; GRASSLAND; IRRIGATION; SALINIZATION.

Global View

The term "dust bowl" has been more recently applied to several semiarid regions throughout the African continent where overgrazing, DEFORESTATION, and poor agricultural practices combined with periodic DROUGHT and winds are making the land vulnerable to EROSION.

EnviroSources

Discovery Channel Online ("Day of the Black Blizzard") Web site: www.discovery.com/area/history/dustbowl/dustbowlopener.html.
Public Television ("The American Experience: Surviving the Dust Bowl") Web site: www.pbs.org/wgbh/pages/amex/dustbowl.
U.S. Library of Congress ("Voices From the Dust Bowl") Web site: lcweb2.loc.gov/ammem/afctshtml/tshome.html.

Dutch elm disease, A devastating fungal DISEASE that affects all SPECIES of elm, ultimately resulting in the deaths of the affected trees. Dutch elm disease is believed to have originated in Asia; however, it was called Dutch elm disease because it was first observed in the Netherlands in the 1930s. Soon after its appearance, this BLIGHT spread throughout Great Britain, continental Europe, and much of North America, where it destroyed entire POPULATIONS of English elm, Dutch elm, and American elm.

The fungus (*Certocystis ulmi*) that causes Dutch elm disease is spread by the elm-bark beetle, an INSECT that lays its eggs beneath the tree's bark. Once introduced, the fungus spreads throughout the xylem (water and mineral transport) tissues of the tree. Initial symptoms of Dutch elm disease include yellowing leaves and wilting. Soon after these symptoms develop, the branches begin to die.

In the 1960s, a second outbreak of Dutch elm disease spread throughout Great Britain following the importation of logs from Canada. This outbreak involved a newer and more virulent strain of *Certocystis ulmi* and killed millions of elms. Other smaller outbreaks of the disease have since occurred throughout North America, resulting in a decline in the elm population. No cure has been developed for Dutch elm disease; however, annual injections of insecticide into elms has been somewhat successful at controlling the disease. When an outbreak of infection is detected, spread of the disease is often prevented by removing large numbers of elms from an area. *See also* BIOLOGICAL CONTROL; FUNGI; GENETIC ENGINEERING; MONOCULTURE; POLYCULTURE; SELECTION CUTTING; TREE FARM.

E

Earle, Sylvia (1935–), A marine scientist, researcher, and author who has devoted her life to exploring the BIODIVERSITY and environmental health of Earth's OCEANS. Earle is one of the world's leading experts in OCEANOGRAPHY and marine science. In the 1990s, she served as chief scientist to the NATIONAL OCEANIC AND ATMOSPHERIC ASSOCIATION, where her responsibilities included monitoring the environmental quality of the nation's waters and evaluating the damage in the Persian Gulf resulting from the burning of Kuwaiti oil fields by Iraqi soldiers at the end of the Gulf War. *(See* oil spill.)

In the 1970s, Earle conducted various studies of the marine ENVIRONMENT. In the first of these, Tektite II, Mission 6, Earle lived in an enclosed HABITAT beneath the ocean's surface for two weeks. The research was conducted to determine how living in a self-contained environment might impact humans over time. Information from this and similar studies was deemed important in the determination of deciding the potential of future colonization of the oceans and space by humans. This expedition, conducted by an all-woman crew, gained worldwide attention for its research potential and has served as a model for studies in space and in the Biosphere 2, which was erected in the Arizona desert in the late-1980s.

In 1977, Earle conducted extensive research on endangered sperm WHALES. Her team followed POPULATIONS of the whales throughout the Pacific and Atlantic Oceans to develop a better understanding of their migration, breeding, and feeding habits to help scientists gain more insight into how these animals live and how human activities threaten their survival. The results of this research were recorded in the film *Gentle Giants of the Sea* (1980).

Earle also has contributed to the technology involved in ocean research. She founded two companies that design and build ocean exploration research vehicles. These vehicles allow marine scientists and oceanographers to study ocean features and life forms at greater depths than previously possible, thus providing scientists with a greater understanding of the ocean ENVIRONMENT. In addition, Earle has written several books and articles to increase public awareness of the importance of the ocean ECOSYSTEM. The first of these was *Exploring the Deep Frontier* (1980), and *Sea Change: A Message of the Oceans* (1995). *See also* COUSTEAU, JACQUES YVES.

EnviroSources

Breton, Mary Joy. *Women Pioneers for the Environment.* Boston, MA: Northeastern University Press, 1998.

Chang, Chris, et al. "Champions of Conservation: *Audubon* Recognizes 100 People Who Shaped the Environmental Movement and Made the Twentieth Century Particularly American." *Audubon* 100, no. 6 (November–December 1998): 120.

Earle, Sylvia. *Dive! My Adventures Undersea in the Deep Frontier.* Washington, DC: National Geographic Society, 1999.

———. *Sea Change: A Message of the Oceans*. New York: Fawcett, 1995.

Earth Day, A day-long, annual global celebration in April that first occurred on April 22, 1970. The first proclamation of Earth Day was made by San Francisco, California (the city of St. Francis, who was the patron saint of animals and ECOLOGY). It was later sanctioned by Secretary General U Thant of the United Nations. The first Earth Day was celebrated by an estimated 20 million people. The event was designed to deepen reverence and responsibility for life on Earth and to serve as the symbol of environmental responsibility and stewardship.

Earth Day evolved from a 1969 proposal by Gaylord Nelson, then a U.S. Senator, for a nationwide "environmental teach-in," modeled after the anti–Vietnam War teach-ins of the time. Earth Day blossomed in grade schools, high schools, and colleges, with grass-roots organization of the events. Representative Pete McCloskey was named cochair and Denis Hayes was hired as national coordinator. The event raised the consciousness of all citizens about the care needed to maintain and enhance life on Earth.

The objectives of the twentieth-anniversary celebration in 1990 included the banning of all logging in ancient FORESTS, the protection of the world's remaining RAINFORESTS, the reduction of SOLID WASTE by 75% by the year 2000, the establishment of RECYCLING and COMPOSTING programs, and a ban on nonrecyclable packaging. Supporters of the movement also hoped to combat ACID RAIN by reducing SULFUR DIOXIDE EMISSIONS by 90% and NITROGEN OXIDES by 75% by the year 2000. In 1990, 200 million people in 141 countries participated in Earth Day events in their communities.

The Earth Day Network made use of the the Internet to plan for Earth Day 2000. They organized Earth Month 2000 for the month of April and involved about one-third of the world's POPULATION in these activities. The Earth Day Network mobilized citizens to demand serious actions to address CLIMATE change and to reform outdated pollution-causing ENERGY systems.

Denis Hayes was involved again in Earth Day 2000 which was held in April 2000. It attracted about 500 million people around the worldwide. *Credit:* Earth Day 2000

Earth Day 2000 had six specific objectives:

- To mobilize and empower citizens around the world who care about the environmental challenges facing us.
- To create global networks that magnify the impact of environmental organizing.
- To facilitate communications and enable citizen groups with similar interests to compare strategies, exchange ideas, and develop potent alliances.
- To highlight innovative strategies and policies that communities are pioneering.
- To catalyze change in countries and institutions on critical millennial issues, including the need for clean, renewable energy.
- To inspire the economic, political, and individual actions necessary for significant cultural shifts to occur.

See also RENEWABLE RESOURCE.

EnviroSource

Earth Day Network, 91 Marion St., Seattle, WA 98104; (206) 264-0114; Web site: http://www.earthday.net/; e-mail: worldwide@earthday.net.

Earth Summit. *See* United Nations Earth Summit.

earthquake, A major NATURAL DISASTER caused by the shifting of rock layers beneath Earth's surface or underground volcanic activity. *(See* volcano.) During an earthquake, the surface of the Earth shifts, shakes, and moves, which can cause destruction of buildings and other structures as well as natural HABITATS. Earthquakes happen because the tectonic plates continue to move and scrape against each other as they have been doing for billions of years. *(See* plate tectonics.) At the plate boundaries, pressure can build up causing a break or crack in Earth's crust called a fault. Most earthquakes are caused by the movements of the plates along the faults, an area also of much volcanic activity.

Much of the activity of earthquakes and volcanoes is located around the edge or rim of the Pacific Ocean—an area extending along the eastern edge of Asia and along the western edge of North and South America. The zone is known as the Ring of Fire or the circum-Pacific belt. About 90% of earthquakes occur in this zone. Another major zone of earthquakes extends from the Mediterranean region, eastward through Turkey, Iran, and northern India. Earthquakes can occur in other areas of the world as well.

Some environmentalists believe that human activity is increasing the potential of earthquakes. They are concerned about activities such as building RESERVOIRS for water supplies, depositing waste deep inside the earth, and pumping fluids into the ground for the secondary recovery of PETROLEUM. Scientists in the United States, Canada, and Japan have documented locations where human activity has induced earthquakes. One of the sites studied was in the Rocky Mountain Arsenal near Denver, Colorado. This area was used to inject fluids into deep wells for waste disposal. *(See* deep-well injection.) In 1967, a series of earthquakes occurred in this location. Studies revealed that there was a link between the fluid injections and the earthquakes. In the next year, fluid injections were discontinued and the earthquakes became less frequent.

Seismologists are scientists who study earthquakes. They use seismographs to record and measure the motion of the ground and the location where the earthquake began. Seismographs are installed in the ground throughout the world and operate as a seismographic network. The strength or magnitude of an earthquake is measured using a scale called a Richter scale. The scale measures the amount of ENERGY from the earthquake in units. Any earthquake movement that registers a value of 3.0 to 3.9 is a minor earthquake. A Richter value of 4.5 to 5.0 would indicate a light to moderate earthquake, and a strong earthquake would have a value of about 6.0 or more. Any tremors registering above a 7.0 would be classified as a major earthquake and would be very dangerous.

Each year 25,000 to 50,000 earthquakes are reported. Most of them are minor and cause no damage; however, some are powerful and damaging earthquakes. In 1985, in central Mexico, 25,000 people were killed resulting from an earthquake that was 8.1 on the Richter scale. In 1990, northwestern Iran experienced an earthquake of 7.7. Approximately 50,000 were killed, 60,000 injured, and 400,000 were homeless. In 1994, an earthquake that registered 6.6 caused the

Earthquake damage to these houses occurred in the famous San Francisco earthquake of 1906. *Credit:* The Bancroft Library, University of California at Berkeley, Roy D. Graves Collection. W.E. Worden, photographer.

death of 60 people and injured about 8,000 in the San Fernando Valley of California. The damages were estimated between $13 billion and $20 billion. In 1995, a powerful earthquake rocked the cities of Osaka, Kyoto, and Kobe, in Japan. The earthquake registered 7.2. More than 5,000 people were killed, 26,000 were injured, and the property damage was about $100 billion. In 1999, the death toll from an earthquake in Turkey registered at 7.2 claimed more than 16,000 lives, and almost 38,000 were reported injured. Thousands were homeless. The earthquake was one of the most powerful recorded this century. There were as many as 250 aftershocks. The worst earthquake in Turkish history killed an estimated 33,000 people in the eastern province of Erzincan in 1939. Besides the tragedy of lost lives, homeless people, and property damage, earthquakes disrupt community life. There is also massive economic loss to cities and states due to earthquake damage that alters landforms and river beds, destroys major roads, bridges, water, SEWER, and telephone lines, and water aqueducts and reservoirs.

Many geologists and other scientists would like to be able to predict earthquakes and to set up an early warning system to alert people before an earthquake appears, but earthquakes cannot be predicted with any reliability. The best option, therefore, is to develop earthquake prevention programs to avoid loss of life. The INTERNATIONAL DECADE FOR NATURAL DISASTER REDUCTION (IDNDR) was launched by the United Nations to encourage countries to implement measures to reduce natural disasters and save lives. The IDNDR recommendations include restricting people from living too close to active fault areas where earthquakes can occur, better building codes, more effective emergency plans for the evacuation of people during a disaster, and quicker government responses for detecting and locating survivors buried under debris. *See also* GEOLOGY; TSUNAMI.

EnviroSources

Bolt, Bruce A. *Earthquakes*. New York: W. H. Freeman and Co., 1993.

Sieh, Kerry E., and Simon LeVay. *The Earth in Turmoil: Earthquakes, Volcanoes, and Their Impact on Humankind*. New York: W. H. Freeman and Co., 1998.

Wilson, Edmund. *The American Earthquake*. New York: Da Capo Press, 1996.

U.S. Geological Survey (Earthquake Information) Web sites: geology.er.usgs.gov/eastern/earthquakes/faq8.html, quake.wr.usgs.gov/, http://geology.usgs.gov/quake.html, geohazards.cr.usgs.gov/earthquake.html, eohazards.cr.usgs.gov/eq/.

Earthwatch, An international nonprofit, nonconfrontational program of the UNITED NATIONS ENVIROMENTAL PROGRAMME, which supports scientific field research. Volunteers and scientists work together to build a sustainable world through an active partnership on field research expeditions in over 100 countries, including the United States, South America, Europe, and Australia. The expeditions provide a unique opportunity to learn scientific investigation methods and learn how fieldwork can contribute to solving real world problems. Major areas of research include world OCEANS and FORESTS, BIODIVERSITY, cultural diversity, learning from the past, global change, and world health.

Earthwatch's mission is to improve human understanding of the planet, the diversity of its inhabitants, and the processes that affect the quality of life on Earth. Volunteers have contributed over $40 million and 6.5 million hours to Earthwatch's efforts. Earthwatch provides scholarships and financial aid to teachers and students. It is funded through grants, private funds, the National Geographic Society, and World Wide Fund for Nature.

EnviroSources

Earthwatch Europe, Belsyre Court, 57 Woodstock Rd., Oxford OX2 GHU, UK.

Earthwatch Institute, P.O. Box 403, 680 Mt. Auburn St., Watertown, MA 02272.

Earthwatch Institute International Web site: www.earthwatch.org.

ecofeminism, A combination of the principles of the feminist and DEEP ECOLOGY movements that has grown in popularity over the past few decades. French feminist Françoise d'Eaubonne introduced the term "*ecofeminisme*" in 1974 to focus attention on women's potential for bringing about an ecological revolution. The ecofeminist philosophy provides a framework for an environmental ethic that seriously considers the historical, experiential, symbolic, and theoretical connections between the domination of women and the domination of nature.

Ecofeminists believe that western patriarchal society has rejected its essential connections with the ENVIRONMENT as it has sought control over women, other peoples, and nature. They believe that there cannot be liberation for women nor a solution to the ecological crisis within a society whose fundamental model of relationships continues to be one of domination. In uniting the demands of the women's movement with those of the ECOLOGY movement, they envision a radical reshaping of the basic socioeconomic structure and underlying values of modern industrial society. They seek to ultimately halt the process of domination and live in harmony with the environment.

Feminist philosophers claim that some of the most important feminist issues are conceptual and relate to philosophical notions such as reason and rationality, ethics, and what it is to be human. Ecofeminists extend this feminist philosophical

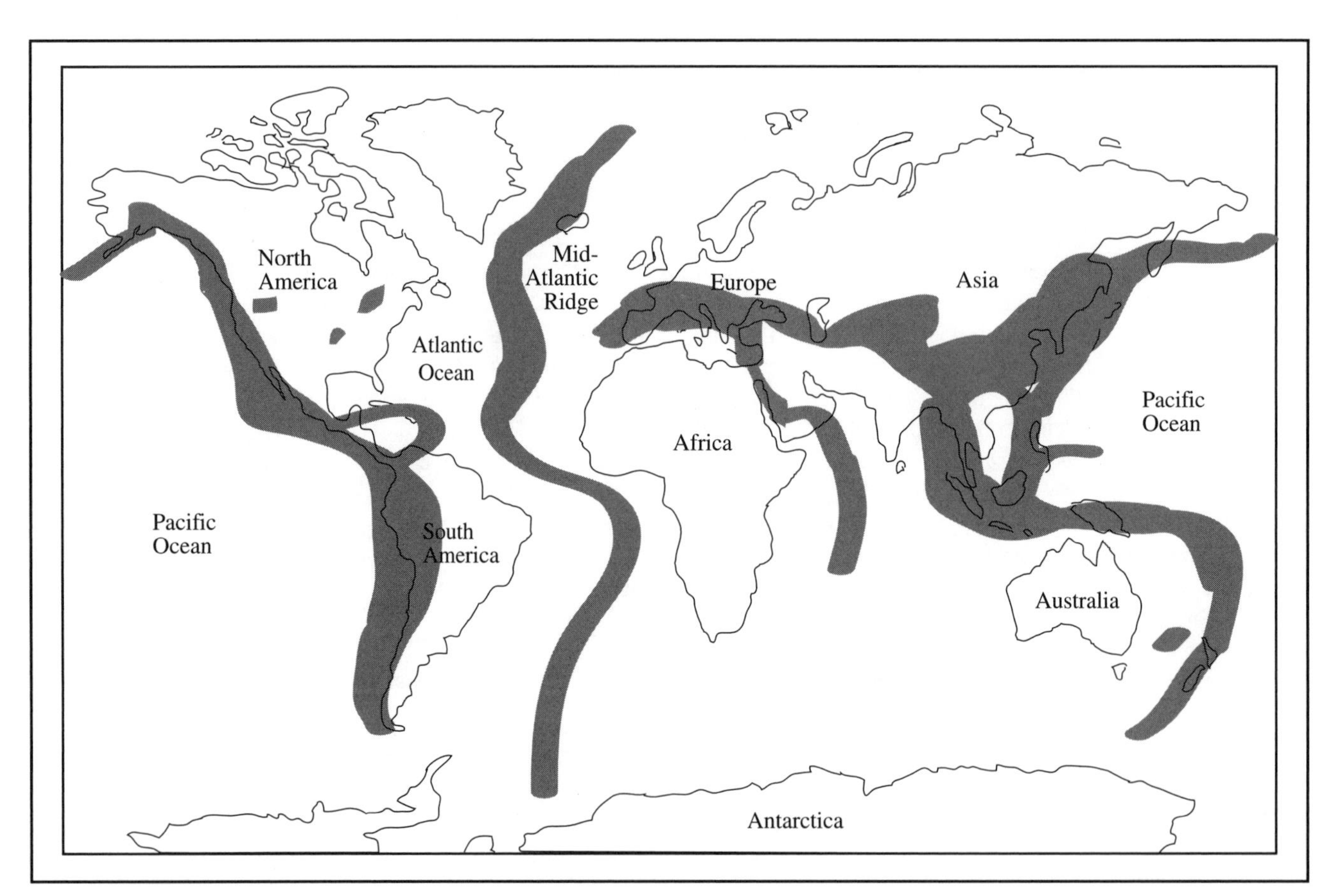

Earthquake Regions of the World. Much of the earthquake and volcano activity, shown in gray, is located around the edge or rim of the Pacific Ocean—an area extending off the eastern section of Asia and along the western edge of North and South America.

concern to nature. They argue that ultimately, some of the most important connections between the domination of women and the domination of nature are conceptual. At the same time, to ecofeminists, the global impact of environmental degradation on women's day-to-day lives—particularly in developing nations—suggests that environmental degradation also is a feminist issue in practical terms. *See also* ENVIRONMENTAL RACISM; GREENBELT MOVEMENT.

EnviroSource

Hypatia: A Journal of Feminist Philosophy Web site: www.indiana.edu/~iupress/journals/hypatia.html.

ecolabeling, Product or package labeling, also known as green labeling, that states the item's environmental benefits. For example, a label on a package could indicate that the product is recyclable or BIODEGRADABLE. In 1982, the U.S. Federal Trade Commission (FTC) set up guidelines for the use of the labels "recyclable," "compostable," and "biodegradable." Other countries, such as Canada, Japan, and India, have environmental labeling certifications for products that pass government tests for environmental benefits such as recyclability. Ecolabeling has come under fire by environmentalists because of false claims and other abuses. They also state that for some products, more research and testing is needed before determining if one formulation is less harmful to the ENVIRONMENT than another. *See also* COMPOSTING; RECYCLING.

EnviroSource

Green Seal: 1730 Rhode Island Avenue, NW, Suite 1050/ Washington, DC 20036. Web site: http://www.greenseal.org.

E. coli. *See Escherichia coli.*

ecological pyramid, A model used by scientists to represent the relative amounts of ENERGY, BIOMASS, or numbers of ORGANISMS present at the different TROPHIC LEVELS within an ECOSYSTEM. The base of an ecological pyramid represents the first trophic level, or PRODUCERS, of an ecosystem. This trophic level contains the greatest amount of energy available to the ecosystem and is assigned a value of 100%. Above the base is a level representing the primary CONSUMERS, those organisms that feed on producers or their products to obtain their food and energy. Because some of the energy present at the producer level is used by the producers to carry out their life processes, less energy is available to the organisms comprising the second trophic level, thus this level is shown as being smaller in size than the producer level. In fact, scientists estimate that only about 10% of the total energy available at one trophic level is passed to the next higher trophic level through

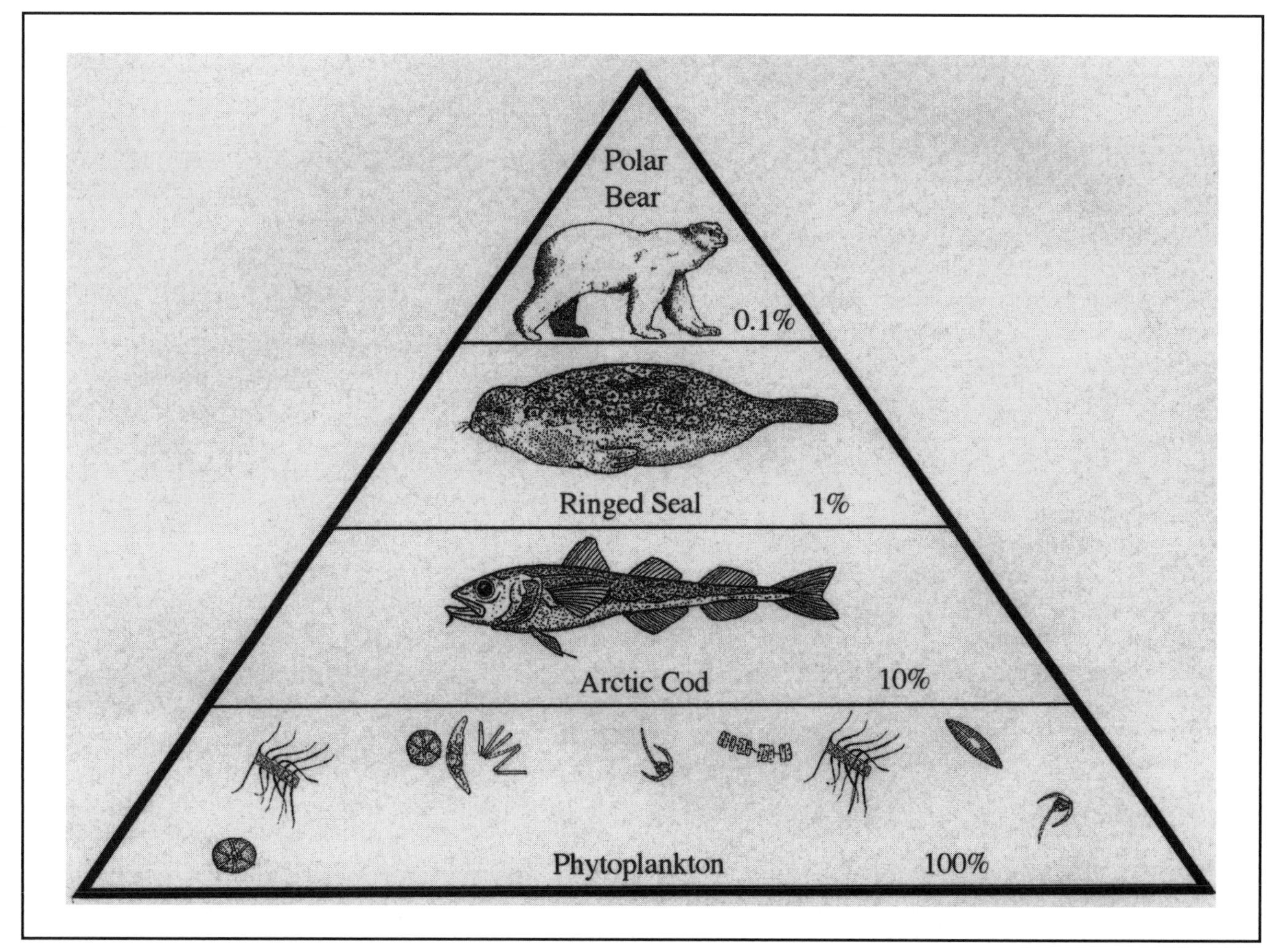

This is an example of an ecological pyramid.

the FOOD CHAIN. The remaining 90% is either used by the organisms at the preceding level or returned to the ENVIRONMENT as heat or in parts of an organism that are not consumed by other organisms (for example, bones, teeth, hooves, and bark.)

The third level of an ecological pyramid is comprised of the secondary consumers—those organisms (CARNIVORES and OMNIVORES) that use primary consumers as food. Only about 1% of the energy originally available to the producers is available to the organisms comprising this feeding level (10% of the energy that was available at the second trophic level). The fourth level of the ecological pyramid is comprised of tertiary consumers—organisms that feed on secondary consumers. At this level much less energy is available (0.1% of the energy present in the producer level).

In addition to showing the amount of energy available at each trophic level, ecological pyramids also are used to show the amount of biomass or the number of organisms present at each trophic level. As with energy, the greatest amount of biomass or greatest number of organisms are present in the producer level that makes up the pyramid's base. A pyramid of biomass follows the same 10% rule observed in the pyramid of energy; each trophic level contains only 10% of the biomass of the preceding level. By contrast, a pyramid of numbers does not strictly follow the 10% rule, but it does show that because less energy is available to each successive trophic level, fewer organisms can be supported at each successive feeding level. *See also* AUTOTROPH; CHEMOAUTOTROPH; COMPETITION; FOOD WEB; HERBIVORE; HETEROTROPH; PHYTOPLANKTON; PLANKTON; PLANTS; POPULATION; PREDATION; ZOOPLANKTON.

ecological succession, A natural process through which one BIOLOGICAL COMMUNITY is slowly replaced by another that has different characteristics in response to changing environmental conditions. Succession occurs partly in response to changes in the ABIOTIC or physical characteristics of an ECOSYSTEM and partly due to changes in the types of ORGANISMS that make up the PRODUCER level of that ecosystem. There are two types of ecological succession: primary succession and secondary succession. Primary succession occurs when a community colonizes an area that was not previously inhabited by organisms, such as an area covered by bare rock. Secondary succession occurs in areas where other biological communities previously lived, but were displaced as a result of a NATURAL DISASTER or through a major change in the area such as the clearing of its vegetation. Secondary succession occurs in an area only if the soil in the area remains intact.

Primary Succession. The first stage in primary succession is the development of soil. Primary succession occurs in areas that lack the soil needed for PLANTS (the main producers of terrestrial ecosystems) to colonize the area. Primary succession is most likely to occur in areas of bare rock recently exposed by a retreating GLACIER or in a cooled lava field following a volcanic eruption. *(See* volcano.)

LICHENS are often the first organisms to colonize an area that will undergo primary succession. A lichen is a SYMBIOTIC association of an alga or cyanobacterium and a fungus. *(See* algae; cyanobacteria; fungi.) As the organisms that comprise a lichen carry out their life processes, they secrete ACIDS and other substances that tend to decompose or weather the surfaces of the rocks upon which they live. *(See* decomposition; weathering.) This breaking down of rock results in the formation of new soil. Lichens are known as PIONEER SPECIES because they are the first organisms to colonize an area during primary succession.

Secondary Succession, Once new soil has formed in an area, the stages of primary and secondary succession become very similar. The presence of soil permits an area to support the growth of small plants such as mosses and grasses. If seeds of such plants are carried into the area by wind, running water, or passing animals, and conditions for growth are favorable, new plants may begin to grow. Without COMPETITION from other organisms, these plants can thrive, allowing a large plant POPULATION to quickly develop. If a lichen population was present (as occurs during primary succession), the grasses and mosses will replace the lichen population.

The mosses and grasses that identify an area undergoing secondary succession may thrive for many generations. As several generations of these small plants continue to go through their life cycles, they also help to build new soil as plants that die are broken down through physical and chemical processes as well as the activities of BACTERIA of decay and other DECOMPOSERS. In time, the soil becomes deep enough and rich enough to support the growth of larger plants such as shrubs. These plants quickly out compete the smaller plants for resources such as sunlight and replace the earlier population. At the same time, animals may begin to colonize the area as plant sources of food become more available. As occurred with the grass population, the shrub and animal populations may thrive for several generations, with deaths of individuals again helping to build and enrich the soil. In time, this soil will become able to support the growth of larger plants, allowing small trees to begin growing in the area.

The communities in the early stages of succession are not very stable and are easily displaced by new populations that outcompete them for resources or by natural disturbances such as fire or FLOOD. If succession continues to advance, the shrub and small tree population will in time be replaced by larger trees, whose thick foliage prevents sunlight from reaching the ground. Without sunlight, many of the ground-level plants will die off and be replaced by the larger trees. In turn, the animal populations of the ecosystem also change as new and diverse HABITATS become available and the major plant food sources in the area changes.

The first tree communities to colonize an area often are coniferous trees, whose roots are shallow and spread out laterally near the surface of the soil. In time, as soil becomes deeper, these trees may be replaced by a community comprised of broad-leaved deciduous trees such as birch, aspen, and maples mixed with larger conifers. In time, a dense forest ecosystem will develop. The forest ecosystem is generally more stable than the earlier communities that inhabited the area. If left undisturbed, this community will thrive in the

Ecological succession following a volcanic eruption in Hawaii. *Credit:* Linda Zierdt-Warshaw

area and not undergo further succession. This final, stable community is referred to as a climax community.

Succession in Aquatic Ecosystems, Succession does not occur only with terrestrial environments. For example, lakes and ponds also undergo succession as they slowly fill with soil that allows plants to begin growing. Soil generally begins to fill in a lake or pond at its edges. In addition, aquatic plants may begin to take root and thrive in the water of the lake or pond. Over many years, the growth of plants and development of new soil that results from their activities inches nearer and nearer the center of the lake or pond, causing the water to become more shallow. In time, the lake or pond community may be replaced by a MARSH or SWAMP.

As the aquatic plants making up a swamp or marsh continue to go through their life cycles, the breakdown of the remains of these plants results in the formation of more soil. Over many years, enough soil may collect in the WETLAND to allow the aquatic plants to be replaced by grasses and mosses; a terrestrial ecosystem may then replace the aquatic ecosystem. Over many years, the grasses may form a meadow, which may continue to undergo succession until it is replaced by a more stable forest community. *See also* ACID RAIN; ADAPTATION; ADAPTIVE RADIATION; BIODIVERSITY; BIOMASS; BIOSPHERE; BIOTIC; BOG; CLEAR-CUTTING; CONIFEROUS FOREST; DECIDUOUS FOREST; DEFORESTATION; EARTHQUAKE; EUTROPHICATION; FIRE ECOLOGY; FOREST FIRE; SLASH AND BURN.

EnviroSources

Bernstein, Leonard, Alan Winkler, and Linda Zierdt-Warshaw. *Environmental Science: Ecology and Human Impact*. Menlo Park, CA: Addison-Wesley, 1995.

Rees, Robin, Sr., ed. *The Way Nature Works*. New York: Macmillan, 1992.

ecology, The branch of science that focuses on studies of the relationships between ORGANISMS and their ENVIRONMENTS. Ecologists, scientists who study ecology, are very concerned with how the BIOTIC (living) parts of the environment interact with each other as well as with the ABIOTIC (nonliving) parts. An understanding of such relationships is important because all organisms derive the substances and materials needed to sustain life from their environment. For example, AIR contains the OXYGEN most organisms require to carry out their respiratory processes. The water all organisms need for life also is derived from the environment. In addition, with the exception of PRODUCERS—organisms capable of synthesizing their food from raw materials obtained from the environment—organisms are dependent upon each other for their food needs.

In addition to obtaining what they need to survive from the environment, organisms also change or alter the nonliving components of their environments. For example, PLANTS take in CARBON DIOXIDE from the air to carry out PHOTOSYNTHESIS (their food-making process). Using the ENERGY of sunlight, this carbon dioxide is combined with water to produce the glucose plants use as food; oxygen also is produced during photosynthesis. This oxygen is released into the ATMOSPHERE, where it becomes available to organisms that obtain their energy from the process of respiration. During respiration these organisms combine oxygen with glucose to obtain the energy needed to carry out a variety of life processes. In addition to energy, respiration produces water and carbon dioxide. These waste products are released back into the environment, where they may again be used by plants or other producers to carry out photosynthesis. This complex cycling of oxygen, carbon dioxide, and water (*see* carbon cycle; water cycle) between the living and nonliving parts of the environment is only one of the many interactions occurring between the biotic and abiotic elements of the environment. By identifying and developing an understanding of such interactions ecologists hope to learn more about organisms and their environments and also to develop an understanding of how the activities of organisms, including humans, may impact the ability of the global environment to continue to sustain life. *See also* ADAPTATION; AUTOTROPH; BACTERIA; BIODIVERSITY; BIOGEOCHEMICAL CYCLE; BIOLOGICAL COMMUNITY; BIOMES; BIOSPHERE; CALCIUM CYCLE; CARRYING CAPACITY; CHEMOAUTOTROPH; CHEMOSYNTHESIS; CLIMATE; COMMENSALISM; COMPETITION; CONSUMER; DECOMPOSER; DECOMPOSITION; ECOSYSTEM; FAUNA; FLORA; FOOD CHAIN; FOOD WEB; FUNGI; GENETIC DIVERSITY; GLOBAL ENVIRONMENT MONITORING SYSTEM; HABITAT; HETEROTROPH; HYDROSPHERE; MICROORGANISM; MUTUALISM; NATURAL RESOURCE; NICHE; NITROGEN CYCLE; POPULATION; PROTIST; SPECIES; SYMBIOSIS.

EnviroSource

Massa, Renato. *The Deep Green Planet: The Breathing Earth.* Austin: Raintree Steck Vaughn, 1997.

ecosphere, A term used by ecologists and environmental scientists to describe the physical space occupied by the BIOTIC and ABIOTIC elements of the ENVIRONMENT. The ecosphere is composed of the ATMOSPHERE, LITHOSPHERE, HYDROSPHERE, and BIOSPHERE. In ECOLOGY, the term "ecosphere" also encompasses the interactions that result in the synthesis and breakdown of all matter existing within this zone, including those that occur naturally and those resulting from human activities. Thus, the processes of nitrification and ammonification that are part of the NITROGEN CYCLE are considered parts of the ecosphere as is the process of pollution. *See also* BIOGEOCHEMICAL CYCLE; CHEMOSYNTHESIS; DECOMPOSITION; NATURAL DISASTER; PHOTOSYNTHESIS; SYMBIOSIS.

ecosystem, All of the BIOTIC and ABIOTIC things present in a given area along with the interactions that occur between them. The biotic parts of an ecosystem are its ORGANISMS—the SPECIES and POPULATIONS that live in that area. The abiotic parts of an ecosystem are its nonliving components, such as its temperature, soil, the amount of sunlight it receives, and its AIR. The living things in an ecosystem constantly interact with the nonliving components and often depend upon these components to sustain life. For example, most organisms use OXYGEN from the air to carry out respiration. In this process, CARBON DIOXIDE and water vapor are given off as wastes. These substances, which also are components of air, are used in turn, by PRODUCERS (PLANTS, CYANOBACTERIA, and some PROTISTS) that meet their food needs through the process of photosynthesis.

Ecosystems vary greatly in size. For example, a puddle that contains both living and nonliving components may be considered an ecosystem. By contrast, the BIOSPHERE—the portion of Earth that sustains all living things—also is an ecosystem. Ecologists—scientists who study the relationships between living things and their environments—often focus their studies on an ecosystem. The size of the ecosystem under study generally is determined by the individual who is studying its components and interactions. In addition, ecosystems often are named for a dominant species. For example, an ecosystem in which the dominant plants are deciduous trees may be identified as a DECIDUOUS FOREST ecosystem, while one in which the dominant species are corals, may be identified as a CORAL REEF ecosystem. Sometimes, scientists identify ecosystems according to a dominant physical feature; bodies of water, for example comprise the major characteristic of a pond or lake ecosystem. *See also* ADAPTATION; AGROECOSYSTEM; ALPINE TUNDRA; ATMOSPHERE; AUTOTROPH; BIODIVERSITY; BIOLOGICAL COMMUNITY; BIOME; BIOREGION; BIOSPHERE RESERVE; BOG; CARARA BIOLOGICAL RESERVE; CARRYING CAPACITY; CHAPARRAL; CHEMOAU-

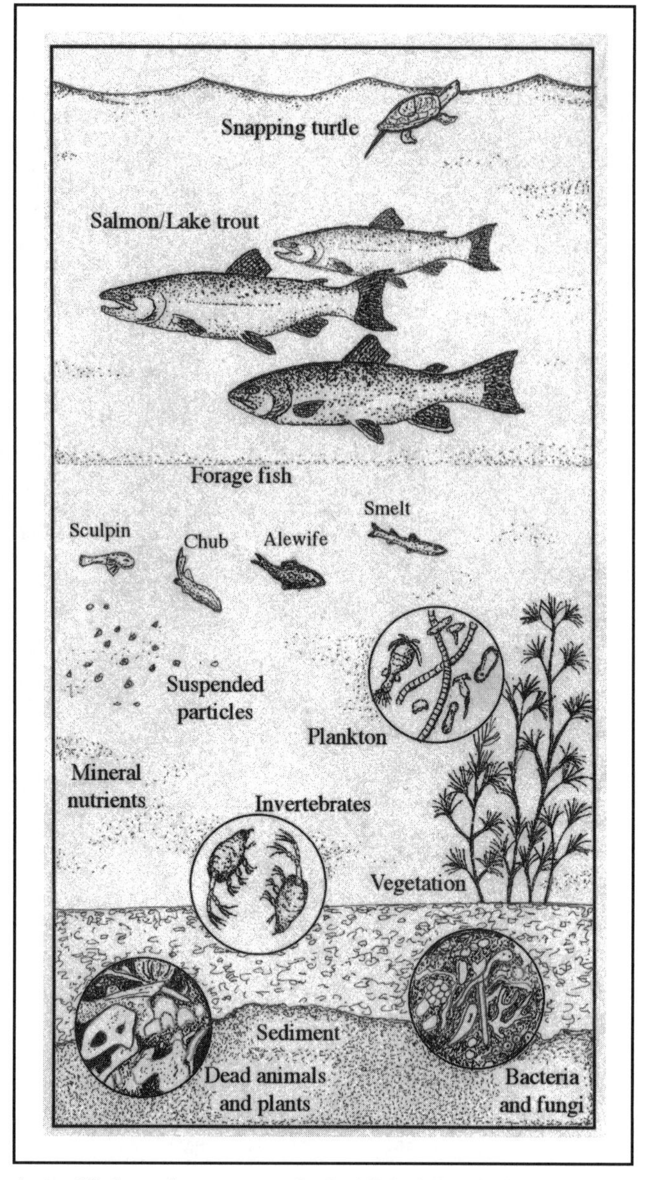

A simplified aquatic ecosystem. *Credit:* Michael Kamrin, Michigan State University, *Ecotoxicology for the Citizen.*

TOTROPH; COMMENSALISM; COMPETITION; CONSERVATION; CONSUMER; DEBT-FOR-NATURE SWAP; DECOMPOSER; DESERT; ECOLOGICAL SUCCESSION; ECOLOGY; ECOSPHERE; ECOTAGE; ECOTONE; ESTUARY; EVERGLADES NATIONAL PARK; GALÁPAGOS ISLANDS; HABITAT; HYDROSPHERE; LITHOSPHERE; MAN AND THE BIOSPHERE PROGRAMME; MARSH; NATIONAL ESTUARY PROGRAM; NATIONAL PARK SERVICE; NATIONAL WILDLIFE REFUGE SYSTEM; NICHE; OCEAN; OLD-GROWTH FOREST; PAMPAS; POPULATION; RAINFOREST; SALT MARSH; SAVANNA; SERENGETI NATIONAL PARK; SWAMP; TAIGA; TEMPERATE DECIDUOUS FOREST; TEMPERATE RAINFOREST; TSAVO NATIONAL PARK; URBAN FOREST; WETLAND; WILDERNESS; WILDLIFE REFUGE; YELLOWSTONE NATIONAL PARK; YOSEMITE NATIONAL PARK.

EnviroSource

Massa, Renato. *The Deep Green Planet: The Breathing Earth.* Austin: Raintree Steck Vaughn, 1997.

ecotage, A radical or aggressive practice used by some environmental groups to discourage such activities as timber harvesting and building roads through national forests. Ecotage practices include inflicting property damage such as sabotaging machines such as bulldozers and other heavy equipment used to build roads to access land for timber and logging activities. Ecotage is more destructive than other protest activities such as sit-ins, blocking roads, and climbing trees to stop logging and other timber operations. *See also* ABBEY, EDWARD; GREENPEACE; FOREMAN, DAVID.

ecotone, An ecological zone or boundary formed in an area where two or more ECOSYSTEMS meet. An ecotone serves as a zone of transition between distinctly different HABITATS that generally have their own characteristic FLORA and FAUNA. An ecotone may have characteristics of each of its adjacent ecosystems or its own unique characteristics. Ecotones commonly exist at the edges of FORESTS or woodlands that abut land areas used for housing or AGRICULTURE. *See also* AGROECOSYSTEM; BIOME; CARARA BIOLOGICAL RESERVE.

ecotourism, A segment of the travel industry that provides people the opportunity to visit places known for their natural beauty or their exotic or abundant WILDLIFE. In addition to providing tourists with opportunities to observe nature and natural phenomena, ecotourism also attempts to make people more aware of how human activities, such as clearing land for construction or the improper disposal of waste materials and GARBAGE, may adversely affect the ENVIRONMENT. Ecotourism is generally beneficial to the environment because it encourages protection of natural environments and wildlife through the promise of economic benefits to local and national communities.

Ecotourism destinations often feature visits to regions known for their natural scenic beauty, FLORA and FAUNA, or an unusual natural phenomenon. For example, YOSEMITE NATIONAL PARK, the Grand Canyon, and Niagara Falls, the geysers of YELLOWSTONE NATIONAL PARK and the VOLCANOES of Hawaii (along with the black sand beaches they produce) are areas of the United States that attract visitors because of their natural beauty and their natural phenomena. Other common destinations of the ecotourism industry offer people opportunities to observe native wildlife in natural settings. Popular sites for such visits include the tropical forests of the Amazon region of South America, the GALÁPAGOS ISLANDS off the coast of Ecuador, and the many game reserves and national parks throughout the African continent. Vacations in such places often center around observing a single SPECIES of animal. For example, in recent years Rwanda and Zaire have begun promoting travel to the Virunga Mountains, within the Parc National des Volcans for the purpose of observing endangered mountain GORILLAS, while visitors to the ADDO NATIONAL ELEPHANT PARK travel there largely to observe their protected POPULATION of endangered African ELEPHANTS. *(See* endangered species.) In the United States, a large ecotourism industry has developed around WHALE watching, while Canada has begun to encourage an ecotourism industry centered around its polar bear and moose populations. *See also* ARCTIC NATIONAL WILDLIFE REFUGE; BAY OF FUNDY; BIOSPHERE RESERVE; CARARA BIOLOGICAL RESERVE; EVERGLADES NATIONAL PARK; MADAGASCAR; NATIONAL AUDUBON SOCIETY; NATIONAL PARK SERVICE; SERENGETI NATIONAL PARK; WILDERNESS.

EnviroSources

Markels, Alex. "The Next Great Eco-Trips." *Audubon* 100, no. 5 (September–October 1998): 66–69.

Perney, Linda, and Pamela Emanoil. "Where the Wild Things Are." *Audubon* 100, no. 5 (September–October 1998): 82–89.

edge effect, An ecological phenomenon in which higher numbers of PLANTS, animals, and other ORGANISMS adapt to the edges of ECOSYSTEMS, as compared to their interiors. The edge effect is a basic phenomenon of ECOLOGY. Birds living along the fringes of FORESTS, or by the edge of the sea, are good examples of the edge effect.

Edges are the narrow zones of overlap between two different ecosystems. In these transition zones, or ECOTONES, SPECIES diversity is high. Many types of plants thrive and grow, using the best qualities of each zone to their benefit. Animals, because they are mobile organisms, constantly move back and forth between the two ENVIRONMENTS in search of food, water, shelter, and other basic needs. Additionally, there are a few organisms found almost exclusively in the ecotone. The edge effect can be explained by the fact that more resources are available to organisms along the edges of ecosystems. Species that are well adapted to the ecotone, for instance, can take advantage of the resources of both adjacent ecosystems, as well as the flows of organisms, ENERGY, and other materials that inhabit the edge.

Wildlife managers must also take the edge effect into account when managing POPULATIONS of animals. Managing includes studying and understanding the food, water, and shelter requirements of the population. The management process may involve building structures that provide cover and protection from predators or the destruction of some part of the HABITAT to encourage or discourage other species. Understanding the edge effect helps wildlife managers to better determine an animal's needs and how best to provide for those needs. *See also* BIODIVERSITY; DEFORESTATION; NICHE; SELECTION CUTTING.

effluent, The DISCHARGE of WASTEWATER from a stationary, or POINT SOURCE, such as a factory or electrical power plant. Industrial effluent is a chief cause of water pollution. Water used by factories such as paper mills often becomes contaminated during the manufacturing process. When the wastewater is discharged as effluent, it ends up in rivers and streams, lakes and ponds, the OCEAN, or GROUNDWATER. Industrial effluent may contain PETROLEUM, HEAVY METALS, ACIDS, salts, ORGANIC substances, and excessive heat. All these substances pollute the ENVIRONMENT and can severely damage aquatic ECOSYSTEMS.

Under the CLEAN WATER ACT of 1972, industries can discharge only certain amounts and types of POLLUTANTS in their effluent. Industries must also obtain a permit before they can legally release any effluent into the environment. *See also* SEWAGE TREATMENT PLANT; SEWER; THERMAL POLLUTION; WASTE MANAGEMENT FACILITY; WATER POLLUTION CONTROL ACT.

Ehrlich, Paul Ralph (1932–), A U.S. biologist who claims that human overpopulation may devastate the ENVIRONMENT. In his book, *The Population Bomb* (1968), Ehrlich suggested that the human POPULATION worldwide is growing larger than Earth's ability to support it with food resources. Thus, the human population is exceeding Earth's CARRYING CAPACITY for humans. Ehrlich predicts that if this trend continues, overpopulation will result in large-scale FAMINE and epidemics of DISEASE.

To help develop awareness of the population crisis, Ehrlich founded the organization Zero Population Growth (ZPG) in 1968 and served as honorary president of the organization. The goal of the ZPG is to curb the population boom as a means of developing a sustainable balance of people, NATURAL RESOURCES, and the environment. In addition to his work with the ZPG, Ehrlich continued his campaign to improve awareness of how the human population affects the environment in his books, *Extinction: The Causes and Consequences of the Disappearance of Species* (1981) and the *Population Explosion* (1991). Ehrlich and DAVID BROWER were nominated for the Nobel Peace Prize in 1998. *See also* STERILIZATION; REPRODUCTIVE; ZERO POPULATION GROWTH.

EnviroSources

De Seiguer, J. Edward. *The Age of Environmentalism*. New York: McGraw-Hill, 1996.

Ehrlich, Paul. *The Population Bomb*. San Francisco, CA: Sierra Club, 1968.

Ehrlich, Paul R., and Anne H. Ehrlich. *Betrayal of Science and Reason: How Anti-Environment Rhetoric Threatens Our Future*. Washington, DC: Island Press, 1998.

El Niño/La Niña, Major climatic events that occur periodically in the eastern Pacific Ocean and produce remarkable WEATHER changes over large portions of Earth. The NATIONAL OCEANIC AND ATMOSPHERIC ADMINISTRATION describes El Niño and La Niña as disruptions of the OCEAN-ATMOSPHERE system in the tropical Pacific that have important consequences for weather around the globe.

Earth's oceans, with their various currents and winds, affect global weather events and patterns. During the last quarter of the twentieth century, climatologists and oceanographers have developed an understanding of how regional changes in the temperatures and directions of OCEAN CURRENTS can create far-reaching effects on global CLIMATE.

El Niño, a Spanish name for the Christ Child, was the name given by Peruvian fishers to an occasional warming of coastal waters off South America, beginning in the winter season, near the time of Christmas. Over many generations, these fishers had observed these periods of ocean warming that resulted in smaller-than-usual catches of fish. Only recently have scientists achieved at an understanding of the causes and effects of El Niño from a global perspective.

The El Niño climate cycle begins in an area of warm water in the eastern Pacific Ocean where high atmospheric pressure causes the prevailing trade winds to blow to the west. The effect of these west-bound trade winds is a higher sea level in Indonesia. Rainfall occurs as air rises over the warm

El Niño

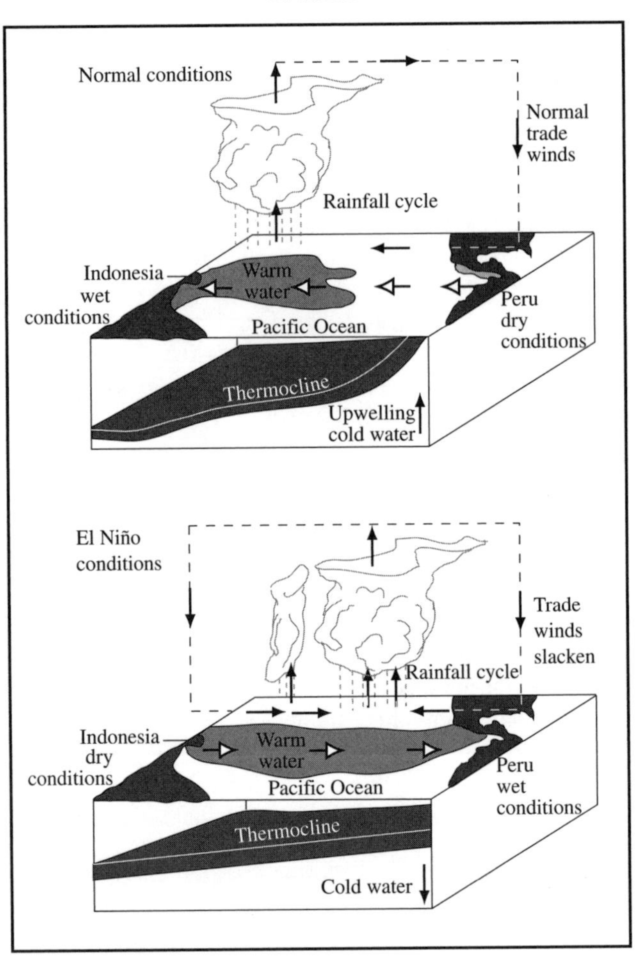

The illustration shows normal and El Niño conditions in the Pacific Ocean. *Source:* NOAA.

water in the western Pacific, while little or no rainfall occurs in the eastern Pacific. Due to the cooler temperatures in the eastern Pacific, there is an UPWELLING of cold water from the bottom of the ocean that is rich with nutrients supporting many diverse marine ECOSYSTEMS and fisheries. However, an occasional slackening of the trade winds will cause the usual westward movement of water to cease and cause instead an eastward flow of water to occur beneath the surface of the ocean. This eastward flow depresses the THERMOCLINE, the layer of cold water in the ocean. As a result, the eastern Pacific Ocean temperatures rise to above normal at the surface. Rainfall follows the eastward movement of the warm surface waters causing flooding conditions in Peru. Due to the lack of rain in the west, dry conditions occur in Indonesia.

La Niña ("the little girl," or sometimes known as El Viejo—"the old man"), the counterpart of El Niño, occurs at the opposite point in the climatic cycle. The El Niño–La Niña cycle consists of a continuously changing balance between two extremes of climatic conditions: as El Niño fades, La Niña emerges. La Niña appears to occur because the deep eastward-flowing water (which causes the thermocline to sink) bounces off the South American coast, back toward Indonesia, and brings the thermocline back to the surface.

El Niño has a significant effect on the jet streams, critical factors in global climate. The warmer eastern Pacific waters cause the Pacific jet stream to develop a more southerly branch that brings greater moisture to southern California, while making the Pacific Northwest region drier. This jet stream change reduces the frequency and intensity of HURRICANES in the Gulf of Mexico also. The far-reaching climatic effects can include DROUGHT in southern Africa and Indonesia and unseasonably warm weather in North America and eastern China.

In turn, severe weather changes may cause DISEASE epidemics in portions of the world. Diseases such as malaria and dengue fever occur with greater frequency during periods of higher rainfall and warmer temperatures because these diseases are spread by mosquitoes, which are more abundant in warm, moist conditions. Similarly, instances of the disease cholera, which is associated with PLANKTON, may increase with warmer ocean temperatures.

Weather impacts attributable to El Niño during 1998 included FLOOD damage in southern California, a mild hurricane season in the CARIBBEAN SEA, severe drought in Australia, and forest fires in Indonesia. During the winter, snow fell in the southwestern United States and in Mexico in much greater quantities than usual. The west coast of South America suffered from heavy rains and snow melt as well, with floods in Peru and mudslides in Ecuador.

The 1998 El Niño episode caused the starvation of thousands of southern sea lions and South American fur seals along the coasts of Chile, Peru, Ecuador, and the Galápagos. *(See* seals and sea lions.) SPECIES such as anchovies and sardines—important food sources for these marine MAMMALS—became scarce because the fish migrated away from the warmer-than-usual coastal waters. The sea lions and seals were then forced to travel much farther from their home territories in search of food. A large proportion of infant animals died.

Scientists generally agree about the mechanism of this climatic cycle but have not reached a widespread understanding of the history and frequency of El Niño occurrences. During the 1990s, scientists studied the dynamics of El Niño more thoroughly than ever before. Several El Niño episodes occurred during the 1990s, whereas formerly they had occurred every two to seven years. Some investigators believe the apparent increase in El Niño events is related to GLOBAL WARMING phenomena. Others contend that El Niño represents a type of normal variation in climate conditions that should be expected. The intensive study of this phenomenon has demonstrated how changes in one critical region of Earth can have widespread and significant consequences over the much of the globe.

EnviroSources

Fagan, Brian M. *Floods, Famines, and Emperors: El Niño and the Fate of Civilizations.* New York: Basic Books, 1999.

Glantz, Michael H. *Currents of Change: El Niño's Impact on Climate and Society.* Cambridge, UK: Cambridge University Press, 1996.

National Center for Atmospheric Research Web site: ww.ncar.ucar.edu.

National Hurricane Center/Tropical Prediction Center Web site: www.nhc.noaa.gov/.

National Oceanic and Atmospheric Administration Web site: www.noaa.gov/.

National Oceanic and Atmospheric Administration (El Niño/El Niña) Web site: www.pmel.noaa.gov/toga-tao/el-nino/nino-home-low.html.

National Oceanic and Atmospheric Administration (La Niña) Web site: www.elnino.noaa.gov/lanina.html

Porter, Henry F., and Eric Wybenga. *Forecast: Disaster; The Future of El Niño.* New York: Dell, 1998.

Scripps Institute of Oceanography Web site: sio.ucsd.edu/supp_groups/siocomm/elnino/elnino.html.

electric vehicle, A motor vehicle powered by ELECTRICITY, or by the combination of an electric motor and a GASOLINE engine, that produces zero or low EMISSIONS. The production of electric-powered vehicles is not a new concept. During the early 1900s approximately 50,000 electric cars powered by BATTERIES were used in the United States; however, the introduction of the internal-combustion engine, the low cost of gasoline and PETROLEUM, and the ability of gasoline-powered engines to travel long distances caused the decline of battery-powered engines. By the 1920s, electric cars had all but disappeared in the United States.

Now electric vehicles (EVs) are back; spurred on by government pressure to reduce harmful vehicle emissions around the world, auto-makers are taking note of the vehicles' advantages. The state of California is mandating that 10% of all new vehicles sold in California in 2003 be zero-emission vehicles. Overseas, countries such as Sweden, France, and Germany are providing tax incentives to customers who purchase EVs. The Swedish postal service is currently switching over to EVs.

EVs are sometimes referred to as zero-emission vehicles because they produce no exhaust pollution and use no FUELS that evaporate. EVs can greatly reduce emissions of CARBON MONOXIDE and SMOG-forming POLLUTANTS in major cities.

An electric vehicle model (EV) built by General Motors. *Credit:* Used with permission of General Motors Media Archives.

In 1996, General Motors (GM) became the first major automaker in the later part of the century to make EVs commercially available in the United States. In 1998, about 2,500 EVs were sold in the United States; most were purchased in California, Arizona, and Michigan.

EVs are under development in all of the European countries as well as in Japan and other countries in Asia. The 1997 estimates for the number of EVs on the road in Europe included 3,500 in France, 2,500 in Switzerland, 2,200 in Germany, and 950 in Italy. Although these numbers seem small, sales of EVs are anticipated to climb much higher beginning in the year 2000. Currently, the prices for an EV range between $20,000 and $35,000 or more.

Most of the EVs on the road in 1999, including minivans and some trucks, use LEAD-ACID batteries similar to those used in gasoline-powered cars. Lead-acid batteries are the most commonly used and least expensive technology. Vehicles that use these batteries need recharging every 100–200 kilometers (60–120 miles). Charging an EV requires a special charging unit that provides the same voltage as is used by a large household appliance such as an electric clothes dryer or a central air conditioner. Some of the units can be installed in home garages, parking lots, and outside buildings. Charging a vehicle usually takes from four to eight hours, depending on how depleted the battery is. Battery life is about three years. Chrysler, Ford, GM, and Toyota vehicles use such battery technology in their EVs.

Nickel–metal hydride (NiMH) batteries are another popular option. General Motors introduced the 1999 model EV-1 electric car with NiMH batteries in California. NiMH batteries are used in laptop computers and in some small appliances. NiMH batteries in EVs offer a greater driving range than vehicles powered by lead-acid batteries but are still expensive because large-scale production is only now being started. NiMH batteries last about 160 kilometers (100 miles) per charge. The life expectancy of this battery is about 100,000 miles. Chrysler, Ford (California only), GM, Honda, and Toyota offer vehicles with NiMH technology.

HYBRID electric vehicles are also being built. The hybrids use two sources of power—an electric motor powered by batteries and a gasoline engine. The gasoline engine provides some power and is used as a generator to recharge the batteries. Toyota was the first to sell a hybrid gasoline-electric car in Japan in 1997; the company then introduced the car to the U.S. market in late 1999. The 1999 Toyota hybrid car has a NiMH and achieves speeds of 160 kilometers (100 miles) per hour when the engine and electric motor are used in combination. In the same year, Honda introduced its hybrid in the American market. GM, Ford, and Chrysler are also working on hybrids. Other hybrid-electric systems being tested include combinations of electric motors powered by FUEL CELLS, gasoline engines, TURBINES, and flywheels.

Although EVs produce zero emissions or low emissions, generating the electricity to charge the EVs from electric power plants using FOSSIL FUELS will produce airborne pollutants and SOLID WASTE at the plants. And there are environmental concerns with the use of batteries for EVs and hybrid cars. They contain TOXIC CHEMICALS and produce some toxic emissions, which can make battery production and disposal a waste issue. Other criticisms include the assertion that today's batteries are still too expensive, store little energy, are heavy, and need to be replaced eventually. The future for EV battery technology looks promising, however, since it represents another option for producing zero- or low-emission vehicles. *See also* AUTOMOBILE; CATALYTIC CONVERTER; GREENHOUSE EFFECT; GREENHOUSE GAS; TOXIC WASTE.

EnviroSources

The Electric Vehicle Association of the Americas, (800) 438-3228; Web site: www.evaa.org

Electric Vehicle Technology Web site: www.avere.org/

Miller, R. K., and M. E. Supnow. *Electric Vehicles*. Future Technology Surveys, 1994.

National Renewable Energy Laboratory, 1617 Cole Boulevard, Golden, CO 80401-3393.

electricity, A form of ENERGY composed of a flow of electrons that can be used to produce light, mechanical energy, or heat. The flow of electrons through a conductor such as a metal wire is referred to as an electric current. The rate at which the electric charge flows in the current is measured in amperes. Current is measured with an ammeter. The complete route of an electric current from a source such as the terminal of a BATTERY, or generator, through an electric-powered mechanism such as a toaster and back to the other terminal is an electric circuit. The electric circuit is a closed path through which electrons can flow.

The rate at which electric energy is converted to another form of energy is called electric power. The most commonly used unit of electric power is the kilowatt (kW), which is the unit of power equal to 1,000 watts (1.34 horsepower) or the energy consumption at the rate of 1,000 joules per second. Electricity utility companies charge customers a rate based on the amount of kilowatts the customers use. The standard measure for large amounts of electricity is the kilowatt hour (kWh), which equals 1,000 watts per hour. The use of ten 100-watt bulbs for one hour would consume one kilowatt. The average home in the United States uses about 9,000 kWh

Electricity Usage of Common Appliances

Appliance	Power (W)	Typical Energy Consumption (kWh/year)
Clock	2	15–50
Electric clothes dryer	4600	900–1,000
Hair dryer	1000	60
Light bulb	100	108
compact fluorescent	18	19
Refrigerator	360	1600
Television	350	300–1,000
Washing machine	700	1,008

in a year's time. Electric utility companies use billing rates based on cents per kilowatt hours.

The following are the average power requirements, or rates of energy used per hour, of some home appliances: a hair dryer, 600–1,000 watts; a microwave, 1,300–1,450 watts; a color television, 200–300 watts; an oven range, 2,600 watts; and a refrigerator, 500–700 watts. Energy costs in a home can be calculated by multiplying the average power in watts of an appliance by the total hours of use for a month. The total will be monthly watt hours, which can be converted to kWh per month. Then to get monthly cost, multiply kWh by the utility rate charged by the electric utility company. Electrical energy use can also be estimated for the year. For example, if a hair dryer has the power of 1,000 watts and is used for a total of 60 hours in one year, the approximate energy used is 60 kWh per year.

EnviroTerms

electric motor: A machine that changes electric energy to mechanical energy.

static electricity: Electricity in the resting state.

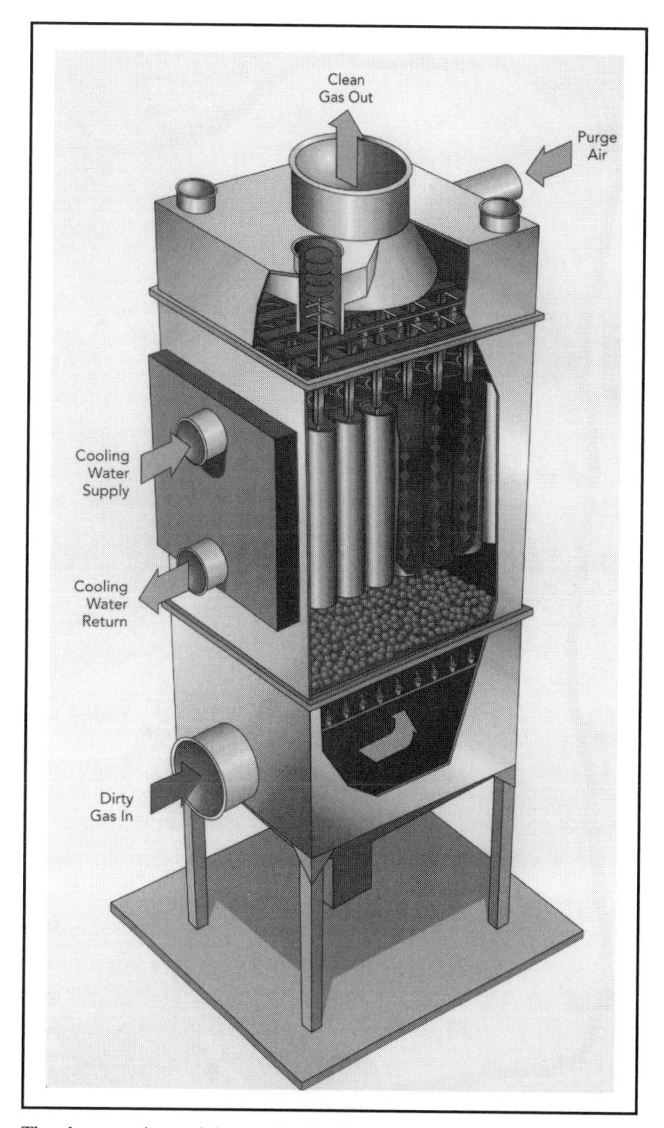

The electrostatic precipitator takes in dirty gas and uses electrostatic forces to remove particulates from gas streams. The particulates settle to the bottom where they are removed. *Credit:* Monsanto

electrostatic precipitator, A pollution-control device that removes very small particles, known as "particulates," from AIR EMISSIONS. In capturing particulates, these devices serve the same purpose as a dust filter but operate differently. Electrostatic precipitators use the physical principle of electrostatic attraction to capture particulates. Small particles can be given a positive or negative electrical charge by creating a sufficient electrical field. Once charged, particulates can be attracted to a collecting device, usually a plate or tube, of the opposite charge. Negatively charged particles will be attracted to a positively charged collector. This process of removing particulates is termed "electrostatic precipitation." Electrostatic precipitators are very effective means of reducing particulates from emissions sources such as incinerators or industrial plants. *See also* INCINERATION; SCRUBBER.

elephants, The largest land animals on Earth and an ENDANGERED SPECIES. There are two distinct types: the African elephant (*Loxodonta Africana*) and the Asian elephant (*Elephas maximus*). Elephants are some of the most studied animals and are considered to be among the most intelligent MAMMALS, along with DOLPHINS AND PORPOISES and higher primates. Groups of elephants have a highly established social order and sophisticated communication. Their lineage can be traced back 55 million years to ancestral Proboscideans.

Elephants have been brought to the brink of EXTINCTION by loss of their HABITAT and POACHING for their IVORY tusks. Male tusks provide the most ivory. The extremely strong tusks are made of dentine—cartilaginous substances and calcium salts. Elephants use them to uproot plants for food, to dig for water, and occasionally to fight. Poachers often disable elephants and leave them to die after severing their tusks. Ivory has been sold to make piano keys, furniture inlays, billiard balls, chess pieces, jewelry, and toilet articles.

During the 1970s and 1980s, a massive upsurge in the illegal slaughter of elephants occurred. Controls to limit the trade in ivory and the killing of elephants forced the practice

The African elephant is the largest land animal on Earth. It is also an endangered species.

underground. The price of ivory soared from $10 per kilogram (2.2 pounds) in 1970 to $300 per kilogram (2.2 pounds) in 1989. By 1999, the estimated POPULATION of African elephants was about 650,000. Asian elephants numbered about 60,000.

Environmentalists led a worldwide effort to shut down the ivory trade. In 1985, the CONVENTION ON INTERNATIONAL TRADE IN ENDANGERED SPECIES OF WILD FLORA AND FAUNA (CITES) introduced an ivory control system. In 1989, the United States and the European Economic Community (the European Union) banned all ivory imports. In January 1990, CITES strengthened laws to prohibit commercial trade in any elephant parts or products. Uncontrolled human POPULATION growth and habitat destruction have also contributed to the decline. The population of elephants has recently increased to a point that population control measures are being discussed in order to preserve their habitats. *See also* ADDO NATIONAL ELEPHANT PARK; AFRICAN WILDLIFE FOUNDATION; ENDANGERED SPECIES LIST; RED LIST OF ENDANGERED SPECIES.

EnviroSources

African Wildlife Foundation Web site: www.awf.org.

Elephant Research Foundation, 106 E. Hickory Grove, Bloomfield Hills, MI 48304.

Stuart Chris, and Tilde Stuart. *Africa's Vanishing Wildlife*. Washington, DC: Smithsonian Institute Press, 1996.

U.S. Fish and Wildlife Service, The Species List of Endangered and Threathened Wildlife Web site: www.fws.gov/r9endspp/lsppinfo.html.

World Wildlife Fund Web site: www.wwf.org.

Elton, Charles Sutherland (1900–1991), A British biologist and NATURALIST who was the leading advocate in animal ECOLOGY—the study of relationships among ORGANISMS and their ENVIRONMENTS. Elton proposed the concept that living things formed an ECOLOGICAL PYRAMID of food levels. At the bottom of the pyramid were large numbers of PRODUCERS and at the top was a small number of CARNIVORES. Elton defined the three kinds of pyramids—a pyramid of BIOMASS, a pyramid of numbers, and a pyramid of ENERGY. Elton wrote several books. In his first book, *Animal Ecology* (1927), Elton reported on the HABITATS and FOOD CHAINS of ARCTIC life. His other books included *Animal Ecology*, *Animal Ecology and Evolution*, and *Moles, Mice, and Lemmings*. *See also* FOOD CHAIN; FOOD WEB; HERBIVORE; OMNIVORE; PREDATION; TROPHIC LEVEL.

EnviroSource

Elton, Charles Sutherland. *The Ecology of Invasions by Animals and Plants*. London: Methuen, 1958.

Emergency Planning and Community Right-to-Know Act, Federal legislation regarding community safety, which was enacted by Congress in 1986. The Emergency Planning and Community Right-to-Know Act (EPCRA) was designed to protect public health, safety, and the ENVIRONMENT in potentially harmful situations such as chemical accidents that release HAZARDOUS SUBSTANCES or TOXIC CHEMICALS into the AIR, water, or soil. The act provides a means for citizens and community health and safety representatives to be alerted when such problems occur. To implement the EPCRA, Congress requires each state to appoint a State Emergency Response Commission (SERC) to work with the U.S. ENVIRONMENTAL PROTECTION AGENCY (EPA) and industries. The SERCs are required to divide their state into emergency planning districts and to name a Local Emergency Planning Committee (LEPC) for each district. Broad representation by firefighters, health officials, government and media representatives, community groups, industrial facilities, and emergency managers ensures that all elements of the planning process are represented.

Under the EPCRA, industries that manufacture toxic chemicals, or use toxic chemicals in the manufacture of their products, must contact state agencies and the EPA when any release of hazardous materials into the environment occurs. Manufacturers are also required to file a MATERIAL SAFETY DATA SHEET on any hazardous material they produce or use. Failure to report releases of hazardous materials into the environment by such companies can result in heavy fines. In addition, all chemical releases must be reported to and published in a TOXIC RELEASE INVENTORY. This annual publication is available free of charge to citizens upon request. The EPCRA is also known as Title III of the Superfund Amendments and Reauthorization Act. *See also* COMPREHENSIVE ENVIRONMENTAL RESPONSE, COMPENSATION, AND LIABILITY ACT; OIL SPILL; SUPERFUND.

EnviroSources

Resource Conservation and Recovery Act, Superfund, Emergency Planning and Community Right-to-Know Act Hotline: (800)

424-9346 or Washington, DC area only (703) 412-9810. Web site: www.epa.gov/epaoswer/hotline.

U.S. Environmental Protection Agency, Toxic Release Inventory Public Data Release Web site: www.epa.gov/opptintr/tri/ttorder.htm.

emigration, The movement of members of a POPULATION out of an area, resulting in a decrease in population size in that area. In wild populations, ORGANISMS often emigrate when resources, such as food, water, and living space, become scarce. Along with BIRTH RATE, DEATH RATE, and IMMIGRATION, emigration is a key factor considered when determining population growth.

In many animal SPECIES, such as caribou and Canadian geese, seasonal emigrations, called migrations, take place. Such species leave their HABITATS in seasons when food and water resources become scarce, and travel, sometimes in great numbers, to areas in which these resources are more abundant. Migratory species generally return to their original habitats when CLIMATE conditions in those habitats again change with the seasons and resources become more plentiful. *See also* CARRYING CAPACITY; COMPETITIVE EXCLUSION; COMPETITION; POPULATION DENSITY.

emission standards, Published requirements that indicate the amount of various substances that may be emitted by sources. The 1990 federal CLEAN AIR ACT (CAA) and the U.S. ENVIRONMENTAL PROTECTION AGENCY (EPA) created AIR quality standards that states must achieve. If an area fails to meet the air quality standards, the CAA requires that EMISSIONS controls be implemented to protect air quality. Emissions standards are typically expressed in units of pounds, kilograms, or metric tons per year.

The CAA established rules that industrial plants and AUTOMOBILE manufacturers must follow to reduce air emissions of many hazardous air POLLUTANTS (HAPs). Under the CAA, industries that emit air pollutants must implement pollution reduction and control technologies. The CAA requires that certain major air emission sources use the maximum achievable control technology to reduce their emissions of HAPs.

Furthermore, new sources of pollutants must use the BEST AVAILABLE CONTROL TECHNOLOGY to meet specific performance standards, which means implementation of systems or air pollution control equipment that achieve the greatest possible reductions of emissions.

The EPA has instituted a program that requires emissions sources to obtain permits and pay fees to operate. Facilities that obtain air emissions permits must demonstrate that their operations meet the emissions standards and performance standards. *See also* CATALYTIC CONVERTER; CRITERIA POLLUTANTS; ELECTROSTATIC PRECIPITATOR; SCRUBBER.

emissions, POLLUTANTS such as PARTICULATE MATTER, gaseous substances, and radioactive material that enter the ENVIRONMENT as a result of activities such as INCINERATION or evaporation. The major sources of emissions include airborne pollutants from motor vehicles and power plants that burn FOSSIL FUELS.

AIR emissions consist principally of chemical compounds that contain CARBON, NITROGEN, or SULFUR in combination with OXYGEN. CARBON DIOXIDE and CARBON MONOXIDE are major carbon emissions. Other significant emissions include HYDROCARBONS and a variety of VOLATILE ORGANIC COMPOUNDS. *See also* EMISSION STANDARDS; RADIOACTIVE WASTE.

endangered species, A SPECIES of animal or PLANT that is in immediate danger of EXTINCTION because of environmental threats, such as CLIMATE change, or as a result of human activities, including purposeful extermination, overhunting, and HABITAT destruction. The term is usually a formal designation made by a governmental agency such as the U.S. FISH AND WILDLIFE SERVICE (FWS) or an organization such as the WORLD CONSERVATION MONITORING CENTRE. Endangered species generally cannot avoid extinction without some form of direct human intervention, such as habitat protection or CAPTIVE PROPAGATION, the breeding of ORGANISMS in captivity for eventual release back into the environment.

Species may become endangered for a variety of reasons, but the most frequent is human-caused habitat loss. Drainage of WETLANDS, conversion of shrub lands for livestock grazing, CLEAR-CUTTING FORESTS (especially tropical forests), urbanization and suburbanization, and highway and DAM construction have significantly reduced available habitats. As habitats are fragmented into "islands," POPULATIONS crowd into smaller isolated areas, where intensive use of available resources may cause further habitat deterioration. Island populations lose contact with other members of their species, thereby reducing GENETIC DIVERSITY and making the species less adaptable to environmental changes. *(See* adaptation.) Small populations are particularly vulnerable to extinction; for some species,

The Oryx is one of many endangered species.

Total Endangered Species in the United States and Worldwide

Group	Endangered U.S.	Endangered Foreign	Threatened U.S.	Threatened Foreign	Total Species
Mammals	61	251	8	16	336
Birds	75	178	15	6	274
Reptiles	14	65	21	14	114
Amphibians	9	8	8	1	26
Fishes	69	11	41	0	121
Clams	61	2	8	0	71
Snails	18	1	10	0	29
Insects	28	4	9	0	41
Arachnids	5	0	0	0	5
Crustaceans	17	0	3	0	20
Animal subtotal	357	520	123	37	1,037
Flowering plants	540	1	132	0	673
Conifers	2	0	1	2	5
Ferns and others	26	0	2	0	28
Plant subtotal	568	1	135	2	706
Grand total	925	521	258	39	1,743

Source: United States Fish and Wildlife Service, 1998.

the fragmented habitats become too small to support a viable population.

Introduced DISEASES, parasites, and EXOTIC SPECIES of predators also have exterminated or greatly reduced the populations of some species. The accidental introduction of a BLIGHT, a fungal disease, for example, eliminated the chestnut tree from DECIDUOUS FORESTS in North America. In addition, control of predators and pests involving the use of PESTICIDES has had unintended effects on nontarget species. Excessive killing of prairie dogs, for example, has nearly eliminated one of their main predators, the black-footed ferret, which is now endangered. *(See* predation.)

Pollution and chemical exposure present additional threats. TOXIC CHEMICALS, especially CHLORINATED HYDROCARBONS such as DICHLORODIPHENYLTRICHOROETHANE (DDT) and POLYCHLORINATED BIPHENYLS (PCBs), most strongly affect species at the top of FOOD CHAINS. Both DDT and PCBs, for example, interfere with calcium metabolism in birds, causing the production of soft-shelled eggs and deformed chicks. Water pollution and elevated water temperatures have eliminated NATIVE SPECIES of fish in several habitats.

Today, more than 475 species of animals native to the United States and other areas of the world are listed by the FWS as endangered, including the FLORIDA PANTHER, several species of WOLF, the CALIFORNIA CONDOR, the whooping crane, several REPTILES, and more than 45 species of fishes. Hunting, trapping, and poisoning predators to protect sheep and cattle are among the greatest threats to predatory MAMMALS; bird populations have suffered great losses because of the use of PESTICIDES. At present, the FWS also has more than 700 species of plants listed as endangered or threatened. Threatened species are those that are abundant in parts of their range, but are declining in total number and are therefore at risk of extinction in the foreseeable future. Recovery plans have been approved for 886 of the total of both endangered and threatened listed plant and animal species. An additional 35 species of animals and 35 species of plants are currently proposed for listing.

The first WILDLIFE protection laws were enacted in the United States in the early 1900s to restrict commercial trade and overhunting. In 1973, the ENDANGERED SPECIES ACT provided mechanisms for the CONSERVATION of the ECOSYSTEMS on which endangered species depend; it also discouraged the exploitation of endangered species in other countries by banning U.S. importation or trade of any endangered species or any product made from such species. The United States also is a party to various agreements with neighboring nations; for example, Canada, Mexico, and the United States signed the Tripartite Agreement on Wetlands, which benefits migratory birds. Other important actions included passage of the NATIONAL ENVIRONMENTAL POLICY ACT of 1969 and the banning of DDT in the early 1970s.

On an international level, the CONVENTION ON INTERNATIONAL TRADE IN ENDANGERED SPECIES OF WILD FLORA AND FAUNA (CITES) a 1973 treaty, now protects more than 600 species of animals and plants, including species of GORILLA and RHINOCEROS. By the early 1990s, some success had been achieved in prohibiting trade in rhinoceros horn, ELEPHANT IVORY, and endangered orchids; however, in many countries, a lack of local law enforcement, the willingness of some individuals to trade in endangered species, and the activities of poachers and traders put the future of many species in jeopardy despite the existence of legal protections. *(See* poaching.)

Direct actions to save endangered species include the propagation of breeding stock for release into the wild, either to restore a breeding population (as for the PEREGRINE FALCON) or to supplement a natural population (as for the whooping crane), and the reintroduction of a species to its historical habitat (such as the North American gray wolf). Another approach involves the determination of critical habitats that must be preserved for endangered species. These habitats may be protected by the establishment of preserves and WILDLIFE REFUGES, though this approach is of limited value if such areas are so small that they result in the creation of island populations. Conservationists have been pressing for the application of stricter land use planning criteria that provide for development without habitat destruction. *See also* ADAPTATION; ADDO NATIONAL ELEPHANT PARK; AFRICAN WILDLIFE FOUNDATION; AMERICAN ZOO AND AQUARIUM ASSOCIATION; ARCTIC NATIONAL WILDLIFE REFUGE; BALD EAGLES; BATS; BIODIVERSITY; BIODIVERSITY TREATY; BIOREGION; BIOSPHERE RESERVE; BISON; CANNED HUNT; CARARA BIOLOGICAL RESERVE; CARRYING CAPACITY; CHEETAHS; COMPETITION; CONSERVATION EASEMENT; CROCODILES; DEBT-FOR-NA-

TURE SWAP; DEFORESTATION; DETRIVORE; DODO BIRD; DOLPHINS AND PORPOISES; DOUGLAS, MARJORY STOKEMAN; DUTCH ELM DISEASE; ECOTOURISM; ENDANGERED SPECIES LIST; GALDIKAS, BIRUTÉ; GOODALL, JANE; GREENPEACE; KEYSTONE SPECIES; LEOPARDS; MANATEES; MARINE MAMMAL PROTECTION ACT; MIGRATORY BIRD TREATY ACT; ORANGUTANS; ORYX; PARASITISM; PARROTS; RAINFOREST; RED LIST OF ENDANGERED SPECIES; THREATENED SPECIES; WHALES; WORLD CONSERVATION UNION.

EnviroSources

Chadwick, Douglas H., and Joel Sartore. *The Company We Keep: America's Endangered Species*. Washington, DC: National Geographic Society, 1995.

Convention on International Trade in Endangered Species of Wild Flora and Fauna (CITES) Web site: www.cites.org

Ehrlich, Paul, and Anne Ehrlich, *Extinction: The Causes and Consequences of the Disappearance of Species*. New York: Random House, 1981.

Luoma, Jon R. "It's 10:00 p.m. We Know Where Your Turtles Are." *Audubon* 100, no. 5 (September–October 1998): 52-57.

Middleton, Susan, and David Littschwager. "The Endangered 100." *Life* (September 1994): 50–63.

U.S. Fish and Wildlife Service, Division of Endangered Species Web site: www.endangered.fws.gov/wildlife.html

Williams, Ted. "Back from the Brink." *Audubon* 100, no. 6 (November–December 1998): 70–76.

Wilson, Edward O. *The Diversity of Life*. Cambridge, MA: Belknap Press, 1992.

World Conservation Monitoring Centre Web site: www.wcmc.org.uk

Endangered Species Act, A federal law enacted in 1973 that is one of the strongest, most far-sighted, and most comprehensive pieces of environmental legislation ever enacted by Congress. The act requires the CONSERVATION of threatened and ENDANGERED SPECIES and the ECOSYSTEMS upon which they depend. Although amendments to the Endangered Species Act (ESA) were enacted in 1978, 1982, and 1988, the overall framework of the 1973 act has remained essentially unchanged.

The U.S. FISH AND WILDLIFE SERVICE (FWS) and the NATIONAL MARINE FISHERIES SERVICE are the two federal agencies charged with implementing the ESA. Their responsibilities include working with partners to conserve species, determining the species that need protection *(see* endangered species list), and restoring listed species to a secure existence (recovery). All federal agencies are required to take into account the conservation of endangered and THREATENED SPECIES and are prohibited from authorizing, funding, or carrying out any action that may jeopardize a listed species or destroy or modify its critical HABITAT.

Earlier endangered species laws passed in 1966 (the Endangered Species Preservation Act) and 1969 (the Endangered Species Conservation Act) raised public awareness about the plight of rare PLANTS and animals; however, it was the 1973 Endangered Species Act that provided the tools needed to help WILDLIFE that were facing EXTINCTION. It combined and considerably strengthened the provisions of its predecessors, while also breaking new ground. For example, the law does not protect only well-known or well-loved species, but seeks to assure healthy and balanced ECOSYSTEMS for all species, humans included. The U.S. List of Endangered and Threatened Wildlife and Plants maintained by the FWS includes listings of both endangered and threatened species, and includes domestic and foreign species. U.S. implementation of the CONVENTION ON INTERNATIONAL TRADE IN ENDANGERED SPECIES OF WILD FLORA AND FAUNA (CITES) also is provided for by the ESA.

By the time a species is deemed to merit the special protection provided by the ESA, it may have been declining for decades and face multiple threats to its continued existence. Often, research is needed to determine the exact causes of its decline and the best means of restoring the species. Typically, species recovery is a gradual process, requiring several generations of successful reproduction before there are sufficient individuals to make up one or more viable POPULATIONS.

The ESA has proved remarkably effective at preventing extinctions and slowing the decline of imperiled species. Nearly half of all species listed for a decade or more are now either stable or improving in status; seven—less than 1%—have been found to be extinct. Preventing the extinction of the remaining 99% of listed species is one of the act's greatest challenges. One of the most noteworthy successes in recent years was the 1999 "graduation" and removal of the PEREGRINE FALCON from the list of endangered and threatened species. Many partner groups, states, and volunteers worked with the FWS over the last two decades to successfully breed and release peregrines into the wild. Today, their numbers have reached 1,593 breeding pairs in 40 states.

Another highlight is the CAPTIVE PROPAGATION program for the endangered black-footed ferret, which in 1998 produced 339 surviving young from a total of 452 born in captivity that spring. As a result, more than 200 ferret kits were placed into reintroduction sites and field breeding programs.

The CALIFORNIA CONDOR reintroduction program also experienced success in 1998, when condors were released in Arizona and California, increasing the population of wild condors from 0 in 1987 to 44 in 1998. The act's "flagship" species, the whooping crane, whose population had fallen to less than 20 individuals in the early 1940s, now is represented by more than 200 individuals in the wild, with 200 more in captive breeding populations. The greenback cutthroat trout, widely believed to be extinct until a remote population was discovered in the 1960s, also is soon to be removed from the list thanks to recovery actions under the ESA. Many more species are well on the road to recovery, too, including the BALD EAGLE and the Aleutian Canada goose.

Despite its successes, the need for the ESA continues. In 1998, the FWS added 57 species to the endangered species list, bringing the total to 1,179. The ESA also has not been free of controversy. Although protection of most species has occurred without much public notice, a few species, such as the snail darter and the northern spotted owl, have caused well-publicized conflict about how people should manage our rivers and streams, FORESTS, and other NATURAL RESOURCES. *(See* owls, northern spotted.) In addition, 70% of all endangered and threatened species inhabit privately owned lands. Thus, the involvement of landowners is crucial to the management of vulnerable species. The FWS works with landowners to conserve species and their habitats, while also providing for economic growth. To date, 243 habitat conservation plans (HCPs) have been established, and about 200 more are in development. HCPs are agreements between the FWS and private companies that establish what percentage of the company's land will be excluded from activities that may threaten the species listed on the endangered species list. The goal of the HCP is to allow companies to benefit from their land, while also setting aside some habitat for listed species. These habitat conservation plans cover approximately 65,000 hectares (6.5 million acres).

Under the ESA, funding also may be provided to state agencies through the Cooperative Endangered Species Conservation Fund to conduct conservation activities for listed and candidate plants and animals. Grants provide states with the resources to participate in various recovery activities ranging from population assessment and habitat restoration to propagation and reintroduction of listed species. States also may use grants to initiate conservation actions before a species is listed and to monitor the status of recovered species. *See also* BIODIVERSITY; BIODIVERSITY TREATY; BIOSPHERE RESERVE; BISON; CACTI; RED LIST OF ENDANGERED SPECIES; WILDLIFE REFUGE; WORLD CONSERVATION UNION.

EnviroSources

Chadwick, Douglas, H. and Joel Sartore. *The Company We Keep: America's Endangered Species*. Washington, DC: National Geographic Society, 1995.

Endangered Species and Wetlands Report: www.eswr.com

U.S. Fish and Wildlife Service ("What You Need to Know . . . about the Endangered Species Act") Web site: www.fws.gov/r9dia/public/esa.pdf.

U.S. Fish & Wildlife Service (Endangered Species) Web site: www.fws.gov/r9endspp/endspp.html.

U.S. Fish and Wildlife Service, Endangered Species Habitat Conservation Planning Web site: endangeredspecies.fws.gov/hcp/hcp.html.

endangered species list, A survey developed and maintained by the U.S. FISH AND WILDLIFE SERVICE (FWS) of PLANTS and animals that are deemed to be either ENDANGERED SPECIES or THREATENED SPECIES. For purposes of listing, the FWS defines endangered species as those at immediate risk of EXTINCTION, which will not likely survive without direct human intervention. Threatened species are those that are abundant in parts of their range, but are declining in total numbers and are thus at risk of extinction in the foreseeable future. The development of this list was initially mandated by the U.S. Congress through its passage of the Endangered Species Conservation Act in 1969. In accordance with provisions of this act, the FWS (through the secretary of the interior) became responsible for maintaining a list of endangered and threatened U.S. species as well as a parallel list of species threatened with worldwide extinction. Although the lists were developed, the effort was widely criticized as being ineffective since listed species generally were limited to VERTEBRATE animals and no major efforts were being made to prevent the animals from extinction. As a result of these criticisms and data obtained from the 1973 CONVENTION ON INTERNATIONAL TRADE IN ENDANGERED SPECIES OF WILD FLORA AND FAUNA, the U.S. Congress in 1973, replaced the Endangered Species Conservation Act with the ENDANGERED SPECIES ACT. This act expanded the goals of the earlier act by requiring inclusion of plants as well as INVERTEBRATES on the endangered species list and also by providing additional resources to the FWS for protection of the HABITATS of those species appearing on the list.

As of March 31, 1999, the FWS included 357 animal species and 568 plant species from the United States on its list of endangered species. The animal species included 60 MAMMALS, 75 birds, 14 REPTILES, 9 AMPHIBIANS, 70 fishes, 61 clams, 18 snails, 28 INSECTS, 5 arachnids, and 17 CRUSTACEANS. U.S. plant species listed as endangered include 540 species of flowering plants, 2 conifers, and 26 species of ferns or other plants. In addition, 121 animal species and 135 plant species are identified as threatened species. The number of foreign species included on the endangered species list includes 520 endangered

Number of Animal Species Included on the Endangered Species List

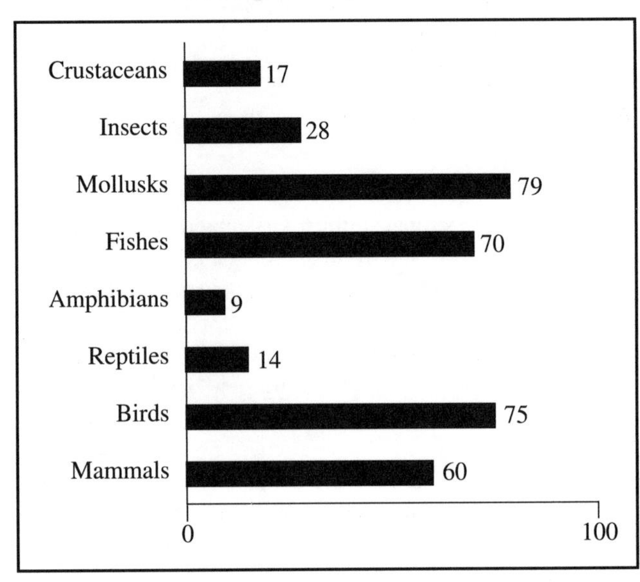

Number of Plant Species Included on the Endangered Species List

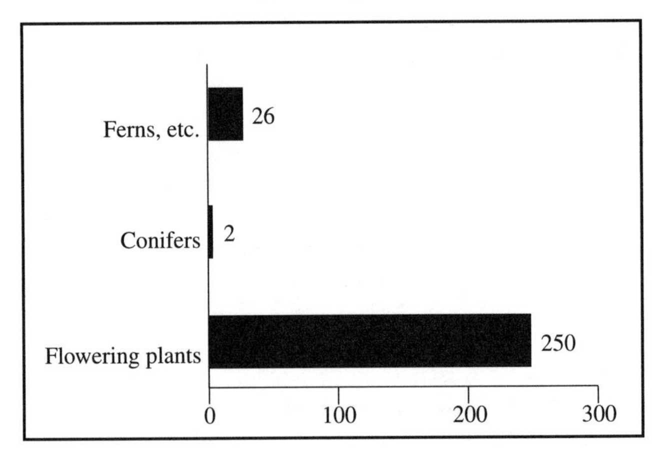

animal species, 37 threatened animal species, 1 endangered plant species, and 2 threatened plant species. A more complete survey of endangered and threatened plant and animal species worldwide is maintained by the WORLD CONSERVATION UNION in its RED LIST OF ENDANGERED SPECIES. *See also* ALLIGATORS; BALD EAGLES; BATS; BIODIVERSITY; BIODIVERSITY TREATY; BIOSPHERE RESERVE; BISON; CACTI; CALIFORNIA CONDORS; CAPTIVE PROPAGATION; CHEETAHS; CONSERVATION; CROCODILES; DODO BIRDS; DOLPHINS AND PORPOISES; ECOSYSTEM; ELEPHANTS; FLORIDA PANTHERS; GORILLAS; GREENPEACE; INTERNATIONAL BIOLOGICAL PROGRAM; INTERNATIONAL CONVENTION FOR THE REGULATION OF WHALING; INTERNATIONAL WHALING COMMISSION; LEOPARDS; MANATEES; MARINE MAMMAL PROTECTION ACT; MIGRATORY BIRD TREATY ACT; ORANGUTANS; ORYX; OWLS, NORTHERN SPOTTED; PARROTS; PASSENGER PIGEONS; PEREGRINE FALCONS; POACHING; RHINOCEROSES; SALMON; SEA TURTLES; SEALS AND SEA LIONS; TIGERS; TRADE RECORDS ANALYSIS OF FLORA AND FAUNA IN COMMERCE; TUNA; VICUNA; WHALES, WILD BIRD CONSERVATION ACT; WILDLIFE REFUGE; WOLVES; WORLD CONSERVATION MONITORING CENTRE; WORLD WILDLIFE FUND; ZEBRAS.

EnviroSources

Chadwick, Douglas H., and Joel Sartore. *The Company We Keep..America's Endangered Species*. Washington, DC: National Geographic Society, 1998.

U.S. Fish and Wildlife Service Web site: http://www.fws.gov.

endemic species. *See* native species.

energy, The capacity for doing work or a force that produces an activity. Energy exists in many forms and can be converted from one form to another. The different forms of energy include chemical, nuclear, electric, thermal, and mechanical. Humans use a variety of energy sources to do work including chemical energy (FOSSIL FUELS and wood), GEOTHERMAL ENERGY, electricity, TIDAL POWER, HYDROELECTRIC POWER, NUCLEAR ENERGY, SOLAR ENERGY, WIND POWER, and FUSION.

The energy a substance or object possesses due to its motion is called kinetic energy. A moving train possesses kinetic energy. Kinetic energy is expressed as $\frac{MV^2}{2}$ —a function of velocity (V) and mass (M).

The energy a substance or object possesses due to its place or location is referred to as potential energy. Water in a reservoir has potential energy. Potential energy can be expressed as the product of the mass (weight) and the height to which the substance is raised; it is expressed as PE= mh. Potential energy is also stored in chemical compounds in fossil fuels, (COAL, PETROLEUM, NATURAL GAS) FUEL WOOD, and BIOMASS.

Energy use is measured in a variety of ways, including British thermal units (Btus), kilowatts (kW), and quads:

- Btu—This is a common measurement for the amount of heat generated or consumed from various types of energy. The Btu is the quantity of heat required to raise the temperature of one pound of water 1°F. One Btu is the energy equivalent of one burning match tip. One Btu equals 252 calories or 1055 joules.

Global Energy Usage per Year

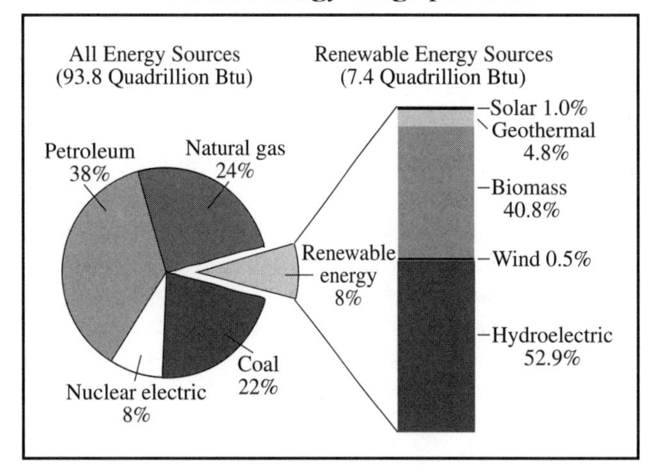

Source: United States Department of Energy.

- Each 3.78 liters (1 gallon) of petroleum produces about 12,500 Btus. Propane has a heating value of 2,500 Btus per cubic foot. One cubic foot of natural gas produces between 900 to 1,200 Btus of energy. About 100 cubic feet of natural gas would contain 100,000 Btus and is referred to as a "therm."
- kilowatt (kW)—This is a common unit in which electricity is measured. It is equal to 1,000 watts or 1,000 joules per second. A kilowatt hour (kWh) is a unit of electrical energy equal to 1,000 watts per hour. One kilowatt hour is equal to 3,413 Btus. Utility companies bill their customers in cents per kilowatt hour. The average home in the United States uses about 9,400 kWh of electricity a year.
- quad— A unit of measurement equal to one quadrillion (1,000,000,000,000,000 or 1,000 trillion) Btus. Quads are used by scientists to measure large quantities of energy. The United States averages about 94 quads of energy per year.

EnviroSource

U.S. Geological Survey (Energy Resources) Web site: energy.usgs.gov

energy pyramid. *See* ecological pyramid.

environment, Everything comprising the surroundings of an ORGANISM. The environment includes all of the external conditions present in the place where an organism lives, including the other organisms, or BIOTIC factors, and the nonliving things, or ABIOTIC factors, with which that organism interacts. The environment is important because all organisms derive the substances and materials they need to sustain life from their environment. For example, most organisms require OXYGEN to carry out respiration—an essential life process that permits the organism to convert matter in food into the ENERGY needed to carry out a variety of processes. For terrestrial organisms, this oxygen is derived from the AIR in the ATMOSPHERE; aquatic organisms may obtain oxygen from air or water. The environment also provides organisms with the food, living space, water, and conditions such as appropriate temperature required for its survival. *See also* ADAPTATION; BIODIVERSITY; COMPETITION; DISSOLVED OXYGEN; ECOSYSTEM; FOOD CHAIN; FOOD WEB; HABITAT; NATURAL RESOURCES; NICHE.

Environmental Education Act, Federal legislation enacted by Congress in 1990 that makes grant money available for environmental education and training programs. Competition for grant money, which has a cap of $250,000 per applicant, is open to state education or environmental agencies, colleges, universities, tribal or local education agencies, nonprofit organizations, and noncommercial educational broadcasting entities. Congress provided authority for determining grant eligibility to the U.S. ENVIRONMENTAL PROTECTION AGENCY. Grant requests for $25,000 or less are determined by regional EPA offices. Those between $25,001 and the maximum $250,000 are decided by the EPAs national headquarters in Washington, DC. *See also* EARTHWATCH.

EnviroSource

U.S. Environmental Protection Agency (Environmental Education Grants) Web site: www.enviroed.grants.html.

environmental impact statement, A written evaluation of how a proposed project is likely to affect the ENVIRONMENT of an area, including its WILDLIFE, water quality, and AIR quality. An environmental impact statement (EIS) is required for any federal project that has the potential to change the quality of the environment of a region. The legislation requiring EIS preparation resulted from the passage of the NATIONAL ENVIRONMENTAL POLICY ACT in 1969, and its amended version in 1979.

Guidelines regarding what information an EIS must include were established by the COUNCIL ON ENVIRONMENTAL QUALITY. The guidelines were also updated in 1979. According to the 1969 and revised 1979 guidelines, a current EIS must include a summary of how the proposed project will affect the environment, a statement of purpose and need for the project, a comparison of alternate means of carrying out the project, a description of the environment that will be impacted by the project, and a prediction of the likely effects the project will have on the environment along with suggestions for alterations to the plan that might lessen adverse affects to the environment. In addition, the 1979 guidelines also require every EIS to include scoping and a record of decision. Scoping is the early identification of likely environmental issues that may arise from a project being carried out. *See also* BUREAU OF RECLAMATION; CONSERVATION; TENNESSEE VALLEY AUTHORITY.

Environmental Justice, The ENVIRONMENTAL PROTECTION AGENCY (EPA) defines environmental justice as the fair treatment for people of all races, cultures, and incomes, regarding the development of environmental laws, regulations, and policies.

Many believe that the environmental justice movement began in the early 1980s in a mostly African American community in Warren County, North Carolina. The county was planning to build a landfill for POLYCHLORINATED BIPHENYL (PCB) wastes. Many groups thought the selection of the site was racially-motivated and posed an environmental health problem for the residents. As a result, there were widespread protests including marches. More than 500 demonstrators were arrested. Although the protestors were unsuccessful in blocking the PCB landfill, they brought national attention to waste facility siting inequities and galvanized African American church and civil rights leaders' support for environmental justice.

Since that time hundreds of environmental justice groups have been established by people of color to center on government and industry practices that create enviromental inequities. The environmental justice movement is comprised of small, democratically run grassroots groups; many led by women of color activists who got involved because of ENVIRONMENTAL RACISM and environmental threats to family, home, and community. The goals of the organizations range from pollution prevention to SUSTAINABLE AGRICULTURE, from land rights and sovereignty to eliminating PESTICIDE poisoning of migrant farm workers.

According to the 2000 directory, *People of Color Environmental Groups*, edited by Robert Bullard there are over 300 people of color environmental groups in the United States, Puerto Rico, Mexico, and Canada; and the list continues to expand.

Concerned that minority POPULATIONS and/or low-income populations bear a disproportionate amount of adverse health and environmental effects, President William J. Clinton issued an executive order in 1994, focusing federal agency attention on these issues. The EPA responded by developing the Environmental Justice Strategy, which focuses on the Agency's efforts in addressing these concerns. Arkansas and Louisiana were the first two states to pass environmental justice laws.

EnviroSources

Angel, Brady. *The Toxic Threat to Indian Lands: A Greenpeace Report.* San Francisco, CA: Greenpeace, 1992.

Bullard, Robert D., ed. *Dumping in Dixie: Race, Class and Environmental Quality.* 2nd edition. Boulder, CO: Westview Press, 1994.

———, ed. *People of Color Environmental Groups.* Flint, MI: Charles Steward Mott Foundation, 1994, 1995, 2000.

———, ed. *Unequal Protection: Environmental Justice and Communities of Color.* San Francisco, CA: Sierra Club Books, 1994.

Environmental Protection Agency (Environmental Justice Hotline), 401 M Street, SW, Mail Code 3103, Washington, DC 20460; e-mail: *environmental-justice-epa@epamail.epa.gov*

Environmental Justice Resource Center, Clark University, Atlanta, GA; Web site: www.ejrc.cau.edu/

environmental medicine, The practice of medicine in which the major focus is identifying and treating illnesses or DISEASES that result from environmental causes. Environmental causes of disease may include such things as exposure of individuals to PATHOGENS or toxic substances present in the

AIR, water, soil, home, or workplace. ALLERGIES are a group of disorders of concern to environmental medicine. Common causes of allergic responses are exposure to pollen and dust, chemicals used in household cleaning products and DETERGENTS, fragrances, SOLVENTS, paints, PESTICIDES, FERTILIZERS, contact with certain PLANTS, and INSECT bites. Symptoms of allergic reactions may range from skin rashes, to eye, nose, or throat irritation, to shock, or even death.

Several major diseases, including several types of CANCER, also have environmental causes. Lung cancer is one of the more common cancers attributed to environmental causes. Among the leading causes of this disease are exposure to tobacco smoke, which contains many chemicals that cause cancer. *(See* carcinogen.) This risk does not affect only the smoker, but also anyone who comes into contact with the smoke. Exposure to RADON gas, a radioactive substance that seeps into the basements of homes overlying certain types of rock also is a cause of cancer. *(See* radiation.) This odorless, colorless gas can collect in enclosed spaces and is easily taken into the lungs with air. Breathing in the gas for long periods of time often results in development of lung cancer. Skin cancer also is attributed to environmental factors. In this case, the major factor is overexposure of the skin to the ultraviolet-B (UV-B) rays of the sun. These rays also have been identified as a cause of premature wrinkling of the skin and damage to the corneas of the eyes.

Many other diseases and disorders are attributed to environmental causes. For example, asbestosis occurs when airborne ASBESTOS fibers are breathed into the lungs. People most at risk for asbestosis include those who became exposed to the fibers through their work, such as construction workers, AUTOMOBILE mechanics, and firefighters. This serious disease led to a ban on asbestos use in the United States and several countries of Europe.

In recent years, indoor environments in homes, office buildings, and other structures have become a major source of illness. Two major problems with such environments are poorly cleaned or maintained ventilation systems, which can serve as a breeding ground for disease-causing BACTERIA and FUNGI, and repeated exposure to chemicals (derived from products such as furnishings, tobacco smoke, paints, perfumes, and other sources) that are cycled in the air in buildings. The problems that can arise from poorly maintained ventilation systems received international attention in the late 1970s, when numerous convention attendees at a hotel in Philadelphia developed a mysterious illness that was given the name "Legionnaire's disease." The cause of the disease was eventually traced to the building's ventilation system, which, left unclean, had provided a perfect breeding ground for bacteria.

More recently, as buildings have become more tightly sealed and fitted with windows incapable of opening to the outside, inadequate ventilation systems that continuously recycle the same air throughout the building have led to the development of a variety of problems that are cumulatively known as SICK BUILDING SYNDROME. Symptoms of this syndrome include such problems as headache, nausea, stomach upset, and increased occurrence of allergies and respiratory disorders such as asthma. A condition, sometimes called MULTIPLE CHEMICAL SENSITIVITY (MCS), also has become more common, although not all doctors believe the illness actually exists. MCS, which is also referred to as "chemical hypersensitivity" or "environmental illness," is a disorder in which a person becomes intolerant of or sensitive to a number of chemicals and other irritants, all at very low concentrations. Such substances may be present in either the indoor or outdoor environment. Although not all physicians agree that MCS is a legitimate disorder, the federal government has recognized MCS as a true illness and provides assistance to those suffering with the disorder through the Americans with Disabilities Act.

Contaminated food and water are two other major sources of environmental illnesses. Among the most common CONTAMINANTS in food are pathogens such as *SALMONELLA* bacteria, listeria bacteria, *ESCHERICHIA COLI* bacteria, and some pesticide residues. Ingestion of foods contaminated by these substances can cause illnesses ranging from minor stomach irritation to death. Water also may become contaminated with chemicals and pathogens. Common chemical contaminants of water include chemicals present in pesticides and fertilizers that run off lawns, farms, and feedlots and chemicals such as those contained in PETROLEUM products, antifreeze, and road-deicing products that are carried into water by storm SEWERS. Other chemicals may enter the water supply as a result of the deliberate dumping into water or on land. For example, MINAMATA DISEASE is a type of MERCURY poisoning that was first observed in people living near Minamata Bay in Japan. This disease was found to be caused by eating fishes that had become contaminated with mercury that was discharged into the water from nearby factories. *(See* discharge.) Among the many pathogens that may be present in water are *GIARDIA LAMBLIA*, CRYPTOSPORIDIUM, and the bacteria that cause cholera and typhoid. These pathogens generally do not pose a problem to people who drink water from a public water supply that makes use of modern water treatment methods including AERATION and CHLORINATION; however, people who live in undeveloped nations and obtain their water directly from lakes, ponds, rivers, or streams are at increased risk of illness from such pathogens. *(See* potable water.) Regions of the world where such contamination is common include many remote areas of Africa and the Latin American countries, including Mexico. People living in these regions can often lessen their risk of disease by boiling the water to kill any pathogens that may be present prior to use. *See also* ACUTE TOXICITY; AEROSOL; AGENT ORANGE; AIR POLLUTION; ALAR; ALDRIN; ALUMINUM; ARSENIC; BENZENE; BERYLLIUM; BIOASSAY; BIOLOGICAL WEAPONS; BLIGHT; BOVINE SPONGIFORM ENCEPHALOPATHY; BRUCELLOSIS; CADMIUM; CARBON MONOXIDE; CARBON TETRACHLORIDE; CARCINOGEN; CHEMICAL WEAPON; CHLORDANE; CHLORINE; CHROMIUM; CHRONIC TOXICITY; COLIFORM BACTERIA; COPPER; DEFOLIANT; DICHLORODIPHENYLTRICHOROETHANE; DIOXIN; DISINFECTANT; DOSE/DOSAGE; ECOLABELING; FALLOUT; HAZARDOUS SUBSTANCE; HEAVY METAL; HEPTACHLOR; HYDROCARBON; INFECTIOUS WASTE; LEAD; LETHAL DOSE–50%; MANGANESE; METHYL MERCURY; MICROORGANISM;

MUTAGEN; NICOTINE; NITROGEN DIOXIDE; NITROGEN OXIDES; NUCLEAR WEAPON; OCCUPATIONAL SAFETY AND HEALTH ADMINISTRATION; ORGANOPHOSPHATES; OXIDATION; OZONE; PARASITISM; PARTICULATE MATTER; PHOSPHATE; PLUTONIUM-239; POLLUTANT; POLYCHLORINATED BIPHENYLS; RADIATION SICKNESS; SELENIUM; SEWAGE; SMOG; STERILIZATION, PATHOGENIC; STRONTIUM-90; SULFATE; SULFUR DIOXIDE; SULFURIC ACID; TEMPERATURE INVERSION; TETRACHLOROETHYLENE; TOLERANCE LEVEL; TOLUENE; TOXIC CHEMICAL; TOXICITY; TOXICOLOGY; TRICHLOROETHANE OR 1,1,1-TRICHLORO-ETHANE; VINYL CHLORIDE; VIRUS; VOLATILE ORGANIC COMPOUND; WARNING LABEL; X-RAYS; XYLENE.

EnviroLink

ecopsychology: A form of therapy that deals with diagnosing and treating depression and other emotional conditions related to environmental pollution.

EnviroSources

Moeller, Dade W. *Environmental Health*. Cambridge: Harvard University Press, 1997.

Pope, Andrew M. *Environmental Medicine: Integrating a Missing Element into Medical Education*. Washington, DC: National Academy Press, 1995.

Vos, J.G. *Allergic Hypersensitivities Induced by Chemicals: Recommendations for Prevention*. Cleveland, OH: CRC Press, 1996.

Environmental Protection Agency, A cabinet-level federal agency created in 1970 when Congress integrated 15 federal programs from 5 executive departments and agencies into a single regulatory body. The Environmental Protection Agency (EPA) was established in response to growing public concern about unhealthy AIR, polluted rivers and GROUNDWATER, unsafe drinking water, ENDANGERED SPECIES, and HAZARDOUS WASTE disposal.

The agency's mission is to "protect public health and to safeguard the natural ENVIRONMENT—air, water, and land—upon which human life depends." Its areas of responsibility include the control of AIR POLLUTION and water pollution, SOLID WASTE management, protection of the drinking water supply, and PESTICIDE regulation. The EPA is divided into ten regions, with field offices for research, investigations, and administration.

The EPA's purpose is to ensure that federal environmental laws are implemented and enforced fairly and effectively; that environmental protection is a part of U.S. policies, including those related to economic growth; that efforts to reduce environmental risk are based on the best available science; that all parts of society have full access to environmental information; that environmental protection contributes to making our communities and ECOSYSTEMS diverse, sustainable, and economically productive; and that the United States plays a leadership role in protecting the global environment.

The EPA coordinates research, monitoring, standard setting, and enforcement with state and local governments, private and public organizations, and educational institutions. The agency also serves as a focal point for all federal agencies whose operations affect the environment.

Congress directed the EPA to implement and enforce an ambitious set of federal environmental laws, including the CLEAN AIR ACT, the CLEAN WATER ACT, the Pollution Prevention Act, the RESOURCE CONSERVATION AND RECOVERY ACT, and the COMPREHENSIVE ENVIRONMENTAL RESPONSE, COMPENSATION, AND LIABILITY ACT (also known as SUPERFUND). The EPA is also beginning to look more closely at new, nonregulatory mechanisms for protecting the environment; such mechanisms build on regulatory requirements but go beyond them by encouraging voluntary actions as well.

The EPA's implementation of environmental statutes has contributed significantly to improvements in environmental quality in the United States. In virtually every American city, the air is cleaner than it was 25 years ago. Since 1990 the number of metropolitan areas that do not meet air quality standards has decreased by more than 50%, from 199 to fewer than 70, and an additional 50 million Americans now live in areas where SMOG no longer exceeds health-based standards. More than 450 million kilograms (992 million pounds) of toxic pollution have been prevented from entering our nation's rivers and streams each year because of WASTEWATER standards put into place over the last 25 years. Today, about 60% of the nation's surveyed inland water bodies are clean enough to meet basic needs for fishing and swimming. From the mid-1970s to 1992, the volume of OIL SPILLS in U.S. waters has decreased by 86%, from 58 million to 8 million liters (15.3 million to 2.1 million gallons) a year. OCEAN DUMPING of SEWAGE SLUDGE, industrial waste, plastic debris, and medical waste has been banned. Hundreds of hazardous waste sites are being cleaned up, and the use of several especially hazardous chemicals has been restricted or banned entirely. Since the early 1970s, the EPA has banned or eliminated the use of over 230 pesticides and 20,000 pesticide products, including DICHLORODIPHENYLTRICHOROETHANE (DDT).

Resources protected by EPA statutes are of critical importance to threatened and endangered species. For example, 85% of all such species utilize WETLANDS and aquatic HABITATS. A preliminary analysis by the Environmental Defense Fund and the WILDERNESS SOCIETY indicates that 52% of the 920 listed and proposed species examined are at least somewhat affected by POLLUTANTS regulated by the EPA or EPA-approved environmental programs.

The agency is responsible for ENVIRONMENTAL JUSTICE in addition to environmental safety. Environmental justice is the fair and equitable application of environmental laws, regulations, and programs in every community. It seeks to address and correct the disproportionate burdens that some environmental hazards impose on economically disadvantaged and minority Americans.

Citizens also have a legal right to know about toxic pollution in their communities. The EPA has expanded both the number of companies that must report about pollution and the number of TOXIC CHEMICALS on which they must report. The agency's TOXIC RELEASE INVENTORY is one of many EPA resources that provides information to the public about pollutants and industrial facilities in their communities.

In the late 1980s, the EPA expanded its mission to include the problems of GLOBAL WARMING and environmental change. The agency created a CLIMATE change division to research the impacts of increased CARBON DIOXIDE and other gases in the ATMOSPHERE. In 1990, the EPA also developed a grant program to improve environmental education nationwide. *See also* CHRONIC TOXICITY; EMERGENCY PLANNING AND COMMUNITY RIGHT-TO-KNOW ACT; INTERNATIONAL REGISTER OF POTENTIALLY TOXIC CHEMICALS; NATIONAL AMBIENT AIR QUALITY STANDARDS; NATIONAL EMISSIONS STANDARDS FOR HAZARDOUS AIR POLLUTANTS; NATIONAL POLLUTANT DISCHARGE ELIMINATION SYSTEM; NOISE CONTROL ACT; NOISE POLLUTION, NOT IN MY BACKYARD; RADIOACTIVE WASTE (NUCLEAR WASTE); RECYCLING; SAFE DRINKING WATER ACT; TOXIC WASTE.

EnviroSources

U.S. Environmental Protection Agency Web site: www.epa.gov.
U.S. EPA, Envirofacts Warehouse Web site: www.epa.gov/enviro.

environmental racism, The inequitable practice of locating a high percentage of HAZARDOUS WASTE sites, incinerators, LEAD smelter operations, paper mills, chemical plants, and other polluting industries in residential areas of minority communities. In 1987 a study conducted by the United Church of Christ Commission for Racial Justice under the leadership of Reverend Benjamin F. Chavis, Jr., he used the term "environmental racism" to describe the unfair distribution of dumps and incinerators in minority neighborhoods. The organization defined environmental racism as racial discrimination in environmental policy-making, and the enforcement of regulations and laws that represent "the deliberate targeting of communities of color for TOXIC WASTE facilities, the official sanctioning of the life-threatening presence of poisons and pollutants in those communities, and the history of excluding people of color from leadership of the environmental movement."

Reverend Benjamin F. Chavis, Jr. was one of many who participated in a protest march against a proposed POLYCHLORINATED BIPHENYL (PCBs) landfill in Warren County, North Carolina; a county of mostly African Americans. Many believed that march was the impetus for the ENVIRONMENTAL JUSTICE movement in the United States. *See also* ECOFEMINISM; NOT IN MY BACKYARD.

EnviroSources

Angel, Brady. *The Toxic Threat to Indian Lands: A Greenpeace Report*. San Francisco, CA: Greenpeace, 1992.
Bullard, Robert D. *Confronting Environmental Racism. Voices from the Grassroots*. Boston: South End Press, 1993.
———, ed. *Dumping in Dixie: Race, Class and Environmental Quality*. 2nd edition. Boulder, CO: Westview Press, 1994.
———, ed. *People of Color Environmental Groups*. Flint, MI: Charles Steward Mott Foundation: 1994, 1995, 2000.
———. *Unequal Protection: Environmental Justice and Communities of Color.* San Francisco: Sierra Club Books, 1994.
United Church of Christ, Commission for Racial Justice, 700 Prospect Avenue, Cleveland, OH, 44115. (216) 736-2168.

environmental science, The study of nature and natural processes. Environmental science is a broad area of science that builds upon information from virtually every other area of science. For example, ECOLOGY—the study of how ORGANISMS interact with their ENVIRONMENTS—is one branch of environmental science. Because ecology deals with the study of organisms, an understanding of biology is essential to understanding ecology. Environmental science also is concerned with natural processes such as the cycling of water, OXYGEN, CARBON DIOXIDE, and NITROGEN through the environment. Understanding these processes requires a knowledge of both chemistry—the branch of science that studies matter and its properties—and physics—the branch of science that studies the interactions between matter and ENERGY.

In addition to studying nature and natural processes, environmental science also is concerned with how humans interact with nature. Such an understanding is important because humans, like other organisms, rely on products of nature, such as oxygen, clean water, and food (PLANTS and animals) for their survival. At the same time, human activities, such as clearing land for housing or burning COAL or wood for energy, often alter the environment in ways that affect other organisms. One goal of environmental science is to understand these interactions to ensure that the health of Earth is maintained in a manner that ensures the survival of human POPULATIONS and those of other organisms. *See also* AGRICULTURE; AGROCLIMATOLOGY; BIOTECHNOLOGY; BOTANY; GENETIC ENGINEERING; HYDROLOGY; METEOROLOGY; TAXONOMY; ZOOLOGY.

environmentally sensitive area, A term used in Great Britain to identify regions of the country that are deemed worthy of protection from damage through agricultural practices. Environmentally sensitive areas (ESAs) were created in 1987 through legislation enacted by the European Commission (EC). To create ESAs, farmers are asked to conserve WILDLIFE HABITAT and maintain the natural features or topography of the landscape by setting aside and not using some lands for AGRICULTURE. Program participation is voluntary; however, once an individual agrees to participate in the program they must do so for a five-year period. In return, participants receive funds from the Ministry of Agriculture as reimbursement for income lost by not cultivating the land. Through this program, about 1 million hectares (2.5 million acres) of land have been set aside for CONSERVATION of wildlife or for its natural beauty. *See also* AGRICULTURAL POLLUTION; AGRICULTURE; EROSION; PESTICIDES.

epiphyte, A PLANT that grows upon and uses another plant for support, without causing harm to its host. The relationship between epiphytes and their hosts is commensal, since the epiphyte benefits while the host neither benefits nor is harmed. *(See* commensalism.) Because epiphytes are not rooted in the ground, they do not obtain their water and nutrients from soil, as do most plants. Instead, epiphytes usually have small, hairlike structures on their leaves and stems through which they absorb needed materials from rainwater or the ATMOSPHERE. The ability of epiphytes to obtain nutrients from the atmosphere has led them to be referred to as AIR plants.

Most epiphytes live in ENVIRONMENTS with warm, wet CLIMATES, such as those in tropical RAINFORESTS. Epiphytes often live high in the FOREST canopy where they are exposed to sunlight, a resource that is scarce at the forest floor, and which is needed for PHOTOSYNTHESIS. Life in the canopy also increases the exposure of epiphytes to rainfall and air.

The vanilla plant, many species of mosses, and many tropical orchids are epiphytes. An epiphyte common to the southeastern United States is Spanish moss. This plant, which is not a true moss, often lives on the branches of oak or cypress trees, or on utility poles and wires. *See also* MUTUALISM; PARASITISM; SYMBIOSIS.

erosion, A natural process in which particles of rock, soil, and sediment produced through WEATHERING, the physical or chemical breaking apart of natural substances, are removed or carried away from an area. Erosion greatly affects the ENVIRONMENT by altering the appearance and physical characteristics of Earth's surface and is generally considered a destructive force. Agents of erosion include wind, moving water (such as RUNOFF, streams, waves, or currents), GLACIERS, and gravity. Together, these agents transport huge amounts of matter from one place on Earth's surface to another. This matter is then dropped in a new location in a process called DEPOSITION.

Wind and moving water are the most common agents of erosion. Together these agents carry away billions of hectares of topsoil from areas around the globe each year. This topsoil carries with it nutrients, such as NITROGEN, PHOSPHORUS, and potassium, that are essential to the growth of PLANTS, and thus erosion has the potential to greatly impact agricultural production. Currently, it is estimated that the United States loses about 4 billion metric tons (4.4 billion tons) of topsoil per year through the erosive activities of wind and moving water. Similar amounts of topsoil are lost from central Belgium, the Gondor region of Ethiopia, the Chao River basin region of Thailand, and the Orinoco River Basin regions of Venezuela and Columbia. Erosion rates in China, the most heavily populated nation in the world, are about 2.5 times greater than those in the United States, providing this country even greater challenges in meeting the food needs of its peoples.

The United States experienced its greatest losses of topsoil from erosion in the late 1930s and early 1940s, during the event that has come to be called the DUST BOWL. The devastation that resulted from this disaster awakened the need for understanding the factors that can lead to increased erosion and led to SOIL CONSERVATION practices that would prevent such losses in the United States and in many other countries throughout the world. Such practices include no-till or low-till farming, planting trees around the edges of a field to slow the wind and lessen its ability to carry away topsoil, and maintaining some ground cover throughout the year to help hold soil in place. Methods such as CONTOUR FARMING, TERRACING, and STRIP CROPPING (alternating strips of plants with plowed fields) also are being implemented throughout the world to help slow wind and water as it moves over land, thus decreasing its ability to carry away topsoil.

Moving water does not only affect topsoil. Each year, billions of metric tons of sand is carried away from beaches by the erosive action of waves. This sand may be deposited farther out in the OCEAN or carried ashore in other areas where it builds new beaches. Such action becomes most intense during severe storms such as CYCLONES and HURRICANES. Beach erosion can greatly impact ORGANISMS that live along the shoreline by destroying their HABITATS. It also affects people who use beaches as recreational areas or who have built homes or other properties very near the shoreline. If enough sand is carried away from populated beaches, homes risk being washed into the sea, resulting in major losses of property and possibly lives.

Glaciers are large, moving masses of snow and ice. As they move, glaciers often pick up rock, soil, and other sediment that becomes imbedded in the frozen mass. These materials can be transported great distances and wear away surfaces they contact through abrasive action. When the glacier begins to melt, the rock and sediment it carries is deposited in a new location, often far from its origin.

Gravity is a major agent of erosion and is responsible for great loss of both human property and lives each year. The erosive action of gravity also can destroy the habitats of numerous organisms in a very brief period of time. Two quickly occurring and potentially destructive examples of the erosive effects of gravity are mudslides and avalanches. Mudslides occur when gravity pulls huge masses of mud that have been formed by heavy rains downhill. These moving masses of mud can be extremely large and can destroy everything in their path. In February 1998, a mudslide in Sonoma County, California required the evacuation of people from 140 homes. Residents were displaced for almost six weeks as mudslides continuously threatened their homes. At the end of the six-week period, 30 homes had been severely damaged enough or left in such potentially dangerous condition that residents were forbidden to return. In May of the same year, mudslides in Kelso, Oregon, threatened to destroy as many as 55 homes in the region. At least five of these homes were completely destroyed. That same month, a mudslide in Naples, Italy, not only destroyed homes but also claimed the lives of at least 135 people. Hundreds of others who were buried in the mud were rescued and treated for injuries ranging in severity.

Avalanches also are caused by gravity. Avalanches involving mass movements of rock often occur in mountainous areas through which a road has been cut. Barriers may be placed near the bottoms of the slopes to capture any loose rocks that slide down the mountain. Avalanches also occur in mountainous regions that receive heavy snowfall. This snow may become suddenly loosened and move down the mountain in a heavy, destructive mass that buries everything in its path. In many regions of the world where avalanches are a continual threat, people now intentionally cause avalanches before snow gets too deep at higher elevations to prevent more severe avalanches from occurring later. This practice is especially common in regions of Switzerland and Italy that host a large tourism industry based on skiing. *See also* DESERTIFICATION; DROUGHT; NATURAL DISASTER; ROCK CYCLE; SALINIZATION; SEDIMENTATION; SILTATION.

EnviroSources

Bernstein, Leonard, Alan Winkler, and Linda Zierdt-Warshaw. *Environmental Science: Ecology and Human Impact.* Menlo Park, CA: Addison-Wesley, 1995.
Geobrugg Web site: www.geobrugg.com/mudslide.html.
Geoindicators Web site: www.gcrio.org/geo/soil.html.

Escherichia coli, A BACTERIUM within the family Enterobacteriaceae that typically represents about 0.1% of the total MICROORGANISMS within the intestines of western adult humans. The gastrointestinal tracts of many animals contain *Escherichia coli* (*E. coli*) within a few hours or days after birth after it is ingested in food or water or transferred from other individuals. *E. coli* was named for Theodor Escherich, who first isolated and characterized the bacterium in 1885.

E. coli and other helpful bacteria are needed by human bodies to develop and function properly; they provide us with several necessary vitamins, for example. These bacteria, however, are helpful only if located in regions of the body that are directly exposed to the outside ENVIRONMENT, such as the intestines. They should not be present within the bloodstream or body tissues. *E. coli* are everywhere in the environment, making contact difficult to avoid; any time humans eat, drink, or touch something that has been a part of, or in contact with animals, there exists the potential to ingest these bacteria.

The presence of the bacteria in water is an indicator of fecal contamination. *(See* coliform bacteria; contaminant.) Although, most strains of *E. coli* are harmless, several are known to produce toxins that can cause urinary tract infections, neonatal meningitis, and intestinal diseases such as gastroenteritis. One potentially deadly *E. coli* strain—O157:H7—can cause severe diarrhea, hemorrhaging, kidney damage, and even death. Its toxin is so potent that the total number of bacteria required for infection may be as low as 10. Symptoms of infection may appear within hours to several days after ingestion.

Foods that have caused well-publicized outbreaks of *E. coli* O157:H7 include undercooked hamburger (possibly contaminated during cattle slaughter or during meat processing), unpasteurized apple juice (contaminated by *E. coli* on windfall apples), and alfalfa sprouts. A waterborne *E. coli* infection, caused by children's soiled diapers in a community swimming pool, also has been documented, as has an outbreak due to WELL water contaminated by livestock.

The Food Safety and Inspection Service of the U.S. DEPARTMENT OF AGRICULTURE regulates and is responsible for the safety of food products in the United States; however, the agency currently can seek only voluntary industry recalls of foods that are unsafe. *See also* BOVINE SPONGIFORM ENCEPHALOPATHY; POTABLE WATER; SEWAGE.

EnviroSources

Centers for Disease Control (CDC) Web site: www.cdc.gov.
National Food Safety Database: www.foodsafety.org
U.S. Department of Agriculture, Food Safety and Inspection Service Web site: www.fsis.usda.gov.
U.S. Food and Drug Administration, Center for Food Safety and Applied Nutrition Web site: vm.cfsan.fda.gov.

estivation, An ADAPTATION of some ORGANISMS that enables them to enter a state of dormancy or torpidity to survive periods of extreme heat or dryness. Estivation is similar to the state of HIBERNATION entered into by some endothermic animals during cold winter months. It is common is some snakes and rodents in summer, especially SPECIES that live in hot DESERT regions. Some terrestrial snails estivate during periods of DROUGHT, when PLANT food supplies are scarce. Estivation helps preserve Earth's BIODIVERSITY by preventing mass EXTINCTIONS of species during periods of severe, adverse CLIMATE conditions. *See also* MAMMALS; REPTILES.

estuarine system. *See* estuary.

estuary, An ECOSYSTEM formed in an area where a region of fresh water, such as a river, meets a body of salt water, such as the OCEAN. Because estuaries are exposed to ocean TIDES and currents, these ecosystems are continually changing. ORGANISMS that make their homes in these ecosystems must be able to adapt to such changes. In addition to changes in water level, salinity levels also vary throughout estuaries, with variations forming three somewhat distinct zones. Typically, water in the part of the estuary nearest the ocean has a salinity very near that of ocean water. Farther upstream, water is BRACK-

Threats to Estuarine Ecosystems

Pollutant	Source	Effects
Nutrients	Fertilizers; sewage	Algal blooms
Chlorinated hydrocarbons pesticides, DDT, PCBs	Agricultural runoff; industrial waste	Contaminated and diseased fish and shellfish
Petroleum hydrocarbons	Oil spills; industrial waste; urban runoff	Ecosystem destruction
Heavy metals (arsenic, cadmium copper, lead, zinc)	Industrial waste; mining	Diseased and contaminated fish
Particulate matter	Soil erosion; dying algae	Smothers shellfish beds; blocks light needed by marine plant life
Plastics	Ship dumping; household waste; litter	Strangles, mutilates wildlife

ISH—composed of a mixture of both fresh water and salt water. Most estuaries also have a tidal river zone that contains water having a salinity roughly equal to that of other freshwater ecosystems. Despite the frequently changing conditions of estuaries associated with changes in tide (water depth) and salinity, estuaries are among the most productive ecosystems in the world.

There are several types of estuaries, the characteristics of which are determined by their physical location, the CLIMATE of the area in which they are located, and the type of PLANTS that are common to the ecosystem. SALT MARSHES, for example, are intertidal communities whose main PRODUCERS are plants that are adapted to salt water and can survive being covered by water during periods of high tide. MANGROVE SWAMPS are among the main types of tropical estuaries. The dominant plant form of such estuaries are mangrove trees, a type of tropical evergreen that is known for its stiltlike prop roots. Mangroves typically grow in dense thickets along tidal shores and are adapted to soils that have a high salinity and are frequently covered with water. Such estuaries are common in the tropical regions of Asia as well as along the Gulf Coast of the United States. Such ecosystems are most often located in areas with temperate climates and contain mostly salt water. The most diverse estuarine ecosystems are the tropical estuaries that form tidal RAINFORESTS. Such estuaries are located along the Amazon River and, like the rainforests of the area, are comprised of very tall trees with dense canopies.

In addition to plants, ALGAE and PHYTOPLANKTON also are major producers in estuaries. These ecosystems also provide HABITAT to a large diversity of animals and MICROORGANISMS. As a rule, the more tropical the climate of the estuary, the more diverse its plant and animal life. Among the HERBIVORES that feed on the producers are a variety of ZOOPLANKTON, MAMMALS, INSECTS, and waterfowl. In addition, many fishes, live in estuary waters, while AMPHIBIANS and REPTILES make their homes along the coastline. Along the floor of an estuary are a variety of shellfishes, such as mussels, clams, and barnacles. These animals provide a source of food to the many fishes and CRUSTACEANS that also make their home in this ecosystem. *See also* FOOD CHAIN; FOOD WEB.

ethanol, The chemical name for ethyl alcohol, or grain alcohol, a substance made from renewable NATURAL RESOURCES that is used as a FUEL or a fuel additive. Ethanol is produced by distilling BIOMASS feedstocks such as sugar cane, beets, grain, and corn. Ethanol can also be produced from GARBAGE and agricultural wastes, such as animal dung.

Ethanol is often used as an additive in unleaded GASOLINE to boost its octane rating and make the fuel burn more efficiently. The property of ethanol that makes it suited to this purpose is that it is an oxygenate—a substance containing OXYGEN. Gasolines containing ethanol produce fewer CARBON MONOXIDE EMISSIONS than do nonoxygenated fuels. The U.S. ENVIRONMENTAL PROTECTION AGENCY (EPA) has determined that ethanol-blended fuels reduce carbon monoxide emissions by as much as 25–30%.

Ethanol is an ALTERNATIVE FUEL. Because it is produced from biomass, ethanol is known as a BIOFUEL. Benefits of ethanol over FOSSIL FUELS include the fact that ethanol is derived from RENEWABLE RESOURCES and the fact that, when burned, ethanol releases little SULFUR in the form of SULFUR DIOXIDE into the ATMOSPHERE. (Sulfur dioxide is a major contributor to the formation of ACID RAIN.) Ethanol is currently used as a primary fuel in Brazil. In the United States, about 8% of ethanol exists as a blend called GASOHOL—a fuel containing about 10% ethanol and 90% gasoline.

About 4% of the corn crop in the midwestern United States is currently used to produce ethanol. Experts believe that new manufacturing processes and technologies will lead to increased production and use of ethanol in the future. The result will be a high-quality fuel that can be made from a variety of plant feedstocks and will also be economically competitive with gasoline, reduce the need for the importation of PETROLEUM from foreign countries, and reduce the emission of harmful substances into the ENVIRONMENT. *See also* AIR POLLUTION; ALTERNATIVE ENERGY RESOURCE.

EnviroSource

U.S. Department of Energy, Energy Efficiency and Renewable Energy Clearinghouse, P.O. Box 3048, Merrifield, VA 22116; e-mail: energyinfo@delphi.com; Web site: www.doe.gov

ethyl alcohol. *See* ethanol.

euphotic zone. *See* oceans.

eutrophication, An increase in the concentration of nutrients such as NITROGEN (in the form of NITRATES) and PHOSPHORUS (in the form of PHOSPHATES) in an aquatic ECOSYSTEM. The presence of nutrients in an aquatic ecosystem is necessary for the survival of its ORGANISMS; however, too much of these substances can be harmful to the ecosystem. As the amount of nutrients in water increases, it promotes the growth of ALGAE and aquatic PLANTS. If nutrients are present in abundance, a POPULATION explosion of algae—called an ALGAL BLOOM—may occur. Once the growing algae and plants consume all the nutrients, they begin to die in great numbers. In turn, BACTERIA and FUNGI begin the process of DECOMPOSITION and release large amounts of ORGANIC matter into the ecosystem. At the same time, the DECOMPOSERS—also experiencing population growth—use up much of the DISSOLVED OXYGEN in the water, making the OXYGEN unavailable to other, larger organisms, such as fishes. Unable to meet their oxygen needs, these organisms also die.

Eutrophication is usually a slow, natural process that often leads to ECOLOGICAL SUCCESSION: In time, the process builds nutrient-rich soil on the ecosystem floor, making the water more shallow, while also providing HABITAT in which plants can begin to grow. The rate of eutrophication is increased by human activities, such as the deliberate dumping of GARBAGE into bodies of water and the use of FERTILIZERS on lawns and farms that can be washed into aquatic ecosystems in RUNOFF.

Many DETERGENTS that contained phosphates also once contributed greatly to eutrophication as they were carried into aquatic ecosystems by storm SEWERS or drains; however, many manufacturers have reformulated these products to remove the phosphates. *See also* BIOGEOCHEMICAL CYCLE; RED TIDE.

Everglades National Park, A national park established in 1947 on the southwestern tip of the Florida peninsula. The Everglades National Park covers 609,658 hectares (1,506,499 acres.) The Everglades National Park makes up about one-fifth of the larger Everglades ECOSYSTEM, one of the most unusual and interesting subtropical WETLAND areas of the world. The northern part of the Everglades consists of shallow water

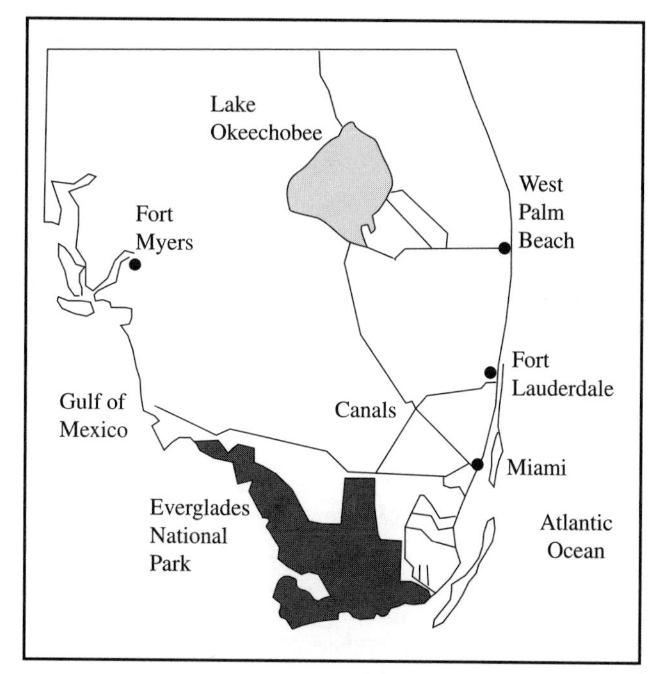

The diverse ecosystems of the Everglades provide habitat for abundant wildlife, including rare birds, manatees, turtles, snakes, alligators, Florida panther, and crocodiles.

and is dominated by sawgrass, a PLANT with sharp, jagged-edged leaves. To the west of this river of grass lies Big Cypress Swamp, a dark, dimly lit SWAMP filled with huge, twisted, moss-covered trees. Near the southern coast, SALT MARSHES and MANGROVE swamps exist. As the largest remaining subtropical WILDERNESS in the continental United States, the diverse ecosystems of the Everglades provide HABITAT for abundant WILDLIFE, including rare and colorful birds, MANATEES, turtles, snakes, ALLIGATORS, FLORIDA PANTHERS, and CROCODILES.

The Everglades National Park was the first national park to be established to preserve purely biological, not geological, resources. Unfortunately, the Everglades National Park is today considered to be one of the most endangered national parks in the United States. Over the last 60 years, a number of environmental problems have developed in the Everglades, including a 90% decrease in the POPULATIONS of wading birds, toxic levels of MERCURY found at all levels of the FOOD CHAIN, and die-off of sea grass in Florida Bay. Most of these problems are caused by changes in the distribution of water throughout the Everglades: Nature no longer controls the flow of water into the Everglades, people do.

Much of the region today is criss-crossed by an elaborate system of dikes, levees, and canals to supply south Florida with fresh water and to control FLOODS. Other concerns include AIR POLLUTION, PESTICIDES, declining wildlife populations, and the proliferation of exotic plant and animal species. Today, an unprecedented Everglades ecosystem restoration effort is under way, involving government agencies at all levels—federal, state, local, and tribal—as well as civic leaders, environmental groups, and business professionals. *See also* DOUGLAS, MARJORY STONEMAN; ENDANGERED SPECIES; EXOTIC SPECIES; NATIONAL PARK SERVICE; NATIVE SPECIES.

Endangered Species in the Everglades National Park

- American crocodile (*Crocodylus acutus*)
- Green turtle (*Chelonia mydas*)
- Atlantic Ridley turtle (*Lepidochelys kempi*)
- Atlantic hawksbill turtle (*Eretmochelys imbricata*)
- Atlantic leatherback turtle (*Dermochelys coriacea*)
- Cape Sable seaside sparrow (*Ammodramus maritima mirabilis*)
- Snail (Everglades) kite (*Rostrhamus sociabilis plumbeus*)
- Wood stork (*Mycteria americana*)
- West Indian manatee (*Trichechus manatus*)
- Florida panther (*Felis concolor coryi*)
- Key Largo wood rat (*Neotoma floridana smalli*)
- Key Largo cotton mouse (*Peromyscus gossypinus allapaticola*)
- Red-cockaded woodpecker (*Picoides borealis*)
- Schaus swallowtail butterfly (*Papilio aristodemus ponceanus*)

EnviroSource

Everglades National Park, 40001 State Road 9336, Homestead, Florida 33034-6733; Web site: www.nps.gov/ever

evolution, The process by which the traits of POPULATIONS of ORGANISMS change over time. Evolution explains the great BIODIVERSITY that exists among Earth's organisms by providing the means by which new SPECIES arise from existing species. The theory of evolution accepted by most scientists was proposed by British NATURALIST CHARLES DARWIN in 1859. This theory provides not only an explanation for how new species arise, but also helps explain why organisms are adapted to life in particular ENVIRONMENTS. *(See* adaptation.)

Darwin's theory of evolution contains four main observations. The first is that all organisms can produce more offspring than can survive. For example, a single female frog may lay thousands of eggs at one time; however, only a small number of these will actually hatch and survive to develop into adult frogs. Many eggs may not be fertilized, while others that are fertilized will be eaten before they hatch; similarly, some tadpoles that do hatch may be eaten by other organisms before they develop into adults. *(See* predation.)

Darwin's second observation held that individuals of the same species had differences, or variations, in their traits. These variations include such characteristics as coloration, running speed, size, and strength, as well as behavioral differences and resistance to DISEASE. Although Darwin did not know it at the time, these differences in traits result from the genes an organism inherits from its parents. *(See* genetic diversity.)

The third observation made by Darwin is generally referred to as natural selection. This observation holds that some traits, or adaptations, of an organism provide the organism with a survival advantage in its environment. Darwin hypothesized that organisms having advantageous traits are more likely to survive and reproduce in their environments than are organisms lacking these traits. Thus, beneficial traits are naturally selected in a population.

Darwin's final observation stemmed from his concept of natural selection. This idea holds that traits which provide a species with advantages for survival in its environment will be passed to the offspring of the individuals having those traits. Thus, the offspring will have the same advantages for survival as do their parents. Through the process of evolution, these advantageous traits become more common in successive generations, resulting in more and more individuals with the advantageous traits. At the same time, traits not selected for in earlier generations will become less common in successive generations.

Scientists use Darwin's theory of evolution to explain why organisms are so well adapted to their environments and why so many different species exist on Earth. For example, when organisms are not well adapted to their environments, they must move to a different environment to which they are adapted or they will die. If the members of a population that lack certain traits move to new areas where they can survive, they will undergo the same process of evolution by natural selection as the population from which they originated. Over time, the traits of the new population may differ enough from those of their ancestral population that a new species emerges. *See also* ADAPTIVE RADIATION; CONVERGENT EVOLUTION; EXTINCTION; GALÁPAGOS ISLANDS; GENE POOL.

EnviroSources

Cain, A. J. *Animal Species and Their Evolution.* Princeton, NJ: Princeton University Press, 1993.

Darwin, Charles. *The Origin of the Species by Means of Natural Selection.* Reprint 1859. Oxford, UK: Oxford University Press, 1998.

Margulis, Lynn, and Karlene V. Schwartz. *Five Kingdoms: An Illustrated Guide to the Phyla of Life on Earth.* Reprint. New York: W.H. Freeman and Company, 1988.

Massa, Renato, Monica Carabella, and Lorenzo Fornasari. *The Deep Green Planet: From the Water to the Land.* Austin: Raintree Steck Vaughn, 1997.

exotic species, Any type of ORGANISM that is present in an ECOSYSTEM as a result of being introduced to that ecosystem through accidental or intentional means. Because they are not native to an area, exotic species sometimes are called alien species, foreign species, or introduced species. Such SPECIES are of concern to environmental scientists because they often disrupt the natural balance of the ecosystem into which they are introduced.

Exotic species often create problems in an ecosystem because they do not have natural predators in their new ENVIRONMENTS. *(See* predation.) This permits the population of the new species to grow, virtually unchecked. The growing population, in turn, may then outcompete NATIVE SPECIES for such essential resources as food, water, and living space. *(See* competition.) When this occurs, the native species often abandons its HABITAT or begins to die off, which can possibly lead to EXTINCTION. *(See* adaptation.)

Exotic species are sometimes intentionally introduced into an environment by humans. For example, the Asian walking catfish is a species of fish that is able to leave the water and actually walk across land for a brief period of time. Several years ago, people imported these unusual fishes into Florida to be kept as pets; however, many pet owners did not realize that the fish grows quite large. Some of the fish were released into the wild, where they have thrived, to the detriment of the native pond fishes they feed on. The problem has been worsened by the catfish's walking ability, which allows it to move from one pond to another when the food source in one pond has been depleted. In a relatively short time, the catfish has created such a problem that many native fish of Florida became threatened with extinction as a result. Another example of how a species introduced to an area by humans has devastated native populations can be seen in the Norfolk Islands of Australia. In the early 1900s, rabbits, which are HERBIVORES, were introduced to the Norfolk Islands by humans. The rabbits had an abundant food source and no natural predators on the islands, and quickly increased in number. Within a few years of their introduction, more than five PLANT species became extinct as a result of the rabbits feeding on them. By 1977, this number had grown to 13 species. To control the problem, people have begun hunting the rabbits in an effort to control their population growth.

Not all exotic species are introduced to an area intentionally; however, even accidental introductions can create problems. During the early colonization of the United States, many species of rats, mice, and other "pest" animals were carried in ships along with human immigrants. The Norway rat, which is a problem in many cities in the northeastern United States, is believed to have originated in Europe. Many weeds and wildflowers were transported into the country when their seeds became mixed with crop seeds imported by early settlers. Such plants compete with crop plants for living space and soil nutrients, and thus become a problem for farmers. In addition, the attempted removal of such plants from the environment has led to an increased use of PESTICIDES that may be potentially harmful to humans or organisms other than their targets. *See also* BIOLOGICAL CONTROL; DODO BIRDS; DUTCH ELM DISEASE; EMIGRATION; IMMIGRATION; INTEGRATED PEST MANAGEMENT.

extinction, The disappearance of a SPECIES as a result of a NATURAL DISASTER, CLIMATE change, overspecialization, unsuccessful COMPETITION for resources, or human-caused stresses such as HABITAT destruction, hunting, and the introduction of EXOTIC SPECIES. Extinction has occurred since the beginning of life on Earth more than four billion years ago; in fact, most of the ORGANISMS that have ever existed are now extinct. Today's increased concern about extinction is not over the fact that it occurs, but that human activities have greatly accelerated the species extinction rate far beyond the natural rate. Conservative estimates of the current rate of extinction indicate that at least one species per day suffers this fate. At a minimum, 100,000 species have died out within the past 100 years. Experts predict that if present trends continue, we are likely to lose half of all living species within the next century.

Of all the species that have lived on Earth since life first appeared, only about 1 in 1,000 is still living today; all others have become extinct, typically within about 10 million years of their first appearance. This extinction rate has had an important influence on the rate of EVOLUTION of life on Earth and on the world's BIODIVERSITY: The POPULATION and repopulation of an ecological NICHE, by species after species, allow for the testing of a wider range of survival strategies than the slower process whereby a species gradually adapts itself to its ENVIRONMENT. *(See* adaptation.) The species alive today also will presumably become extinct within the next 10 million years or so, and thus make way for successors.

Several mass extinctions—sudden disappearances of great numbers of species within short periods of time—are recorded in the fossil record. Such large-scale extinctions may have been caused by catastrophic agents, such as meteorite impacts, or terrestrial agents, such as massive volcanic eruptions, ICE AGES, SEA LEVEL RISES, global climate changes, and changes in OCEAN OXYGEN or salinity levels.

Paleontologists, scientists who study the past based on fossils, have been able to recognize patterns within and between such extinction events. Mass extinction has struck both marine and terrestrial species. On land, animal species have tended to suffer most, while PLANT species have been more resistant. There have been frequent disappearances of tropical forms of life during mass extinctions, and the fossil record shows a tendency for certain groups of animals, such as trilobites and ammonoids, to experience extinctions of member species relatively often.

Mass extinctions have occurred about every 26 million years; scientists recognize 5 to 10 major episodes in the geologic record. Dinosaurs are the most well known victims of a major mass extinction that occurred 65 million years ago, at the boundary between the Cretaceous and Tertiary periods; however, the world's largest mass extinction took place about 240 million years ago, at the end of the Paleozoic era. Scientists estimate that during that event, 80 to 96 percent of all species disappeared from Earth.

The overall pattern of extinction within human time, about the past two million years, generally follows the movement of humans from Africa and Asia to Europe, and their subsequent migration to the Americas and the world's islands. North American species that became extinct during the prehistoric era (circa 11,000 years before present) due, directly or indirectly, to human overhunting, included the mammoth, mastodon, woolly rhinoceros, dire wolf, and sabre-tooth cat. Other well-known species from around the world that have become extinct in historical time, also because of human activities, include Steller's sea cow, the great auk, the Carolina parakeet, the DODO BIRD, and the PASSENGER PIGEON.

The extinctions occurring in the world today involve complex issues that are in many cases relatively new to science, and that have global implications. A complete inventory of species threatened with extinction would include every type of organism: MICROORGANISMS, FUNGI, plants, and animals. Because we know so little about the variety of life on Earth, it is likely that many species have already died out during our history without our knowing of their existence. One method that scientists use to better understand and explain mass extinction is to focus research on an isolated region, such as an island, or on a particular type of organism to determine what environmental pressures are acting on that species. Using information gathered from such specific studies, scientists hope to assemble a clearer picture of worldwide extinction patterns. *See also* BIODIVERSITY TREATY; CARRYING CAPACITY; COMPETITIVE EXCLUSION; ECOLOGICAL SUCCESSION; ENDANGERED SPECIES; ENDANGERED SPECIES ACT; ENDANGERED SPECIES LIST; FISH AND WILDLIFE SERVICE; GENETIC DIVERSITY; GLACIER; GLOBAL WARMING; GREENPEACE; INDICATOR SPECIES; RAINFOREST; RED LIST OF ENDANGERED SPECIES; THREATENED SPECIES; UNITED NATIONS EARTH SUMMIT; VOLCANO; WORLD CONSERVATION MONITORING CENTRE; WORLD CONSERVATION UNION.

EnviroSources

Endangered Species Recovery Council Web site: www.esrc.org.
U.S. Fish and Wildlife Service, Division of Endangered Species Web site: http://www.fws.gov/r9endspp/endspp.html.
Williams, Ted. "Back from the Brink." *Audubon* 100, no. 6 (November–December 1998): 70-76.
World Conservation Union Web site: www.iucn.org.
World Wildlife Fund Web site: www.panda.org.

***Exxon Valdez* oil spill,** One of the greatest and most environmentally damaging OIL SPILLS in history. The *Exxon Valdez* oil spill occurred in Prince William Sound, off the coast of Alaska, on March 24, 1989. The OIL TANKER *Exxon Valdez*, carrying a cargo of 1.2 million barrels of crude oil, hit a reef, began leaking, and gradually released approximately 260,000 barrels of oil into the sea. The remaining oil aboard the *Exxon Valdez* was pumped into another oil tanker. But the damage was done. The extent of the oil spill destroyed thousands of sea birds and sea otters, along with WHALES. The fishing grounds of SALMON, cod, and herring were devastated.

The crude oil, when spilled onto the OCEAN, spread and became distributed through the action of wind and waves. For the most part, the spilled crude oil, being less dense than sea

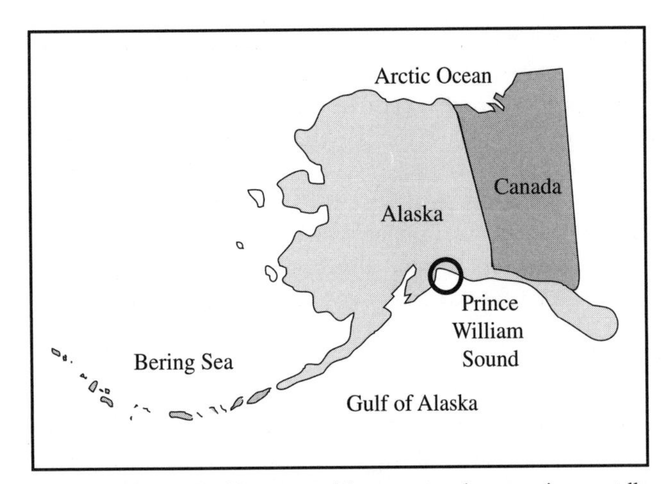

The *Exxon Valdez* oil spill was one of the greatest and most environmentally damaging oil spills in history. The oil spill occurred in Prince William Sound, off the coast of Alaska, on March 24, 1989.

water, floated on the surface of the ocean. The wave and wind action mixed much of the oil with sea water to create a substance known as "mousse," which contains too much water to be ignitable. Thin layers of unmixed oil, or "sheens," covered much of the sound and began to spread with the movement of the waves.

Oil, because it is a complex mixture of HYDROCARBONS, contains many different constituents with different chemical properties. Of the spilled crude oil from the *Valdez*, approximately 35%, consisting of the lighter and more volatile ingredients, evaporated from the surface of the spill.

About 40% of the oil washed ashore in various portions of Prince William Sound, where it formed oil slicks on the beaches. The remaining 25% of the oil was distributed throughout the Gulf of Alaska, some of it reaching the coastline. The oil reached approximately 727 kilometers (452 miles) of the Prince William Sound shoreline and another 1,600 kilometers (1,000 miles) of Alaskan coast to the southwest.

The U.S. Coast Guard, the Exxon Company, and the Alaska Department of Environmental Conservation conducted the initial emergency response to the oil spill. The response effort consisted of removing unspilled oil, salvaging parts of the vessel, skimming floating oil off the ocean surface, and placing floating oil-absorbent booms in locations that were sensitive to oil-related environmental damage. In addition, emergency workers tracked the path of the oil, cleaned up oiled beaches, and rescued WILDLIFE affected by contact with oil. Response activities, including physical and biological treatment methods, continued until June 1992, when the government officially declared the cleanup program complete. The cleanup activities cost about $2.5 billion. Environmentalists, many in the fishing industry, local residents, and others were not satisfied with the cleanup and believe that many marine HABITATS are still suffering from the disaster. *See also* *AMOCO CADIZ* OIL SPILL; CARTAGENA CONVENTION; OIL SPILL EQUIPMENT; PETROLEUM.

fallout, RADIOACTIVE dust and particles emitted into the ATMOSPHERE from a nuclear explosion, caused by a NUCLEAR WEAPON or accident, and then settling on Earth's surface. Radioactive particles of iodine-131, CESIUM-137, STRONTIUM-90, and PLUTONIUM-239 can be picked up by winds and swept thousands of kilometers before falling to Earth. During the CHERNOBYL nuclear accident in 1986, radioactive particles fell on more than 20 countries outside of Russia with the largest amounts dumped in eastern and southern Europe, Scandinavia, and the United Kingdom including Scotland, Wales, and Ireland. The fallout caused hundreds of deaths and injuries.

Radioactive materials also contaminate the soil. PLANTS take in the radioactive material, and then it is passed along the FOOD CHAIN. Humans injest the particles into their bodies along with food. CANCER and other illnesses can occur.

The effect of the fallout can linger in the environment for long periods of time. As an example, four years after the Chernobyl accident, the soil on hundreds of hectares in Wales was so contaminated with cesium-137 from the fallout, that there were restrictions in selling sheep grazed on those lands. *See also* NUCLEAR REACTOR; NUCLEAR WINTER.

famine, A severe food shortage in an area that causes widespread hunger and malnutrition. Famines result from both natural causes and human activities. Natural causes of famine include DROUGHT, FLOODS, EARTHQUAKES, and DISEASES that decimate PLANTS and livestock. Human activities responsible for famine include wars, government seizures of food, the deliberate destruction of crops, crop failures resulting from improper farming techniques that leave land unarable, and overpopulation. *(See* carrying capacity.)

People living in developing nations are most at risk of experiencing famine, especially those resulting from natural causes. When food shortages begin, many people try to escape the problem by migrating to other areas. Such was the case in the 1840s when the potato famine of Ireland, which was caused by a fungus, resulted in mass migrations of that country's POPULATION to other areas. *(See* blight.) More recently, mass migrations out of Ethiopia and Somalia occurred, as civil wars within those countries led to food shortages. *(See* Time Capsule.)

Those unable to leave a region experiencing famine suffer from health problems associated with various forms of malnutrition, including significant weight loss, retarded growth, dehydration, weakness, fatigue, and a diminished immune system. In time, these conditions worsen and people die either from starvation or from diseases that spread unchecked throughout the weakened population.

To reduce the likelihood of famine, many countries have made attempts to curb population growth, increase food production, or both. To help slow its population growth, China has mandated family planning practices. India also has provided incentives to its people to achieve this goal. India has simultaneously increased crop production by adopting many of the farming practices associated with the GREEN REVOLUTION. In many parts of Africa, lands cleared of native vegetation for use in AGRICULTURE have become unsuitable for this purpose as exposure of the soil to sunlight and periodic drought have led to DESERTIFICATION of the land. To reverse this effect and help increase soil fertility of such land, many African nations are replanting trees in areas that were once FOREST. *(See* greenbelt movement.) *See also* AGROECOLOGY; AGROFORESTRY; DEFORESTATION; EMIGRATION; FOOD AND AGRICULTURE ORGANIZATION OF THE UNITED NATIONS; FUNGUS; IMMIGRATION; IRRIGATION; KWASHIORKOR; SALINIZATION.

EnviroSources

Botkin, Daniel B., and Edward A. Keller, *Environmental Science: Earth as a Living Planet.* 2d ed. New York: John Wiley and Sons, 1995.

Chiras, Daniel. *Environmental Science: A Framework for Decision Making.* Menlo Park, CA: Addison-Wesley, 1989.

Operation United Shield Web site: www.fas.org.

UMCOR (United Methodist Comittee on Relief) Web site: gbgm-umc.org.

farming. *See* agriculture.

Time Capsule
Major Famines of the World

Year	Area Affected/Results
1846–1847	As many as one million people in Ireland die from a famine caused by a potato blight. Many others emigrate from the country to avoid starvation.
1878	More than 10 million people in China die as a result of a famine brought about by two years of drought.
1928–1929	Almost three million people living in northern China die from famine caused by drought.
1932–1934	About five million people die in the Soviet Union following a famine that results when Stalin collectivizes farms and seizes grain and livestock in the Ukraine and Caucasus regions.
1941–1943	Famine resulting from World War II activities causes the deaths of more than 40,000 people in Warsaw, Poland. During the same period, another 450,000 people living in Greece, Poland, and Yugoslavia experience famine when German soldiers block wheat shipments from the Ukraine and north Caucasus.
1943	A famine in Rwanda results in the deaths of between 35,000 and 50,000 people.
1968	A civil war in Biafra causes a famine that kills almost one million people.
1968–1974	A famine resulting from a drought in the Sahel region of Africa (located south of the Sahara) causes the deaths of 500,000 people.
1973	Famine resulting from drought is responsible for the deaths of 100,000 in Ethiopia.
1982–1984	Ethiopia is again stricken with a severe famine, this time caused by the combined effects of drought and civil war. Deaths from the famine number near 800,000. Another 1.5 million flee the country in search of food.
1992–1999	Somalia experiences severe famine, leading to the deaths of millions of people and displacement of many more. The famine is the result of both adverse climate conditions and an intense civil war, which by 1999 had divided the country into territories under the leadership of five different warring factions.
1995–1999	In North Korea, poor harvests resulting from a combination of economic and environmental factors during the years 1989–1994 are followed by severe flooding and massive landslides during the summers of 1995 and 1996. The problem is compounded by severe drought during the following growth season. As many as 24 million North Koreans are suffering from malnutrition and starvation as a result of these repeated food shortages.

fauna, A general term often used by biologists, ecologists, and environmentalists to refer to the animals or animal life of a particular region or ECOSYSTEM. For example, emus, koalas, and kangaroos are all part of the fauna of Australia. Fauna may also refer to the animals common to a particular time in geologic history. The term fauna contrasts with FLORA, which refers to PLANTS or plant life. *See also* BIOME.

fecal coliform. *See* coliform bacteria.

Federal Emergency Management Agency, A federal agency in charge of helping people before, during, and after a disaster. The Federal Emergency Management Agency (FEMA) is charged with building and supporting the nation's emergency management system. FEMA reports to the president of the United States; the governor of a state where a disaster has occurred must ask the president for help so that funds can be distributed. Disasters are declared after HURRICANES, TORNADOES, FLOODS, EARTHQUAKES, or other similar events. FEMA coordinates the federal response to a disaster.

FEMA helps disaster victims find temporary housing and repair private and public buildings. FEMA also trains firefighters and emergency workers and runs a flood and crime insurance program. The agency helps communities to build stronger and safer buildings that will withstand disasters.

Funding for FEMA's budget is provided by Congress with money set aside for disasters. *See also* NATURAL DISASTER.

EnviroSource

Federal Emergency Management Agency: Federal Center Plaza, 500 C St. S.W., Washington, D.C. 20472, (800) 480-2520; Web site: www.fema.gov.

Federal Energy Regulatory Commission, An independent regulatory agency within the U.S. DEPARTMENT OF ENERGY. The commission is responsible for overseeing the country's ENERGY industries including: electric utilities, HYDROELECTRIC POWER facilities, and NATURAL GAS and PETROLEUM pipelines. It seeks to balance consumers' and suppliers' needs. Its actions are governed by the Energy Policy Act of 1992. The commission focuses its efforts on environmental issues and compliance. It strives to make regulations work better through improved automation and efficiency.

EnviroSource

Federal Energy Regulatory Commission Web site: www.ferc.fed.us.

Federal Insecticide, Fungicide, and Rodenticide Act, A law that provides for federal control of PESTICIDE distribution, sale, and use. The federal government first regulated pesticides when Congress passed the Insecticide Act of 1910. This law was intended to protect farmers from adulterated or mislabeled products. Congress broadened the federal government's control of pesticides by passing the original Federal Insecticide, Fungicide, and Rodenticide Act (FIFRA) of 1947. The FIFRA required the U.S. DEPARTMENT OF AGRICULTURE (USDA) to register all pesticides prior to their introduction into interstate commerce. In 1970, Congress transferred the administration of the FIFRA to the newly created ENVIRONMENTAL PROTECTION AGENCY (EPA).

The EPA was given authority under the FIFRA not only to study the consequences of pesticide use but also to require users (farmers, utility companies, and others) to register when purchasing pesticides. Through later amendments to the law, users also must take exams for certification as applicators of pesticides. All pesticides used in the United States must be registered (licensed) by the EPA. Registration includes approval by the EPA of the pesticide's label, which must give detailed instructions for its safe use. The EPA must classify each pesticide as either "general use," "restricted use," or both. General-use pesticides may be applied by anyone, but restricted-use pesticides may be applied only by certified applicators or persons working under the direct supervision of a certified applicator. Registration assures that pesticides will be properly labeled and that, if used in accordance with specifications, will not cause unreasonable harm to the ENVIRONMENT. *See also* AGENT ORANGE; CARSON, RACHEL LOUISE; DEFOLIANT; FUNGICIDE; WARNING LABEL.

EnviroSource

U.S. Environmental Protection Agency, Office of Pesticide Programs, Pesticide Information Network: (703) 305-7499.

fertilizer, A substance that is added to soil for the purpose of enhancing PLANT growth. In 1998 the major users of fertilizers were China, the United States, and India; altogether, these three countries used a little more than 50% of all manufactured fertilizer. China is the world leader.

Fertilizers may be classified as either ORGANIC or synthetic. Both types supply plants with the three most important plant nutrients—NITROGEN, potassium, and PHOSPHORUS. They may also contain elements, called secondary nutrients, that plants need in smaller amounts, such as calcium, magnesium, and SULFUR, and materials that plants require in trace amounts, such as boron, MANGANESE, COPPER, cobalt, iron, and zinc. Fertilizers come in the form of solids, semisolids, suspensions, liquids, and gases. Organic and artificial fertilizers are primarily used on farms, lawns, and in greenhouses to enhance the natural fertility of the soil and to replace soil nutrients used up by previous crops.

Synthetic fertilizers are basically mixtures of inorganic substances that supply plants with large amounts of the three most essential nutrients. Most synthetic fertilizers come in different grades indicated by a code listed on the package. The code—known as the N-P-K code—represents the percent by mass of the essential nutrients contained in the fertilizer. A fertilizer designated with a 8-32-16 grade, for instance, contains 8% nitrogen (N), 32% phosphorus (P), and 16% potassium (K) by weight. The main value of synthetic fertilizers is that they provide plants with abundant nutrients quickly, in the exact proportion needed. Soil-testing kits can be used to evaluate the nutrient content of the soil; appropriate fertilizers can then be added to the soil according to which nutrients are deficient.

The use of synthetic fertilizer is an extremely important and valuable practice in modern AGRICULTURE; however, when fertilizers end up in lakes, rivers, and streams, as a result of RUNOFF from farms and lawns, they can cause a number of environmental problems, the most devastating of which is EUTROPHICATION. Eutrophication often results when too many nutrients, especially nitrogen and phosphorus, enter an aquatic ECOSYSTEM. The excess nutrients often lead to rapid growth in POPULATIONS of ALGAE and BACTERIA. The growing populations quickly deplete the water of DISSOLVED OXYGEN, causing the suffocation deaths of fish and other aquatic animals.

Synthetic fertilizers often readily release NITRATES, chemical compounds that are high in nitrogen. Plants need nitrates as their source of nitrogen, but the chemicals are harmful to animals, including humans. Today, there is worldwide concern about nitrates accumulating in drinking water, in both GROUNDWATER and SURFACE WATER, as a result of farm runoff. The body changes nitrates into similar chemicals called nitrites *(see* nitrate), which interfere with the body's ability to carry OXYGEN and are also known to cause headaches, low blood pressure, nausea, and diarrhea. Nitrites can also combine with other chemicals in the body to form nitrosamines, which have been associated with stomach CANCER.

By contrast organic fertilizers are natural substances made of naturally occurring organic matter, for example: animal manure, GREEN MANURE, and compost. Animal manure refers to the dung and urine of various animals, usually cattle, sheep, horses, chickens, and BATS. Animal manures are high in NITRATES. Green manure, also known as plant manure, consists of plants that have been plowed into the soil. Typically, a green-manure crop is grown in the fall; LEGUMES are most commonly used because their roots bear nodules containing bacteria that are capable of NITROGEN FIXATION. In the spring, the plants are then plowed under before the summer crop is grown. The main value of green manure is that it provides the soil with abundant nitrogen, the single most important plant nutrient. Compost, sometimes called "synthetic manure," consists of organic matter made from decomposed leaves, stems, grass clippings, leftover household food, animal wastes, SEWAGE SLUDGE, and other organic remains.

Organic fertilizers contain all or most of the essential nutrients plants need for healthy growth, but in very small amounts; however, they are used exclusively in the practice of organic farming, which is becoming much more widespread. Organic farming refers to the practice of growing crops without reliance on synthetic fertilizers or synthetic PESTICIDES. Globally, people are becoming increasingly concerned about how artificial fertilizers and pesticides affect the ENVIRONMENT, the foods they eat, and thus their health. These concerns have sparked the commercial growth of the organic foods market. Supermarkets today offer consumers a wide variety of organically grown fruits and vegetables. Many people are willing to pay higher prices for these foods if they believe they are safer for the environment and human health.

International concern exists about the management of waste byproducts of synthetic fertilizer manufacturing. For instance, considerable dumping of chemical wastes from fertilizer plants has occurred in the MEDITERRANEAN SEA, adversely affecting on the marine environment.

During the late twentieth century, AQUACULTURE has been used increasingly to produce fish-based fertilizers. China, Peru, and Chile are the largest producers and exporters of fish meal, some of which is processed into fertilizer. Some analysts have observed that the large-scale production and use of fish-based fertilizer may not be an efficient or sustainable ecological practice. *See also* AGROECOLOGY; CALCIUM CYCLE; COMPOSTING; CARCINOGENS; NITROGEN CYCLE; SUSTAINABLE AGRICULTURE.

filtration, The process of passing water or another liquid through a porous medium to remove suspended solid materials or particulates and thus clean the liquid, a practice often used to purify drinking water. Filtration is one step in the treatment of WASTEWATER and SEWAGE. The pore size of the filter used to trap particles varies according to the types of particles to be removed from the liquid. For example, screens may be used to separate very large debris, such as leaves or branches, from wastewater derived from storm RUNOFF before the water is discharged to a stream. *(See* discharge.) For smaller particles, layers of sand, gravel, and/or charcoal are often used as the filter media. These layers of materials have smaller spaces between their particles than do screens and thus are useful for removing small particles from water. *See also* ADSORPTION; PRIMARY, SECONDARY, AND TERTIARY TREATMENTS; REVERSE OSMOSIS; SEWAGE TREATMENT PLANT.

fire ecology, The study of how naturally caused fires are important to a GRASSLAND, SAVANNA, or FOREST ECOSYSTEM. Fire ecology is the relationship between natural communities and fire. Fire is often a main factor in determining the type of PLANT and animal communities in an area. In many forests, plant and animal SPECIES are adapted to live, reproduce, and tolerate fires.

Under ideal conditions where there are no fires, trees will grow, mature, die, and decompose. At each stage of this cycle the forest will support a wide variety of plant and animal HABITATS. Fire interrupts this cycle in a number of different ways to create conditions that favor some species at the expense of others. Fires remove forest LITTER, such as fallen branches and leaves. The burning and removal of tall grasses or bushes that prevent sunlight from reaching the ground can improve growing conditions for small PLANTS and can maintain open spaces for WILDLIFE habitat. Fires generally improve plant growth conditions by releasing nutrients such as calcium, magnesium, potassium, and PHOSPHORUS into the soil. The seeds of the lodgepole pine require exposure to fire to germinate. Fires also assist root growth in aspen and many shrubs. In grasslands, fire may be essential to the survival of native grasses and the animal or INSECT life they support.

Fires can have negative effects, too. The plants and trees of RAINFORESTS and alpine areas are sensitive to high temperatures and may be damaged and killed by fire. And if fires occur too frequently it will significantly reduce the HERBIVORES by exposing them to predators. *(See* predation.) Fires also reduce INVERTEBRATE POPULATIONS and have an effect on nesting birds and young MAMMALS born in the spring. The loss of available food can have adverse effects on wildlife as well. *See also* PRESCRIBED BURNING.

first law of thermodynamics. *See* laws of thermodynamics.

Fish and Wildlife Service, An agency within the U.S. DEPARTMENT OF THE INTERIOR whose mission is to conserve, protect, and enhance fish and WILDLIFE SPECIES and their HABITATS. The major areas of responsibility of the Fish and Wildlife Service (FWS) are migratory birds, ENDANGERED SPECIES, certain marine MAMMALS, and fresh-water and anadromous fish. The FWS provides expert biological advice to other federal agencies, states, industry, and members of the public concerning the CONSERVATION of fish and wildlife habitat that may be affected by development activities.

The origin of the FWS was in 1871, when Congress established the U.S. Fish Commission to study and reverse the decline in the nation's food fish POPULATIONS. In 1885, Congress created an Office of Economic Ornithology within the U.S. DEPARTMENT OF AGRICULTURE (USDA) to study migratory birds. The Fish Commission was placed under the Department of Commerce in 1903 and renamed the Bureau of Fisheries. In 1905, the Office of Economic Ornithology was renamed the Bureau of Biological Survey. In addition to studying the abundance, distribution, and habits of birds and mammals, the responsibilities of the Bureau of Biological Survey included managing the nation's first WILDLIFE REFUGES, controlling predators *(see* predation), enforcing wildlife laws, and conserving dwindling populations of migratory birds. Both bureaus were transferred to the Department of the Interior in 1939; one year later, they were combined as the Fish and Wildlife Service. Further reorganization came in 1956 when the Fish and Wildlife Act created the U.S. Fish and Wildlife Service (FWS). Its Bureau of Commercial Fisheries was ultimately transferred to the Department of Commerce in 1970 and is now known as the NATIONAL MARINE FISHERIES SERVICE. Finally, in 1993, the research activities of the FWS were transferred to a new Interior Department agency—the National Biological Survey.

Today, the FWS maintains headquarters in Washington, DC, seven regional offices, and nearly 700 field units and installations, including national wildlife refuges, fish hatcheries, ecological field offices, and law enforcement offices. A major function of the FWS is the identification and recovery of endangered species. The FWS leads the federal effort to protect and restore animals and PLANTS that are in danger of EXTINCTION, both in the United States and worldwide. Using the best scientific information available, the FWS identifies species that appear to be endangered or threatened; species that meet the criteria of the ENDANGERED SPECIES ACT are placed on the Interior Department's official List of Endangered and Threatened Wildlife and Plants. *(See* endangered species list.) In recent years, the FWS has placed increased emphasis on two provisions of the Endangered Species Act: Habitat Conservation Plans and Special 4(d) Rules. These provisions are designed to avoid or resolve conflicts between private development projects and the protection of endangered species. The FWS also consults with other federal agencies and renders opinions on the effects of proposed federal projects on endangered species, recommending ways for developments to avoid harm to endangered species.

The U.S. NATIONAL WILDLIFE REFUGE SYSTEM is the world's largest and most diverse collection of lands specifically set aside for wildlife. The refuge system was initiated in 1903 when President THEODORE ROOSEVELT designated Florida's 1.2-hectare (3-acre) Pelican Island as a sanctuary for pelicans and herons. Today, 500 national wildlife refuges have been established, ranging in location from the Arctic Ocean to the South Pacific, and from Maine to the CARIBBEAN SEA. Varying in size from 0.2-hectare (0.5-acre) parcels to thousands of square kilometers, they encompass more than 37 million hectares (92 million acres) of prime wildlife habitats. The vast majority of these lands are located in Alaska, with the rest spread across the United States and several U.S. territories.

Most refuges were created to protect the more than 800 species of migratory birds that travel along the four major north-south flyways. As specified in various international treaties (with Canada, Mexico, Japan, and the former Soviet Union), as well as U.S. legislation such as the Migratory Bird Conservation Act of 1929, the United States has responsibilities to protect migratory birds. The FWS also cooperates with state wildlife agencies and the Canadian Wildlife Service to regulate migratory bird hunting. Presently, the FWS also is increasing efforts to identify nongame bird species that may be declining and to undertake efforts to restore them. Through the Partners for Wildlife Program, the FWS provides technical and financial assistance to private landowners wishing to restore wildlife habitat on their properties, primarily WETLANDS, RIPARIAN HABITAT, and native PRAIRIE. To date, nearly 11,000 landowners have participated in Partners for Wildlife, restoring approximately 85,000 hectares (210,000 acres) of habitat.

Restoring nationally significant fisheries that have been depleted by OVERFISHING, pollution, or other habitat damage is also a responsibility of the FWS. Currently, FWS fishery specialists are devoting much of their efforts to helping four important fish groups: lake trout in the upper Great Lakes, striped bass of the CHESAPEAKE BAY region and the Gulf Coast, Atlantic SALMON of New England, and chinook, coho, and steelhead salmon of the Pacific Northwest. The FWS's fishery program also works to compensate for losses of fishery resources caused by federal water projects and to improve fishery resources on Indian reservations and federal lands.

Two laws administered by the FWS, the Federal Aid in Wildlife Restoration Act and the Federal Aid in Sport Fisheries Restoration Act, have created some of the most successful programs in the history of fish and wildlife conservation. These programs provide federal grant money to support specific projects carried out by state fish and wildlife agencies. The money comes from federal excise taxes on sporting arms and ammunition, archery equipment, and sport-fishing tackle. In 1984, the basic Sport Fisheries Restoration law was supplemented by new provisions that increased revenue for sport-fish restoration. The FWS also works with other nations to preserve their NATIVE SPECIES through the CONVENTION ON INTERNATIONAL TRADE IN ENDANGERED SPECIES OF WILD FLORA AND FAUNA. *See also* BALD EAGLES; CALIFORNIA CONDORS; LACEY ACT; MIGRATORY BIRD TREATY ACT; THREATENED SPECIES.

EnviroSources

Chadwick, Douglas H., and Joel Sartore. *The Company We Keep: America's Endangered Species.* Washington, DC: National Geographic Society, 1995.
National Fish and Wildlife Foundation Web site: www.nfwf.org.
U.S. Fish and Wildlife Service: 1849 C St., NW Washington, DC 20240; (202) 208-5634; Web site: www.fws.gov

fish farm. *See* aquaculture.

fishing. *See* commercial fishing.

fission, A nuclear CHAIN REACTION in which a neutron splits the nucleus of a PLUTONIUM or URANIUM atom releasing ENERGY, nuclei of other elements, and neutrons that split more nuclei. These nuclei emit more neutrons and so on resulting in a chain reaction that releases great amounts of heat and energy. Fission is the process that produces energy in NUCLEAR REACTORS and is also the process that detonates NUCLEAR WEAPONS. Plu-

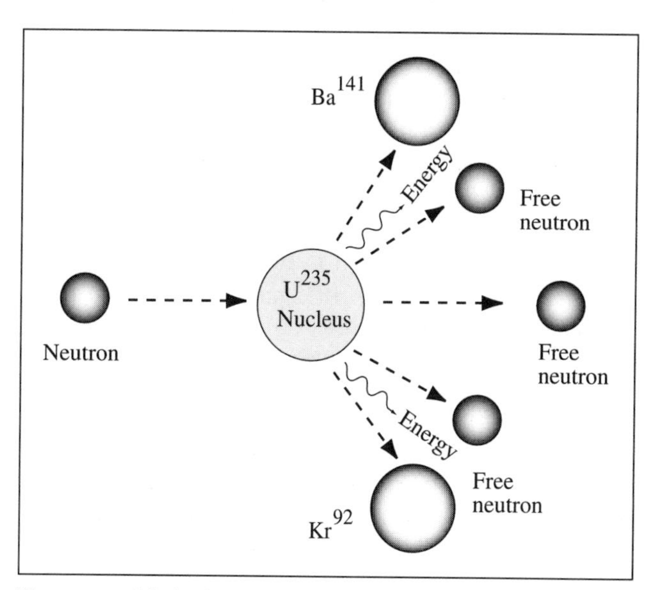

The process of fission in a nuclear reactor is used to produce energy and in detonating nuclear weapons.

tonium-239 and uranium-235 are elements that undergo the process of fission. *See also* RADIATION.

EnviroTerms
proton: The positive-charged particle of an atom located in the nucleus.
neutron: The particle of an atom in the nucleus that has no charge.
electron: The negative-charged particle of an atom that orbits the nucleus.

floc. *See* flocculation.

flocculation, A process by which biological or chemical action causes clumps of solids, called floc, to PRECIPITATE through, or settle out of, water or SEWAGE. Flocculation takes place in the secondary sewage treatment processes in which MICROORGANISMS decompose ORGANIC matter in WASTEWATER. The floc settles to the bottom of a wastewater tank where it can be removed as SLUDGE. *See also* DECOMPOSITION; PRIMARY, SECONDARY, AND TERTIARY TREATMENTS; WASTEWATER TREATMENT.

flood, Water that flows over the top of a river bank and then spreads out over a FLOODPLAIN, becoming a NATURAL DISASTER. Floods can occur when any relatively high stream overflows the natural or artificial banks in any reach of a stream. A flood is also a relatively high flow of water as measured by either gage height or DISCHARGE quantity. Floods, a natural occurrence in all rivers, begin when soil and vegetation in a particular area cannot absorb rain, when water runs off the land so quickly that it cannot be diverted by the natural channels, or when ponds and lakes cannot contain excess water. Some floods occur during the rainy season or when heavy rains are caused by severe storms. Other floods, called "flash floods" because of their sudden and unpredictable occurrence, are caused by heavy localized rainfall. Drowning is the most common cause of flood-related deaths.

All parts of the United States are threatened by floods, although they occur most frequently in low-lying or OCEAN front areas. Each year, approximately 300,000 Americans are driven from their homes by floods. In September 1999, Hurricane Floyd dumped up to 50 centimeters (20 inches) of rain on eastern portions of the state of North Carolina. It was the most expensive natural disaster in North Carolina history, topping the $6 billion price tag from 1996's Hurricane Fran. Agricultural losses alone in North Carolina were expected to top $1 billion, far more than the $872 million in damage caused by Hurricane Fran, three years earlier. In North Carolina, floodwaters destroyed or heavily damaged 3,000 homes and forced 42,500 people to apply for state and federal assistance. In 1997 severe floods and flash floods caused $2 billion worth of damage in the West Coast and interior plains of the United States. And one of the worst years of flood-related deaths in the United States was in 1988–1989 when more than 120 people died. In the United States the FEDERAL EMERGENCY MANAGEMENT AGENCY (FEMA) helps disaster victims find temporary housing and helps repair private and public buildings. FEMA also trains firefighters and emergency workers and runs a flood and crime insurance program. The agency helps communities to build stronger and safer buildings that will withstand disasters.

Some of the worst flooding occurs in Europe, Africa, and Asia. In 1996 severe flooding in the Yangtze and Yellow River floodplains caused the deaths of more than 2,000 people and left 2 million homeless. The cost of the damages was more than $20 million. Flood conditions linked to EL NIÑO killed more than 1,000 people and left 200,000 homeless in the eastern part of Africa in 1997. Severe floods have caused billions of dollars worth of damage in central Europe.

In 1999, torrential rains caused horrific flooding conditions in the state of Vargas, north of the capital of Caracas, Venezuela. When the flood subsided, the death toll reached more than 20,000. About 6,000 people were reported missing and presumed dead and another 200,000 people were homeless. It was the country's worst natural disaster in the twentieth century.

The kind of damage that a flood causes is related to a number of variables. They include the effectiveness of WEATHER forecasting to alert citizens, the land use of the floodplain, season of the year, depth and velocity of the water, and how long the flooding lasts. Efforts to control flooding include building levees, reservoirs, channels, DAMS, and flood walls. Channelization is also used to enlarge the size of the stream or river channel to divert the water away from populated areas. However, most experts agree, although these structures and mechanisms can help prevent some flooding, floodplain regulation is the best method to minimize flood damage. Constructing homes on steep, poorly graded hillsides should also be avoided if possible. *See also* ARMY CORPS OF ENGINEERS; EROSION; HURRICANE.

floodplain, A long flat area of land along either side of a river made up of soil deposits carried by river water. The soil is deposited when the river overflows its banks. The floodplain is an important part of the natural river system. Many civilizations of the world were established on floodplains because of the rich alluvial soil deposits that supported farming, access to fresh-water resources, and the readily available transportation opportunities the river provided to other areas. Some major floodplains that are home to large POPULATIONS of people and major industries are those of Yangtze, Ganges, Indus, Tigris-Euphrates, Nile, Mississippi, and Rhine. One of the most densely populated areas in the world is the southern section of the Yellow River floodplain.

The rapid expansion of buildings, including residential, farm, and industrial structures, in high-risk floodplains has caused extensive FLOOD damages and disasters. In fact, many insurance companies have stopped issuing premiums on properties that are constantly vulnerable to flood damage. Most environmentalists and hydrologists agree that one of the best ways of reducing flood damage is to manage and regulate the land use of floodplains more carefully. *See also* HYDROLOGY.

flora, A general term used by biologists, ecologists, and environmentalists to refer to the PLANTS or plant life of a particular region or ECOSYSTEM. For example, mosses, ferns, oaks, and maples are flora of a DECIDUOUS FOREST ecosystem. Flora may also refer to the plants common to a time in geologic history such as the Carboniferous period. Flora contrasts with FAUNA, which refers to animals or animal life. *See also* BIOME; ECOLOGY.

Florida panthers, An endangered subspecies of the cougar (mountain lion) family that lives in the subtropical FORESTS of the EVERGLADES NATIONAL PARK in southern Florida. The Florida panther (*Felis concolor coryi*) is a predatory animal that generally feeds upon the deer, raccoons, and wild hogs that share its ECOSYSTEM, helping to keep POPULATIONS of these animals in balance. (*See* predation.) As the top predator in its FOOD CHAIN, the panther's only natural enemies are humans and, occasionally, other panthers.

The Florida panther lives in the subtropical forests of EVERGLADES NATIONAL PARK in southern Florida.

The Florida panther population once numbered in the thousands and ranged throughout the southeastern United States; however, current population studies show that as few as 30–50 adult animals now exist in the wild. The animal's decline is attributed to human activities and to genetic abnormalities that have arisen in the subspecies as a result of inbreeding. The human activities threatening the panther population include HABITAT destruction and the DISCHARGE of TOXIC CHEMICALS, including MERCURY, resulting from industrial wastes and PESTICIDES into the water supply. Genetic abnormalities resulting from inbreeding include heart defects, infertility, and a lowered resistance to infectious DISEASES.

The Florida panther has been on the ENDANGERED SPECIES LIST since March 1967. Efforts to prevent its EXTINCTION have been under way for many years. Until 1995, these efforts focused on preserving the panther's habitat by disallowing further clearing and filling of WETLAND areas in and around the Everglades National Park; creating WILDLIFE REFUGES and preserves, such as the Florida Panther National Wildlife Refuge, the Big Cypress National Preserve, and the Fakahatchee Strand State Preserve; and by restoring some wetland areas. Despite such efforts, the Florida panther population has continued its decline and breeding groups have become more isolated, encouraging greater inbreeding of the animals. To prevent additional declines, the U.S. FISH AND WILDLIFE SERVICE (FWS) changed its Florida Panther Recovery Plan in 1995, proposing that measures be taken to restore GENETIC DIVERSITY in the remaining population. To accomplish this goal, the FWS introduced eight female Texas cougars—the closest relatives of the Florida panthers—into the panther's range, hoping that the cougars and panthers would mate and produce fertile, healthy offspring. Within its first year, the program met with some success when two of the introduced female cougars gave birth. The program will be deemed truly successful when the resulting HYBRID offspring reproduce and begin increasing the size of the Florida panther population. Additional efforts are also being made to work with area farmers, businesses, and industry to develop CONSERVATION efforts that will restore the Florida panther's habitat. *See also* DOUGLAS, MARJORY STONEMAN; ENDANGERED SPECIES; ENDANGERED SPECIES ACT; GENE POOL; HYBRIDIZATION; THREATENED SPECIES.

EnviroSources

Chadwick, Douglas H., and Joel Sartore. *The Company We Keep: America's Endangered Species*, Washington, DC: National Geographic Society, 1995.

La Pierre, Yvette. "On the Edge." *National Parks* 71, nos. 11–12: (November–December 1997): 44.

National Wildlife Federation: 8925 Leesburg Pike Vienna, Virginia 22184-0001; Web site: www.nwf.org.

flue gas, The gaseous material emitted from a chimney following combustion of FOSSIL FUELS in a COAL-fired burner or incinerator. The gases can include fly ASH, NITROGEN OXIDES, carbon oxides, water vapor, SULFUR oxides, PARTICULATE MATTER, and many chemical POLLUTANTS. A process, called flue gas desulfurization uses limestone to remove SULFUR DIOXIDE from the gases emitted by power plants that burn fossil fuels. *See also* CARBON; CARBON DIOXIDE; CARBON MONOXIDE; COAL; INCINERATION; SCRUBBER.

fluidized bed combustion, A low-polluting technology, sometimes referred to as atmospheric fluidized bed combustion (AFBC), for burning low-grade COAL in a boiler that traps SULFUR DIOXIDE EMISSIONS before being emitted into the ATMOSPHERE. The technology was created through research and development sponsored by the U.S. DEPARTMENT OF ENERGY. The U.S. government was awarded the first patent for SULFUR retention in AFBC boilers.

In the AFBC process, excess AIR is blown in from underneath the boiler. A mixture of pulverized coal and limestone is forced into the boiler where it "floats" on the air while it burns. The calcium and some magnesium from the limestone absorb the sulfur dioxide from the SULFUR materials in the coal. This state-of-the-art technology is quite successful in both U.S. and overseas operations. Besides burning coal, AFBC is being considered as an option for burning SOLID WASTES such as SLUDGE or REFUSE. *See also* INCINERATION.

EnviroTerm

flue gas decombustion: A current state-of-the-art technology that uses lime and limestone to remove SULFUR DIOXIDE from the gases produced in a COAL-fired furnace.

fluorine, A nonmetal element that is a green-yellow poisonous gas at room temperature. Fluorine is used in the manufacturing of CHLOROFLUOROCARBONS (CFCs) and FLUOROCARBONS. At one time fluorocarbons and CFCs were used as propellants for domestic AEROSOLS in the United States. Now they are present mainly in coolants and some industrial processes. Fluorine is present in rocks and soils and as a trace element in the form of fluoride compounds. Fluoride compounds are used in toothpaste and the fluoridation of water supplies to help prevent tooth decay; however, excessive amounts of fluoride in drinking water can cause bone and teeth DISEASE. *See also* GLOBAL WARMING; GLOBAL WARMING POTENTIAL; OZONE DEPLETION.

fluorocarbon, Any of a number of ORGANIC compounds similar to HYDROCARBONS in which one or more HYDROGEN atoms are replaced with FLUORINE. A fluorocarbon is a compound of fluorine and CARBON. At one time, fluorocarbons were used as propellants for domestic AEROSOLS in the United States. Now they are mainly used in coolants and some industrial processes.

Fluorocarbons containing CHLORINE, called CHLOROFLUOROCARBONS (CFCs), cause depletion of the OZONE layer in the stratosphere, thereby allowing more harmful solar RADIATION to reach Earth's surface. *See also* OZONE DEPLETION.

fly ash. *See* ash.

Food and Agriculture Organization of the United Nations, The largest autonomous agency within the United Nations. Since its inception in 1945, the Food and Agriculture Organization (FAO) has worked to alleviate poverty and hunger by promoting agricultural development, improved nutrition, and the pursuit of food security. The FAO acts as an international forum for debate on food and AGRICULTURE issues and plays a major role in dealing with food and agricultural emergencies.

The FAO offers direct development assistance; collects, analyzes, and disseminates information; and provides policy and planning advice to governments. The FAO also deals with land and water development, PLANT and animal production, FORESTRY, fisheries, investment, nutrition, economic and social policy, food standards, and commodities and trade.

The FAO gives practical help to developing countries, drawing on local expertise and ensuring a cooperative approach to development. The FAO has an average of 1,800 field projects operating at any one time. *See also* DEPARTMENT OF AGRICULTURE; DUST BOWL; FAMINE; UNITED STATES.

EnviroSource

Food and Agriculture Organization of the United Nations Web site: www.fao.org.

Food and Drug Administration, A federal public health agency within the Department of Health and Human Services with the mission of protecting consumers by enforcing the Federal Food, Drug, and Cosmetic Act of 1938 and related public health laws. The U.S. Food and Drug Administration (FDA) is one of the oldest consumer protection agencies in the United States. Major environmental concerns of the agency include DISEASES spread in food, especially those resulting from improper handling, processing, storage, or packaging. To carry out its mandate of consumer protection, the FDA employs more than 1,100 investigators and inspectors to scrutinize the nation's 95,000 FDA-regulated businesses. The FDA also operates the National Center for Toxicological Research, which studies the biological effects of widely used chemicals, and the Engineering and Analytical Center, which tests medical devices and radioactive drugs and other products. *(See* radiation.)

A century ago, conditions in the U.S. food and drug industries were primitive. The use of chemical additives and toxic colorants was virtually uncontrolled. Sanitary practices and standards for food storage and handling were minimal, and ice remained the principal means of refrigeration. Thousands of so-called patent medicines, often containing drugs such as opium, morphine, heroin, and cocaine, were sold without restriction, in packages that did not disclose their contents. Some otherwise harmless preparations were labeled for the cure of every conceivable disease or symptom. What scant information the public did receive, frequently resulted from negative experiences with a particular product.

There was strong opposition to federal food and drug regulation from whiskey distillers and patent medicine firms—at the time the largest commercial advertisers in the nation. Many thought they would be put out of business under federal control. On the other side of the dispute were agricultural organizations, including many food packers, state food and drug officials, and health professionals. It was American women who made the difference in the establishment of federal regulation of foods and drugs. In 1906, the Food and Drug Act was passed, resulting in the formation of the Food, Drug, and Insecticide Administration, which was renamed the Food and Drug Administration in 1931. In 1940, to prevent recurring conflicts between producer interests and consumer interests, the FDA was transferred from the U.S. DEPARTMENT OF AGRICULTURE (USDA) to the Federal Security Agency, which, in 1953, became the Department of Health, Education, and Welfare. Today, the department is known as the Department of Health and Human Services.

FDA investigators and inspectors visit more than 15,000 facilities each year, to ensure that products are made correctly and labeled truthfully. As part of their inspections, they annually collect about 80,000 domestic and imported product samples for label checks or examination by FDA scientists. If a company is found to be in violation of any laws that the FDA enforces, the agency may encourage the firm to voluntarily correct the problem or recall a faulty product from the market. A recall usually is the fastest and most effective way to protect the public from an unsafe product. About 3,000 products each year are found to be unfit for consumers and are withdrawn from the marketplace, either by voluntary recall or by court-ordered seizure. For example, in 1998 and again in 1999, numerous beef products and pork products were recalled because of evidence of ESCHERICHIA COLI *(E. coli)* and listeria contamination at the plants where the products were processed and packaged, respectively. In addition, about 30,000 unacceptable import shipments are detained at U.S. ports of entry each year. In deciding whether to approve new drugs, the FDA does not itself conduct research, but rather examines the results of studies done by drug manufacturers. The agency must then determine that a new drug produces the benefits it claims, without causing side effects that outweigh those benefits.

Another major FDA mission is to ensure the safety and wholesomeness of foods. The agency's scientists test samples to see if substances, such as PESTICIDE residues, are present in unacceptable amounts. If CONTAMINANTS are identified, the FDA takes corrective action, such as requiring the recall of contaminated products and/or closing down the facility where the product was produced until the source of contamination is identified and eliminated. The agency also sets labeling standards to educate consumers about the ingredients and nutritional values of the foods they buy and ensures that medicated animal feeds and other livestock drugs do not jeopardize the health of consumers. All dyes and other additives used in drugs, foods, and cosmetics are subject to FDA scrutiny.

The safety of the nation's blood supply is another FDA responsibility. The agency's investigators routinely examine blood bank operations, from record keeping to contaminant testing. The FDA also ensures the purity and effectiveness of biological preparations (medicinal compounds made from living ORGANISMS and their products), such as insulin and vaccines. *(See* biotechnology.) *See also* ACUTE TOXICITY; CHRONIC TOXICITY.

EnviroSource

U.S. Food and Drug Administration Web site: www.fda.gov.

food chain, A model used by ecologists to trace the flow of nutrients and ENERGY through an ECOSYSTEM based on the feeding patterns of its ORGANISMS. A food chain traces only one pathway of food through an ecosystem; thus, a single ecosystem generally has many different food chains. All food chains have some elements in common. For example, they all begin with PRODUCERS—organisms that synthesize their nutrients from materials in the ENVIRONMENT. The producers at the base of the food chain are followed by one or more levels of CONSUMERS, organisms that obtain their nutrients by eating other organisms. The food chain concludes with DECOMPOSERS—organisms that obtain their nutrients by feeding on the wastes or remains of other organisms. *(See* decomposition.)

Because the sun is the major source of energy for most ecosystems, the producers at the base of a food chain are usually organisms that derive their nutrients through the process of PHOTOSYNTHESIS. In terrestrial ecosystems, the producers are usually PLANTS, whereas in aquatic ecosystems photosynthetic producers are often ALGAE. In some ecosystems, such as thermal vent communities in the deep OCEAN, the producers are BACTERIA that obtain their nutrients by synthesizing inorganic materials using the process of CHEMOSYNTHESIS.

Each feeding level in a food chain is known as a TROPHIC LEVEL. Because they make their own food, producers do not comprise a trophic level. Primary consumers make up the first trophic level of a food chain; they are HERBIVORES or OMNI-

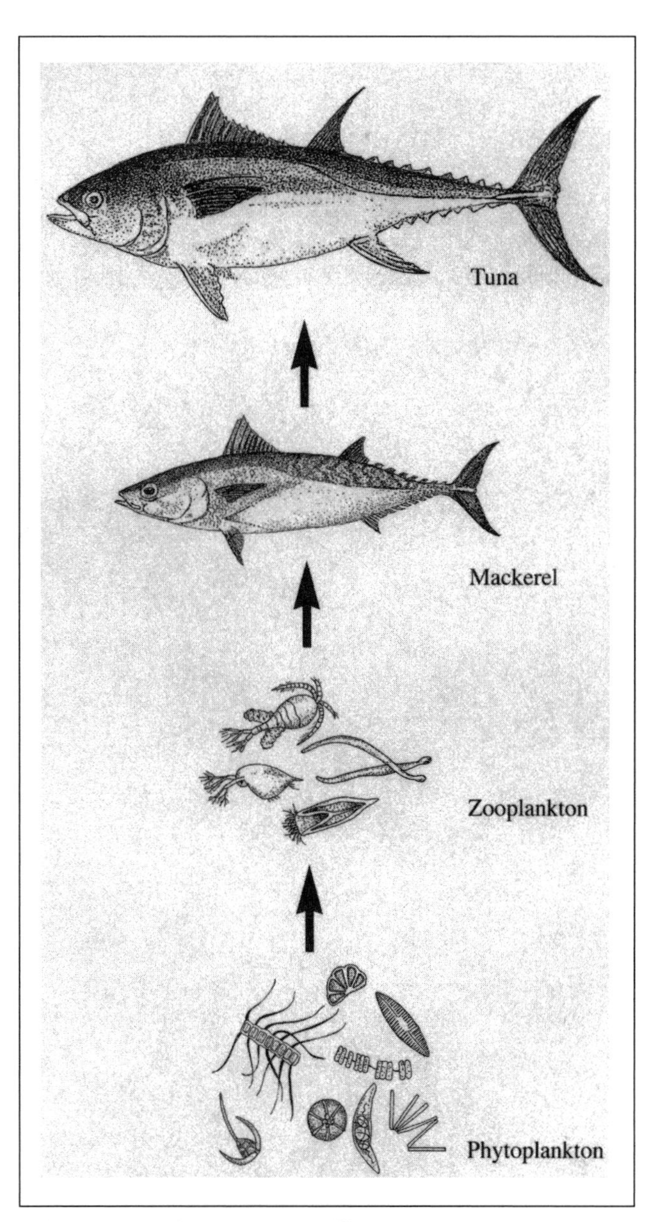

A food chain showing producer and different consumers.

VORES that feed directly on producers. As they feed, the primary consumers obtain both nutrients and energy stored in the tissues of the producers. A primary consumer may then be eaten by a secondary consumer (a CARNIVORE or omnivore) located in the second trophic level of the food chain. As in the first trophic level, the secondary consumer obtains its nutrients and energy from the primary consumer it eats. A food chain may contain one or more additional levels of consumers above the secondary consumer, before ending with a top predator. This predator generally does not serve as prey to any other consumers in the ecosystem (*see* predation); however, when this predator dies, it may become a source of food for a SCAVENGER or be consumed directly by decomposers (bacteria and FUNGI), as would any other organism in the food chain.

Food chains are often represented in visual models by arrows showing the movement of nutrients and energy from one organism to the next. A possible food chain for an ocean ecosystem is shown. This food chain is comprised of PHYTOPLANKTON (producer), ZOOPLANKTON (primary consumer), mackerel (secondary consumer), and TUNA (tertiary consumer). *See also* ECOLOGICAL PYRAMID; ECOLOGY; FOOD WEB.

food web, A model composed of several overlapping FOOD CHAINS that is used by ecologists to trace the flow of nutrients and ENERGY through an ECOSYSTEM by analyzing the feeding relationships among its ORGANISMS. All ecosystems have some CONSUMERS—organisms that obtain their food and energy by eating other organisms. Often, consumers use more than one kind of organism as a food source. Thus, many consumers are involved in more than one food chain. A food web provides ecologists with a mechanism for simultaneously tracing multiple pathways involving the transfer of food and energy.

A food web includes all the same TROPHIC LEVELS, or feeding levels, as a food chain; however, unlike a food chain, a food web illustrates how a single organism may feed at more than one trophic level. For example, a field mouse is an OMNIVORE that sometimes feeds on seeds and sometimes feeds on INSECTS. When feeding on seeds, the mouse is feeding directly on a PRODUCER, and thus is a primary consumer eating at the first trophic level; however, if this same mouse eats a grasshopper (which feeds on PLANTS), it is a secondary consumer feeding at the second trophic level. A food web can show both these feeding relationships at the same time.

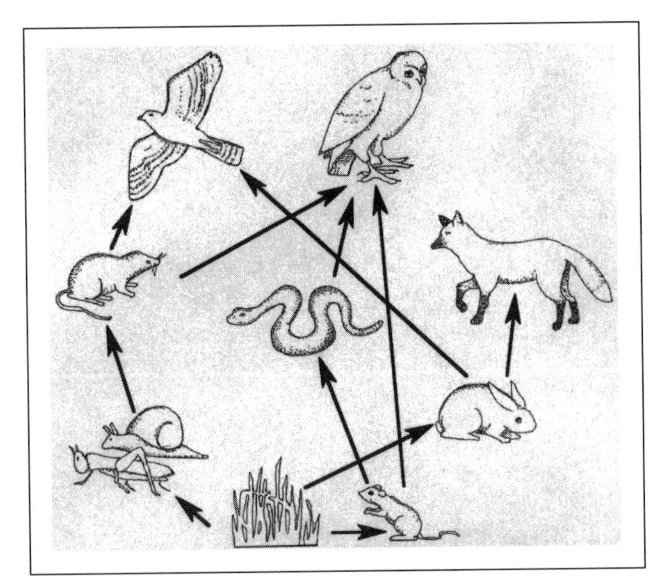
This food web shows several feeding patterns involving organisms of neighboring grassland and deciduous ecosystems.

Food webs are represented in visual models that use arrows to show the movement of nutrients and energy from one organism to the next. The diagram shows a possible food web for the overlapping areas of a GRASSLAND and a DECIDUOUS FOREST ecosystem. In the food web shown, grass (producer) is the food source of a rabbit (primary consumer) which may then be eaten by a fox (secondary consumer)—a top predator in this ecosystem. *(See* predation.) In addition, a mouse (primary consumer) may be eaten by a snake (secondary consumer), which is then eaten by the owl (tertiary consumer)—also a top predator in this ecosystem. These overlapping food chains provide scientists with more information about the interactions among the organisms in this ecosystem than does a single food chain. This food web also shows several other feeding relationships that may involve these same organisms. *See also* AUTOTROPH; CARNIVORE; DECOMPOSER; DETRIVORE; ECOLOGICAL PYRAMID; HERBIVORE; HETEROTROPH; INSECTIVORE; SCAVENGER.

EnviroSource

Bernstein, Leonard, Alan Winkler, and Linda Zierdt-Warshaw. *Environmental Science: Ecology and Human Impact.* Menlo Park, CA: Addison-Wesley, 1995.

foreign species. *See* alien species.

Foreman, David (1946–), An environmentalist who was a WILDERNESS SOCIETY lobbyist, the national director of the SIERRA CLUB from 1995–98, and the founder of Earth First!–a group that is periodically accused of engaging in "ecoterrorism," radical activities to prevent damage to the environment that often result in damage to both equipment and the people operating the equipment. *(See* ecotage.) David Foreman authored the 1985 book, *Ecodefense: A Field Guide to Monkeywrenching*, which provides detailed instructions about how to sabotage equipment, industrial establishments, roads, and vehicles in the name of environmental defense. Among the techniques outlined in *Ecodefense* is "tree spiking," the insertion of long nails into trees as a means of deterring loggers from felling them. This sabotage technique has resulted in serious injury to timber industry workers who have unknowingly attempted to use saws on spiked trees.

Some of Foreman's detractors have alleged that he founded Earth First! in 1980 with the financial backing of the Sierra Club and the Wilderness Society. According to these critics, Earth First! was specifically created to present an extreme environmental agenda, as a means of making more mainstream environmental groups appear reasonable by comparison. In 1989, Foreman and three other members of Earth First! were arrested in Arizona by the Federal Bureau of Investigation (FBI), and charged with conspiracy involving the sabotage of nuclear power facilities. Foreman left Earth First! the following year.

Foreman is currently chairman of the Wildlands Project, an ambitious CONSERVATION program whose mission is to bring together grassroots environmental activists and conservation biologists to help protect and restore North America's ecological richness and native BIODIVERSITY through the establishment of a connected system of reserves. *See also* ABBEY, EDWARD; BIOREGION; BIOSPHERE; BIOSPHERE RESERVES; BROWER, DAVID ROSS; DEEP ECOLOGY; ECOLOGY.

EnviroSources

Chang, Chris. "Champions of Conservation: *Audubon* Recognizes 100 People Who Shaped the Environmental Movement and Made the twentieth Century Particularly American." *Audubon* 100, no. 6 (November–December 1998): 120.

———. *Ecodefense: A Field Guide to Monkeywrenching.* Chico, CA: Abbzug Press (1993.)

Earth First! Web site: www.enviroweb.org/ef.

Foreman, David. *Confessions of an Eco-Warrior.* New York: Crown Publishers, Inc., (1993.)

Wall, Derek. *Earth First! and the Anti-Roads Movement: Radical Environmentalism and Comparative Social Movements.* New York: Routledge, 1999.

Wildlands Project Web site: www.wildlandsproject.org.

forest, A type of biome dominated by trees, with a unique variety of PLANTS, animals, MICROORGANISMS, soils, and CLIMATE. Natural forests are composed largely of native tree SPECIES, while plantation forests often are managed as MONOCULTURES of particular commercial species. *(See* tree farm.)

Trees in forests are critical to the welfare of our planet and play a vital role in controlling climate and WATER CYCLES. They help keep the AIR clean by filtering POLLUTANTS and reduce the risk of GLOBAL WARMING by absorbing CARBON DIOXIDE and other GREENHOUSE GASES. The majority of Earth's species are dependent on the survival of trees. People who live in and near forests depend on them for much of their food, medicine, clothing, and timber; most of these products come from plants and trees that grow best in natural forests. As WATERSHEDS, forests absorb rainfall and slowly release it into streams and rivers, moderating both FLOODS and DROUGHTS, and regulating water flows. Soil EROSION is also kept in check, which prevents SILTATION of waterways and damage to CORAL REEFS, as well as protecting fisheries and spawning grounds. Forests are vibrant ecosystems with complex symbiotic relationships that have assured their survival for over 180 million years. *(See* symbiosis.)

Forests exist in almost every part of the world. Only the North and South Poles, some mountains, the DESERTS, and some PRAIRIES are bare of forests. Different trees have adapted to live in different parts of the world, and a variety of forest types have evolved. Forests can be grouped by location, climate, or the type of hardwood or softwood trees most common in them. Forests can also be described in terms of the uses made of them. Commercial forests, for example, are lands used for growing successive crops of trees for products. *(See* sustained yield.) WILDERNESS preserves, on the other hand, are areas where no harvesting is allowed.

Forests do not only contain trees. They also contain millions of plants, animals, and microorganisms, whose existences are all closely interlinked. Tropical forests alone are home to some 90% of the world's terrestrial species.

One-fourth of the medicines available today are derived from plants. Along with timber and timber products, forests are also an abundant source of industrial commodities such as rubber, waxes, and fibers and some foods such as nuts.

Worldwide, forests are being destroyed and are thus disappearing at an alarming rate, with 50% of the world's original forest cover already having been lost to the pressures of logging, FOREST FIRES, and land clearing. *(See* deforestation.) Over 8 billion hectares (about 19 billion acres) of forest existed in the world 8,000 years ago; today, only 3–4 billion hectares (about 10 billion acres) remain. The pace of forest destruction has accelerated in the 1990s and continues to rise; currently, more than 400,000 hectares (950,000 acres) of forest are cleared or degraded every week. Increasing numbers of people are concerned about forest loss and FOREST MANAGEMENT. Recent surveys show that the prospect of forest loss has overtaken RECYCLING and chemical use as a key environmental issue for the general public.

More than 8,750 of the 80,000 to 100,000 tree species known to science have been found to be threatened with EXTINCTION, and 77 are already extinct. Almost 1,000 species are believed to be critically endangered; the POPULATIONS of some species have been so depleted that only one individual or a handful of individuals now exist. Fewer than one-quarter of these threatened tree species are protected by CONSERVATION measures. Only 12% of these species are recorded in protected areas, and only 8% are known to be in cultivation. The major threats to tree species include felling for timber and FUEL WOOD, clearing land for AGRICULTURE, and expansion of human settlements, uncontrolled forest fires, introduction of invasive EXOTIC SPECIES, and unsustainable forest management. With more than 1,000 tree species threatened as a result of felling, the sustainable management of forests is a top priority for conservation organizations.

Key solutions advocated include sustainable forest management, protection of forest HABITAT, and control of invasive alien species, supplemented by off-site conservation in botanical gardens, arboreta, and seed banks. *(See* germ plasm bank.) If a forest is to be used, the best way to keep its natural balance is to harvest it sustainably—that is, take only as much of any product as the forest can afford to give.

One way to reduce pressures is for international trade organizations to acknowledge the importance of sustainable harvesting. Currently, businesses that have adopted sustainable forest management practices can lose in the marketplace because they produce less timber than their competitors. For the survival of forests, it is therefore important for trade systems to be altered to distinguish between sustainably and unsustainably produced timber, and to favor the former.

The WORLD WILDLIFE FUND (WWF) now believes that certification by the Forest Stewardship Council (FSC) is a reliable way to promote improved forest management. This certification is the only independent global guarantee that forest products come from well-managed forests. The FSC provides companies with a simple and clear way to demonstrate to consumers and investors that they are committed to the highest international standards in forest management. *See also* CLEAR-CUTTING; FOREST SERVICE; FORESTRY; NATIVE SPECIES; OLD-GROWTH FOREST; SELECTION CUTTING; SHELTERWOOD HARVESTING; SUSTAINABLE DEVELOPMENT.

EnviroSources

American Forest, P.O. Box 2000, Washington, DC 20013.

Fornasari, Lorenzo, and Renato Massa. *The Deep Green Planet: The Temperate Forest.* Austin: Raintree Steck Vaughn, 1997.

Greenpeace International (forests) Web site: www.greenpeace.org/~forests.

Massa, Renato. *The Deep Green Planet: The Tropical Forest.* Austin: Raintree Steck Vaughn, 1997.

Massa, Renato, and Monica Carabella. *The Deep Green Planet: The Coniferous Forest.* Austin: Raintree Steck Vaughn, 1997.

Rainforest Alliance, 65 Bleeker St., New York, NY 10012.

Save America's Forests, 4 Library Court, SE, Washington, DC 20003.

Society of American Foresters Web site: www.safnet.org.

U.S. Forest Service Web site: www.fs.fed.us.

World Conservation Monitoring Centre Web site: www.wcmc.org.uk.

World Resources Institute Forest Frontiers Initiative Web site: www.wri.org/ffi.

World Wildlife Fund, Worldwide Fund for Nature (Forests for Life Campaign) Web site: www.panda.org/forests4life.

forest fire, A natural or human-induced FOREST phenomenon that is integrally linked to the health and dynamics of many PLANT communities and animal POPULATIONS. Fire serves ECOSYSTEMS by RECYCLING nutrients, regulating plant succession and WILDLIFE HABITAT, maintaining BIODIVERSITY, reducing BIOMASS, enriching soils, and controlling the populations of INSECTS and other DISEASE-causing ORGANISMS. *(See* ecological succession.) Fire is a natural, necessary, and inevitable part of our ENVIRONMENT, that helps to maintain a healthy equilibrium in ecosystems. Fire is beneficial to some SPECIES and detrimental to others, is sometimes unpredictable and potentially destructive, and has shaped ecosystems throughout time. Its effects on wildlife are complex and often indirect, affecting habitat more than individuals.

Some tree species are fire adapted or even fire dependent (that is, they need fires to thrive successfully), with thick barks and tall, limbless trunks. For example, the longleaf pine forests of the southeastern coastal plain of the United States are among the world's most fire-dependent forest types. Naturally occurring fires regularly burn most original longleaf pine forests every two to seven years. Exclusion of fire from forest ecosystems can cause a shift toward species that are not adapted to fire and may not even be native.

In the United States, today's forests differ significantly from virgin forests, primarily because of humans' elimination of fire. Successful fire exclusion over the past 60–70 years has disrupted fire cycles, contributing to greater stand densities and an increase in the potential for devastating crown fires. Fire exclusion increases the incidence of tree diseases because the forest understory becomes unnaturally dense, competing with taller trees for space, water, and nutrients, which causes stress on them and decreases their disease resistance. As a result of such long-term suppression efforts, large wildfires since the 1980s in dead and dying western U.S. forests have accelerated the rate of forest mortality, and threatened people, property, and NATURAL RESOURCES.

Fire serves ecosystems by recycling nutrients, regulating plant succession and wildlife habitat, maintaining biodiversity, reducing biomass, enriching soils, and controlling insect and disease organism populations. *Credit:* National Interagency Fire Center, Bureau of Land Management

Fire can be friend or foe, depending on whether or not it is under control. Currently, an average of 12–13 million hectares (about 28 million acres) burn out of control worldwide each year. In 1988, the great Yellowstone fires alone burned 569,000 hectares (1.4 million acres.) Catastrophic fires such as these result from suppression of low-intensity fires, which allows a buildup of fuel. If the amount of FUEL loading is high, the resultant fire intensity can cause severe ecological damage to the point of destroying the forest.

The arrangement of dead and living vegetation affects the way fires burn. For example, too much dead, fallen material creates a continuous horizontal arrangement of fuel, causing surface fires to ignite more quickly, burn with greater intensity, and spread more rapidly and extensively. An increase in the density of small understory trees creates a continuous vertical fuel arrangement, which may allow a fire that would normally stay at the ground level to spread upward and become a crown fire. Older trees can survive ground-level fires, but cannot live through flames that climb up densely packed small trees to engulf treetops. This type of crown fire can also travel much farther up slopes than ground-level fires.

Unfortunately, several late twentieth-century forest fires destroyed valuable woodland in areas where the forests were already threatened by pressure from human activity. During 1992, the WEATHER phenomenon known as EL NIÑO created conditions that favored the development of forest fires. One of the worst forest fires in the history of Zimbabwe occurred in 1992, following an extended DROUGHT. More than 5,000 hectares (12,000 acres) of eucalyptus forests burned beyond the control of firefighters. Also in 1992, a forest fire destroyed approximately 10,000 hectares (25,000 acres) near Ratibor, Poland, after unusually hot weather and a prolonged drought in Europe. Several concurrent fires burned nearby in Poland, the Ukraine, and Russia during this period. Before the Ratibor fire, this forest had functioned as a natural filter to remove POLLUTANTS that had been emitted into the AIR from industrial sources in neighboring Czechoslovakia.

In 1994, a severe drought in Southeast Asia produced extensive forest fires throughout the Philippines, affecting reforested areas, fruit plantations, and primary virgin forest. Large-scale forest fires that occurred on the Indonesian islands of Borneo and Sumatra produced thick smoke that disrupted airline flights and caused severe AIR POLLUTION. At least 8,000 hectares (20,000 acres) of natural forest were lost in these events.

Today, proper FOREST MANAGEMENT dictates that fire be included as part of a forest ecosystem's natural forces. PRESCRIBED BURNING is a tool used by land managers to reintroduce fire to ecosystems. This method attempts to mimic natural fires, although prescribed fires are often too small to be effective and are sometimes conducted at an unnatural time of year—in the wet season rather than the dry season. *See also* ECOLOGICAL SUCCESSION; ECOLOGY; FIRE.

EnviroSource

National Wildland Fire Page Web site: www.nifc.gov

forest management, The application of scientific, economic, and social principles to the administration of a FOREST for specific objectives. Historically, the goals and objectives of forest management have reflected the changing needs and values associated with forests. Traditional objectives revolved around timber production; by contrast, many present-day forest managers apply the principles of MULTIPLE USE and SUSTAINED YIELD, balancing CONSERVATION of BIODIVERSITY, maintained or increased productivity, and preservation of recreational values.

Constraints that may influence implementation of forest management objectives include the presence of ENDANGERED SPECIES or sensitive areas, such as steep slopes or RIPARIAN HABITATS, which require special attention. Ease of transportation also affects how a forest is managed; the cost of road-building and maintenance influences which practices can be carried out and where. Ownership and size of the forest in question are additional factors. A small forest or a forest that falls under numerous jurisdictions must be considered in the context of the overall landscape rather than political boundaries. *See also* BIOREGION; DEFORESTATION; FORESTRY; FOREST SERVICE; GREENBELT MOVEMENT; OLD-GROWTH FOREST; OWL, NORTHERN SPOTTED; SILVICULTURE, SUSTAINABLE DEVELOPMENT; TREE FARM.

EnviroSources

National Forest Foundation Web site: www.nffweb.org.
U.S. Forest Service Web site: www.fs.fed.us.

Forest Service, The largest agency of the U.S. DEPARTMENT OF AGRICULTURE, with the responsibility for national CONSERVATION leadership in the management, protection, and use of America's FORESTS and rangelands. The Forest Service's mission is summarized as "caring for the land and serving people." The agency manages NATURAL RESOURCES within a MULTIPLE-USE, SUSTAINED-YIELD concept to achieve quality land management and meet people's diverse needs. Multiple use and sustained yield require the management of resources within the best combination of uses to benefit the American people, ensure the productivity of the land, and protect the ENVIRONMENT.

Congress established the Forest Service in 1905 to provide water and timber for the nation's benefit. When the nation's resource needs changed, Congress directed the agency to manage national forests for additional multiple uses and benefits, and for the sustained yield of RENEWABLE RESOURCES such as water, WILDLIFE, and recreation.

The Forest Service has four branches: the National Forest System; International Forestry; Research; and State and Private Forestry. The National Forest System has jurisdiction over national forests and GRASSLANDS. There are 155 national forests and 20 grasslands in 44 states, Puerto Rico, and the Virgin Islands. The national forests comprise 77.3 million hectares (191 million acres), an area the size of Texas.

The International Forestry branch establishes partnerships with other nations to exchange technical and administrative expertise on conservation and FOREST MANAGEMENT. The Institute for Tropical Forestry is an important component of the Forest Service's international program, providing U.S. support for protection and sound management of the world's forest resources. *(See* rainforest.) The International Forestry branch works closely with other federal agencies, such as the U.S. Agency for International Development, and with nonprofit developmental, wildlife, and international assistance organizations.

Forest Service Research is concerned with developing new FORESTRY-related technology and information. Such research is conducted through a network of forest and range experiment stations and the Forest Products Laboratory. The U.S. Forest Service is the largest forestry research organization in the world.

The State and Private Forestry branch provides financial and technical assistance to state and local governments, Indian tribes, and private organizations and landowners. This branch cooperates with other land managers in the management, protection, and development of forest land in nonfederal ownership. Cooperative activities include fire management and URBAN FOREST initiatives. *(See* forest fire.)

The National Forest Foundation was created by Congress to be the official nonprofit partner of the U.S. Forest Service. Its mission is to help the Forest Service to care for the nation's forests for the benefit of future generations. The National Forest Foundation receives funding made available by Congress and also solicits funds from the private sector. While the Forest Service is prohibited by law from soliciting outside funding, the National Forest Foundation has been expressly designated to fulfill that function. *See also*

DEFORESTATION; FOREST MANAGEMENT; NATURAL PUBLIC SERVICE; OLD-GROWTH FOREST.

EnviroSources

National Forest Foundation Web site: www.nffweb.org.
U.S. Forest Service Web site: www.fs.fed.us.

forestry, The development and application of scientific principles and practices related to FORESTS, forest products, and the integrated management, protection, and CONSERVATION of forested lands. Forestry addresses a complex issue: how to use trees, forests, and related resources to improve people's economic, social, and environmental conditions while ensuring that resources are conserved to meet the needs of future generations.

The science of forestry, adapted by GIFFORD PINCHOT was established in the United States at the beginning of the twentieth century, when vast areas of American forests had already been cut down with little thought for the future. Since that time, forestry has become an international science, and many forestry subspecialties have been developed to address the particular characteristics, requirements, and potential productivity of diverse forest ECOSYSTEMS, from tropical RAINFORESTS to URBAN FORESTS. *See also* AGROFORESTRY; FOREST MANAGEMENT; MUIR, JOHN; FOREST SERVICE.

EnviroSources

Robinson, Gordon. *The Forest and the Trees: A Guide to Excellent Forestry*. Washington, DC: Island Press. 1988.
Society of American Foresters Web site: www.safnet.org.
United Nations Food and Agricultural Organization, Forestry Programme Web site: www.fao.org/waicent/faoinfo/forestry.
U.S. Forest Service Research Web site: www.fs.fed.us/links/research.shtml.

Fossey, Dian (1932–1985), An American biologist and WILDLIFE activist whose life and work with the endangered mountain GORILLAS of Rwanda and Zaire were chronicled in her 1983 book, *Gorillas in the Mist*, and a 1986 film of the same name. Largely as a result of her work, mountain gorillas are now protected by the government of Rwanda and their wild POPULATIONS are recovering slowly.

Dian Fossey became interested in Africa and made a six-week trip there in 1963. At Olduvai Gorge, she met anthropologist Louis Leakey who impressed upon her the importance of doing research on great apes; this meeting inspired her to study mountain gorillas. In 1966, Fossey obtained support from the National Geographic Society and the Wilkie Foundation for a research program in Zaire (then called the Belgian Congo). A civil war there forced her to move across the border to Rwanda, where in 1967 she established the Karisoke Research Center in the Virunga Mountains, within the Parc National des Volcans. There, she used many of the methods JANE GOODALL employed in her chimpanzee research, such as naming each animal. In 1970, her efforts to acclimate the gorillas to her presence succeeded when an adult male touched her hand, the first friendly gorilla-to-human contact ever documented.

Intense observation over thousands of hours enabled Fossey to earn the trust of the wild gorilla groups she studied and brought forth new knowledge concerning gorilla behavior. When poachers attacked and killed Digit, a gorilla of whom Fossey was particularly fond, she initiated an intense public

Dian Fossey was an American biologist and wildlife activist who lived and worked with the endangered mountain gorillas. *Credit:* The Dian Fossey Gorilla Fund International

campaign against gorilla POACHING. *National Geographic* assisted her by featuring a cover story on the gorillas' plight. Monetary contributions poured in from around the world, allowing Fossey to establish the Digit Fund (renamed the Dian Fossey Gorilla Fund International in 1992) and to dedicate the rest of her life to the protection of the mountain gorillas.

In 1980, Fossey accepted a position at Cornell University and began writing *Gorillas in the Mist*. Its publication in 1983 brought her worldwide attention and helped to again focus international attention on the mountain gorillas, whose numbers had by then dwindled to 250, from an estimated 500 only 13 years earlier. She then returned to Karisoke to continue her campaign for the animals' survival.

Dian Fossey was murdered in her cabin at Karisoke on December 26, 1985. Many believe the crime was committed in retaliation for her antipoaching efforts. Her work has been continued by dedicated researchers and by the Rwandan staff at Karisoke. Today, the mountain gorilla population is making steady gains in the Virunga area and elsewhere, a trend that can be substantially attributed to the achievements of the Dian Fossey Gorilla Fund International and its supporters. *See also* ADAMSON, JOY GESSNER; ECOTOURISM; ENDANGERED SPECIES; GALDIKAS, BIRUTÉ; ORANGUTANS.

EnviroSources

Chang, Chris. "Champions of Conservation: *Audubon* Recognizes 100 People Who Shaped the Environmental Movement and Made the Twentieth Century Particularly American." *Audubon* 100, no. 6 (November–December 1998): 121.
Dian Fossey Gorilla Fund International Web site: www.gorillafund.org.
Fossey, Dian. *Gorillas in the Mist*. New York: Houghton Mifflin, 1983.
———. "The Imperiled Mountain Gorilla." *National Geographic Magazine* (April. 1981) 159: 501–23.

———. "Making Friends with Mountain Gorillas." *National Geographic Magazine* (January 1970) 137: 48–67.

———. "More Years with Mountain Gorillas." *National Geographic Magazine.* (October 1971) 140: 574–585.

Montgomery, Sy. *Walking with the Great Apes.* New York: Houghton Mifflin, 1991.

Mowat, Farley. *Woman in the Mists: The Story of Dian Fossey and the Mountain Gorillas of Africa.* New York: Warner Books, 1991.

Mountain Gorilla Protection Project Web site: deathstar.rutgers.edu/projects/gorilla/gorilla.html.

fossil fuel, A naturally occurring, nonrenewable ENERGY resource, including COAL, PETROLEUM, and NATURAL GAS, formed in Earth's crust over millions of years by the chemical and physical alteration of PLANT and animal residues. The United States currently is reliant on fossil fuel for about 85% of the energy it consumes; worldwide, fossil fuel also supplies most of the energy used. Because of increased energy demands, Earth's known natural gas and petroleum reserves are expected to last 50–60 years, while coal reserves will last longer, about 250–300 years.

Coal, the most economically important solid fossil fuel, is used in developed nations primarily for production of ELECTRICITY, with smaller amounts required for heating and for metallurgical processes. The principal liquid fossil fuels are the refined products of crude petroleum, including GASOLINE, kerosene, and diesel oil. Gasoline is by far the major petroleum product because of its widespread use in AUTOMOBILES and other vehicles. The most important gaseous FUEL is natural gas, which contains light HYDROCARBON compounds, mostly METHANE and ethane, and is used for home heating and cooking and for industrial heating. Gaseous fuels are convenient to use because they can be readily turned on and off, produce no smoke, and leave no ASH behind; however, they generally are not as easily transported as other fuels.

There is growing anxiety that oil and natural gas are being used up quickly, which has led to a search for ALTERNATIVE ENERGY RESOURCES. Oil shale, a type of sedimentary rock with embedded oils, is one such alternative. *(See* rock cycle.) Although there has been some commercial extraction of oil from shale, the process is difficult and costly. Large-scale oil extraction from shale is not presently economical, although eventually it may become a major source of liquid fuel. Even more complex is the extraction of the heavy, tar-like hydrocarbons from tar sands (sedimentary rocks). Present technology is not adequate for large-scale, economical extraction of fuels from tar sands, though rising petroleum prices and limited availability of other fossil fuels eventually may make tar sands an attractive fuel source.

Until the late 1940s, the United States was a net exporter of fuels, primarily oil and coal. Since 1952, however, the nation has had to import fuel to meet its needs; by 1981, the cost of fuel imports exceeded $75 billion per year. While the situation improved slightly with the drop in oil prices during 1982–83, and more recently in the late 1990s, the long-range outlook is for an increase in fuel costs. Exploration for oil and natural gas is becoming more expensive as WELLS are drilled deeper and in less accessible locations, such as deep-sea basins. Areas that are currently under legal protection from MINERAL activities, such as the ARCTIC NATIONAL WILDLIFE REFUGE, are now being considered as potential areas for new oil and gas exploration. Coal MINING expenses will also increase as mines become deeper and/or yield coal of poorer quality.

When large volumes of fossil fuels are burned, the GREENHOUSE GASES and other byproducts that are released into the AIR cause serious environmental problems such as SMOG, ACID RAIN, and an enhanced GREENHOUSE EFFECT. Scientists generally believe that the combustion of fossil fuels and other human activities are the primary reason for the increased concentration of CARBON DIOXIDE in the ATMOSPHERE. The combustion of fossil fuels to run cars and trucks, heat homes and businesses, and power factories is responsible for about 80% of the world's carbon dioxide EMISSIONS, about 25% of U.S. methane emissions, and about 20% of global nitrous oxide emissions.

Industrial processes that form greenhouse gases also produce a host of other air POLLUTANTS, including airborne PARTICULATE MATTER that may have immediate public health impacts. Air pollutants from the combustion of fossil fuels have global impacts because such pollutants can be transported thousands of kilometers from their source by wind and air currents, as occurred during the Kuwaiti oil fires. *See also* AIR POLLUTION; ALTERNATIVE FUEL; CARBON MONOXIDE; GLOBAL WARMING; NITROGEN OXIDES; NONRENEWABLE RESOURCES; GASOHOL; VOLATILE ORGANIC COMPOUNDS.

EnviroSources

U.S. Department of Energy, Energy Information Administration Web site: www.eia.doe.gov.

U.S. Department of Energy, Office of Fossil Energy Web site: www.fe.doe.gov.

U.S. Geological Survey Energy Resources Program Web site: energy.usgs.gov/index.html.

Freedom of Information Act, National legislation passed by Congress in 1966 to allow any person to request and obtain information from the federal government of the United States. Major amendments to the Freedom of Information Act (FOIA) were made in 1986 and 1996. The FOIA provides citizens with a means of obtaining information gathered by the government about various issues, including documentation regarding environmental problems, such as spills of toxic substances; fines imposed on violators of environmental law; efforts to protect the ENVIRONMENT; and information about uses and activities involving PUBLIC LANDS, such as national parks. The position of Congress in passing the FOIA was that government work is "for and by the people" and that government information should be available to everyone. Citizens who request government information under the FOIA are not required to identify themselves or explain why they want the information they have requested. All branches of the federal government are required to adhere to the provisions of the FOIA with certain restrictions for works in progress (early drafts), confidential information, classified documents, and national security information. *See also* EMERGENCY PLANNING AND COMMUNITY RIGHT-TO-KNOW ACT; ENVIRONMENTAL PROTECTION AGENCY; TOXIC RELEASE INVENTORY.

Friends of the Earth, A nonprofit global organization founded by DAVID Brower in 1970, dedicated to the CONSERVATION, restoration, and rational use of Earth's resources. Friends of the Earth (FOE), whose slogan is "The Earth needs all the friends it can get," focuses on protecting the planet from environmental degradation; preserving biological, cultural, and ethnic diversity; and empowering citizens to have an influential voice in decisions affecting the quality of their ENVIRONMENT and their lives. Areas of interest include GLOBAL WARMING, tropical DEFORESTATION, GROUNDWATER and drinking water contamination, preserving marine BIODIVERSITY, NUCLEAR ENERGY, AIR POLLUTION, and East-West cooperation.

The FOE is a federation of autonomous environmental organizations from all over the world. Members in more than 50 countries campaign on the most urgent environmental and social issues of the day. Friends of the Earth sponsors many projects, such as the Community Support Project, begun in 1996. Through this effort, the FOE provides assistance to grassroots groups and individuals who are working for local environmental improvement. The Clean Steel Coalition, a subsidiary of the FOE, is a consortium of environmental, ENVIRONMENTAL JUSTICE, and labor groups working to promote pollution prevention and lessen the adverse impacts of iron and steel production on health and the environment.

The FOE's Healing the Atmosphere Campaign was started in the 1980s to push for the phase-out of OZONE-depleting chemicals. The FOE has led efforts in this area on a national and international basis. Another FOE project is the River Restoration Project, in which the FOE works with environmentalists, Native American tribes, labor groups, and fishing organizations to restore historic SALMON runs and damaged RIPARIAN HABITATS in the Pacific Northwest.

The FOE's new publication, *Close to Home*, provides news, information, contacts, and ideas to activists working on state and local environmental protection issues. *See also* BIODIVERSITY; GREENPEACE; WORLD WILDLIFE FUND.

EnviroSources

Friends of the Earth, 1025 Vermont Ave., NW, Suite 300, Washington, DC 20005-6303; (202) 783-7400; fax: (202) 783-0444; Web site: www.foe.org.

Friends of the Earth-Northwest, 4512 University Way, NE, Seattle, Washington 98105; (206) 633-1661; fax: (206) 633-1935

frugivore. *See* herbivore.

fuel, Any material that can be chemically converted to ENERGY to provide power for motor vehicles, appliances, and other machines as well heat and ELECTRICITY. The most common fuels are FOSSIL FUELS, which contain CARBON compounds and, when burned, combine with OXYGEN in a chemical reaction that produces heat. This heat can then be converted into mechanical or electrical energy. Worldwide, nonrenewable fossil fuels provide people with most of their electricity, heat, and the energy source for modern forms of transportation.

ALTERNATIVE FUELS cause less environmental damage than traditional fossil fuels do. All alternative fuels reduce OZONE-forming tailpipe EMISSIONS in motor vehicles. The U.S. DEPARTMENT OF ENERGY (DOE) currently recognizes several substances as alternative fuels. METHANOL and ETHANOL are fuel mixtures that contain no less than 70% alcohol. Methanol is produced from NATURAL GAS, COAL, and BIOMASS, and ethanol is produced from sources such as corn and other grains. Other alternative fuels include compressed natural gas and liquefied NATURAL GAS, both produced from underground reserves, and liquefied PETROLEUM gas, a byproduct of petroleum refining and natural gas processing. Other alternative fuels include HYDROGEN, coal-derived liquid fuels, fuels derived from BIOMASS, and SOLAR ENERGY.

The energy in fuel is expressed as a heating value measured in British thermal units (Btus) per pound of fuel or kilocalories per kilogram. One Btu is the amount of heat required to raise the temperature of 1 pound of water by 1°F. *See also* HYDROCARBONS; NONRENEWABLE RESOURCES; RENEWABLE RESOURCES.

EnviroSources

National Alternative Fuels Hotline: (800) 423-1363.

U.S. Department of Energy, Energy Information Administration Web site: www.eia.doe.gov.

U.S. Department of Energy, Office of Fossil Energy Web site: www.fe.doe.gov.

U.S. Department of Energy Web site: www.afdc.doe.gov.

U.S. Department of Energy Web site: www.fleets.doe.gov.

fuel cell, A device that generates ELECTRICITY by electrochemically combining HYDROGEN and OXYGEN without combustion, producing water and heat as its only waste products. A fuel cell consists of two electrodes sandwiched around an electrolyte. Oxygen passes over one electrode, hydrogen over the other. This activity results in the flow of electrons known as electricity. Unlike a BATTERY, a fuel cell does not "run down" or require recharging; it generates ENERGY as long as FUEL is supplied. In addition, fuel cells are quiet, flexible, and operate at low temperatures.

There are many types of fuel cells, including alkaline fuel cells (AFCs), phosphoric ACID fuel cells (PAFCs), proton exchange membrane fuel cells (PEMs), solid polymer fuel cells (SPEs), molten carbonate fuel cells (MCFCs), and solid oxide fuel cells (SOFCs). The different types of fuel cells use different electrolytes, operate at different temperatures, and are suited to different uses. For example, PEMs are considered best suited for use in AUTOMOBILES (still in development); MCFCs are suited for use with gas TURBINES; and AFCs have been used to produce electricity aboard both Russian and U.S. space vehicles.

Engineers are currently experimenting with using fuel cells to power automobiles and buses. The vehicles are now tested in Chicago, and Vancouver, British Columbia. California is also testing nonpolluting buses and fuel-cell cars. Fuel-cell automobiles are at an earlier stage of development than are the battery-powered cars; however, use of fuel cells offers several advantages over battery-powered vehicles, such as the ability of fuel cells to be refueled quickly and to travel greater distances between refuelings. They do not need constant recharging like battery-powered automobiles. In addition, automobiles powered by fuel cells are likely to produce fewer GREENHOUSE GASES than are the HYBRID cars powered by both GASOLINE and batteries.

Although fuel-cell technology is at an early stage, many scientists and environmentalists believe fuel-cell driven vehicles can cut down on much of the current urban pollution from vehicle exhausts. *See also* ALTERNATIVE ENERGY RESOURCES; GLOBAL WARMING; GREENHOUSE EFFECT.

Time Capsule

The first FUEL CELL was built in 1839 by Welsh judge and scientist Sir William Grove; however, the use of fuel cells as a practical generator began in the 1960s, when the National Aeronautics and Space Administration (NASA) chose fuel cells over NUCLEAR POWER and SOLAR ENERGY to power the *Gemini* and *Apollo* spacecraft. Today, fuel cells provide ELECTRICITY and water on the space shuttles.

Tech Watch

Many automobile experts believe that fuel cell technology will be the automotive propulsion system of the future. So several auto companies are working on this technology. In 1999 the Daimler Chrysler automobile company introduced a fuel cell car in the United States. The electric-powered car, the NECAR 4, was also chosen at the "1999 International Engine of the Year Awards" as the best engine concept for the future.

According to the automaker, the NECAR 4 is based on the Mercedes-Benz A-class cars and operates on liquid hydrogen. The fuel cells generate the electrical energy required for powering the vehicle from a chemical reaction between hydrogen and oxygen. NECAR 4 is a zero-emission vehicle which can cover about 448 kilometers (280 miles) on one tank filling, reaches a top speed of 144 kilometers (90 miles) per hour and provides ample room for five occupants and their luggage. The NECAR 4 is powered by liquid hydrogen stored in a cryogenic cylinder resembling a large thermos at the rear of the vehicle. The fuel is then processed by a Proton Exchange Membrane Fuel Cell (PEMFC.) Inside the PEMFC, a platinum-coated membrane separates hydrogen into protons and electrons and combines them with oxygen in the air to form water. This surplus and deficit of electrons and protons creates positive and negative terminals that, when connected, produce electricity, which in turn, powers the vehicle. Daimler Chrysler plans to have fuel cell vehicles in limited production by 2004.

EnviroSources

Crest's Guide to the Internet's Alternative Energy Resources Web site: solstice.crest.org/online/aeguide/aehome.html.
Department of Energy Alternative Fuels Data Center at Web site: www.afdc.nrel.gov.
Department of Energy Website: www.doe.gov.
Fuel Cells and Their Applications, Gunter Simader, and Karl Kordesch. New York: John Wiley & Sons, 1996.
U.S. Department of Energy, National Alternative Fuels Hotline, P.O. Box 12316, Arlington, VA 22209, 1-800-423-1363

fuel rod, A long metal cylinder or tube containing URANIUM oxide pellets used as a FUEL in a NUCLEAR REACTOR. The tubes are made of stainless steel or other materials such as magnesium oxide alloy, zirconium, or carbide/graphite. Fuel rods are joined together to make a reactor core. Once the fuel is used up, the fuel rods are taken out of the reactor and stored. Because the SPENT FUEL is highly radioactive and produces a considerable amount of heat, it must be cooled and shielded. The fuel rods are stored in pools near the reactor site or placed in special dry-storage casks made of metal or concrete. The casks are then stored on a concrete pad above ground. *See also* CONTROL ROD; FISSION; RADIATION; RADIOACTIVE WASTE (NUCLEAR WASTE).

fuel wood, Harvested wood, also referred to as firewood, that is commonly used as a source of ENERGY for cooking and heating by as many as three billion people on Earth. For thousands of years, wood was the world's only source of energy utilized by humans; it has only been in the last few hundred years, since the Industrial Revolution, that people have used other sources of energy, such as FOSSIL FUELS.

Countries with Fuel Wood Shortages

Africa	Asia	Latin America
Botswana	Afghanistan	Haiti
Cape Verde	China	Peru
Ethiopia	India	Bolivia
Kenya	Nepal	
Mali	Pakistan	
Rwanda		
Sudan		

As late as the 1850s, wood supplied over 90% of U.S. energy requirements. Many developing nations still use wood as their primary FUEL. At the end of the twentieth century, half of the energy used in the continent of Africa, for example, came from fuel wood, and one-third of the world's POPULATION relied on fuel wood as a significant energy source. According to the United Nations, 1.3 billion metric tons (1.4 billion tons) of wood was used in 1990, either directly as fuel wood or in other energy production, representing half of all wood consumption.

There are three basic sources of wood for the production of energy: existing FORESTS, wastes from the forest products industry, and fuel wood plantations. In much of the developing world, existing forests are the major source of fuel wood. In most cases there is no FOREST MANAGEMENT, and wood may be harvested more rapidly than it can be replaced, sometimes resulting in DEFORESTATION. Dwindling fuel wood resources in many developing countries both cause hardship to rural populations and have damaging effects on the ENVIRONMENT. Overall, the use of wood as a fuel has fewer negative impacts on the environment than the use of fossil fuels if it is harvested sustainably and used with properly designed stoves, or other equipment. *See also* ALTERNATIVE ENERGY RESOURCES; ALTERNATIVE FUEL; BIOFUEL; BIOMASS; CARBON; CARBON CYCLE; GREENBELT

MOVEMENT; NONRENEWABLE RESOURCES: RENEWABLE RESOURCES; SUSTAINABLE DEVELOPMENT.

fungi, A kingdom of ORGANISMS, consisting of about 100,000 SPECIES, that include mushrooms, yeasts, rusts, smuts, molds, mildews, and puffballs. As a group, fungi include both single-celled and multicellular species. *(See* cell.) Most fungi are DECOMPOSERS that derive their nutrition by absorbing nutrients

The coconut scented milkcap is a fungus.

obtained through the DECOMPOSITION, or breakdown, of ORGANIC matter, such as the wastes or remains of other organisms. Many fungi also derive their nutrition from other living organisms, and thus are parasites. *(See* parasitism.) In addition to their roles as decomposers and parasites, fungi are of environmental importance because they can both cause and be used in the treatment of DISEASE, and because they provide a source of food to many other types of organisms.

Fungus Ecology. Fungi are widely dispersed throughout the world and exist in virtually every type of ENVIRONMENT. In most ECOSYSTEMS, fungi serve a vital role as decomposers—a role that benefits the environment in two ways. First, fungi help cleanse ecosystems of wastes that would otherwise build up on Earth's surface. In addition, as they feed, fungi break down complex organic matter into simpler organic and inorganic substances. Once broken down, matter that is not used by the fungi is returned to the environment, where it can be reused by PLANTS and other organisms. In this way, fungi, along with BACTERIA of decay, serve a vital role in Earth's BIOGEOCHEMICAL CYCLES.

Most fungi thrive in HABITATS that are moist, dark, and warm. Such habitats include both soil and water. Many fungi are able to live in polluted habitats, where they derive nutrition by breaking down the organic matter present in the POLLUTANTS. For example, fungi have been found living in ponds and other aquatic habitats that are polluted with SEWAGE. Fungi that live in soil include mushrooms, puffballs, and the water molds and downy mildews. Other fungi such as bracket fungi and various types of mushrooms live on decomposing plant matter such as rotting logs, where they decompose the cellulose and proteins contained in the plant and contribute to the formation of the nutrient-rich soil known as HUMUS.

Some fungi live in a symbiotic association with ALGAE in which they form a type of organism known as a LICHEN. *(See* symbiosis.) Lichens are able to colonize some harsh environments that most fungi cannot colonize by themselves; for example, in the arctic TUNDRA and TAIGA, lichens often grow on bare rocks or on tree trunks. These lichens provide a major source of food for animals, including the large moose that are common in arctic and taiga environments of North America. Another symbiotic association involves MYCORRHIZAL FUNGI, which grow among the roots of PLANTS, providing some benefit to the plants.

Other Uses of Fungi. In addition to aiding in decomposition and manufacture of a source of food to many organisms, humans have found fungi useful in the manufacture of a variety of products. For example, the single-celled yeasts, which obtain their energy through a process known as fermentation, are used in the production of breads and some alcoholic beverages. Other fungi, particularly those classified as molds, have been used for many years by peoples around the world to make a variety of cheeses, including blue cheese and brie. The mold that commonly causes bread to spoil (the black bread mold) is used in industry to make resins.

Fungi also produce a variety of substances that have medicinal value. Among the most well known is the antibiotic penicillin, which has been used worldwide to treat a variety of bacterial infections since the 1940s. Cyclosporine, a drug that is used to suppress the immune system following organ transplant surgery, also is derived from a fungus. Another fungus that is used by the medical community is ergot, a fungus

The fungi known as mushrooms are commonly observed growing in moist, shady areas. *Credit:* Linda Zierdt-Warshaw

that has long been known to cause humans to hallucinate when ingested. More recently, a substance obtained from ergot has been used to induce uterine contractions during childbirth.

Harmful Fungi. Not all activities of fungi benefit other organisms. Many fungi are extremely toxic to humans and other organisms when ingested. Other fungi are PATHOGENS, or agents of disease. For example, DUTCH ELM DISEASE is caused by fungi and has decimated many of the elm trees in North America and Europe. Thrush is a human fungal disease common in small children and in persons with diminished immune systems, such as those suffering from AIDS. A few fungal species are parasitic and often harm the organisms on which they live and feed. For example, ringworm is a fungal disease caused by a parasitic fungi that affects humans as well as some other MAMMAL species. The potato BLIGHT that decimated the potato crop in Ireland in the mid-1800s, bringing about that country's well-known FAMINE, also was caused by a parasitic fungi—in this case, a water mold. Other plant diseases caused by fungi include potato wart, rust and smut in corn, and clubroot in cabbage. These diseases, and others, destroy millions of dollars worth of crops throughout the world each year. *See also* AGRICULTURAL POLLUTION; AGRICULTURE; ALLERGY; ANAEROBIC DECOMPOSITION; BIODIVERSITY; BIOLOGICAL CONTROL; BIOTECHNOLOGY; BLIGHT; COMPOSTING; DETRITUS; ECOLOGICAL SUCCESSION; PESTICIDES; PIONEER SPECIES.

EnviroSources

Margulis, Lynn and Karlene V. Schwartz. *The Five Kingdoms: An Illustrated Guide to the Phyla of Life on Earth.* 2d ed. New York: W.H. Freeman and Co., 1996.

Rees, Robin, ed. *The Way Nature Works.* New York: Macmillan, 1998.

fungicide. *See* pesticides.

fusion, A nuclear reaction that occurs when two or more atoms combine to form a larger atom of a different element. The mass of the new element is less than the mass of the two atoms. The lost mass is converted to ENERGY. The sun and other stars produce energy through the process of fusion.

Helium is formed by the fusion of HYDROGEN ISOTOPES—deuterium and tritium. Temperatures greater than 1,000,000°C are required to initiate a fusion reaction. *See also* NUCLEAR ENERGY.

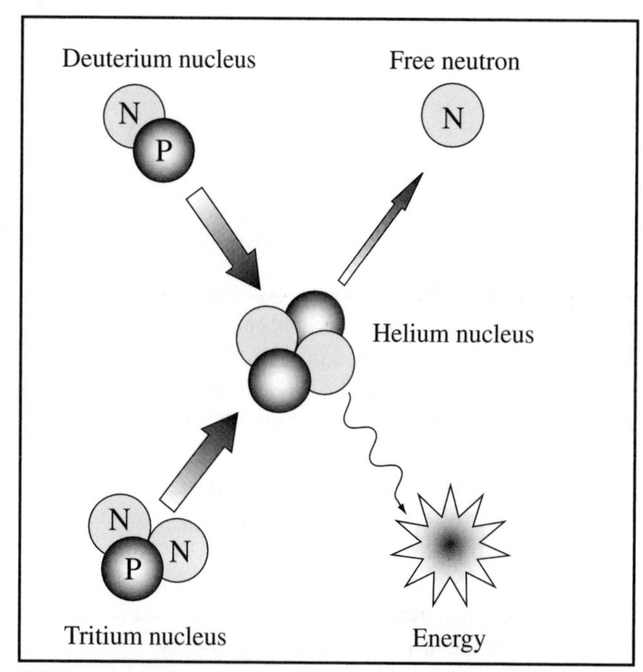

Fusion is a nuclear reaction that occurs when two or more nuclei combine to form a single, heavier nucleus releasing a large amount of energy.

G

Galápagos Islands, A region, also known as the Archipelago de Colon, that is an Equatorial island chain in the Pacific Ocean located about 1,000 kilometers (1,609 miles) off the South American coast. There are 13 large islands, 6 smaller ones, and 107 islets, with a total land area of about 8,000 square kilometers (4,971 square miles).

The Galápagos Islands had no native peoples and were not officially discovered by humans until 1535. The Galápagos were annexed by Ecuador in 1832, and small colonies gradually were established on several of the islands. Many of the present inhabitants moved to the islands from the Ecuadorian mainland over the last half-century, and the POPULATION is currently increasing at an annual rate of more than 8%. Five of the islands are inhabited by people, with a total population of around 15,000.

The Galápagos Islands were formed by volcanic activity three to five million years ago and have remained volcanically active. (*See* volcano.) The islands never were connected to the South American mainland; thus, long-distance dispersal brought life to the Galápagos Islands. AIR currents, OCEAN CURRENTS, and birds dispersed seeds, INSECTS, and other life forms.

Today, the islands are famous as a living laboratory of EVOLUTION in which scientists study processes related to the formation of new SPECIES. In 1835, the British research ship H.M.S. *Beagle* visited the Galápagos as part of a five-year voyage to make navigational charts for the Royal Navy. On board was CHARLES DARWIN, a young NATURALIST. The *Beagle* spent just five weeks in the archipelago, during which time Darwin visited four of the islands—Chatham, Charles, Albermarle, and James—collecting and observing the islands' FLORA and FAUNA.

Darwin accumulated extensive collections of Galápagos PLANTS and animals and was impressed by the fact that closely related species existed on different islands. He did not realize the significance of this observation at the time, however, and often neglected to label specimens with their island of origin. After years of additional research, he published *The Origin of Species by Means of Natural Selection* in 1859, in which he proposed his theory of evolution by natural selection. Darwin understood that the variations he had observed on Galápagos were nonrandom ADAPTATIONS to the ENVIRONMENT, which could be passed on to successive generations.

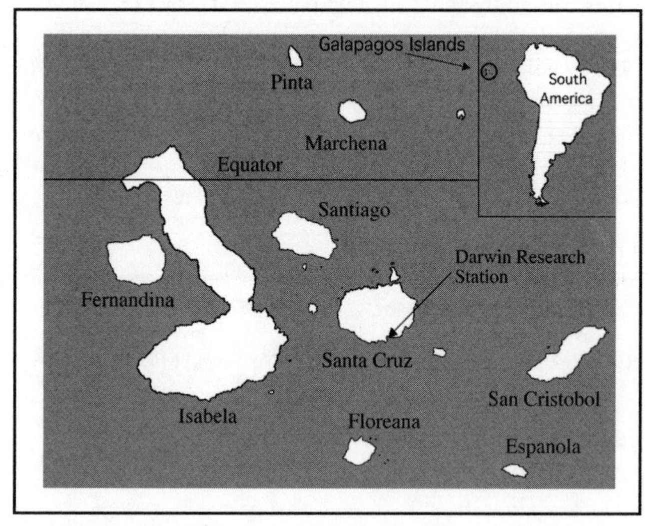

The Galapágos Islands are located off the South American coast.

Galápagos is home to many unique animals, most of which lack natural predators. (*See* predation.) The islands' isolation, confined HABITATS, and relatively small number of species have resulted in many instances of ADAPTIVE RADIATION. One of the best-known Galápagos species, the giant tortoise, has evolved into 14 distinct forms on the different islands of the archipelago. The 14 species of small, brown Galápagos finches—known collectively as Darwin's finches because they were central to the development of his theory—are each adapted to different foods. Other endemic Galápagos birds include four species of mockingbird, a flightless cormorant, and the only penguin species living in the Northern Hemisphere. The only native terrestrial MAMMALS are rice rats and two species of BATS. Sea lions and fur seals are present around the coasts while DOLPHINS, WHALES, and SEA TURTLES are common marine species. More than 300 species of fishes from the rich OCEAN waters around the archipelago also have been identified and described.

The plants of Galápagos are equally diverse and include endemic tree ferns, bromeliads, and orchids. Many kinds of plants, particularly those belonging to the daisy family, have evolved on the different islands into whole arrays of NATIVE SPECIES, providing scientists with classic examples of adaptive radiation in plants.

The uniqueness of the islands' flora and fauna has led to development of a strong ECOTOURISM industry in the region. In 1959, Ecuador designated 97% (800,000 hectares, or 1.98 million acres) of the land area of Galápagos as a national park. In 1986, the 7-million-hectare (17.2-million-acre) Galápagos Marine Resources Reserve was established, protecting the waters around the archipelago. The islands also have been recognized internationally as a BIOSPHERE RESERVE and a United Nations Educational, Scientific, and Cultural Organization (UNESCO) World Heritage Site. The Charles Darwin Research Station, operated by the international Charles Darwin Foundation since 1964, carries out scientific research and works in cooperation with the U.S. National Park Service.

The unique flora and fauna of the Galápagos Islands face many serious CONSERVATION problems today. EXOTIC SPECIES of introduced animals are the greatest threat to the native plants and animals of Galápagos. Today, only 2 out of 14 major islands are untouched by introduced mammals, and new introductions still occur. The aggressive Norway rat, for example, introduced around 1983, is present on at least two islands and is rapidly expanding its range. Goats compete for grazing resources with native HERBIVORES, such as tortoises and iguanas, and on some islands have devastated the natural vegetation. Feral dogs and cats prey on iguanas and sea bird chicks; rats and pigs eat the eggs of tortoises, turtles, and sea birds. For nearly 100 years on the island of Pinzon, for example, rats have killed every giant tortoise hatchling, leaving only an aging adult tortoise population. Some exotic plant species also seriously threaten the survival of native species.

Since 1961, personnel from the U.S. NATIONAL PARK SERVICE and the Charles Darwin Research Station have worked to eradicate these foreign species, but the process is difficult and expensive, especially on the larger islands. Hunting requires great effort because of the rough lava terrain and dense vegetation. Targeted trapping and poisoning also are difficult; extreme caution must be used to ensure that native animals are not harmed. Fencing is costly and not effective against smaller animals. The use of BIOLOGICAL CONTROLS involves careful research and poses the risk of creating new problems on the islands.

There have been notable successes, however. Researchers are breeding ENDANGERED SPECIES in captivity and then reintroducing them to boost wild populations. Goats have been eradicated from six of the smaller islands thanks to organized hunts, which may eliminate thousands of goats in a two-week period. Black rats are also gone from a few of the small islands. Wild dog packs that once threatened colonies of iguanas, fur seals, and penguins on the island of Isabela are now controlled. A yearly trapping and poisoning program for wild cats and rats around the nesting sites of the endangered dark-rumped petrel on Floreana and Santa Cruz has greatly increased the species' hatchling survival rate.

More recently, illegal fisheries for sharks, lobsters, and sea cucumbers have been operating in the marine park, taking stock for export to countries in the Far East. The effects of this industry have highlighted the need for effective patrols of the national park and Marine Resources Reserve. *See also* BIODIVERSITY; CAPTIVE PROPAGATION; COMMERCIAL FISHING; SEALS AND SEA LIONS; SPECIES DIVERSITY.

EnviroSources

Boyce, Barry. *The Traveler's Guide to the Galápagos Islands.* Edison, NJ: Hunter, 1998.
Charles Darwin Foundation for the Galápagos Islands, P.O. Box 17-01-3891, Quito, Ecuador, South America.
Charles Darwin Research Station: Web site: fcdarwin.org.ec.
Darwin, Charles. *The Origin of the Species by Means of Natural Selection.* Oxford, U.K.: Oxford University Press, 1998.
De Roy, Tui. *Galápagos: Islands Born of Fire.* Toronto, Canada: Warwick, 1998.
Galápagos Conservation Trust Web site: www.gct.org.
Virtual Galápagos Web site: www.terraquest.com/Galápagos.
World Wildlife Web site: www.panda.org.

Galdikas, Biruté (1948–), German-born Canadian primatologist and conservationist who has spent most of her adult life working in Indonesia, studying and fostering CONSERVATION of ORANGUTANS. Orangutans are anthropoid (humanlike) apes that inhabit tropical RAINFORESTS of Borneo and Sumatra. The animals have shaggy, red-brown coats, very long arms, no tail, and, unlike other primates, are known for being shy and leading solitary lifestyles.

Galdikas began her studies of orangutans, called the Orangutan Research and Conservation Project, in 1971; she has remained in Borneo, working within the Tanjung Puting National Park, ever since. Like DIAN FOSSEY, who studied mountain GORILLAS, and JANE GOODALL, who devoted her life's work to chimpanzees, Galdikas's research began under the guidance of noted anthropologist Louis Leakey with funding for the research provided at different times by the WORLD WILDLIFE FUND, the National Geographic Society, the Leakey Foundation, the Chicago Zoological Society, the New York Zoological Society, and EARTHWATCH.

In her work, Galdikas has observed orangutans to learn about their lifestyle and behavior and has become the world's leading authority on the SPECIES, which has consistently appeared on the ENDANGERED SPECIES LIST maintained by the U.S. FISH AND WILDLIFE SERVICE since 1970. In addition, Galdikas has spent much of her time raising orphaned animals whose mothers were killed by poachers (*see* poaching), treating sick or injured individuals, and rehabilitating orangutans that were kept in captivity as pets for release back into the wild.

For her work with orangutans, Galdikas has received a Humanitarian Award from People for the Ethical Treatment of Animals (1991), the CHICO MENDES Award of the SIERRA CLUB (1992), and a United Nations Global 500 Award (1993), and was recognized as an Eddie Bauer Hero of the Earth (1991). In 1995, she recounted her work in her book, *Reflec-*

tions of Eden: My Years with the Orangutans of Borneo. In addition, a documentary film featuring Galdikas and American actor Julia Roberts was produced in 1999 to help increase awareness of the problems faced by orangutans in the wild. *See also* ANIMAL RIGHTS; CONVENTION ON INTERNATIONAL TRADE IN ENDANGERED SPECIES OF WILD FLORA AND FAUNA; ENDANGERED SPECIES ACT; RED LIST OF ENDANGERED SPECIES; WORLD CONSERVATION UNION.

EnviroSources

Galdikas, Biruté. *Reflections of Eden: My Years with the Orangutans of Borneo.* Boston: Little, Brown & Company, 1995.

Montgomery, Sy. *Walking with the Great Apes.* Boston: Houghton Mifflin, 1991.

gamma ray. *See* radiation.

garbage, Disposable, BIODEGRADABLE food waste material from household and industrial sources. The U.S. ENVIRONMENTAL PROTECTION AGENCY, describes garbage as food waste from animals or PLANTS resulting from the handling, storage, packaging, sale, preparation, cooking, and serving of foods. Garbage is generally disposed in SANITARY LANDFILLS or burned in incinerators (*see* incineration); however, in some countries garbage is dumped in open pits where food wastes are exposed to INSECTS and rodents. Each year, millions of people get sick or die from illnesses resulting from the unsanitary disposal of garbage.

Garbage is a major problem for many urban areas in the world. The United States wastes the most. According to the NATIONAL AUDUBON SOCIETY, in 1988, the United States generated about 150 million metric tons (165 million tons) of commercial and residential waste a year. Americans generate about 1.5 kilograms (3.3 pounds) of garbage per person per day.

Several cities are initiating programs to deal with the garbage problem. One unique campaign, called the Garbage Purchase Program, takes place in Curitiba, Brazil, a city with a POPULATION of about two million people: Because the narrow streets in certain sections of the city prevent garbage trucks from picking up the garbage on many routes, residents who participate in the program come to the garbage trucks waiting along outlying streets. In exchange for the garbage, the residents receive compensation in the form of surplus food, bus tokens, and school supplies. The program is well received by the city dwellers. More than 35,000 families participate in this program. Other city programs make use of SOURCE SEPARATION, RECYCLING, and COMPOSTING to reduce some of the garbage being transported to landfills. *See also* LITTER; REFUSE; SEWAGE; SEWAGE TREATMENT PLANT; SEWER; SOLID WASTE; WASTE MANAGEMENT FACILITY.

gasohol, The trade name for a blend of GASOLINE (90% by volume) and ETHANOL (10% by volume) that can be substituted for 100% gasoline, serving to stretch gasoline supplies. The ethanol in the mixture acts as an oxygenate by adding chemically bonded OXYGEN. Gasohol has higher octane, or antiknock, properties than 100% gasoline and burns more completely, resulting in reduced EMISSIONS of some POLLUTANTS, including CARBON MONOXIDE and some GREENHOUSE GASES. Ethanol is the only motor FUEL additive that does not contribute to the GREENHOUSE EFFECT. According to the U.S. ENVIRONMENTAL PROTECTION AGENCY, a 10% ethanol blend reduces carbon monoxide emissions by 25–30% over pure gasoline; however, it also vaporizes more readily than straight gasoline, potentially aggravating OZONE pollution in warm weather.

In the United States, inexpensive gasoline supplies, a large demand for gasoline, and other parameters have historically limited investment in gasohol; however, political pressures and the large rural constituency of corn-growers particularly in the midwest prompted many advances in ethanol production during the 1980s. Although the initial fervor has been subdued by lackluster public interest and political concerns, use of such ALTERNATIVE FUELS has continued to rise in the 1990s. *See also* AIR POLLUTION; ALTERNATIVE ENERGY RESOURCES; AUTOMOBILE; ENERGY; FOSSIL FUELS.

EnviroSource

U.S. Department of Energy, Alternative Fuels Data Center Web site: www.afdc.nrel.gov.

gasoline, A light, volatile, highly flammable mixture of HYDROCARBONS obtained in the fractional distillation of PETROLEUM, SHALE OILS, or COAL and used as a FUEL for internal-combustion engines and as a SOLVENT. Gasoline is a complex mixture, containing hundreds of different hydrocarbons, most with 3 to 12 CARBON atoms per molecule, but varying widely in structure. Perhaps the most widely used product refined from petroleum, gasoline is useful as an AUTOMOBILE fuel because it easily evaporates to a gas, which when burned, releases a great deal of ENERGY.

The antiknock quality of gasoline used in engines is rated by octane number; isooctane has a value of 100. To increase octane rating, additives containing LEAD were widely used until the late 1960s; however, because of the health and environmental hazards posed by lead and the harmful effects it has on pollution control devices, in the 1970s, manufacturers began to change automobile designs and gasoline composition to exclude lead. In 1990, the CLEAN AIR ACT forced major compositional changes in gasoline; lead additives are now banned in the United States. Compounds such as methyl tertiary butyl ether (MTBE), which raises octane ratings and promotes more thorough combustion, and ETHANOL, used in GASOHOL, are now added to reduce pollution. However, MTBE is becoming an environmental problem from gasoline leaking into GROUNDWATER from buried gasoline storage tanks. Alternatives to gasoline as a vehicular fuel, such as NATURAL GAS and ELECTRICITY from storage BATTERIES, have also been used on a small scale, particularly in urban areas, where travel distances are shorter and AIR POLLUTION problems greater.

There are several known TOXINS in gasoline, some of which are confirmed human CARCINOGENS. The most well known of these toxins are lead and BENZENE, and both are regulated. Other aromatic hydrocarbons and some toxic olefins are also controlled. About 20% of all U.S. GREENHOUSE GAS EMISSIONS are attributable to motor vehicle gasoline consumption, and some 60% of the total weight of POLLUTANTS discharged into the atmosphere of the United States originates from this source. *See also* ALTERNATIVE ENERGY RESOURCES; ALTERNATIVE FUEL; CARBON MONOXIDE; NONRENEWABLE RESOURCES; VOLATILE ORGANIC COMPOUNDS.

EnviroSource

U.S. Department of Energy, Energy Information Administration Web site: www.eia.doe.gov.

gene pool, The total collection of genes that exists among a POPULATION of ORGANISMS. The traits of an organism are controlled by its genes, sections of deoxyribonucleic acid (DNA) that occupy a specific place on a particular chromosome. For organisms that reproduce asexually (reproduction involving only one parent), there is no variation in the genetic makeup of parent and offspring, unless a MUTATION occurs that alters the genes of the offspring. The diversity in such a gene pool is therefore low. In organisms that reproduce sexually (reproduction involving the union of CELLS from two parents), genes may combine in different ways to produce greater diversity among offspring. These variations explain why all offspring in the same litter are not identical to each other. As variation in a population's gene pool increases, the GENETIC DIVERSITY of the population also increases. Genetic diversity provides a population some survival advantages in a changing ENVIRONMENT by increasing the likelihood that some individuals will be better able to adapt to changes that occur. *(See* adaptation.) These individuals, in turn, will then pass the traits that permitted their survival onto their offspring, thus increasing their chances for survival. *See also* BIODIVERSITY; EVOLUTION; EXTINCTION; GENOME; HYBRIDIZATION.

General Mining Law of 1872, A law passed by the U.S. Congress and signed by President Ulysses S. Grant in May 1872, which regulates the MINING of hardrock MINERALS (nonfuel minerals such as gold, COPPER, and silver) in the United States. It was enacted, in part, to promote settlement of the western United States by prospectors and their families. The law transformed miners from trespassers into legitimate occupants of the PUBLIC LANDS. Since 1872, the law has never been updated.

Environmentalists believe that the mining law is outdated. Under the law, anyone who discovers a "valuable mineral deposit" on open public lands has a right to claim the land and mine it, no matter what other nonmineral values exist. Individuals were limited to 8 hectares (20 acres), while associations or groups were limited to 66-hectare (160-acre) claims. Claimants can purchase the land and minerals for $5 an acre or less. They do not have to pay a royalty for the value of ores taken from public lands, and there are no restrictions or requirements included in the law to protect the ENVIRONMENT during mining operations.

Multinational mining corporations have constructed huge mining operations that environmentalists believe could not have been anticipated by the original authors of the 1872 Mining Law. Massive machinery and new mining technologies and processes have devastated the landscape. Environmental impacts include contamination of precious water resources, destruction of WILDLIFE HABITAT, and the creation of vast amounts of TOXIC WASTE. Environmentalists would like to see the law changed and updated to ensure that future mining for gold, silver, copper, and other minerals on federal lands will incorporate environmental projections. *See also* CONSERVATION; NONRENEWABLE RESOURCES.

EnviroSource

Bureau of Land Management Web site: www.blm.gov.

genetic diversity, The variety of differences that exist among individuals of the same SPECIES as a result of their genes. Genes are the portions of chromosomes that carry the information for specific traits, such as leaf shape, flower color, hair or fur color, or body size. Differences in the traits of individual members of a species result from differences in their genes, or their genetic diversity.

ORGANISMS that reproduce sexually have the greatest genetic diversity, since each member of the POPULATION receives a different set of genes from each parent. Genetic diversity is a benefit to organisms because it increases the likelihood that some members of the species will survive changes or alterations in their ENVIRONMENT. *(See* adaptation.) Organisms that have traits that allow them to survive in a changing environment may pass these beneficial traits to their offspring, thus increasing the likelihood that their offspring will survive in a similar environment. By contrast, a species that lacks genetic diversity, such as those which reproduce asexually (reproduction involving only one parent), is less likely to have traits that allow individuals to adapt to a changing environment. Such species are thus more likely to become extinct. *(See* extinction.)

Scientists are applying their knowledge of genetic diversity to AGRICULTURE as a means of increasing crop yields. For example, many species of PLANTS grown as food crops, such as wheat and rice, are now being cross-bred with wild species to help increase the genetic diversity in the offspring. This cross-breeding, known as HYBRIDIZATION, has helped produce varieties of crop plants that have a greater resistance to changing environmental conditions such as DROUGHT, freezing temperatures, INSECTS and other parasites, and DISEASE. Hybridization also is used to produce traits in animals that people find appealing. For example, the great differences in appearance among the many breeds of domestic dogs and domestic cats has resulted from the genetic diversity that exists among their species. *See also* BIODIVERSITY; EVOLUTION; GENE POOL; GENETIC ENGINEERING; GENOME; HYBRID.

genetic engineering, A scientific practice, also known as genetic modification and genetic manipulation, that involves the artificial manipulation of the genetic material of living ORGANISMS, particularly by transferring the genes (sections of deoxyribonucleic acid [DNA]) that occupy a specific place on a particular chromosome) of one organism into another, rearranging the genes within an individual, or cloning (duplication) of an individual. Genetic engineering falls into the broader category of BIOTECHNOLOGY.

Breeding is the natural process of sexual reproduction within a SPECIES. In this process, the hereditary information carried in the genes of both parents is combined and passed on to the offspring. When sex cells combine, the genes from the sex cells may be exchanged between similar chromosomes, but the genes always remain at a precise position and order on the chromosomes, unless a MUTATION or other accident occurs. Generally, only organisms of the same species breed to produce offspring; however, it is sometimes possible for closely related species, such as a horse and a donkey, to interbreed, resulting in a new type of individual called a hybrid. As a rule, HYBRID offspring are infertile, or incapable themselves of breeding to produce offspring.

The field of genetic engineering can be divided into three major areas: gene splicing (the transfer of genetic material from one species to another), cloning, and GENOME studies, which involve the determination of DNA sequences in chromosomes.

Gene Splicing. Gene splicing is the artificial gene transfer from one species to another to improve or alter the functions of the recipient species. A great deal of research occurred in the 1990s, as part of commercial and scientific programs, to create genetically engineered, or transgenic, animals. Private enterprises have been actively testing this method on animals and PLANTS as a means of obtaining "super products," such as freeze-resistant tomatoes and citris fruits and DROUGHT-resistant varieties of wheat. Mice are often bred to carry human genes so that drugs can be tested on them.

Cloning. A sheep named Dolly was the first animal to be genetically duplicated via cloning using a complete set of chromosomes from an adult; a mouse named Cumulina, also was cloned. The first successful cloning of a male animal, a mouse named Fibro, took place in June 1999, suggesting that animals can be cloned from any CELL in the body, and not just from reproductive or fetal cells.

Genome Studies. Most research to determine the sequence of chromosomes and genes in organisms is supported largely by the federal government. Many private industrial enterprises also are extensively involved, because the patenting of biotechnological breakthroughs usually is lucrative.

Genetic engineering is not without controversy. Proponents of genetic engineering believe that if genetic information in the chromosomes is decoded and genetic mechanisms are fully understood, we can potentially control and improve human health, quality of life, and the BIOCHEMICAL processes in our bodies. We also might improve the genes of other animals and plants to enable them to better serve human needs. Many people, including ANIMAL RIGHTS activists, however, object to all genetic engineering, particularly on moral and ethical grounds. In the case of genetically engineered foods, such as transgenic tomatoes, many people are concerned about consumer safety. Furthermore, opponents of genetic engineering are concerned about the potential for accidental introduction of new TOXINS, allergens (*see* allergy), and PATHOGENS into the ENVIRONMENT. They also are concerned about how genetically engineered species released into the environment might impact ECOSYSTEMS and the NATIVE SPECIES and other WILDLIFE living in the region. To address such concerns, most countries have established some laws to regulate genetic engineering and its uses. For example, the United States and many countries throughout Europe have enacted laws that prohibit the cloning of humans, largely on moral and ethical grounds. Other laws have been passed regarding transgenic organisms designed for use as food. Such organisms generally must undergo testing and be qualified as safe for use before they may be made available to the public. In the United States, the FDA mandates that such products must carry labels identifying them as products of trangenesis to allow consumers opposed to the practice to avoid their purchase. *See also* GENE POOL; GENETIC DIVERSITY; HYBRIDIZATION.

EnviroSources

Council for Responsible Genetics Web site: www.gene-watch.org.
Genetic Engineering News Web site: www.genengnews.com.
National Institutes of Health, National Human Genome Research Institute Web site: www.nhgri.nih.gov.

genome, All the genes contained in a single, or haploid, set of chromosomes. The traits of all ORGANISMS are determined by their genome. In organisms that reproduce sexually, the genome of each parent is contained in its reproductive CELLS—sperm or egg. When these cells unite in fertilization, the genome of each parent is passed to the resulting zygote (the cell produced through the union of sperm and egg).

Scientists worldwide are working to identify the number, location, and functions of each of the genes contained in the genomes of various types of organisms. Such information can help scientists better understand the traits of organisms. The data may also lead to the use of GENETIC ENGINEERING techniques that can prevent organisms from developing certain hereditary DISEASES or disorders that could be passed on to offspring and lead to the EXTINCTION of their SPECIES. *See also* GENE POOL; GENETIC DIVERSITY; HYBRIDIZATION.

Geological Survey, An agency in the DEPARTMENT OF THE INTERIOR that was established by the U.S. Congress in 1879, to collect, analyze, interpret, publish, and disseminate information about the NATURAL RESOURCES of the nation's PUBLIC LANDS. An important part of that mission includes publishing earth science information needed to understand, to plan the use of, and to manage the nation's ENERGY, land, MINERAL, and water resources.

The U.S. Geological Survey is the nation's largest civilian map-making agency and the primary source of data on the nation's SURFACE WATER and GROUNDWATER resources. The agency assesses the energy and mineral potential of U.S. land and offshore areas, and issues warnings about impending

EARTHQUAKES, volcanic eruptions, landslides, and other geologic and hydrologic hazards. The agency also studies the geologic features, structure, processes, and history of the other planets of the solar system. *See also* GEOLOGY.

EnviroSource

U.S. Geological Survey Web site: www.usgs.gov; mapping.usgs.gov/www/products/mappubs.html.

geology, The branch of science that studies the origin, history, structure, and chemical and physical composition of Earth's crust. In geology, PLANT and animal fossils are studied to understand more about a given location and its history. The major principles of geology include superposition and uniformitarianism. Superposition is the process whereby younger rocks are deposited on older rocks. *(See* deposition; rock cycle.) Geologists use this principle to determine the age of the rock layers. Uniformitarianism is the assumption that the geological processes that formed rocks in the past are the same processes that are at work today. These processes include EARTHQUAKES, VOLCANOES, WEATHERING, EROSION, and the faulting and folding of rocks layers.

Environmental geology is the study of how physical conditions such as weathering and erosion, flooding, earthquakes, volcanoes, and landslides interact and impact human health, life, and property. It is also a study of how human activity affects the ENVIRONMENT. In a broad sense, environmental geologists study and analyze data from these NATURAL DISASTERS to suggest programs for minimizing or avoiding the risk of injury to human life and property. These proposals might include better zoning laws to prevent overbuilding in earthquake zones or constructing homes on FLOODPLAINS and hills where there are frequent FLOODS and landslides. Environmental geologists also initiate CONSERVATION methods to preserve NATURAL RESOURCES.

geothermal energy, The use of natural heat ENERGY from steam, hot water, and hot dry rocks in the interior of Earth. Geothermal energy is an ALTERNATIVE ENERGY RESOURCE that can be used for the direct heating of buildings or for generating ELECTRICITY. Geothermal energy is not always listed as a renewable energy source because the depletion rate of sources such as hot water springs can be higher than the rate of replenishing or recharging those sources. Italy, Iceland, New Zealand, Russia, Japan, France, and the United States, use geothermal energy. Other countries include the Philippines, Indonesia, and Mexico, and various countries in Central and South America, eastern Africa, and eastern Europe.

The Geysers Geothermal Field, located in northern California, is the largest geothermal electric power plant in the world. It produces about 1,300 megawatts of electricity, enough to satisfy the residential electricity needs of 1.7 million Californians. This is more than the combined POPULATIONS of San Francisco, Oakland, and Berkeley in the California Bay Area. According to the U.S. Energy Information Agency, geothermal energy has the potential to provide the United States with 12,000 megawatts of electricity by the year 2010 and 49,000 megawatts by 2030.

Geothermal: A Hot Water Hydrothermal System

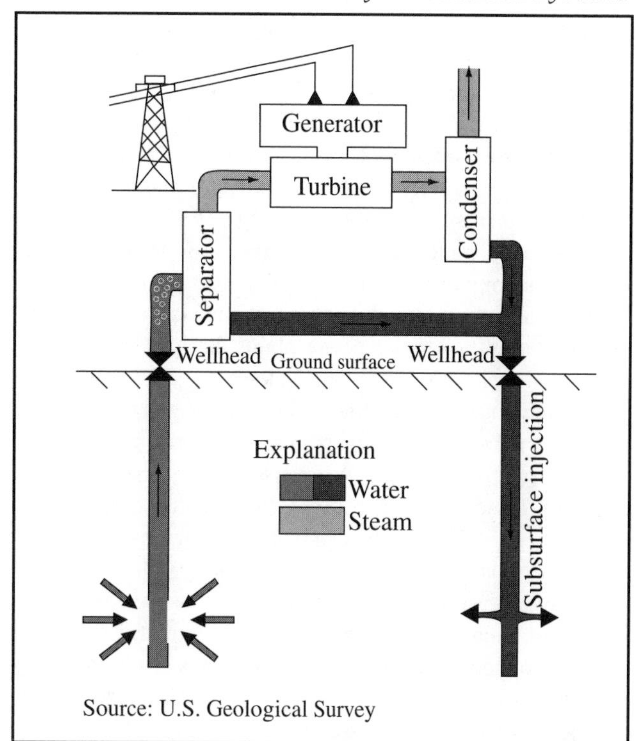

The diagram illustrates how electricity is generated from a hot-water hydrothermal system. The part of hydrothermal water that becomes steam is separated and used to drive a turbine generator. Wastewater from the separator and condenser is injected back into the subsurface to help replace the groundwater used in the hydrothermal system. *Credit:* Geothermal Education Office. Marilyn Nemzer.

The geothermal process begins when rainwater seeps into the ground near hot igneous rocks. *(See* rock cycle.) The GROUNDWATER is heated to form naturally occurring hot water and steam. These resources are tapped by WELL-drilling methods in order to generate electricity or to produce hot water for direct use.

For producing electricity, hot water is brought from the underground RESERVOIR to the surface through production wells where steam is separated from the liquid and fed to a TURBINE engine, which turns a generator. Spent geothermal fluid is injected back into parts of the reservoir to help maintain reservoir pressure. Geothermal heat sources were first used for electrical power production at the Larderello geothermal field in Italy, in 1903; electrical power is still being produced there.

Heated water from geothermal resources can be circulated by pipes through a home or building in order to provide heat or can be used to dry vegetables or heat greenhouses. In Iceland, for example, the entire city of Reykjavik is heated by geothermal energy.

Geothermal energy can also be produced by a "hot dry rock" system. In this technique, holes are drilled in the rocks to a depth of more than 7 kilometers (25,000 feet) where the rock temperature can exceed more than 250°C (480°F.) Surface water is injected into holes or cracks where hot rocks are located. The water is heated by the rocks and rises to the surface in the form of steam, which can be used to drive a turbine to generate electricity. Many geothermal energy specialists are enthusiastic about this technique since it poses few if any environmental concerns. Geothermal energy systems have proven to be extremely reliable and flexible. Geothermal electric power plants are online an average of 97% of the time, whereas nuclear and COAL plants average less online time. One of the most important advantages of geothermal energy is that it does not pollute the ATMOSPHERE with SULFUR DIOXIDE and NITROGEN EMISSIONS as do power plants that use FOSSIL FUELS. NITROGEN OXIDE emissions are much lower in geothermal power plants than in fossil fuel power plants. Nitrogen oxides combine with HYDROCARBON vapors in the atmosphere to produce ground-level OZONE, a gas that causes adverse health effects and crop losses as well as SMOG. Geothermal power plants require very little land and can be constructed quickly.

Geothermal energy has some disadvantages. It can be more expensive to use than other fossil fuels. THERMAL POLLUTION can occur from the WASTEWATER unless it is treated. The construction of geothermal plants in RAINFORESTS can destroy sensitive ecological HABITATS. Drilling wells can also disrupt underground faults and fissures, which can lead to seismic activity and landslides. *(See* earthquake.)

The future of using geothermal energy is still cloudy. Presently, it is not a major player in the overall energy needs of the world. Advocates believe that geothermal energy has great potential as an energy source, but more research and development has to be initiated by countries.

TechWatch

Geothermal Heat Pumps (GHP) are being used today to produce heat and air conditioning in several areas of the United States. More than 100,000 electrically-powered GHPs had been installed in homes and buildings in the United States by the mid-1990s. The heat pump operates on the same principal as the home refrigerator, which is a one-way heat pump. The GHP, however, can move heat in either direction. In the cold season, heat is removed from Earth and delivered into a building or home. In the warm months, heat is removed from the buildings or homes and delivered back to Earth to be stored for use during the cold months.

The GHP unit sits inside the home or building at the site of a normal gas furnace. In a typical installation, a loop of plastic pipe is placed in a nearby vertical drill hole at a depth of 30 meters to more than 100 meters (100 to several hundred feet.) The hole is backfilled with clay. A water/antifreeze solution is circulated through the loop and through the heat pump for removing heat from or transferring heat to the ground. In this system, there is no use of GROUNDWATER, nor is there any contact between the solution in the plastic pipe and surrounding soil, rocks, or groundwater. Installation conforms to local construction codes. Typical loop installations are warranted for 50 years.

EnviroSources

Center for Renewable Energy and Sustainable Technology Web site: solstice.crest.org.

Duffield, W.A., J.H. Sass, and M.L. Sorey, "Tapping the Earth's Natural Heat." U.S. Geological Survey Circular 1125.

———, 1994. (U.S. Geological Survey, Branch of Information Services, P.O. Box 25286, Denver, CO 80225.)

Energy and Geoscience Institute, University of Utah, 423 Wakara Way, Salt Lake City, UT 84108; Web site: www.egi.utah.edu.

Graham, Ian. *Geothermal and Bio-Energy* (Energy Forever.) Chatham, NJ: Raintree/Steck Vaugh, 1999.

Geothermal energy information Web site: geothermal.marin.org.

Geothermal database worldwide Web site: www.geothermal.org.

International geothermal Web site: www.demon.co.uk/geosci/igahome.html.

germ plasm bank, A repository for seeds and PLANT parts (germ plasm), such as cuttings, from which plants can be grown. Germ plasm banks, or seed banks, provide a great resource to AGRICULTURE and research. They also provide a means by which the genetic components of rare, threatened, and ENDANGERED SPECIES of plants can be maintained.

The first germ plasm banks were opened following World War II. Currently, the largest plant gene bank in the world is the National Plant Germ Plasm System of the United States. This system contains seeds or germ plasm from about 520,000 plant SPECIES. In addition, the system maintains a database of information about the genetic traits of each species along with its ADAPTATIONS for survival in different ENVIRONMENTS. Thus far, the United States has shared its germ plasm resources with more than 100 different countries throughout the world, helping these nations improve their crop yields by providing them with plants that are more suited to the natural environment and CLIMATE conditions of their region or providing them with plants that could be cross-bred with other varieties to produce hardier plants.

Using products from germ plasm banks, scientists have successfully developed varieties of crop plants that are resistant to certain types of DISEASE, parasites, and climate conditions, such as DROUGHT and freezing temperatures. For example, a variety of wheat that is resistant to the fungal disease called wheat bunt has been successfully produced. *See also* BIOTECHNOLOGY; BLIGHT; FUNGI; GENETIC ENGINEERING; GENOME; HYBRID; HYBRIDIZATION; PARASITISM.

Giardia lamblia, A protozoan CONTAMINANT in the feces of humans and animals that is associated with the DISEASE giardiasis. *(See* protist.) Ingestion of this protozoan in contaminated drinking water can cause severe gastrointestinal diseases that may last for weeks or months. Symptoms of the diseases include diarrhea, fatigue, and stomach cramps. Unless the drinking water is treated, *Giardia lamblia* can live in contaminated water from one to three months. *See also* COLIFORM BACTERIA; CRYPTOSPORIDIUM.

gill net, A long fishing net that is suspended vertically in the water to catch fish by their gills. The mesh of the net allows the head of the fish to go through but not the fins. Some environmentalists state that the use of the nets has caused a reduction of fish POPULATIONS and HABITATS and has destroyed other marine ORGANISMS as well. Millions of nontargeted fish or other organisms are killed each year when they become entangled in the nets and die.

The waters of the Sea of Cortés, off the western coast of Mexico, are breeding grounds for many SPECIES of fish. The Sea of Cortés produces as much protein per cubic meter of water as any OCEAN in the world. In the late 1970s, Mexican fishers began to use gill nets. The 130-meter (400-foot) gill nets could increase a boat's catch tenfold. By the mid-1980s, the number of fishing vessels with large nets increased and as a result huge quantities of fish were harvested. During the peak fishing years in the Sea of Cortés, there were more than 20,000 gill nets in use every day, capable of killing several metric tons of fish in one night. Within a few years, a dramatic decrease in the fish populations of the region occured, especially in species such as Cabrilla, snapper, yellowtail, grouper, and Jewfish. *See also* COMMERCIAL FISHING; DRIFTNETS; OVERFISHING.

EnviroSource

SeaWatch, 3939 N. Marine Drive, Slip 12, Portland, OR 97217, (619) 678-0235.

glacial deposit, Soil and rock laid down by the action of a GLACIER, (a large, naturally occurring, moving body of ice that exists for many years). Glacial deposits can provide important data about former CLIMATE and vegetation because their contents represent thousands of years of Earth's history. The chemical composition of old sediments may reveal information about the ATMOSPHERE in the past. Pollen buried in glacial deposits can indicate the types of PLANT communities that existed as the deposits formed. Geologists can estimate historical lake and sea levels from some glacial deposits. This information can help scientists to understand the possible effects of present global climate conditions.

Glacial deposits exist all over Earth, particularly in high latitudes (nearer Earth's poles) and in high mountainous regions where glaciers exist today or existed several thousand years ago. Glaciers have the ability to scrape, scour, and break up large quantities of rock and transport this material to new locations where the material is deposited in various forms. (*See* deposition.) From huge building-size boulders to microscopic clay particles, the materials that comprise glacial deposits have a vast range of sizes.

Large glacial deposits formed on the edges and at the front of glaciers are called moraines. Much of the land of Long Island (New York) and Cape Cod (Massachusetts) is made up of glacial moraines. Glacial tills are types of glacial deposits that consist of many different sizes of clay, sand, and gravel particles mixed together, usually densely packed. Other important glacial deposits include areas of sand or gravel deposited by glacial streams or rivers and clay deposits formed in the bottoms of glacial lakes. *See also* ICE AGE.

glacier, A large, moving mass of ice and snow that is melting away and retreating due to warming temperatures on Earth. A glacier is formed on land from layers of compressed snow that continually builds up year after year. Glaciers move slowly, pulled by gravity less than 1 meter (3.3 feet) a day. Glaciers

Before and after photos of glacier shrinkage in Glacier National Park, Montana. The glacier cave in 1932 and in 1988. *Credit:* top, George Grant, GNP Archives; bottom, Jerry DeSanto, photographer.

can range in length between 3 and 130 kilometers (2–75 miles) and can be as wide as several kilometers, with a thickness of 340 meters (1,000 feet) or more. Glaciers are different in appearance from ice sheets. Glaciers are usually long and narrow and are located in mountainous areas such as the Alps, while continental ice sheets, such as in ANTARCTICA, are larger in volume and have a more oval shape.

At one time ice sheets and glaciers covered about 30% of all land area in the world. Today, ice covers about 10% of Earth's land surface. Of the total amount of ice, the Antarctica ice sheets contains over 90% of the world's ice; the rest is contained mostly in glaciers. The last ICE AGE ended about 10,000 years ago, but there was a period called the Little Ice Age that lasted from about 1500 A.D. to 1900 A.D. During those years ice sheets and glaciers were advancing in some areas of Earth. Since the end of the Little Ice Age, glaciers and ice sheets have been retreating and shrinking. Environmentalists are now concerned that glaciers and ice sheets are disappearing too quickly because of warming conditions on Earth. Many scientists are convinced that GLOBAL WARMING over the last 60 years has led to the drastic decline of glaciers.

In 1999, researchers reported that two Antarctica ice shelves had lost nearly 1,700 square kilometers (1,100 square miles) of volume in one year. Such ice shelves are breaking up more quickly than were predicted. Scientists have reported that parts of Antarctica have warmed up by as much as 2.4°C (4.5°F) in the last 50 years. For example, the glaciers of Glacier National Park, in Montana, like glaciers all over the world, are shrinking. Some of the park's glaciers have already lost more than half of their volume, and the number of glaciers in the park has decreased from an estimated 150 in 1850 to approximately 50 in 1999. Scientists are concerned that the disappearing glaciers will have an adverse effect on the park's ECOSYSTEMS. If nothing is done to reduce global warming, park scientists predict that by the year 2030 there may not be a single glacier left in Glacier National Park.

Ice sheets in the Andes have been disappearing since the early 1960s, and glaciers in the Alps have lost about 50% of their total volume since 1850. Many countries and the United Nations are developing programs to monitor shrinking ice sheets and glaciers. The EARTHWATCH program of the UNITED NATIONS ENVIRONMENT PROGRAMME established the GLOBAL ENVIRONMENT MONITORING SYSTEM to link up and coordinate existing monitoring systems such as the World Glacier Monitoring Service (WGMS). Since the world's glaciers and ice sheets act as a thermometer of climatic change as they shrink or advance, the WGMS now has over 750 stations that monitor glaciers in 21 countries.

EnviroSources

Sierra Club Global Warming and Energy Team: (202) 547-1141; e-mail: information@sierraclub.org.
Sierra Club Public Information Center: (415) 923-5653.
U.S. Geological Survey (climate change and history) Web site: geology.usgs.gov/index.shtml.

Global Environment Monitoring System (GEMS), A special program established in 1975 by the EARTHWATCH program of the UNITED NATIONS ENVIRONMENT PROGRAMME (UNEP). UNEP established GEMS to link up and coordinate existing monitoring systems to research and analyze global ATMOSPHERE and CLIMATE conditions, environmental pollution, RENEWABLE RESOURCES, and environmental data. Over 140 nations and thousands of scientists participate in GEMS.

GEMS is one of the four main components of the Earthwatch program. The other components include the Global Resource Information Database, the INTERNATIONAL REGISTER OF POTENTIALLY TOXIC CHEMICALS, and the global information system called INFOTERRA. All of these components were brought together because the UNEP believed that in the past many scientists, governments, and decision-makers worked in isolation on environmental problems. Scientific inaccuracies and inadequate data were the result. The UNEP believed that with state-of-the-art technology, such as satellite remote sensing, OCEAN-going research ships, ground observation stations, computers, and telecommunications, a better way could be found to facilitate, monitor, manage, and disseminate environmental data at a national, regional, and global scale. Global monitoring can keep nations informed with the most up-to-date and accurate data on such issues as DEFORESTATION, GLOBAL WARMING, and POPULATION increases in various places of the world.

Climate data within the framework of GEMS has assisted in international action to save the OZONE layer and to study the effects of climate change. Two of the main networks involved in GEMS are the World Glacier Monitoring Service and the World Meteorological Organization UNEP Background Air Pollution Monitoring Network. Since the world's GLACIERS and ice sheets act as a thermometer of climate change as they shrink or advance, there are now over 750 stations that monitor glaciers in 21 countries.

GEMS produces the biennial *UNEP Environmental Data Report*, which brings together information at regional, national, and local levels in order to illustrate all major environmental issues, including environmental pollution, climate, NATURAL RESOURCES, population/settlements, human health, ENERGY, transportation/tourism, wastes, NATURAL DISASTERS, and international cooperation.

EnviroSource

United Nations Environment Programme, Global Environment Monitoring System, Monitoring and Assessment Research Center: King's College London, University of London, The Old Coach House, Campden Hill, London W8 7AD United Kingdom.

Global Ocean Ecosystems Dynamics (GLOBEC), A U.S. research program organized by oceanographers and other scientists to learn more about how global CLIMATE change may affect the sustainability of animals in the sea. GLOBEC scientists are using circulation models to predict how OCEAN CURRENTS respond to global climate changes such as GLOBAL WARMING. The research focuses on how climate changes will impact the distribution, abundance, and productivity of marine animals—especially ZOOPLANKTON and fish POPULATIONS. Zooplankton (tiny aquatic CONSUMERS) are a main focus of this work because they are the key link in the FOOD CHAIN between PHYTOPLANKTON (aquatic PRODUCERS) and higher levels of FOOD CHAINS and FOOD WEBS. Many fishes such as cod, haddock, and sardines and benthic INVERTEBRATES such as oysters, scallops, and sea urchins depend upon the survival of zooplankton and fish larvae.

GLOBEC is a component of the U.S. Global Change Research Program, with support from the U.S. National Science Foundation Division of Ocean Sciences and the U.S. NATIONAL OCEANIC AND ATMOSPHERIC ADMINISTRATION Coastal Ocean Program Office. The GLOBEC program currently has major research efforts under way in the Georges Bank, in the northwest Atlantic region, and in the northeast Pacific. *See also* EL NIÑO/LA NIÑA.

EnviroSource

GLOBEC Web site: cbl.umces.edu/fogarty/usglobec/misc/education.html.

global warming, The relatively recent, ongoing elevation in global surface AIR temperatures primarily resulting from

Global Warming

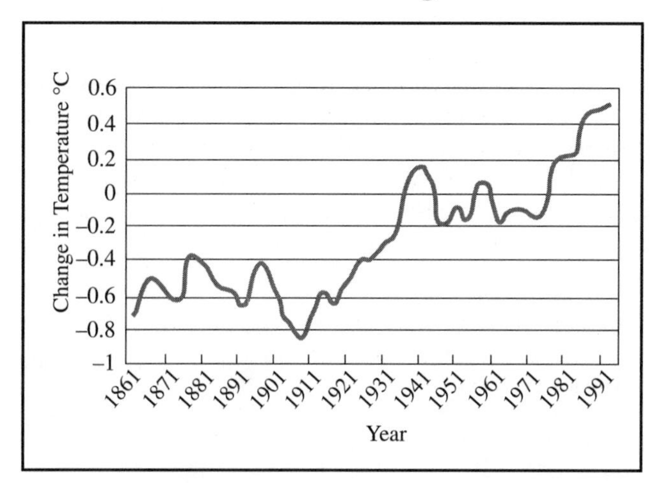

Source: United Nations Intergovernmental Panel on Climate Change, 1995.

human-caused increases in the concentrations of GREENHOUSE GASES in the lower ATMOSPHERE. CARBON DIOXIDE the most abundant greenhouse gas, is produced when FOSSIL FUELS are used to generate ENERGY and when FORESTS are cut down and burned. METHANE and nitrous oxide are emitted from agricultural activities and other sources. FLUOROCARBONS (including CHLOROFLUOROCARBONS), CARBON TETRACHLORIDE, and methyl chloroform are released during industrial processes, while OZONE in the lower atmosphere is generated indirectly by AUTOMOBILE exhaust fumes. Like the glass panes of a greenhouse, these gases trap the sun's heat, and control the flow of natural energy through the world's CLIMATE system. It is estimated that 76% of global warming is caused by carbon dioxide alone.

Approximately half the carbon dioxide released into the atmosphere from the combustion of fossil fuels is quickly absorbed by the OCEANS or PLANTS, which use the gas in the process of PHOTOSYNTHESIS. The other half remains in the atmosphere for many decades. The average concentration of carbon dioxide has increased from about 275 parts per million (ppm) before the Industrial Revolution, to 315 ppm in 1958, (when precise monitoring stations were set up), to 361 ppm in 1996. The change in concentration of carbon dioxide has increased the amount of energy striking Earth's surface by an amount equivalent to about 1% of the energy in sunlight.

Global average temperatures have remained relatively stable over the last 10,000 years. But since 1880, when reliable temperature data was first recorded worldwide, the global average temperature has risen by nearly 0.5°C (0.9°F). Snow cover in the Northern Hemisphere and floating ice in the Arctic Ocean have decreased, and cold-season precipitation has increased in the high latitudes. Globally, sea level has risen 10–25 centimeters (4–10 inches) over the past century, and worldwide precipitation over land has risen by about 1%. Earth's northern latitudes have become about 10% greener since 1980, due to more vigorous plant growth associated with warmer temperatures and higher levels of atmospheric carbon dioxide.

Starting in the mid-1700s, human activities began to alter the composition of the atmosphere. Vast supplies of charcoal, and later COAL and PETROLEUM, were burned to power the Industrial Revolution; the CARBON stored in these FUELS was released into the air as carbon dioxide. Since then, atmospheric concentrations of carbon dioxide have increased by nearly 30%, methane concentrations have more than doubled; and nitrous oxide concentrations have risen by about 15%. Today, for each of the more than 6 billion people on Earth, nearly 5.4 metric tons (6 tons) of carbon dioxide are spewed into the air annually. It is notable that with only 5% of the world's POPULATION, the United States accounts for 22% of world's greenhouse gas EMISSIONS from fossil fuel use.

Climate change as a result of global warming is likely to have a significant impact on global ENVIRONMENTS. According to the United Nations Intergovernmental Panel on Climate Change (IPCC), climate models indicate that unless the world takes steps to reduce emissions of greenhouse gases, global temperature is projected to rise another 0.7–3.0°C (1°–4°F) by the year 2100. Other projections are higher. However, an increase of just a few degrees in the global average temperature can translate into a major climate change.

At current emissions rates, mean sea level is projected to rise 15–95 centimeters (6–38 inches) by the year 2100, causing flooding of low-lying areas and other damage, particularly along the U.S. coast. Studies by the U.S. ENVIRONMENTAL PROTECTION AGENCY and others estimate that along the Gulf and Atlantic Coasts, a rise of 30 centimeters (11 inches) is likely by 2050, and could occur as soon as 2025. Nationwide, such a SEA LEVEL RISE could eliminate 17–43% of American WETLANDS, with more than half the loss taking place in Louisiana alone.

Patterns of rainfall and snowfall are also expected to change; many plant and animal SPECIES may not be able to adjust to such shifts. Climate zones, ECOSYSTEMS, and AGRICULTURE zones could migrate 150–550 kilometers (90–340 miles) toward the poles in the mid-latitude regions. Human society will face new risks and pressures, including issues related to food security, water resources, economic activities, and human health. The changing climate could alter forests, crop yields, and the distribution of water supplies. It could also threaten human health and harm birds, fish, and many types of ecosystems.

Carbon dioxide increases could have some positive impacts on plants—a beneficial fertilization effect and enhancement of water-use efficiency. In colder areas, warmer temperatures would lengthen the growing season, and might also make it possible to cultivate new lands, potentially increasing agricultural production by 0.2–1.2%.

Extreme temperatures can directly cause loss of human life, particularly for those with cardiac or respiratory problems. Global warming may also increase the prevalence of some infectious DISEASES, particularly those that only appear in warm areas. Diseases such as malaria and dengue fever, which are spread by INSECTS, could become more prevalent if warmer temperatures enable those insects to extend their ranges.

Stabilization of atmospheric concentrations of greenhouse gases will demand a major, international effort. The IPCC was established by the UNITED NATIONS ENVIRONMENT PROGRAMME and the World Meteorological Organization in 1988 to assess information related to all significant components of global climate change, including its consequences for society and the environment. With its capacity to report on climate change, the IPCC also functions as the official advisory body to the world's governments on the state of climate-change science.

Another international effort is the United Nations Framework Convention on Climate Change (FCCC), finalized in 1992 at the UNITED NATIONS EARTH SUMMIT in Rio de Janeiro, Brazil. The 165 signatory nations agreed to nonbinding goals for reducing emissions of greenhouse gases by developed nations to 1990 levels by the year 2000. To fulfill its agreement, the United States established a Climate Change Action Plan, a variety of programs through which American business and industry instituted voluntary efforts at reducing emissions.

When signatory nations met in 1995 to review the FCCC progress at the first such conference in Berlin, they outlined a process for establishing mandatory limits on greenhouse gas emissions after the year 2000, even though voluntary programs were just getting under way. One of the most critical elements of the Berlin mandate was a provision that exempted developing nations from limits on greenhouse gas emissions while at the same time imposing restrictions on the United States and other developed countries. Thus, nations such as China, India, Korea, Mexico, and Indonesia were exempted, despite the fact that these countries now account for about half of the world's greenhouse gas emissions. Because of their expanding populations, economies, and demands for ELECTRICITY and transportation, their share is expected to approach 75% within the next 50 years.

At the second conference in 1997 in Kyoto, Japan, the signatories agreed to reduce greenhouse gas emissions by harnessing the forces of the global marketplace to protect the environment. The Kyoto Protocol's key points, including emissions targets, timetables for industrialized nations, and market-based measures for meeting targets, reflect proposals advanced by the United States. To become binding, the Kyoto Protocol had to be ratified by at least 55 countries, accounting for at least 55% of the total 1990 carbon dioxide emissions of developed countries. As of September 1998, 58 states had signed the Kyoto Protocol.

A central feature of the Kyoto Protocol is a set of binding emissions targets for developed nations. The specific limits vary from country to country, though those for the key industrial powers—the European Union, Japan, and the United States—are similar: 6–8% below 1990 emissions levels. Under the protocol's Clean Development Mechanism, developed countries are able to use certified emissions reductions from project activities in developing countries to contribute toward their compliance with greenhouse gas reduction targets. The protocol represents only a start on securing the meaningful participation of developing countries, which remains a core U.S. goal. On the bright side, many global warming and climate change researchers believe that the world can realize energy efficiency gains of 10–30% over the next 20–30 years at no net cost, using current knowledge and today's best technologies. In the longer term, increases in energy efficiency may make it possible to move close to a zero-emissions industrial economy. *See also* GLOBAL ENVIRONMENT MONITORING SYSTEM; GLOBAL WARMING POTENTIAL; NATIONAL OCEANIC AND ATMOSPHERIC ADMINISTRATION; NITROGEN OXIDES.

EnviroSources

Environmental Defense Fund (global warming) Web site: www.edf.org/pubs/Brochures/GlobalWarming.
Environmental Defense Fund (global warming) Web site: www.enviroweb.org/edf.
Greenpeace International (climate) Web site: www.greenpeace.org/~climate.
United Nations Environment Programme Web site: www.ipcc.ch.
U.S. Environmental Protection Agency (global warming) Web site: www.epa.gov/globalwarming.

Information Sheets, Brochures, and Shorter Documents

State Climate Change Impacts Information Sheets, no. 230-F-97-008. 1997–8.
United Nations Framework Convention of Climate Change. Kyoto Protocol. Text from the UNFCCC Third Conference of the Parties, (Adobe Acrobat file.) 1997.
U.S. Department of State. "The Kyoto Protocol on Climate Change Fact Sheet." 1998.
U.S. Environmental Protection Agency. "Climate Change and Public Health." Publication no. 236-F-97-005 (Adobe Acrobat file.) 1997.
U.S. Environmental Protection Agency. "Cool Facts about Global Warming." Brochure. Publication no. 320-F-97-001 (Adobe Acrobat file.) 1997.

global warming potential, A number that is used to indicate the amount of GLOBAL WARMING caused by a specific substance. The global warming potential (GWP) of a substance represents a ratio that compares the amount of global warming caused by that substance to the amount of global warming caused by an equivalent mass of CARBON DIOXIDE, which has been assigned a GWP of 1.

CHLOROFLUOROCARBONS (CFCs) have GWPs of between 4,000 and 11,700. As a group, these compounds are the most damaging ozone-depleting substances. HYDROCHLOROFLUOROCARBONS (HCFCs) have GWPs between 93 and 12,100, ratings which indicate that they, too, are very damaging to OZONE. Another group of ozone-depleting substances, the HALONS, have GWPs of around 5,600. Because of their ozone-depleting qualities, production and use of all these substances has been stopped in the United States and 92 other countries under the terms of the MONTREAL PROTOCOL. *See also* AIR POLLUTION; FLUOROCARBON; GREENHOUSE EFFECT; GREENHOUSE GAS; OZONE DEPLETION; OZONE HOLE.

Goodall, Jane (1934–), A British primatologist known worldwide for her extensive studies of chimpanzees in the Gombe Stream Reserve on the coast of Lake Tanganyika in Tanzania, Africa, and for her efforts to provide protection to this animal SPECIES. Goodall's methodologies and scientific discoveries revolutionized the field of primatology and influenced such other primatologists as DIAN FOSSEY, who worked with mountain GORILLAS, and BIRUTÉ GALDIKAS, who works with ORANGUTANS. Goodall distinguished individual chimpanzee personalities, giving the animals names instead of numbers. She also documented chimpanzees making and using tools, a skill once believed to be exclusive to humans. Her scientific articles have appeared in many issues of *National Geographic* magazine as well as in internationally known scientific journals. Such articles have provided a decades-long picture of the development, social behavior, ranging patterns, feeding habits, and life cycle of wild chimpanzees. Goodall's research was detailed in her books *Through a Window: My Thirty Years with the Chimpanzees of Gombe* (1990) and *In The Shadow of Man* (1971).

On a trip to Kenya in 1957, Jane Goodall met renowned anthropologist and paleontologist Louis B. Leakey. To gain insight into the evolutionary history of humans, Leakey suggested that Goodall initiate a long-term field study on wild chimpanzees (because of their close genetic ties to humans). Although Goodall had no formal training, her patience and dedication to the understanding of animals prompted Leakey to choose her for the study. In the summer of 1960, she arrived in Tanzania, marking the beginning of the longest continuous field study of animals in their natural HABITAT.

In the beginning, studying the Gombe chimpanzees was difficult and frustrating for Goodall. The chimpanzees fled from her in fear; often, Goodall could observe them only through binoculars from a peak overlooking the FOREST. It took many months before she could successfully approach them. Gradually, the chimpanzees became accustomed to her presence and grew to trust her.

In 1965, Goodall earned a Ph.D. in ethology (the study of animal behavior) from Cambridge University, after which she returned to Tanzania to establish the Gombe Stream Research Centre. To provide ongoing support for chimpanzee research, Goodall founded the Jane Goodall Institute (JGI) in 1977. In 1984, Goodall received the J. Paul Getty Wildlife Conservation Prize for helping to build global awareness of the need for WILDLIFE CONSERVATION. She has received countless other awards as well.

Research at Gombe continues today, conducted primarily by a trained team of Tanzanians. Moral and philosophical support from the Tanzanian government is a chief reason that the program has endured for so long. Since the 1980s, Goodall has spent much of her time traveling around the world, lecturing on her experiences at Gombe and speaking to school groups about "Roots and Shoots," a JGI environmental education and humanitarian program for children. *See also* COUSTEAU, JAQUES-YVES; EARLE, SYLVIA.

EnviroSources

Goodall, Jane. *In the Shadow of Man.* Reprinted, 1971. New York: Houghton Mifflin, 1988.

———. *Through a Window: My Thirty Years with the Chimpanzees of Gombe.* Reprint, 1990. New York: NY: Houghton Mifflin, 1991.

Jane Goodall Institute Web sites (United States): wwww.janegoodall.org; (Canada): www.janegoodall.ca.

Montgomery Sy. *Walking with the Great Apes.* New York: Houghton Mifflin, 1991.

gorillas, The largest and rarest of the great apes, which live only in equatorial regions of east and west Africa. There are three distinct subspecies of gorilla, each of which lives in a different geographical region. The eastern lowland gorilla (*Gorilla gorilla grauere*) is the only gorilla native to eastern Africa and makes its home in dense FORESTS. The western lowland gorilla (*Gorilla gorilla gorilla*) and the mountain gorilla (*Gorilla gorilla beringel*) both make their homes in dense forests of western Africa. The lowland gorilla lives in low mountainous areas, while the mountain gorilla, as its name implies, lives at higher ALTITUDES. The greatest concentration of the mountain gorilla is found in the Parc National des Volcans of Rwanda. All three subspecies of gorilla are recognized as ENDANGERED SPECIES.

Gorillas were once widely distributed throughout equatorial forests of Africa; however, hunting and destruction of HABITAT greatly reduced gorilla POPULATIONS and reduced suitable ranging areas. Habitats were destroyed when forests were cleared to create land for AGRICULTURE and for commercial logging. Scientists estimate that fewer than 126,000 animals remain in the wild. Currently, hunting gorillas is prohibited throughout Africa, and trade in gorillas or their parts is prohibited under the CONVENTION ON INTERNATIONAL TRADE IN EN-

The mountain gorilla is an endangered species.

Gorillas At a Glance	
Height	When standing, between 125 cm (50 in) and 175 cm (69 in); males generally larger than females.
Arm span	200 – 275 cm (78 in – 108 in)
Weight	Females—70 –140 kg (154 – 308 lbs); Males—135 to 275 kg (298 – 606 lbs)
Life Span	50 years
Diet	Fruits, leaves, herbs
Family Groups	Typically 5 to 10 individuals that include 1 fully adult male, 2 – 4 adult females, and 2 – 4 young

DANGERED SPECIES OF WILD FLORA AND FAUNA; however, POACHING gorillas to obtain body parts such as hands, which are used as ashtrays, or heads, which are mounted as trophies, continues to be a problem.

Several CONSERVATION programs have been established to protect remaining gorilla populations and to encourage increases in their numbers. Rwanda has three programs in place for the protection of the mountain gorillas residing in the Parc National des Volcans. They include the Karisoke Research Center, which was established by DIAN FOSSEY in 1967, the International Gorilla Conservation Programme, and the Virunga Veterinary Centreform. The government works in conjunction with these programs and in return benefits economically from an ECOTOURISM industry that has evolved around interest in this SPECIES. The Nigerian Conservation Foundation has established a program to educate the people of Nigeria about the need to conserve WILDLIFE through decreased hunting and preserving habitat. In addition, the International Gorilla Conservation Programme, a partnership of the AFRICAN WILDLIFE FOUNDATION and Fauna Flora International, also is working to get international cooperation among Zaire, Uganda, and Rwanda to ensure protection of the eastern lowland and mountain gorillas that reside near the common borders of these nations. *See also* GALDIKAS, BIRUTÉ; GOODALL, JANE; ORANGUTANS; RED LIST OF ENDANGERED SPECIES; VOLCANO; WORLD CONSERVATION MONITORING CENTRE; WORLD CONSERVATION UNION.

EnviroSources

Bourne, Joel. "Gorillas in Our Midst." *Audubon* 100, no. 5 (September–October 1998): 70–72.

Fossey, Dian. *Gorillas in the Mist*. New York: Houghton Mifflin, 1983.

Montgomery, Sy. *Walking with the Great Apes.* New York: Houghton Mifflin, 1991.

Mowat, Farley. *Woman in the Mists: The Story of Dian Fossey and the Mountain Gorillas of Africa.* New York: Warner Books, 1987.

World Conservation Monitoring Centre Web site: www.wcmc.org.

granite. *See* rock cycle.

granulated activated carbon treatment, A filtering system often used in small water systems and individual homes to remove ORGANIC CONTAMINANTS that cause taste and odor in the water. Some units can remove CHLORINATION byproducts and PESTICIDES but are not efficient in removing some metals such as LEAD and COPPER. Granulated activated carbon treatment can be highly effective in removing elevated levels of RADON from water. *See also* CARBON.

grassland, A dry, flat BIOME in which shrubs and grasses are the dominant PLANTS. Grasslands represent about one-third of Earth's land mass, although most have been converted to farmlands or livestock grazing lands. The rich, fertile soil and the availability of water from underground sources has produced vast farm and ranching enterprises.

Today, many grasslands are suffering from overgrazing, wind and water EROSION of topsoil, the overuse of PESTICIDES, the depletion of AQUIFERS, and the invasion of nonnative or EXOTIC SPECIES of plants. Overuse of grassland for farming and grazing has caused DESERTIFICATION in many parts of Africa and was the cause of the 1930s DUST BOWL in the United States.

Rhea

There are several major types of grasslands. The grasslands of central North America are called PRAIRIES and they extend from south-central Canada to Texas, and from the Rocky Mountains to Ohio. In central South America, the grasslands are called PAMPAS. Grasslands located across central Eurasia are called STEPPES. The vast steppes stretch from southern Russia, through Kazakstan and southern SIBERIA, west to Mongolia, and south to western China. The steppe is dominated by grassy, flat plains with few, if any trees. The tropical grasslands of central and southern Africa and parts of South America are called SAVANNAS. Savannas consist mainly of grasses, bushes, and scattered small trees and are located at the edges of FORESTS and dry regions.

Precipitation in the grasslands is irregular and varies season to season during the year. There are times of heavy rainfall and periods of dry WEATHER. Precipitation averages between 25 and 100 centimeters (10–40 inches) a year. In grasslands with high precipitation, grasses can grow as high as 3 meters (9 feet) tall. In the drier regions, grasses grow to between 25 and 50 centimeters (10–20 inches) tall.

Grasslands are HABITATS for a wide variety of animals and plants. Depending on the location of the grassland, there may

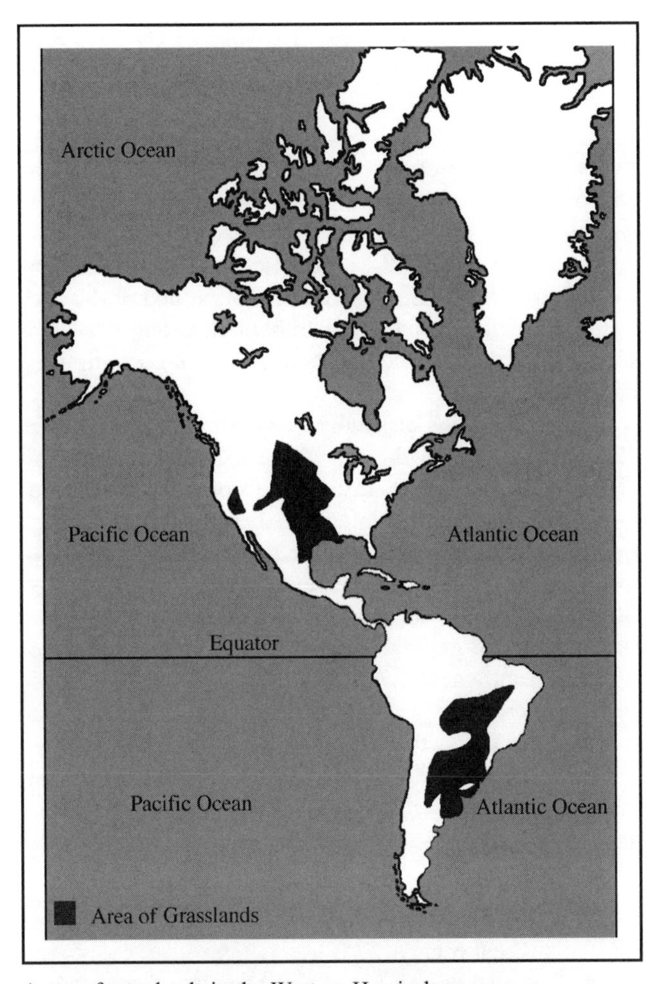

Areas of grasslands in the Western Hemisphere.

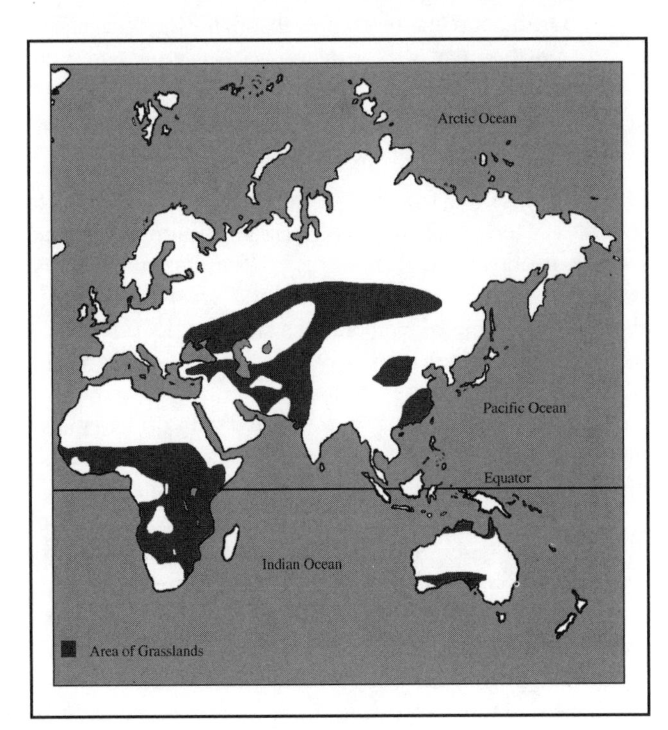

Areas of grasslands in the Eastern Hemisphere.

be BISON, coyotes, eagles, deer, and bobcats, as well as CHEETAHS, jackals, lions, and wildebeest. Migrating birds such as Canada geese and wild ducks, can be seen on grasslands as they make their way north or south. *See also* ECOSYSTEM.

EnviroSources

Brown, Lauren. *Grasslands.* New York: Alfred A. Knopf, 1985.

Knapp, Alan K., et. al., eds. *Grassland Dynamics: Long-Term Ecological Research in Tallgrass Prairie.* Long-Term Ecological Research Network Series, no. 1 Oxford, UK: Oxford Univ Press, 1998.

Manning, Richard. *Grassland: The History, Biology, Politics and Promise of the American Prairie.* New York: Penguin, 1997.

Pipes, Rose. *Grasslands* (World Habitats.) Chatham, NJ: Raintree/ Steck Vaugh, 1998.

University of California, Berkeley (world biomes: grasslands) Web site: www.ucmp.berkeley.edu/glossary/gloss5/biome/ grasslan.html.

U.S. Geological Survey, (postcards from the prairie) Web site: www.nrwrc.usgs.gov/postcards/postcards.html.

Van Dyne, G. M. *Grasslands, Systems Analysis and Man.* Cambridge, UK: Cambridge University Press, 1980.

Waterlow, Julia. *Grasslands (Habitats.)* Thomson Learning, 1996.

World Wildlife Fund (grasslands and their animals) Web site: www.panda.org/kids/wildlife/idxgrsmn.html.

Great Barrier Reef. *See* coral reef.

green manure, PLANTS grown instead of commercial crops for the purpose of controlling weeds, increasing soil fertility, and to minimize soil EROSION. Some green manure plants include clover (a LEGUME), rye, sunflower, and buckwheat. The plants are grown in fields until they are needed again for growing commercial crops. Then the plants are cut, crushed, and mixed with the soil, and allowed to decompose. During DECOMPOSITION, important nutrients are released into the soil making it fertile for the new crop. *See also* COMPOSTING; COVER CROPS; NITROGEN CYCLE.

green revolution, Term used to describe the attempt to significantly increase crop production and diversity in certain developing nations (particularly Mexico, the Philippines, India, and Pakistan) from the 1960s–1980s. The green revolution stressed the planting of high-yielding HYBRID grains, along with increased use of FERTILIZERS, PESTICIDES, and IRRIGATION.

greenbelt movement, An organized effort to restore environmental quality by planting trees in regions suffering the effects of DEFORESTATION, EROSION, and soil nutrient depletion. The greenbelt movement had its official start in June 1977 on World Environment Day, an annual celebration sponsored by the UNITED NATIONS ENVIRONMENT PROGRAMME. The movement was the brainchild of Kenyan zoologist and conservationist Wangari Muta Maathai. In its first year, the greenbelt movement involved only a few women who planted a total of seven trees in Maathai's backyard. Less than two decades later, greenbelt volunteers and employees numbered more than

60,000 and had planted an estimated 17 million trees throughout Kenya and several other African nations.

Maathai, who was born and raised in Africa, traveled to the United States to pursue her undergraduate studies. When she returned to her homeland, she was horrified to see how much of the FORESTS of the Kenyan landscape were gone. Like many other areas of the world, Kenya's forests had been largely clear-cut by the timber industry or to make land available for AGRICULTURE. As a result, fruits and nuts derived from forest trees and FUEL WOOD—the main source of ENERGY for cooking—was no longer readily available. Forced to change their diets, many Kenyans were now suffering the effects of malnutrition. In addition, deforestation left the barren soil vulnerable to erosion. Thousands of metric tons of topsoil were washed off the mountains and into the Indian Ocean each year. SEDIMENTATION of this soil in the OCEAN buried and killed coastal marine ORGANISMS, thus depleting another source of food for local peoples.

Maathai began the greenbelt movement with two goals: to restore the health of the Kenyan ENVIRONMENT and to provide better economic lives to poor rural women by providing them with both employment and a means of providing food to their families. At the time the project began, Maathai was a member of the National Council of Women in Kenya (NCWK). With the aid of women from this group, Maathai set up a tree nursery on the NCWK office grounds. She started the nursery with seeds she collected from established trees and with seedlings donated by the Kenyan Department of Forestry. Volunteers nurtured the seedlings until they were ready for planting, and then began raising new seedlings. New chapters of greenbelt volunteers have since formed, and the number of nurseries throughout Kenya has increased to more than 3,000.

As a result of the greenbelt movement, much of the once barren soil in Kenya is now covered by trees. Children, who once were hungry, now readily eat fruit from the trees. Fuel wood is once again available, along with medicinal substances derived from the trees. In addition, by planting trees on land also used for agriculture, soil quality and crop harvests have improved.

Maathai and the greenbelt movement have been well recognized for their successes. In 1987, the greenbelt movement was presented the Global 500 Award by the United Nations Environment Programme. Maathai was presented the Right Livelihood Award in 1984, the Goldman Environmental Prize in 1991, and the Golden Ark Award from Prince Bernhard of the Netherlands in 1994. In addition, Maathai has worked with nongovernmental organizations of other countries to help them establish greenbelt movements. Through this effort, such movements now exist in more than 30 nations throughout the world. *See also* AGROFORESTY; CLEAR-CUTTING; COMPOSTING; COVER CROPS; MUIR, JOHN; NITROGEN CYCLE; SLASH AND BURN; SUSTAINABLE DEVELOPMENT; TREE FARM.

EnviroSources

Breton, Mary Joy, *Women Pioneers for the Environment,* Boston: Northeastern University Press, 1998.

Seymour, John, and Herbert Girardet. *Blueprint for a Green Planet: Your Practical Guide to Restoring the World's Environment,* New York: Prentice-Hall, 1987.

greenhouse effect, A naturally occurring phenomenon wherein certain atmospheric gases, such as water vapor and CARBON DIOXIDE, regulate the radiant ENERGY balance of Earth, thus making it habitable. So-called GREENHOUSE GASES allow SOLAR ENERGY to pass through the ATMOSPHERE and be absorbed by Earth's surface, but they trap most of the radiant heat emitted from Earth's surface, preventing it from escaping back into space.

The wavelengths of sunlight falling on the surface of a planet are primarily within the visible and ultraviolet portions of the spectrum; light energy emitted from the surface tends to be longer-wavelength infrared RADIATION, or radiant heat. On Earth, much infrared radiation is absorbed by carbon dioxide and water vapor molecules in the atmosphere and reflected back to the surface as heat. (Because of their molecular structures, these gases are essentially transparent to visible light but readily absorb infrared radiation.) The greenhouse analogy refers to the fact that greenhouse gases function much like the glass panels that trap heat inside a greenhouse.

The process leading to the greenhouse effect may be summarized as follows:

- Light energy penetrates Earth's atmosphere as short-wave radiation.
- This energy is absorbed by Earth's surface and changed into heat energy.
- The heat energy is radiated back into Earth's atmosphere as long-wave radiation.
- Greenhouse gas molecules absorb the long-wave, heat radiation that strikes them on its way back to outer space.
- In the troposphere, greenhouse gas molecules release the captured heat energy.

Energy resulting from the greenhouse effect is essential to a number of terrestrial processes, including heating the ground surface, melting ice and snow, the evaporation of water, and PHOTOSYNTHESIS in PLANTS. Without the greenhouse effect, Earth's average global temperature would be –18°C (–0.4°F), rather than its current 15°C (59°F). Even the OCEANS would be frozen under such conditions.

Current scientific concern associated with GLOBAL WARMING and CLIMATE change focuses on the sensitivity of Earth's climate to an *enhanced* greenhouse effect resulting from ongoing human-induced increases in atmospheric concentrations of greenhouse gases. Scientists fear that, once started, global warming becomes self-perpetuating. For example, a warmer Earth would increase evaporation and result in more atmospheric water vapor. Because atmospheric water vapor is a significant greenhouse gas, this could lead to further warming. Increased water vapor and changes in atmospheric circulation also could create changes in cloud cover, although more clouds would, on average, have a cooling effect. Similar feedbacks occur as a result of marine, terrestrial, and sea-ice pro-

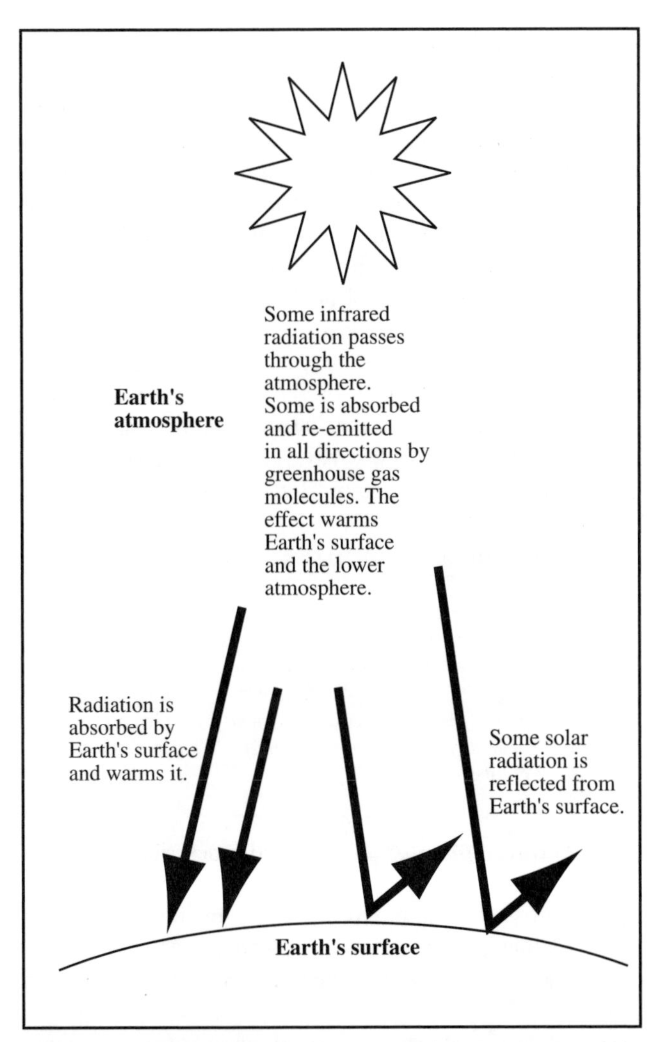

Without the greenhouse effect, Earth's average global temperature would be -18°C (-0.4°F), rather than its current 15°C (59°F).

cesses, many of which are highly temperature dependent. An increased understanding of the extent to which these feedbacks affect Earth's climate will help scientists to better estimate the environmental, economic, and human health risks associated with an enhanced greenhouse effect.

A "runaway" greenhouse effect also is a possible long-term result of accelerated accumulation of greenhouse gases. Earth contains huge amounts of water vapor and carbon dioxide, although they are primarily bound up in oceans and carbonate rocks. If an enhanced greenhouse effect causes the temperature of the atmosphere to approach the boiling point of water (100°C or 212°F), assuming that no other counterbalancing effects intervene, the oceans may begin to convert to water vapor. The water vapor will accelerate the greenhouse effect, causing the temperature to rise even further, thus causing the oceans to evaporate faster. When the oceans are gone, the atmosphere will finally stabilize at a much higher temperature and at a much higher density, because all Earth's water will be contained in the atmosphere.

A further runaway stage might be reached if the temperature rose so high that chemical reactions drove the carbon dioxide from the rocks into the atmosphere. This would again accelerate greenhouse heating, which, in turn, would accelerate the transfer of carbon dioxide from the rocks to the atmosphere in another positive feedback loop. The atmosphere would finally stabilize at a still higher temperature and pressure after all the carbon dioxide had been driven from the rocks. In fact, if this runaway sequence were to occur on Earth, the temperature and pressure of the resulting atmosphere would be similar to that of present-day Venus: The atmospheric temperature would be hundreds of degrees celsius, and the pressure would be many times greater than it is today.

EnviroSources

Christianson, Gale E. *Greenhouse: The 200-Year Story of Global Warming.* New York: Walker and Co., 1999.

Edmunds, Alex. *The Greenhouse Effect.* Brookfield, CT: Copper Beech Books, 1997.

Hocking, Colin, Cary Sneider, and Lincoln Bergman, eds. *Global Warming and the Greenhouse Effect.* Berkeley, CA: University of California, Berkeley, Lawrence Hall of Science (Gems), 1992.

Mabey, Nick. *Arguments in the Greenhouse: The International Economics of Controlling Global Warming.* New York: Routledge, 1997.

U.S. Environmental Protection Agency (global warming) Web site: www.epa.gov/globalwarming.

greenhouse gas, Any natural or artificial gas that contributes to the GREENHOUSE EFFECT. The greenhouse effect is a natural phenomenon caused by the layer of gases in Earth's ATMOSPHERE. Acting like the glass walls of a greenhouse, these gases prevent heat that reaches Earth from escaping back into space. Scientists recognize five important greenhouse gases: CARBON DIOXIDE (CO_2), water vapor, CHLOROFLUOROCARBONS (CFCs), METHANE, and NITROGEN OXIDES (NO_x).

The greenhouse effect is vital to life on Earth; without the greenhouse effect, Earth would be much too cold to sustain life. Today, however, environmental scientists are concerned that many human activities are releasing excessive amounts of greenhouse gases into the atmosphere. The result could be a much warmer Earth in the future. Studies of carbon dioxide levels in the atmosphere support this view. Scientists have documented that when atmospheric carbon dioxide levels rise, global temperatures also rise.

Carbon dioxide and nitrogen oxides are released into the ENVIRONMENT every time FOSSIL FUELS, such as PETROLEUM and GASOLINE, are burned for ENERGY by AUTOMOBILES, power plants, and factories. CFCs also gradually enter the AIR from refrigerators and air conditioners, and methane is given off in fairly large amounts from farms, landfills, and other sources. Natural occurrences such as volcanic activities, FOREST FIRES, and PHOTOSYNTHESIS and respiration also produce greenhouse gases.

Environmentalists have advanced a number of recommendations for reducing greenhouse gas EMISSIONS. To reduce CO_2 emissions, they suggest using ALTERNATIVE ENERGY SOURCES, such as SOLAR ENERGY, WIND POWER, or TIDAL POWER, that produce fewer emissions than fossil fuels. CONSERVATION agencies are also looking at ways to reduce the destruction of tropical RAINFORESTS, which adds billions of tons of CO_2 into the atmosphere each year. *See also* AIR POLLUTION; ALTERNATIVE ENERGY RESOURCES; CHLORINATED HYDROCARBONS; DEFORESTATION; FLUOROCARBONS; GLOBAL WARMING; SLASH AND BURN.

EnviroSources

Greenpeace International (climate) Web site: www.greenpeace. org/~climate.

Pratt, Robert, ed. *Global Climate Change Digest: A Guide to Current Information on Greenhose Gases and Ozone Depletion.* (1988– .)

United Nations Intergovernmental Panel on Climate Change Web site: www.ipcc.ch.

U.S. EPA (global warming) Web site: www.epa.gov/global warming.

Greenpeace, An international organization dedicated to bringing an end to activities that harm the ENVIRONMENT and protecting ENDANGERED SPECIES. Greenpeace originally formed in 1971 in Vancouver, British Columbia, as the Don't Make a Wave Committee with the goal of opposing nuclear testing at Amchitka Island in Alaska. It later changed its name to Greenpeace to more accurately reflect its goal of achieveing a "greener" environment through peaceful means.

To achieve its goals, Greenpeace members often engage in direct confrontations with businesses and corporations engaged in environmentally harmful activities. For example to prevent the killing of WHALES, Greenpeace volunteers often position themselves in small watercraft between the harpoon guns of whalers and the whales they are trying to kill. Members also have plugged pipes of companies guilty of discharging TOXIC WASTE into bodies of water. *(See* discharge.) Such actions often bring Greenpeace media attention that helps to increase public awareness of the environmental problems they are trying to correct. For example, Greenpeace's alleged involvement in the 1985 bombing of the ship *Rainbow Warrior* was widely publicized. At the time of the bombing, the ship was docked in Auckland Harbour, New Zealand, awaiting travel to the Moruroa atoll, where Greenpeace members planned a protest of atmospheric NUCLEAR WEAPONS tests being conducted by France. It was later discovered that French intelligence agents were responsible for bombing the ship and the death of a Greenpeace photographer who was on board. The international attention resulting from this event led to the dismissal of the head of France's intelligence service and the resignation of its defense minister.

Since its inception, Greenpeace has grown from a small group into the world's largest environmental organization. The group is supported mostly by volunteers and has branches in many countries, including the United States, Canada, Argentina, Australia, Costa Rica, Japan, Germany, Belgium, Switzerland, the Netherlands, the Russian Federation, and Denmark, among others. In 1987, the U.S. chapter of Greenpeace, established a second branch known as Greenpeace Action, which carries out public education programs and attempts to influence the legislative process by gaining grassroots support and running various political campaigns. To achieve its goals, Greenpeace Action has initiated a variety of nonpolitical campaigns including the Ocean Ecology Campaign, which focuses on protecting marine MAMMALS and their HABITATS; the Atmosphere and Energy Campaign, which seeks to decrease the EMISSIONS of POLLUTANTS responsible for GLOBAL WARMING by encouraging countries to switch from dependence on FOSSIL FUELS to the use of ALTERNATIVE ENERGY RESOURCES; the Tropical Forests Campaign, which strives to reduce DEFORESTATION by encouraging sustainable use of FOREST resources and the preservation of areas such as BISOPHERE RESERVES or national parks; the Toxics Campaign, which stresses pollution prevention through reduction of the use of toxic materials, RECYCLING, and the development of safer technologies for waste disposal; and the Nuclear Campaign, which strives to end nuclear weapons testing and eliminate the construction and possible use of nuclear weapons, as well as to stop the dumping of RADIOACTIVE WASTE at sea. *See also* CONSERVATION; DEBT-FOR-NATURE SWAP; DEEP ECOLOGY; EARTH DAY; FOREST MANAGEMENT; FRIENDS OF THE EARTH; GREENBELT MOVEMENT; INTERNATIONAL WHALING COMMISSION; MARINE MAMMAL PROTECTION ACT; MARINE PROTECTION, RESEARCH, AND SANCTUARIES ACT; NATIONAL AUDUBON SOCIETY; NATIONAL WILDLIFE REFUGE SYSTEM; NUCLEAR WASTE POLICY ACT; OCEAN DUMPING; SIERRA CLUB; WHALING; WORLD CONSERVATION UNION.

EnviroSources

Greenpeace International Web site: www.greenpeace.org.

Lanier-Graham, Susan D. *The Nature Directory: A Guide to Environmental Organizations.* New York: Walker and Company, 1991.

Seymour, John, and Herbert Girardet. *Blueprint for a Green Planet: Your Practical Guide to Restoring the World's Environment.* New York: PrenticeHall, 1987.

gross national product, A term used to describe a country's economic wealth and a measure of a nation's economic activity. The gross national product (GNP) is the total output of a country's final goods or services valued at market prices, including net exports, during a specific period of time such as a year.

A country produces goods such as AUTOMOBILES, paper goods, GASOLINE, food products, chemicals, homes, and buildings. These goods are produced by manufacturing companies and other industries such as AGRICULTURE, wood harvesting, MINING, and oil drilling. Service industries generate services rather than goods or products. Those services include banking, retail and wholesale sales, marketing, and professional services such as education and government services.

Environmental economists are critical of the GNP because it does not take into account the losses and depreciation of NATURAL RESOURCES used for mining, fishing, logging, farming activities, or industrial uses. They would prefer more environmental accounting in the GNP. As an example, the accounting in a GNP statement would include a country's sale of wheat minus the depreciated costs of farm equipment and buildings to grow and store the wheat; however, the GNP does not adjust for any loss of natural resources such as the amount of water used to grow the crops or the depletion of fertile soil due to soil EROSION. Nor does the GNP take into account other debits such as AIR and water pollution. As a result of these concerns, members attending the UNITED NATIONS EARTH SUMMIT in 1992 proposed measures to use economic and environmental accounting to provide a better indicator of a country's wealth. *See also* SUSTAINED YIELD.

EnviroTerm

gross domestic product (GDP): The total market value of goods and services produced by a country but does not include any income from investments outside the country.

groundwater, Water that occurs beneath Earth's surface at depths of a few centimeters to more than 300 meters (900 feet). Groundwater collects in spaces between rock and soil layers beneath Earth's surface. Groundwater concerns include the LEACHING of POLLUTANTS such as ARSENIC and MTBE into the water making it unfit for human consumption. The leaching of buried wastes can also pollute groundwater resources.

Most groundwater derives from precipitation that percolates downward through soil layers. *(See* percolation.) Some of the water collects in pores in rocks and the spaces between soil particles; however, much of the water collects in the space above a layer of impermeable rock. Groundwater moves through layers of porous materials called AQUIFERS or groundwater resources. The area where water in an aquifer leaves the aquifer is known as an outflow. If the groundwater is under pressure, its natural outflow may be a spring, a region where groundwater breaks through the surface. Other natural outflows for groundwater are ponds, lakes, and rivers.

As much as 20% of the water from precipitation becomes groundwater. *(See* water cycle.) As this water passes through layers of rock and soil, chemical substances and MINERALS present in the rock and soil may dissolve in the water and be carried into the groundwater supply. In some cases, these substances are pollutants produced in MINING, AGRICULTURE, or other human activities. Such substances become a problem because they may be ingested by humans and other ORGANISMS that derive their drinking water from groundwater sources through WELLS or springs and because they can be transported into aquatic ENVIRONMENTS.

The importance of groundwater to humans is evidenced by the OGALLALA AQUIFER of the United States. This aquifer underlies 12 states in the central United States and supplies a great deal of the water used for drinking and agricultural purposes by people living in these states. Because the aquifer is dependent upon precipitation for maintaing its water level, the water level of the aquifer is greatly subject to CLIMATE conditions such as severe DROUGHT. In addition, since much of the water in the central plains region of the United States is derived from this aquifer, concerns have also arisen about the potential for OVERDRAFT—the use of more water than is replaced by nature.

In the United States, state environmental regulations usually define standards of groundwater quality, based on the chemical properties of the groundwater that indicate whether it may be contaminated or otherwise unfit for consumption. Often, states also delineate and map areas of groundwater protection (aquifer protection) and known or suspected groundwater degradation. *See also* ACID MINE DRAINAGE; ARTESIAN WELL; DISCHARGE; DRAINAGE BASIN; FILTRATION; HYDROLOGY; HYDROSPHERE; IRRIGATION; LEACHING; LEACHING FIELD; NATURAL RESOURCE; PERCOLATION POND; PERMEABILITY; POTABLE WATER; RESERVOIR; SEEPAGE; SINK; SINKHOLE; SUBSIDENCE; SURFACE WATER; ZONE OF AERATION; ZONE OF DISCHARGE; ZONE OF SATURATION.

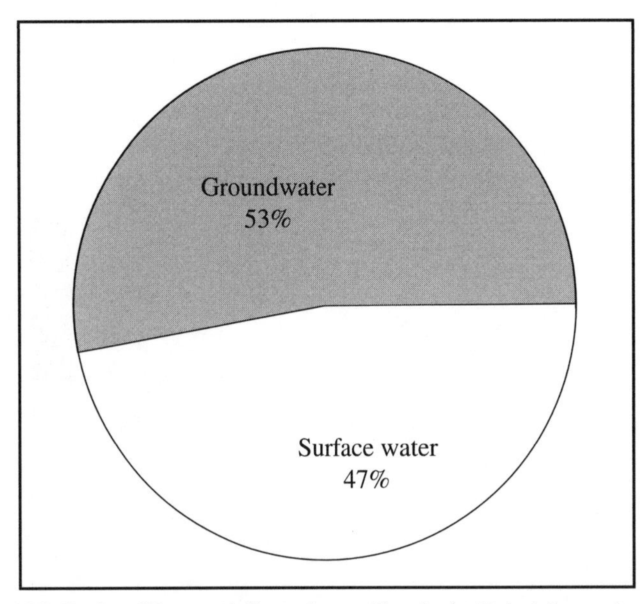

U.S. Surface Water and Groundwater Use. In the United States, the percentage of the population using surface water and groundwater is almost even. However, in many countries, most populations get their potable water from groundwater. *Source:* United States Geological Survey

EnviroSources

Environmental Protection Agency Web site: www.epa.gov/swerosps/ej/.

The Groundwater Foundation, P.O. Box 22558, Lincoln, NE 68542-2558, (800) 858-4844.

United States Geological Survey Web site: water.usgs.gov.

growth rate, A measure of the increase or decrease in the size of a POPULATION over time. Growth rate may be used by environmental scientists as an indicator of the overall health of an ECOSYSTEM. For example, an increasing growth rate suggests that environmental conditions are favorable for the population under study. An unchanging growth rate is an indicator of a balanced ecosystem. By contrast, a decreasing growth rate may indicate that a population is suffering from changing environmental pressures such as increased COMPETITION, PREDATION, PARASITISM, or DISEASE.

Growth rate is determined by subtracting the number of deaths per 1,000 members of a population from the number of births per 1,000 members during the same period. This figure is then adjusted to account for net migration. The total derived from these calculations is expressed as a percentage. *See also* EMIGRATION; IMMIGRATION; POPULATION DENSITY.

EnviroLink

The term *growth rate* is sometimes used to refer to the speed at which individual ORGANISMS increase in size. For colonial MICROORGANISMS, the term may be applied to the speed at which the colony increases in size.

Gulf Stream. *See* ocean currents.

H

habitat, The ENVIRONMENT, or place, where an ORGANISM normally lives. The habitat of an organism provides the organism with all the BIOTIC and ABIOTIC FACTORS—food, water, shelter, proper temperature, and mates—it needs to sustain life, as well as to ensure the survival of its SPECIES.

A habitat can be thought of as an organism's home. Thus, the habitat of an individual organism may be described in very specific terms, such as a particular tree in a particular FOREST. When discussing a species, however, habitat is often identified in broader terms that describe an entire ECOSYSTEM or BIOME. These broader terms are generally descriptive of the dominant PLANTS or physical features of an area. For example, a hardwood forest is a habitat in which the dominant plants are broad-leaved, deciduous trees such as oak, maple, and birch. Such a forest provides habitat not only to the trees, but also to a variety of other plant species as well as animals, FUNGI, PROTISTS, and BACTERIA. The presence of water, as in a BOG, lake or pond, hot spring, or OCEAN, is a physical feature that may be used to identify a habitat. Such environments provide habitat not only to the organisms living in the water, but also to terrestrial organisms living nearby that interact with the water or its organisms. For example, many migratory birds stop over in water environments such as bogs, ponds, or lakes to obtain drinking water and food—plants, INSECTS, or fish. They may also rest either in the water or in surrounding shrubbery.

Destruction and loss of habitat are major causes of EXTINCTION. Each species is adapted to life in a particular habitat. *(See* adaptation.) Thus, destruction or loss of habitat threatens the survival of the species if organisms are unable to adapt to such changes or move to a new habitat. Natural occurrences that can result in habitat loss include such events as FLOODS, volcanic eruptions, and FOREST FIRES. Human activities, such as the release of POLLUTANTS into AIR, water, and soil; diversion of water from rivers and streams to other areas for use as drinking water or for crop IRRIGATION; and clearing land to develop roads, farmland, grazing pastures, or for housing developments or shopping malls, also destroy or alter habitats to the extent where organisms can no longer survive there. Environmentalists are concerned that as the human POPULATION continues growing and altering environments to meet its needs, more organisms will become extinct as a result of habitat loss, thus decreasing Earth's BIODIVERSITY. To help save some species from extinction, particularly those included on the ENDANGERED SPECIES LIST, some areas have been set aside as WILDLIFE REFUGES or preserves; however, such efforts provide only limited protection to organisms and their habitats. *See also* BIOREGION; BIOSPHERE RESERVE; CARARA BIOLOGICAL RESERVE; CARRYING CAPACITY; DEBT-FOR-NATURE SWAP; DEFORESTATION; ECOLOGICAL SUCCESSION; ECOTONE; ENDANGERED SPECIES ACT; ENVIRONMENTAL IMPACT STATEMENT; ENVIRONMENTALLY SENSITIVE AREA; FISH AND WILDLIFE SERVICE; NATIONAL WILDLIFE REFUGE SYSTEM; NICHE; WILDERNESS.

habitat loss. *See* habitat.

half-life, The length of time it takes for 50% of a RADIOISOTOPE to decay into another element. After a period equal to 10 half-lives, the radioactivity of the element will have decreased to about 0.1% of its original value. The half-life of radioactive substances may vary in time from a fraction of a second to thousands of years. Radioactive carbon-14, for example, has a half-life of approximately 5,760 years, while POLONIUM-218 has a half-life of about three minutes. *See also* RADIOACTIVE DECAY; RADON; URANIUM.

halon, Any of a family of halogen compounds that cause OZONE DEPLETION and GLOBAL WARMING when released into the ENVIRONMENT. Halons are generally composed of CARBON, bromine, and other halogen elements such as CHLORINE and FLUORINE. In the past, halon gases were widely used in fire extinguishers; however, because of their OZONE-depleting ability, governments worldwide agreed to discontinue halon production and use as part of the 1990 amendment to the MONTREAL PROTOCOL. *See also* CHLOROFLUOROCARBONS, GLOBAL WARMING POTENTIAL.

Hanford Nuclear Waste Site, A facility that originally was a U.S. Department of Defense site for the production of PLUTONIUM FUEL used for developing NUCLEAR WEAPONS. The Hanford Nuclear Waste Site is an area of approximately 1,450 square kilometers (560 square miles) located in the southeastern part of Washington State near the Columbia River.

Nuclear weapons production began in the 1940s, during World War II, when little consideration was given to the management of nuclear wastes. Such activities continued for over 40 years. Between 1944 and 1989, the Hanford Nuclear Site produced over 45 metric tons (50 tons) of weapons-grade plutonium, which represented more than half of the U.S. supply. During this time much RADIOACTIVE WASTE was released into the ENVIRONMENT in and around the Hanford facility.

EnviroSources

National Research Council, Board on Radioactive Waste Management Web site: www4.nas.edu/brwm/brwm-res.nsf.
U.S. Department of Energy, Office of Civilian Radioactive Waste Management Web site: www.rw.doe.gov.
U.S. Environmental Protection Agency (mixed-waste) Homepage: www.epa.gov/radiation/mixed-waste.
U.S. Nuclear Regulatory Commission, Radioactive Waste Page Web site: www.nrc.gov/NRC/radwaste.

Hazard Ranking System, The principal screening tool used by the U.S. ENVIRONMENTAL PROTECTION AGENCY (EPA) to evaluate the risks to public health and the ENVIRONMENT associated with abandoned or uncontrolled HAZARDOUS WASTE sites. The Hazard Ranking System (HRS) uses a formula to develop a score that determines if a site should be placed on the NATIONAL PRIORITIES LIST (NPL). The factors that determine the score are grouped into three categories: the likelihood that a site has released or has the potential to release HAZARDOUS SUBSTANCES into the environment, the characteristics of the waste (such as TOXICITY and waste quantity), and the proximity of human POPULATIONS or sensitive environments that might be affected by the release.

Once a site is placed on the NPL, its HRS score can also be used to assess its hazard and cleanup priority in comparison to other sites. *See also* COMPREHENSIVE ENVIRONMENTAL RESPONSE, COMPENSATION, AND LIABILITY ACT; SUPERFUND; TOXIC RELEASE INVENTORY.

EnviroSources

U.S. Environmental Protection Agency (Hazard Ranking System) Web site: www.epa.gov/superfund/programs/npl_hrs/hrsint.html.
U.S. Environmental Protection Agency (Superfund): Web site www.epa.gov/superfund; Web site: www.pin.org/superguide.html.

hazardous material. *See* hazardous substance.

Hazardous Materials Transportation Act, Federal legislation that was enacted in 1975 to establish regulations for the transportation of HAZARDOUS WASTES and procedures for enforcement of those regulations. The primary goal of the Hazardous Materials Transportation Act (HMTA) is to eliminate the illegal dumping of HAZARDOUS SUBSTANCES. According to the HMTA, all vehicles that carry hazardous wastes must have conspicuously displayed stickers indicating the kind of waste they are transporting.

In 1990 the HMTA was strengthened with passage of the Hazardous Materials Transportation Uniform Safety Act. The new laws ensure better enforcement of the HMTA by increasing the number of safety inspectors, increasing the criminal penalties for violations of the laws, and helping states develop better response plans for hazardous waste accidents. *See also* HAZARD RANKING SYSTEM; TOXIC CHEMICALS.

EnviroSources

U.S. Department of Transportation Web site: www.dot.gov.

hazardous substances, A substance containing materials or having characteristics that could cause harm to people or the ENVIRONMENT if the materials are not managed properly. The COMPREHENSIVE ENVIRONMENTAL RESPONSE, COMPENSATION, AND LIABILITY ACT (CERCLA) (also known as SUPERFUND) included a list of materials designated as hazardous substances. For each hazardous substance, CERCLA has established a maximum amount that can be released without being reported to the ENVIRONMENTAL PROTECTION AGENCY (EPA) or the state environmental agency. Any amounts released that exceed this maximum are considered reportable quantities.

Hazardous substances include chemicals in solid, liquid, or gaseous form as well as organic, and inorganic chemicals. Some of the following properties make hazardous substances potentially harmful:

corrosive Capable of burning the skin, eyes, or other body surfaces. Strong ACIDS, such as NITRIC ACID and SULFURIC ACID are corrosive.
reactive Capable of reacting to produce poisonous gases. Compounds containing cyanide may be reactive hazardous substances. HYDROGEN SULFIDE is reactive.
ignitable Capable of catching on fire easily. Propane is an ignitable gas.
toxic If eaten, breathed, or absorbed through the skin is capable of causing harmful short-term or long-term health effects. Toxic substances include the metal LEAD and many PESTICIDES.
reactive Causing a dangerous reaction, such as an explosion, when combined with water. The element sodium is highly reactive.
etiological Containing biological matter, such as BACTERIA OR VIRUSES, that may cause disease.

Proper management of hazardous substances entails labeling to identify each substance and its hazards, maintaining secure storage of substances in appropriate containers, the segregation of substances that could cause a harmful reaction if accidentally mixed, and the preparation of emergency plans to deal with leaks or spills. "Hazard communication" refers to the process of alerting people to the hazards of various substances, providing information about the safe handling of substances, and explaining how to respond to accidents involving such substances.

A hazardous substance is not equivalent to a HAZARDOUS WASTE, which is defined as hazardous material that is discarded, abandoned, or disposed of. A hazardous substance can become a hazardous waste if it is released into the ENVIRONMENT or improperly managed.

In addition to CERCLA, other laws and regulations that govern the management of hazardous substances include the HAZARDOUS MATERIALS TRANSPORTATION ACT (HMTA), the U.S. Department of Transportation (DOT) regulations, the OCCUPATIONAL HEALTH AND SAFETY ADMINISTRATION (OSHA) regulations, and the EMERGENCY PLANNING AND COMMUNITY RIGHT-TO-KNOW ACT (EPCRA). The HMTA and the DOT regulate the labeling and shipping of hazardous materials. OSHA regulates the management of chemicals and hazardous materials in the workplace and enforces the "Worker Right-to-Know" rule. The EPCRA deals with issues of community planning and communication of information about hazardous chemicals. *See also* HAZARDOUS WASTE TREATMENT; TOXIC CHEMICAL.

EnviroSources

U.S. Environmental Protection Agency Superfund Web site: www.epa.gov/oerrpage/superfund/programs/er/hazsubs/index.html.

U.S. Occupational Safety and Health Administration (OSHA) Web site: www.osha-slc.gov/sltc/hazardoustoxicsubstances/index.html.

Vermont SIRI MSDA Archive Material Safety Data Sheets Web site: www.siri.uvm.edu/msds/grep/g2.cgi.

hazardous waste, Any SOLID WASTE, or combination of solid wastes, that poses a potential risk to the health of humans, other ORGANISMS, or the ENVIRONMENT. The RESOURCE CONSERVATION AND RECOVERY ACT (RCRA) identifies hazardous waste as any solid waste or combination of solid wastes, that, because of its quantity, concentration, or physical, chemical, or infectious characteristics, may cause or significantly contribute to DEATH RATES or the risk of serious irreversible or incapacitating reversible illnesses, or may pose a substantial present or future threat to human health or the environment when improperly treated, stored, disposed of, or otherwise managed. The U.S. ENVIRONMENTAL PROTECTION AGENCY (EPA) determines which wastes are classified as hazardous. The RCRA uses a CRADLE-TO-GRAVE manifest to keep track of a waste from its origin to its site of final disposal. *See also*; ACUTE TOXICITY; CHRONIC TOXICITY; HAZARDOUS MATERIALS TRANSPORTATION ACT ; HAZARDOUS SUBSTANCES; HAZARD RANKING SYSTEM ; TOXIC CHEMICAL; TOXIC SUBSTANCES CONTROL ACT; TOXIC WASTE.

Some hazardous wastes include

aldrin	mercury
asbestos	methane
benzene	plutonium-239
boron	propane
lead	

EnviroSources

Blackman, William C., Jr. *Basic Hazardous Waste Management.* Boca Raton, FL: Lewis, 1996.

Citizen's Clearinghouse for Hazardous Wastes: P.O.Box 6806, Falls Church, VA , 22040.

Clean Sites: 1199 North Fairfac Street, Suite 400, Alexandria, VA 22314.

Environmental Industry Associations, Hazardous Waste Ombudsman Program: 4301 Connecticut Avenue, NW, Suite 300, Washington, DC 20008.

Environmental Protection Agency: 401 M St., SW, Mail Code 5101, Washington, DC 20460.

Haun, J. William, and Bill Haun. *Guide to the Management of Hazardous Waste: A Handbook for the Businessman and Concerned Citizen.* Philadelphia: North American Press, 1991.

U.S. Environmental Protection Agency, Office of Solid Waste and Emergency Reponse Web site: www.epa.gov/swerosps/ej.

U.S. Environmental Protection Agency Web site: es.epa.gov/oeca/hazsol.html.

Wagner, Travis P. *The Complete Guide to Hazardous Waste Regulations: RCRA, TSCA, HMTA, OSHA, and Superfund.* New York: John Wiley and Sons, 1999.

Hazardous Wastes Present in Landfills

Hazardous waste	All landfilled hazardous waste (%)
Electroplating wastewater treatment sludge	16.3
Lead	5.9
Chromium	5.9
Electric steel furnace sludge	4.4
Petroleum refinery wastes	3.8

Hazardous includes any solid waste, or combination of solid wastes, that poses a potential risk to the health of humans, other organisms, or the environment. *Source:* U.S. EPA.

hazardous waste treatment, Waste treatment methods used to destroy HAZARDOUS WASTES or make them less hazardous. Methods of hazardous waste treatment include INCINERATION, stabilization, and neutralization. These methods are used to treat hazardous wastes that are in containers or capable of being placed in containers easily. Wastes that have been spilled or released into the ENVIRONMENT (usually soil or water) often require special techniques to control or treat the hazardous waste, known as hazardous waste "remediation" methods.

Incineration is the process of burning a hazardous waste to destroy it. Incineration often may produce additional waste in the form of smoke (or gases) and ASH, which must be properly controlled and managed to prevent pollution. The heat produced by incineration can provide the energy to drive a TURBINE and produce ELECTRICITY. Hazardous waste incinerators are specially designed and controlled devices.

Stabilization of a hazardous waste is a process that prevents the harmful components of the waste from escaping into the environment. Usually, in stabilization, a waste is treated chemically to make it more stable (or inert) and less likely to enter the AIR or water. Once stabilized, a waste may be considered safe to use for some other purpose or to place in a SANITARY LANDFILL.

Neutralization, a treatment often applied to corrosive wastes, refers to a chemical change that makes the waste neutral and less harmful. Usually other chemicals must be added to the hazardous waste to neutralize it. For instance, acidic wastes often are treated by combining them with a basic, or alkaline, substance to neutralize the ACID. *(See* alkali.) Sometimes neutralization is used to control an accidental chemical spill or leak.

Many remediation methods involve chemical, physical, or biological treatments that alter the waste material by changing or removing its hazardous constituents. Several of these methods, known as "transfer technologies," involve transferring CONTAMINANTS from one form (medium) to another, such as removing VOLATILE ORGANIC COMPOUNDS from water by transferring the compounds to the air. Various filtering systems are used to filter air or water, which transfers the hazardous waste from the air or water to the filter medium. Special treatments can break apart hazardous chemicals into smaller, less harmful chemicals. In addition, biological treatments exist that can break down chemicals in soil or water. The U.S. ENVIRONMENTAL PROTECTION AGENCY and the RESOURCE CONSERVATION AND RECOVERY ACT have established special permits and procedures for hazardous waste treatment.

When hazardous waste is to be treated, the methods of treatment selected depend on the type of waste, the amount of waste, and the cost of the treatment. In some cases, more than one treatment will be necessary to manage the waste. For instance, incineration may destroy part of the waste, but the ash that is left over may require a second treatment such as stabilization.

Hazardous waste disposal is not considered a type of hazardous waste treatment because the process of disposal does not necessarily destroy the waste or make the waste nonhazardous, but places the waste in another, usually more secure, location.

EnviroSources

Federal Remedial Technologies Roundtable Web site: www.frtr.gov
Hazardous Waste Cleanup Information Web site: www.clu-in.org.

heat island, An effect that takes place when there is an increase of stagnant warm AIR over an urban area due to the combustion of FOSSIL FUELS from apartment buildings, homes, schools, office complexes, and industrial plants. A heat island is a dome-like cover over the center of a city of trapped POLLUTANTS and other PARTICULATE MATTER that have absorbed heat. The sun's rays are also absorbed and retained by the surface area of large buildings, streets, and other structures. Building materials such as slate, roofing shingles, asphalt, granite, marble, concrete, and bricks act as SOLAR ENERGY collectors. As a result, the temperatures in urban areas are generally warmer than the surrounding, less populated, suburban and rural towns. During the year, cities will probably have warmer days, less sun, more fog and SMOG, and rainier conditions than those living in the nearby surrounding areas.

heavy metal, A metallic element that is hazardous and toxic, even at low concentrations. Heavy metals can cause adverse health conditions in humans and other ORGANISMS when they accumulate in the FOOD CHAIN. These metals remain in the ENVIRONMENT because they cannot be broken down. Among the common heavy metals are MERCURY, CHROMIUM, LEAD, ARSENIC, COPPER, zinc, and CADMIUM.

Heavy metals are widespread in the environment and exist as PARTICULATE MATTER in the ATMOSPHERE, dissolved in the waterways or soil. Heavy metals are released into the environment as a result of the combustion of leaded GASOLINE, in EMISSIONS from manufacturing plants, foundries, incinerators, SMELTING operations, and via LEACHING from landfills. Most of the contamination of soil comes from heavy metal wastes from SEWAGE SLUDGE, WASTEWATER, and ASH.

Humans, who are at the top of the food chain, are at great risk for health hazards associated with heavy metals because of the process of BIOACCUMULATION. Since heavy metals are soluble in water, they can be easily absorbed in the bodies of humans and other organisms. As the metals progress along the food chain, their concentration in each organism's tissue increases. By the time a human consumes the metal in food, the concentration is very high and can easily cause health problems.

ACUTE TOXICITY can result from the inhalation of AIR or the ingestion of liquids containing metals in very high concentrations. Inhalation of high concentrations of metals is irritating and may cause severe damage to the respiratory tract. Symptoms following the ingestion of excessive amounts of metals are similar to those of food poisoning; they include nausea, vomiting, and abdominal pain. In excessive amounts, cadmium can cause renal (kidney) and neural (nervous system) damage, as well as pulmonary (lung), skeletal, and testicular problems. Other chronic effects of cadmium in humans may include heart damage, decreased fertility, and changes in physical appearance or behavior. Chronic exposure to lead, particularly from lead paint, can be dangerous for young children. Other illnesses associated with lead include hypertension, reproductive disorders, and neurological problems. Too much arsenic in drinking water also can cause a variety of illnesses including forms of CANCER. METHYL MERCURY poisoning causes irreversible damage to the central nervous system; furthermore, some children who exhibit severe cerebral palsy–like brain damage were born to mothers who had been exposed to mercury.

EnviroSources

Foulkes, E.C., ed., *Biological Effects of Heavy Metals: Metal Carcinogenesis*. Cleveland, OH: CRC Press, 1990.
Salomons, W. et al., eds. *Heavy Metals: Problems and Solutions*. New York: Springer-Verlag, 1995.
Vernet, J.P., ed., *Heavy Metals in the Environment*. Elsevier Science Publishers, 1991.

U.S. Environmental Protection Agency, Office of Pollution Prevention and Toxics Web site: www.epa.gov/opptintr.

heptachlor, An insecticide belonging to the CHLORINATED HYDROCARBON group that may be harmful to the ENVIRONMENT and human health. Exposure to heptachlor and heptachlor epoxide, a breakdown product, happens most commonly from eating contaminated foods and milk or by skin contact with contaminated soil. In high concentrations, these compounds can cause damage to the nervous system.

Heptachlor is a manufactured chemical that does not occur naturally. Trade names include Heptagran, Basaklor, Drinox, Soleptax, Termide, and Velsicol 104. Heptachlor was used extensively in the past to kill INSECTS in homes, buildings, and on food crops, especially corn. Use of the PESTICIDE slowed in the 1970s, and in 1988 the U.S. ENVIRONMENTAL PROTECTION AGENCY (EPA) banned the sale of all heptachlor products and restricted the use of heptachlor to the control of fire ants in power transformers.

The EPA has established a maximum acceptable concentration of 0.4 parts of heptachlor per billion parts of drinking water. *(See* parts per billion and parts per million.) The U.S. FOOD AND DRUG ADMINISTRATION (FDA) limits the amount of heptachlor that may be present on raw food crops and on edible seafood.

The UNITED NATIONS ENVIRONMENT PROGRAMME (UNEP), along with other international organizations, convened in 1997 and 1998 to seek resolutions to the problem of persistent ORGANIC POLLUTANTS (POPs), pollutants that do not readily break down into harmless substances. The list of POPs of greatest concern included heptachlor because of its TOXICITY and persistence in the environment. Other POPs considered most damaging to the environment included the pesticides DICLORODIPHENYLTRICHOROETHANE (DDT), toxaphene, CHLORDANE, endrin, mirex, ALDRIN, and dieldrin. While the UNEP discussions included proposals to phase out the use of certain POPs, they also recognized the importance of pesticide use in the maintenance of world food production. *See also* AGENT ORANGE; AGRICULTURAL POLLUTION; DIOXIN.

EnviroSource

Agency for Toxic Substances and Disease Registry, Division of Toxicology: 1600 Clifton Road NE, Mailstop E-29, Atlanta, GA 30333; (800) 447-1544; fax: (404) 639-6315; Web site: www.atsdr1.atsdr.cdc.gov:8080/atsdrhome.html.

herbicide. *See* pesticides.

herbivore, Any ORGANISM that derives its nutrition by feeding only on PLANTS or plant products such as seeds, nuts, or fruits. Herbivores, also known as primary CONSUMERS, feed directly on PRODUCERS and make up the first TROPHIC LEVEL in all ECOSYSTEMS. Depending on the type of plant food they eat, herbivores are classified into several groups. For example, herbivores that feed primarily on grains and grasses or seeds are classified as granivores. This group of herbivores includes grazing animals such as cattle, horses, ZEBRAS, BISON, and sheep; seed-eating MAMMALS such as mice; and birds such as cardinals, grosbeaks, and longspurs. Herbivores that feed mostly on fruits are classified as frugivores. This group includes oilbirds, some BATS, and many primates that cannot feed on herbaceous (soft, green) plant parts because of their inability to digest cellulose. Another group of herbivores are the browsers. These include animals such as deer and ELEPHANTS that feed on woody vegetation. This group also includes many xylophagus organisms, such as termites, elephant beetles, unicorn beetles, and rhinoceros beetles, which feed on or in wood, often causing great damage to human property. *See also* CARNIVORE; DECOMPOSER; ECOLOGICAL PYRAMID; FOOD CHAIN; FOOD WEB; OMNIVORE; SCAVENGER.

The kangaroo is a herbivore.

heterotroph, An ORGANISM that is incapable of synthesizing its own food and must therefore obtain its nutrients by feeding on other organisms. Matter ingested by heterotrophs is broken down during the digestive process to provide the nutrients and ENERGY needed by the organism. Because they obtain nutrition by feeding on other organisms, all heterotrophs are CONSUMERS. All animals and FUNGI and most BACTERIA and PROTISTS are heterotrophs. *See also* AUTOTROPH; CARNIVORES; CHEMOAUTOTROPH; CHEMOSYNTHESIS; DECOMPOSER; DETRIVORE; ECOLOGICAL PYRAMID; FOOD CHAIN; FOOD WEB; HERBIVORE; OMNIVORE; PARASITISM; PHOTOSYNTHESIS; PREDATION; PRODUCER; SCAVENGER; TROPHIC LEVEL.

hibernation, An ADAPTATION of some endothermic animals that allows them to enter a dormant or torpid state to survive periods of extreme cold. During hibernation, an animal becomes inactive and enters a deep sleep as its body temperature drops to a lower-than-normal level. During this period, the animal's metabolic functions, respiration, and heart rate slow, allowing the animal to conserve ENERGY.

Examples of animals that are true hibernators are marmots and ground squirrels. These animals can enter a dormant state at the first sign of winter and remain in this state until spring, even during periods when temperatures rise above freezing. Brown bears and some rodents enter states of partial or temporary hibernation during the winter. In these animals, metabolic processes, respiratory rates, and heart rates slow only while the animal is in a sleeplike state. When outdoor conditions become more favorable, these animals may awaken briefly and leave their dens or nesting sites to obtain food or water. *See also* ESTIVATION.

high-level radioactive waste, Dangerous RADIOACTIVE WASTE materials that include FUEL or materials that remain after reprocessing SPENT FUEL from NUCLEAR REACTORS and NUCLEAR WEAPONS production facilities. High-level radioactive wastes are more hazardous than LOW-LEVEL RADIOACTIVE WASTES, and usually will remain dangerous for thousands or even millions of years. These wastes contain a combination of transuranic radioactive elements (those elements with atomic numbers and atomic masses greater than uranium) and FISSION products in concentrations high enough to require permanent isolation from humans and other ORGANISMS. In the United States, the federal government is responsible for disposing of high-level radioactive waste. *See also* HAZARDOUS WASTE; NUCLEAR REGULATORY COMMISSION; PLUTONIUM-239; RADIOISOTOPE OR RADIOACTIVE ISOTOPE; URANIUM; VITRIFICATION.

EnviroSources

International Atomic Energy Agency ("Managing Radioactive Waste" fact sheet) Web site: www.iaea.org/worldatom/inforesource/factsheets/manradwa.html.
National Research Council, Board on Radioactive Waste Management Web site: www4.nas.edu/brwm/brwm-res.nsf.
U.S. Department of Energy, Office of Civilian Radioactive Waste Management Web site: www.rw.doe.gov.
U.S. Environmental Protection Agency (Mixed waste) Web site: www.epa.gov/radiation/mixed-waste.
U.S. Nuclear Regulatory Commission (radioactive waste) Web site: www.nrc.gov/NRC/radwaste.

host. *See* parasitism.

humus, ORGANIC matter present in soil that is derived from the DECOMPOSITION of PLANT and animal remains. Humus forms through the actions of soil ORGANISMS, such as the BACTERIA, PROTISTS, and FUNGI that obtain their nutrition through decomposition and worms, beetles, and other INSECTS that are DETRITUS feeders. *(See* detrivore.) Humus is a dark-colored material that aids in soil fertility by returning various salts, proteins, and other organic nutrients to the soil as it decomposes. These nutrients are then used by plants for their growth and repair processes and passed through the FOOD CHAIN when animals feed on plants.

The amount of humus in soil affects the physical properties of the soil. For example, soil structure, texture, water-holding capacity, and color all are impacted by the humus content of the soil. Because humus aids in plant growth, it is desirable in soils used for the cultivation of plants. In areas used for AGRICULTURE, humus may be depleted from soil due to the successive growth of crops with similar nutrient needs. To maintain and restore humus to such soils, it is necessary to periodically add organically rich materials such as manure, compost, organic MULCH, and PEAT moss to soil that will form humus as they decay. *See also* BIOGEOCHEMICAL CYCLE; COMPOSTING; DECOMPOSER; FERTILIZER; SOIL CONSERVATION.

hurricane, A violent storm that is generated over a tropical area of the OCEAN and which causes destruction and loss of life in primarily coastal areas. Hurricanes are actually tropical CYCLONES: in the eastern Pacific Ocean, these storms are called TYPHOONS.

Hurricanes frequently occur in the Caribbean West Indies, usually during the period from August to October. Hurricanes sustain wind speeds of at least 150 kilometers (90 miles) per hour, with some reaching 300 kilometers (180 miles) per hour. The high winds associated with hurricanes cause huge waves that batter coastal areas, leading to flash FLOODS, beach EROSION, and often significant damage to marinas and shoreline homes. When hurricanes move inland, their powerful winds and torrential rains can destroy homes, buildings, and powerlines, and cause extensive damage to crops and NATURAL RESOURCES. Drowning is the major cause of deaths due to hurricanes. The development of accurate hurricane warning systems and forecasting has decreased the number of lives lost during hurricanes; however, continuing POPULATION growth along coastal areas, unsatisfactory building codes on shoreline structures, and little or no planning of evacuation routes by some local governments are still significant concerns.

The most disastrous hurricanes in the 1990s include Hurricane Andrew, in 1992, which swept along a path through the northwestern Bahamas, the southern Florida peninsula, and south-central Louisiana. Damage in the United States was near $25 billion, making Andrew the most expensive NATURAL

The Path of a Typical Hurricane

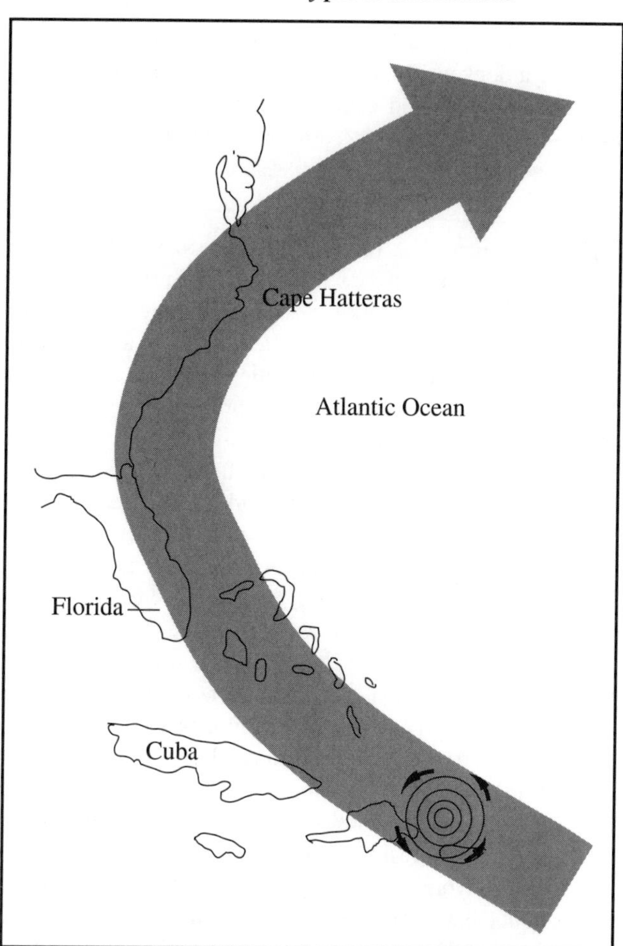

DISASTER in United States history at that time. In 1994, Hurricane Gordon followed an unusual, erratic path from the western CARIBBEAN SEA and then on to Florida and the southwestern Atlantic Ocean. Its torrential rains caused much loss of life in Haiti and extensive agricultural damage in south Florida. In 1998, Hurricane Mitch brought massive devastation and loss of life across Central America. Officials estimate that Mitch killed more than 10,000 people across Central America, with an additional 13,000 missing and presumed dead and a staggering 2.8 million people left homeless. The storm hit hardest in Honduras and Nicaragua. *See also* EL NIÑO/LA NIÑA; FEDERAL EMERGENCY MANAGEMENT AGENCY; NATURAL DISASTERS; WEATHER.

EnviroSources

Elsneer, J. B. *Hurricanes of the North Atlantic: Climate and Society.* Oxford, UK: Oxford University Press, 1999.
National Hurricane Center Web site: www.nhc.noaa.gov.

hybrid, An offspring produced by the breeding of ORGANISMS of different SPECIES, breeds, or varieties. People often intentionally produce hybrid PLANTS and animals that exhibit traits that are more desirable than those of either parent organism. For example, to produce the hybrid known as a mule, a female horse is bred with a male donkey, resulting in an offspring that is generally more resistant to DISEASE (especially hereditary diseases) and has a longer life span than either parent species. The mule also is generally stronger and better suited for use as a pack animal or for farmwork than either a horse or a donkey. The disadvantage of a hybrid is that they are usually sterile and cannot reproduce. *See also* FLORIDA PANTHERS; GENETIC DIVERSITY; GENETIC ENGINEERING; GERM PLASM BANK; HYBRIDIZATION; INTEGRATED PEST MANAGEMENT.

hybridization, The breeding of ORGANISMS of different breeds, SPECIES, or varieties to produce offspring with traits that differ from the parents. Organisms resulting from hybridization are often stronger, have better resistance to DISEASE, or produce better yields than the varieties from which they were formed, a phenomenon known as hybrid vigor. An offspring resulting from hybridization is called a HYBRID or crossbreed.

Hybridization can occur in nature; however, it is often carried out by breeders, agricultural workers, and genetic engineers to produce PLANTS or animals with distinct traits or to develop new varieties of food. For example, many hybrid flower varieties, such as roses and tulips, are created to attain flowers of specific colors or sizes. The mule is a hybrid farm animal that is stronger and generally has a longer life span than either the male donkey or female horse that are its parents. The loganberry is an edible fruit produced by crossing the red raspberry with the western dewberry.

Hybridization has been used successfully to produce heartier crop plants that are more DROUGHT resistant, disease resistant, and frost resistant than traditional varieties. For example, many hybrid varieties of wheat and corn are now grown throughout the world to meet the food demands of both people and farm animals. Continued development of such crops may increase food production, decrease waste, and make it possible to grow some crops in ENVIRONMENTS where they could not otherwise be grown. *See also* BORLAUG, NORMAN ERNEST; FLORIDA PANTHERS; GENETIC DIVERSITY; GENETIC ENGINEERING; GERM PLASM BANK; INTEGRATED PEST MANAGEMENT; SUSTAINABLE AGRICULTURE.

hydrocarbons, Any combustible ORGANIC compound that contain only the elements HYDROGEN and CARBON. Hydrocarbons are used primarily as FUELS. They are primary POLLUTANTS and have an adverse effect on PLANTS and animals. Hydrocarbons can be synthetic or natural substances and are found in PETROLEUM, COAL, NATURAL GAS, SOLVENTS, paints, and printing materials. Some common hydrocarbons include METHANE (CH_4), propane (C_3H_8), acetylene (C_2H_2), and butane (C_4H_{10}). Hydrocarbon EMISSIONS caused by the incomplete combustion of FOSSIL FUELS used by AUTOMBOBILES and power plants can react with sunlight to form photochemical SMOG.

hydrochloroflurocarbons, Any of a group of chemicals used as a replacement for CHLOROFLUOROCARBONS (CFCs) in refrigeration and in some aerosols. Hydrofluorocarbons (HCFCs) are less damaging to the OZONE layer than CFCs.

hydroelectric power, An ALTERNATIVE ENERGY RESOURCE and RENEWABLE RESOURCE, sometimes referred to as hydropower, that uses flowing water to drive TURBINES to generate ELECTRICITY. The amount of electric ENERGY produced by the turbine depends on the pressure and the volume of the water that flows through it.

Hydroelectric power accounts for about 22% of the world's electricity. The largest hydroelectric power producers include Canada, the United States, Brazil, Norway, Russia, and China. About 10% of all U.S. electricity is produced by hydropower. By contrast, hydroelectric power accounts for less than 2% of the electricity produced in the United Kingdom.

The Itaipú hydroelectric power plant, located on the border between Brazil and Paraquay, is the largest development of its kind in operation in the world. Built between 1975 and 1991, Itaipú represents the efforts and accomplishments of the two neighboring countries. In 1995, Itaipú alone provided 25% of the energy supply in Brazil and 78% in Paraguay.

A major benefit of hydroelectric power is that it is nonpolluting: It produces no harmful EMISSIONS such as CARBON DIOXIDE, SULFUR DIOXIDE (SO_2), or NITROGEN OXIDES (NO_x) and no liquid or SOLID WASTES. However, there are concerns about and opposition to building large DAMS for hydroelectric power because they have negative effects on the ENVIRONMENT and may damage archaeological sites or historic lands.

Building a large dam requires a large area behind the dam to be flooded to form a RESERVOIR. The flooding destroys natural HABITATS, WETLANDS, and farmlands. People living in areas designated for flooding have to be relocated. Over time reservoirs fill up with silt and sediments, burying the spawning areas of fish and other aquatic ORGANISMS. Stagnant pools can

Site of a Hydroelectric Power Plant in South America

The Itaipu Hydroelectric Power Plant is the largest of its kind in the world. In 1995, Itaipú provided 25% of the energy supply in Brazil and 78% in Paraguay.

become breeding grounds for DISEASE-causing INSECTS and waterborne PATHOGENS

Large dams such as the Aswan High Dam in Egypt and the Tarbela in Pakistan have produced severe impacts on ECOSYSTEMS and the health of the people nearby. Because of environmental concerns and the high cost of building large dams, interest in hydroelectic power has decreased and in some countries has ceased; however, other countries, such as Brazil, India, and China, view hydoelectric power as one option to reduce their dependency on imported PETROLEUM, preferable to the use of COAL-fired power plants that pollute the ATMOSPHERE.

Building small, rather than large, hydroelectric power systems may be the trend for the future. Today, small-scale hydroelectric power systems, called "minihydro" or "microhydro" systems are being used on rivers and tributaries and in remote areas where construction is difficult. Such small-scale systems do not require the damming of rivers. These minihydro systems are used in China, the United States, and several smaller countries, including Indonesia, Nepal, Sri Lanka, and Zaire. Small-scale hydropower stations provide the advantages of hydropower without the problems of large-scale hydropower plants. They have less impact on the environment and are less costly to build and maintain than are large hydroelectric systems. Small-scale hydropower can be used locally in remote villages and towns to generate electricity for businesses, farming needs, local lighting, and for pumping water. *See also* AIR POLLUTION; FOSSIL FUEL; JAMES BAY HYDROPOWER PLANT.

EnviroSources

Escheverria, J.D., et al. *Rivers at Risk: The Concerned Citizens's Guide to Hydropower.* Washington, DC: Island Press, 1990.
McCutheron, S. *Electric Rivers: The Story of the James Bay Project.* St. Paul, MN: Black Rose Books, 1991.
U.S. Bureau of Reclamation hydropower information) Web site: www.usbr.gov/power/edu/edu.html.
U.S. Geological Survey Web site: wwwga.usgs.gov/edu/hybiggest.html.

hydrogen, An element that is a colorless and odorless gas existing as a H_2 molecule that makes up 75% of the universe's mass. Its ISOTOPES include deuterium and tritium. Hydrogen is found on Earth only in combination with other elements such as OXYGEN, CARBON, and NITROGEN. Research is being conducted to separate hydrogen from other elements so it can be used as a potential RENEWABLE RESOURCE.

Today, hydrogen is used primarily in ammonia manufacturing, PETROLEUM refining, and METHANOL synthesis. It is also used in National Aeronautic and Space Administration's (NASA's) space program as FUEL for the space shuttles and in FUEL CELLS that provide heat, ELECTRICITY, and drinking water for astronauts. Fuel cells are devices that directly convert hydrogen into electricity. In the future, hydrogen could be used to fuel vehicles and aircraft and to provide power for homes and offices. When hydrogen is used as an ENERGY source, it generates no EMISSIONS and its only byproduct is water, which is recycled to make more hydrogen. *(See* recycling.)

As part of the U.S. DEPARTMENT OF ENERGY'S (DOE) Hydrogen Program, the National Renewable Energy Laboratory (NREL) conducts research on advanced technologies to produce, store, and safely transport and use hydrogen. The goal is to use hydrogen in quantities large enough, and at low-enough costs, as a fuel.

Researchers must overcome several obstacles if hydrogen is to become a major energy resource, however. Hydrogen is currently more expensive than traditional energy sources, the production efficiency—or the amount of energy or feedstock used to produce hydrogen—must improve, and the infrastructure to efficiently transport and distribute hydrogen must be developed.

Future uses of hydrogen as an energy source would primarily be transportation, where it would help reduce pollution. Internal combustion engines can be fueled with pure hydrogen, or hydrogen blended with NATURAL GAS. Vehicles can also be powered with hydrogen fuel cells, which are three times more efficient than GASOLINE-powered engines. Fuel cells can also supply heat and electricity for homes and buildings. The overall goal of the DOE's Hydrogen Program is to replace the use of conventional energy in 2–4 million households with hydrogen by the year 2010. Its goal in 2030 is to replace conventional energy in 10 million households. *See also* FOSSIL FUEL; NUCLEAR POWER; WIND POWER.

EnviroSources

National Renewable Energy Laboratory Web site: www.nrel.gov/lab/pao/hydrogen.html.

U.S. Department of Energy, Hydrogen InfoNet Web site: / www.eren.doe.gov/hydrogen/infonet.html.

hydrogen fuel cell. *See* hydrogen.

hydrogen sulfide, A colorless, corrosive gas composed of HYDROGEN and SULFUR that has a characteristic "rotten egg" odor. The gas is produced under ANAEROBIC CONDITIONS. Hydrogen sulfide (H_2S) is a major AIR POLLUTANT. Inhaling high concentrations of hydrogen sulfide dulls the sense of smell, making it difficult to detect over long periods. This can lead to asphyxiation. The corrosive nature of hydrogen sulfide also makes the compound damaging to PLANTS and aquatic ORGANISMS.

Hydrogen sulfide is produced by natural sources such as BOGS and SWAMPS as well as from human sources such as WASTEWATER treatment plants, SMELTING and PETROLEUM plants, and SEWERS. *See also* AIR POLLUTION, CHEMOAUTOTROPHS, SULFUR DIOXIDE.

hydrologic cycle. *See* water cycle.

hydrology, A science concerned with all aspects of SURFACE WATER on Earth. Two fundamentally important areas of hydrology are the occurrence of water and the movement of water over the surface of Earth. Hydrologists study the ways in which surface water occurs to form streams, rivers, ponds, and lakes and how water flows through drainage channels, streams, and rivers.

Much of hydrology deals with the effects of precipitation (such as rain and snow) on surface water bodies. For instance, hydrologic studies often evaluate the amount of surface water flow or flooding that could occur in an area as a result of a storm of a certain size. Hydrologists classify storms according to how frequently storms of various sizes would be expected to occur in the area being studied. A "10-year storm" would be expected to occur once every 10 years.

Hydrologic investigations are necessary to plan many construction projects, evaluate water supplies, and design FLOOD control structures. Examples of important hydrologic data include the total amount of precipitation that occurs in a storm (inches or centimeters), the rate of flow (feet or meters per second), and the volume of flow (cubic feet or cubic meters per second). *See also* FLOODPLAIN; RUNOFF; WATER CYCLE.

hydroponics, A method of growing PLANTS in a nutrient solution rather than in soil, which provides regions that have nonarable soil or lengthy periods of DROUGHT with a means for growing crops. In hydroponics, materials such as glass wool, sand, gravel, and PEAT are used to anchor plants. Liquid nutrient solutions containing NITROGEN, PHOSPHORUS, magnesium, potassium, and SULFUR—materials generally obtained from soil—are provided to the plants at regular intervals. Trace amounts of zinc, COPPER, molybdenum, iron, and MANGANESE may also be provided.

Hydroponics is successfully used in areas that lack arable soil or sufficient water for the cultivation of crops using traditional agricultural methods. For example, hydroponics is used to grow plants in DESERT regions of the southwestern United States and in several countries of the Middle East. Hydroponics also is used in many nurseries to grow plants out of season or to grow certain types of plants in areas where the CLIMATE is not suited to their year-round growth. *See also* AGRICULTURE; AGROCLIMATOLOGY; IRRIGATION.

hydropower. *See* hydroelectric power.

hydrosphere, The part of Earth's ECOSPHERE that contains or is covered by water, including WETLANDS, OCEANS, rivers, streams, lakes, ponds, seas, GROUNDWATER, and atmospheric water vapor. The hydrosphere includes portions of the ATMOSPHERE and the LITHOSPHERE.

hydrothermal ocean vents, A hot spring containing MINERAL-rich water flowing up from cracks in the bottom of the Pacific OCEAN floor in an area where there is ocean floor spreading at the boundary of tectonic plates. ORGANISMS living near the vents produce nutrients through CHEMOSYNTHESIS, not PHOTOSYNTHESIS. Inorganic SULFUR compounds, nutrients for the organisms, flow out of the vents, which are as deep as 2,500 meters (8,200 feet) at temperatures of about 340°C (630°F). Organisms that live in this unique ENVIRONMENT include clams, mussels, white tube worms, fish, and even octopi. Most experts believe that large-scale explorations of minerals will take place at these hydrothermal vents where there are large deposits of MANGANESE oxides and sulfides. Other deposits include COPPER, ZINC, LEAD, gold, and silver. *See also* CHEMOAUTOTROPH; LAKE BAIKAL; PLATE TECTONICS.

I

ice age, A time in Earth's history during which an extended period of atmospheric cooling causes much of Earth's surface to be covered by large sheets of snow and ice, called GLACIERS. Ice ages, or glacial ages, have occurred several times over the past two billion years and have had effects on SPECIES distribution and EXTINCTIONS. Ice ages consist of fluctuating glacial periods and warmer interglacial periods. Earth last entered an ice age during the Pleistocene epoch about two million years ago. Known as the Quaternary Ice Age, about 27% of Earth's land surface, including much of Europe and North America, was covered by thick sheets of ice. Although the ice sheets withdrew from much of these areas about 11,000 years ago, many scientists believe that Earth is still in an interglacial period of the Quaternary ice age.

Glaciers can drastically reshape land. Some parts of the world are still covered by glaciers today; other areas show scars from previous glaciation. North America's Great Lakes, for example, were carved by a continental glacier that extended as far south as St. Louis, Missouri, during the Quaternary Ice Age.

The causes of ice ages are a subject of controversy. According to many scientists, gradual drifting of the continents in conjunction with minor variations in Earth's orbit around the sun may play a role. Another argument that has gained credibility in recent years suggests that the rotation of our solar system within the larger Milky Way galaxy may be involved. *See also* CONTINENTAL DRIFT; PLATE TECTONICS.

igneous rock. *See* rock cycle.

immigration, The movement of individual members of a POPULATION into an area. Some SPECIES of ORGANISMS move from their primary HABITAT into new ECOSYSTEMS when seasonal changes such as ice, snowfall, or DROUGHT make resources such as food and water scarce. Such regular movement of organisms from one ENVIRONMENT to another is called migration.

Other environmental pressures such as habitat loss or NATURAL DISASTERS may force populations to move into a new area in order to survive. When immigration occurs, the overall size of a population increases, affecting resource availability by increasing COMPETITION. Along with BIRTH RATE, DEATH RATE, and EMIGRATION, immigration is one of the key factors scientists must consider when determining the population growth of a species in a particular area. *See also* CARRYING CAPACITY; LIMITING FACTOR; POPULATION DENSITY.

***in situ* vitrification,** A process used to treat and safely dispose of LOW-LEVEL RADIOACTIVE WASTES (LLWs) using on-site burial at the waste processing facility. *In situ* vitrification differs from the technology known simply as VITRIFICATION because *in situ* vitrification uses an electric current to heat the soil surrounding buried LLWs on site. The electric current is sent a few centimeters into the ground, where it raises the temperature of the surrounding soil to about 2,000°C (3,632°F). This temperature is hot enough to melt the soil and wastes into a vitrified mass that cools to form a solid glass block.

In situ vitrification provides several advantages over other disposal methods for RADIOACTIVE WASTES. When using this technology, for example, the waste material does not have to be transported to a separate treatment site. The cost of the process is lower than other methods used for treatment and diposal of radioactive wastes. In addition, the glass blocks created during *in situ* vitrification immobilize the wastes they contain for about one million years. The Oak Ridge National Laboratory in Tennessee currently makes use of the *in situ* vitrification technology to treat its LLWs. *See also* RADIATION; RADIOACTIVE DECAY.

EnviroSource

U.S. Department of Energy Web site: www.em.doe.gov/fs/fs3m.html.

incineration, A burning process using controlled high temperatures to reduce waste materials to noncombustible ASH,

CARBON DIOXIDE, and water. Incineration is used for municipal SOLID WASTE and HAZARDOUS WASTE. Most municipal solid waste incinerators burn such wastes as GARBAGE, paper, and cardboard after other materials, rubber, metal, and glass are removed. These incinerators can reach about 500°C (915°F). Some of these incinerators that process large amounts of solid waste per day can be used to produce steam to be used for local heating or to be piped to electric power plants to produce ELECTRICITY.

To burn TOXIC WASTES, particularly ORGANIC SOLVENTS, POLYCHLORINATED BIPHENYLS (PCBs), and CHLORINATED HYDROCARBONS, high-temperature incinerators are used. Temperatures during the combustion process can reach 1,000°C (1,830°F) or higher. There are several HAZARDOUS WASTE TREATMENT incinerators in the United States.

One of the advantages of incineration is that it reduces the volume of original waste by 50 to 75% and vaporizes the liquid wastes. The remaining ash is periodically removed, transported, and deposited in SANITARY LANDFILLS. The disadvantages of incineration include the EMISSION of toxic ash, SULFUR, and NITROGEN OXIDES from chimney stacks into the ATMOSPHERE that contribute to AIR POLLUTION and ACID RAIN. Due to the CLEAN AIR ACT legislation, many modern incinerators have SCRUBBERS or ELECTROSTATIC PRECIPATORS to remove and trap POLLUTANTS before they are emitted into the atmosphere. Another environmental concern is that the ash residue, usually disposed of in landfills, can contain hazardous materials such as HEAVY METALS that can leak into the GROUNDWATER.

Most environmental groups report that incineration is not a viable method for disposing of hazardous wastes because DIOXIN emissions are produced. In 1998, three French municipal waste incinerators were ordered to close after elevated levels of DIOXINS were discovered in milk from cows grazing near one of the plants. Environmental groups emphasized that source reduction and RECYCLING are better alternatives to waste reduction than by incineration.

TechWatch

Plasma Energy Applied Technology, a company located in Huntsville, Alabama, has developed a process called thermal destruction and recovery (TDR), which, they claim completely destroys all types of medical and hospital waste and leaves nothing remaining to be shipped to a landfill.

The TDR process, which destroys ORGANIC materials, is a combination of pyrolysis and gasification. All organic materials, including pharmaceuticals, infectious ORGANISMS, surgical waste, and contaminated materials such as bandages, syringes, tubing, sponges, and so on, can be completely converted to a FUEL gas by a combination of infrared and ultraviolet ENERGY EMISSIONS from a plasma torch. Virtually all organic materials are reduced to basic elements and simple gases. The high-energy, oxygen-starved process converts the solid and liquid wastes into a gas, composed mainly of HYDROGEN and CARBON MONOXIDE which is then filtered and cleaned of particulates and gases. The gas is then oxidized to CARBON DIOXIDE and water before being released into the ATMOSPHERE. The process has been certified for treatment of medical waste in California as an alternative to incineration.

EnviroLink

quench tank: A water-filled tank used to cool incinerator residues or hot materials during industrial processes.

EnviroTerm

source reduction: The layering of amount of waste such as packaging requiring disposal through a reduction in the amount of waste material produced such as garbage. Source reduction conserves resources and reduces pollution.

indicator species, An ORGANISM that scientists monitor to gauge the health of an ECOSYSTEM or the changes affecting an ecosystem. KRILL and LICHENS are indicator SPECIES. Krill are a tiny shrimp-like CRUSTACEANS that are the main source of food for many SPECIES of marine WILDLIFE. Scientists studying the hole in the OZONE layer above ANTARCTICA have observed that changes in the size of the krill POPULATION correspond with changes in the size of the OZONE HOLE. Because a dramatic decrease in the krill population would devastate the marine FOOD CHAIN near Antarctica, scientists now monitor the size of the krill population to determine the overall health of this ecosystem. Lichens (organisms formed through the association of CYANOBACTERIA or ALGAE with FUNGI) are a major food source for moose and other large MAMMALS that live in TUNDRA and TAIGA ecosystems and some other FORESTS. Scientists have discovered that lichens are extremely sensitive to some chemicals, such as SULFUR DIOXIDE (SO_2). Sulfur dioxide is a common AIR POLLUTANT that derives primarily from electric power plants that burn COAL as a FUEL. By examining the lichens in an ecosystem, scientists often are able to determine whether AIR POLLUTION is increasing or decreasing in a particular area. They also may be able to determine how far AIR pollutants are carried from their source by wind and air currents. *See also* BACTERIA; FOOD CHAIN; KEYSTONE SPECIES; SYMBIOSIS; ZOOPLANKTON.

EnviroLink

biological indicator; indicator; index species: Synonyms for the term "indicator species."

EnviroSource

Luomo, Jon R. "Vanishing Frogs." *Audubon* 99, no. 3: (May–June 1997) 60–69.

indoor air pollution. *See* sick building syndrome.

infectious waste, A term used to describe any hazardous medical waste having infectious characteristics. Such wastes include contaminated animal waste, human blood and blood products, isolation waste, pathology waste, and discarded used sharp instruments, such as needles, scalpels, or broken medical instruments. *See also* DISEASE; PATHOGEN.

infiltration, The SEEPAGE or downward movement of water or another liquid through soil, rock, or other sediments. Gravity provides the driving force for infiltration, moving rainwater or meltwater from snow or ice from Earth's surface through permeable soil and rock layers below. *(See* permeability.) When the water reaches an impermeable layer, it collects as GROUNDWATER in an area known as the ZONE OF SATURATION. Some water from infiltration also moves through soil to feed lakes, ponds, streams, and rivers.

In many parts of the world, groundwater provides people with water for drinking, cooking, crop IRRIGATION, and other needs; however, many groundwater supplies are becoming polluted because of CONTAMINANTS contained in rainwater or substances present in the soil that dissolve in water as it moves through soil layers. For example, agrochemicals such as PESTICIDES and FERTILIZERS can be carried into groundwater as infiltration occurs in agricultural regions. HEAVY METALS, ACIDS, and MINERALS often enter groundwater supplies as water moves through the soil in areas used for MINING. Thus, in addition to replenishing groundwater supplies, infiltration occurring in some areas also contributes to groundwater pollution. *See also* ACID MINE DRAINAGE; AGRICULTURAL POLLUTION; AQUIFER; LEACHING; POTABLE WATER; SUBSIDENCE.

infrared satellite imaging. *See* Landsat.

injection well. *See* deep-well injection.

inorganic. *See* organic.

insecticide. *See* pesticide.

insectivore, A member of a small order of MAMMALS that feed largely, although not exclusively, on INSECTS, thus helping to keep insect POPULATIONS in check. Other food sources for insectivores include some PLANTS, small VERTEBRATES, and small INVERTEBRATES, such as earthworms. Because their diet includes both plant and animal products, insectivores are OMNIVORES.

Most insectivores are small and nocturnal (active at night). They generally have small and poorly developed eyes and ears, which are often partially hidden by their fur or skin. Their most well developed sense is their sense of smell, which is used to detect food as they scurry over or burrow through soil.

Insectivores include shrews, moles, hedgehogs, and the lesser-known solenodons, tenrecs, and desmans. Members of this group are distributed worldwide, except in ANTARCTICA, Australia, and the southernmost regions of South America. Moles and shrews are the only insectivores native to North America. Spiny hedgehogs are sometimes imported into North America from Africa and kept as pets. Of the SPECIES native to the United States, the Dismal Swamp southeastern shrew, which is native to Virginia and North Carolina, is the only insectivore currently on the ENDANGERED SPECIES LIST of the U.S. FISH AND WILDLIFE SERVICE. Of the species living outside North America, two—the Cuban solenodon and the Haitian solenodon—are endangered. *See also* CARNIVORE; FOOD CHAIN; FOOD WEB; HERBIVORE; RED LIST OF ENDANGERED SPECIES; TROPHIC LEVEL; WORLD CONSERVATION UNION.

insects, INVERTEBRATE animals that, as adults have three pairs of legs, a segmented body comprised of a head, thorax, and abdomen, and usually two pairs of wings. Additionally, all insects are arthropods—invertebrate animals that have jointed appendages and an exoskeleton composed of chitin. Insects are the single largest animal group and comprise almost 95% of all animals.

The gypsy moth is an insect pest that strips the leaves of trees in orchards and forests.

Beneficial Insects. Insects are an extremely diverse group of animals and are found living in virtually every type of ENVIRONMENT. As a group, these animals are of great environmental importance because of the diverse NICHES they occupy. For example, many insects such as grasshoppers, butterflies, and honeybees are HERBIVORES that feed on either the herbaceous (fleshy, green parts) of PLANTS or the nectars produced by plants. These insects are primary CONSUMERS and thus feed on PRODUCERS—ORGANISMS that produce their own nutrients. Other insects, such as ladybug beetles, are secondary consumers that feed on plant-eating insects such as aphids. These insects, in turn, serve as a food source for animals feeding at higher levels of the FOOD CHAIN. *(See* trophic level.)

Many insects help to decompose ORGANIC matter in the environment by feeding on DETRITUS or other organic remains of organisms. For example, the larvae of flies, commonly called maggots, are SCAVENGERS that feed on the remains of dead animals and on organic wastes in GARBAGE. Such insects serve a vital role to ECOSYSTEMS by helping to cleanse the environment of wastes that would otherwise build up on Earth's surface. They also help to return nutrients they do not use as food to the environment, thus serving an important role in Earth's BIOGEOCHEMICAL CYCLES.

Many insects are beneficial to humans. For example, many people throughout the world raise bees for the honey they produce. Several types of moth caterpillars (the larval stage of the moth) are raised for the silk fibers they produce as they spin their cocoons. For example, the *Bombyx mori* moth, which is native to Asia, produces a silk that is of great ecomonic importance to the clothing industry. This silk is known for its high luster and great strength. In some parts of the world, such as Mexico, some insects are used as food by people. Insects, such as a variety of ants and the hissing cockroach of MADAGASCAR, also are kept by people as pets.

Harmful Insects. Many of the plants on which insects feed are important for their use as food or in the making of other products. Insects that feed on such plants cause billions of dollars of crop losses worldwide each year and thus are considered pest organisms. In some areas of the world, insects that feed on plant-eating insects have become useful in AGRICULTURE, where they are used to help keep pest insect POPULATIONS in check. Insects used in this way are known as BIOLOGICAL CONTROLS—a key component of the SUSTAINABLE AGRICULTURE practice known as INTEGRATED PEST MANAGEMENT (IPM). Many insects also benefit agriculture when they serve as agents of pollination. Pollination is the transfer of pollen (male gametes) from one plant to another and is essential for reproduction in many flowering and nonflowering plants.

In addition to destroying food crops, many insects also are considered destructive for other reasons. For example, termites feed on wood. Carpenter bees and carpenter ants destroy wooden structures as they dig into wood (to make their home or lay their eggs), causing it to break apart. These insects cause billions of dollars in damage to homes and other wooden structures each year.

Many insects are environmentally important because of the roles they play in the transmission of DISEASE. For example, the numerous SPECIES of mosquitoes are common to all environments except the extreme ARCTIC. Mosquitoes are parasitic insects that require substances contained in the blood of MAMMALS, including humans, or birds to carry out their reproductive functions. *(See* parasitism.) Throughout the world, mosquitoes are responsible for the transmission of a variety of diseases. For example, an estimated 400 million people each year who live in tropical CLIMATES become infected with malaria when they are bitten by a mosquito carrying the PROTIST that causes the disease. In more temperate regions, some mosquitoes transmit a form of encephalitis, known as West Nile encephalitis, between birds and humans. In the summer of 1999, outbreaks of this disease occurred during the late summer in the northeastern United States. As a result, several states sprayed PESTICIDES into the AIR in mosquito-infested areas to kill the pests. Other areas discharged BACTERIA into bodies of standing water—common HABITAT for mosquitoe larvae; this bacteria would disrupt the maturation of mosquito larvae.

In addition to causing disease, some insects produce substances that are toxic to other organisms. For example, several types of bees produce a substance that can cause an extremely severe allergic reaction in some humans who are sensitive to the substance. (*See* allergy.) Other insects produce toxins that are designed to paralyze the organisms they use as food. *See also* BIODIVERSITY; FOOD WEB; NEUROTOXIN; PREDATION; WILSON, EDWARD OSBORNE; ZOOPLANKTON.

EnviroSources

Saign, Geoffrey C. *Green Essentials: What You Need to Know about the Environment.* San Francisco, CA: Mercury House, 1994.

Seymour, John, and Herbert Girardet. *Blueprint for a Green Planet: Your Practical Guide to Restoring the World's Environment.* New York: Prentice-Hall Press, 1987.

integrated pest management, A sustainable, long-term approach to the regulation of pests such as INSECTS, weeds, MICROORGANISMS, rodents, and PLANT DISEASES using methods that are effective, economical, and the least harmful to the ENVIRONMENT. Pest control strategies in an integrated pest management (IPM) program extend beyond the application of PESTICIDES to include structural and procedural modifications that reduce the food, water, shelter, and access used by pests. Some plants are capable of producing their own insecticides, and advances in BIOTECHNOLOGY are offering new ways to enhance this natural ability to repel insect pests, diseases, and weeds. IPM is a process that is redefined for each particular situation; in AGRICULTURE, for example, IPM practices vary from place to place according to CLIMATE, growing conditions, and soil types.

IPM reduces reliance on pesticides by using a range of practices that control pests effectively. For example, during a growing season, farmers routinely count pests in their fields to estimate how rapidly they are accumulating. Often, they find that POPULATIONS of beneficial insects (ones that prey upon pests) are growing fast enough to prevent a potential problem, thus circumventing the use of chemical applications altogether. Although pesticides are often essential parts of IPM strategies, they are typically applied only after such field checks or other evidence indicate that their use is necessary to prevent extensive crop damage. The practice of IPM has reduced pesticide use in some crops by as much as 70%.

IPM applies to pests in many types of environments, including those in businesses, homes, gardens, crops, and even on animals. IPM for the garden, for example, includes the use of plant varieties that naturally resist pests; the maintenance of plant health to increase resistance to insects and diseases; fostering pests' natural enemies, such as ladybugs and spiders; and using pesticides that are less toxic and degrade quickly, and only when necessary.

Agricultural producers pioneered the IPM approach in the United States in the 1920s and 1930s. Initially, progress in IPM was slow, mainly because inexpensive synthetic pesticides were abundant and there was only limited knowledge of the long-term effects of pesticides on ORGANISMS and the environment; however, sustained use of these chemicals eventually had serious negative impacts on farm profits, leading to increased interest in IPM programs. Three primary factors in this philosophical shift were pesticide resistance, the effects of pesticides on nontarget organisms, and increased governmental PESTICIDE REGULATION.

Working with land-grant colleges, agencies of the U.S. DEPARTMENT OF AGRICULTURE (USDA), and private organizations, farmers began to develop IPM methods for the production of practically every food product. In some of the earliest IPM practices, for example, cotton farmers destroyed cotton stalks immediately after harvesting to reduce boll weevil populations by removing the insects' source of food. Other effective practices included scheduling planting and harvesting when pest populations were low and using fast-maturing crops that had limited exposure time to pests.

The USDA Integrated Pest Management Initiative, an outcome of the 1992 National IPM Forum, combines IPM resources into a single, cooperative effort, involving farmers, land-grant universities, federal agencies, private consultants, and industry, that addresses important pest control problems. One goal of the initiative is to achieve IPM implementation on 75% of U.S. crop acres by the year 2000. The IPM initiative will provide farmers with the tools and knowledge needed to implement more comprehensive IPM strategies. The USDA program is important to American farmers and consumers to ensure the future profitability, sustainability, and competitiveness of U.S. agriculture by reducing crop losses due to pests. A unique feature of this program is that it is based on grower-identified research and voluntary participation. The program also addresses implementation of IPM in homes and other structures, urban landscapes, RANGELANDS, FORESTS, parks, and other PUBLIC LANDS.

The early focus of IPM programs was farm profitability, and in almost every measured case, IPM programs have reached profitability goals. The newer philosophy of IPM retains some of the same objectives as the traditional approach, but also incorporates social welfare and environmental sustainability considerations into an ecologically based approach. Interdisciplinary scientific teams work with consultants and agricultural producers to incorporate effective pest management into production systems, addressing issues such as timing and efficacy of pest control techniques, the increasing problem of pesticide resistance in insects, worker safety, agrochemical pollution of soils and water, and the threat of pesticide residues in foods. *See also* AGRICULTURAL POLLUTION; AGROECOLOGY; FEDERAL INSECTICIDE, FUNGICIDE, AND RODENTICIDE ACT; SUSTAINABLE AGRICULTURE.

EnviroSources

National Science Foundation, Center for Integrated Pest Management Web site: ipmwww.ncsu.edu/cipm.
Pesticide Environmental Stewardship Program Web site: www.pesp.org.
University of Minnesota (Radcliffe's IPM World Textbook) Web site: ipmworld.umn.edu/textbook.html.
U.S. Department of Agriculture, National Integrated Pest Management Network Web site: www.reeusda.gov/nipmn.

International Atomic Energy Agency, A United Nations agency, headquartered in Austria, Vienna, and founded in 1957, that promotes the safe, peaceful use of NUCLEAR ENERGY. It was formed a few years after U.S. President Dwight D. Eisenhower proposed the creation of an international atomic energy agency in his historic speech before the United Nations General Assembly.

The International Atomic Energy Agency (IAEA) provides technical cooperation and training, food and agricultural programs, and life science and physical science programs offered to UN members. Advice on nuclear energy and the nuclear FUEL cycle, RADIOACTIVE WASTE management, international nuclear information, nuclear safety programs, and international safeguards is also available.

The agency maintains offices in Switzerland, New York, and Japan and operates laboratories in Austria and Monaco. It also supports a research center in Italy, that is administered by the United Nations Educational, Scientific, and Cultural Organization (UNESCO).

The IAEA Secretariat is headed by the director general, who is the chief administrative officer and is appointed for a term of four years. The director general is assisted by six deputy director generals who head six separate departments. *See also* NUCLEAR REGULATORY COMMISSION.

EnviroSources

IAEA, 1400 Vienna, POB 100, Wagramerstr 5, Austria; Web site: www.iaea.org.
International Atomic Energy Agency ("Managing Radioactive Waste" fact sheet) Web site: www.iaea.org/worldatom/inforesource/factsheets/manradwa.html.

International Biological Program (1964–1974), An international program founded to coordinate the application of biological sciences to human welfare. The program brought together and fostered communication among scientists concerned with the world ENVIRONMENT and human interaction with the environment. Its purpose was to develop the rational use and CONSERVATION of NATURAL RESOURCES of the BIOSPHERE and to improve the global relationship between people and the environment. The program advanced the concept of the ECOSYSTEM as a unit for ecological investigation and management of the world's resources. Research was performed in the areas of aerobiology, marine MAMMALS, the conservation of PLANT GENETIC materials and ecosystems, the biosocial ADAPTATION of migrant and urban POPULATIONS, and nutritional adaptation to the environment.

The program had an operational budget of $40–50 million, and hundreds of scientists from 57 countries participated. The participants represented diverse disciplines all working to obtain an understanding of the structure and function of a highly complex ecological system.

The program was promoted by the International Union of Biological Sciences, one of the unions of the International Council of Scientific Unions. Out of the International Biological Program grew UNESCO's the United Nations Educational, Scientific, and Cultural Organization, MAN AND THE BIOSPHERE PROGRAMME. *See also* AGE STRUCTURE; BIOREGION; BIOSPHERE RESERVE; EMIGRATION; GERM PLASM BANK; IMMIGRATION.

EnviroSources

Big Biology. The US/IBP, W. Frank Blair; Dowden, Hutchinson and Ross, Inc., 1977.
Division of Ecological Sciences: 7 place de Fontenoy, F-75352, Paris 07SP, France.

International Convention for the Regulation of Whaling, An International agreement, signed in 1946, intended to ensure the CONSERVATION of WHALES. WHALING is an important commercial industry in some parts of the world. Whales are hunted for meat and for the valuable oils produced from their blubber. Whale oil is used in a variety of products, including perfume, soap, cosmetics, candles, and cooking oil. The International Convention for the Regulation of Whaling (ICRW) was the first attempt by nations of the world to protect and conserve whales in international waters. Signed by 40 nations, this agreement seeks to safeguard for future generations the great NATURAL RESOURCES represented in whales by limiting the hunting of those whale SPECIES in danger of EXTINCTION

The International Convention and Regulation of Whaling was established in 1946 to protect and conserve whales in international waters such as the fin whale (pictured here). *Credit:* Linda Zierdt-Warshaw.

and by collecting and distributing scientific information about whales.

In 1949, the INTERNATIONAL WHALING COMMISSION (IWC) was established to monitor and enforce the ICRW agreement. The IWC is an international committee of representatives from each of the signatory nations of the ICRW, as well as scientific advisors and industry representatives. The IWC publishes annual reports on whale science as well as catch and related data on whaling operations. One of the first important decisions made by the IWC set a limit on the number of whales that could be caught. They also voted that whales could only be hunted within a nation's own territorial waters; however, some nations, including France, Norway, Iceland, and Japan protested and elected not to follow the IWC's rulings. Because no laws had yet been passed to limit which whales could be caught, many species were hunted to the brink of extinction despite the efforts of the IWC.

During the 1970s and 1980s, public pressure to "save the whales" urged the IWC to create new whaling laws. In 1982, the IWC enacted a worldwide ban on commercial whaling that was in effect from 1986 to 1992. During this six-year period, the IWC carefully monitored and studied whale POPULATIONS throughout the world. Whaling was allowed to continue after the ban only in certain areas and for particular whale species. Today, most nations, including the United States, have agreed to maintain the moratorium on commercial whale hunting. Other nations, including Norway, Japan, Peru, and the former Soviet Union, refused to uphold the ban. The IWC has since decided that these countries could continue whaling for scientific purposes only. In 1994, a permanent sanctuary for whales was created in ANTARCTICA, the largest feeding ground for whales in the world. Japan and a few other nations that are severely critical of the IWC's worldwide whaling ban and the establishment of the Southern Ocean Sanctuary still continue to harvest whales today under the guise of scientific whaling. *See also* BIODIVERSITY; ECOTOURISM; ENDANGERED SPECIES; ENDANGERED SPECIES ACT; GILL NET; GREENPEACE; MARINE MAMMAL PROTECTION ACT.

EnviroSource

International Whaling Commission Web site: www.ourworld.compuserve.com/homepages/iwcoffice.

International Council for Local Environmental Initiatives, An international association of local governments dedicated to the prevention and solution of local, regional, and global environmental problems through local action. Worldwide, more than 300 cities, towns, counties, and their associations are members of the council; their philosophy is that local action can have a global impact. The Internation Council for Local Environmental Initiatives (ICLEI) was established in 1990 with the UNITED NATIONS ENVIRONMENT PROGRAMME and has offices in Toronto, Canada; Frieburg, Germany; Tokyo, Japan; and Harare, Zimbabwe. One of the ICLEI's programs, the Urban CO_2 Reduction Project, includes a network of 14 cities worldwide that work together to develop local initiatives and plans to reduce CARBON DIOXIDE EMISSIONS. These cities have proposed a number of ways to reduce emissions by using such measures as retrofitting municipal buildings, improving public transportation and waste management, and switching to more cost-efficient FUELS to generate ELECTRICITY. Even simple measures such as insulating pipes and repairing leaks, implementing better building codes, and imposing local ENERGY taxes have been initiated by the cities in the program. *See also* ALTERNATIVE ENERGY RESOURCES.

EnviroSource

ICLEI Web site: www.iclei.org.

International Decade for Natural Disaster Reduction, An international movement declared by the United Nations General Assembly in December 1989 to foster worldwide cooperation in reducing the global effects of NATURAL DISASTERS. In 1994, a world conference on natural disaster reduction was held in Yokohama, Japan. The conference was an important milestone in the awareness-building process to help reduce the loss of life and property and the economic disruption that can result from natural disasters. Within the movement, special attention is given to assisting developing nations in assessing disaster damage potential, establishing early warning systems, and in building disaster-resistant structures. *See also* CYCLONE; DUST BOWL; EL NIÑO/LA NIÑA; FLOOD; HURRICANE; MOUNT PINATUBO; TORNADO; VOLCANO.

International Register of Potentially Toxic Chemicals, A United Nations program based in Nairobi, Kenya, that provides information on more than 60,000 chemicals that can affect human health and are hazardous to the ENVIRONMENT. The International Register of Potentially Toxic Chemicals (IRPTC) was established by the UNITED NATIONS ENVIRONMENT

PROGRAMME (UNEP) in 1976. For the first time, the world had an institution to collect and process information on HAZARDOUS SUBSTANCES and to disseminate that information to all member states. The IRPTC is intended to manage chemicals throughout their life, including disposal, thereby reducing the risks to human health and the environment posed by chemicals and HAZARDOUS WASTES.

The IRPTC is also a record of data used to assess the hazards to human health and the environment posed by various chemicals. The UNEP is moving toward making the IRPTC available on personal computers and electronic networks which should be of much benefit to global scientists. *See also* ACID; AGENT ORANGE; ALKALI; CADMIUM; COMPREHENSIVE ENVIRONMENTAL RESPONSE, COMPENSATION, AND LIABILITY ACT; CHLORDANE.

EnviroSources

International Register of Potentially Toxic Chemicals Web site: www.unep.org/unep/program/hhwb/chemical/irptc/home.html.

International Union for the Conservation of Nature and Natural Resources. *See* World Conservation Union.

International Whaling Commission, A body created by the INTERNATIONAL CONVENTION FOR THE REGULATION OF WHALING in 1946 to conserve the world's WHALE stocks and provide for the orderly and SUSTAINABLE DEVELOPMENT of the WHALING industry. Forty nations are members of the commission.

The International Whaling Commission (IWC) has used various means to regulate commercial whaling, including the designation of open and closed seasons, open and closed areas, protected SPECIES, size limits for each species, and limits on whale catches in any season. Past actions by the IWC include the establishment of whale sanctuaries in the Indian Ocean and in the southern oceans (most waters below 40° south latitude), a moratorium on all commercial whaling as of the beginning of the 1985–86 whaling season, and the adoption of a separate and distinct management scheme for aboriginal subsistence whaling.

At its 1997 annual meeting, the IWC approved a maximum limit on the total number of bowhead whales killed by the Eskimos living in Alaska and Russia. The Alaska Eskimos have been conducting aboriginal subsistence hunts with approval of the IWC since the commission began regulating such hunts in the 1970s. The IWC also allows aboriginal subsistence hunts of an average of four nonendangered gray whales a year by the Makah Indian tribe and an average annual harvest of 120 gray whales by Russian natives of the Chukotka region.

The IWC maintains its moratorium on commercial whaling; however, Norway lodged a timely objection to the 1986 moratorium decision and therefore is not bound by that decision. Thus, Norway continues to harvest minke whales from the northeast Atlantic. In 1997, as it has done in previous years, the IWC passed a resolution condemning Norwegian whaling outside the parameters set by the commission. *See also* ENDANGERED SPECIES ACT; MARINE MAMMAL PROTECTION ACT.

Member Nations of the International Whaling Commission

Antigua and Barbuda	India	Russian Federation
Argentina	Ireland	Saint Kitts and Nevis
Australia	Italy	Saint Lucia
Austria	Japan	Saint Vincent and Grenadines
Brazil	Kenya	Senegal
Chile	Republic of Korea	Solomon Islands
The People's Republic of China	Mexico	South Africa
Costa Rica	Monaco	Spain
Denmark	Netherlands	Sweden
Dominica	New Zealand	Switzerland
Finland	Norway	United Kingdom
France	Oman	United States
Germany	Peru	Venezuela
Grenada		

EnviroSources

International Whaling Commission: The Red House, Station Road, Histon, Cambridge, CB4 9NP, United Kingdom. NOAA Web site: www.nmfs.gov.

intertidal zone. *See* ocean.

introduced species. *See* exotic species.

inversion. *See* temperature inversion.

invertebrates, An animal lacking a vertebral column, or backbone. Invertebrates constitute the majority of the animals on Earth and are the most diverse group of ORGANISMS. Invertebrates range from the simple sponges to more advanced animals such as INSECTS and cephalopod mollusks. Examples of invertebrates are jellyfish; sea anemones; corals; fan, tube, and bristle worms; and echinoderms such as sea stars, sea urchins, sand dollars, and sea cucumbers.

Arthropods are the largest group of invertebrates, with 750,000 described SPECIES. This group's members all have jointed legs and include CRUSTACEANS, spiders, insects and several less familiar organisms. Invertebrates are important in the efficient DECOMPOSITION of ORGANIC matter, the pollination of PLANTS, and as vital food resources in the FOOD CHAIN.

During the 1980s, Terry Erwin of the National Museum of Natural History studied insect POPULATIONS in the canopies of tropical RAINFORESTS. He concluded that probably 15–30 million insect species live on Earth—compared to the previous estimate of around 1 million. The vast majority of these species live in the tropics. Because rainforests are being destroyed, countless species are becoming extinct before scientists can even discover them. Invertebrates took 3.5 billion years to evolve; scientists estimate that they are now vanishing at a rate of 1 species every 12 minutes. *See also* BACTERIA; EXTINCTION; PROTISTS; VERTEBRATES; WILSON, EDWARD OSBORNE.

The octopus is one of the biggest and most intelligent of all the invertebrates.

EnviroSource

U.S. Fish and Wildlife Service, Species List of Endangered and Threathened Wildlife Web site: www.fws.gov/r9endspp/lsppinfo.html.

ion exchange treatment, A common water-softening method often found at water purification plants and homes. The treatment removes hard-water MINERALS such as calcium and magnesium ions (atoms that carry an electrical charge) as they flow through the water-softener tank. Some units can remove radium and barium from water. Special ion exchange treatment systems use activated alumina to remove flouride and ARSENIC compounds from water. *See also* WASTEWATER TREATMENT.

ionosphere. *See* atmosphere.

irradiation, Exposure to rays such as gamma, X-RAYS, and ultraviolet RADIATION; also a process of preserving food by using gamma rays to destroy BACTERIA or mold in the food. Irradiated food, such as milk, can be stored without refrigeration or chemical preservatives. Irradiation is also used to kill INSECT pests and to sterilize medical instruments.

In 1980 a joint committee of the FOOD AND AGRICULTURE ORGANIZATION OF THE UNITED NATIONS (FAO), the INTERNATIONAL ATOMIC ENERGY AGENCY (IAEA), and the WORLD HEALTH ORGANIZATION (WHO) declared that a dose of radiation within prescribed limits was safe for food products. Several countries use irradiation for specific purposes, such as to prevent sprouting in potatoes and onions and to increase the shelf life of strawberries and other fruits. In 1997, the U.S. FOOD AND DRUG ADMINISTRATION approved irradiation for use on raw ground beef to kill foodborne PATHOGENS such as *ESCHERICHIA COLI*. One of the drawbacks of irradiated food is that irradiation can mask the taste, smell, and even the appearance of food that is spoiling. *See also* SALMONELLA.

irrigation, The process of artificially supplying water to agricultural land used for crop cultivation. According to the Food and Agricultural Organization of the United Nations, about 17% of the world's cropland used some form of irrigation in 1994. This land, which represents almost 250 million hectares (618 million acres), produces almost 40% of the world's food. Worldwide, five main types of irrigation are in use. They include FLOOD irrigation, furrow irrigation, sprinkler systems, drip (or trickle) irrigation, and subirrigation. The method used on a particular farm is determined by such factors as crop type, topography, water source, soil drainage, and cost.

Flood Irrigation. In flood irrigation, farmland is flooded with water. Flood irrigation is most often used where land is relatively flat, the water source is nearby, and the crops being grown require large amounts of water. Flood irrigation is among the least expensive irrigation methods, because it does not require much machinery to carry and deliver water to PLANTS; however, flood irrigation must be used in areas that have good soil drainage to avoid water-logging and SALINIZATION of the soil. Salinization occurs slowly as water evaporates from land leaving behind salts and MINERALS that accumulate in the soil. A drawback of flood irrigation is that it wastes great amounts of water because much of the water goes to land not occupied by plants. In addition, more than 50% of water may be lost to evaporation.

Furrow Irrigation. The delivery of water to crops via small channels, or furrows, dug between crop rows is called furrow irrigation. Furrow irrigation is used in areas with fairly flat topography. Because it delivers water nearer to plants, furrow irrigation is more efficient than flood irrigation; however, furrow irrigation is also more expensive than flood irrigation because it requires the use of machines or farm laborers to dig the furrow between crop rows. To offset the higher costs, this irrigation method is used for crops with high market values, such as cotton and vegetables. The main disadvantage of furrow irrigation is that it requires a lot of water, much of which is lost to evaporation. The high evaporation rate may decrease soil fertility through salinization.

Sprinkler Systems. Use of sprinkler systems to deliver water to crops is a type of overhead irrigation useful with most types of topography. Sprinkler systems are more expensive than flood or furrow irrigation because they require the purchase of equipment to deliver water from above, simulating rainfall; however, the system allows water to be directed to where it is needed and wastes less water than flood or furrow irrigation. Thus, sprinkler systems are useful in regions with limited water supplies. A drawback to sprinklers is that their efficiency is decreased by strong winds, which divert water away from crops.

Pivot irrigation, a type of sprinkler system, is used on a cotton crop in Mississippi. *Credit:* United States Department of Agriculture. Photo by Tim McCabe.

Drip Irrigation. The technology for drip (or trickle) irrigation was first developed in Israel and is today widely used throughout the United States, Israel, and Australia. Drip irrigation delivers water through narrow tubes directly to the root area of individual plants at frequent intervals and in small amounts. The slow, frequent release of water has several advantages over other irrigation methods. First, it reduces the total amount of water needed to irrigate crops, making this irrigation method useful in regions with limited water supplies. Second, because water is released slowly and over a small area, little water is lost to the AIR through evaporation or to the soil via PERCOLATION. At the same time, fewer salts are deposited in the soil, helping to preserve soil quality. Third, drip irrigation can be used on almost all lands, regardless of their topography. The main disadvantage of drip irrigation is that the cost involved in setting up the system is higher than that of other irrigation methods; however, this disadvantage is generally offset by its benefits.

Subirrigation. As its name implies, subirrigation involves delivering water to plants from beneath the soil's surface. Subirrigation can be used only in places where the WATER TABLE is near Earth's surface. In subirrigation, water is placed in ditches and allowed to naturally percolate through soil into the water table. As a result of the added water, the level of the water table is raised to a point where it reaches plant's roots. An advantage to subirrigation is that it allows plant parts located above the surface to remain dry, lessening the likelihood of DISEASE caused by FUNGI and damage caused by rot. This irrigation method is most often used to raise low-growing fruit crops, such as strawberries, or to raise vegetables.

Problems with Irrigation. Use of irrigation is declining in many regions because of problems such as salinization of soil and water shortages. Mineral deposits build up during salinization and render the soil infertile.

The water used for irrigation comes from a variety of sources. In many parts of the world, SURFACE WATER is diverted from lakes, ponds, or streams; however, such water is not available in sufficient supply to allow for crop irrigation in many areas. More than 40 countries in the Middle East and Africa have water supplies that are insufficient to meet the needs of factories and homes, in addition to those of farmers. In addition, the diversion of water from such sources often significantly alters HABITAT, making an area unable to support its NATIVE SPECIES. Impacts resulting from water diversion are evident in California's MONO LAKE, Florida's EVERGLADES NATIONAL PARK, and Russia's ARAL SEA. The construction of DAMS to create RESERVOIRS also can alter habitat and threaten ECOSYSTEMS.

GROUNDWATER, such as that contained in the OGALLALA AQUIFER of the United States, is also sometimes used for irrigation or to meet the needs of homes and industry. Use of such water greatly increases the amount of water available for irrigation; however, water usage from AQUIFERS must be closely monitored to avoid OVERDRAFT, a condition that results when water is removed from a source more quickly than it can be naturally replenished. When overdraft occurs, lands above an aquifer may sink to fill the space once occupied by the water. *(See* subsidence.) Aquifers located near coastlines also risk contamination with salt water from the sea, a process known as SALT-WATER INTRUSION.

Another major problem with irrigation is that water that runs off farmland can carry with it dissolved substances such as PESTICIDES and FERTILIZERS. These substances can become POLLUTANTS when they enter lakes, ponds, and streams. Once in aquatic ecosystems, fertilizers can stimulate the growth of aquatic PRODUCERS, resulting in an ALGAL BLOOM. As these producers die, the population of DECOMPOSERS that break down their remains will increase and can use such large amounts of DISSOLVED OXYGEN in the water that other aquatic ORGANISMS may be unable to carry out respiration. Pesticides may serve as pollutants and harm the organisms they contact directly or be harmful to their offspring. In addition, if irrigation water runs off land at a fast enough speed, it can carry away topsoil needed for the growth of crops. This soil may be deposited in new land areas or be dropped into bodies of water where it alters the ecosystem by changing water depth or becomes harmful to organisms that ingest or inhale the particles. *See also* AGRICULTURAL POLLUTION; DEPOSITION; EROSION; RUNOFF; SEDIMENTATION; SILTATION.

EnviroSources

Abromovitz, Janet, et al. *State of the World, 1998.* New York: Norton, 1998.

Brown, Lester R., et al. *Vital Signs, 1997.* New York: Norton, 1997.

isotope, An atom of an element that differs from another atom, or other atoms, of the same element because it has a different number of neutrons in its nucleus. An isotope has the same number of electrons and protons, chemical properties, and atomic number as other atoms of the element, but has a different atomic mass. For example, one isotope of HYDROGEN is deuterium,which has an atomic mass twice that of hydrogen. It is called a heavy hydrogen since it has one proton and one neutron in its nucleus. An ordinary hydrogen atom has only one proton in the nucleus. *See also* RADIOISOTOPE OR RADIOACTIVE ISOTOPE.

ivory, The tusks and teeth of certain MAMMALS, most notably the ELEPHANT and the walrus. Ivory is valued for its beauty and has been used to make art and jewelry since ancient times. Ivory also was once used to make billiard balls, piano and organ keys, and decorative handles for tableware and furniture.

The most valuable ivory comes from elephant tusks. The use of firearms to kill these animals for their tusks has devastated wild elephant POPULATIONS. To combat this problem, in 1977, African and Asian elephants were listed by the CONVENTION ON INTERNATIONAL TRADE IN ENDANGERED SPECIES OF WILD FLORA AND FAUNA as one of the most ENDANGERED SPECIES. Under this agreement, it became illegal to trade in elephant ivory. Although this ban has been honored by most countries, some countries, most notably China, continue to trade in ivory.

POACHING of African elephants to obtain their tusks is a problem in many African nations; however, successful CONSERVATION and protection measures have allowed wild elephant populations in Botswana, Namibia, and Zimbabwe to increase. As a result of these increases, some international trade in ivory derived from African elephants resumed, with limitations, in 1999. The limitations allow for export of a predetermined amount of ivory only to Japan, which has a long history of domestic ivory use. Ivory derived from African nations other than Botswana, Namibia, and Zimbabwe and from Asian elephants continues to be prohibited.

Walrus tusk ivory is less valuable than that of elephants, and walruses are not considered an endangered species. In the United States, however walruses are protected under the MARINE MAMMAL PROTECTION ACT of 1972. *See also* ADDO NATIONAL ELEPHANT PARK; ENDANGERED SPECIES LIST; RED LIST OF ENDANGERED SPECIES; SEALS AND SEA LIONS; WORLD CONSERVATION MONITORING CENTRE.

J–K

James Bay Hydropower Plant, North America's largest planned HYDROELECTRIC POWER project, located in the James Bay region of northern Quebec. Started in 1971, it is still incomplete as of 1998.

Phase one of the project, on the La Grande River, was completed in 1985 at a cost of $13.7 billion Canadian. Construction of phase one, which has the capacity to generate 10,300 megawatts, required the flooding of 11,500 square kilometers (4,480 square miles) of WILDERNESS that is home to native Cree Indians and Inuits. The flooding has also caused MERCURY to leach from rock layers in the RESERVOIR, contaminating fish. Another DAM on the La Grande River is under construction.

Other dams that are part of the James Bay project are on the Great Whale, Nottaway, and Rupert Rivers. Together, these dams, will divert a total of nine free-flowing rivers and flood an area the size of Belgium. The entire project is planned to have a capacity of 27,000 megawatts and has an estimated cost of $63 billion Canadian. Some of the power from the James Bay project was to be sold to the U.S. states of New York, New Hampshire, Maine, and Vermont. In 1992, then-Governor Mario Cuomo of New York directed the New York Power Authority to cancel its contract with Hydro-Quebec in favor of ENERGY CONSERVATION and to purchase power from other sources. Because of the lack of a market for this hydroelectricity, the Great Whale Complex was shelved. See also ASWAN HIGH DAM; THREE GORGES DAM; TENNESSEE VALLEY AUTHORITY.

EnviroSource

Environmental Defense Fund Web site: www.edf.org.

J-shaped curve, A line plotted on a graph showing the negative growth pattern of an unsuccessful POPULATION. The graph plots the size of the population against time. The J-shaped curve is formed when the line on the graph swings up and then down. It displays a rise in the growth of the population and the drop point—the point at which the population begins dying off. The "crash" occurs in the population when the number of individuals exceeds the CARRYING CAPACITY of a particular ENVIRONMENT. WEATHER changes, food depletion, predators, and DISEASES can cause changes in population GROWTH RATE. If the reproduction rate increases and can surpass the DEATH RATE, then the population can recover; if not, the population will disappear. *See also* BIRTH RATE; COMPETITION; POPULATION DENSITY, PREDATION; S-SHAPED CURVE; ZERO POPULATION GROWTH.

Typical J-shaped Curve

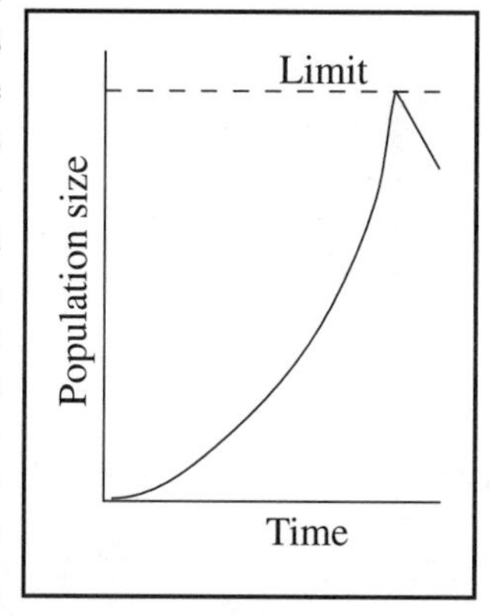

A J-shaped curve plotted on a graph shows the negative growth pattern of an unsuccessful population. It displays a rise in the growth of the population and the drop point, the point at which the population is dying off.

keystone species, A SPECIES whose presence or absence in an ECOSYSTEM significantly impacts the other species in that ecosystem. Removal of a keystone species from an ecosystem can significantly reduce its BIODIVERSITY. For example, the sea otter that inhabits the kelp forests of the Pacific Ocean off the coasts of the United States and Canada has been identified as a keystone species. *(See* algae.) Kelp forests are areas of dense kelp growth that provide HABITAT or breeding grounds for numerous fishes and other WILDLIFE. During the early twentieth century, sea otters were greatly overhunted, decreasing their POPULATION size. In addition, scientists observed a significant decline in the kelp that comprised the ecosystem in which these animals had thrived. Studies of the region indicated that decimation of the kelp resulted from a population explosion of the sea urchins on which the sea otters fed. Without a predator to keep their numbers in check, the sea urchin population thrived and exploded, while the kelp

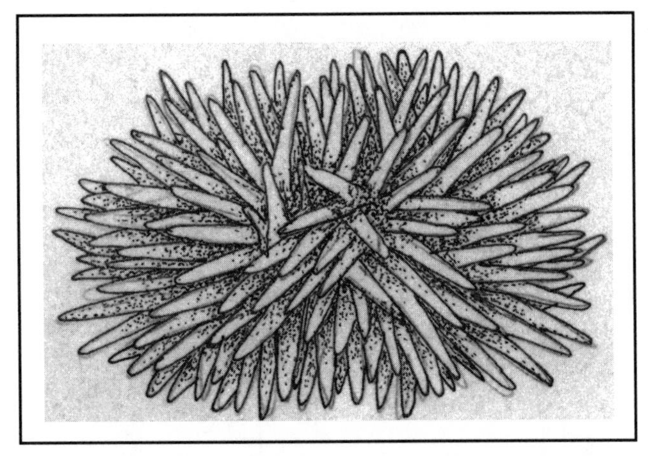

The overpopulation of sea urchins can decimate kelp that other animals depend on for food and habitat.

and seaweed populations on which the sea urchins fed greatly declined. The thinning of the kelp population not only removed a major food source, but destroyed the habitat of many ORGANISMS, forcing them to move out of the area. Some organisms that could not meet their food needs died. This once diverse undersea ecosystem was transformed into a territory barren of life. To reverse the problem, the sea otter was reintroduced to the region and was provided protected status under the MARINE MAMMAL PROTECTION ACT of 1972. Since its reintroduction, the kelp forests have recovered and are once again a thriving ecosystem. *See also* BIOREGION; ENDANGERED SPECIES; EXOTIC SPECIES; FOOD CHAIN; FOOD WEB; MARINE PROTECTION, RESEARCH, AND SANCTUARIES ACT; PREDATION; PROTECTED AREA.

kilowatt hour. *See* electricity.

Komi Republic oil spill. *See* oil spills.

Krakatoa, A small island of Indonesia that was virtually destroyed by a volcanic explosion in 1883. *(See* volcano.) Krakatoa is located off southwestern Indonesia between Sumatra and Java. The island was approximately 47 square kilometers (18 square miles) in size prior to the eruption; however, the impact of the explosion was so severe, the island was reduced to its current size of only 16 square kilometers (6 square miles).

Krakatoa provides evidence of how devastating a volcanic eruption can be to the surrounding ENVIRONMENT. In addition to almost destroying the island on which it stood, the eruption of Krakatoa also resulted in the local EXTINCTION of virtually all island WILDLIFE. In addition, thousands of people living along the coasts of Sumatra and Java also were killed by the waves resulting from a TSUNAMI brought on by the eruption and an accompanying EARTHQUAKE. *See also* AEROSOL; ECOLOGICAL SUCCESSION; MOUNT PINATUBO; NATURAL DISASTER.

krill, Small, shrimplike CRUSTACEANS of vital importance to OCEAN FOOD CHAINS and FOOD WEBS. Krill grow in abundance throughout the world's oceans, where they feed on PLANKTON that float near the ocean's surface. Although krill do not usually grow to more than 2.5 centimeters (1 inch) in size, they are the major food source for Earth's largest animal—the blue WHALE—as well as for other whale SPECIES, SEALS AND SEA LIONS, and many other marine ORGANISMS.

Krill is major food source for Earth's largest animal—the blue whale—as well as for other whale species, seals, sea lions, and many other marine organisms.

Scientists studying the environment in ANTARCTICA often use krill as an INDICATOR SPECIES to monitor changes in global CLIMATE related to the hole in the OZONE layer above Antarctica. Increases in the size of this hole are expected to accelerate GLOBAL WARMING as the amount of ultraviolet RADIATION reaching Earth increases. A temperature increase would likely devastate the krill POPULATION near Antarctica and thus threaten the survival of populations that feed upon krill. Changes in these population sizes are used to indicate the environmental health of Antarctica and may forecast global climate changes as well. *See also* CONSUMERS; ZOOPLANKTON.

Krutch, Joseph Wood (1893–1970), American NATURALIST, college professor, and writer, whose books on nature themes helped to inspire awareness of concerns about protecting natural ENVIRONMENTS. Krutch spent much of his adult life teaching at Columbia University in New York City. Following his retirement in 1952, he moved to a DESERT community near Tucson, Arizona. During his retirement, Krutch became a leading conservationist and nature writer, with more than 20 books to his credit. His works have been read by millions of people worldwide, motivating many to become more involved in protecting natural environments. Among his books are *The Desert Year* (1952), *The Voice of the Desert* (1955), *The Great Chain of Life* (1957), *The World of Animals* (1961), and *Herbal* (1965). *See also* ABBEY, EDWARD; CACTI; CARSON, RACHEL LOUISE; DARWIN, CHARLES ROBERT; DOUGLAS, MARJORY STONEMAN; LEOPOLD, ALDO; MUIR, JOHN; RAY, DIXY LEE.

EnviroSources

Krutch, Joseph. *The Desert Year.* Tucson: University of Arizona Press, 1985.

———. *Forgotten Peninsula: A Naturalist in Baja California.* Tucson: University of Arizona Press, 1986.

———. *More Lives than One.* Tucson: University of Arizona Press, 1962.

kwashiorkor, A nutritional disorder resulting from a severe dietary protein deficiency. Kwashiorkor is a form of malnutrition that occurs when protein-rich foods such as meats, dairy products, and beans are lacking in the diet. Because kwashiorkor is nutrient-specific, it can occur even when an individual's overall caloric intake is sufficient.

Kwashiorkor is most often observed in young children who live in developing nations of Latin America, Africa, and Asia. Sometimes, the disorder develops in children who have been weaned from breast milk too early; however, large outbreaks of kwashiorkor may appear in areas experiencing severe FAMINE as a result of natural events, such as DROUGHT, or human activities, such as civil wars. In the 1950s, the civil war between North and South Korea resulted in the deaths of many children from kwashiorkor. During the 1980s, large-scale outbreaks of kwashiorkor affected children and adults in the African nations of Ethiopia and Somalia. Many Somalians continue to suffer from kwashiorkor as a result of famine that is tied to the civil wars that continue to plague this nation.

Early symptoms of kwashiorkor include lethargy, apathy, and irritability. Later symptoms include slowed growth, hair loss, and an accumulation of fluids in the abdominal region. The fluids in the abdominal region cause the belly to take on a greatly enlarged appearance, while other parts of the body become so emaciated as to make the skeletal features very apparent. *See also* AGRICULTURE; BLIGHT.

L

Lacey Act, A federal law of the United States enacted in 1900 for the purpose of protecting birds from being overhunted to the point of EXTINCTION. At the time of its creation, many people were becoming concerned about decreases in several wild bird POPULATIONS, including the PASSENGER PIGEON and the snowy egret. Most declines in bird populations were attributed to overhunting for commercial sale of birds to restaurants and the sale of feathers or plumes to clothing manufacturers. Passage of the Lacey Act prohibited illegal game obtained in one state from being sold in another state that did not prohibit the hunting of that SPECIES. Although the act slowed hunting of some bird species, it was largely ineffective due to a lack of enforcement and the large amounts of money available to poachers who violated the law. *(See* poaching.) To correct its weaknesses, Congress passed the Weeks-McLean Law in 1913, which made commercial hunting and interstate shipment of most migratory birds illegal. This law was later replaced by the more comprehensive MIGRATORY BIRD TREATY ACT of 1918. *See also* BALD EAGLES; CALIFORNIA CONDORS; DODO BIRDS; ENDANGERED SPECIES ACT; ENDANGERED SPECIES LIST; FISH AND WILDLIFE SERVICE; RED LIST OF ENDANGERED SPECIES; NORTH AMERICAN WETLANDS CONSERVATION ACT; WILD BIRD CONSERVATION ACT.

EnviroSource

U.S. Fish and Wildlife Service Web site: www.fsw.org.

land trust, A private or government organization that acquires land as a means of conserving or protecting natural areas, especially those deemed as having scenic or historic significance or those encompassing critical WILDLIFE HABITAT. The Nature Conservancy is among the leading Environmental organizations attempting to conserve wildlife habitat in the United States, Canada, Costa Rica, Dominican Republic, and Ecuador. In addition, many smaller organizations are forming in rural communitites throughout the United States to help prevent land now used for farming and ranching from being sold for other types of development, such as construction of shopping malls or housing developments. In suburban areas, groups also are forming to purchase open or forested areas for the purpose of preserving these regions as green areas.

Land trust properties are generally purchased by the trust, donated by private citizens, or acquired through CONSERVATION EASEMENTS. In some cases, land trusts do not purchase the lands they are trying to preserve, but rather help to develop restrictions on land use or development of the property. *See also* CONSERVATION; DEBT-FOR-NATURE SWAP; NATIONAL PARK SERVICE.

EnviroSource

Nature Conservancy: (703) 841-5300; Web site: www.tnc.org.

landfills. *See* sanitary landfills.

Landsat, A program of satellites (craft that orbit Earth) that produce images of Earth's surface using thermal sensors to measure solar RADIATION reflected from Earth. The first Landsat satellite was launched in 1972. The most recent, Landsat 7, was launched on April 15, 1999. The instrument on board Landsat 7 is the Enhanced Thematic Mapper Plus. At launch, the satellite, including the instrument and FUEL, weighed approximately 2,200 kilograms (4,800 pounds.) It is about 4.3 meters (14 feet) long and 2.8 meters (9 feet) in diameter. The Landsat Program is managed by the National Aeronautics and Space Administration (NASA) and the U.S. Geological Survey.

The Landsat Program was approved by Congress in October 1992, when it passed the Land Remote Sensing Policy Act. The act allows a continuous collection and utilization of land remote sensing data from space in order to study and understand human impacts on the global ENVIRONMENT and to better manage Earth's NATURAL RESOURCES. The instruments on the Landsat satellites have acquired millions of images, which are archived in the United States and at Landsat receiving stations around the world. The images provide a unique resource for global change research and have applications in AGRICULTURE, GEOLOGY, FORESTRY, regional planning, education, and national security. Landsat satellites have been used to monitor timber losses in the U.S. Pacific Northwest, to map the extent of winter snow pack, to measure FOREST cover, to locate MINERAL deposits, to monitor STRIP MINING, and to assess natural changes due to fires and INSECT infestations.

Current research using images from the Landsat program is in the following areas:

- **Growth patterns of urban sprawl**. Geographers are using Landsat data to study the amount of land area used by increasing human POPULATIONS. Landsat imagery can show where the growth is occuring and help geographers evaluate how different urban planning programs effect population growth and land use.
- **Farming and land management.** Important agricultural factors such as PLANT health, plant cover, and soil moisture can be monitored from space. Such data provide a much bigger picture of the land surface that can be combined with other technologies to help cut costs and increase crop yields.
- **Water pollution.** One university is using Landsat imagery to help predict the extent, duration, and impacts the formation of the thermal bar in the Great Lakes—the narrow band of warm water close to the shore that contains a concentration of salt, sediment, FERTILIZERS, and chemical POLLUTANTS from RUNOFF.
- **Rainforest deforestation**. Landsat images of forest loss obtained over a decade from the mid-1970s to mid-1980s found that an average of 16,120 square kilometers (6,200 square miles) of rainforest were lost each year in the Brazilian Amazon. The annual rate in the smaller Southeast Asia region was estimated to be 12,400 square kilometers (4,800 square miles). Using Landsat imagery, scientists are performing multiyear studies in the Amazon basin to determine the year-to-year changes in DEFORESTATION rates.
- **Mapping wildfire hazards.** Landsat imagery is being used to identify different types of dry BIOMASS, matter derived from organisms on the ground. (*See* detritus.) Dry biomass can act as fuel that feeds fires. Knowing the amount and condition of the dry mass can be an important factor in predicting possible wildfire hazards. *See also* FIRE ECOLOGY; GLOBAL ENVIRONMENT MONITORING SYSTEM; PLUME.

EnviroSources

Earth Observing System Project Science Office: Code 900, NASA Goddard Space Flight Center, Greenbelt, MD 20771

NASA Earth Science Enterprise: Code Y, NASA Headquarters, Washington, DC 20546.

National Aeronautics and Space Administration (Landsat Gateway) Web site: landsat.gsfc.nasa.gov/main.html.

U.S. Geological Survey: EROS Data Center, Sioux Falls, SD 57198.

Langelier index, A measuring system used to indicate the ability of water to deposit or dissolve calcium carbonate in pipes and therefore to affect water flow. To determine this ability, the Langelier index, which is also called the stability index, makes use of measurements of water temperature, PH, dissolved salts, and calcium concentrations. Langelier measurements with a negative value indicate that the water is likely to dissolve calcium carbonate deposits; measurements with a positive value indicate that the water is likely to form calcium carbonate deposits.

The Langelier index was developed in 1949 by Dr. Wilfred H. Langelier. Deposits of calcium carbonate scale and accumulations of SLUDGE can disrupt water flow. The Langelier index provides a means of determining if DEPOSITION is occurring. The index is most often used by industry when monitoring pipes in an effort to prevent equipment failure caused by insufficient water flow or deposits of matter within the equipment. *See also* LAWS OF THERMODYNAMICS; SEWAGE TREATMENT PLANT; SEWER.

laws of thermodynamics, Scientific principles that describe the behavior of heat or thermal ENERGY in the ENVIRONMENT. The laws describe the interrelationships between energy, particularly heat energy, and work. There are three laws of thermodynamics; however, it is the first two laws that are of the greatest concern to ENVIRONMENTAL SCIENCE.

The first law of thermodynamics is sometimes called the Law of Conservation of Energy and was first stated in 1847 by German physicist Hermann Ludwig Ferdinand von Helmholtz. The law states that in a closed system, energy cannot be created or destroyed but can be changed from one form to another. The principle of this law is illustrated by the combustion of COAL or wood. The coal and wood contain stored chemical energy which, when ignited, is changed to heat energy that can be used to heat a home or for cooking. Light energy is also produced in the combustion process.

The second law of thermodynamics was stated in 1850 by German physicist Rudolf Clausius. This law holds that when energy is used to do work, the conversion of energy to work is not 100% efficient because some energy is lost as heat. This principle is observed when heat is given off by a running AUTOMOBILE engine. The energy lost as heat results from the friction between the moving parts of the engine, and directly impacts the FUEL efficiency of the automobile. The second law of thermodynamics explains why a machine cannot run indefinitely under its own power.

The third law of thermodynamics is often called the Law of Absolute Zero (0K or –273° C or –459°F). Absolute zero is the temperature at which all molecular motion stops. The law states that at absolute zero, all matter has the same entropy. Because the particles of all matter cease to move at absolute zero and movement of particles is required for work to be done, no work can be accomplished at the temperature of absolute zero. The Law of Absolute Zero was formulated in 1906 by German physical chemist Walter H. Nernst. *See also* ALTERNATIVE ENERGY RESOURCES; ALTERNATIVE FUEL; BIOFUEL; BIOMASS; ECOLOGICAL PYRAMID; ELECTRICITY; GEOTHERMAL ENERGY; SOLAR ENERGY; THERMAL POLLUTION; WIND POWER.

leachate. *See* leaching.

leaching, A process in which a liquid (usually water) flows through a material and extracts substances from the material as it passes through. Leaching naturally occurs in soils as rainfall or other water soaks into the ground surface and works its way deeper into the soil. Substances that leach from soil include metals, nutrients, and other chemicals that dissolve in water. The natural leaching processes help to develop soils of various types and to make nutrients available to the roots of PLANTS; however, leaching from some waste materials can have harmful effects on SURFACE WATER, soil, or GROUNDWATER if harmful constituents of wastes are removed and transported elsewhere by water flowing through the waste. "LEACHATE" is the term for water that has flowed through waste material (such as a landfill) and contains waste CONTAMINANTS. *See also* ACID MINE DRAINAGE; PERCOLATION; RUNOFF; SANITARY LANDFILL.

leaching field, The part of a septic system in which pipes in the ground safely DISCHARGE WASTEWATER into a bed of soil. A LEACHING field usually consists of a network of horizontal pipes constructed in a bed of gravel placed in the ground above the WATER TABLE. The pipes often consist of perforated polyvinyl chloride (PVC) pipes that allow the SEWAGE water to percolate into the gravel and surrounding soil. (*See* percolation.) A leaching pit, a similar structure, is usually a precast concrete cylinder or box that is constructed with holes to allow water to flow out. Leaching pits, like leaching fields, are constructed below ground in gravel or crushed stone.

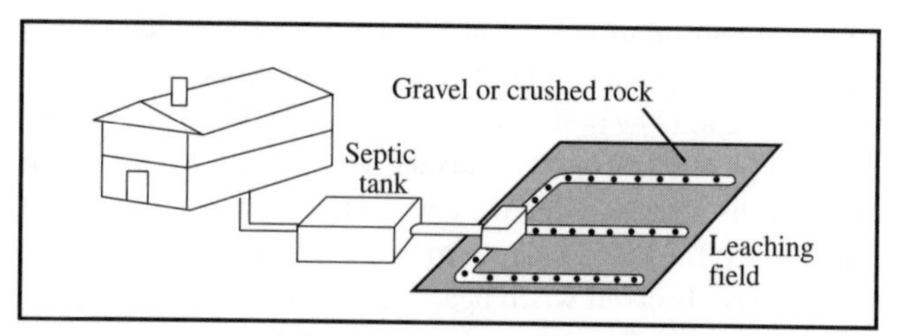

A leaching field usually consists of a network of horizontal pipes constructed in a bed of gravel placed in the ground above the water table.

Leaching fields or pits should be used to discharge only sanitary wastewater, not industrial wastewater, which may contain hazardous materials that could pollute GROUNDWATER. Appropriately designed and constructed leaching fields or leaching pits allow the liquid portion of sanitary sewage to infiltrate the soil but not the groundwater. When a leaching structure functions properly, MICROORGANISMS in the soil will degrade and filter the sewage as it percolates through the surrounding soil so that the discharged water becomes cleaner as it flows to the water table. *See also* SEPTIC TANK; SEWER.

lead, A soft, tasteless, naturally occuring HEAVY METAL found in small amounts in all parts of the ENVIRONMENT including AIR, food, water, and soil. Compounds of lead are HAZARDOUS SUBSTANCES and air POLLUTANTS. Exposure to lead occurs mostly from breathing workplace dust, eating contaminated foods, and ingesting lead in house paints and house dusts, especially by children who are most vulnerable to the effects of lead.

In adults, high levels of lead exposure may decrease reaction time and possibly affect memory. Such lead exposure may also cause weakness in the fingers, wrists, or ankles and may increase blood pressure. At even higher levels of exposure, lead can severely damage the brain and kidneys in adults and children. In the United States, lead is regulated by both the CLEAN WATER ACT and the CLEAN AIR ACT.

Lead has been used in paints, glazes, and enamels since about 1000 B.C. The ancient Romans used lead to build water pipes. Today, lead is used in the manufacturing of GASOLINE, paints, plumbing supplies, roofing supplies, PESTICIDES, and BATTERIES. Lead is also used in RADIATION shields that protect against X-RAYS and in the manufacturing of surgical equipment and computer circuit boards. The most common sources of lead pollution are industrial activities such as lead-SMELTING operations and BATTERY RECYCLING plants.

Leaded Gasoline. Gasoline burned in internal-combustion vehicle engines has been the single largest source of lead in the ATMOSPHERE. In the United States, the ENVIRONMENTAL PROTECTION AGENCY (EPA) phased out the use of leaded gasoline in the 1970s. Less than 35% of the lead released into the air now comes from gasoline.

Although use of leaded gasoline was banned in the United States and eight other countries, it is still a health risk to children in developing countries, where it is sold and used. Average blood levels of lead in children in central and eastern Europe range from 15 micrograms per deciliter to 40 micrograms per deciliter while in Canada blood levels average 5.3 micrograms per deciliter in children. At one time, Bangkok, Thailand, ranked among the highest in the world in blood lead levels in children and adults. In 1996, Thailand made the use of unleaded gas mandatory for vehicles, an action which may help in reducing lead levels.

Many countries throughout the world are switching to low-lead or no-lead FUELS as well as NATURAL GAS, liquefied natural gas, compressed natural gas, and ETHANOL. Some countries have imposed taxes on vehicles that use leaded gasoline that are higher than the taxes imposed on vehicles that use low-leaded fuel.

Leaded Paint. The U.S. Department of Housing and Urban Development reports that lead hazards exist in 500,000 U.S. homes occupied by young children. The lead found in painted walls of older homes can be very harmful to young children, as the paints may contain very large amounts of lead. The paint in these houses often chips off and mixes with dust and dirt. Some old paint (when it is dry) contains 5–40% lead. Exposure to lead is often increased among preschool age children because they put many things into their mouths. Their hands, toys, and other items may have lead-containing dirt on them. Like unborn children, preschool age children are more sensitive to its effects. Lead exposures may decrease intelligence quotient (IQ) test scores and stunt the growth of young children. A federal law requires that individuals have access to certain information related to prior use of lead paint before renting, buying, or renovating pre-1978 housing.

Lead Use in Plumbing. Sources of lead in drinking water include lead pipes, faucets, and solder used in plumbing. In

1986 lead pipe was banned for use in new construction in the United States; however, lead-containing plumbing may be found in public drinking water systems, in old houses, apartment buildings, and public buildings. Experts believe that lead solder is the major lead CONTAMINANT in U.S. homes today. If lead levels in water are high, health experts suggest that the old pipes be removed. In general, however very little lead is found in drinking water. More than 99% of all drinking water contains less than 0.005 part of lead per million parts of water. (*See* parts per million and parts per billion.)

Lead is one of the five most common types of HAZARDOUS WASTES sent to landfills. Landfills contain lead waste from lead ore MINING, smelters, ammunition manufacturing, and from other industrial activities such as battery recycling plants. If lead enters the atmosphere, it may travel thousands of kilometers. In time, the particles fall to the ground or into SURFACE WATER. Once lead particles deposit on the soil, they usually adhere to soil particles. Lead-contaminated soil can then be ingested by young children who eat contaminated soil, eat unwashed plants that have grown in such soil, or who accidentally ingest soil particles deposited on their hands.

People living near landfills can be exposed to lead and chemicals that contain lead by breathing air, drinking local water, eating locally grown foods, or swallowing or touching dust or dirt that contains lead. For people who do not live near hazardous waste sites, most exposure to lead occurs by eating foods that contain lead, occupationally in brass/bronze foundries, or in buildings where leaded paints exist. Foods such as fruits, vegetables, meats, grains, seafood, soft drinks, and wine may have lead in them. Cigarettes also contain small amounts of lead. Because of health concerns, the content of lead in gasoline, paints, ceramic products, power and communication cable casings, and pipe solder has been drastically reduced in recent years. *See also* AIR POLLUTION; ALTERNATIVE ENERGY RESOURCES; AUTOMOBILES; ELECTRIC VEHICLES; FUEL CELLS.

EnviroSources

Alliance to End Childhood Lead Poisoning: (202) 543-1147.

National Conference of State Legislatures. *State Lead Poisoning Prevention Directory*. 3rd edition. 1994.

National Lead Information Center's Clearinghouse: (800) 424-LEAD.

U.S. Environmental Protection Agency. Lead in Your Home: A Parent's Reference Guide. Booklet, no. 747-B-98-002, 1998.

U.S. Environmenal Protection Agency Web site: www.epa.gov/lead/.

legumes, PLANTS that have nodules on their roots in which NITROGEN-fixing BACTERIA live. Examples of legumes include peas, alfalfa, peanuts, clover, soybeans, and other kinds of beans. Planting legumes benefits the ENVIRONMENT by preventing NITROGEN depletion in soil. Soil fertility improves when crops, such as corn, are rotated with legumes from year to year.

Bacteria living in the nodules, or swellings on the roots of legumes, secrete NITRATES—chemical compounds that are soluble in water, and can be absorbed by plants through their roots. Once inside a plant, nitrates are converted into a variety of nitrogen-containing compounds that are useful to ORGANISMS. Decay of root nodules enriches the soil by freeing nitrogen compounds for use by other plants and organisms. Eventually the nitrogen compounds contained in plants are used to make proteins in the animals and humans who use the plants as food. *See also* NITROGEN CYCLE; NITROGEN FIXATION.

Legumes

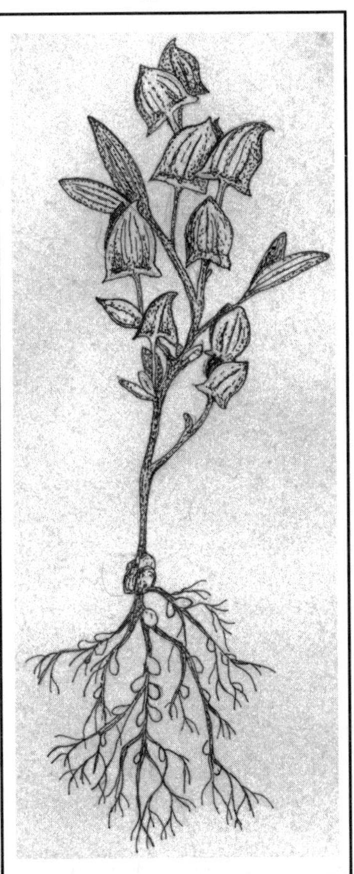

The planting of legumes benefits the environment by preventing nitrogen depletion in soil

leopards, Several species and subspecies of the cat family including two on the ENDANGERED SPECIES LIST. Endangered and threatened leopards are protected by the CONVENTION ON INTERNATIONAL TRADE IN ENDANGERED SPECIES OF WILD FLORA AND FAUNA (CITES), the ENDANGERED SPECIES ACT, and other laws. An estimated 3,000–10,000 are left in the wild, and about 370 are in captivity.

The HABITAT of the leopard ranges from DESERT to FOREST and lowland plains to mountains. They can be found in Asia, the Middle East, Africa, and the United States. Leopards are either buff or tawny with black spots. Some, known as panthers, are entirely black.

Snow leopards (*Uncia uncia*) have been hunted for their skins and are on the endangered species list. The snow leopard lives on the mountain slopes and in forests in Russia, Pakistan, India, Afghanistan, and the Himalayas east to China and North Korea. The snow leopard's prey includes antelopes, ibexes, markhors, wild sheep and goats, boars, and ground-dwelling birds. When food is scarce, snow leopards will attack domestic livestock. The subspecies is protected in India

The habitat of the leopard ranges from desert to forest and lowland plains to mountains.

and Russia but is still hunted illegally for its fur. There are perhaps 6,000 snow leopards left in the wild.

The Amur leopards (*Panthera pardus orientalis*), a subspecies of the panther genus, live in remote areas of China, Russia, and North Korea and are at risk of EXTINCTION. The International Union for the Conservation of Nature placed these leopards on their Critically Endangered Red List. (*See* World Conservation Union.) There may only be less than 100 in the wild.

EnviroSources

International Snow Leopard Trust: 4649 Sunnyside Ave. N, Suite 325, Seattle, WA 98103.

U.S. Fish and Wildlife Service, The Species List of Endangered and Threathened Wildlife Web site: www.fws.gov/r9endspp/lsppinfo.html.

Leopold, Aldo (1887–1948), An internationally respected scientist, NATURALIST, and scholar who is considered the founding father of WILDLIFE ECOLOGY and CONSERVATION as a result of his observations on land-use management approaches and his appeals for a land ethic. He is perhaps best known for his posthumously published book, *A Sand County Almanac* (1949), often acclaimed as the twentieth century's literary landmark in conservation. He also wrote the 1933 text, *Game Management*, which defined fundamental skills and techniques for managing and restoring wildlife POPULATIONS. Techniques outlined in this book are still in use today. During his lifetime, Leopold also wrote more than 350 articles, mostly on scientific and policy matters. He also served as a conservation advisor to the United Nations.

Aldo Leopold was born in Burlington, Iowa. As a boy, he developed a keen interest in ornithology (the study of birds) and natural history. Later, he attended the Sheffield Scientific School at Yale and the Yale FORESTRY school—the first graduate school for forestry in the United States. In 1909, he joined the U.S. FOREST SERVICE, and by 1912 he was supervisor of the 0.4 million-hectare (1 million-acre) Carson National Forest in what is now New Mexico. During his tenure there, he began to see the land as a living ORGANISM and to refine the concept of the BIOLOGICAL COMMUNITY. This concept became the foundation of his approach. In 1924, Leopold accepted the position of associate director of the U.S. Forest Products Laboratory in Madison, Wisconsin, the principal research institution of the Forest Service at that time. In 1933, he was appointed chairperson of the newly created Department of Game Management at the University of Wisconsin, a position he held until his death.

Leopold was inducted into the National Wildlife Federation's Conservation Hall of Fame, and in 1978, the John Burroughs Memorial Association awarded him the John Burroughs Medal for his life's work, in particular for *A Sand County Almanac*. *See also* ADAMSON, JOY GESSNER; AUDUBON, JOHN JAMES; BROWER, DAVID ROSS; CARSON, RACHEL LOUISE; DOUGLAS, MARJORY STONEMAN; FOSSEY, DIAN; GALDIKAS, BIRUTÉ; MACARTHUR, ROBERT H.; MARSH, GEORGE PERKINS; MUIR, JOHN; PETERSON, ROGER TORY; ROOSEVELT, THEODORE; WILSON, EDWARD OSBORNE.

EnviroSources

The Aldo Leopold foundation Web site: www.aldoleopold.org.

Leopold, Aldo. *A Sand County Almanac*. New York: Oxford University Press, 1949.

Watkins, Thomas H. "Voice of the Land Ethic: Aldo Leopold." *Audubon* 100, no. 6 (November–December 1998): 85.

Leopold Education Project Web site: www.lep.org.

lethal dose–50%, The amount of a TOXIC CHEMICAL that will kill 50% of the ORGANISMS exposed to the DOSE. The lower the lethal dose–50% (LD50) the more acutely toxic the chemicals. The higher the LD50, the less acutely toxic the chemical. Thus, the LD50 is a measure of the relative TOXICITY of different chemicals: NICOTINE for example, is much more acutely toxic than aspirin. It is more specifically a measure of acute toxicity—effects that occur soon after exposure to one or very few doses.

The Lethal Dose–50% for Various Substances

Substance	LD_{50} (milligrams of substance per kilogram of body weight)
Ethyl alcohol	10,000
Sodium chloride (salt)	4,000
Lindane (a pesticide)	130
Nicotine	1
Dioxin (a toxic waste product)	0.001

Source: National Institute of Occupational Safety and Health

EnviroTerm

lethal concentration 50%: The concentration of a POLLUTANT that kills 50% of the organisms exposed to the substance.

lichens, Mutualistic associations between FUNGI and green ALGAE or fungi and CYANOBACTERIA in which the ORGANISMS are so interdependent they are considered to be a single organism. In a lichen, the algae or cyanobacteria makes the food needed by both organisms through PHOTOSYNTHESIS. The fungi provide protection, moisture, and needed MINERAL salts to the algae or cyanobacteria. Lichens have been used in the manufacture of dyes, perfumes, and medicinal products. Lichens also serve as a food source for humans and other animals.

Lichens make ACIDS that neither fungi nor algae alone can produce. This acid often breaks down the bare rock or other surface colonized by the lichens, aiding in SOIL formation. (*See* weathering.) Lichens are PIONEER SPECIES in many ECOSYSTEMS because they can establish a foothold where no other living things can grow. When a NATURAL DISASTER such as a volcanic eruption or a FOREST FIRE occurs and destroys all vegetation in an area, lichens often become the first organisms to return. Eventually, SPECIES of moss gain a foothold in the rocky areas. As more soil forms, PLANTS begin to establish themselves in the moss mats and the soil. Cover gradually increases and the BIOLOGICAL COMMUNITY continues to grow, expand, and diversify. In time, some of the community may undergo ECOLOGICAL SUCCESSION to the point where a FOREST—a climax community—can form.

Species of lichens whose algal or bacterial components are NITROGEN-fixers add to the level of usable nitrogen compounds in their ENVIRONMENT by producing nitrogen-rich soil for neighboring plants. Lichens that exist as EPIPHYTES in trees provide the trees with a supply of nitrogen in a form that is useful to plants, which is why lichens are so advantageous to FOREST growth, particularly on older timber. Thus, destruction of OLD-GROWTH FORESTS often robs the remaining environment of nitrogen-fixing lichens.

Lichens can live almost anywhere—from the coldest to the hottest places on Earth. Their ability to grow in many places makes them very good indicators of pollution. Lichens absorb minerals, nutrients, water, and other substances from both solid and liquid substances and from the air around them; however, lichens cannot excrete the unnecessary substances, some of which are TOXINS from the air, water, and soil they absorb. The absorption of these substances in high amounts can cause the deterioration and the breakdown of the photosynthesizing unit of the lichens. One such toxin is SULFUR DIOXIDE, a major air pollutant. Many lichens are too sensitive to exist in areas where sulfur dioxide is present in the air, or may exist only in small POPULATIONS. Therefore, lichens can be used as INDICATOR SPECIES to monitor and gauge AIR POLLUTION levels in many areas. *See also* MUTUALISM; NITROGEN FIXATION; TAIGA; TUNDRA.

life-cycle assessment, A scientific evaluation of the potential environmental impact of a new product or chemical at each stage of its life. A thorough life-cycle assessment has three phases: an inventory, an environmental impact analysis, and recommendations for the future. The life-cycle inventory lists the amounts and types of polluting substances in a product at every stage of its life. The environmental impact analysis assesses the short- and long-term effects of each substance in the inventory. When these first two critical phases are completed, manufacturers are given suggestions (recommendations for the future) for improving the product and limiting potential environmental dangers. *See also* CRADLE-TO-GRAVE; ENVIRONMENTAL IMPACT STATEMENT.

lignite. *See* coal.

limestone. *See* rock cycle.

liming, The process of adding powdered limestone or other alkaline materials such as calcium carbonate into lakes to reduce the levels of acidity. (*See* alkali.) Liming raises the PH level of the water, thus bringing it back to normal levels. Lime can either be added directly to the lake or to streams or waterways that drain into the lake. The limestone is completely harmless to fish and WILDLIFE and instantly neutralizes acidic water as it settles. (*See* acid.)

Upstate New York; Ontario, Canada; and Sweden use liming programs in many of their lakes that have high acidic levels due to ACID RAIN; however, some studies have indicated that although liming programs have been successful in Canada and Scandinavia, not all of the lakes of upstate New York have

Liming is used to treat acidic water in a lake. *Credit:* Sweetwater Technology, Teemark Corporation

responded well to liming. Research data from some lakes revealed short periods of pH increases were followed by periods of reacidification. Liming does not restore the original water chemistry of the lake nor does it restore its BIODIVERSITY. It is expensive and requires the continuous use of liming to ensure normal acidity levels. Although liming is only a temporary solution to the problem, it can be a viable option for many acidic lakes. So, unless there is a decrease in EMISSIONS that cause acid rain, many lakes will continue to be acidic. *See also* ACID MINE DRAINAGE; LEACHING.

TechWatch

Sweetwater Technology, a division of TEEMARK Corporation in Minnesota, has developed a precise and efficient method for applying ground limestone to acid lakes. A specially designed barge is used to apply the limestone. The limestone is delivered to the site as a dry powder by pneumatic tank trucks. It is blown through hoses into the barge's hold where it is mixed with lake water in a dust-free operation to form a 50–70% slurry. Continuous mixing maintains a uniform slurry while transferring, transporting, and applying the mixture.

During distribution, the barge covers the lake with the slurry much as a farmer would fertilize a field. A computerized satellite navigational system is used to guide the operator back and forth across the lake in parallel paths. This guarantees a no-skip, no-overlap coverage and eliminates the need to drop and retrieve buoys to mark zones and paths. The DISCHARGE of the limestone slurry is controlled by a patented process using a computer. The computer continually monitors the water depth and barge speed and automatically adjusts the valves to deliver a precise DOSE of limestone to the water.

EnviroSource

Sweetwater Technology: HC 7 BOX 14 – T, Aikin, MN, 56431,(218) 927-2200; fax (218) 927-2333.

limiting factor, Any variable or combination of variables in an ENVIRONMENT that can inhibit or encourage the growth of ORGANISMS and POPULATIONS. An understanding of limiting factors is important to ecologists because the CARRYING CAPACITY—the number of individuals an ECOSYSTEM can support—is determined by its limiting factors. Limiting factors exist in all ecosystems and may be BIOTIC or ABIOTIC. For example, the types and numbers of PLANTS in an ecosystem is a biotic limiting factor that determines the types and numbers of animals the ecosystem can support. Rainfall and temperature are examples of abiotic limiting factors.

The effects of some limiting factors are regulated, in part, by the sizes of the populations living in that ecosystem. Such limiting factors are known as density-dependent limiting factors and include availability of food, water, and living space. These factors are density-dependent because COMPETITION for these resources becomes more intense as the number of organisms that require these resources increases. DISEASE, PREDATION, and PARASITISM are other density-dependent limiting factors; each of these conditions is most likely to thrive among dense populations.

Limiting factors that are not affected by population size are known as density-independent limiting factors. Such limiting factors include NATURAL DISASTERS, human disturbances, and CLIMATE. These factors impact a population in the same way regardless of its size. For example, a TORNADO that cuts a path through a FOREST may destroy an entire ecosystem; it does not selectively limit its destruction to one type of resource or to one type of organism as do parasites and some diseases. Similarly, the clearing of an area by humans affects all populations within the cleared area; it does not affect only those populations that exist in the greatest numbers. *See also* ADAPTATION; CONSERVATION; ECOLOGICAL SUCCESSION; HABITAT; NATURAL RESOURCE.

liner. *See* sanitary landfill.

lithosphere, The rocky, outer layer of Earth that is solid. The lithosphere is comprised of all of Earth's lithic (land) areas, including those underlying the OCEANS, and therefore provides HABITAT to many of the world's ORGANISMS. In addition, the lithosphere is an important RESERVOIR for COAL, PETROLEUM, NATURAL GAS, rocks, MINERALS, and most chemical elements, the simplest substances of which all matter is made. The lithosphere is also where many geological events, such as EARTHQUAKES and VOLCANOES, occur.

The lithosphere is composed of Earth's thin outer crust and a portion of the underlying mantle. On average, the lithosphere is about 100 kilometers (62 miles) thick. According to the theory of PLATE TECTONICS, Earth's lithosphere is divided into several sections or "plates," which fit together like pieces of a jigsaw puzzle. These plates float and move above the molten material of the mantle.

By studying the activities and forces within Earth's lithosphere, scientists can learn about geologic events. For instance, when the plates of the lithosphere rub and grind against one another, ENERGY builds up along plate boundaries. It is here that most mountains form, volcanoes erupt, and earthquakes occur. For example, movement of the Pacific plate against the North American plate has resulted in frequent earthquakes in California. Scientists also study processes within the lithosphere to learn more about Earth's history. Natural processes, such as WEATHERING, EROSION, compaction, cementation, melting, and cooling, for instance, all occur within the lithosphere and are responsible for the formation of the world's different rock types. (*See* rock cycle.) *See also* ATMOSPHERE; BIOSPHERE; CONTINENTAL DRIFT; GEOLOGY; HYDROSPHERE; NATURAL DISASTER.

litter, Materials scattered around in disorder, especially rubbish. According to research by the organization Keep America Beautiful, there are seven primary sources of litter: household trash placed at the roadside curb for collection; dumpsters used by business, at loading docks, and construction and demolition sites; and street litter from trucks with uncovered loads, pedestrians, and motorists.

Street litter is often swept away with rainwater into storm drains and then makes its way to the OCEAN where it washes up on beaches. Such litter is often a problem for marine animals that get entangled in the plastic litter, such as the discarded rings that hold canned beverages together, or that try to eat plastic, such as that used in plastic bags.

The Rates of Degradation of Types of Litter

Type of Litter	Period of Degradation
Piece of paper	2–4 weeks
Rolled newspaper	2–6 weeks
Candy wrapper	1–3 months
Cotton rag	1–5 months
Degradable plastic bag	2–3 months
Unpainted wood stake	1–4 years
Railroad cross tie	30 years
Tin or steel can	100 years
Aluminum can	200–500 years
Plastic six-pack holder	450 years
Glass bottle	1 million years

A correlation between litter and crime in city neighborhoods has been established by experts in urban planning, government, and sociology. "The Broken Window Theory" was developed by political scientist James Q. Wilson and criminologist George L. Kelling. It hypothesizes that neighborhoods deteriorate when even one broken window is left unrepaired, starting a chain reaction that fuels further deterioration of the area. Organizations such as Keep America Beautiful have formed to combat the problem of litter. In addition, many cities have written ordinances establishing fines for littering. *See also* GARBAGE; OCEAN DUMPING; REFUSE.

EnviroSource

Keep America Beautiful Web site: www.kab.org.

littoral zone. *See* ocean.

Love Canal, A residential community near Niagara Falls, New York, that became the nation's first recognized environmental disaster area in 1980. The Love Canal community was developed on land that previously had held a canal that was used as a dump site by the Hooker Chemical Company. In the 11-year period between 1942 and 1953, the company disposed of about 20,000 metric tons (22,000 tons) of chemical waste in the canal.The practice ceased when the city of Niagara Falls began legal proceedings, which led to condemnation of the site and a threat by the city to seize the company's property. To avoid further litigation and to ensure a release from future liability for problems resulting from the dumping, the company gave its land to the city to use for the construction of a school and residential community.

Once acquired by the city, the canal and its contents were covered with clay and soil; however, workers removed the clay cap during construction of the school, which allowed buried chemicals to ooze to the surface. This continued for the next two decades, exposing community residents (especially schoolchildren) to toxic and carcinogenic chemicals. (*See* carcinogen.) The problem worsened in the early 1970s when a season of heavy rainfall led to widespread flooding of basements with water and a thick, noxious chemical SLUDGE.

Over the next few years, residents began complaining to state officials about the foul-smelling material oozing from the soil and their basement walls. Tests later conducted on the AIR, water, and soil showed high levels of toxic and carcinogenic chemical CONTAMINANTS, including DIOXIN. Additional tests conducted by the New York State Health Department revealed that Love Canal residents showed higher-than-expected rates of certain DISEASES, including some types of CANCER, as well as miscarriages and birth defects. In response to these findings, the state evacuated 240 families from their homes. Later that year, more than 700 additional families were evacuated from Love Canal and adjacent areas. Soon after, the federal government became involved in assessing the damage to the area, and President Jimmy Carter advised thousands of other families to leave the area to allow a comprehensive cleanup program to be undertaken.

In the decade following the evacuation of Love Canal, New York State and the federal government spent more than $42 million to relocate residents, research damage to the area, and clean up and reclaim the area for use by people. Eight years after cleanup began, 236 of the evacuated homes were declared safe for occupancy by the state. Two years later, the U.S. ENVIRONMENTAL PROTECTION AGENCY (EPA) gave its approval for the homes to be sold to new owners. Despite the concerns of many environmentalists the low-cost homes of the community which was renamed the Black Creek Village community quickly attracted buyers. *See also* COMPREHENSIVE ENVIRONMENTAL RESPONSE, COMPENSATION, AND LIABILITY ACT; EMERGENCY PLANNING AND COMMUNITY RIGHT-TO-KNOW ACT; HANFORD NUCLEAR WASTE SITE; HAZARD RANKING SYSTEM; HAZARDOUS WASTE; LEACHING; RECLAMATION; SANDOZ CHEMICAL SPILL; SANITARY LANDFILL; SUPERFUND; TIMES BEACH SUPERFUND SITE; TOXIC RELEASE INVENTORY.

EnviroSources

Colten, Craig E., and Peter N. Skinner. *The Road to Love Canal: Managing Industrial Waste Before EPA.* Austin: University of Texas Press, 1996.

Gibbs, Lois Marie. *Love Canal: The Story Continues.* Gabriola Island, B.C.: New Society, 1998.

Mazur, Allan. *A Hazardous Inquiry: The Rashomon Effect at Love Canal.* Cambridge: Harvard University Press, 1998.

United States Environmental Protection Agency (Love Canal) Web site: www.epa.gov/r02earth/superfnd/site_sum/0201290c.html.

low-level radioactive waste (LLW), Any waste material such as tools, clothing, rags, papers, filters, equipment, soil, and construction rubble that is contaminated with low levels of radioactivity. Such materials generally originate from hospitals and research centers or are byproducts of the MINING industry. Most low-level radioactive wastes (LLWs) involve small amounts of radioactivity dispersed in large amounts of material.

LLWs are less hazardous than HIGH-LEVEL RADIOACTIVE WASTES (HLWs) because the radioactivity of LLWs diminishes to harmless levels—through RADIOACTIVE DECAY—in only a few years. Approximately 3% of LLWs require shielding during handling and transportation operations. *See also* HALF-LIFE.

EnviroSources

International Atomic Energy Agency ("Managing Radioactive Waste" fact sheet): www.iaea.org/worldatom/inforesource/factsheets/manradwa.html.

National Research Council, Board on Radioactive Waste Management Web site: www4.nas.edu/brwm/brwm-res.nsf.

Low-level Radioactive Waste Policy Act, Legislation passed in 1980 that makes each state responsible for managing and disposing of the LOW-LEVEL RADIOACTIVE WASTES (LLWs) within its borders. These materials include only low-level wastes, such as discarded protective clothing and equipment from power plants or hospitals and other medical facilities. The act also allows groups of states to jointly develop low-level radioactive disposal facilities. *See also* HAZARDOUS WASTE; HIGH-LEVEL RADIOACTIVE WASTE; RADIATION.

M

Maathai, Wangari Muta. *See* greenbelt movement.

MacArthur, Robert H. (1930–1972), A Canadian-born American biologist and ecologist, who did pioneering work in evolutionary ECOLOGY combining his skills in biology and mathematics. MacArthur had a special interest in birds, and much of his work dealt with bird POPULATIONS. His conclusions transformed population biology and biogeography, the patterns of distributions of ORGANISMS across Earth's surface. He earned the Mercer Award of the Ecological Society of America for his study of warblers and their specialized roles in their shared HABITAT. (*See* niche.)

MacArthur received an undergraduate degree from Marlboro College, a master's degree from Brown University, and a doctorate in 1957 from Yale University. He studied ornithology (birds) at Oxford University in England the following year. He then accepted a position as assistant professor of biology at the University of Pennsylvania in 1958.

MacArthur is the co-author, with EDWARD O. WILSON, of the book *The Theory of Island Biogeography*, published in 1967. He also co-authored *Geographic Ecology: Patterns in the Distribution of Species* (1972). *See also* AUDUBON, JOHN JAMES; EHRLICH, PAUL RALPH; EVOLUTION.

mad cow disease. *See* bovine spongiform encephalopathy.

Madagascar, An island located in the southwest Indian Ocean that lies about 400 kilometers (250 miles) off the southeast coast of Africa. Madagascar is the world's fourth-largest island; it measures approximately 1,570 kilometers (976 miles) from north to south and 571 kilometers (355 miles) at its widest point from east to west; it boasts almost 3,990 kilometers (2,480 miles) of coastline. Madagascar is of great interest to environmental scientists largely because of its BIODIVERSITY. The island provides HABITAT for numerous PLANT and animal SPECIES—many of which are endangered or threatened—found nowhere else in the world.

Madagascar is located almost entirely within the tropics; however, its CLIMATE conditions vary widely. The eastern coast has a humid, tropical climate and is covered by MANGROVE forests in regions of SWAMPS and tropical RAINFORESTS farther inland; the rainforests contain many valuable hardwoods. This area is prone to heavy rains and FLOODS that accompany tropical CYCLONES originating in the Indian Ocean. By contrast, the southwestern portion of Madagascar is dry and desertlike, and is bordered by SAVANNA GRASSLANDS and woodlands. Between the two coasts lies a central mountainous plateau dotted by VOLCANOES and with a temperate climate. This region is used largely for the cultivation of crops, which provides employment to almost 80% of the people who live on the island. The main food crop of the island people is rice. Vanilla, coffee, sugar cane, and cloves are grown for export.

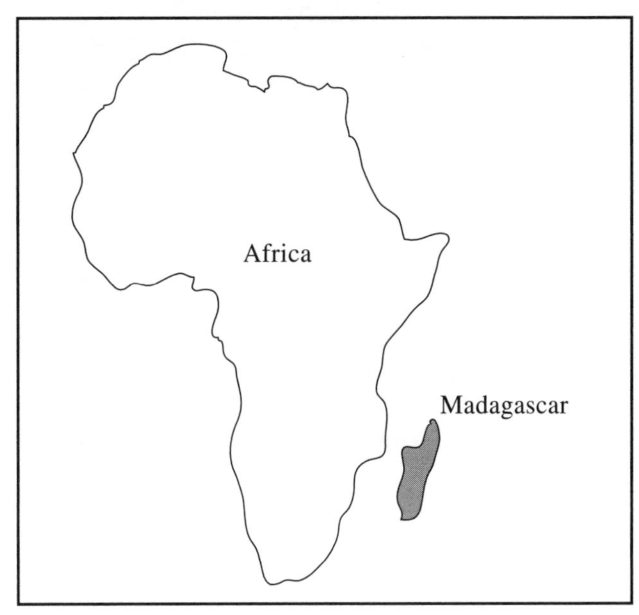

Madagascar is the world's fourth largest island. The island provides habitat to numerous plant and animal species—many of which are endangered or threatened—found nowhere else in the world.

FORESTS—coniferous and deciduous—once covered most of Madagascar; however, most of the trees have been cut to provide island inhabitants with FUEL WOOD and building materials. This activity destroyed the habitat of many of the island's unique WILDLIFE, including that of many species of lemurs and various tree-dwelling lizards of the chameleon family. Although almost 40 species of lemurs exist on the island, many are endangered and threatened with EXTINCTION. Another important island species is the rosy periwinkle (also called the Madagascar periwinkle), a plant that grows only in India and Madagascar. Like the lemurs, this flowering plant is now an ENDANGERED SPECIES, largely due to habitat destruction. The rosy periwinkle is toxic to most animals; however, its survival is of concern to many scientists because the plant contains substances that are useful in the treatment of CANCER. *See also* CONIFEROUS FOREST; DECIDUOUS FOREST; DESERT; ECOTOURISM; GALÁPAGOS ISLANDS; RED LIST OF ENDANGERED SPECIES; THREATENED SPECIES; WORLD CONSERVATION MONITORING CENTRE; WORLD CONSERVATION UNION.

EnviroSources

Garbutt, Nick. *Mammals of Madagascar.* New Haven: Yale University Press, 1999.

Greenway, Paul, and Deanna Swaney. *Lonely Planet Madagascar and Comoros.* Oakland, CA: Lonely Planet, 1997.

mahogany, A family of tropical hardwood trees that are harvested for making furniture, musical instruments, and paneling. Environmentalists believe that harvesting mahogany is one of the most destructive activities in the RAINFOREST: For every mahogany tree cut down, many other trees are damaged and destroyed in the process.

Environmentalists believe that the harvesting of mahogany is one of the most destructive activities in the rainforest.

Mahogany, a tall evergreen tree, is native to South America, particularly Brazil, Africa, and the East Indies. Most mahogany wood is obtained from several SPECIES of trees in the mahogany family.

The CONVENTION ON INTERNATIONAL TRADE IN ENDANGERED SPECIES OF WILD FLORA AND FAUNA (CITES) has proposed the placement of some species of mahogany that are endangered or threatened with EXTINCTION on their list of protected species. CITES is also seeking trade restrictions on exports and imports of mahogany.

FRIENDS OF THE EARTH (FOE) has called for a complete ban on imports of Brazilian mahogany to the United Kingdom. The organization has asked the European Union to ban mahogany imports under WILDLIFE trade laws. *See also* ENDANGERED SPECIES ACT; ENDANGERED SPECIES LIST; RED LIST OF ENDANGERED SPECIES; THREATENED SPECIES.

EnviroSource

Friends of the Earth e-mail: webmaster@foe.co.uk.

mammals, A large class of endothermic (warm-blooded) VERTEBRATES, whose offspring are nourished by milk from the mother. Other characteristics of mammals include body hair, live birth in all but one group of mammals, and outer ears. Mammals also differ from other animals in that they have a flat muscle, called a diaphragm that separates the chest from the abdominal cavity.

Mammals probably evolved more than a 150 million years ago but did not become the dominant form of life until the Cenozoic period, approximately 65 million years ago. Mammals total about 4,500 SPECIES and are widely distributed today in varied ENVIRONMENTS that include DESERTS, FORESTS, GRASSLANDS, and TUNDRA as well as in the warm and cold waters of the OCEAN. Mammals are classified into three main groups: There are egg-laying mammals, or monotremes, which include the platypus and spiny echidnas; marsupials are mammals with pouches in which the young develop, and include the kangaroo and opposum. All other mammals are placental mammals—animals whose young develop inside the body of the female and are born alive. The placental mammals are further classified into orders that include flying mammals such as BATS; hoofed mammals such as RHINOCEROSES, horses, and deer; and mammals that live in the ocean such as blue WHALES, and MANATEES among others.

Scientists estimate that about 19 orders of mammals have become extinct and that many POPULATIONS of species are now threatened or faced with EXTINCTION. Loss of HABITAT, AIR and water pollution, POACHING, OVERFISHING and overhunting, and COMPETITION with other species for food, are just some of the causes. Fortunately, many local and international environmental groups and organizations, such as the INTERNATIONAL WORLD CONSERVATION OF UNION, the INTERNATIONAL WHALING COMMISSION, the Nature Conservancy, and the WORLD WILDLIFE FUND

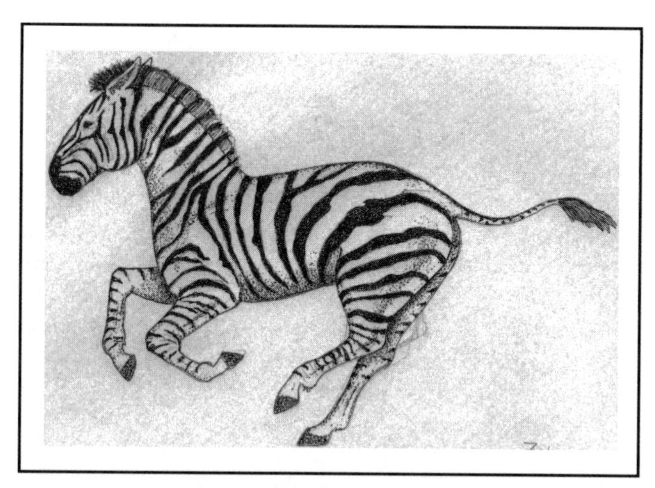

The zebra is a mammal, a large class of endothermic (warm-blooded) vertebrates, whose offspring are nourished by milk from the mother.

are working hard to maintain and protect these animals. *See also* ADDO ELEPHANT NATIONAL PARK; AFRICAN WILDLIFE FOUNDATION; AMERICAN ZOO AND AQUARIUM ASSOCIATION; ARCTIC NATIONAL WILDLIFE REFUGE; BIODIVERSITY; BISON; CARARA BIOLOGICAL RESERVE; DOLPHINS AND PORPOISES; ORANGUTAN; ORYX; RHINOCEROS; TIGERS; WOLVES; ZEBRAS.

Examples of Threatened or Endangered Mammalian Species

Common Name	*Latin Name*	*Countries*
Anthony's woodrat	*Neotoma anthonyi*	Mexico
Arabian oryx	*Oryx leucoryx*	Reintroduced in Jordan, Oman, Saudi Arabia
Bowhead whale	*Balaena mysticetus S5*	Atlantic Ocean
Eastern chimpanzee	*Pan troglodytes schweinfurthi*	Burundi, Rwanda, Sudan, Tanzania, Uganda, Zaire
Giant-striped mongoose	*Galidictis grandidieri*	Madagascar
Giant panda	*Ailuropoda melanoleuca*	China
Indiana bat	*Myotis sodalis*	United States
Javan leopard	*Panthera pardus melas*	Indonesia
Maned sloth	*Bradypus torquatus*	Brazil
Northern marsupial mole	*Notoryctes caurinus*	Australia
Northern sea lion	*Eumetopias jubatus*	Canada, Japan, Russia, United States
Riverine rabbit	*Bunolagus monticularis*	South Africa

EnviroSource

U.S. Fish and Wildlife Service Web site: www.fws.gov/r9endspp/lsppinfo.html.

Man and the Biosphere Programme, An international, interdisciplinary research and training program established in 1971 by the United Nations Educational, Scientific, and Cultural Organization (UNESCO) to provide information about NATURAL RESOURCES and solutions to environmental issues worldwide. The Man and the Biosphere (MAB) Programme is composed of a World Network of BIOSPHERE RESERVES established to promote a balanced relationship between humans and the BIOSPHERE and to respond to the challenges posed by increasingly complex and globalized CONSERVATION and development issues. The MAB is further divided into regional networks, including MABNetAmericas, EuroMAB, and AfriNet. Program activities are conducted in more than 100 countries under the direction of their MAB national committees or focal points.

Each biosphere reserve is intended to fulfill three basic objectives: conservation, SUSTAINABLE DEVELOPMENT, and support for research, education, and information exchange. The program is intended to promote for the rational use and conservation of the resources of the biosphere and to improve the relationship between humans and the ENVIRONMENT. As an intergovernmental program, the MAB also presents an opportunity for international cooperation and serves as a focus for the coordination of related programs for improving the management of natural resources and the environment.

The overall program is guided by the MAB International Coordinating Council, consisting of 34 member states elected by the UNESCO General Conference. The council generally meets every two years. In addition, other UNESCO member states that are not members of the council may send representatives as observers, as may other United Nations agencies and nongovernmental organizations.

The Guidelines or Statutory Framework of the World Network of Biosphere Reserves stipulates that each biosphere reserve is subject to a review by representatives from MAB nations committees every ten years—a process designed to improve the functioning of the reserves. Should any nation decide that one of its biosphere reserves is unable to meet the criteria set forth in the Statutory Framework, the periodic review makes it possible for that nation to request removal of its site from the world network. In this way, the network as a whole maintains quality and credibility. *See also* ADDO NATIONAL ELEPHANT PARK; AGENDA 21; ARCTIC NATIONAL WILDLIFE RESERVE; BIODIVERSITY TREATY; BIOME; CARARA BIOLOGICAL RESERVE; COASTAL ZONE MANAGEMENT ACT; ECOSYSTEM; ENVIRONMENTALLY SENSITIVE AREA; EVERGLADES NATIONAL PARK; GALÁPAGOS ISLANDS; MADAGASCAR; YELLOWSTONE NATIONAL PARK; YOSEMITE NATIONAL PARK.

EnviroSource

United Nations Educational, Scientific, and Cultural Organization; Man and the Biosphere Programme Web site: www.unesco.org/mab.

manatees, A group of PLANT-eating marine MAMMALS that live in ESTUARIES, bays, and slow-moving rivers. Manatee SPE-

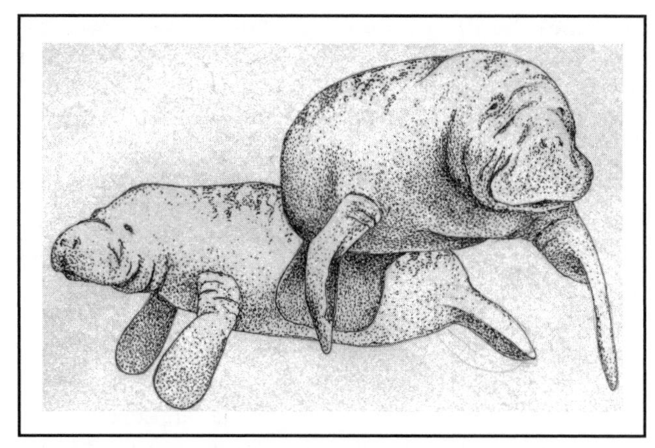

Manatees are protected at the federal level by the Marine Mammal Protection Act of 1972 and the Endangered Species Act of 1973.

CIES are all listed as either endangered or threathened as a result of pollution, boating and fishing accidents, POACHING, and the destruction of the sea grass beds that are a large part of their diet.

The three species of manatees are the Amazonian manatee (*Trichechus inunguis*), which is found in the Amazon River and had been listed as an ENDANGERED SPECIES since 1970; the West African manatee (*Trichechus senegalensis*), which lives in the rivers and coastal areas of West Africa and was included on the THREATHENED SPECIES list in 1979; and the West Indian manatee (*Trichechus manatus*), which lives in the CARIBBEAN SEA and along the east coasts of tropical North America (Florida) and South America and was placed on the endangered list in 1967.

Among the major threats to manatees are the gill nets used in COMMERCIAL FISHING. The manatees accidentally become entangled in the nets and suffocate when they are unable to surface to get air. Poaching of manatees still occurs in some areas. TOXINS in polluted rivers and estuaries have also been linked to manatee deaths. Another significant threat to manatees in areas with heavy pleasure boat or personal watercraft activity, such as in south Florida, is propellers which sometimes inflict gruesome wounds that often cause death. Administrative efforts by the state of Florida, the U.S. FISH AND WILDLIFE SERVICE, and local governments, combined with increased public awareness of this problem, has helped to reduce such accidents.

Manatees are protected at the federal level by the MARINE MAMMAL PROTECTION ACT of 1972 and the ENDANGERED SPECIES ACT of 1973. The Endangered Species Act makes it a violation to harass, harm, pursue, hunt, shoot, wound, kill, capture, or collect endangered species. The Marine Mammal Protection Act provides the same protection to marine mammals, including manatees. *See also* DOLPHINS AND PORPOISES; OCEAN DUMPING; WHALES.

EnviroSources

Save the Manatee Club: 500 N. Maitland Ave., Maitland, FL 32751; (800) 432-JOIN; fax: (407) 539-0871.

Save the Manatees Web site: www.savethemanatee.org.

Sea World (Manatees) Web site: www.seaworld.org/manatee/sciclassman.html.

U.S. Fish and Wildlife Service, Species List of Endangered and Threathened Wildlife Web site: www.fws.gov/r9endspp/lsppinfo.html.

manganese, A silver-colored metallic element that is present in many types of rocks. Small amounts of manganese are essential for the lives of humans and other ORGANISMS; however, in larger amounts, most manganese compounds are highly toxic. In humans, high levels may be carcinogenic and teratogenic (harmful to a developing fetus). Although manganese can react to form a number of compounds, it does not break down in the ENVIRONMENT. Low levels of manganese are present in lakes, ponds, streams, and particles suspended in AIR. *(See* particulate matter.)

In industry, manganese is combined with iron to make steel and other alloys. Manganese compounds are also used to make a variety of products, including PESTICIDES, FERTILIZERS, ceramics, paints, dry cells (BATTERIES), DISINFECTANTS, antiseptics, and a dietary supplement that helps the body use vitamin B_1. *See also* ACUTE TOXICITY; BIOACCUMULATION; CANCER; CARCINOGEN; CHRONIC TOXICITY; HEAVY METAL.

mangrove, An ECOSYSTEM that occurs in the intertidal zone along coastlines in tropical and subtropical areas, and is disappearing because of human activities. Worldwide, more than 100 countries have mangrove FORESTS; the most extensive ones are located in Indonesia, Nigeria, Australia, Mexico, Malaysia, and Brazil. The largest mangrove forest on Earth is located in Bangladesh.

Mangrove trees are adapted to grow in the BRACKISH waters of lagoons, SWAMPS, and creeks. The trees grow special breathing roots from their branches which take in OXYGEN from the AIR. This feature enables the trees to live in water-logged

More than 100 countries have mangrove forests, but the most extensive ones are found in Bangladesh, Indonesia, Nigeria, Australia, Mexico, Malaysia, and Brazil.

soil that is deprived of oxygen. Their roots trap silt and mud, creating a more solid, drier ENVIRONMENT over time. Mangroves, like ESTUARIES, are important ecosystems and provide breeding grounds for fishes and shellfishes. Many SPECIES of CROCODILES, ALLIGATORS, birds, and small MAMMALS live in mangrove forests.

Estimates indicate that more than 50% of the world's mangroves have been destroyed. The major causes of the losses are human activities such as timber harvesting, the conversion of mangrove forests to AQUACULTURE pools in Southeast Asia, and the clearing of the forests for farming activities and housing developments. Use of the DEFOLIANT AGENT ORANGE during the Vietnam War also led to significant loss of mangrove forests in that part of the world. Mangrove forests are also subject to excess buildup of SEDIMENTATION due to soil EROSION from nearby cleared land and RUNOFF from farms. *See also* AGRICULTURAL POLLUTION.

EnviroSource

Tropical Silverculture, Mangrove Forest Management Web site: www.metla.fi/conf/iufro95abs/dlpap135.htm.

Manhattan Project, A top-secret program of the U.S. government conducted between 1942 and 1946 for the purpose of developing an atomic (nuclear) bomb for use during World War II. To accomplish this goal, several secret laboratories, overseen by the U.S. Army, were established. Primary among these were the Metallurgical Lab in Chicago, Illinois, which conducted NUCLEAR REACTOR research (now known as the Argonne National Laboratory-East); the Clinton Engineer Works, which researched ISOTOPE separation (now known as the Oak Ridge National Laboratory); the Hanford Reservation of Washington, which produced weapons-grade PLUTONIUM (now the HANFORD NUCLEAR WASTE SITE); and the Los Alamos National Laboratory, which manufactured bomb assemblies. As a result of the Manhattan Project, three atomic bombs (Trinity Test, Fat Man, and Little Boy) were produced at a cost of more than $2 billion. The production and use of the bombs brought an end to World War II and marked the beginning of NUCLEAR WEAPONS construction and use, as well as the beginning of the search for other uses for NUCLEAR ENERGY. Use of the weapons in Japan also provided scientists with a first-hand account of the devastation to life, property, and the environment that detonation of nuclear weaponry could produce. *See also* BREEDER REACTOR; FALLOUT; FISSION; NUCLEAR REGULATORY COMMISSION; RADIATION, RADIATION SICKNESS; RADIOACTIVE WASTE.

marine biome. *See* ocean.

Marine Mammal Protection Act, U.S. legislation passed in 1972 to protect and manage marine MAMMAL SPECIES by making it illegal to take, import, possess, buy, sell, or offer to buy or sell marine mammals, their parts, or their products. Before the passage of the Marine Mammal Protection Act (MMPA), many of the species protected by this legislation were in decline and facing the possibility of EXTINCTION. The protection of walruses, polar bears, dugongs, sea otters, marine otters, and MANATEES is managed by the U.S. FISH AND WILDLIFE SERVICE (FWS). Management of seals, sea lions, and all WHALE species including DOLPHINS AND PORPOISES, is the responsibility of the NATIONAL MARINE FISHERIES SERVICE (NMFS). (*See* seals and sea lions.)

Humpback whales are one of many marine mammals that receive protection under the Marine Mammal Protection Act. *Credit:* Linda Zierdt-Warshaw

The MMPA allows for the capture of marine mammals in some circumstances. For example, permits that allow the importation of marine mammals for scientific research or public display may be issued by the secretaries of interior and commerce. In addition, Aleuts, Indians, and Eskimos who reside in Alaska are permitted to hunt marine mammals for subsistence purposes or for use in the manufacture and sale of native hand crafts.

Violators of the MMPA are subject to fines, imprisonment, or both. Individuals who violate the act may be fined as much as $100,000 and receive up to one year's imprisonment. Organizations may receive fines of up to $200,000. Violation of the MMPA may also result in the seizure of property such as water craft and aircraft used when the law was violated, including seizure of cargo.

Enforcement of the MMPA is limited to the jurisdiction of the United States; however, many marine mammal species receive additional protection worldwide through the enforcement of the CONVENTION ON INTERNATIONAL TRADE IN ENDANGERED SPECIES OF WILD FLORA AND FAUNA as well as the efforts of the INTERNATIONAL WHALING COMMISSION (IWC) and CONSERVATION groups such as GREENPEACE and the International Marine Mammal Project. *See also* BIODIVERSITY TREATY; COUSTEAU, JACQUES YVES; EARLE, SYLVIA; ENDANGERED SPECIES; KEYSTONE SPECIES; MIGRATORY BIRD TREATY ACT; RED LIST OF ENDANGERED SPECIES; WORLD CONSERVATION MONITORING CENTRE; WORLD CONSERVATION UNION.

EnviroSources

Greenpeace International Web site: www/greenpeace.org.
World Conservation Monitoring Centre Web site: www.wcmc.org.

Marine Protection, Research, and Sanctuaries Act, Legislation passed by the U.S. Congress in 1972 to prohibit the dumping of material into the OCEAN that would unreasonably degrade or endanger the health of humans, marine ORGANISMS, or the marine ENVIRONMENT. The Marine Protection Research, and Sanctuaries Act (MPRSA) is sometimes called the Ocean Dumping Ban Act. In 1988, the MPRSA was amended to include a ban on the dumping of industrial wastes and SEWAGE SLUDGE into the oceans. Virtually all OCEAN DUMP-

ING occurring today involves dredged material—sediment removed from the ocean bottom to maintain navigation channels and ship berthing areas.

Ocean dumping is illegal unless a permit is issued under the MPRSA. The U.S. ENVIRONMENTAL PROTECTION AGENCY (EPA) is charged with developing the criteria for evaluating permit applications. The decision to issue a permit is made by the U.S. ARMY CORPS OF ENGINEERS, using the criteria developed by the EPA and only with agreement from the EPA. *See also* CLEAN WATER ACT; COASTAL ZONE MANAGEMENT ACT; DREDGING; OCEAN DRILLING PROGRAM; OIL SPILL.

EnviroSources

U.S. Environmental Protection Agency: 401 M St., SW, Washington, DC 20460; (202) 260-8448.

U.S. Environmental Protection Agency, Office of Water Web site: www.epa.gov/OWOW.

U.S. Environmental Protection Agency, "Proposal to Clarify Ocean Dumping Testing Regulations." Fact sheet, no. 842-K-96-001.)

marsh, A type of ECOSYSTEM characterized by flooded soils and the growth of grass PLANTS. Marshes are disappearing due to human POPULATION growth, urban sprawl, and conversion to farmland.

A marsh is one type of WETLAND ecosystem, an area of land where water is the dominant ABIOTIC factor in the ENVIRONMENT. All wetlands, including SWAMPS, BOGS, and marshes, are transitional areas between land and water. In a wetland, water covers or saturates the soil all year or for varying periods of time during the year. Thus, all plants that inhabit marshes are adapted to changing water levels and can grow in water-logged soils.

Marshes can be distinguished from other wetlands in that the dominant plants are grasses, grasslike sedges, and reeds. Swamps, on the other hand, are characterized by the growth of trees and other woody plants. Marshes occur both inland and along coasts. Inland marshes are fresh-water environments that generally form on low-lying or depressed areas of land with poorly drained soils. In most fresh-water marshes the primary sources of water are rainfall and overflow from rivers, streams, and lakes. Fresh-water marshes are also common at the mouths of rivers where sediments are deposited to form deltas. Delta soils are poorly drained but rich in nutrients, a perfect environment for the growth of marsh grasses. In the United States, the most extensive delta marshes occur at the mouth of the Mississippi River where it opens into the Gulf of Mexico.

Another kind of coastal marsh is the SALT MARSH. Salt marshes occur along coasts, in bays, and in lagoons, and are influenced, not by rainfall, but by the daily rise and fall of the tides. Salt marsh plants must adapt to the constantly fluctuating water levels as well as the high levels of salt deposited by sea water. Spartina, or salt marsh cordgrass, for instance, is a common salt marsh plant with stiff, pointed leaves and specialized glands that help it excrete excess salt absorbed by its roots.

Marshes and other wetlands are extremely important ecosystems to WILDLIFE as well as people. Many inland fresh-water marshes, for instance, serve as feeding, breeding, and nesting areas for ducks, geese, and other migratory waterfowl. Salt marshes and other coastal wetlands are also spawning grounds for commercially important shellfish and game fish species. In addition to providing food, marshes also help improve the quality of the environment. Inland marshes minimize flooding by storing excess water from heavy rainfalls and also help maintain water quality of lakes, rivers, and streams by filtering out sediments and nutrients that might otherwise enter and harm the environment. (*See* flood.) Coastal marshes also protect the land by slowing waves and protecting coastlines from EROSION. *See also* CHESAPEAKE BAY; ESTUARY, EVERGLADES NATIONAL PARK; NORTH AMERICAN WETLANDS CONSERVATION ACT; MANGROVE; SEDIMENTATION.

EnviroSources

U.S. Environmental Protection Agency, Environmental Protection Agency's Office of Wetlands, Oceans, Watershed Web site: www.epa.gov/owow/wetlands/wetland2.html.

U.S. Environmental Protection Agency, Wetlands Hotline: (800) 832-7828, e-mail: wetlands-hotline@epamail.epa.gov.

U.S. Fish and Wildlife Service, North American Waterfowl and Wetlands Office Web site: www.fws.gov/r9nawwo.

U.S. Fish and Wildlife Service, North American Wetlands Conservation Act Web site: www.fws.gov/r9nawwo/nawcahp.html.

U.S. fish and Wildlife Service, North American Wetlands Conservation Council Web site: www.fws.gov/r9nawwo/nawcc.html.

Marsh, George Perkins (1801–1882), An American conservationist, lawyer, teacher, writer, diplomat, and scholar who was born in Vermont. Environmentalists consider him to be the father of the CONSERVATION movement. In 1864, Marsh wrote *Man and Nature*, which for the first time reported how human activity was destroying FORESTS and causing EROSION that was devasting the land. Marsh also believed that it was important to protect and preserve the WILDERNESS.

Marsh was appointed as the first U.S. Minister to Italy in 1860, in which capacity he served for the remaining 21 years of his life. In addition, he researched and was published in numerous fields. The George Perkins Marsh Institute was founded at Clark University to enhance the understanding of nature-society relationships through research and development. The insitute is internationally known for its work on the human dimensions of global change. *See also* FORESTRY; GREENBELT MOVEMENT; MARSHALL, ROBERT.

EnviroSources

Curtis, Jane. *The World of George Perkins Marsh, America's First Conservationist and Environmentalist: An Illustrated Biography.* Woodstock, VT: Billings Farm and Museum, 1982.

Lowenthall, David. *George Perkins Marsh: Versatile Vermonter.* New York: Columbia University Press, 1980.

University of Vermont Web site: sageunix.uvm.edu:6336/dynaweb/dwebdoc/marsh/@Generic_BookTextView/103;cd=3.

Marshall, Robert (1901–1939), A devoted conservationist and cofounder, with ALDO LEOPOLD and Benton MacKaye, of the WILDERNESS SOCIETY. Marshall's 1930 article, "The Problem of the Wilderness," which appeared in *Scientific Monthly*, was an important work in the development of a national WILDERNESS policy. An avowed socialist, Marshall believed that forests must be publicly owned to ensure their protection. Marshall was among the first to suggest that large tracts of land in Alaska be preserved. The WILDERNESS ACT OF 1964 was the crowning achievement of Marshall's Wilderness Society; the law marked the beginning of a new environmental "preservation" movement characterized by government regulation of CONSERVATION measures.

Marshall's first major efforts to inventory roadless areas began in the mid-1920s; at the time, assessments were also being conducted by the U.S. FOREST SERVICE. In 1926, the Forest Service found that the largest roadless area in the United States was 2.8 million hectares (7 million acres) in size. By 1936, Marshall had identified and overseen mapping of 48 forested areas that were each larger than 138,000 hectares (300,000 acres) in size and 29 DESERT areas, each over 230,000 hectares (500,000 acres) in size.

Marshall served first as the assistant silviculturist at the Northern Rocky Mountain Experiment Station from 1925 to 1928 before engaging in exploration, ecological studies, and anthropological research in northern Alaska from 1929 to 1931. (*See* silviculture.) From 1933 to 1937, Marshall served as director of FORESTRY for the U.S. Department of the Interior's Bureau of Indian Affairs. In 1937, he was appointed chief of the Division of Recreation and Lands for the U.S. Forest Service. Marshall was also active in many environmental and professional organizations, such as the Society of American Foresters, the Ecological Society, the Society of Plant Physiologists, and the Society of Anthropologists.

During the 1930s, Marshall succeeded in setting aside over 2 million hectares (4.9 million acres) of wilderness. In 1941, two years after his death, the U.S. Forest Service immortalized him in naming the Bob Marshall Wilderness Area in the Flathead and Lewis and Clark National Forests in Montana in recognition of his work in the development of the agency's system of wilderness areas. In addition, there are three other geographic features in the United States named in his honor: Marshall Lake in the Brooks Range, Alaska; Mount Marshall in the Adirondack Mountains; and the Bob Marshall Recreation Camp in the Black Hills National Forest, South Dakota. In the 1970s, the addition of the contiguous Great Bear and Scapegoat Wildernesses to the Bob Marshall Wilderness Area in Montana created a roadless area ranking among the largest and finest in the world, and which is sometimes referred to as "Bob Marshall Country." *See also* ECOLOGY; MUIR, JOHN; MARSH, GEORGE PERKINS.

EnviroSources

Glover, James M. *A Wilderness Original: The Life of Bob Marshall.* Seattle: Mountaineers Books, 1986.
Marshall, Robert. *Alaska Wilderness: Exploring the Central Brooks Range.* 2d ed. Berkeley: University of California Press, 1970.
Bob Marshall Foundation Web site: www.bobmarshall.org.
Wilderness Society Web site: www.wilderness.org.

mass extinction. *See* extinction.

material safety data sheet, Published information that is circulated to workers concerning a HAZARDOUS SUBSTANCE. Material safety data sheets were established for employee safety by the OCCUPATIONAL SAFETY AND HEALTH ADMINISTRATION. The information is given to employees who may be exposed to hazardous materials in the workplace, and includes the identity of a hazardous chemical, any health and physical hazards, exposure limits, and precautions. It also includes safe handling recommendations, firefighting techniques, and proper disposal of a hazardous material. *See also* CRADLE-TO-GRAVE; HAZARDOUS MATERIALS TRANSPORT.

EnviroSources

Interactive Learning Paradigms Incorporated Material Safety Data Sheets Web site: www.ilpi.com/mads/index.html.
Vermont Safety Information Resources, Inc., Material Safety Data Sheets Web site: www.siri.org/msds.

maximum contaminant level, The maximum level of a CONTAMINANT such as BACTERIA or chemicals that is permitted in drinking water. Contaminant levels that exceed this maximum are known to have adverse health effects on human health. Maximum contaminant levels (MCLs) are typically expressed as micrograms of substance per liter (μg/L) of water. For example, the MCL for chemicals linked to CANCER would be set at 0mg/L. MCLs are determined by the U.S. ENVIRONMENTAL PROTECTION AGENCY, as set forth in the federal SAFE DRINKING WATER ACT and cannot be exceeded in public water systems. MCLs are published in the NATIONAL PRIMARY DRINKING WATER REGULATIONS. *See also* CLEAN WATER ACT; HEAVY METALS; PRIMARY, SECONDARY, AND TERTIARY TREATMENTS; POLLUTION; SEWAGE; WASTEWATER.

medical waste. *See* infectious waste.

Mediterranean Sea, A sea that separates the European and African continents, extending east from the Atlantic Ocean to the Near East. The Suez Canal, in Egypt, connects the Mediterranean to the Red Sea. A host of water pollution problems plagues the Mediterranean, which features many industrialized ports and a great volume of commercial maritime traffic. Industries that DISCHARGE WASTEWATER, sometimes untreated, into the sea include paper mills, refineries, and chemical FERTILIZER plants. In addition, urban and agricultural RUNOFF reaches the Mediterranean through storm water outflows and waterways that empty into the sea. SOLID WASTE and sediment dumped in the Mediterranean have caused further degradation of its water quality.

Beginning in 1975, a group of Mediterranean nations adopted an action plan aimed at reducing pollution in the Mediterranean. Among the initiatives encouraged by the Mediterranean Action Plan are the protection of the sea against pollution from land-based sources, the establishment of specially protected areas, the cooperation of nations in the cleanup of OIL SPILLS and chemical pollution, and the protection against the effects of marine MINERAL and PETROLEUM exploration. *See also* AGRICULTURAL POLLUTION; BLACK SEA; NORTH SEA.

meltdown, The melting of FUEL in a NUCLEAR REACTOR that occurs when a CHAIN REACTION goes out of control. During the meltdown dangerous RADIATION is released. The overheating of the fuel is caused only when there is a problem in removing heat from the FUEL RODS. The meltdown can occur due to a loss of coolant in the reactor cooling system followed by a failure of the emergency protection system to operate and shutdown the reactor. There were partial meltdowns at THREE MILE ISLAND in Pennsylvania and CHERNOBYL in Ukraine. *See also* CONTROL ROD; HIGH-LEVEL RADIOACTIVES WASTES.

Mendes, Francisco (Chico), A Brazilian environmental activist who spoke out about the DEFORESTATION of the Brazilian RAINFOREST. Chico Mendes was born in Xapuri, Brazil, in the state of Acre. Acre is a large area of cattle ranches and plantations in the western part of Brazil. Mendes and his family lived in a community of Seringeiros or rubber tappers, who gathered latex from the seringera trees in northwestern Brazil. The Seringeiros did not own the land but leased it from landowners. Chico Mendes organized the Seringeiros to fight for independence from the landowners and the cessation of the unnecessary destruction of the rainforests by cattle ranchers. For many years, ranchers cut and burned the FOREST so that the land would be more suitable for cattle raising. (*See* slash and burn.) The ranchers were also destroying the livelihood of the Seringeiros. The rising smoke from the burning forest covered a large area. Photos taken of the fires alerted many environmentalists of the massive destruction that was taking place in the rainforest. The damage of the rainforest and Mendes's advocacy to preserve it caused constant problems between ranchers and government officials. In 1988, after years of conflict between the ranchers and Seringieros, Mendes was shot and killed at his home in Xapuri. A local rancher and his son were prosecuted for the murder. After Mendes's death, the Brazilian government established the Reserva Extractavisitas (extractive reserves) a program that set aside several forest preserves in northwestern Brazil. These reserves were established to protect the livelihood of the Seringieros and to save and preserve the trees from any further deforestation. One of the reserves was named the Chico Mendes Extractive Reserve, an area that covers 1 million hectares (2.5 million acres). *See also* ENVIRONMENTALLY SENSITIVE AREA; FORESTRY; SILVICULTURE.

EnviroSource

Environmental Defense Fund, The Chico Mendes Sustainable Rainforest Campaign Web site: www.edf.org/programs/internaitonal/chico/index.html.

mercury, A metallic chemical element that occurs naturally in the ENVIRONMENT as part of several stable compounds. Because of its high TOXICITY, mercury is regulated by the U.S. ENVIRONMENTAL PROTECTION AGENCY (EPA) under the SAFE DRINKING WATER ACT and the CLEAN AIR ACT. The most common source for mercury is the MINERAL cinnabar, which is a mercury sulfide. In its elemental form, mercury is a shiny, silver-white, odorless liquid. It can combine with other elements, such as CHLORINE, CARBON, SULFUR, or OXYGEN, to form mercury compounds.

Mercury and its compounds have many uses. They are often used to make thermometers, barometers, BATTERIES, lamps, skin care products, and medicinal products. Mercury compounds are also used to make neon lights, fungicides, paints, plastics, electrical equipment such as switches, and in the production of paper and paper goods. Mercury is also used in dental fillings.

Symptoms of mercury poisoning in humans were first observed in workers in the 1800s who used mercury NITRATE to cure beaver skins for hats. Mercury nitrate was also used as a felting agent for rabbit fur. The workers in the hat-making industry suffered from nervous conditions such as loss of memory, incoherent speech, and other illnesses. The expression "mad as a hatter" referred to those workers.

Mercury can enter the AIR from the burning of FUELS or GARBAGE, as a result of MINING activities, and in EMISSIONS from industries that use mercury in the manufacture of paper or steel or other metal products. The EPA estimates that combustion of mercury-containing material accounts for 86% of the mercury emissions in the United States. This combustion occurs primarily in coal-fired electric utility boilers, which contribute approximately 33% of emissions; municipal waste incinerators, which contribute around 19%; coal and oil-fired commercial/industrial boilers, which contribute about 18%; medical waste incinerators, which contribute approximately 10%; HAZARDOUS WASTE incinerators, which contribute approximately 4%; and residential boilers, which contribute approximately 2%. Another 10% of the total emissions come from manufacturing such as chlor-ALKALI, Portland cement, and pulp and paper manufacturing. The remaining 3% comes from sources such as laboratory uses and dental preparations.

ORGANIC forms of mercury can enter the water and remain there for a long time, particularly if there are particles in the water to which the mercury can attach. Mercury in water can enter the FOOD CHAIN as ORGANISMS ingest matter contaminated with mercury. Once in the food chain, mercury concentrations in tissues of organisms can increase through the processes of BIOACCUMULATION and BIOMAGNIFICATION. Fishes living in some contaminated waters have been found to have such high levels of mercury in their bodies that their consumption by people has been prohibited. Despite the prohibitions, contaminated fish have become a major source of exposure to organic mercury for people.

During the 1950s and through the 1970s, many people living near Minamata Bay, Japan, became ill because of mercury poisoning. Chemical analysis indicated that the fishes and shellfishes of the bay contained appreciable amounts of mercury in their tissues. Hundreds of people who ate the fishes and shellfishes developed symptoms of mercury poisoning, including a loss of peripheral vision, weakness, excessive salivation and perspiring, tremors, mental disturbances, and coma.

It was reported in 1999 that some people in the fishing communities of the remote Amazon RAINFOREST had high levels of METHYL MERCURY in their bodies. The mercury is present in fish in the region. But there is confusion about whether the mercury comes from gold MINING or is LEACHING from soils following DEFORESTATION.

Human exposure to mercury may also occur near hazardous waste sites. Such exposure may result from breathing contaminated air, having contact with contaminated SOIL, or drinking contaminated water. Most work-related exposure to mercury occurs from breathing air that contains mercury or absorbing mercury through the skin. Such exposure is most common in medical, dental, and health services, in chemical and metal processing plants, and in industries that make electrical equipment, automotive parts, and building supplies. Short-term exposure to high levels of metallic mercury in the air can cause skin rashes. It can also have adverse effects on the lungs and eyes. Long-term exposure to high levels of either inorganic or organic mercury can cause permanent damage to the brain and kidneys, and can also be harmful to a developing fetus. *See also* ACUTE TOXICITY; CHRONIC TOXICITY; HEAVY METAL; MINAMATA DISEASE; TUNA.

EnviroSources

National Institute of Environmental Health Sciences: 100 Capitol Drive, Suite 108, Durham, NC 27713; (800) 643-4794.

U.S. Department of Health and Human Services, National Institute for Occupational Safety and Health Center for Disease Control: 4676 Columbia Parkway, Cincinnati, OH 45226; (800) 356-4674

metalimnion. *See* thermocline.

metamorphic rock. *See* rock cycle.

meteorology, The branch of earth science that studies Earth's ATMOSPHERE and all WEATHER and CLIMATE conditions. Scientists who study the atmosphere are called meteorologists. They study the composition of gases in the atmosphere and how they interact with Earth's water supply and its solid surface. Meteorologists study wind conditions, distribution of temperature, precipitation, humidity, cloud formation and movement, thunderstorms, AIR POLLUTION, and NATURAL DISASTERS such as HURRICANES, TORNADOES, and CYCLONES. One of the major goals in meteorology is to be able to make long-term weather forecasts. Forecasts for a period of one to three days can be fairly accurate; however, forecasting for longer periods of time is not as accurate. *See also* AGROCLIMATOLOGY; EL NIÑO/LA NIÑA; FLOOD; HURRICANE; NATURAL DISASTER; NATIONAL WEATHER SERVICE.

methane, A colorless, odorless, gaseous HYDROCARBON formed by the thermal DECOMPOSITION or ANAEROBIC DECOMPOSITION of ORGANIC matter. Methane (CH_4) is the simplest, lightest, most abundant hydrocarbon. It occurs naturally as the chief component of NATURAL GAS, in association with COAL beds, and as the MARSH gas released by the ANAEROBIC bacterial decomposition of vegetable matter buried in WETLAND soils. Methane gas also is emitted as a waste product of digestion in some animals such as cattle and sheep.

Methane is combustible and can form explosive mixtures with AIR at concentrations between 5 and 14%; explosions of such mixtures have been the cause of many coal mine disasters. As a component of natural gas, methane is used for FUEL, and also in the manufacture of SOLVENTS and certain freons, a group of chemical compounds used as coolants and solvents.

Methane is a GREENHOUSE GAS whose concentration in the ATMOSPHERE has increased sharply as a result of human activities; in fact, the increase of global methane has essentially kept pace with increases in world POPULATION. Large quantities of methane are believed to be released by rice paddies, where vegetation rots in the water-logged soils. The digestive systems of cattle and other grazing livestock are also another major source emitting methane gas as a waste. It is estimated that the large global cattle population and increasing land areas covered by rice paddies now account for almost 50% of the global release of methane; another 20% is produced by the burning of wood and other vegetation. Lesser amounts of methane are produced by the MINING of coal and production of natural gas, which release methane trapped with these deposits. *See also* AIR POLLUTION; BACTERIA; CLIMATE; FOSSIL FUELS; GLOBAL WARMING; GREENHOUSE EFFECT; METHANOGEN; VOLATILE ORGANIC COMPOUNDS.

methanogen, ANEROBIC BACTERIA that are present in SWAMPS and MARSHES, where they break down, or decompose, PLANT material, releasing METHANE gas or marsh gas in the process. These bacteria are also found in the digestive system of many MAMMALS, including cattle, sheep, and other ruminant livestock. Ruminant livestock are those which must regurgitate and re-chew their food as part of their digestive process.

methanol, A colorless liquid chemical that evaporates easily. It is naturally present in wood, decaying ORGANIC materials, and volcanic gases. Methanol (CH_3OH), known also as methyl alcohol or wood alcohol, is produced by the destructive distillation of coal or wood. It is used as an ingredient in wall paints, carburetor cleaners, car windshield washer products, and as a GASOLINE additive.

In 1999, according to a study conducted by the University of California, methanol-based additives (MTBE) used in reformulated gasoline has affected drinking water in thousands of GROUNDWATER sites. MTBE is used in gasoline in several states to reduce SMOG-causing EMISSIONS from car exhausts. MTBE molecules can penetrate very quckly through soil and into groundwater once gasoline seeps into the ENVIRONMENT. Spills can occur through leaks in underground tanks as well as while refueling motor vehicles and boats. As a result of the study, California has banned MTBE in gasoline.

Methanol is not likely to be toxic at levels normally found in the environment. Its potential TOXICITY to humans depends on the frequency and intensity of exposure. Most direct releases of methanol are to the air, especially in the workplace.

Workers repeatedly exposed to high levels of methanol have experienced adverse effects such as headaches, sleep disorders, gastrointestinal problems, and optic nerve damage. Exposure to large amounts of methanol-contaminated liquids can lead to death.

Occupational exposure to methanol is regulated by the OCCUPATIONAL SAFETY AND HEALTH ADMINISTRATION. The exposure limit is 200 parts per million parts of air. This must not be exceeded during any eight-hour workshift. *See also* PARTS PER BILLION AND PARTS PER MILLION.

EnviroSources

Agency for Toxic Substances and Diseases Registry, Division of Toxicology, 1600 Clifton Road NE, Mailstop E-29, Atlanta, GA 30333.

United States Environmental Protection Agency (chemical summary) Web site: www.epa.gov/opptintr/chemfact/s_methan.txt.

methyl bromide, An agrochemical that is used to sterilize soil prior to the planting of many fruits and vegetables and is being eliminated from use primarily because it contributes to OZONE DEPLETION. To apply methyl bromide, farmers inject the chemical into the soil, and then cover the soil with plastic sheeting. Within 24 hours, the chemical kills parasites, the seeds of weeds, rodents, and other soil ORGANISMS that can harm crop PLANTS. Methyl bromide is so effective in controlling pests that many countries refuse to import produce that has not been grown in soil treated with the chemical. Other countries, such as Japan, require methyl bromide fumigation of fruit prior to export as a means of controlling pests.

Although it is a highly effective pest control agent, methyl bromide can cause environmental problems. One such problem is that methyl bromide is highly toxic to all organisms, not only to organisms that harm plants. Another problem is that methyl bromide is a gas that easily enters the ATMOSPHERE, where it reacts with other atmospheric substances to contribute to ozone layer depletion. Because of these negative effects, the U.S. ENVIRONMENTAL PROTECTION AGENCY (EPA) has issued a ban on methyl bromide use that must be met by January 1, 2005. This date places the United States five years ahead of other countries that have issued bans that take effect in 2010 under an international agreement.

Environmentalists and concerned citizens generally favor the ban on methyl bromide; however, many farmers believe that when the methyl bromide ban takes effect, they will lose hundreds of millions of dollars each year in export sales. Other methods used to sterilize soil prior to planting are much less effective than methyl bromide treatment. For example, a soil sterilization method that involves sprinkling hot water from truck-mounted boilers provides only short-term relief from pest organisms. The method also uses a great deal of ENERGY and water, and if overused can expose soil to the problem of SALINIZATION. Nonchemical soil sterilization methods, such as the use of pinebark compost or BIOLOGICAL CONTROLS, are currently being tested. *See also* FEDERAL INSECTICIDE, FUNGICIDE, AND RODENTICIDE ACT OF 1972; FUNGI; GLOBAL WARMING; INSECTS; INTEGRATED PEST MANAGEMENT; MONTREAL PROTOCOL; OZONE; OZONE HOLE; PARASITISM; PESTICIDE .

TechWatch

Leesch and GFK Consulting, a company in San Clemente, California, has developed an industrial gas mask that fits over the vents of fumigation chambers. The mask traps methyl bromide in activated carbon filters (*see* adsorption) and may reduce methyl bromide EMISSIONS by as much as 95%. Methyl bromide not captured by the mask still threatens the ozone layer, but the reduction in emissions is an improvement over current methods (which allow as much as 80% of the gas to escape to the atmosphere).

U.S. plant pathologists at the Agricultural Research Services Appalachian Fruit Research Station in Kearneysville, West Virginia, have discovered that the natural oil found in peach seeds can kill fungi and other pests in the soil. They believe that the extract could be a possible alternative to methyl bromide fumigation. If their tests prove successful, it could be a significant breakthrough in eliminating the use of methyl bromide.

EnviroSources

U.S. Department of Agriculture, Agricultural Research Services Web site: www.ars.usda.gov/.

U.S. Department of Energy, Toxicology Information Response Center, Oak Ridge National Laboratory: 1060 Commerce Park, Oak Ridge, TN 37830.

methyl chloroform. *See* thrichlorethane.

methyl mercury, The most common ORGANIC form of MERCURY. Methyl mercury is produced mainly by BACTERIA and small ORGANISMS in the water and soil. Methyl mercury enters the FOOD CHAIN and undergoes BIOACCUMULATION and BIOMAGNIFICATION as it builds up in the tissues of organisms and moves up the food chain. People who eat fish or seed grains that contain high levels of methyl mercury can develop permanent damage to the brain, kidneys, and growing fetuses. Methyl mercury was the cause of the MINAMATA DISEASE that developed in Minamata, Japan beginning in the 1950s. The FOOD AND DRUG ADMINISTRATION has set a maximum permissible level of 1 part of methyl mercury in 1 million parts of seafood. *See also* PARTS PER BILLION AND PARTS PER MILLION.

EnviroSource

Agency for Toxic Substances and Disease Registry, Division of Toxicology: 1600 Clifton Road NE, Mailstop E-29, Atlanta, GA 30333.

methyl tertiary butyl ether. *See* MTBE.

methylene chloride, A colorless liquid chemical that evaporates easily. People may be exposed to methylene chloride in contaminated AIR, water, food, or consumer products such as paint strippers. Exposure occurs when the vapors given off by these products are inhaled by humans. Breathing methylene chloride at high levels for short periods of time can cause mild deafness and slightly impaired vision. At very high concentrations, it can lead to unconsciousness and death.

Methylene chloride, also known as dichloromethane, is made from METHANE gas or wood alcohol. It is widely used as an industrial SOLVENT and as a paint stripper as well as in the manufacture of photographic film. The chemical may be found in a variety of household products including spray paints, automotive cleaners, and some PESTICIDE formulations. The highest and most frequent exposures to methylene chloride usually occur in workplaces where the chemical is made or used.

The U.S. ENVIROMENTAL PROTECTION AGENCY (EPA) has determined that methylene chloride is a probable human CARCINOGEN. The EPA has set a maximum drinking water level of 5 parts of methylene chloride per 1 billion parts of water. The FOOD AND DRUG ADMINISTRATION has established limits on the amounts of methylene chloride that can remain in spices, hops extract, and decaffeinated coffee. *See also* VOLATILE ORGANIC COMPOUNDS.

EnviroSources

Agency for Toxic Substance and Disease Registry, Division of Toxicology: 1600 Clifton Road NE, Mailstop E-29, Atlanta, GA 30333.

Mike Kamrin, Institute for Environmental Toxicology: C231 Holden Hall, MSU, kamrin@pilot.msu.edu

U.S. Environmental Protection Agency (chemical summary) Web site: atsdr1.atsdr.cdc.gov:8080/tfacts19.html.

micron. *See* particulate matter.

microorganisms, Any living thing so minute that is unable to be observed without the use of a microscope. Microorganisms include all BACTERIA, many PROTISTS, and many FUNGI. These ORGANISMS are important to ENVIRONMENTAL SCIENCE because they fill a variety of environmental NICHES. For example, microorganisms play important roles in the FOOD CHAINS and FOOD WEBS of all ECOSYSTEMS as PRODUCERS, CONSUMERS, and DECOMPOSERS. Microorganisms also play major roles in Earth's BIOGEOCHEMICAL CYCLES by helping to cycle substances such as OXYGEN, CARBON DIOXIDE, water, and NITROGEN through the ENVIRONMENT. Many microorganisms are PATHOGENS, or agents of DISEASE in PLANTS, animals, and other organisms. Other microorganisms are used to make medicines such as antibiotics or vaccines that are used to treat or prevent diseases. Additional uses for microorganisms have been discovered in the food-making industry, the textile industry, and even in the cleanup of oil and TOXIC WASTES released into the environment. *See also* ALGAE; ANAEROBIC; ANAEROBIC DECOMPOSITION; BIODIVERSITY; BIOLOGICAL CONTROL; BIOLOGICAL WEAPON; BIOREMEDIATION; BIOTECHNOLOGY; BLIGHT; BOVINE SPONGIFORM ENCEPHALOPATHY; BRUCELLOSIS; CANCER; CARCINOGEN; CELLS; CHEMOAUTOTROPH; COLIFORM BACTERIA; CRYPTOSPORIDIUM; CYANOBACTERIA; DECOMPOSITION; DENITRIFICATION; DISINFECTANT; DUTCH ELM DISEASE; ENVIRONMENTAL MEDICINE; *ESCHERICHIA COLI*; EUTROPHICATION; GENETIC ENGINEERING; *GIARDIA LAMBLIA*; HETEROTROPH; LEGUMES; LICHENS; METHANOGEN; MYCORRHIZAL FUNGI; NITROGEN CYCLE; NITROGEN FIXATION; OPPORTUNISTIC ORGANISM; PARASITISM; PESTICIDE; PHYTOPLANKTON; PHYTOREMEDIATION; PIONEER SPECIES; PLANKTON; POTABLE WATER; PRIMARY, SECONDARY, AND TERTIARY TREATMENTS; RED TIDE; *SALMONELLA*; SEWAGE; SICK BUILDING SYNDROME; STERILIZATION, PATHOGEN; WORLD HEALTH ORGANIZATION; ZOOPLANKTON; ZOOXANTHELLAE.

migration, *See* emmigration; immigration.

Migratory Bird Treaty Act, A federal law enacted in 1918 to protect most SPECIES of common wild birds of the United States and therefore prevent the EXTINCTION of some species. The Migratory Bird Treaty Act was created, in part, to address issues not covered in the LACEY ACT (1900) and the Weeks-McLean Law (1913), federal laws that dealt with the hunting and shipment of birds used for commercial purposes. The act also helped affirm the commitment of the United States to international conventions established with Canada, Mexico, Japan, and Russia in 1916 dealing with the protection of birds whose migratory routes were shared by these countries.

The Migratory Bird Treaty Act makes it unlawful to kill, capture, collect, possess, buy, sell, trade, ship, import, or export any migratory bird or its eggs, nests, or parts, including feathers. The provision regarding bird parts includes both parts that are found and parts obtained by intentionally killing a bird. Restrictions concerning feathers were made to protect birds from overharvesting by the millinery industry, a practice that was decimating POPULATIONS of some bird species. The law also applies to craftspeople who use feathers to decorate their work.

Several common birds are not protected under the Migratory Bird Treaty Act. Among them are the starling, the house sparrow, the feral (wild) pigeon, and species identified as resident game birds. Resident game bird protection is managed by individual states and applies to grouse, quail, wild turkey, and pheasant, among others. Some species protected by the Migratory Bird Treaty Act receive additional protection through the Bald Eagle Protection Act (*see* bald eagles), the Waterfowl Depredations Prevention Act, the Fish and Conservation Act, the WILD BIRD CONSERVATION ACT, and the ENDANGERED SPECIES ACT.

The snowy egret is protected by the Migratory Bird Treaty Act.

Year	Law
	Migratory Bird Treaty Act
1959	The ANTARCTIC TREATY includes measures adopted by the Third Antarctic Treaty Consultative Meeting that are designed to protect native birds as well as the other FAUNA and FLORA of Antarctica.
1973	The CONVENTION ON INTERNATIONAL TRADE IN ENDANGERED SPECIES OF WILD FLORA AND FAUNA (CITES) was signed by 80 nations to prevent international trade (importing and exporting) of specific PLANTS and animals identified in the appendix of the CITES agreement.
1983	The Convention on the Conservation of Migratory Species of Wild Animals, also known as the Bonn Convention, is an international agreement involving 49 nations from Africa, Central America, South America, Asia, Europe, and Oceania intended to ensure the CONSERVATION of WILDLIFE and wildlife HABITATS of migratory SPECIES throughout the world.

Permits that allow possession of bird parts protected by the Migratory Bird Treaty Act are sometimes issued to groups or individuals. The permits make exceptions for birds or bird parts used for scientific collecting, falconry, or taxidermy. In addition, Native Americans can obtain permits allowing them to possess eagles, eagle parts, and migratory bird feathers for use in religious ceremonies. (Similar exceptions for aboriginal peoples and indigenous peoples of Alaska were made as amendments to the Migratory Bird Convention agreed to by the United States and Canada in 1995.) To obtain such parts, Native Americans holding permits must contact the National Eagle Repository located near Denver, Colorado. The repository serves as a collection center for all dead eagles acquired by the U.S. FISH AND WILDLIFE SERVICE.

Prohibitions in the Migratory Bird Treaty Act are intended to eliminate any commercial market for wild birds or their parts. Violation of the act is a criminal offense with punishments including fines and imprisonment. In addition, the government can seize property, such as vehicles and weapons, used to commit crimes from individuals found guilty of violating the law. *See also* WETLANDS.

EnviroSources

U.S. Fish and Wildlife Service Web site: www.fws.gov.html.
U.S. Fish and Wildlife Service, International Treaties Deal for Migratory Birds Web site: www.fws.gov/9mbmo/intrnlt/intreat.html.
World Conservation Monitoring Centre Web site: www.iucn.org/info_about_iucn/index.html

millirem. *See* radiation; sievert.

Minamata disease, A DISEASE of the nervous system resulting from ingestion of aquatic ORGANISMS contaminated with a form of MERCURY used in industrial processes. Minamata disease received its name because it was first observed in people living near Minamata Bay, Japan. People ingested the mercury by eating bay organisms contaminated with mercury that had been discharged from a nearby plastics factory. (*See* discharge.) This incident alerted people worldwide to the potential danger of exposure to mercury through the consumption of marine organisms. Prior to the events at Minamata Bay, mercury poisoning of humans via the FOOD CHAIN was unknown.

Once it was determined that sick people living near Minamata Bay shared a common illness, a cause for the disease was sought. Chemical analyses showed that many kinds of fishes and shellfishes of Minamata Bay contained appreciable amounts of METHYL MERCURY in their tissues. People who ate these organisms developed symptoms of the disease. Hundreds of poisonings occurred from the 1950s through the 1970s: More than 40 people died; approximately 100 people were permanently disabled; and some babies were born with birth defects due to the ingestion of mercury by the mothers during their pregnancies. Even cats and sea birds that fed on the fish showed symptoms of disease. Eventually, it was determined that the mercury had come from WASTEWATER that was dumped into the bay by the plastics factory. The factory stopped discharging mercury and the company closed its plant in 1968.

Outbreaks of Minamata disease have not been limited to Japan. Residents living near the Songhua River of China also have shown symptoms of the disease. More recently fishes removed form lakes in Tanzania and the Amazon region of South America also have been identified as the source of Minamata disease outbreaks.

Symptoms of Minamata disease include a loss of peripheral vision, weakness, excessive salivating and perspiring, tremors, mental disturbances, and coma. Recovery from the

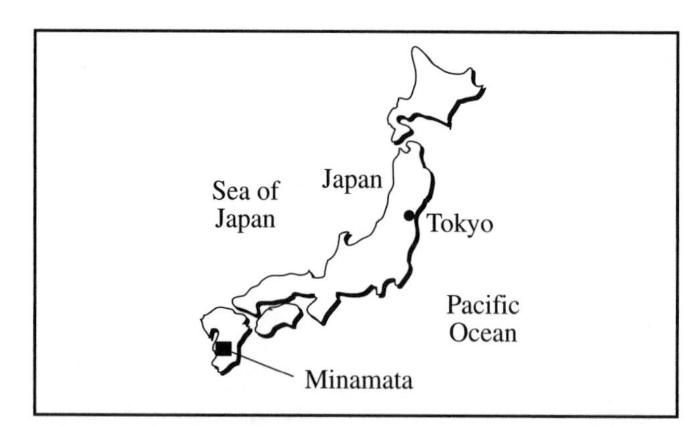

People living in Minamata Bay contacted a disease of the nervous system resulting from ingestion of aquatic organisms contaminated with a form of mercury used in industrial processes.

disease is unlikely; most often, the disease results in death. *See also* BIOACCUMULATION, CHRONIC TOXICITY.

EnviroSource

Minamata Disease Municipal Museum in cooperation with the National Institute for Minamata Disease,Ten Things to Know About Minamata Disease Web site: www.fsinet.or.jp/~soshiba/10tisiki/10tisiki-e.html.

mineral, A naturally occurring solid having a crystal structure and a definite chemical composition. Minerals are widely distributed in Earth's crust and are excavated by people for a variety of uses, including construction, the manufacture of various products, as a FUEL source, and for ornamentation. Worldwide use of minerals is of concern to environmental scientists because most minerals are being used at a faster rate than they can be replaced by nature. Thus, minerals are nonrenewable NATURAL RESOURCES. In addition, the excavation of minerals often harms the ENVIRONMENT by destroying HABITATS, making land more vulnerable to EROSION, and by releasing POLLUTANTS into nearby AIR, water, and soil. Minerals also are of interest to environmental scientists because many minerals are needed by ORGANISMS, including humans, for growth and proper functioning of body systems, and because some minerals are toxic to organisms if taken into the organism in too great an amount.

Quartz is a common mineral. *Credit:* United States Department of Agriculture. Photo by Ken Hammond.

In nature, most minerals are present as ores or compounds, but a small number of minerals consist of a single element. Minerals generally form through inorganic processes (processes not involving organisms); however, some ORGANIC substances such as COAL, graphite, and diamond (which are formed from carbon), also are considered minerals. Minerals are classified into several major groups according to their chemical composition and the anion (group of atoms that together carry a negative charge) it contains. Silicates, compounds of silicon and OXYGEN, are the most commonly occurring minerals because silica is the most abundant element in Earth's crust. This group of minerals is currently very important to the computer and electronics industries because of their uses in making computer chips. Another important mineral group is the hydroxides, minerals that form ACIDS or bases in the presence of water. This group of minerals includes lime, a strong ALKALI, and lye, an acid. Some other major mineral groups include the native elements such as SULFUR and COPPER, sulfides (iron pyrite and galena), oxides (iron oxide), halides (halite), carbonates (calcium carbonate), NITRATES, SULFATES, and PHOSPHATES. *See also* ACID MINE DRAINAGE; AGRICULTURAL POLLUTION; ALGAL BLOOM; ASBESTOS; BUREAU OF LAND MANAGEMENT; CALCIUM CYCLE; CONSERVATION; DETERGENT; FERTILIZER; FOSSIL FUEL; GENERAL MINING LAW OF 1872; LEACHING; MINING; NONRENEWABLE RESOURCES; PESTICIDE; SURFACE MINING; SURFACE MINING CONTROL AND RECLAMATION ACT; TAILINGS; UNITED NATIONS CONVENTION ON THE LAW OF THE SEA.

EnviroTerm

ore: A natural mixture of MINERALS from which elements or compounds can be extracted.

mineral cycles. *See* biogeochemical cycle.

mining, The process of extracting MINERALS from Earth's crust. Minerals may be located very near Earth's surface or may lie at great depths below the surface. The location of the minerals to be extracted determines what mining method will be used to obtain the minerals. The two most common mining methods are STRIP MINING and open-pit mining. Strip mining is used to obtain minerals located near the surface and involves using bulldozers or similar heavy equipment to "strip" away or clear the overburden (rock and soil) that lies just over the mineral deposit. Once the overburden is cleared, access to the mineral is achieved. Open-pit mining is used to obtain minerals located deeper in the Earth. This mining method involves digging large holes deep into the Earth to provide access to its minerals. Once access to a deposit is made, shafts may be dug perpendicular to the surface to provide even greater access by both people and machinery. Because both mining methods involve the removal of rock, soil, and vegetation from an area, both have the potential to greatly impact the ENVIRONMENT.

Mining is of great economic importance to people; however, the effects of mining on the environment and the health of ORGANISMS is of great concern to many environmental scientists. Another concern is that because minerals are NONRENEWABLE RESOURCES—they are not quickly replaced through natural processes—the availability of mineral reserves for future generations is greatly diminished by their use.

Habitat Destruction. Because mining operations involve the removal of some soil, rock, and vegetation from an area, all mining operations are at least somewhat disruptive to natural ECOSYSTEMS. As vegetation is removed, FOOD CHAINS and FOOD WEBS may be disrupted. In addition, the HABITAT of many SPECIES may be destroyed. As a result of these changes, organisms may be forced to move out of the area in order to meet their needs for survival. Those organisms unable to move may die if unable to meet their needs.

Pollution. Not all damage to natural habitats from mining results from direct destruction of the habitat. The release of POLLUTANTS into the environment also is a major concern in many locations where mining occurs. Depending upon the material being mined, some pollutants may be released directly into the environment. As soil and rock are cleared from an area, silt, dust, and other small pieces of matter may become airborne and be carried great distances by the wind. Such substances pose threats to organisms that inhale such PARTICULATE MATTER.

Water pollution also commonly occurs as a result of mining practices. As rainwater washes through overburden or mine TAILINGS, substances contained in this matter can dissolve in the water and be carried to new locations. (*See* leaching.) For example, water moving over mine tailings from COAL mines frequently combines with iron sulfide in the tailings to form SULFURIC ACID. This ACID is extremely caustic and can result in severe damage to the CELLS and tissues of organisms that contact the acid. The iron from the iron sulfide also may combine with water to form an alkaline substance called ferric hydroxide. (*See* alkali.) Like sulfuric acid, this substance also is harmful to the cells and tissues of organisms.

Iron sulfide is not the only substance that can combine with rainwater to form potentially harmful pollutants. Rainwater also may pick up and carry away aluminum, magnesium, MANGANESE, calcium, and a variety of HEAVY METALS. Once dissolved in water, these materials can be carried to new areas, where they are deposited in soil, streams and lakes, or enter the GROUNDWATER supply. Soil contaminated with heavy metals can devastate PLANT communities. Because plants form the base of most terrestrial food chains, such devastation can greatly impact the other organisms living in the environment, including the FUNGI and BACTERIA that aid in DECOMPOSITION.

Health Problems. Mining is a dangerous occupation. People engaged in open-pit mining, in particular, are often at risk of being buried alive as vibrations caused by the movements of heavy machinery overhead or natural movements of Earth's crust cause a cave-in of soil and rock layers. Water that moves through mines also may weaken soil and rock structures; as a result, SUBSIDENCE may occur.

Another potential danger faced by miners is exposure to a variety of TOXIC CHEMICALS, including poisonous gases that may collect in underground mines. The types of chemicals and gases to which miners are likely to be exposed varies according to the substance being mined. Often, as with ASBESTOS, LEAD, MERCURY, and BERYLLIUM, the substance being sought is itself the material that places miners at risk. For example, asbestos is easily inhaled into the lungs, where it causes the formation of scar tissue that can impede breathing. Mercury and lead, both of which are toxic substances, can be absorbed into the body through the skin as well as through the lungs, resulting in a variety of illnesses, ranging from headaches or tremors to death. To prevent exposure to such substances, miners are encouraged to wear special clothing and breathing apparatus.

Subsidence. When groundwater enters and collects at the bottom of a mine, pumps are used to raise this water to the surface, where it is discharged into the environment. This pumping of water out of the mine can lower the WATER TABLE in a region. As the water level in underground passageways drops, overlying layers of rock and soil that are no longer supported by the water pressure may drop to fill the available space. This can result in the formation of large holes at the surface called SINKHOLES. A sinkhole may develop rapidly and without warning, leading to the destruction of any property lying atop the location where the sinkhole develops. The formation of sinkholes has resulted in extensive property damage in many parts of the United States, especially in the eastern states of Florida, Georgia, Alabama, Tennessee, and Pennsylvania. *See also* ACID MINE DRAINAGE; ACUTE TOXICITY; AIR POLLUTION; ARSENIC; BUREAU OF LAND MANAGEMENT; BUREAU OF RECLAMATION; CADMIUM; CALCIUM CYCLE; CANCER; CARCINOGEN; CESIUM-137; CHROMIUM; CHRONIC TOXICITY; CONSERVATION; CONTAMINANT; CORE SAMPLE; DEEP-WELL INJECTION; DISCHARGE; DOSE/DOSAGE; DREDGING; EARTHQUAKE; EFFLUENT; EROSION; FOSSIL FUEL; GEOLOGY; HAZARDOUS SUBSTANCE; HAZARDOUS WASTE; HIGH-LEVEL RADIOACTIVE WASTE; HYDROGEN SULFIDE; LETHAL DOSE–50%; LOW-LEVEL RADIOACTIVE WASTE; METHANE; METHYL MERCURY; MULTIPLE USE; NATURAL GAS; NATURAL RESOURCE; NATURAL RESOURCES CONSERVATION SERVICE; NITRIC ACID; OCEAN DRILLING PROGRAM; OVERDRAFT; PHOSPHATE; PHOSPHOROUS CYCLE; POTABLE WATER; PYRITE; PYROLYSIS; RADIATION; RADIOACTIVE WASTE (NUCLEAR WASTE); RADON; RECLAMATION; RESTORATION ECOLOGY; SEEPAGE; SOIL CONSERVATION; SULFATE; SULFUR; SURFACE MINING; SURFACE MINING CONTROL AND RECLAMATION ACT; TAILINGS; TOXICITY; TRACE METAL; URANIUM; URANIUM MILL TAILINGS RADIATION CONTROL ACT; WORLD RESOURCES INSTITUTE.

EnviroSources

Saign, Geoffrey C. *Green Essentials: What You Need to Know about the Environment.* San Francisco: Mercury House, 1994.

Seymour, John, and Herbert Girardet. *Blueprint for a Green Planet: Your Practical Guide to Restoring the World's Environment.* New York: Prentice-Hall, 1987.

moneran. *See* bacteria.

Mono Lake, A saline, or salt-water, lake in California that has been adversely affected by diversion of water from its feeder streams. Mono Lake formed in a volcanic crater and is fed by streams carrying meltwater from the Sierra Nevadas. Beginning in 1941, a system of aqueducts began diverting water from four of the streams feeding Mono Lake to the city of Los Angeles, located about 650 kilometers (405 miles) away, to help meet the water needs of its residents. Diversion of this water has significantly altered the Mono Lake ECOSYSTEM. By 1989, the lake's water level dropped as much as 10 meters (33 feet). This drop caused many tufa spires—calcium carbonate deposits—to emerge. As the water level dropped, the salinity of the lake increased, creating an ENVIRONMENT unsuitable for fresh-water ORGANISMS. Currently, the main PRODUCERS of the lake are ALGAE that can survive the saline environment. The algae provide food for brine flies and brine

In 1941, a system of aqueducts began diverting water from several streams feeding Mono Lake to the city of Los Angeles, located about 650 kilometers away, to help meet the water needs of its residents. Diversion of this water has significantly altered the Mono Lake ecosystem.

shrimp, which in turn, provide food for various migratory birds and California gulls.

In 1989, concern about the future of the Mono Lake ecosystem resulted in a court order that forbade further diversion of water from its feeder streams; however, in 1993, the state legislature appropriated monies to allow for further usage of feeder stream water, while also providing protection to the ecosystem and surrounding WILDLIFE HABITAT. Despite the provision for protection of the lake ecosystem, many ecologists fear that the salinity of Mono Lake will increase so much that the lake will be unable to support the organisms that now live in its water or the migratory birds that rely on these organisms for food. *See also* AQUIFER; ARAL SEA; BAIKAL, LAKE; BRACKISH; IRRIGATION; OVERDRAFT; SALINIZATION.

EnviroTerms

saline: Term used to describe a solution that contains salt; for example, OCEAN water is a saline solution.

salinity: A measure of the degree of saltiness of a substance, often expressed in terms of density or as a percent; for example, ocean water has a salinity of 35,000 parts per million.

monoculture, An agricultural practice that involves the growth of a single PLANT SPECIES in an area year after year. Monoculture may also refer to the development of a plantation devoted to a single tree species. (*See* tree farm.) Monoculture often adversely effects the ENVIRONMENT. For example, each plant species removes specific nutrients from soil. Over time, as nutrients are depleted, soil fertility decreases. To combat this problem, many farmers must rely on FERTILIZERS to replenish nutrients. These substances can wash into nearby aquatic ECOSYSTEMS and promote excessive algal growth, which in time decreases the OXYGEN content of water, resulting in the deaths of other aquatic ORGANISMS. (*See* algal bloom.) Monoculture also increases the vulnerability of plants to DISEASE, PREDATION, and PARASITISM by providing abundant hosts or food for pest organisms, thus encouraging increases in pest POPULATIONS. To combat this problem, farmers often must use chemical PESTICIDES that can adversely affect NATIVE SPECIES. In addition, in response to heavy pesticide use, some species have developed a resistance to these chemicals.

Monoculture leads to a general decrease in the BIODIVERSITY in an area. In recent years, many farmers have begun to abandon monoculture practices for the more sustainable practice of POLYCULTURE—the growth of multiple plant species in an area. This practice is one of the components of the pest control method known as INTEGRATED PEST MANAGEMENT. *See also* AGRICULTURAL POLLUTION; AGRICULTURE; AGROFORESTRY; BIOLOGICAL CONTROL; EUTROPHICATION; SOIL CONSERVATION; SUSTAINABLE AGRICULTURE.

monofill, A LANDFILL disposal site that accepts waste only from a single source. There are monofills set up specifically for ASH produced and removed from waste-burning incinerators. (*See* incineration.) Wood waste landfills are often monofills containing primarily bark and scrap wood mixed with dirt and sawdust. Other monofills may receive just wood ash. Studies conclude that the disposal of ash in an appropriately designed monofill greatly minimizes the potential for release of any harmful leachable material into the environment. *See also* DEEP-WELL INJECTION; HAZARDOUS WASTE TREATMENT; LEACHING; SANITARY LANDFILL.

monsoon, A major wind system that changes its direction during a seasonal change. During one season, the winds blow from sea to land. During another season, the winds blow from land to sea. The monsoon is caused by differences in annual temperature and AIR pressure between land and sea. Monsoons appear along the Gulf Coast of the United States and in central Europe, but the most extreme ones occur in South Africa, northern Australia, and Southeast Asia, particularly in India.

In India, for example, there are two distinct monsoon seasons. In the summer, the monsoons move over the Indian Ocean and blow southwest over the land. The summer monsoons bring heavy winds and plenty of precipitation in the form of rain. This is an important growing season. In the winter, the monsoon blows across the land in a northeast direction out to the Indian Ocean. Winter monsoons, by contrast, are cold and dry with very little precipitation. The dry harsh winds can cause DROUGHTS and damage and destroy crops. The summer monsoons can also pose problems. Any disruption of the summer monsoon can be a disaster to heavily populated towns and cities in the monsoon belt: Too much rain can cause massive flooding conditions, and too little rain can impact the growth of crops. *See also* FLOOD; WATER CYCLE.

Montreal Protocol, A 1987 international treaty agreement approved by 29 countries for phasing out the production of

CHLOROFLUOROCARBONS (CFCS), which deplete the OZONE LAYER in the ATMOSPHERE. The protocol came into force in 1989 and was updated and amended in 1990. That year the agreement was strengthened to eventually prohibit all members from producing CFCs. Members of the Montreal Protocol are forbidden to purchase CFCs or products containing them from nations that have not agreed to the treaty. The treaty was amended in 1992 to phase out CFCs by 2000.

Without the Montreal Protocol and its amendments, the continuing use of CFCs and other compounds would have tripled the stratospheric abundances of CHLORINE and bromine by the year 2050. HALONS and CARBON TETRACHLORIDE are to be phased out by 2000 and methyl chloroform by 2005. The outright bans were based on scientific theory and evidence suggesting that, once emitted to the atmosphere, these compounds could significantly deplete the stratospheric ozone layer that shields the planet from damaging ultraviolet-B RADIATION. Under the Montreal Protocol, global consumption of CFCs dropped from 1.3 billion kilograms (2.9 billion pounds) in 1988 to some 510 million kilograms (1,124 million pounds) in 1993. The Monreal Protocol agreement has also addressed the use of HYDROCHLOROFLUROCARBONS (HCFCs), the replacement chemical for CFCs. Although HCFCs are less harmful to the ozone layer than CFCs, they will be banned by the year 2040 or earlier. *See also* OZONE DEPLETION; OZONE HOLE.

EnviroSource

United Nations Environment Programme. *Montreal Protocol Handbook.* Center for International Earth Science Information Network. Ozone Depletion and Global Environmental Change.

mortality rate. *See* death rate.

Mount Pinatubo, A VOLCANO in the Philippines that erupted in 1991, producing a large AEROSOL cloud that lowered the mean global temperature during the Northern Hemisphere winter by about 0.5°C (0.9°F) as the cloud became evenly distributed throughout the stratosphere. The eruption of Mount Pinatubo emitted particles into the ATMOSPHERE over a wide area, resulting in brilliant sunsets and blocked sunlight. Scientists believe that large decreases in the amount of OZONE in the atmosphere in 1992 and 1993 were related, in part, to the eruption of Mount Pinatubo. The eruption produced large amounts of stratospheric SULFATES that temporarily accelerated OZONE DEPLETION caused by human-made CHLORINE and bromine compounds. During 1992, Mount Pinatubo erupted several more times, releasing ASH that buried entire villages and destroyed roads and bridges. More than 850 people died; many thousands of people were left homeless. More lives would have been lost had it not been for the predictions made by scientists who had been studying this once-dormant volcano. *See also* KRAKATOA; NATURAL DISASTERS

EnviroSource

U.S. Geological Survey Web site: geology.usgs.gov/whatsnew.html.

MTBE (methyl tertiary butyl ether), An oxygenate additive in GASOLINE that was developed by PETROLEUM companies in the 1970s to replace LEAD additives and reduce AUTOMOBILE exhaust EMISSIONS. The use of MTBE is one of several methods used to reformulate gasoline for lower emissions.

MTBE has improved gasoline combustion and has reduced SMOG and AIR POLLUTION in large cities where there are an exceptional number of automobiles, such as Los Angeles County, California. However, although MTBE is making the air cleaner, it is becoming an environmental problem in GROUNDWATER that supplies many municipalities with drinking water. The danger to humans is unknown, but studies have linked MTBE to liver and kidney tumors in mice.

Water officials in several states have reported that MTBE leaking from gas stations' underground FUEL tanks has been detected in varying amounts in groundwater. As a result of the MTBE contamination, several public drinking WELLS have been closed.

MTBE has an easily detectable odor—it smells like turpentine. Local, state, and federal investigations have initiated studies into the problem. Many environmentalists and water quality experts are asking for a ban on the use of MTBE. As of early 2000, some bans have been established. California banned MTBE, and New York state will ban MTBE in 2004. *See also* ALTERNATIVE FUELS, FUEL CELLS.

Muir, John (1838–1914), A Scottish-American NATURALIST and conservationist, who helped found the SIERRA CLUB and was largely responsible for the establishment of many national parks in the United States, including Sequoia National Park, YOSEMITE NATIONAL PARK, Rainier National Park, and Grand Canyon National Park. John Muir was born in Dunbar, Scotland, and moved with his family to the United States in 1849.

Muir's work as a naturalist began in 1868 after he moved to California's Yosemite Valley. One of his great passions was hiking, during which he studied wild animal and PLANT life. He was interested also in the waterways and spectacular rock formations he saw in Yosemite. Muir became convinced that these unusual land forms were carved by GLACIERS. Although Muir's ideas were controversial at the time, most scientists today accept his view of glacial action. (*See* erosion.)

Muir's glacial studies were his main scientific contribution. The rest of his life was devoted primarily to travel, writing, and environmental activism. In 1889, Muir helped focus public attention on the negative impacts of sheep herding and grazing in Yosemite Valley. Several of his articles published in *Century* magazine that year helped pressure Congress into establishing Sequoia and Yosemite National Parks in 1890. In 1892, Muir founded the Sierra Club, a nonprofit CONSERVATION and outdoors organization dedicated to the preservation of America's WILDERNESS and WILDLIFE.

Muir's environmental activism continued through his later years. In 1903, Muir and President THEODORE ROOSEVELT went on a camping trip to the Yosemite region. Muir convinced

John Muir was a Scottish-American naturalist and conservationist who helped found the Sierra Club and was largely responsible for the establishment of many national parks in the United States. *Credit:* The Bancroft Library, University of California at Berkeley California Faces Collection.

Roosevelt about the need for conservation throughout the country. Later that year, Roosevelt added millions more acres to the list of federally protected FORESTS and established what would become the NATIONAL WILDLIFE REFUGE SYSTEM. In 1908, the U.S. government honored Muir by establishing the Muir Woods National Monument in Marin County, California. *See also* ABBEY, EDWARD; AUDUBON, JOHN JAMES; BENNETT, HUGH HAMMOND.

EnviroSources

Browning, Peter, ed. *John Muir in His Own Words: A Book of Quotations.* Lafayette, CA: Great West Books, 1988.

Fox, Stephen. *John Muir and His Legacy: The American Conservation Movement.* Boston: Little Brown, 1981.

Muir, John. *John Muir: The Eight Wilderness Discovery Books.* Seattle: Mountaineers Books, 1992.

Turner, Frederick. *Rediscovering America: John Muir in His Time and Ours.* New York: Viking, 1985.

Wadsworth, Ginger. *John Muir: Wilderness Protector.* Minneapolis: Lerner, 1992.

Wilkins, Thurman. *John Muir: Apostle of Nature.* (Oklahoma Western Biographies, vol. 8.) Norman, OK: University of Oklahoma Press, 1996.

mulch, Any inorganic or ORGANIC material that is placed on the ground to protect the roots of PLANTS from extreme moisture or temperature changes that could harm the plant. The spreading of mulch also reduces soil EROSION and prevents the growth of weeds—unwanted plants that compete with cultivated plants for resources such as sunlight, nutrients, and living space. (*See* competition.)

Common inorganic mulches include rocks, vermiculite, and polyethylene film (a plastic). Fine loose soil, called dust mulch, is used to protect plant roots from drying out through evaporation. Plants use the nutrients provided through the decay of organic mulches to meet their growth needs. The leaf litter (fallen leaves and branches) common on FOREST floors is a natural organic mulch. As this material decays, it forms a nutrient-rich soil called HUMUS. Other organic mulches include sawdust, grass and leaf clippings, PEAT moss, straw, animal manure, and shredded tree bark or wood chips. These mulches are sometimes called stubble mulch. *See also* AGRICULTURE; COMPOSTING; DECOMPOSER; DECOMPOSITION; DETRITUS; DROUGHT.

multiple chemical sensitivity, A condition, also referred to as "chemical hypersensitivity" or "environmental illness," in which a person is intolerant of or sensitive (as distinct from allergic) to a number of chemicals and other irritants, all at very low concentrations. These irritants may include both recognized POLLUTANTS, such as tobacco smoke and formaldehyde, and other substances ordinarily considered innocuous, such as mold. Symptoms of multiple chemical sensitivity (MCS) generally involve more than one body organ, and manifestations are subjective, with no clinical evidence of organ damage. One listing of symptoms for MCS cites more than 100 common problems, ranging from headaches to dizziness to sleep disturbances.

Some who accept the validity of MCS believe it may explain some chronic conditions such as certain forms of arthritis and colitis, in addition to widely recognized types of hypersensitivity reactions, such as rashes, sneezing, and watering eyes. MCS has become more widely known and increasingly controversial as more patients receive the label. Definition of the condition is elusive; its pathogenesis has not been confirmed; and diagnostic criteria have not been established.

There are differing views among medical professionals about the existence, causes, diagnosis, and treatment of MCS. Some practitioners believe the condition has a purely psychological basis. For example, one study reported a 65% incidence of current or past clinical depression, anxiety disorders, or similar illnesses in subjects diagnosed with MCS, compared with 28% in the control group. Other researchers contend that MCS itself may cause such problems, since affected individuals can no longer lead normal lives, or that these conditions stem from the effects of MCS on the nervous system. Current consensus is that in cases of claimed or suspected MCS, complaints should not be dismissed as psychogenic and a thorough physical workup is essential. Primary-care health professionals should determine that the individual does not have an underlying physiological problem and should consider consultation with allergists and other specialists, since controlling primary ALLERGIES can also reduce reactions to multiple irritants. *See also* DISEASE; SICK BUILDING SYNDROME.

EnviroSources

Moeller, Dade W. *Environmental Health.* Cambridge, MA: Harvard University Press, (1997.)

Pope, Andrew M. *Environmental Medicine: Integrating a Missing Element into Medical Education.* Washington, DC: National Academy Press, 1995.

Vos, J.G. *Allergic Hypersensitivities Induced by Chemicals: Recommendations for Prevention.* Cleveland, OH: CRC Press, 1996.

U.S. Environmental Protection Agency (Indoor air quality) Web site: www.epa.gov/iedweb00.

multiple use, A land management policy that permits a combination of diverse recreational activities such as camping, fishing, and hiking along with logging, MINING, and scientific and historical exploration. According to the BUREAU OF LAND MANAGEMENT, multiple use takes into account long-term needs of future generations for renewable and NONRENEWABLE RESOURCES. *See also* NATURAL RESOURCE; RENEWABLE RESOURCE.

mutagen, Any agent—matter or ENERGY—that can cause a permanent change in the genes or chromosomes of an ORGANISM, sometimes resulting in new traits, new SPECIES, or DISEASE. A change caused by a mutagen is called a MUTATION. Because mutagens affect the genetic material of an organism, changes they cause can be passed to the offspring of the affected organism, as well as to subsequent generations of offspring.

Understanding mutagens is important to environmental scientists and ecologists because of the potential effects of mutagens on individuals and entire species. Some mutagens cause harmful mutations to an organism and its offspring. If such changes are significant, they can alter POPULATIONS enough to bring about the EXTINCTION of a species. Changes that are significant and beneficial to an organism and its offspring can lead to the EVOLUTION of a new species through natural selection.

Changes caused by mutagens are usually observable only when they affect the physical appearance or physical structures of an organism or alter its biochemistry in a way that changes the organism's physiology or behavior. One way in which mutagens may affect an organism's physiology is by causing disease. For example, many chemical substances and forms of energy that have been identified as mutagens are carcinogenic, or CANCER-causing. (*See* carcinogen.) DIOXIN, a byproduct formed during the manufacturing of various industrial and chemical substances such as herbicides and plastics, is a mutagenic chemical that has been linked to birth defects and cancer in animals. Many forms of RADIATION also act as mutagens. For example, the ultraviolet-B radiation emitted by the sun is a mutagen that has been linked to skin cancer. Other forms of radiation, such as X-RAYS, also have been linked to the development of cancers and birth defects. *See also* ADAPTATION; AGENT ORANGE; BIODIVERSITY; ENVIRONMENTAL MEDICINE; LOVE CANAL; MERCURY; PESTICIDE; RADIATION SICKNESS; VIRUS.

mutation, An alteration in the genetic material (genes or chromosomes) of an ORGANISM that has the potential to change traits of that organism or its offspring. Most mutations are neutral or harmful to an organism. Mutations are neutral when they do not cause any observable alterations in an organism's traits. Some mutations may provide an organism traits with that are beneficial. When beneficial mutations occur, they may be passed to offspring and over time be spread throughout a POPULATION, possibly leading to the development of new SPECIES. (*See* adaptive radiation.) Causes for mutations include environmental factors such as exposure to forms of RADIATION, certain chemicals, or extremely high temperatures. Mutations may also result from errors during the replication of an organism's deoxyribonucleic acid (DNA) prior to CELL division. DNA is the chemical substance present in the cells of an organism that is responsible for the organism's traits. The material also is responsible for carrying traits from parent to offspring.

Mutations may affect either the genes or the chromosomes of an organism. Mutations affecting genes are more likely to be harmful to the organism, and if the organism survives and reproduces, such mutations are more likely to be passed to offspring. Most gene mutations are recessive (expressed only when present on both genes that carry that specific trait). Thus, for such mutations to be expressed, an organism must inherit the mutation from each of its parents. Such an occurrence is most likely to take place during inbreeding, the mating of closely related organisms.

Chromosome mutations often occur when a nucleotide (the fundamental compound of which DNA is made that is composed of a nitrogen base linked to a sugar) is entirely lost or added during DNA replication. Other mutations in chromosomes occur when a section of the chromosome detaches from its strand, becomes inverted, and then reattaches to the original chromosome strand (a condition known as chromosome inversion), or attaches to a different part of the original chromosome or an entirely different chromosome (a condition known as translocation). Another type of chromosome mutation occurs when a portion of a chromosome is lost from one of the two chromosomes of a homologous (matching) pair and gained by the other chromosome. The chromosome that is incomplete is described as having a deficiency; the chromosome with the additional piece has a duplication. In many cases, chromosome deficiencies and duplications result in conditions in organisms that prevent their survival.

Another common type of chromosome mutation results when homologous chromosomes do not separate during meiosis, the process by which gametes (sperm or egg) are formed in organisms that reproduce sexually. When this happens, the resulting gametes, as well as any offspring that result from them, either lack a chromosome or have an extra chromosome. For example, humans normally have 23 chromosomes in their gametes and 23 matching pairs of chromosomes in their body cells. If the chromosomes in the gametes do not replicate properly, the cell formed when a sperm and egg unite may have only one chromosome of the matching pair—a condition called monosomy—or may gain an extra chromosome—a condition known as trisonomy. Monosomy and trisonomy generally result in offspring that have serious health problems. For example, in humans, Down's syndrome is a common condition resulting from trisonomy, in this case an extra twenty-first chromosome.

Mutations may also result when chromosomes replicate, but then do not separate during meiosis. When this occurs, a gamete with twice the normal number of chromosomes is produced. If this gamete fuses with a gamete that contains the normal number of chromosomes during fertilization, the resulting offspring will have three homologous sets of chromosomes rather than the normal two. If two gametes having twice the normal number of chromosomes fuse, the resulting offspring will have four homologous sets of chromosomes. Organisms that have additional sets of chromosomes are known as polyploids. When polyploidy occurs, a new species may arise in only one generation. Polyploidy most often results in flowering PLANTS or INVERTEBRATES that are capable of self-fertilization (hermaphrodites). Because they have traits consistent with their parent organisms, polyploids generally are adapted to their environments and often are fertile. Thus, they are capable of producing new polyploid offspring. In plants, polyploids often are larger and hardier than the diploid (organisms having two complete sets of chromosomes) plants from which they formed. The characteristics of polyploids make them desirable in AGRICULTURE. Today, many species of oats, potatoes, wheat, tobacco, and cotton grown on farms throughout the world are polyploids. *See also* ADAPTATION; CANCER; CARCINOGEN; EVOLUTION; GENETIC ENGINEERING; GENOME; HYBRIDIZATION; MUTAGEN; SPECIESISM.

EnviroTerm

hermaphrodite: An ORGANISM that has functional male and female reproductive structures.

mutualism, A symbiotic association between two ORGANISMS of different SPECIES through which both species derive some benefit. (*See* symbiosis.) The relationship between the mutualistic species, called symbionts, is often, though not always, obligatory. An obligatory relationship exists when the symbionts cannot survive without the interaction their mutualism provides. Thus, if one of the organisms is removed from the ENVIRONMENT, the other perishes as well. An example of such a relationship is seen in LICHENS. Lichens are an association between FUNGI and either green ALGAE or CYANOBACTERIA. In this relationship, the fungi provide the algae or cyanobacteria with shelter and the raw materials (CARBON DIOXIDE and water) needed for PHOTOSYNTHESIS. The algae or cyanobacteria provides the fungi with a source of nutrients. Because the association between the organisms in a lichen is obligatory, many scientists consider the lichen a single organism.

Many PLANTS have a mutualistic relationship with FUNGI. MYCORRHIZAL FUNGI benefit by absorbing nutrients made by the plant through PHOTOSYNTHESIS; the plants benefit as the fungi decompose ORGANIC matter in soil, which the plants absorb as nutrients. (*See* decomposition.) In addition, the hyphae (threadlike structures) of the fungi increase the surface area of the plant's roots, enhancing the plant's ability to absorb water and nutrients in soil. Plants having mycorrhizae on their roots generally grow larger than those lacking the association. The digestive enzymes produced by the fungi also seem to make such plants more resistant to DISEASE than plants lacking the association. *See also* COMMENSALISM; PARASITISM.

mycorrhizal fungi, Certain types of FUNGI that live in a mutualistic association with PLANTS, which provides benefit to both ORGANISMS. (*See* mutualism.) The fungi, which live on the plant's roots are provided with nutrients released by the plant into the soil as the plant carries out PHOTOSYNTHESIS. In this association, the hyphae, or threadlike structures that form the "body" of the fungus, become intertwined with the plant's roots, increasing their surface area. As a result, access of the roots to soil water and the dissolved nutrients needed by the plant for growth and maintenance also increase. Studies have shown that plants living in association with mycorrhizae tend to grow more rapidly, and to be stronger and more DISEASE-resistant than those that lack the association. *See also* COMMENSALISM; COMPETITION; PARASITISM; PREDATION; SYMBIOSIS.

nanometer, A distance of one billionth of a meter (SI unit of the length 10^{-9}). Nanometers (nm) are used to measure wavelengths of ultraviolet (UV) RADIATION, which is in sunlight. The entire UV radiation spectrum falls into four categories: UV-A (320–400 nm), UV-B (280–320 nm), UV-C (200–280 nm), and UV-D (less than 200 nm). UV-B radiation is the most dangerous because it is not blocked by the OZONE layer. Skin CANCER may result if humans are overexposed to sun.

Nanometers are also used to measure wavelengths of light or other electromagnetic radiation. For example, green light has wavelengths of about 500–550 nm, while violet light has wavelengths of about 400–450 nm.

National Ambient Air Quality Standards, Measurable levels of POLLUTANTS permissible in the outside AIR throughout the United States as established by the U.S. ENVIRONMENTAL PROTECTION AGENCY (EPA). The National Ambient Air Quality Standards (NAAQS) was a provision of the 1970 amendments to the CLEAN AIR ACT. As part of the NAAQS, the EPA identified and established standards designed to protect human health and welfare for six pollutants known as CRITERIA POLLUTANTS. The six pollutants are OZONE, CARBON MONOXIDE, SULFUR DIOXIDE, LEAD, NITROGEN OXIDE, and suspended particulates. (*See* particulate matter.) All these pollutants are common EMISSIONS of power plants that use FOSSIL FUELS to generate ELECTRICITY, AUTOMOBILES, and industries that manufacture products from PETROCHEMICALS.

Monitoring areas to determine whether they are meeting the NAAQS is carried out by the Office of Air and Radiation (OAR), a division of the EPA, and the data are published annually in the *National Air Quality and Emission Trends Report.* Under the provisions of the EMERGENCY PLANNING AND COMMUNITY RIGHT-TO-KNOW ACT of the United States, this publication is available to the public upon request. *See also* ACID RAIN; AEROSOL; AFTERBURNER; AIR POLLUTION; AIR POLLUTION CONTROL ACT; ASH; GLOBAL WARMING; GLOBAL WARMING POTENTIAL; GREENHOUSE EFFECT; GREENHOUSE GAS.

EnviroSource

U.S. Environmental Protection Agency, Office of Air and Radiation Air Quality Trends Analysis Group: Research Triangle Park, North Carolina 27711 Web site: www.epa.gov/docs/oar/oarhome.html.

National Audubon Society, A nonprofit, national organization founded in 1905 to conserve and restore natural ECOSYSTEMS, focusing on birds and other WILDLIFE, for the benefit of humanity and to maintain Earth's BIODIVERSITY. The National Audubon Society is named for JOHN JAMES AUDUBON, renowned ornithologist, explorer, NATURALIST, and wildlife artist. It has 550,000 members, with 508 chapters in North America, and maintains 100 sanctuaries and nature centers nationwide. The largest sanctuary is the 10,525 hectare (26,000 acre) Paul J. Rainey Sanctuary in Louisiana, which was acquired in 1924.

Wildlife faced a bleak future at the beginning of the twentieth century. There were no laws to control the hunting of birds and other animals. Birds were slaughtered by the millions, their plumes used as hat decorations, their nests robbed for eggs, and many SPECIES hunted for food. Entire species, such as the PASSENGER PIGEON, the Carolina parakeet, and the great auk, were driven to EXTINCTION. In the late 1800s, citizens banded together in outrage against the mass slaughter of these animals, eventually forming state Audubon Societies. By 1899, there were 17 such societies. That same year, Frank Chapman, an ornithologist with the American Museum of Natural History, began publishing *Bird Lore* magazine, which became a unifying forum for the Audubon movement. In 1901, some of the Audubon Societies formed a loose alliance called the National Committee of the Audubon Societies; in 1905, they incorporated into the National Association of Audubon Societies for the Protection of Wild Birds and Animals. Eventually, the name was shortened to its present form.

In the early 1900s, Audubon members worked to have important CONSERVATION laws passed, including the New York State Audubon Plumage Law of 1910, which banned the sale of plumes of all NATIVE SPECIES of birds in the state, and the MIGRATORY BIRD TREATY ACT of 1918. The National Audubon

Society also encouraged the federal government to protect critical HABITATS by including them in a NATIONAL WILDLIFE REFUGE SYSTEM.

The society has acted as a leader within the overall twentieth-century environmental movement in the United States. In the 1930s, it began sponsoring scientific research projects on ENDANGERED SPECIES of birds, such as the ivory-billed woodpecker and the roseate spoonbill. During the 1970s and 1980s, the National Audubon Society was a prominent presence when the nation's landmark environmental laws were enacted. For example, Audubon staff and members worked for passage of the CLEAN AIR ACT, the CLEAN WATER ACT, the WILD AND SCENIC RIVERS ACT, and the ENDANGERED SPECIES ACT.

Today, Audubon scientists use sanctuaries to study the effects of environmental degradation on ECOSYSTEMS. The society is still guided by the ethic that citizens must work toward preserving their natural heritage. To help achieve this goal, the organization publishes the bimonthly magazine *Audubon,* which provides a diversity of articles about various species that face extinction and those whose POPULATIONS are growing, as well as updates on the society's preservation activities, among other issues. The organization sponsors environmental education programs for 500,000 children a year through its Audubon Adventures program. Other major initiatives for the 1990s included protecting OLD-GROWTH FORESTS in the Pacific Northwest, preventing PETROLEUM development in the ARCTIC NATIONAL WILDLIFE REFUGE, preserving WETLANDS, and reauthorizing the Endangered Species Act. *See also* ADAMSON, JOY GESSNER; ADAPTATION; ADDO NATIONAL ELEPHANT PARK; AFRICAN WILDLIFE FOUNDATION; ALASKA NATIONAL INTERESTS LANDS CONSERVATION ACT; AMERICAN ZOO AND AQUARIUM ASSOCIATION; ANIMAL RIGHTS; BALD EAGLES; BIODIVERSITY TREATY; BIOMONITORING; BIOREGION; CALIFORNIA CONDORS; CAPTIVE PROPAGATION; CARARA BIOLOGICAL RESERVE; CARSON, RACHEL LOUISE; DICHLORODIPHENYLTRICHOROETHANE; DODO BIRDS; DOUGLAS, MARJORY STONEMAN; ENDANGERED SPECIES LIST; EVERGLADES NATIONAL PARK; FISH AND WILDLIFE SERVICE; GREENPEACE; LACEY ACT; LEOPOLD, ALDO; MAN AND THE BIOSPHERE PROGRAMME; MUIR, JOHN; OWLS, NORTHERN SPOTTED; PARROTS; PEREGRINE FALCONS; RED LIST OF ENDANGERED SPECIES; SIERRA CLUB; SPECIES DIVERSITY; WILD BIRD CONSERVATION ACT; WILDLIFE REFUGE; WORLD CONSERVATION MONITORING CENTRE; WORLD CONSERVATION UNION.

EnviroSource

National Audubon Society: 700 Broadway, New York, NY 10003; (212) 979-3000; Web Site: www.audubon.org.

National Emissions Standards for Hazardous Air Pollutants, Guidelines set forth by the U.S. ENVIRONMENTAL PROTECTION AGENCY under the provisions of the CLEAN AIR ACT that set limits on the amounts of specific POLLUTANTS that may be released into the AIR. Exposure to the air pollutants regulated under this provision is known or suspected to result in serious health problems. These pollutants include both chemical agents and radioactive substances that are derived from natural sources such as soil and human sources such as motor vehicles and factories. *See also* AIR POLLUTION; NATIONAL AMBIENT AIR QUALITY STANDARDS; RADIATION.

EnviroSource

U.S. Environmental Protection Agency, Office of Air and Radiation Web site: www.epa.gov/radiation/neshaps.

National Environmental Policy Act, Legislation enacted by Congress in 1969 that requires the government to consider how projects it may undertake could affect the quality of the ENVIRONMENT. The National Environmental Policy Act (NEPA) was the first general law of the United States that required government to consider the environmental consequences of construction involving federal funds. To achieve this consideration, the NEPA requires that ENVIRONMENTAL IMPACT STATEMENTS (EIS) be prepared and filed for any project funded with federal dollars. In addition, the NEPA also established the COUNCIL ON ENVIRONMENTAL QUALITY (CEQ), an independent agency that is charged with the responsibility of advising the president of the United States on matters regarding environmental policy. The CEQ also helps to establish guidelines intended to support the NEPA, such as those dictating what information must be included as part of an EIS, and aids in the preparation of an annual report on the quality of the environment of the United States. *See also* ENVIRONMENTAL JUSTICE; ENVIRONMENTAL PROTECTION AGENCY.

National Estuary Program, Formed by an amendment to the CLEAN WATER ACT to identify, restore, and protect nationally significant ESTUARIES of the United States. Founded in 1987, the program is administered by the U.S. ENVIRONMENTAL PROTECTION AGENCY (EPA). Estuaries are selected for inclusion through a nomination process. The governor of the state where the estuary is located submits a nomination to the EPA.

The program focuses on maintaining the chemical, physical, and biological properties along with economic, recreational, and aesthetic values of estuaries. The program is managed by representatives from federal, state, and local government agencies organized into local programs to manage each of the 28 estuaries. The objectives of the programs are designed to identify problems in each estuary, develop specific actions, and create and implement a formal management plan to restore and protect the estuary. Some of the primary environmental problems of an estuary might include conventional POLLUTANTS, human POPULATION growth, HABITAT loss or alteration, loss or decline in SPECIES, introduced pest species, problems with the quantity of fresh-water inflow, red or brown tides, and insufficient, polluted, or saline drinking water.

At the local level, the structure of each NEP program usually includes a management committee to oversee routine operations of the program, a policy committee, a technical advisory committee, and a citizens advisory committee. The committees develop a comprehensive conservation and management plan, which addresses the whole range of environmental problems facing the estuary. *See also* NATIONAL PARK SERVICE.

National Estuary List of Programs

Albemarle-Pamlico Sounds, North Carolina
Barataria-Terrebonne Estuarine Complex, Louisiana
Barnegat Bay, New Jersey
Buzzards Bay, Massachusetts
Casco Bay, Maine
Charlotte Harbor, Florida
Lower Columbia River Estuary, Oregon and Washington
Corpus Christi Bay, Texas
Delaware Estuary, Delaware, New Jersey, and Pennsylvania
Delaware Inland Bays, Delaware
Galveston Bay, Texas
Indian River Lagoon, Florida
Long Island Sound, New York and Connecticut
Maryland Coastal Bays, Maryland
Massachusetts Bays, Massachusetts
Mobile Bay, Alabama
Morro Bay, California
Narragansett Bay, Rhode Island
New Hampshire Estuaries, New Hampshire
New York-New Jersey Harbor (Harbor Estuary Program), New York and New Jersey
Peconic Bay, New York
Puget Sound, Washington
San Francisco Estuary, California
San Juan Bay, Puerto Rico
Santa Monica Bay, California
Sarasota Bay, Florida
Tampa Bay, Florida
Tillamook Bay, Oregon

EnviroSource

U.S. Environmental Protection Agency, National Estuary Program Web site: www.epa.gov/nep/nep.html.

National Marine Fisheries Service (NMFS), A part of the NATIONAL OCEANIC AND ATMOSPHERIC ADMINISTRATION (NOAA) that supports the CONSERVATION and management of living marine resources. One of their services is to protect SPECIES and HABITATS. In 1999, for example, the National Marine Fisheries Service (NMFS) added nine POPULATIONS of SALMON and steelhead in Washington and Oregon, including metropolitan Portland and Seattle, to the ENDANGERED SPECIES LIST. This was the first time federal protection has been extended to salmon living in streams in heavily populated areas of the Pacific Northwest.

The agency, originally called the independent Office of the Commissioner of Fish and Fisheries, was founded in 1871 by President Ulysses S. Grant. The commission was charged with studying and recommending solutions to the declining fish populations in lakes and along the coasts. Fish were important food products. The first commissioner was Spencer Fullerton Baird, an internationally acclaimed scientist. Baird initiated the country's first marine ecological studies, which analyzed the temperature of the OCEAN water at different depths, as well as its varying transparency, density, CHEMICAL composition, percentage of saline matter, surface, and undercurrents. Later on, Commissioner Baird supervised construction of the first federal fishery research laboratory at the Woods Hole Oceanographic Institute of Massachusetts in 1885.

The NMFS also provides services and products to support fisheries management operations, trade and industry assistance activities, enforcement of the protection of species, and the scientific aspects of NOAA's marine fisheries program.

EnviroSources

National Marine Fisheries Service Web sites: www.nmfs.gov.; kingfish.ssp.nmfs.gov.

National Oceanographic and Atmospheric Administration (NOAA), National Marine Fisheries Service (history of National Marine Fisheries Service) Web site: www.wh.whoi.edu/125th/125th.html.

National Oceanic and Atmospheric Administration (NOAA), An organization of the U.S. Department of Commerce founded in the 1970s and composed of several organizations such as the NATIONAL WEATHER SERVICE; the National Environmental Satellite, Data, and Information Service; the NATIONAL MARINE FISHERIES SERVICE; the National Ocean Service; and the Office of Oceanic and Atmospheric Research. All of these organizations provide a vast number of services.

National Marine Fisheries Service is a major force in protecting marine SPECIES around the globe. As an example, endangered right WHALES that live in the Atlantic Ocean off New England are individually tracked by satellite to help maintain these fragile stocks. To help ensure sustainable fish yields, NOAA scientists study the life history, stock size, and ECOLOGY of economically important fishes, and the effects of CLIMATE and OCEAN processes on their POPULATIONS.

The NOAA issues warnings regarding impending HURRICANES, heat waves, TORNADOS, winter storms, FLOODS, and wildfires. It develops global climate models to help scientists study climates and provides an EL NIÑO web site on the Internet. The NOAA has provided coastal navigation services, revolutionizing marine transportation to help move increased cargoes and passengers safely and efficiently.

The organization employs meteorologists, hydrologists, cartographers, oceanographers, fishery managers, physicists, computer scientists, engineers, technicians, law enforcement agents, support personnel, and others. These people work around the world on ships and aircraft, in undersea HABITATS, and at laboratories and WEATHER stations. The organization's goal is to make all environmental data easily accessible. The possible benefits of the dissemination of environmental information include saving billions of dollars, facilitating an easy transition to a global free market, and increasing knowledge at all levels. *See also* HYDROLOGY; METEOROLOGY; OCEANOGRAPHY.

EnviroSources

National Oceanographic and Atmospheric Administration Web site: www.noaa.gov.

National Oceanographic and Atmospheric Administration (climate) Web site: www.cdc.noaa.gov/Seasonal.
National Oceanographic and Atmospheric Administration (El Niño) Web site: www.pmel.noaa.gov/toga-tao/el-nino/nino-home-low.html.
National Oceanographic and Atmospheric Administration (recover protected species) Web site: www.noaa.gov/nmfs/recover.html.
National Oceanographic and Atmospheric Administration (safe navigation) Web site: www.anchor.ncd.noaa.gov/psn/psn.htm.

National Park Service, A federal bureau within the U.S. DEPARTMENT OF THE INTERIOR (DOI) that is charged with the responsibility of managing the historic and natural scenic areas and WILDLIFE included within the national park system. The National Park Service (NPS) was established in 1916 through passage of the National Park Service Act and was immediately placed in charge of managing the natural historic objects and wildlife within the lands set aside as national parks of the United States. Among these parks were YELLOWSTONE NATIONAL PARK, which had been established in 1872, and Yosemite and Sequoia National Parks, which had been designated as national parks several years later.

Today, more than 350 different areas fall under the management jurisdiction of the NPS, including national parks, national wildlife preserves, national seashores, national lakeshores, and those memorials, battlefields, and cemeteries that have been deemed to have national historic value. The areas support a large tourism industry. Almost 300 million tourists per year visit these areas because of their historic significance, natural scenic beauty, or WILDLIFE attractions. (*See* ecotourism.) As part of their mission, the NPS is charged with encouraging such visitation, while also overseeing recreational activities to ensure public safety and preservation of the natural ENVIRONMENT, including protection of native FLORA and FAUNA. To meet these challenges, the NPS has begun to initiate policies designed to help lessen the effects of AIR, water, land, and NOISE POLLUTION within its park areas by limiting the number of AUTOMOBILES and other motor vehicles that may enter park grounds and banning the use of some recreational vehicles such as motor boats, personal watercraft, all terrain vehicles, and snowmobiles within some park environments. In addition, logging, MINING, and hunting has been banned in some park areas, while monitoring livestock grazing and fishing has increased in park areas that permit such activities. *See also* ALASKA NATIONAL INTERESTS LANDS CONSERVATION ACT; ARCTIC NATIONAL WILDLIFE REFUGE; BIOSPHERE RESERVE; BISON; CACTI; CALIFORNIA CONDORS; COASTAL ZONE MANAGEMENT ACT; CONSERVATION; DOUGLAS, MARJORY STONEMAN; EVERGLADES NATIONAL PARK; FLORIDA PANTHERS; LEOPOLD, ALDO; MAN AND THE BIOSPHERE PROGRAMME; MARSH, GEORGE PERKINS; MUIR, JOHN; NATIONAL ESTUARY PROGRAM; NATIONAL WILDLIFE REFUGE SYSTEM; ROOSEVELT, THEODORE; TRAGEDY OF THE COMMONS; WILD AND SCENIC RIVERS ACT; WILDERNESS; WILDERNESS ACT OF 1964; WILDLIFE REFUGE; YOSEMITE NATIONAL PARK.

The Old Faithful geyser is in Yellowstone National Park. *Credit:* National Park Service.

EnviroSources

National Parks and Conservation Association: 1776 Massachusetts Ave., NW, Washington, DC 20036.
National Park Service Web site: www.nps.gov.

National Pollutant Discharge Elimination System, A program used to protect human health and the ENVIRONMENT under a provision of the Federal Water Pollution Control Act, also know as the CLEAN WATER ACT. The CLEAN WATER ACT requires that all POINT SOURCES discharging POLLUTANTS into waters of the United States must obtain an National Pollutant Discharge Elimination System (NPDES) permit. By point sources, the ENVIRONMENTAL PROTECTION AGENCY (EPA) means discrete conveyances such as pipes or human made ditches.

Although individual households do not need permits, facilities such as WASTEWATER plants must obtain permits if their DISCHARGES go directly to SURFACE WATERS. A pollutant includes any type of industrial, municipal, and AGRICULTURAL waste discharged into water. Some pollutants that may threaten public health and the nation's waters are: human wastes, ground-up food from sink disposals, laundry and bath waters, TOXIC CHEMICALS, oil and grease, metals, and PESTICIDES. Other examples of pollutants include SEWAGE SLUDGE, agricultural wastes, and incinerator residue. The permit will contain limits on what the facility can discharge and other provisions to ensure that the discharge does not hurt water quality or people's health.

EnviroSource

Environmental Protection Agency Web site: www.epa.gov/OW.

National Primary Drinking Water Regulations, Part of the SAFE WATER DRINKING ACT of 1974 that established limits on the amounts of various chemicals and forms of ENERGY permitted to be present in public water systems. These limits are intended to protect the public from CONTAMINANTS that pose potential health risks. The National Primary Drinking Water Regulations (NPDWRs) established MAXIMUM CONTAMINANT LEVELS (MCLs) for a variety of substances such as LEAD, MERCURY, etc., that may be present in the drinking water supply or recommended water treatment techniques that can minimize the effects of substances for which MCLs are not feasible. The NPDWRs were established by the U.S. ENVIRONMENTAL PROTECTION AGENCY (EPA), but enforcement of the standards is delegated to individual states.

Included among the substances for which NPDWRs have been established are CADMIUM, cyanide, SELENIUM, inorganic mercury, NITRATES and nitrites (both of which are determined through measures of NITROGEN in water), acrylamide, BENZENE, radionuclides, and MICROORGANISMS, including COLIFORM BACTERIA and *ESCHERICHIA COLI*. All of these substances are known to pose health risks to humans and other ORGANISMS. For example, in relatively small amounts, benzene is highly toxic and carcinogenic, and may cause reproductive disorders. In addition, cadmium and cyanide are both TOXIC CHEMICALS and are known to corrode pipes in household plumbing systems. *See also* ADSORPTION; AGRICULTURAL POLLUTION; ARSENIC; CANCER; CARCINOGEN; CHLORINATION; CLEAN WATER ACT; CRYPTOSPORIDIUM; LANGELIER INDEX; PATHOGEN; PRIMARY, SECONDARY, AND TERTIARY TREATMENTS; RUNOFF; SEWAGE.

EnviroSources

Cronin, John, and Robert F. Kennedy, Jr. *The Riverkeepers: Two Activists Fight to Reclaim Our Environment as a Basic Human Right*. New York: Scribner, 1997.

U.S. Bureau of Reclamation Web site: www.usbr.gov/laws/safedrin.html.

U.S. Environmental Protection Agency, Office of Water Web site: www@pa.gov.OW.

National Priorities List, A list of the most serious uncontrolled or abandoned HAZARDOUS WASTE sites that have been identified for possible long-term remedial action under the COMPREHENSIVE ENVIRONMENTAL RESPONSE, COMPENSATION, AND LIABILITY ACT (also known as SUPERFUND). The responsibility for maintaining the National Priorities List (NPL), as well as developing and managing the cleanup process at sites on the list, rests with the U.S. ENVIRONMENTAL PROTECTION AGENCY (EPA). Before a site can obtain money from the Superfund trust for remedial action, it must be placed on the NPL. The inclusion of a contaminated area on the NPL is based primarily on the score the site is assigned using the HAZARD RANKING SYSTEM. To be listed on the NPL, a site must meet all three of the following requirements:

1. The AGENCY FOR TOXIC SUBSTANCES AND DISEASE REGISTRY of the U.S. Public Health Service has issued a health advisory that recommends evacuating people from the site;
2. The EPA has determined the site poses a significant threat to public health;
3. The EPA has anticipated it will be more cost effective to use its remedial authority (available only at NPL sites) than to use its emergency removal authority to respond to the site.

The EPA is required to update the NPL each year, focusing primary effort on the most serious hazardous waste sites. As of May 1997, 10,653 sites were included on the NPL in the United States and its territories. An additional 30,305 sites were identified as archived sites of the NPL, sites at which no further remedial action was planned. *See also* ARMY CORPS OF ENGINEERS; BROWNFIELDS.

The EPA added 15 sites to the NPL as of July 1999, including the following:

Site Name	*Location*
Alameda Naval Air Station	Alameda, California
Eastland Woolen Mill	Corinna, Maine
Emmell's Septic Landfill	Galloway Township, New Jersey
Former Nansemond Ordnance Depot	Suffolk, Virginia
Hanlin-Allied-Olin	Moundsville, West Virginia
Hart Creosoting Company	Jasper, Texas
Hudson Refinery	Cushing, Oklahoma
Kim-Stan Landfill	Selma, Virginia
Martin Aaron, Inc.	Camden, New Jersey
Mountain Pine Pressure Treating, Inc.	Plainview, Arkansas
Norfolk Naval Shipyard	Portsmouth, Virginia
North Belmont PCE	North Belmont, North Carolina
United States Avenue Burn	Gibbsboro, New Jersey
Vasquez Boulevard and I-70	Denver, Colorado
Vega Baja Solid Waste Disposal	Vega Baja, Puerto Rico

EnviroSource

U.S. Environmental Protection Agency (Superfund) Web site: www.epa.gov/superfund.

National Weather Service, The U.S. WEATHER forecasting agency, which is a division of the NATIONAL OCEANIC AND ATMOSPHERIC ADMINISTRATION (NOAA) and is located in Maryland. The concept of a national weather service was first voiced on February 9, 1870, when President Ulysses S. Grant signed a joint resolution of Congress authorizing the secretary of war to establish such a service; it was initially known as the Weather Bureau. In 1940, President Franklin D. Roosevelt transferred the Weather Bureau to the Department of Commerce. Then in 1970, the name of the bureau was changed to the National Weather Service, and the agency became a component of the Commerce Department's newly created National Oceanic and Atmospheric Administration.

The National Weather Service (NWS) is the primary source of weather and CLIMATE forecasts for the United States and its territories for the protection of life and property. NWS data form a national information database and infrastructure, which can be used by other governmental agencies, the private sector, the public, and the global community. The NWS issues warnings and provides forecasts of hazardous weather, including thunderstorms, FLOODS, HURRICANES, TORNADOES, winter weather, TSUNAMIS, and climate events. The NWS is the sole official U.S. voice for issuing warnings during life-threatening weather situations.

The NWS relies on its partners in community emergency management organizations and the media to issue severe warnings and keep communities safe. The NWS broadcasts continuous weather forecasts, data, and severe weather warnings over a nationwide network of more than 480 NOAA Weather Radio stations. The audio weather forecasts heard over many cable television systems are NOAA Weather Radio broadcasts.

Due to the work of the NWS, today's 3-day and 4-day forecasts are as accurate as the 2-day forecast was 15 years ago. The NWS is working to make the 6- to 10-day forecast as accurate as the forecast for tomorrow. While the NWS uses many state-of-the-art technologies, meteorologists still depend heavily on some traditional sources for important weather data, including radiosondes carried by weather balloons and manual river observations. *See also* METEOROLOGY.

EnviroSource

National Oceanic and Atmospheric Administration, National Weather Service Web site: www.nws.noaa.gov.

National Wildlife Refuge System, A network of federally owned sanctuaries for PLANTS, animals, and other WILDLIFE. Today, the National Wildlife Refuge System of the United States embraces more than 500 such refuges on 37 million hectares (92 million acres) scattered across 50 states and several overseas territories. The primary focus of the National Wildlife Refuge System, which is administered by the U.S. FISH AND WILDLIFE SERVICE, is to conserve and protect the breeding and wintering grounds for migratory waterfowl such as ducks and geese and to preserve HABITATS for ENDANGERED SPECIES. In many places, including the PRAIRIE pothole region, each year thousands of hectares are planted with native grasses, thus improving waterfowl habitat.

In 1903, President THEODORE ROOSEVELT created the first WILDLIFE REFUGE, Florida's Pelican Island, to save herons, egrets, pelicans, and other birds from overhunting. Since then, concerned citizens, private organizations, and government agencies have worked cooperatively to acquire, protect, restore, and manage millions of hectares of land set aside for the CONSERVATION of wildlife POPULATIONS and their habitats. The National Wildlife Refuge System has been a successful program. In the 1930s, it was discovered that the whooping crane, a large and beautiful migratory bird, had been hunted to near EXTINCTION. Scientists had estimated that there were fewer than 50 birds left in both Canada and the United States. In 1937, the Arkansas National Wildlife Refuge was created in Texas to protect the wintering grounds of whooping cranes. Working with the Canadian government, wildlife managers were also able to establish protection for the birds on their breeding grounds in northern Canada and along their 4,000-kilometer (2,500-mile) migration route through North America. Whooping cranes remain an endangered species in North America today; however, their population is slowly growing.

The main purpose of a wildlife refuge is to provide sanctuary for wildlife; however, over the years, Congress has allowed so many different uses of these sanctuaries—including grazing, farming, logging, and MINING—that critics fear some parts of the system have become too public. The most immediate issue confronting the refuge system today is: What are refuges for? As long as the goal of wildlife conservation was not compromised, other uses have always been a part of the refuge formula. The Fish and Wildlife Service has had no quarrel with hunting and fishing or other outdoors recreation on refuges where these uses are appropriate. Maintaining a protected habitat for animals, while at the same time allowing the land to be used for human activities, presents wildlife managers with the problem of balancing the needs of the economy and the needs of the environment. *See also* DOUGLAS, MARJORY STONEMAN; ENDANGERED SPECIES ACT; MIGRATORY BIRD TREATY ACT; MULTIPLE USE.

EnviroSource

U.S. Fish and Wildlife Service, National Wildlife Refuge System Web site: www.refuges.fws.gov/NWRSHomePage.html.

Nationwide 26, A permit program of the U.S. ARMY CORP OF ENGINEERS that allowed up to 10 acres of WETLANDS to be filled without an individual permit application and without advance public notice. Environmentalists believe that this permit program allowed the destruction of thousands of hectares of valuable wetlands each year. Because of protests from environmental groups and wetland activists, the Army Corps of Engineers agreed to rescind Nationwide 26 in 1998.

native species, A SPECIES that evolved, or originated, in a particular area. Species that live in certain locations for long periods of time become well adapted to their surroundings. Through the process of natural selection, species evolve ADAPTATIONS that help them survive within their BIOLOGICAL COMMUNITIES and interact with the ABIOTIC factors of their ENVIRONMENT. EXOTIC SPECIES, species that are introduced to new locations, can sometimes disrupt stable ECOSYSTEMS by introducing new conditions (such as COMPETITION or DISEASE) to an ecosystem to which native species are not adapted. These conditions may make it impossible for native species to coexist with newly introduced species that might be better equipped to gather important resources. *See also* ADAPTIVE RADIATION; BIOLOGICAL CONTROL; DARWIN, CHARLES ROBERT.

natural disaster, A geological or WEATHER-related event, such as a FOREST FIRE, FLOOD, or EARTHQUAKE, that kills and injures people and other ORGANISMS, destroys personal property, damages the ENVIRONMENT, and disrupts a country's economy. Natural disasters can occur at any time and on any continent. More life is lost per disaster in poor countries than in wealthy countries.

When natural disasters occur, scientists try to learn from them so they can identify the factors that caused them. Instruments, such as weather satellites, computers, and ground sensors are used to gather data about conditions in the ATMOSPHERE and on Earth. Modern technologies such as these enable scientists to better predict when natural disasters may strike.

Floods are the most common type of natural disaster and usually occur when heavy rains cause rivers and streams or lakes to overflow. Floods can kill people and cause extensive destruction to homes, AUTOMOBILES, buildings, and other structures by submerging them under water. In addition, floods may disrupt ECOSYSTEMS by destroying the HABITATS of ORGANISMS or changing the conditions of the habitat.

Forest fires most often occur during periods of DROUGHT when conditions are very dry. Fires often devastate ecosystems by burning trees and other vegetation that provide habitat or food for animals and other organisms. Such fires also release POLLUTANTS such as CARBON DIOXIDE, soot, and ASH into the ATMOSPHERE. While the damaging effects of forest fires are clear, there are some ecological benefits of fire. For example, the Jack pine, a common conifer in North America, releases seeds only after it has been exposed to the heat from a forest fire. Routine fires also help clear dead wood, brush, and other debris that, if allowed to accumulate, might lead to larger, less controllable fires. (*See* detritus.)

Earthquakes occur when the tectonic plates that make up Earth's crust move and scrape against each other. (*See* plate tectonics.) The movements of the crustal plates are themselves very slight; however, over time they can be significant enough to cause the surface to shake, often violently. Earthquakes are not particularly damaging to ecosystems, but they can be devastating to people and their possessions. Strong earthquakes (having a magnitude of 8 on the Richter scale) can collapse buildings, roads, bridges, and other structures.

A VOLCANO is an opening in Earth's surface through which molten rock called lava and gases are released into the environment. Today, there are about 600 active volcanoes around the world. Volcanic eruptions are impressive, yet destructive, events. People are not often greatly affected by volcanic eruptions because cities and towns tend to be built away from active volcanoes; however, damage to the environment within the vicinity of a volcano can be quite extensive, with entire ecosystems being destroyed by a single eruption.

A HURRICANE is a violent tropical storm with sustained wind speeds of at least 150 kilometers (90 miles) per hour. The high winds associated with hurricanes cause huge waves that batter coastal areas, leading to beach EROSION and often significant damage to buildings and structures as well. When a hurricane moves over land, powerful winds and intense rains can destroy homes, businesses, automobiles, and just about everything else in its path. *See also* AEROSOL; CYCLONE; DAM; DUST BOWL; EL NIÑO/LA NIÑA; KRAKATOA; LITHOSPHERE; MOUNT PINATUBO; TORNADO; TSUNAMI.

EnviroSource

Deblieu, Jan. "Whirling Hurricanes" *Audubon,* 101: 5 (September-October 1999): 38–43.

natural gas, A natural mixture of flammable gases, usually composed of 70–80% METHANE; other HYDROCARBON constituents include ethane and propane, which are nonrenewable FUELS. Natural gas was formed in Earth's crust over millions of years by the chemical and physical alteration of ORGANIC matter. Its hydrocarbon molecules have one to four atoms of CARBON each; at Earth's surface, these compounds exist as gases.

Residential and commercial uses consume the largest proportion of natural gas in North America and western Europe; industry is the next-largest consumer, and electric-power generation a distant third. Natural gas is also used to fuel some automotive vehicles, though currently on a very limited basis and only in some countries.

The composition of natural gas varies according to locality; minor components may include CARBON DIOXIDE, NITROGEN, HYDROGEN, CARBON MONOXIDE, and helium. It is the cleanest-burning of the FOSSIL FUELS, yielding little more than carbon monoxide, carbon dioxide, and water as combustion byproducts. Although some natural gases can be used directly from the WELL without treatment, most must first be processed to remove undesirable constituents, such as HYDROGEN SULFIDE and other SULFUR compounds.

Often found in solution with PETROLEUM, natural gas also occurs apart from it in sand, sandstone, and limestone deposits, and in COAL beds. In order for gas to accumulate, it must be trapped (i.e., the underground RESERVOIR must be sealed at the top by an impermeable stratum or cap rock, such as clay or salt, to prevent gas from leaking to the surface). Gas accumulations are mostly encountered in the deeper parts of sedimentary basins.

Natural gas was first used to light the town of Fredonia, New York, in 1821; however, the fuel's use remained localized over the next century because long-distance transportation of gases was difficult. Methods of pipeline transportation were developed in the 1920s; between World War II and the 1980s, there was a period of tremendous residential and com-

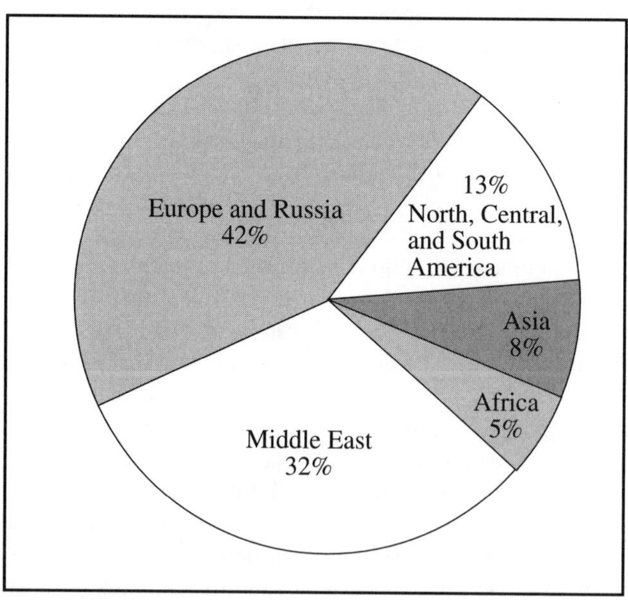

The chart shows the approximate world natural gas reserves as of 1989.

mercial expansion that relied increasingly on the use of pipeline transportation of gas. North American gas pipelines now extend from Texas and Louisiana to the northeast coast, and from the Alberta gas fields to the Atlantic seaboard.

In 1999, many taxi and bus drivers in Cairo, Egypt, converted their vehicles to run on compressed natural gas (CNG)—a fuel that is cleaner burning than GASOLINE. Cairo has the most privately owned natural-gas powered motor vehicles of any city in the world, and now the country's bus and taxi companies are coming aboard. Egypt has abundant natural gas reserves and can offer CNG to car owners. CNG vehicles emit about 80% less carbon monoxide and fewer hydrocarbons than gasoline-powered vehicles. And CNG costs less than gasoline in Egypt. A cubic meter of CNG is 50% less expensive than the equivalent amount of gasoline. By 2010, Egypt will have more than 25 stations to service CNG vehicles.

According to the U.S. GEOLOGICAL SURVEY, as of 1993 worldwide discovered reserves totaled 145,500 billion cubic meters of gas and undiscovered reserves were estimated at 132,600 billion cubic meters of gas. Among the largest accumulations of natural gas are those of Urengoy in SIBERIA, the Texas Panhandle in the United States, the Slochteren-Groningen area in the Netherlands, and Hassi R'Mel in Algeria. *See also* ENERGY; GREENHOUSE EFFECT; GREENHOUSE GAS; NONRENEWABLE RESOURCES; VOLATILE ORGANIC COMPOUNDS.

EnviroSources

American Gas Association Web site: www.aga.org.
Oil and Gas Journal Online Web site: www.ogjonline.com.
U.S. Department of Energy, Energy Information Administration Web site: www.eia.doe.gov.
U.S. Department of Energy, Office of Fossil Energy Web site: www.fe.doe.gov.
U.S. Geological Survey ("Changing Perceptions of World Oil and Gas Resources as Shown by Recent USGS Petroleum Assessments," fact sheet) Web site: www.greenwood.cr.usgs.gov/pub/fact-sheets/fs-0145-97/fs-0145-97.html.
U.S. Geological Survey, Energy Resources Program Web site: www.energy.usgs.gov/index.html.

natural resource, Any matter or ENERGY derived from the ENVIRONMENT that is used by living things. Examples of natural resources include such things as AIR, water, soil, sunlight, MINERALS, FOSSIL FUELS, and other ORGANISMS. Many of these resources are essential to living things for their survival. For example, PLANTS cannot live without sunlight because they rely on the energy of the sun to carry out PHOTOSYNTHESIS, the process by which they make their food. Similarly, organisms require air and water to carry out their respiratory processes and to maintain a proper chemical balance in their CELLS and tissues. Not all natural resources are essential to living things for their survival. For example, fossil fuels are natural resources that have little value to organisms other than humans; however, humans rely heavily on fossil fuels as a source of energy.

Natural resources are classified into two groups: RENEWABLE RESOURCES and NONRENEWABLE RESOURCES. Renewable resources are those natural resources that are regularly replenished through natural processes and thus have the potential to last indefinitely. Examples of such resources include water, which is naturally replenished through the WATER CYCLE; OXYGEN and CARBON DIOXIDE, which are naturally replenished through the oxygen-carbon cycle; and those resources, such as timber, plants, animals, BACTERIA, FUNGI, and PROTISTS, which are replenished through the reproductive processes of organisms. Nonrenewable resources are those resources that exist on Earth in fairly fixed amounts and thus have the potential of being used by organisms faster than they are replaced by nature. Such resources include available land space, fossil fuels (oil, COAL, NATURAL GAS), and MINERALS. These resources are replenished by natural processes; however, these processes occur very slowly, sometimes requiring millions of years to be completed. Thus, availability of these resources is limited and they have the potential of being depleted. *See also* ABIOTIC; AGENDA 21; ALTERNATIVE ENERGY RESOURCE; ALTERNATIVE FUEL; ALUMINUM; ASBESTOS; ATMOSPHERE; BIODIVERSITY; BIODIVERSITY TREATY; BIOFUEL; BIOGEOCHEMICAL CYCLE; BIOMASS; BIOSPHERE; BIOTIC; CADMIUM; CALCIUM CYCLE; CARBON; CARBON CYCLE; CARRYING CAPACITY; CHEMOSYNTHESIS; CHLORINE; CHROMIUM; CLEAN AIR ACT; CLEAN WATER ACT; COMPETITION; CONSERVATION; COPPER; DECOMPOSITION; DESERT; DETRITUS; EARTH DAY; ECOLABELING; ECOSYSTEM; ENVIRONMENTAL PROTECTION AGENCY; ESTUARY; EXTINCTION; FAUNA; FISH AND WILDLIFE SERVICE; FLORA; FORESTRY; FUEL WOOD; GENERAL MINING LAW OF 1872; GROUNDWATER; HABITAT; HUMUS; HYDROGEN; HYDROLOGY; LAND TRUST; LEAD; MAHOGANY; MANGANESE; MERCURY; METHANE; MINING; MULTIPLE USE; NITRATE; NITROGEN; NITROGEN CYCLE; NITROGEN FIXATION; OCEAN; OZONE; PEAT; PETROLEUM; PHOSPHATE; PHOSPHORUS; RECYCLING; ROCK CYCLE; SOIL CONSERVATION; SOLAR ENERGY; SULFATE; SULFUR; SUSTAINED YIELD.

EnviroSources

Bernstein, Leonard, Alan Winkler, and Linda Zierdt-Warshaw. *Environmental Science: Ecology and Human Impact.* Menlo Park, CA: Addison-Wesley, 1995.
U.S. Department of Agriculture, Natural Resources Conservation Service Web site: www.nrcs.usda.gov.

Natural Resources Conservation Service, A CONSERVATION agency within the U.S. DEPARTMENT OF AGRICULTURE (USDA) whose mission is to help private landowners to conserve, improve, and sustain the nation's NATURAL RESOURCES and ENVIRONMENT. Among other services and programs, the National Resources Conservation Service (NRCS) provides soil surveys for privately owned lands. Its National Soil Survey Center provides scientific expertise to maintain a uniform system for soil resource mapping and assessment, enabling comparison of information from different locations.

The NRCS (formerly the Soil Erosion Service and then the Soil Conservation Service) was created in response to the DUST BOWL catastrophe of the mid-1930s. HUGH H. BENNETT, who would become the agency's first chief, convinced the U.S. Congress that soil EROSION was a national menace, that a permanent agency was needed within the USDA to call landowners' attention to their land stewardship responsibilities, and that a nationwide partnership of federal agencies with local communities was needed to help farmers and ranchers conserve their land. The nationwide establishment of SOIL CONSERVATION projects was accelerated with the passage of the SOIL CONSERVATION AND DOMESTIC ALLOTMENT ACT in 1935.

The NRCS relies on many partners to help set CONSERVATION goals and to work with people on the land. Its partners include conservation districts (local agencies), state and federal agencies, NRCS Earth Team volunteers, agricultural and environmental groups, and professional societies. The nation's 3,000 conservation districts—one in almost every county—link the NRCS with local priorities for soil and water conservation. By law and secretarial agreement, NRCS county technical services are delivered through local conservation districts. Nearly 75% of the technical assistance provided by the agency helps farmers and ranchers develop conservation systems (methods for planting and irrigation) suited to their lands and their individual approaches to doing business. The agency also provides assistance to rural and urban communities to reduce erosion, conserve and protect water resources, engineer FLOOD control systems, and solve other resource problems. The NRCS also assists agricultural producers in complying with the provisions of the Food Security Act of 1985; the Food, Agriculture, Conservation and Trade Act of 1990; the Federal Agriculture Improvement and Reform Act of 1996; and the CLEAN WATER ACT. *See also* AGRICULTURE; CONTOUR FARMING; SOIL CONSERVATION.

EnviroSource

U.S. Department of Agriculture, Natural Resources Conservation Service Web site: www.nrcs.usda.gov.

natural selection. *See* evolution.

naturalist, A person who studies nature, especially by direct observation of PLANTS, animals, and Earth. Generally, anyone with an interest or appreciation in nature can be considered to be a naturalist. This can include nature artists, WILDLIFE photographers, bird watchers, or anyone else that enjoys seeing animals and plants in their natural HABITATS.

Professional naturalists typically are employed in national parks, FOREST preserves, nature museums, and botanical gardens. The work of a professional naturalist usually involves leading groups on walks to observe and identify plants, animals, birds, fish, and INSECTS; explaining the interactions among these organisms; lecturing at schools, children's groups, garden clubs, retiree associations, and state and local CONSERVATION groups; and designing exhibits, trail booklets and displays to depict the natural history of a park or preserve. *See also* AUDUBON, JOHN JAMES; CARSON, RACHEL LOUISE; DARWIN, CHARLES ROBERT; ECOTOURISM; ENVIRONMENTAL SCIENCE; MUIR, JOHN.

Nature Conservancy. *See* land trust.

neritic zone. *See* ocean.

neurotoxin, A chemical that can poison, damage, and destroy the human nervous system. Neurotoxins include HEAVY METALS such as LEAD and MERCURY, PESTICIDES, POLYCHLORINATED BIPHENYLS (PCBs), and other chemicals. Neurotoxins can be produced naturally by animals, PLANTS, and other ORGANISMS. For example, the dinoflagelletes (PROTISTS) that are responsible for RED TIDES produce neurotoxins that are harmful to other marine life. Many snakes also produce neurotoxins in their venom and are passed to other organisms in their bite. Symptoms of exposure to neurotoxins include vision loss, low IQ (intelligence quotient) in young children, and the improper development of the fetus. Parkinson's disease is also linked to neurotoxins.

niche, A term used by ecologists to define the role of an ORGANISM in its ECOSYSTEM. A niche also may describe the typical physical location of the organism in its ENVIRONMENT. For example, an oak tree may occupy the niche referred to simply as "PRODUCER," while the niche of grasses might be described as "ground-level producers" as a means of distinguishing their place in the ecosystem from that of taller PLANTS and trees.

All organisms are adapted to the environment in which they live. (*See* adaptation.) For example, each organism lives within a certain part of its ecosystem, eats a certain type of food, and carries out certain functions. Together, all of these elements determine the role, or niche, of the organism. Organisms of the same SPECIES generally have the same niche in an ecosystem; however, organisms of different species cannot occupy the exact same niche at the same time. When this happens, one species generally will outcompete the other, forcing the less adapted species to adapt to a slightly different niche within the same ecosystem, move to another ecosystem where its niche is not already occupied, or risk EXTINCTION.

Often, the niche of an organism is closely related to the feeding level it occupies in the FOOD CHAIN and the specific type of food it eats. For example, two species of *Anolis* lizards that live in the same parts of the tropics feed exclusively on INSECTS; however, each species feeds on insects of different sizes. *Anolis* lizards with small jaws feed only on small insects, while similar lizards with large jaws feed exclusively on large insects. Because they have slightly different food sources, the niches of the two types of lizards are different, even though they share the same HABITAT.

In the case of the *Anolis* lizards, the large-jawed lizard could feed on insects of all sizes. Thus, its fundamental niche, the niche it is able to occupy, is composed of both large and small insects. By not feeding on small insects, however, the large-jawed *Anolis* reduces COMPETITION for food with the small-jawed *Anolis,* and thus the large-jawed lizard feeds almost exclusively on large insects. Feeding on large insects is the realized niche of the large-jawed *Anolis*. The tendency for two similar species to divide a niche between them is sometimes called niche-splitting or niche differentiation. This tendency reduces competition and helps each species meet its needs. *See also* ADAPTIVE RADIATION; CARNIVORE; COMMENSALISM; COMPETITIVE EXCLUSION; CONSUMER; DECOMPOSER; DETRIVORE; ECOLOGICAL PYRAMID; FOOD WEB; HERBIVORE; MUTUALISM; OMNIVORE; PARASITISM; SYMBIOSIS; TROPHIC LEVEL.

nicotine, A colorless, oily liquid derived from the dried leaves of the tobacco PLANT. In AGRICULTURE, nicotine is sprayed on plants as an insecticide. Pure nicotine is toxic to many ORGANISMS, including humans. When tobacco is smoked or chewed, small amounts of nicotine are absorbed into the body. Scientific evidence supports the idea that nicotine is highly addictive and promotes the use of tobacco products. In small DOSES, nicotine is a mild stimulant of the central nervous system. In larger doses, however, nicotine can paralyze the nervous system by preventing the transmission of nerve impulses. The high TOXICITY of nicotine results in the deaths of many small children each year; deaths most often result from children ingesting nicotine from discarded cigarette butts they place in their mouths. In some adults, nicotine can also cause the formation of gastric ulcers.

nitrate, Any chemical compound, typically salts, that contains the chemical elements NITROGEN and OXYGEN. Nitrates (NO_3) are found naturally in PLANTS and forms from ammonia in animal wastes and FERTILIZERS in soil. Nitrogen is perhaps the most essential chemical to living things because it is used in the formation of deoxyribonucleic acid (DNA) and proteins. Plants obtain nitrates from the soil by absorbing them through their roots. Animals obtain the nitrogen they need to make DNA and proteins by feeding on plants and on other animals. (*See* food chain.)

Nitrates are formed naturally by soil BACTERIA as part of Earth's NITROGEN CYCLE. The nitrogen cycle starts with the element nitrogen in the AIR. About 78% of the ATMOSPHERE is made up of nitrogen gas (N_2). In the form of a gas, nitrogen is highly unreactive, making it useless to plants and animals. To become useful, the nitrogen must be first converted into a different form.

Lightning in the atmosphere is one way that gaseous nitrogen can be converted, or fixed, into another form. The tremendous ENERGY released by these electrical discharges breaks the strong bonds between nitrogen atoms, causing them to react with oxygen and form a new chemical compound. Most of the nitrogen in the air, however, becomes part of biological matter through the actions of soil BACTERIA (and CYANOBACTERIA in aquatic environments) through a process known as NITROGEN FIXATION. Many of these nitrogen-fixing bacteria live in nodules on the roots of peas, clover, alfalfa, soybeans, and other members of the LEGUME family.

In nitrogen-fixation, nitrogen is taken from the air and converted into usable ammonia (NH_3). But since ammonia is directly toxic to most plants, it must also be changed into other forms. This is accomplished by nitrifying bacteria, primarily in the bacterial genera *Nitrosomonas*, *Nitrosococcus,* and *Nitrobacter* through the process of nitrification. In nitrification, ammonia is converted first into nitrites (NO_2), and then into the nitrates (NO_3) that plants absorb. Ammonia is abundant in animal wastes and is released from the decaying plants and bodies of animals. Again, nitrifying bacteria convert this ammonia into nitrates. Plants now have a usable form of nitrogen. Some plants can use ammonia directly. These plants do not have to wait until the ammonia has been converted into a nitrate before they absorb it.

Nitrogen that is fixed by the processes described above is eventually returned to the atmosphere by the DENITRIFICATION process. Denitrification is the reduction of nitrates to gaseous nitrogen. Denitrifying bacteria perform almost the reverse of the nitrogen-fixing bacteria. The released nitrogen is returned to the atmosphere to complete the nitrogen cycle.

Because nitrates are so essential to plant growth, they are also incorporated into synthetic fertilizers in the form of ammonium nitrate for use on farms. Ammonia can be made synthetically through an industrial process. Nitrogen and HYDROGEN are combined under great pressure and temperature in the presence of a catalyst. The newly manufactured ammonia may be applied directly to farm fields as fertilizer. When synthetic fertilizers, containing nitrates, end up in lakes, rivers, and streams as a result of RUNOFF from farms, livestock feedlots, and lawns they can cause a number of environmental problems, the most important of which is EUTROPHICATION. In this natural phenomenon, excess nutrients such as nitrate in the water lead to rapid growth in POPULATIONS of ALGAE and bacteria. As a result of the population explosion, the water is quickly depleted of DISSOLVED OXYGEN, causing the suffocation deaths of fish and other aquatic animals. There is also a concern about nitrate levels in drinking water, especially drinking water from shallow WELLS and GROUNDWATER. Once nitrates are ingested, the human body can change the nitrates into nitrites. Nitrate and nitrite compounds used as preservatives for meat, such as bacon, can present health problems. If nitrate and nitrites are ingested in the body and combine with amino acids, CANCER-causing compounds such as nitrosamine can form. However, the health risk is small. Nitrites can interfere with the body's ability to carry oxygen and are also known to cause headaches, low blood pressure, nausea, and diarrhea. Nitrites can also combine with other chemicals in the body to form nitrosamines, which can cause stomach cancer. *See also* BIOGEOCHEMICAL CYCLE.

nitric acid, An ACID formed naturally in the ATMOSPHERE when nitrogen dioxide and other NITROGEN OXIDES react with water vapor. Nitric acid also can be formed in the laboratory. Nitric acid is one of the main CONTAMINANTS in ACID RAIN. The colorless and highly corrosive toxic liquid is a common laboratory reagent and an important industrial chemical used in the manufacturing of explosives, FERTILIZERS, dyes, and other chemicals. *See also* NITROGEN.

nitric oxide. *See* nitrogen oxides.

nitrite. *See* nitrate.

nitrogen, A naturally occurring element important for the growth, reproduction, and metabolism of living ORGANISMS. Nitrogen is one of the most abundant elements and accounts for about 78% of all gases in Earth's ATMOSPHERE.

Nitrogen is present in all organisms and throughout the ENVIRONMENT. At standard temperature and pressure, elemental nitrogen exists as a diatomic gas (N_2). Before most living things can use this nitrogen, however, it must be changed in form, or "fixed" in compounds that are accessible and usable by PLANTS—a process called NITROGEN FIXATION—in a process by which some BACTERIA and ALGAE convert the gaseous nitrogen (N_2) into ammonia (NH_3). Some plants can use the ammonia; other plants depend on bacteria in the soil to convert into nitrates and nitrites.

Besides its importance as a biological chemical, nitrogen can also be a dangerous POLLUTANT in certain chemical compounds, such as NITROGEN OXIDE. Nitrogen oxides, including NITROGEN DIOXIDE (NO_2), are primarily released into the atmosphere in EMISSIONS from power plants, motor vehicles, and factories that burn FOSSIL FUELS. In the atmosphere, nitrogen oxides can react with other substances to form SMOG and ACID RAIN. *See also* ACID; DECOMPOSERS; DECOMPOSITION; DETRITUS; NITRIC ACID; NITROGEN CYCLE.

nitrogen cycle, The continuous flow of NITROGEN throughout the BIOSPHERE as nitrogen moves between living things and the ENVIRONMENT. Nitrogen is a chemical substance present in all ORGANISMS and is essential to many life processes, including PHOTOSYNTHESIS, growth, and reproduction. Nitrogen is also a critical component of deoxyribonucleic acid (DNA), the complex molecule that controls CELL processes in all organisms.

Nitrogen is abundant in Earth's ATMOSPHERE and comprises about 78% of all atmospheric gases. Except for a few SPECIES of BACTERIA, organisms cannot use nitrogen in its gaseous form. To enter living systems, gaseous nitrogen must first be fixed (combined with OXYGEN or HYDROGEN) into compounds that PLANTS can utilize, such as NITRATE (NO_3) or ammonia (NH_3). The high energies provided by lightning and solar RADIATION convert some atmospheric nitrogen into nitrates; however, most NITROGEN FIXATION is performed by species of bacteria that live in soil or in root nodules of leguminous plants, such as peas and beans. (*See* legumes.)

Nitrogen that has been "fixed" as ammonia or nitrates can be taken up directly by plants and incorporated into tissues as proteins. Animals and other CONSUMERS get their nitrogen by eating plants or plant-eating animals. When living things die and produce wastes, nitrogen in the form of ammonia is again returned to the soil. Some of this ammonia is taken up by plants. The rest is converted by bacteria into nitrates and nitrites through a process called nitrification. (*See* EnviroTerm.) Nitrates may be stored in HUMUS or eroded from the soil and carried by moving water to rivers and lakes. They may also be converted to nitrogen gas by denitrifying bacteria and returned to the atmosphere.

In undisturbed ECOSYSTEMS, nitrogen lost from the soil through denitrification and EROSION is replenished by nitrogen fixation. Various human activities, however, can alter the nitrogen cycle; as a result not enough nitrogen may be cycled or too much nitrogen may overload the system. For example, large-scale cultivation and harvesting of crops and CLEAR CUTTING of FORESTS all lead to a loss of nitrogen in the soil. In AGRICULTURE, farmers and gardeners often try to replenish their soils with nitrogen-containing FERTILIZERS. On the other hand, the LEACHING of nitrogen from overfertilized soils can result in an overabundance of nitrogen in aquatic ecosystems, resulting in reduced water quality and the excessive growth of ALGAE or aquatic plants. This process, known as EUTROPHICATION, sometimes leads to massive die-offs of fishes and other small aquatic organisms. As these dead organisms become food for DECOMPOSERS, oxygen depletion occurs in the water. (*See* biochemical oxygen demand.)

Humans impact the nitrogen cycle in other ways, too. AUTOMOBILES, power plants, and factories that burn FOSSIL FUELS release nitrogen-containing POLLUTANTS, such as NITROGEN DIOXIDE, into the atmosphere. Here, the pollutants can contribute to the formation of SMOG or mix with water vapor and other substances to form ACID RAIN. *See also* AIR POLLUTION; ALGAL BLOOM; BIOGEOCHEMICAL CYCLE; DENTRIFICATION; NITRIC ACID.

EnviroTerm

nitrification: The process in the nitrogen cycle in which nitrifying bacteria use OXYGEN to change ammonia molecules in WASTEWATER into nitrites and then to nitrates.

nitrogen dioxide, An AIR POLLUTANT produced from the combustion of FOSSIL FUELS, such as GASOLINE, oil, and COAL, which has adverse effects on human health. The main sources of nitrogen dioxide EMISSIONS in the air are AUTOMOBILES, factories, and electrical power plants. Nitrogen dioxide contributes to the formation of ACID RAIN.

Nitrogen dioxide is one of several nitrogen-containing air pollutants known as NITROGEN OXIDES. Nitrogen oxides make up approximately 11% of all primary air pollutants and are associated with several environmental problems. (*See* primary and secondary air pollutants.) Near highly populated and industrialized areas, nitrogen dioxide is one of the major components of SMOG, the brownish haze often seen over large cities such as Los Angeles and Mexico City. Smog can be a health hazard to humans, causing eye and lung irritation, and possibly CANCER.

In the ATMOSPHERE, nitrogen dioxide can also combine with water and other pollutants to form NITRIC ACID, which can fall to Earth as acid rain. Acid rain, rainfall with a PH of 5.0 and below, often occurs in rural areas surrounding large cities and industrialized regions. All forms of acid precipitation (rain, snow, fog, or dew) are known to seriously damage crops and WILDLIFE. Acid precipitation can also dissolve the calcium carbonate in common building materials such as limestone and concrete. Many historic monuments, statues, bridges, and buildings today are gradually being eaten away by acid precipitation.

Several environmental laws have been passed to help reduce nitrogen dioxide EMISSIONS. Under the CLEAN AIR ACT, the main body of legislation regulating AIR POLLUTION in the United States, all power plants and factories must install SCRUBBERS in their smokestacks to control the amounts of nitrogen dioxide released into the air. New automobiles must also be equipped with CATALYTIC CONVERTERS, which convert nitrogen oxides into less harmful substances. *See also* AIR POLLUTION CONTROL ACT; SICK BUILDING SYNDROME.

nitrogen fixation, The process in which atmospheric NITROGEN (N_2) is changed into other nitrogen compounds, such as ammonia (NH_3) or NITRATE (NO_3) by the actions of certain BACTERIA, lightning, solar RADIATION, or through other chemical means. Nitrogen fixation is an important process because most of the nitrogen on Earth exists as a gas that is unusable by most ORGANISMS. Nitrogen accounts for approximately 78% of all atmospheric gases. The process of nitrogen fixation helps convert this nitrogen gas into compounds that are accessible to PLANTS. Animals obtain this nitrogen by ingesting plants or other CONSUMERS.

Various ABIOTIC factors, such as lightning, ultraviolet radiation, and electrical equipment, play a minor role in nitrogen fixation. The products, ammonia and nitrates, are then washed into soil by rainfall and other types of precipitation where they may be taken up by plants. Biological fixation of nitrogen is much more important. Only prokaryotes (BACTERIA and CYANOBACTERIA) can fix nitrogen. Two kinds of nitrogen-fixing bacteria are known: free living and symbiotic. In marine and fresh-water ECOSYSTEMS, free-living SPECIES, including many types of cyanobacteria, help convert nitrogen gas into ammonia. On land, free-living soil bacteria make some contribution, but it is the bacteria that live within the root nodules of leguminous plants, such as clover, alfalfa, beans, and peas, that produce most of the fixed nitrogen in the form of ammonia. The root nodules, or swellings, house colonies of nitrogen-fixing bacteria (usually of the genus *Rhizobium*), which live in a symbiotic relationship with the plant. Unlike free-living nitrogen fixers that fix what they need for their own purposes and release the fixed nitrogen only upon death, bacteria within root nodules release up to 90% of the nitrogen they fix to the plant and excrete some of the rest into the soil, making it available to other organisms. This is one reason why many farmers alternate crops, periodically planting clover or alfalfa to increase the content of useful nitrogen in the soil.

Ammonia can also be artificially produced by humans through chemical means, using a technique called the Haber-Bosch process. Artificially produced ammonia is often used in the production of nitrogenous FERTILIZERS, which are important to AGRICULTURE. *See also* DENITRIFICATION; LEGUMES; NITROGEN CYCLE; SYMBIOSIS.

nitrogen oxides, A group of gases containing NITROGEN and OXYGEN that are introduced into the AIR primarily in EMISSIONS of AUTOMOBILES and power plants that burn PETROLEUM and COAL. Nitrogen oxides are primary air POLLUTANTS that irritate the lungs, nose, and throat, and also cause a variety of respiratory illnesses. NITROGEN DIOXIDE (NO_2), one type of nitrogen oxide, is one of the main pollutants of the ATMOSPHERE. It is a visible yellow-brown gas that is a major component of photochemical SMOG. Another form of nitrogen oxide is nitric oxide. *See also* AIR POLLUTION; PRIMARY AND SECONDARY AIR POLLUTANTS.

Sources of Emissions from Nitrogen Oxides in the United States

Source	%
Motor vehicle exhausts	51%
Burning soft coal	46%
Nitric acid industries	1%
Solid waste disposal	1%
Miscellaneous	1%

The table illustrates the approximate amount of emissions from nitrogen oxides from various sources as of 1992.

nitrogen-fixing bacteria. *See* nitrogen fixation.

Noise Control Act, Legislation passed in 1972, giving the U.S. ENVIRONMENTAL PROTECTION AGENCY (EPA) authority to establish noise control regulations. The EPA established the Office of Noise Abatement and Control, which was responsible for publishing noise emissions standards, requiring product labeling, coordinating federal noise reduction programs, and assisting state and local efforts by promoting noise education and research; however, funding for the Office of Noise Abatement and Control was terminated in 1982. In February 1997 some members of Congress wanted to reestablish the office. They cited studies showing that many Americans suffer from noise from aircraft, vehicular traffic, and a variety of other sources. Their report also stated that nearly 20 million Americans are exposed to noise levels that can lead to psychological and physiological damage, and another 40 million people are exposed to noise levels that cause sleep or work disruption. Chronic exposure to noise has been linked to increased risk of cardiovascular problems, strokes, and nervous disorders as well as hearing loss. Excessive noise also causes sleep deprivation and task interruptions, both of which pose untold costs on society in diminished worker productivity.

The congressmembers pointed out that the EPA still remains legally responsible for enforcing regulations issued under the Noise Control Act of 1972 even though funding for these activities were terminated, and since the act prohibits state and local governments from regulating noise sources in many situations, noise abatement programs across the country lie dormant. Therefore, they concluded that the Office of Noise Abatement and Control should be reestablished.

Most environmentalists agree that as POPULATION growth and AIR and vehicular traffic continue to increase, offensive NOISE POLLUTION will continue to be a significant problem in the future.

EnviroSource

Noise Pollution Clearinghouse: P.O. Box 1137, Montpelier, VT 05601-1137; (888) 200-8332 Web site: www.nonoise.org.

noise pollution, Any unwanted or unpleasant sound present in the ENVIRONMENT. Sounds and noises are measured in units of decibels (dBA). The noise level in an average house is about 40–45 dBA, while AUTOMOBILE traffic is about 55–70 dBA. Exposure to sounds greater than 120 dBA for a length of time may cause hearing loss in humans. Exposure to excessive noise levels of greater than 80 dBA for a long period of time can have both physiological and psychological effects on the body that result in such conditions as anxiety, stress, and fatigue.

Since 1978, when Congress passed the Quiet Communities Act, more than 25 states have created nearly 1,000 noise-related ordinances. The Department of Urban Development uses a guideline of 65 dBA for acceptable levels of day and night noise pollution. Some European countries are far ahead of the United States in developing quieter machines and tools, and in finding creative ways to reduce noise levels; however, despite such efforts, expanding POPULATIONS and industrial growth are likely to cause offensive noise levels to continue to be a problem. The WORLD HEALTH ORGANIZATION (WHO) reported that 100 million people are exposed to heavy traffic noise in excess of 65 dBA. This figure is above the maximum level of 55 dBA recommended by the WHO. *See also* NOISE CONTROL ACT.

Decibel Levels of Common Sounds

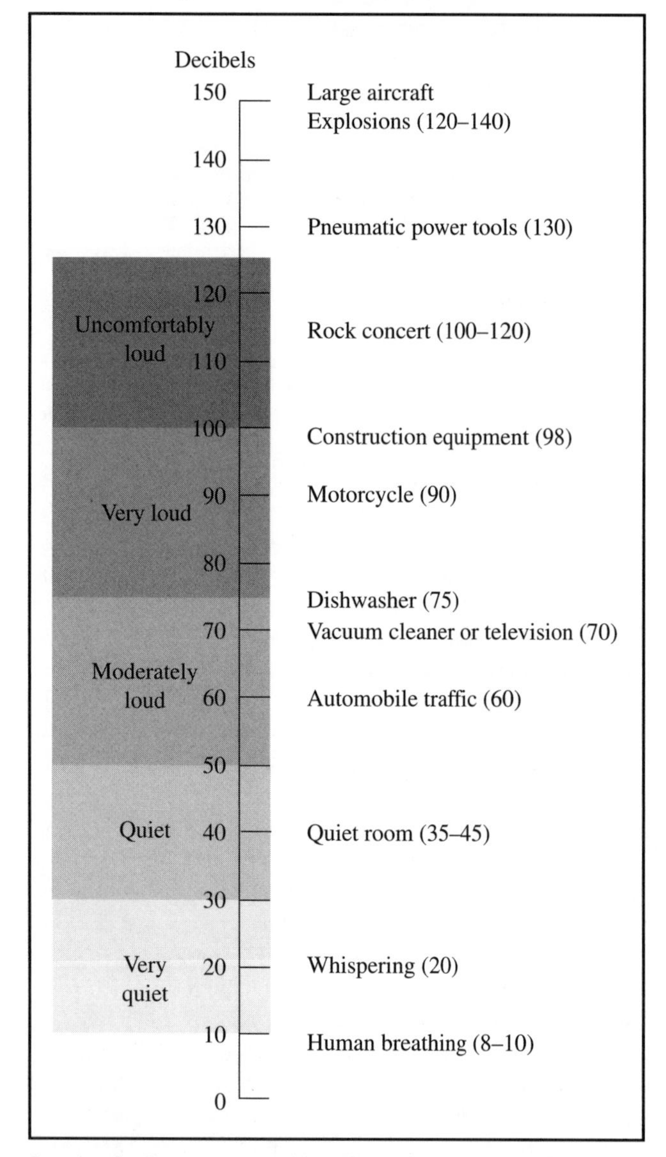

Sound and noise are measured in units of decibels (dBA). Exposure to excessive noise levels of greater than 80 dBA for a long period of time can have both physiological and psychological effects on the body.

nonattainment areas, Areas of the United States that have not met the AIR-quality standards specified in the NATIONAL AMBIENT AIR QUALITY STANDARDS for human health as dictated by the CLEAN AIR ACT. Some states have established nonattainment areas for GROUNDWATER or SURFACE WATER quality, too, where the water quality has been degraded by a history of industrial activity or other impacts. *See also* ATTAINMENT AREA.

nonbiodegradable, A term used to describe ORGANIC wastes that do not readily break down in the ENVIRONMENT through natural processes such as DECOMPOSITION and mineralization. Most often, nonbiodegradable materials are synthetic, or human made. Examples of such materials include many kinds of plastics, styrofoam, and synthetic rubber. Because they are not easily broken down, nonbiodegradable materials remain in the environment for extremely long periods of time, and may remain in landfills for many years, where they take up the limited space available for discarded waste products. To combat this problem, environmentalists encourage people to reuse, recycle, and reduce their use of products made from nonbiodegradable materials. *See also* BIODEGRADABLE; COMPOSTING; DECOMPOSER; DETRITUS; GARBAGE; RECLAMATION; RECYCLING; REFUSE.

nonpoint source, Naturally occurring processes or human activities that cause pollution, such as sediment, nutrients, and inorganic or toxic substances, originating from land use activities to be carried to lakes and streams. Nonpoint sources include surface RUNOFF from agricultural and urban areas, MINING activities, and FORESTRY. Nonpoint source pollution can exist in AIR, soil, and GROUNDWATER. Nonpoint sources are a major factor in the pollution of streams and rivers.

EnviroSource

U.S. Environmental Protection Agency, Nonpoint Source Pollution Control Program: Web site: www.epa.gov/OWOW/NPS.

U.S. Nonpoint Source Pollution

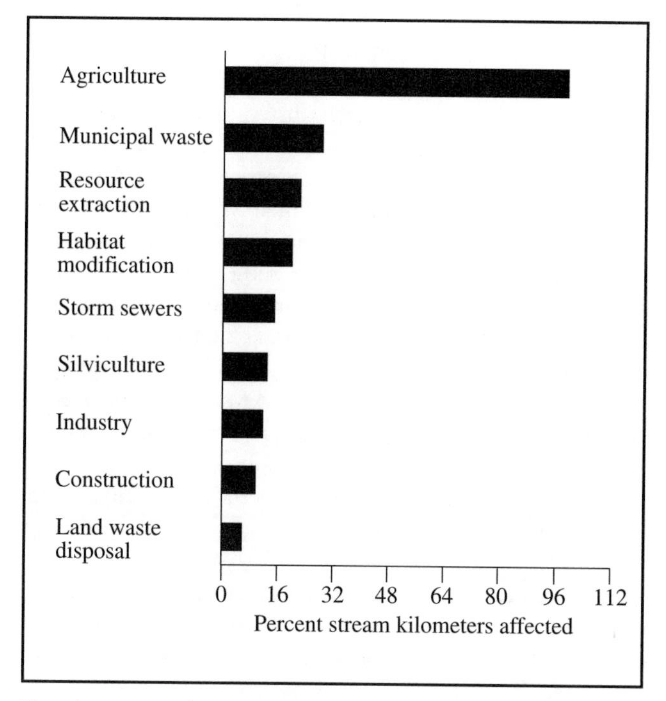

Nonpoint sources of water pollution are a major factor in the pollution of steams and rivers. *Source:* United States Environmental Protection Agency, 1990.

nonrenewable resource, A NATURAL RESOURCE that has the potential of being exhausted because it is being used by people at a faster rate than it can be replaced by nature. Nonrenewable resources exist in fairly fixed amounts throughout the world and include FOSSIL FUELS (COAL, PETROLEUM, and NATURAL GAS) and MINERALS. Fossil fuels are currently the main source of ENERGY for the United States and most other industrialized nations. The combustion of these FUELS to generate ELECTRICITY is the main source of AIR POLLUTION globally. Petroleum also is the main source of fuel for AUTOMOBILES, airplanes, boats, and many other transportation vehicles—all of which also are major contributors to AIR pollution. The major concern over fossil fuels, however, is not that these fuels are a major source of pollution, but that they require millions of years to form, and thus have the potential of being completely used up in a relatively short period of time.

Like fossil fuels, many minerals also form very slowly and exist in fairly fixed amounts on Earth. Minerals are used in the manufacture of a variety of products used by people throughout the world. For example, gypsum is a mineral used to make the dry wall that is used for the interior walls of homes and other buildings. URANIUM and plutonium are minerals used as fuel sources for NUCLEAR ENERGY as well as in the construction of NUCLEAR WEAPONS. Extracting and processing minerals is often costly and time-consuming, is very damaging to the HABITATS of ORGANISMS, and is a major source of pollution. *(See* mining.) As nonrenewable resources, mineral reserves, like those of fossil fuels, have the potential of being depleted by humans. Thus, finding ways to ensure that these minerals will be available for use by future generations is a major concern to the people of many nations throughout the world.

Mineral Conservation. The CONSERVATION of minerals focuses primarily on developing methods for reducing, reusing, and RECYCLING these vital resources. Reducing the use of minerals involves using mineral products that are made wisely and in a manner that is not wasteful. Often, this involves finding substances that can replace minerals in various products used by humans. For example, instead of packaging products in metal cans, many products today are packaged in paper (a RENEWABLE RESOURCE) or plastic (a synthetic resource). These alternatives help reduce the need to mine new minerals. In addition to reducing the use of minerals, many people are now finding ways to reuse products made using minerals. For example, instead of discarding a metal can, it might be reused as a storage container for nails or other household items. Such uses may also help eliminate the need to mine new minerals. In addition, reuse of products made from minerals helps to eliminate the need to find additional landfill space for disposal of such matter. Recycling involves the reprocessing of materials to reclaim their minerals for additional uses. For example, ALUMINUM beverage cans may be recycled to reclaim the aluminum they contain. Once reclaimed, this aluminum may be used to make other products such as automobile parts, lawn furniture, new aluminum beverage cans, or a variety of other products. Many other metals, such as COPPER, LEAD, gold, silver, and iron, also are recycled for use in new materials, helping to preserve these vital resources.

Recycling minerals not only benefits the ENVIRONMENT by reducing the need to mine more of these materials; it also helps to reduce the use of fossil fuels and pollution. For example, recycling the aluminum in a beverage container uses up to 95% less energy than processing aluminum from ore. This helps reduce the amount of fossil fuels that are burned to generate the heat needed to process the aluminum, while also eliminating the polluting EMISSIONS that result from the combustion of these fuels.

Fossil Fuel Conservation. The conservation of fossil fuels also employs the methods of reducing, reusing, and recycling; however, the major emphasis on fossil fuel conservation involves finding suitable ALTERNATIVE ENERGY RESOURCES. For example, in many parts of the world SOLAR ENERGY (the energy of sunlight) is now being used to replace oil or natural gas to meet the heating and electrical needs of homes. Using solar energy for this purpose conserves fossil fuels by reducing their use and also makes use of a RENEWABLE RESOURCE that is nonpolluting. Other alternative energy sources include WIND POWER, GEOTHERMAL ENERGY, and HYDROELECTRIC POWER. Use of these energy sources is often limited by the location of homes and businesses; however, like solar energy, their use can be relatively more beneficial to the environment compared to fossil fuels because they make use of renewable resources and are less polluting.

In addition to finding alternatives to fossil fuels, much of the effort on fossil fuel conservation has focused on making more efficient use of such fuels. For example, engineers have employed a variety of methods to make automobiles more fuel efficient, such as building cars that are more aerodynamic and have smaller engines. These adaptations can allow automobiles to drive greater distances on less fuel, thus helping to decrease the amount of fuel used by each individual. People also have been encouraged to car pool (share rides with others) or make use of mass transportation when possible. These methods help to reduce the total number of automobiles on the roads, reducing traffic congestion while conserving fuel and reducing the amount of pollutants released into the air.

In addition to making use of energy alternatives and increasing energy efficiency, scientists also are working on ways to develop new fuels that can be used in place of fossil fuels. For example, in many parts of the world, the combustion of BIOMASS is used to meet the cooking and heating needs of homes. Other scientists are reducing the use of fossil fuels by mixing these fuels with other materials to create various blends. For example, GASOHOL (a mixture of petroleum and alcohol) can be used in some automobiles instead of gasoline. Since alcohol is made from renewable resources, use of this fuel can help to extend the life of petroleum reserves.

As the global POPULATION continues to grow, meeting the needs of people for products and energy will continue to be a growing challenge; however, if conservation of nonrenewable resources is not encouraged and practiced by a large number of people throughout the world, such resources will not be available for future generations. *See also* ACID MINE DRAINAGE; AEROGENERATOR; AIR POLLUTION CONTROL ACT; ALTERNATIVE FUEL; ASBESTOS; ASWAN HIGH DAM; ATOMIC ENERGY ACT; BERYLLIUM; BIOCONVERSION; BIOFUEL; BREEDER REACTOR; CATALYTIC CONVERTER; CHROMIUM; COGENERATION; DAM; ELECTRIC VEHICLE; EMISSIONS STANDARDS; FUEL CELL; FUEL WOOD; INTERNATIONAL ATOMIC ENERGY AGENCY; JAMES BAY HYDROPOWER PLANT; LAWS OF THERMODYNAMICS; MANGANESE; MERCURY; METHANE; METHANOL; NUCLEAR ENERGY.

EnviroSources

Bernstein, Leonard, Alan Winkler, and Linda Zierdt-Warshaw. *Environmental Science: Ecology and Human Impact.* Menlo Park, CA: Addison-Wesley, 1996.

Saign, Geoffrey C. *Green Essentials: What You Need to Know about the Environment.* San Francisco, CA: Mercury House, 1994.

Seymour, John, and Herbert Girardet. *Blueprint for a Green Planet: Your Practical Guide to Restoring the World's Environment.* New York, NY: Prentice-Hall, 1987.

North American Wetlands Conservation Act, A 1989 U.S. statute that provides funding and administrative direction for implementation of the North American Waterfowl Management Plan and the Tripartite Agreement on Wetlands among Canada, Mexico, and the United States. The goals of the law are to protect, enhance, restore, and manage an appropriate distribution and diversity of WETLAND ECOSYSTEMS and other HABITATS for migratory birds, fishes, and WILDLIFE in North America; to maintain current or improved distributions of migratory bird POPULATIONS; and to sustain an abundance of waterfowl and other migratory birds. It also established the North American Wetlands Conservation Council, the purpose of which is to recommend wetlands CONSERVATION projects to the Migratory Bird Conservation Commission. The director of the U.S. FISH AND WILDLIFE SERVICE is a permanent member of the nine-member council. The act also directs the secretary of the U.S. DEPARTMENT OF THE INTERIOR to maintain and implement a wetlands conservation strategy and to report to Congress on project implementation and assessment.

Congress passed this law because of new findings about the significant functional values of North American wetland ecosystems. At the time the law was enacted, more than 50% of the original wetlands in the United States had been lost. Wetlands destruction, loss of nesting cover, and degradation of migration and wintering habitat had contributed to long-term downward trends in populations of migratory bird SPECIES. In particular, the law recognizes that populations of migratory birds in North America depend on wetland ecosystems in Canada, the United States, and Mexico; wetland ecosystems provide substantial FLOOD and storm control values and can reduce the need for humans to devise control measures; wetland ecosystems make a significant contribution to water resource availability and quality, recharge GROUNDWATER, filter surface RUNOFF, and provide WASTEWATER treatment; and wetland ecosystems provide aquatic areas that are important for recreational and aesthetic purposes. *See also* BOG; DOUGLAS, MARJORY STONEMAN; EVERGLADES NATIONAL PARK; MIGRATORY BIRD TREATY ACT; SWAMP; SWAMPBUSTER PROVISION.

EnviroSources

U.S. Fish and Wildlife Service, Migratory Bird Commission Web site: www.fws.gov/r9realty/mbcc.html.

U.S. Fish and Wildlife Service, North American Waterfowl and Wetlands Office Web site: www.fws.gov/r9nawwo.

U.S. Fish and Wildlife Service, North American Wetlands Conservation Act Web site: www.fws.gov/r9nawwo/nawcahp.html.

U.S. Fish and Wildlife Service, North American Wetlands Conservation Council Web site: www.fws.gov/r9nawwo/nawcc.html.

North Sea, Part of the Atlantic OCEAN, located between the eastern coast of Great Britain and the continent of Europe, which has been threatened by a variety of pollution problems including EUTROPHICATION, OIL POLLUTION, pollution by HAZARDOUS SUBSTANCES and chemicals, and radioactive substances.

The widest part of the North Sea is about 645 kilometers (about 400 miles); its greatest length is about 965 kilometers (about 600 miles); and its area is about 575,000 square kilometers (about 222,000 square miles). Its greatest depth is off the coast of Norway. Its TIDES are very irregular because two tidal waves enter it, one from the north and one from the south. The North Sea contains some of the busiest shipping routes in the world. Its coastal zone is also used for recreation, such as swimming, sailing, and fishing.

The sea supports a huge diversity of animal and PLANT life. Hundreds of SPECIES of fish, shellfish, WHALES, DOLPHINS, SEALS, otters and sea birds live there. Fishing has long been an important activity in all countries bordering the North Sea with total annual landings around 2.5 million metric tons (2.75 million tons). Environmental groups, such as GREENPEACE, however, are concerned that OVERFISHING now poses the greatest threat to the marine BIODIVERSITY in the North Sea. According to Greenpeace, "Reclamation of parts of the sea for farming, the DISCHARGE of chemicals and FERTILIZERS into the

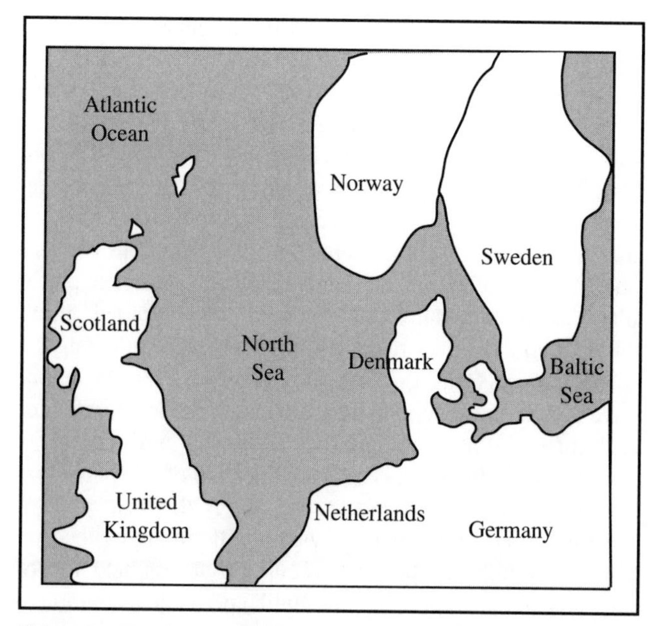

Fishing has long been an important activity in all countries bordering the North Sea; however, pollution is threatening various species and habitats.

sea and rivers that feed into it, and the operation of the oil and gas industry are just some of the causes of environmental degradation to the North Sea."

To address the pollution problems, the First International Conference on the Protection of the North Sea was held in Bremen in 1984 with participation from Belgium, Denmark, France, Germany, the Netherlands, Norway, Sweden, the United Kingdom, and the European Commission. Since then, there have been a number of meetings. In 1995, the Fifth International Conference on the Protection of the North Sea addressed the following issues:

1. the protection of species and HABITATS;
2. the impact of fisheries on the commercially important fish stocks (*see* commercial fishing), on other fish stocks, and on the marine ECOSYSTEM;
3. the prevention of pollution by hazardous substances;
4. further reduction of nutrient inputs;
5. the prevention of pollution from ships;
6. the prevention of pollution from offshore installations;
7. the management of radioactive substances, including waste.

Progress has been made in some areas. As an example, there has been a reduction of DISCHARGES of certain PESTICIDES into the North Sea. Several pesticides have been phased out. The dumping and INCINERATION of waste in the North Sea, which was common in the 1980s, has now ceased. There has been a substantial reduction (about 50%) of nutrients discharged into the sea from agricultural production.

The next North Sea conference is expected to be held in Norway sometime during 2000–02.

EnviroSources

Fifth International Conference on the Protection of the North Sea Web site: www.odin.dep.no/nsc/background.

North Sea Commission, Business and Development Office: Skottenborg 26, DK-8800, Viborg, Denmark; Web site: www.northsea.org.

not in my backyard, A community attitude that opposes any construction of facilities such as waste treatment plants, nuclear facilities, and incinerators near their neighborhood, especially if such construction may lead to increased pollution or decreased property values. Their attitude is that waste should be treated and stored someplace else. Not-in-my-backyard (NIMBY) opposition also includes the construction of airports, chemical plants, and even shopping centers.

EnviroTerm

not in my term of office, A term applied to politicians who will not advocate or sponsor any controversial legislation, and instead say, "Not in my term of office."

nuclear energy, The ENERGY stored in the nuclei of atoms, which may be released—sometimes in the form of RADIATION—during FISSION or FUSION reactions. In fission, a nucleus absorbs a neutron and splits forming nuclei of elements having lower atomic masses. A fusion, or thermonuclear, reaction occurs when two light nuclei combine to form a single, heavier nucleus. Both types of nuclear reactions produce much more energy per unit of FUEL weight than is produced with conventional materials. When 1 kilogram (2.2 pounds) of URANIUM undergoes fission, for example, it releases energy equal to that released by the combustion of 6,000 metric tons (6,600 tons) of COAL or 18,000 metric tons (19,800 tons) of TNT explosive.

The basic fuel used in nuclear reactors is uranium-235 (U-235). In nature, U-235 occurs along with uranium's much more abundant form, U-238. After laboratory separation from U-238, U-235 is bombarded with neutrons. This impact causes

World Usage of Nuclear Energy

Rank	Country	Electricity from nuclear generators (% of total)
1	Lithuania	77%
2	France	76%
3	Belgium	55%
4	Sweden	46%
5	Ukraine	45%
6	Slovakia	44%
7	Bulgaria	42%
8	Republic of Korea	41%
9	Switzerland	41%
10	Slovenia	38%

U-235 to release neutrons, making a CHAIN REACTION possible. In a NUCLEAR REACTOR, this chain reaction is controlled to maintain a steady reaction rate.

The first fission reaction was achieved in 1939. In 1942, it was experimentally proven that a self-sustaining chain reaction could be produced using U-235 and PLUTONIUM-239. Because these discoveries were made during World War II, nuclear energy was used first for destructive purposes—that is, the creation of the atomic bomb, in which fission progresses rapidly to produce an explosion. Many months were required to produce the materials to build the bomb; in August 1945, two atomic bombs were dropped on the cities of Hiroshima and Nagasaki, Japan. The explosive power of each bomb was approximately equal to that of 18,000 metric tons (19,800 tons) of TNT.

Deuterium, a HYDROGEN RADIOISOTOPE that fuels fusion reactions, is available in large amounts; however, temperatures greater than 1 million°C (1.8 million°F) are required to initiate a fusion reaction.

In the hydrogen bomb, such temperatures are provided by the detonation of a fission bomb. Sustained fusion reactions, however, require the containment of nuclear fuel at extremely high temperatures long enough to allow the reactions to take place. In 1994, U.S. researchers used deuterium and tritium (another hydrogen ISOTOPE) to achieve a one-second fusion reaction that generated about 10.7 million watts of power. Fusion reactions result in a greater conversion of mass to energy than occurs in a fission reaction. A thermonuclear explosion, for example, releases thousands of times more energy as does an atomic bomb.

After World War II, a major effort was made to apply nuclear energy to peacetime uses. Nevertheless, five additional nations—the Soviet Union, the United Kingdom, France, China, and India—soon demonstrated the capability to explode nuclear devices. In 1970, the Treaty on the Nonproliferation of Nuclear Weapons went into effect. Signatory nations without nuclear weapons agreed not to develop them in exchange for the provision of nonnuclear materials and technology from the nations that already had nuclear weapons. In a major effort to limit the nuclear arms race between the United States and the Soviet Union, negotiations such as the Strategic Arms Limitation Talks (SALT) were pursued during the 1980s. The INTERNATIONAL ATOMIC ENERGY AGENCY attempts to ensure that weapons proliferation does not occur.

Nuclear reactors are single- or multiunit facilities in which heat produced by nuclear fission is used to boil water, which produces steam that drives TURBINES, which in turn produces ELECTRICITY. The reaction is initiated, controlled, and sustained at a specific rate. BREEDER REACTORS are designed to produce both power and new fuel at the same time.

When nuclear fission of U-235 occurs, the atom may produce hundreds of fission products, which decay and release radiation. Exposure to such radiation can be hazardous to human health, and repeated exposures have a cumulative effect, damaging reproductive CELLS and causing genetic MUTATIONS that may produce physical defects in future generations.

Development of nuclear energy for the generation of electricity in the United States has been slowed by issues related to safety and waste disposal. A 1979 accident at the THREE MILE ISLAND nuclear reactor near Harrisburg, Pennsylvania, raised questions about the safety of nuclear power. The potential health and environmental effects of the radiation released during this event are still being monitored. The extensive media coverage given the event contributed to public concern about reactor safety, and the construction of new nuclear power plants slowed dramatically in the wake of the accident. International concern over reactor safety increased following the Soviet Union's CHERNOBYL MELTDOWN in April 1986. No deaths have thus far been conclusively attributed to the operation or malfunction of any U.S. commercial nuclear power plant.

Another issue of concern is the question of where to store and dispose of RADIOACTIVE WASTE, largely the SPENT FUEL of reactors, some components of which will remain radioactive almost indefinitely. The waste is currently being held at geologically stable, temporary sites until a permanent solution to the problem can be found. In the 1980s, it was reported that radioactive wastes from such sites had begun to leak into the ENVIRONMENT. At present, the most promising solution to the problem of waste storage involves converting waste material into a glassy substance in a process called VITRIFICATION.

In the United States, control of nuclear-energy activities is the responsibility of the NUCLEAR REGULATORY COMMISSION. The commission grants licenses for the construction and operation of nuclear reactors and for the use of nuclear materials. Among its other duties is the establishment of procedures to protect the health and safety of the public. The construction and operation of nuclear reactors have also come under increased scrutiny by involved state and local governments. *See also* ALTERNATIVE ENERGY RESOURCES; ATOMIC ENERGY ACT; CADMIUM; COOLING POND; COOLING TOWER; FALLOUT; FUEL ROD; HIGH-LEVEL RADIOACTIVE WASTE; MANHATTAN PROJECT; NUCLEAR WASTE POLICY ACT; NUCLEAR WEAPONS; NUCLEAR WINTER; RADIATION SICKNESS; RADIOACTIVE SERIES.

EnviroTerms

${}_0n^1$ = one neutron
${}_{36}Kr^{90}$ = krypton-90
${}_{56}Ba^{142}$ = barium-142
${}_1H^2$ = hydrogen-2 (deuterium)
${}_1H^3$ = hydrogen-3 (tritium)
${}_2He^4$ = helium-4

EnviroSources

American Nuclear Society Web site: www.ans.org.
Hodgson, Peter E. *Nuclear Power, Energy and the Environment.* River Edge, NJ: Imperial College Press, 1999.
Murray, Raymond L. *Nuclear Energy: An Introduction to the Concepts, Systems, and Applications of Nuclear Processes.* Woburn, MA: Butterworth-Heinemann, 2000.
Nuclear Energy Institute Web site: www.nei.org.
Nuclear Information and Resource Service Web site: www.nirs.org.
Ramsey, Charles B., and Mohammad Modarres. *Commercial Nuclear Power: Assuring Safety for the Future.* New York: John Wiley and Sons, 1998.
U.S. Department of Energy, Office of Nuclear Energy, Science, and Technology Web site: www.ne.doe.gov.

U.S. Nuclear Regulatory Commission Web site: www.nrc.gov.
Winnacker, Karl, *Nuclear Energy in Germany.* La Grange Park, IL: American Nuclear Society, 1979.
Wolfson, Richard. *Nuclear Choices: A Citizen's Guide to Nuclear Technology*. Rev. ed. Cambridge, MA: MIT Press, 1993.

nuclear power. *See* nuclear energy.

nuclear reactor, The main structure, sometimes referred to as a vessel, of a nuclear power plant where heat is produced by nuclear FISSION to convert boiling water into steam to drive TURBINES to generate ELECTRICITY. The design of nuclear reactors vary but the general features include a thick, reinforced concrete and steel structure that contains FUEL RODS, pumps, moderators, coolants, and CONTROL RODS. As of 1995, there were approximately 430 nuclear reactor power plants operating worldwide. South Korea, Taiwan, and China are planning to construct nuclear facilities within the next 10 years. France is the only western European country constructing nuclear reactors. The last nuclear plant built in the United States was in 1996 and presently there are no plans to build any additional reactors.

The FUEL used in nuclear reactors is natural URANIUM oxide or uranium–235 (U-235). The enriched fuel is made into pellets and placed inside fuel rods (stainless steel tubes). The fuel rods are joined together to make the reactor core. The U-235 is then bombarded with neutrons to cause a fission reaction.

This releases several neutrons per atom of U-235, making a CHAIN REACTION possible. The chain reaction is regulated by control rods made of CADMIUM or boron to absorb neutrons and maintain a steady reaction rate. The control rods in the reactor core are raised to speed up fission and lowered into the core to slow down the fission.

The moderator in the nuclear reactor is used to slow down the neutrons so that the right speed is maintained for a steady fission rate. The moderators contain a variety of materials such as pure water, heavy water or deuterium, or graphite. Coolants are piped into and out of the reactor core to remove excessive amounts of heat that build up in the reactor. The coolants can be pure water, heavy water, graphite, CARBON DIOXIDE, sodium, or helium. Sometimes the coolant is also the moderator. The heated water is pumped into COOLING TOWERS or in nearby waterways.

The reactor that is most commonly used throughout the world is the light water reactor. These reactors use fuel rods containing low enriched U-235 (3% U-235 and 97% U-238) and pure water as a moderator and a coolant. Light water reactors need to be shut down each year for about six weeks for refueling.

The Canadian designed Candu, or Canada Deuterium Uranium, reactor is similar to the light water reactor but uses heavy water (deuterium) as a moderator and coolant. Unlike the light water reactors, the Candu does not need to be shut down for refueling. Advanced gas reactors use graphite moderators and

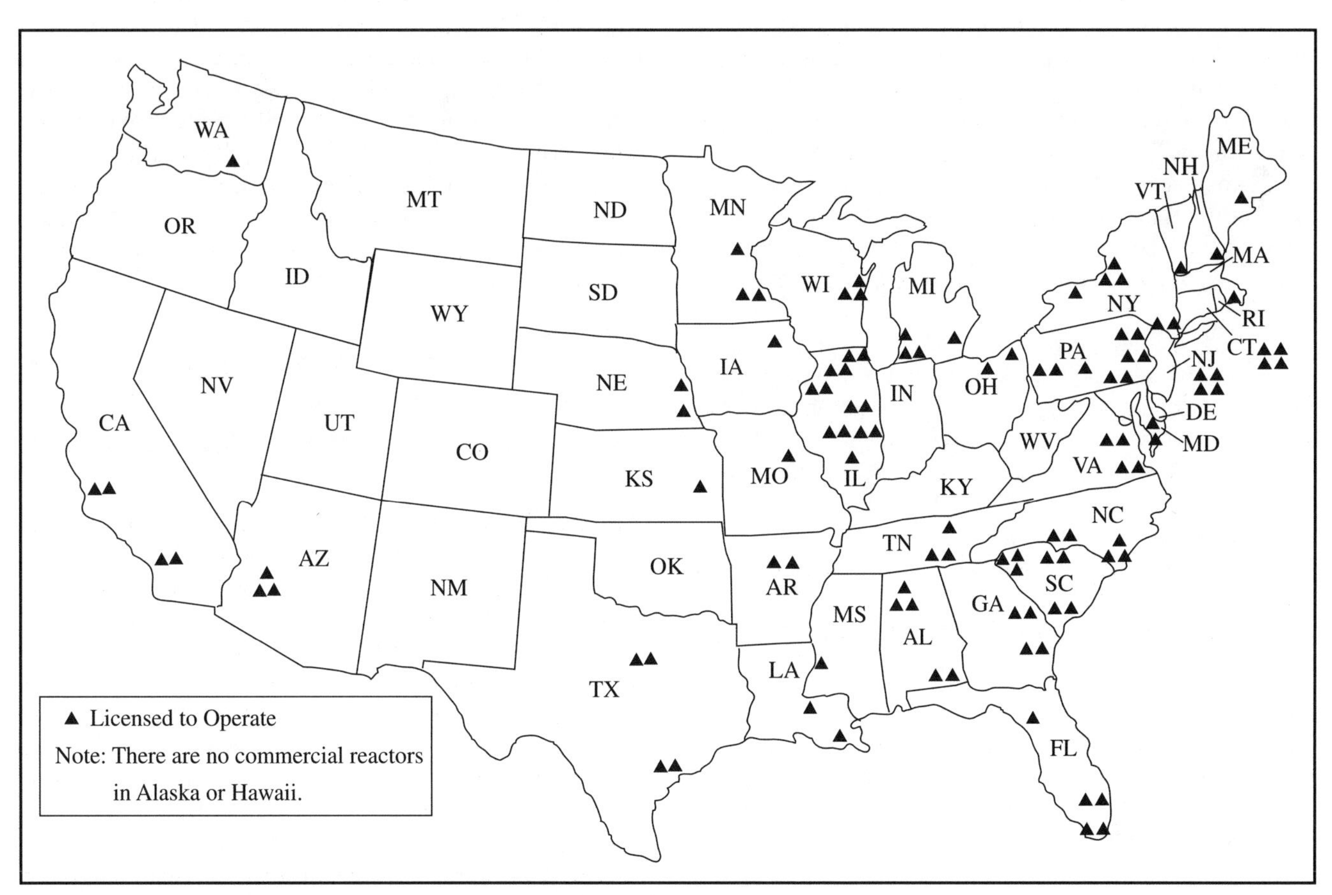

Nuclear Power Plants. The approximate number of commercial nuclear power plants in the United States as of 1999.

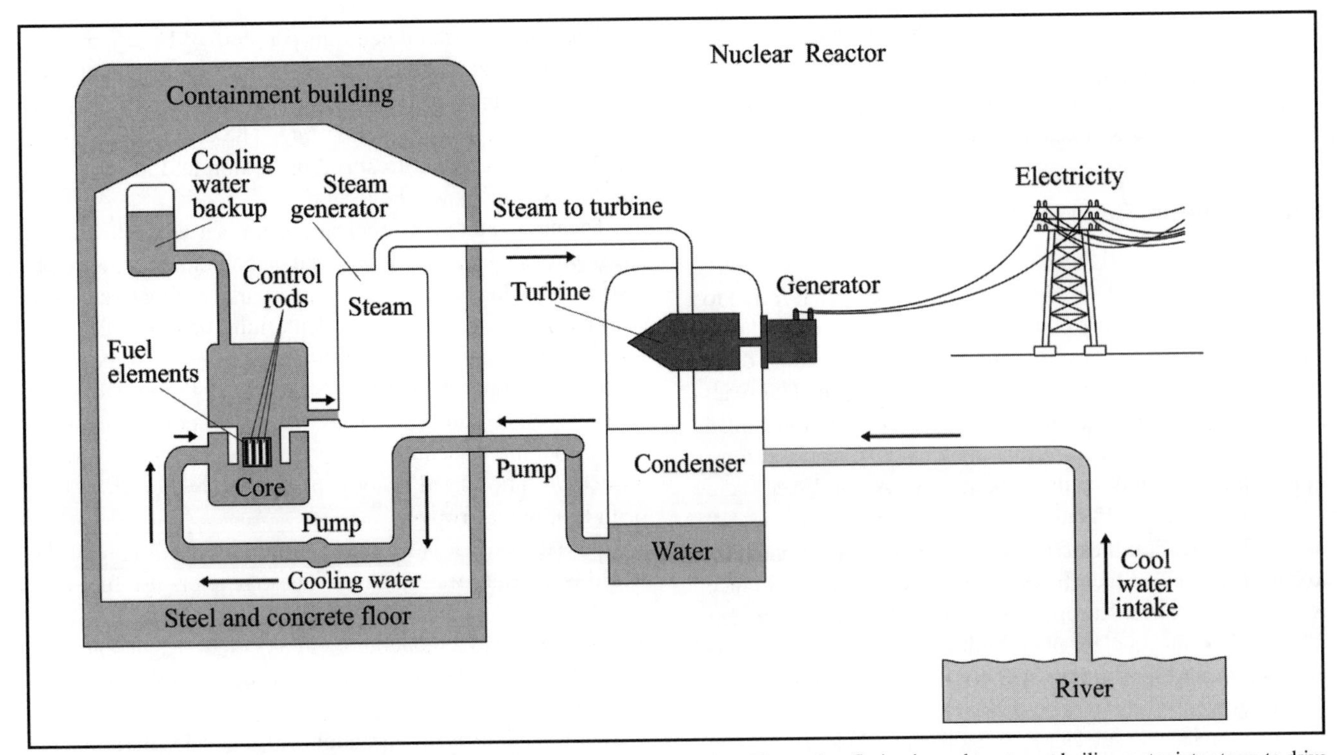

The nuclear reactor is the main component of a nuclear power plant where heat produced by nuclear fission is used to convert boiling water into steam to drive turbines to generate electricity.

carbon dioxide or helium coolants. These reactors are used mostly in Europe.

The fast BREEDER REACTOR is designed to produce both power and new fuel at the same time. A breeder reactor produces fissionable PLUTONIUM-239 (Pu-239) from nonfissionable U-238. Liquid sodium, or a combination of sodium and potassium, is used as a coolant in breeder reactors. The United States, Russia, and France have developed breeder reactors.

Approximately 22% of the U.S. electricity is generated by about 100 operating nuclear reactors in 32 states. Six states rely on nuclear power for more than 50% of their electricity. Thirteen additional states rely on nuclear power for 25–50% of their electricity.

The average nuclear reactor has a life span of about 20 years before it must be retired. In 1999, a 1,000-metric-ton (1,100 ton) reactor, the largest U.S. nuclear power plant ever to be shut down, was shipped to the HANFORD NUCLEAR WASTE SITE in eastern Washington, where it is buried 15 meters (45 feet) below the surface. It is estimated that about 100 nuclear reactors throughout the world will be put out of service by 2000. Since these nuclear reactors are highly radioactive and will remain so for a long time the disposal situation poses a major problem to nuclear-power countries. *See also* NUCLEAR ENERGY.

EnviroSources

Nuclear Information and Resource Service Web site: www.nirs.org.

U.S. Department of Energy, Office of Nuclear Energy, Science and Technology Web site: www.ne.doe.gov.

U.S. Nuclear Regulatory Commission Web site: www.nrc.gov.

Nuclear Regulatory Commission, An agency established by the U.S. Congress under the Energy Reorganization Act of 1974 to replace the Atomic Energy Commission. The Nuclear Regulatory Commission (NRC) ensures adequate protection of the public health and safety, the common defense and security of the nation, and the ENVIRONMENT in the use of nuclear materials in the United States. The NRC's scope of responsibility includes regulation of commercial NUCLEAR REACTORS: nonpower research; test and training reactors; medical, academic, and industrial uses of nuclear materials; and the transport, storage, and disposal of nuclear materials and waste. *See also* ATOMIC ENERGY ACT.

EnviroSource

U.S. Nuclear Regulatory Commission Web site: www.nrc.gov.

nuclear waste. *See* radioactive waste (nuclear waste.)

Nuclear Waste Policy Act, Federal legislation enacted in 1982 and amended in 1987 that authorizes the U.S. DEPARTMENT OF ENERGY (DOE) to design and construct geologic facilities or repositories for the disposal of HIGH-LEVEL RADIOACTIVE WASTES (HLWs) and SPENT FUEL from commercial NUCLEAR REACTORS. In 1987, Yucca Mountain in southern Nevada was designated as a potential site to be studied by the DOE. Yucca Mountain is composed of dense, compacted ASH and is located in an extremely dry area with less than 15 centimeters (6 inches) of precipitation each year. The site has a very deep WATER TABLE, or thick vadose zone. There is no GROUNDWATER above 525 meters (1,680 feet), so the repository could be constructed above the groundwater supply. Despite these features, environmental groups and local citizens have expressed concerns about the site. The DOE and the U.S. GEOLOGICAL SURVEY are conducting studies to evaluate EARTHQUAKE activity, ways to ensure HABITAT preservation, and the socioeconomic impact of constructing such a facility there. If opposition to the site can be resolved, the repository could begin operation by the year 2010. *See also* DEEP-WELL INJECTION; *IN SITU* VITRIFICATION; LOW-LEVEL RADIOACTIVE WASTES; RADIOACTIVE WASTES; VITRIFICATION.

EnviroSources

American Nuclear Society Web site: www.ans.org.
Nuclear Energy Institute Web site: www.nei.org.

nuclear weapon, Any type of weapon, such as a bomb, missile, or torpedo, that is designed to release NUCLEAR ENERGY. Such weapons have the potential to kill great numbers of people with a single explosion and also result in great damage to both human-made and natural structures. In addition, such weapons release great amounts of RADIATION to the ENVIRONMENT, which can have both short-term and long-term effects on humans and other ORGANISMS. Most explosive devices derive their power from the rapid combustion or DECOMPOSITION of chemical compounds. Nuclear explosives, on the other hand, take advantage of the huge amounts of ENERGY stored in the nucleus of an atom.

The first nuclear weapons were developed during World War II as a goal of the MANHATTAN PROJECT, a research effort sponsored by the U.S. government and conducted by American scientists and engineers. The Manhattan Project produced the world's first atomic bomb (nuclear bomb), which was exploded in a remote desert in New Mexico in 1945. This first bomb generated an explosive power equivalent to 13,640–18,180 metric tons (15,000–20,000 tons) of dynamite (TNT). The next month, two more atomic bombs were produced. These were the nuclear bombs dropped on Hiroshima and Nagasaki, Japan, which effectively ended the war.

The first nuclear weapons were FISSION weapons. Fission weapons derive their explosive energy from nuclear fission reactions, which involve the splitting of atoms. The two most common chemical substances used in fission reactions are URANIUM and PLUTONIUM-239.

Another type of nuclear weapon is the thermonuclear, or FUSION, weapon. These weapons generate and release energy through nuclear fusion, or the joining of atomic nuclei. Fusion weapons are much more powerful than fission weapons. Generally, ISOTOPES of HYDROGEN (tritium and deuterium) are used, which is why fusion weapons are sometimes called hydrogen bombs (H bombs). The first H bombs were developed by the United States in the early 1950s. The first successful H bomb was tested in 1952. This bomb, which generated an explosion equivalent to 13.6 million metric tons (15 million tons) of TNT, created a glowing fireball more than 4.8 kilometers (3 miles) in diameter.

When a nuclear weapon explodes, a very high-pressure pulse, or shock wave, is created. As with other explosives, most of the damage to buildings and other structures results from the effects of the initial blast. Nuclear explosions also create fireballs of extremely high temperatures that can instantly incinerate organisms and objects and reduce buildings to rubble. Another dangerous effect associated with nuclear weapons is radioactive FALLOUT. Exposure to high levels of radiation from the blast can cause CANCER and other illnesses in people. Radiation can also damage genes in ways that allow harmful MUTATIONS to occur in offspring. Many scientists also fear that a large-scale nuclear war in which many nuclear weapons are used could lead to widespread changes in CLIMATE, known as NUCLEAR WINTER. *See also* RADIATION SICKNESS; RADIOACTIVE DECAY; RADIOACTIVE WASTE.

nuclear winter, A term used to describe the potential environmental and climatic effects resulting from a large-scale nuclear exchange between nations. The nuclear winter theory was first proposed in 1983 in a paper published in the journal *Science*. In the article, scientists hypothesized that the explosion of just one-half of the combined NUCLEAR WEAPONS in the United States and the former Soviet Union would throw billions of metric tons of dust, soot, smoke, and ASH into the ATMOSPHERE. This, the scientists argued, could produce a blanket of AIR POLLUTION so thick that it would have the potential to block more than 80% of the sunlight now reaching the Northern Hemisphere. As a result, they claimed, severe worldwide climatic changes could occur, including prolonged periods of darkness, below-freezing temperatures, and violent windstorms. The combination of cold temperatures, dryness, and lack of sunlight would also cripple agricultural production and destroy ECOSYSTEMS, putting the majority of the world's POPULATION at risk of starvation.

The nuclear winter theory has been the subject of some controversy. In 1984, the U.S. National Research Council publicly stated that it agreed with the ideas advanced in the *Science* article; however, in 1985 the U.S. Department of Defense issued a report saying that while the nuclear winter theory might be valid, it would not change defense policies. *See also* CLIMATE; FALLOUT; RADIATION; RADIATION SICKNESS.

nutrient cycle. *See* biogeochemical cycle.

Occupational Safety and Health Administration, A U.S. government agency that establishes protective standards for more than 100 million working men and women and their 6.5 million employers. Its mission is to save lives, prevent injuries, and protect the health of the country's workers. The Occupational Safety and Health Administration (OSHA) works diligently in the areas of fire safety, ergonomics, ASBESTOS, preventing needlestick injuries, hearing damage related to NOISE POLLUTION, maintaining confined spaces, and workers' respiratory wellness, and provides legal assistance for workers needing help in getting employment laws enforced.

Founded in 1970 by the Occupational Safety and Health Act of 1970, OSHA has approximately 2,100 inspectors, plus complaint discrimination investigators, engineers, physicians, educators, standards writers, and other technical and support personnel. Workplace inspections are one of OSHA's principal activities.

States can develop and operate their own programs under Section 18 of the Occupational Safety and Health Act of 1970. OSHA approves and monitors state plans. It also provides up to 50% of an approved plan's operating costs. States' job safety and health standards must be at least as effective as comparable federal standards. States can also cover hazards not addressed by federal standards. There are currently 23 states and jurisdictions with complete state plans and two (Connecticut and New York) with a program covering public employees only.

OSHA provides consultation services, voluntary protection programs, training for federal agencies, assists small businesses in setting up programs, and maintains a list of approved blood LEAD laboratories.

EnviroSource

Occupational Safety and Health Administration Web site: www.osha.gov.

ocean, A vast BIOME of Earth's saline waters that provides various HABITATS to more than 250,000 SPECIES of ORGANISMS.

Salts in Ocean Water	
Magnesium sulfate	9.4%
Magnesium chloride	6.4%
Calcium chloride	3.2%
Potassium chloride	2%
Sodium bicarbonate	0.6%
Other	1%
Sodium chloride	77.4%

Oceans help regulate CLIMATE and WEATHER by contributing to the WATER CYCLE and by distributing heat through OCEAN CURRENTS. The oceans also yield an enormous amount of food for humans and are a reservoir for many NATURAL RESOURCES. Humans also rely on oceans and their coastlines for recreation and for transporting goods throughout the world.

The world's oceans cover about 361 million square kilometers (140 million square miles), or about 71% of Earth's total surface area. The three largest oceans are the Atlantic Ocean, the Pacific Ocean, and the Indian Ocean. There are also several smaller oceans, including the ARCTIC Ocean, and the MEDITERRANEAN SEA, the CARIBBEAN SEA, and the BERING SEA, which are enclosed by land or surrounded by island systems.

Ocean Structure. An ocean can be imagined as a giant swimming pool. At the sides of the "pool" are the vast continental shelves, the submerged edges of the continental land masses. Continental shelves extend out to sea an average distance of about 75 kilometers (43 miles), usually sloping gradually downward. At a depth of approximately 200 meters (660 feet), the shelf abruptly gives way to a much steeper zone known as the continental slope, which descends about 3,700 meters (12,000 feet) to the ocean floor. The aphotic zone of the ocean, an area where no light penetrates, is below 800 meters (2,400 feet).

In general, sea-floor features are similar in all oceans. Much of the sea floor is flat and smooth, forming extensive areas called abyssal plains. In some areas, this flatness is broken by mid-ocean ridges, trenches, seamounts, and HYDROTHERMAL OCEAN VENTS.

The sea floor is spreading at the Mid-Atlantic ridge located in the Atlantic Ocean. This process is due to PLATE TECTONICS. Sea-floor spreading occurs when two oceanic plates diverge, move apart from one another. The boundary between the two plates is called a divergent boundary. A mid-ocean ridge forms whenever divergent plates continue to separate. Hot plastic-like material in Earth's crust is forced upward to the surface.

Ocean trenches are long, narrow, steep-sided depressions in the ocean floor. Most trenches are located in the Pacific Ocean. The deepest ocean trench is the Marianas Trench in the Pacific. At 11 kilometers (6.8 miles) deep, it is the deepest place on Earth. A deep ocean trench is formed when two ocean plates collide. A trench can also be formed when one ocean plate (more dense) sinks under another.

Hydrothermal vents, or deep-sea vents, occur at cracks in the ocean floor where the plates of Earth's crust are spreading. Like hot springs, hot MINERAL-rich sea water seeps through the vents. The water is quickly cooled to a temperature of about 23°C (73°F). In the late 1970s, scientists studying the Pacific Ocean floor unexpectedly discovered vibrant ECOSYSTEMS living near these vents. These ecosystems are supported by the food producing activities of chemosynthetic BACTERIA. *(See* chemosynthesis.) Growing on rocks in this warm, fertile ENVIRONMENT are clumps of bacteria that make ENERGY from the HYDROGEN SULFIDE in the water that seeps through the vents. Feeding upon these bacteria are a number of CONSUMERS, including giant clams, lobsters, crabs, and tube worms.

Ocean Water. Ocean water contains a variety of dissolved chemicals, including sodium, CHLORINE, silica, and calcium, of which sodium and chlorine are the most abundant. Ocean water contains on average about 3.5% salt. In areas that receive a lot of fresh water from rivers or precipitation, the salinity level is slightly lower. In drier and more inland regions, such as the Persian Gulf and the Red Sea, the salt content can reach about 4.2%.

Ocean temperature also tends to vary according to location because the sun is the main supplier of heat for Earth's oceans. In places that do not receive much sunlight, such as near the poles, temperature at the ocean surface can be as low as –1.4°C (29.5°F), which is the freezing point of sea water. Near the equator, the area of Earth that receives the most sunlight, ocean temperatures may reach as high as 30°C (86°F). Generally, temperatures decrease with increasing depth, except in polar waters which show the reverse pattern.

Life in the Oceans. The world's oceans contain more animals and PLANTS than anywhere else on Earth. Although oceans are vast, most life is concentrated near the surface and along the coast (the area where ocean water and land meet.) In this area, sunlight is plentiful and water temperature is tolerable.

Marine ECOSYSTEMS can be divided into three distinct groups, depending upon distance from the shore and depth. The neritic zone is the region that exists within a few kilometers of shore. Here, the shallow, warmer waters and the large amounts of nutrients washed from the land promote the growth of PHYTOPLANKTON and other photosynthetic organisms that trap the energy in sunlight to make food. In turn, this large amount of food attracts a variety of animal species, including many types of fishes and birds, sea otters, DOLPHINS AND PORPOISES, MANATEES, and a great variety of INVERTEBRATES, such as jellyfishes, corals, crabs, shrimp, and sea stars.

There is much less BIODIVERSITY in the open ocean, or oceanic zone. Since sunlight can penetrate to only about 180 meters (590 feet) in ocean water, life is concentrated close to the surface. The region of the ocean comprised of water through which sunlight penetrates is called the photic zone. Floating at the surface of the oceanic zone ecosystem are the PRODUCERS, the phytoplankton. The abundant phytoplankton attracts many smaller animals, such as fishes, which in turn draw larger predators, such as larger fishes, dolphins, sharks, and killer WHALES.

The continental slope and the deepest parts of the ocean floor make up the benthic zone. The benthic zone is sparsely populated due to the very cold temperatures and overall lack of producers. Many of the organisms that live in the deep, dark, benthic zone, such as sea cucumbers, shrimp, clams, sea stars, and sea urchins, are SCAVENGERS and DECOMPOSERS that rely on the constant downward flow of decaying ORGANIC matter from the surface. Many organisms that live in the benthic zone are referred to as BENTHOS. Because sunlight does not penetrate into the benthic zone, some deep-dwelling species possess special light-generating organs for attracting live food. The angler fish, for example, dangles a luminous lure over its forehead. When small animals bite at the lure, the angler fish swallows them whole.

Threats to Oceans. Oceans supply the world's human POPULATION with a great amount of food. In the United States alone, seafood is a $10 billion per year industry. Fish and other marine species are also used to produce soaps, medicines, cosmetics, animal feed, FERTILIZER, and other commercial products. The oceans also contain many valuable resources, including PETROLEUM, magnesium, sodium chloride, and bromine. Unfortunately, oceans are being adversely affected by many human activities. For example, dolphins, many species of fishes, crabs, shrimp, clams, and other marine species, are being caught and harvested in record numbers. OVERFISHING threatens some species with EXTINCTION.

Unintentional DISCHARGE of oil from tankers and OCEAN DUMPING also represent significant threats to oceans. Most sources of ocean pollution, however, are land-based. Oceans become polluted when factories and towns discharge SEWAGE and industrial wastes into rivers, streams, and harbors. A significant amount of ocean pollution also comes from NONPOINT SOURCES, such as water RUNOFF from farms, cities, and construction sites. Non-point sources discharge POLLUTANTS such as PESTICIDES, fertilizers, oils, paints, and other chemicals, which collect along shores and poison organisms.

Saving Ocean Ecosystems. In the United States, a number of laws protecting oceans have been passed as public awareness of ocean pollution has grown. The 1977 CLEAN WATER ACT and the MARINE PROTECTION, RESEARCH, AND SANCTUARIES ACT passed in 1988 are important laws that authorize the U.S. ENVIRONMENTAL PROTECTION AGENCY to regulate ocean pollution. The 1972 MARINE MAMMAL PROTECTION ACT protects dolphins, whales, sea otters, and other marine MAMMALS in U.S. waters by regulating the types of nets fishing boats use for their catch.

On an international level, the International Convention for the Prevention of Pollution from Ships was developed in 1973 to help control ocean pollution. THE LAW OF THE SEA CONVENTION, developed in 1982, is another important international agreement to control ocean pollution. Today, the convention has been signed by more than 160 nations. *See also* COMMERCIAL FISHING; CORAL REEF; EL NIÑO/LA NIÑA; OCEAN DRILLING PROGRAM; OCEANOGRAPHY; TSUNAMI; VOLCANO.

EnviroTerms

nekton: Free-swimming animals that live in aquatic ECOSYSTEMS.

pelagic zone: An area where animals live suspended in water.

EnviroSource

Center for Marine Conservation: 1725 DeSales Street, SW, Suite 500, Washington, DC 20036; (202) 429-5609.

Knight, Deborah. "Underwater Wilderness." *Audubon* 102:1 (January-February 2000): 62–69.

Longhurst, Alan R. *Ecological Geography of the Sea.* Boston, MA: Academic Press, 1998.

National Oceanographic and Atmospheric Administration Web site: www.noaa.gov.

National Oceanographic and Atmospheric Administration (safe ocean navigation) Web site: anchor.ncd.noaa.gov/psn/psn.htm.

Talen, Maria. *Ocean Pollution.* Lucent Overview Series. San Diego, CA: Greenhaven Press, 1991.

Van Dyke, Jon M., et al. *Freedom for the Seas in the Twenty First Century: Ocean Governance and Environmental Harmony.* Washington, DC: Island Press, 1993.

Whitfield, Peter. *The Charting of the Oceans: Ten Centuries of Maritime Maps.* Beverly Hills, CA: Pomegranate, 1996.

ocean current, A mass movement or flow of OCEAN water. Ocean currents carry water in various directions. Surface currents—the movement of only the upper few hundred meters of sea water—are the main type of ocean current. Powered by the wind, they drive ocean water in huge circular patterns all around the world. Surface currents are important because they help distribute heat from equatorial regions to other areas of Earth, thus influencing global CLIMATE.

Surface currents are related to the general circulation of surface winds on Earth. Surface winds are influenced by the Coriolis effect, the deflection of AIR and water due to the rotation of Earth. Because of the way Earth rotates, surface currents north of the equator, such as the Gulf Stream, move to the right. Currents south of the equator are deflected to the left. The continents also influence ocean currents. For instance, currents moving west in the Pacific Ocean are deflected by the continents of Asia and Australia. Then the currents, influenced by the Coriolis effect, move eastward until they are deflected once more by North and South America.

Sailors use ocean currents to quickly navigate from place to place. But the primary importance of ocean currents to life on Earth is the way they influence climate. Warmer waters flowing in ocean currents away from the equator transfer heat to the ATMOSPHERE. Because water heats up and cools down more slowly than land, many coastal areas are warmer in the winter and cooler in the summer than inland areas of similar latitude. Iceland, for example, is located far north in the Atlantic Ocean; however, because the Gulf Stream, a warm ocean current that flows north along the eastern edge of North

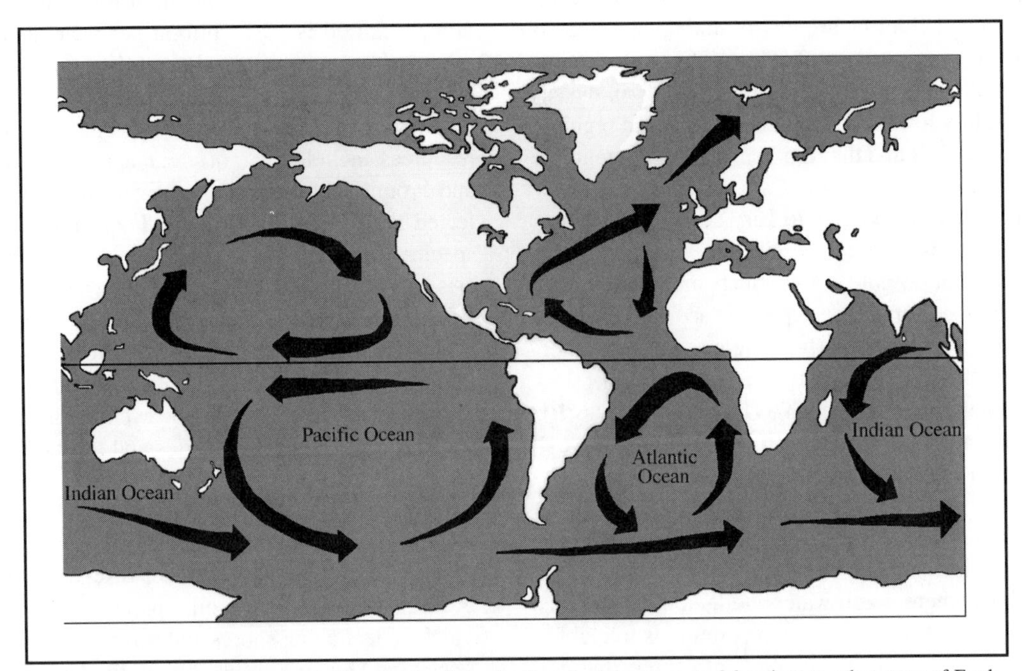

Ocean surface currents are important because they help distribute heat from equatorial regions to other areas of Earth, thus influencing global climate.

America, flows past Iceland, the country has a surprisingly mild climate.

Wind-driven surface currents affect only the upper layers of Earth's oceans. In the depths of the ocean, water circulates not because of wind but because of density differences. Density currents occur when more dense sea water sinks under less dense sea water. Density differences in sea water are due to changes in salt content, pressure, and, primarily, temperature. Cold water is more dense than warm water. Very salty oceans, such as the Red Sea in the Middle East, for example, have a higher density than other oceans. Density currents move much more slowly than surface currents. One important density current occurs in Antarctica where the most dense ocean water forms. As sea water freezes, forming ice on the surface, the salt is left below in the unfrozen water, increasing its density. This dense water sinks and slowly spreads along the sea floor toward the equator, forming a density current.

A third type of ocean current is known as an UPWELLING. An upwelling is a current that brings deep, cold water up to the ocean surface. Found along some coasts, upwellings occur when wind blowing offshore pushes water away from the land. When surface water is pushed away, cold water from the deep ocean rises. Upwellings are important because the cold, deep water brings high concentrations of nutrients with it. The nutrients supply fish, shrimp, crabs, and other marine organisms with plenty of food. This is why some coastal areas of Oregon, Washington, and Peru are prime fishing locations. *See also* AQUACULTURE; COMMERCIAL FISHING; OCEANOGRAPHY.

EnviroSources

National Oceanographic and Atmospheric Administration Web site: www.noaa.gov.

National Oceanographic and Atmospheric Administration (safe navigation) Web site: www.anchor.ncd.noaa.gov/psn/psn.htm.

Ocean Drilling Program, An international partnership of scientists and government agencies investigating Earth's history by studying the structure and history of the sea floor. The Ocean Drilling Program (ODP) is funded primarily by the National Science Foundation with substantial contributions from many international partners, including Germany, France, Japan, and the United Kingdom.

The ODP consists of basic research into the history of the OCEAN and the nature of the crust beneath the ocean floor. Every year, scientists cruise the world's oceans on board the drill ship *JOIDES Resolution.* (JOIDES is an acronym for Joint Oceanographic Institutions for Deep Earth Sampling, and *Resolution* is in honor of the HMS *Resolution,* a ship commanded by the English Captain James Cook over 200 years ago, which explored the Pacific Ocean and the Antarctic.) During these excursions, which last approximately eight weeks, holes are drilled into the sea floor to penetrate millions of years of Earth's geologic past. From these holes, cores—slender cylinders made of sediment and rock—are retrieved. The cores, each approximately 9.5 meters (31 feet) in length, reveal many clues about Earth's history. Scientists from a variety of disciplines—including GEOLOGY, METEOROLOGY, biology, paleontology, and OCEANOGRAPHY—examine these cores to learn about Earth's basic processes, including CONTINENTAL DRIFT, the EVOLUTION of marine life, and the changes over time of global CLIMATE, OCEAN CURRENTS, sea level, and Earth's magnetic field.

Investigating Earth's history by ocean drilling would not be possible without a drill ship uniquely outfitted for this kind of work. The *JOIDES Resolution* is 144 meters (471 feet) long and 21 meters (70 feet) wide. On board are 28 scientists, about 20 ODP engineers and technicians, and a crew of 62. The ship is able to maintain its location, even in heavy seas, by means of computer-controlled thrusters. In addition, the ship contains seven levels of laboratories and other scientific facilities for studying the physical properties of the core samples, as well as paleomagnetism, paleontology, chemistry, x-ray analysis, photography, and geophysics. Advanced computing equipment is also available to aid the scientists in capturing and processing data electronically. Since 1985, *JOIDES Resolution* has recovered nearly 200 kilometers (124 miles) of CORE SAMPLES. *See also* PLATE TECTONICS; SEA LEVEL RISE.

ocean dumping, The deliberate dumping or accidental DISCHARGE of garbage, oil, industrial wastes, SEWAGE, and other POLLUTANTS into the world's OCEANS. Ocean pollution kills marine animals, poisons CORAL REEFS and other aquatic ECOSYSTEMS, and causes health problems in people who consume seafood contaminated with pollutants.

For centuries, oceans have been a favorite dumping ground for waste from human activity. Surprisingly, however, most ocean pollution comes from land-based sources. RUNOFF from farms, lawns, cities, and towns is the most significant source of ocean pollution. When pollutants, such as industrial wastes, sewage, discarded oil from cars, PESTICIDES, and FERTILIZERS, enter rivers as runoff, the rivers may carry the polluted water to the ocean. Coastal development—MINING and the construction of homes, hotels, DAMS, and canals—is another major source of land-based ocean pollution. Coastal development not only contributes to runoff pollution, but also destroys HABITATS and breeding grounds for fish, shellfish, and other aquatic ORGANISMS.

Pollutants are also dumped directly into the oceans. Millions of gallons of oil, for instance, are either accidentally or illegally dumped into the oceans each year. Most OIL POLLUTION comes from the illegal and accidental discharge of oil from tankers, oil drilling platforms, ships, and recreational boats. Large OIL SPILLS, such as the 1989 *EXXON VALDEZ OIL SPILL*, which released more than 41.6 million liters (11 million gallons) of oil into Prince William Sound, Alaska, are rare however, and account for only 5% of the oil polluting the oceans.

GARBAGE, particularly plastic, is also a significant ocean pollutant because it does not break down easily. *(See* nonbiodegradable.) Each year, millions of tons of plastic six-pack rings, garbage bags, nylon fishing nets, fishing line, construction materials, and other forms of garbage are dumped illegally into the oceans. Many animals are suffocated to death when they become entangled in the nets or other plastic debris. Turtles and other animals may also eat clear plastic bags that resemble jellyfish and die from suffocation or blockage of the digestive system. The U.S. Office of Technology Assessment estimates that discarded plastic alone kills more than 1 million birds and more than 100,000 SEALS and SEA LIONS, otters, WHALES, DOLPHINS and porpoises, sharks, and turtles annually.

Until the 1970s, the world's oceans were largely considered by industry and certain governments as convenient "out-of-site, out-of-mind" dumping grounds. Since then, however, public and political perceptions have changed about ocean dumping. People and governments worldwide now recognize that potential polluters should deal with their own wastes rather than dump them. Today, there are a number of laws making it illegal to dump wastes into the ocean. In the United States, the most significant law protecting against ocean dumping is the Ocean Dumping Ban Act. (*See* Marine Protection, Research, and Sanctuaries Act.) Passed in 1972, this important law bans ocean dumping of nuclear, chemical, and biological warfare agents and HIGH-LEVEL RADIOACTIVE WASTE. Amendments in 1988 extended this ban to sewage SLUDGE, industrial waste, and medical wastes. The Ocean Dumping Ban Act also authorizes research on the effects of ocean pollution, OVERFISHING, oil spills, and other human-induced problems. Provisions added in 1992 established a national coastal water quality monitoring program to evaluate the health and quality of ocean waters and the pollution sources that affect them.

On an international level, the London Convention protects ocean waters worldwide. According to the London Convention, all countries shall individually and collectively promote the effective control of all sources of pollution of the marine ENVIRONMENT and pledge themselves to take all practical steps to prevent the pollution of the sea by the dumping of waste and other matter that is liable to create hazards to human health, to harm living resources and marine life, to damage amenities, or to interfere with other legitimate uses of the sea. Also enacted in 1972, the London Convention has now been signed by more than 70 countries, including the United States, Canada, Japan, France, Germany, and China. *See also* CLEAN WATER ACT; NONPOINT SOURCE; POINT SOURCE; UNITED NATIONS CONVENTION ON THE LAW OF THE SEA.

Ocean Dumping Ban Act. *See* Marine Protection, Research, and Sanctuaries Act.

ocean thermal energy conversion, An ALTERNATIVE ENERGY RESOURCE that uses the natural temperature differences between various layers of OCEAN water to produce ELECTRICITY. The idea of using ocean thermal energy conversion (OTEC) to produce electricity is not new. A small OTEC plant was built off the coast of Cuba in the 1930s. The plant produced electricity for the island country until it was destroyed by a HURRICANE. Another plant was built in 1956 off the coast of Africa. This plant was later replaced by a DAM that generates HYDROELECTRIC POWER at a lower cost than did the OTEC plant.

OTEC systems work best in the tropical waters of the central Pacific Ocean, the Indian Ocean, and in the Gulf of Mexico region of the Atlantic Ocean. In these regions, temperature differences between warm surface waters and colder water, at depths of 1,000 meters (3,280 feet) or more, is sufficient to generate electricity.

OTEC plants can be installed on land or in the ocean. One kind of OTEC system is the closed-cycle system. The closed-cycle OTEC plant consists of pipes arranged in a closed loop extending down into the ocean. A liquid chemical with a low boiling point is placed inside the pipes. The loop is connected to the TURBINE of an electric generator. Warm surface sea water at the top of the loop is circulated around the pipes causing the liquid inside the pipes to be heated and changed to a gas. Movement of this gas through the pipes then causes the turbine to rotate and thus generate electricity. After passing through the turbine, the gas flows downward into the bottom part of the loop, where cold water pumped from the deep ocean is circulating around the pipes. The cold water absorbs heat ENERGY from the loop, causing the gas to condense back into a liquid. The cycle then repeats over and over.

Another kind of OTEC system is the open-cycle system. In this type of plant, warm ocean water is boiled within a vacuum chamber. As the water evaporates, it produces low-pressure steam that is used to generate electricity. Cold ocean water is then used to condense the steam into fresh water, which can be pumped to communities for use as drinking water or to agricultural regions for use in IRRIGATION.

OTEC research and feasibility studies are currently being conducted by the state of Hawaii and by the Japanese government. The attraction of the OTEC system is that it may be useful in generating both electricity and fresh water. Water supplied by the systems can also be used to raise seafood in commercial AQUACULTURE projects, for air conditioning, and for refrigeration. OTEC systems do have some drawbacks. One is that the systems are useful only in tropical areas, which are subject to seasonal NATURAL DISASTERS such as hurricanes and TYPHOONS. Such storms can completely destroy an OTEC plant. Another drawback is that electricity produced using OTEC systems is more costly than electricity generated by methods such as hydroelectric power and the combustion of FOSSIL FUELS. Unlike fossil fuels, however, OTEC systems do not release harmful POLLUTANTS into the ATMOSPHERE. A major concern about the use of OTEC systems is that they may alter ocean water temperatures in the areas where they are used. If water temperatures are altered too much, it can affect the ability of the region to support sea life or cause a change in the SPECIES diversity in the area. Construction of OTEC plants and laying pipes in coastal waters may cause localized damage to reefs and near-shore marine ecosystems. Limited OTEC research continues in other countries, especially Japan, Canada, Great Britain, France, and Taiwan. *See also* BIODIVERSITY; DESALINATION; THERMOCLINE.

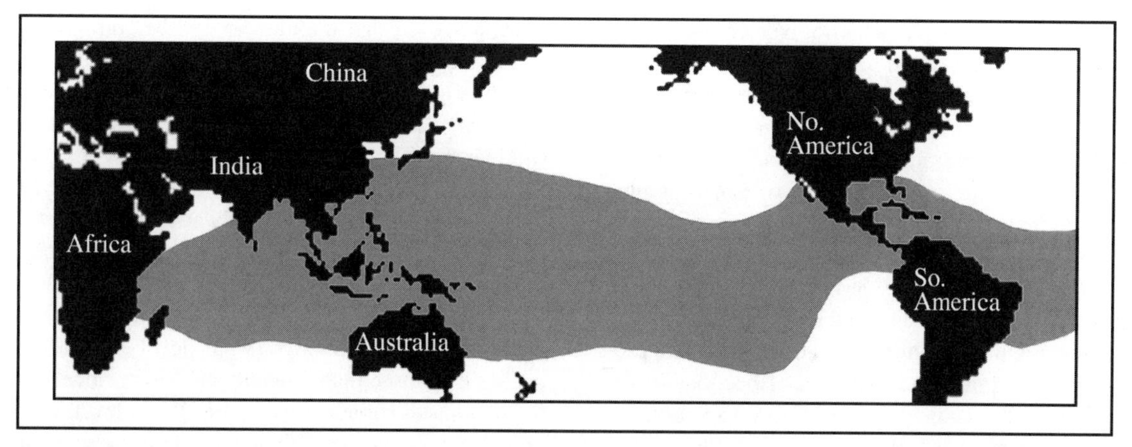

Ocean thermal energy conversion systems work best in the tropical waters of the central Pacific Ocean, the Indian Ocean, and in the Gulf of Mexico region of the Atlantic Ocean.

EnviroSource

National Renewable Energy Laboratory: 1617 Cole Boulevard, Golden, CO 80401; Web site: www.nrelinfo.nrel.gov.

Natural Energy Laboratory of Hawaii Web site: www.bigisland.com/nelha/index.html.

U.S. Department of Energy, Energy Information Administration, National Energy Information Center: 1000 Independence Ave., SW, Forrestal Building, Washington, DC 20585; (202) 586-8800.

oceanography, A scientific discipline dealing with the physical, chemical, geological, and biological properties of the ENVIRONMENT of the OCEANS. The science of oceanography consists of four subdisciplines: physical oceanography, which is concerned with the physical properties of ocean water (temperature, pressure, density), as well as its movement in waves, TIDES, and OCEAN CURRENTS; chemical oceanography, which deals with the chemical makeup of ocean water; geological oceanography, which deals with the structures and features of the ocean floor; and marine ECOLOGY, or biological oceanography, which involve the study of PLANTS, animals, and other ORGANISMS living in the sea. *See also* OCEAN DUMPING.

Odum, Eugene (1913–), Environmental advocate known as the father of modern ECOSYSTEM ECOLOGY. Eugene Odum was one of the most influential figures in the history of twentieth-century ENVIRONMENTAL SCIENCE. After years of retirement, Odum continued to challenge scientific assumptions about the ENVIRONMENT. Odum took a broad view of the environment, pioneering ecosystem ecology as a new integrative science. He explored connections between pristine WATERSHEDS, PLANTS, animals, CLIMATE, and WEATHER. His groundbreaking 1953 text, *Fundamentals of Ecology*, influenced an entire generation of ecologists. In this book, Odum distinguished the concept of the ecosystem from that of the BIOLOGICAL COMMUNITY for the first time. In his view, ecosystems were able to store and transform ENERGY and recycle nutrients, comprising a more complex set of functions than biological communities alone. For 10 years, his was the only ecosystem ecology textbook available worldwide. As a result, his influence in the field of environmental science was enormous. Former U.S. President Jimmy Carter suggested that Odum's ideas changed the way people viewed the natural world and their place in it.

Odum believed that ecosystem theory provided a common denominator for humans and nature and that the "goods and services" of both were intertwined. Before Odum, ecology had been studied only on a small scale within other scientific disciplines. In fact, many scientists doubted that it could be studied on a large scale, such as through an examination of how one natural system interacted with another. Odum, however, made "ecosystem" a household word. Odum began his teaching career at the University of Georgia (UGA) as a ZOOLOGY professor in 1940. He founded the Institute of Ecology there in 1961, which he directed until his retirement in 1984, after which he continued to serve as Director Emeritus of the institute. In 1954, Odum established UGA's Marine Institute on Georgia's Sapelo Island, the southern portion of which had been donated by tobacco magnate R.J. Reynolds for the study of coastal MARSH ecosystems. Odum also was responsible for the founding of the Savannah River Ecology Laboratory, operated by the university under contract with the U.S. DEPARTMENT OF ENERGY (DOE). *(See* Savannah River Site.) This 483-square-kilometer (300-square-mile) area has become one of the largest outdoor science classrooms in the world. *See also* AGROECOSYSTEM; BIOGEOCHEMICAL CYCLE; BIOME; BIOREGION; BIOSPHERE; BIOSPHERE RESERVE; ECOTONE; OCEAN.

EnviroSource

Eugene Odum: An Ecologist's Life. Videotape, 30 minutes. University of Georgia Center for Continuing Education, Athens, Georgia 30602-3603; (800) 359-4040; Web site: www.ssl.gactr.uga.edu/tv/videocatalog.

Office of Surface Mining Reclamation and Enforcement, An office created by the U.S. DEPARTMENT OF THE INTERIOR, that has the responsibility of protecting the ENVIRONMENT during COAL MINING. The Office of Surface Mining Reclamation and Enforcement (OSMRE) ensures that land used for coal mining is reclaimed after coal has been removed as required by the SURFACE MINING CONTROL AND RECLAMATION ACT. As an example, one of the largest producing coal mines in the country, Kerr-McGee's Jacobs Ranch Mine near Gillette, Wyoming, has reclaimed thousands of hectares of mined land and has returned that land to its previous use—cattle grazing—and to WILDLIFE HABITAT for herds of deer and antelope.

The OSMRE is responsible for publishing the rules and regulations needed to enforce the law. The rule-making process includes discussions with coal industry representatives, citizen groups, tribal groups, and state regulators to obtain their input and suggestions. The OSMRE strives to maintain consistency among state programs and to ensure compliance with the law and regulations through evaluation of state programs. The OSMRE plans and conducts inspections, independent reviews, and technical analyses, and also provides technical assistance to coal states, tribes, and industries to improve the effectiveness of the law.

In 1994, the OSMRE started the Appalachian Clean Streams Initiative to develop cleanup strategies to reduce and eliminate ACID MINE DRAINAGE from the nation's streams and rivers. The principle source of acid mine drainage in the Appalachian area is from abandoned coal mines. The OSMRE along with more than 100 government agencies, environmental groups, WATERSHED organizations, coal producers, and local individuals have joined together to create acid mine drainage projects. More than 15 projects have been funded and a growing number of citizen groups have become involved with the stream restoration program.

The OSMRE maintains the Abandoned Mine Land Inventory System. The office generates reports on abandoned mines land accomplishments, and problems that still require RECLAMATION. There are more than 10,000 areas, such as watersheds, that contain one or more abandoned mine problems. Hazardous abandoned mine problems include open shafts; unstable, cliff-like high walls; rusting machinery; defective explosives; and dangerous flooded areas. The OSMRE warns against entering into any abandoned mine. And if an abandoned mine is discovered the location should be reported to the agency.

EnviroSources

Office of Surface Mining Reclamation and Enforcement: 1951 Constitution Ave., N.W., Washington, DC 20240; (202) 208-2719; Fax: (202) 501-0549; Web site: www.osmre.gov.

Office of Surface Mining Reclamation and Enforcement (Appalachian Clean Streams Initiative) e-mail: majordomo@osmre.gov.

Ogallala aquifer, The largest AQUIFER and GROUNDWATER source in North America whose water level has been declining due to an OVERDRAFT of groundwater used primarily for IRRIGATION. The Ogallala aquifer is located in the midwestern United States, beneath portions of South Dakota, Nebraska, Colorado, Kansas, Oklahoma, New Mexico, Wyoming, and northern Texas. The underlying aquifer is approximately 440,000 square kilometers (170,000 square miles) of which most of the volume of water is under Nebraska. Also known as the High Plains Aquifer, the aquifer supplies domestic water and irrigation water for approximately 4,005,000 hectares (10 million acres) of farmland. Some environmentalist believe, however, that the aquifer will be completely depleted by 2020 due to the amount of water being withdrawn.

The Ogallala Aquifer has an average thickness of about 61 meters (200 feet), consisting largely of gravel, sand, and silt deposits. The quantity of water stored within the Ogallala Aquifer is estimated to equal the volume of Lake Huron, one of the Great Lakes. The groundwater level beneath the aquifer has been dropping since the 1940s, which indicates that the amount of water withdrawn from the aquifer is greater than the amount of water entering (or replenishing) it. In parts of Texas and New Mexico, water levels in the aquifer have declined by more than 30 meters (90 feet) since water-pumping operations started in the 1940s. Water levels in the aquifer underlying Kansas and Oklahoma have also declined. As a result of the depletion of water in the aquifer, farming areas in parts of Texas and New Mexico have cut back on using the aquifer for irrigation purposes. *See also* POTABLE WATER; ZONE OF DISCHARGE.

oil. *See* petroleum.

oil spill, Any accidental or intentional DISCHARGE of PETROLEUM into the ENVIRONMENT from POINT SOURCES such as pipelines, OIL TANKERS, and land- and off-shore drilling oil rigs. Oil is sometimes spilled from buried GASOLINE and petroleum tanks, and this is becoming a widespread and expensive problem. Oil is also discharged via NONPOINT SOURCES, such as recreational power boat usage and in RUNOFF from the surface pavements of roads, parking garages, and gasoline stations. Most of this runoff finds it way into waterways, often by way of storm SEWERS, causing water pollution. Runoff from rivers and other waterways represents about 30% of all oil pollution sources in the OCEAN.

Large OIL SPILLS sometimes occur in ocean waters when tankers run aground or leak due to collisions. Some estimates indicate that tanker accidents are the source of about 3–5% of oil pollution in the ocean; however, most oil pollution in the ocean results from less spectacular events, such as accidental discharges of oil when oil tankers are loaded and unloaded and minor spillages from tankers as they transport oil across the ocean.

Oil pollution can have devastating effects on WILDLIFE and ECOSYSTEMS. Oil is toxic and directly kills small animals, such as fishes, shrimp, crabs, and other shellfishes. Other animals are poisoned indirectly when they feed upon oil-soaked prey. In addition, birds, sea otters, and other large animals are harmed by oil when it forms a slimy coating over their feathers or fur. Oil, a thick, black liquid, does not wash off easily in water and destroys the insulating properties of feathers and fur. As a result, oil-contaminated animals develop hypothermia and often freeze to death.

Today, laws help protect against oil pollution in the oceans. For example, the Oil Pollution Act of 1990 is a federal law of the United States that increases the legal liability of oil tanker owners by imposing strict regulations regarding oil transport.

Ogallala Aquifer. The Ogallala aquifer in Midwestern United States is the largest aquifer and groundwater source in North America.

Sources of Oil Pollution in Oceans	
Runoff that enters rivers and waterways	31%
Tanker activity: loading, unloading, etc.	20%
Sewage plants and refineries	13%
Underwater seepage from cracks in ocean floor	9%
Pleasure boots, fishing vessels, ferries, etc.	9%
Oil tanker accidents	3%–5%

The amount of oil released into the environment each year may exceed 1.893 billion liters (500 million gallons), most of which ends up in lakes, streams, or the ocean. Most of this oil, more than 1.325 billion liters (350 million gallons), derives from improper disposal of used motor oil, oil leaks from AUTOMOBILES, or from improperly disposed industrial wastes. Oil from these sources is either deliberately dumped into drains or carried off land and into bodies of water with runoff. The second-largest source of oil occurs through routine ship maintenance, including the cleaning of bilges. Such maintenance releases more than 473 million liters (125 million gallons) of oil into the environment each year.

Much of the oil released into the environment is released into the air as EMISSIONS from automobiles and industry. Once airborne, particles containing petroleum byproducts (HYDROCARBONS) may settle out of the ATMOSPHERE or be washed back to Earth by rain or other forms of precipitation. *(See* water cycle.) The types of oil discharge that receive the most attention are from oil spills involving large OIL TANKERS. Such spills actually account for only about 3%–5% of all oil spills, but a single, large spill can devastate the area in which it occurs for years. (*See* Time Capsule.) Similar devastation may result from petroleum discharges resulting from offshore drilling sites.

Four other sources account for most other releases of oil into the environment: A fire or explosion at a refinery, where crude oil is processed, may release oil onto land or into water, but most of the oil is released into the air as smoke and particulates. Leaks in pipelines that transport oil across land (either above or below ground) can allow large amounts of oil to be discharged into soil. The potential for such an occurrence has continued to be a major concern of environmentalists, especially with regard to the ALASKA PIPELINE, which transports oil more than 1,300 kilometers (808 miles) across the state of Alaska, from Prudhoe Bay in the north to the Port of Valdez in the south. Accidents involving trucks or trains that transport oil above ground also can result in large discharges onto land or in water. Finally, some of the oil that pollutes Earth's oceans results from natural processes: Estimates suggest that more than 227 million liters (60 million gallons) of oil each year may seep from the ocean floor as oil-containing sedimentary rocks are eroded. *See also* *AMOCO CADIZ* OIL SPILL; ARCTIC NATIONAL WILDLIFE REFUGE; *EXXON VALDEZ* OIL SPILL; OIL SPILL EQUIPMENT; TIMES BEACH SUPERFUND SITE.

Time Capsule Major Oil Spills	
Year	**Spill/Effects**
1967	***Torey Canyon* Oil Spill.** Torey Canyon runs aground off Cornwall, England, spilling about 175 tons of crude oil.
1969	**Santa Barbara Oil Well Blowout.** An oil well blowout near Santa Barbara, California, leaks about 2,700 tons of crude oil into the Pacific Ocean.
1973	***Corinthos–Edgar M. Queeny* Collision.** In a collision in the Delaware River in Marcus Hook, Pennsylvania, the port anchor of the *Edgar M. Queeny* penetrates the hull of the *Corinthos,* resulting in an explosion. The 20,000 tons of chemical cargo from the *Edgar M. Queeny*, which includes gasoline, methanol, phenol, vinyl acetate, and styrene monmer, and the 272,000 barrels of crude oil carried on the *Corinthos* are released into the air as a result of the fire or into the waters of the Delaware River .
1976	***Argo Merchant* Oil Spill. Argo Merchant** runs aground off Nantucket, Massachusetts, releasing almost 25,000 tons of fuel oil into the Atlantic Ocean.
1976	***Hawaiian Patriot* Oil Spill.** *Hawaiian Patriot* catches fire, releasing almost 100,000 tons of oil into the Pacific Ocean.
1977	**Ekofisk Oil Well Blowout.** An oil well blowout results in nearly 27,000 tons of crude oil being spilled into the North Sea.
1978	***Amoco Cadiz* Oil Spill.** *Amoco Cadiz* runs aground off Portsall, Brittany, spilling 226,000 tons of oil into the ocean.
1979	***Atlantic Empress–Aegean Captain* Oil Spill.** A collision between the *Atlantic Empress* and the *Aegean Captain* spills more than 370,000 tons of oil into the Carribbean.
1983	**Iran-Iraq War Oil Spill.** Iraq attacks wells of Nowuz oil field in Iran, resulting in 600,000 tons of oil being released into the Persian Gulf.
1988	**Ashland Oil Spill.** Collapse of a storage tank releases nearly 2,500 tons of oil into the Monongahela River near Pittsburgh, Pennsylvania.
1989	***Exxon Valdez* Oil Spill.** The *Exxon Valdez* runs aground in Prince William Sound off the Alaskan coast, releasing about 37,000 tons of oil into the sound.
1989	***Kharg 5* Oil Spill.** The *Kharg 5* oil tanker catches fire off the Canary Islands, releasing 75,000 tons of oil into the surrounding water.
1991	**Persian Gulf War Oil Spills.** Iraqi troops deliberately spill oil stored at Sea Island Terminal in Kuwait and set the oil fields ablaze. Much of the oil enters the water of the Persian Gulf or spreads over the surrounding land. That which does not seep into water or soil is released into the air through the fires that burn for well over a year.
1994	**Komi Republic Oil Spill.** A dike constructed to contain oil leaking from a pipeline near Usinsk in northern Russia (just below the Arctic Circle) collapses releasing nearly 102,000 tons of oil onto the Siberian tundra.
1999	***New Carissa* Oil Spill.** The *New Carissa* runs aground in Coos Bay off the coast of Oregon. The tanker carries more than 1,500 tons of fuel oil, some of which leaks into the water over a period of days. Some remaining oil is pumped from the tanker to containers on shore. Next, the tanker is set ablaze, believing release of hydrocarbons into the air will pose less of a threat than its release into the water. Finally the tanker is towed into the ocean, where it is expected to rest 1,825 meters (6,000 feet) below the surface. Temperatures on the ocean floor are believed to be cold enough to keep the oil in a solid state that will pose little environmental threat.

EnviroSources

Luoma, Jon R. "Spilling the Truth." *Audubon* 101, no. 2 (March – April): 52–55.
NASA (ocean planet) Web site: www.seawifs.gsfc.nasa.
U.S. Environmental Protection Agency Oils Spill Program Web site: www.epa.gov/oilspill/overview.html.
Williams, Ted. "Fatal Attraction." *Audubon* 99, no. 5 (September–October 1997): 24–31.

oil spill equipment, Tools, vessels, and other products that are used for controlling and cleaning up OIL SPILLS. Some of the equipment includes booms, skimmers, vessels, and absorbents. Booms are structures that are placed in water to confine a spill. They may be used in rivers, ESTUARIES, and protected harbors. Booms can be used at the water's edge to protect MARSHES and other low-water-level areas from approaching oil spills.

Oil spill response vessels similar to the one pictured here are used for the collection of oil, floating debris, and floating tarballs. *Credit:* National Response Corporation

Skimmers, equipment used to collect oil that floats on the water's surface, have been used throughout the world in oil spill cleanup operations and have proved to be a highly effective and reliable recovery device. Skimmers can recover a wide range of oils, from light products to heavy crude oil in both coastal and offshore locations. Some skimmers can recover oil at a rate of 70 metric tons (77 tons) per hour, depending on oil type, viscosity, temperature, and sea state.

Oil spill response vessels (small boats) are used for the collection of oil, floating debris, and floating tarballs. The equipment aboard these vessels can recover a wide range of oil viscosities, at a maximum rate of 40 metric tons (44 tons) per hour.

Oil sorbents are a cost effective way to deal with the cleanup and removal of oil where mechanical equipment is no longer economically feasible. The special absorbents contain special materials that can absorb up to 25 times their own weight of HYDROCARBONS. Sorbents remain floating even after becoming saturated with oil.

During an oil spill cleanup operation, it is necessary to temporarily store large volumes of recovered material. Different kinds of storage tanks are used depending on the location of the cleanup. Floating, towable oil barges and floating storage bladders are for use at sea, while other tanks are designed for land storage. *See also* BIOREMEDIATION; *EXXON VALDEZ* OIL SPILL.

oil tanker, An ocean-going ship that transports oil from oil refineries to ports having bulk oil terminals or other oil storage facilities. Some oil tankers are about 240 meters (800 feet) long and can carry 15 to 30 million barrels of crude oil in one year. During each trip, such tankers can transport more than 500,000 barrels of oil cruising at about 16 knots (nautical miles per hour) from port to port. A tanker carries a crew of about 30 members.

Single-hulled oil tankers have been associated with several major OIL SPILLS and will be phased out in U.S. fleets by 2015. At that time, only double-hulled oil tankers will be permitted access to U.S. ports. Presently, several double-hulled vessels are now in service in the Conoco fleet. The new oil tankers have 2.8 meters (8 feet) of space between the outer and inner hulls of the vessel. This type of tanker carries approximately 880,000 barrels of crude oil. The double hull provides additional protection against leakage resulting from accidental collisions and soft groundings, and thus is intended to help control oil spillages. To reduce EMISSIONS, the double-hulled tankers use low-emission engines that will burn only low-SULFUR FUELS. The tankers also have the ability to change ballast water at sea to avoid transporting MICROORGANISMS from one port to another. Black boxes similar to the ones on aircraft will record data including the ship's location, its speed, and WEATHER conditions during the entire voyage.

One of the double-hulled tankers has already been tested against accidental collisions. In 1997, one of these tankers survived a collision with a tug boat. A gash 33 meters (100 feet) long and 1 meter (3 feet) wide was sustained on the outer hull of the tanker; however, no spillage occurred because the inner hull remained intact.

It is important to recognize that catastrophic oil spills from tanker ships, though highly publicized and locally damaging, constitute a relatively small part of the human impact on Earth's OCEANS when compared with such factors as OVERFISHING, marine HABITAT destruction, coastal development, and land-derived water pollution. A more significant impact that ocean tankers have on the marine ENVIRONMENT is the pollution that may result from tank cleaning, which may involve DISCHARGE of oily wastes and oil-contaminated ballast. A 1998 international study by the GREENPEACE organization estimated that shipping was responsible for an estimated 568,000 metric tons (625,000 tons) of oil entering the marine environment annually.

The UNITED NATIONS CONVENTION ON THE LAW OF THE SEA, which became effective in 1994, contained provisions that permitted coastal states to establish, within their territorial seas, specially designated sea lanes to confine the passage of tankers and other vessels transporting potentially hazardous

Presently, there are several double-hulled vessels, like this one built in 1994, transporting oil. *Credit:* OSG Ship Management, Inc.

cargoes. This type of planning and regulation of tankers is expected to reduce the incidence and environmental impact of oil spills from tankers.

old-growth forest, A late stage in FOREST ECOLOGICAL SUCCESSION in which there are many large, mature trees and often several canopy layers. These forests are dominated by tree SPECIES that are able to regenerate in the shade created by the canopy of older trees. Old-growth forests contain a variety of tree species and trees of many sizes; there are decadent old trees, as well as standing snags and fallen logs. Old-growth forests are ecologically mature and have been subjected to only negligible unnatural disturbances, such as logging, road construction, or CLEAR CUTTING. These mature forests exist in the United States, Canada, Russia, Mexico, and Central America.

A classic old-growth forest contains giant redwoods, cedars, Douglas fir, hemlock, or spruce. It has at least 20 large trees per hectare (2.5 acres) that are older than 300 years, or are more than 1 meter (3 feet) in diameter at breast height. In Washington's Mount Rainier National Park, for example, many trees are 500 to 1,000 years old. BIODIVERSITY is another hallmark of old-growth forests. There are hundreds of species of ferns, vines, mosses, LICHENS, shrubs, and understory trees beneath towering canopy trees.

A classic old-growth forest contains giant redwoods (above), cedars, Douglas fir, hemlock, or spruce.

The tree species that dominates an old-growth forest is determined by local topography, elevation, soil, CLIMATE, GEOLOGY, and GROUNDWATER conditions. Throughout the United States, there is a great deal of variety exhibited among old-growth forests. Perhaps the most well known old-growth forests are the TEMPERATE RAINFORESTS of the Pacific Northwest, which have dense, closed canopies of redwood and Douglas fir, a fern-covered forest floor, and large moss-covered dead trees decaying on the ground. In the arid West, by comparison, ponderosa pines grow to large sizes in relatively open conditions; southern pine forests exhibit similar characteristics. These two ECOSYSTEMS often are dependent on frequent, light-intensity ground fires to thin out competing vegetation. In the northern Rocky Mountains, there are old hemlock forests with dense hemlock canopies and little vegetation underneath except for hemlock seedlings and some shade-tolerant PLANTS, such as orchids. Alaska's Tongass National Forest is one of the world's largest tracts of temperate old-growth forest, as well as a unique ecosystem type: the coastal temperate rainforest.

Even forests that are not in a late successional stage can exhibit old-growth characteristics. For example, aspen is short-lived and considered to be an early- to mid-successional species. At 50–100 years old, aspen forests can be dominated by large trees nearing the end of their life, with scattered dead trees, both standing and on the ground. These forests are considered to be in an old-growth condition even though they are not late successional.

Across the United States, many old-growth forests were generated centuries ago as a result of some intense disturbance, such as a large wildfire. It is estimated that between 6–7.6 million hectares (14.8–18.8 million acres) of old-growth forest blanketed Oregon and Washington before European settlement. The U.S. FOREST SERVICE estimates that by the mid-1980s, however, only between 500,000 and 700,000 hectares (1.24 and 1.73 million acres) remained. Other estimates put the area at slightly less than 400,000 hectares (988,000 acres). Today, only 5% of the ancient groves the Europeans first found still exist. About one-third of the remaining old-growth forest in the Pacific Northwest is protected in national WILDERNESS areas. Much of what remains is fragmented or otherwise compromised by proximity to roads and areas of clear-cutting, which cause radical microclimate shifts.

Even protected areas are vulnerable to activities around them. Of all remaining old growth in the Pacific Northwest, 37% occurs in patches smaller than 158 hectares (395 acres), and for every 10 hectares (25 acres) of old growth that has been clear cut, an additional 14 hectares (35 acres) is degraded because of the detrimental effects of fragmentation. Degradation takes the form of loss of diversity, exposure to wind, and dramatic changes in temperature, humidity, and light.

The case of the northern spotted owl illustrates why today's FOREST MANAGEMENT should be focused on ecosystem values. Studies in Oregon show that this owl's POPULATION declined by one-third between 1976 and 1987. (See owls, northern spotted.) The Forest Service estimates that only 3,000–6,000 pairs remain in North America, each requiring 1,600–3,600 hectares (about 4,000 to 9,000 acres) for hunting and foraging. These owls can nest only in the broken tops of dead, old-growth firs. Because this owl was declared an ENDANGERED SPECIES in 1990, a court injunction halted most timber harvesting in national forests where old-growth firs were found.

The Forest Service was also required to ensure sufficient owl HABITAT to maintain viable populations.

The importance of the old-growth forest ecosystem extends far beyond the preservation of this one species, however. Old-growth forests provide habitat for plants and animals that cannot live anywhere else. At least 118 known VERTEBRATE species live primarily in old-growth forests; 41 of them cannot nest, breed, or forage anywhere else. In contrast, new-growth forests are managed as MONOCULTURES, containing only a few species of trees that were planted at the same time; they do not offer habitat diversity. For example, only 9 MAMMAL species make their home in second-growth forests of young firs and hemlocks, compared to 25 in old-growth forests of the same tree species. In 1976, the National Forest Management Act required the Forest Service to maintain viable populations of all vertebrate species throughout their ranges.

Loss of forests throughout the world has led to a greater awareness of the value of the remnants of older natural forests that are relatively free of modern human disturbance. People prize them for their great age, the unique relationships among the plants and animals within them, and their highly complex structures.

EnviroSources

Flicker, John. "The Audobon View: Our Fight for Forests." *Audubon* 102.1 (January–February 2000): 8.

Greenpeace International (forests) Web site: www.greenpeace.org/~forests.

Norse, Eliot A. *Ancient Forests of the Pacific Northwest*. Washington, DC: Island Press, 1990.

World Resources Institute, Forest Frontiers Initiative Web site: www.wri.org/ffi.

Olmsted, Frederick Law (1822–1903), An American architect who designed Central Park in New York City, one of the first city parks in the United States. Olmstead has been recognized as the founder of the profession of landscape-design architecture because of his application of environmental concepts to the design of public parks, recreation areas, and parkways. Although he is most well known for his work on Central Park, Olmsted also designed the landscapes for the U.S. Capitol, the White House, and the Jefferson Memorial. His other landscape works include the West Point Military Academy, Great Smoky Mountains National Park, and Niagara Falls Reservation. All of his work incorporated the use of NATIVE SPECIES rather than nonnative ones.

In 1963, Olmsted's home and office in Brookline, Massachusetts became a national historic site. The Olmsted archives are the most widely researched collections of the NATIONAL PARK SERVICE. Many researchers, including park and city planners, visit the site each year to review the thousands of landscape architectural drawings, plans, photographic prints, letters, and records. The archives contain records for nearly 5,000 projects, including national, state, and city parks; arboretums; and school and college campuses. Olmsted's plans include entire park systems for the cities of Seattle, Chicago, Baltimore, and Buffalo.

EnviroSources

Beveridge, Charles E., et al., eds. *Frederick Law Olmstead: Designing the American Landscape*. New York: Rizzoli Bookstore, 1998.

Frederick Law Olmsted National Historic Site: 99 Warren Street, Brookline, MA 02146; Web site: fredericklawolmsted.com.

omnivore, An animal that feeds on both PLANTS and animals, or products derived from plants and animals such as seeds or eggs. Examples of omnivores are humans, some SPECIES of bears, and some species of birds, such as macaws. (*See* parrots.) All omnivores are CONSUMERS. Because their diet consists of both plant and animal products, omnivores feed at the second TROPHIC LEVEL or above. Thus, omnivores are always secondary or tertiary consumers. *See also* CARNIVORE; DECOMPOSER; DETRIVORE; ECOLOGICAL PYRAMID; FOOD CHAIN; FOOD WEB; HERBIVORE; INSECTIVORE; PARASITISM; PREDATION; SCAVENGER.

Examples of omnivores are humans, bears, and some species of birds, such as macaws. The polar bear occasionally eats eggs and vegetation.

opacity, A measure of the quality or amount of light that is able to come through a PLUME of PARTICULATE MATTER. Changes in opacity can indicate changes in the performance of particulate matter control systems. Opacity can be measured in percentage. As an example, window glass, which is transparent, has 0% opacity while a solid wall, through which light cannot penetrate, is 100% opaque.

OPEC. *See* Organization of Petroleum Exporting Countries.

open-pit mine. *See* mining.

opportunistic organism, A normally harmless, often commensal ORGANISM, such as certain FUNGI and BACTERIA, that may cause DISEASE—also called opportunistic diseases—in persons with immune systems that have been compromised by a primary PATHOGEN or immunosuppressive drug therapy. (*See* commensalism.) Opportunistic organisms are of particular danger to organ-transplant recipients, CANCER victims, the elderly, and people with advanced HIV (human immune deficiency) infection. Among the latter group, common opportunistic diseases that occur in people living in developed nations are *Pneumocystis carinii* pneumonia, *Mycobacterium avium* complex, and cytomegalovirus infection, although the incidence of these diseases has declined because of the widespread use of preventive regimens and more potent medicines.

In 1995, the U.S. Public Health Service (USPHS) and the Infectious Diseases Society of America (IDSA) published comprehensive, disease-specific guidelines to consolidate information pertaining to the prevention of opportunistic infections in HIV-positive persons. The positive response to the 1995 guidelines suggested that they were a valuable reference against which local policies on prevention of opportunistic infections could be compared. The USPHS and IDSA also determined that preventive strategies needed to be developed and applied to a wider spectrum of opportunistic infections. The guidelines have been approved by several medical and health organizations, including the American College of Physicians and the National Foundation for Infectious Diseases.

EnviroSources

Centers for Disease Control Web site: www.cdc.gov.
Infectious Diseases Society of America Web site: www.idsa.org.
National Foundation for Infectious Diseases Web site: www.nfid.org.
U.S. Department of Health and Human Services, Public Health Service Web site: www.dhhs.gov/phs.

orangutans, A highly intelligent, arboreal great ape whose common name derives from the Malay phrase meaning "old man of the forest." There are two subspecies of orangutan: one in Sumatra, the other in Borneo. Orangutans *(Pongo pygmaeus)* are the largest primates existing in Asia; they may grow to a height of 1.5 meters (5 feet) and a weight of 100 kilograms (220 pounds.) An orangutan's coat is normally a deep red-brown and shaggy, but often thin.

The orangutan's HABITAT is primarily tropical FORESTS of varying elevation, where they live in lower and mid-canopy areas 6–30 meters (20–100 feet) off the ground. Higher POPULATION densities exist in SWAMP forests; lower densities are present in mountainous areas. Unlike other primates, orangutans spend their time in small groups rather than within complex social structures.

Prehistorically, orangutan numbers were probably in the hundreds of thousands and their range extended from southern China throughout Southeast Asia. Today, their total global population is 20,000–27,000 individuals. They are now an ENDANGERED SPECIES, primarily because their habitat continues to be altered or destroyed by logging, AGRICULTURE, and FOREST FIRES, forcing populations into smaller areas that cannot support them. Large tracts of potential habitat remain unoccupied by orangutans because of impassable natural obstructions, such as mountains and rivers. Under ideal conditions, these animals roam the forests in search of widely distributed food sources. To a lesser degree, populations are also being reduced because mother orangutans are killed to obtain infants or juveniles for the live animal trade. This practice is a problem because the orangutan's reproductive rate is very low; an adult female may reproduce only every eight to nine years in the wild, producing only three or four offspring during her reproductive lifetime. Though totally protected by law in Indonesia, Malaysia, and internationally, enforcement is extremely difficult in most areas. *See also* CLEAR-CUTTING; CONVENTION ON INTERNATIONAL TRADE IN ENDANGERED SPECIES OF WILD FLORA AND FAUNA; ENDANGERED SPECIES LIST; GALDIKAS, BIRUTÉ; RED LIST OF ENDANGERED SPECIES; WORLD CONSERVATION MONITORING CENTRE; WORLD CONSERVATION UNION.

Orangutans are the largest primates in Asia; they may grow to a height of 1.5 meters (5 feet) and a weight of 100 kilograms (220 pounds). Infants or juveniles (above) are sold illegally for the live animal trade.

EnviroSources

Galdikas, Biruté. *Reflections of Eden: My Years with the Orangutans of Borneo.* Boston, MA: Little Brown and Company, 1995.
Montgomery, Sy. *Walking with the Great Apes.* New York: Houghton Mifflin, 1991.
Orangutan Foundation International: (800) ORANGUTAN; Web site: www.ns.net/orangutan.
U.S. Fish and Wildlife Service, Division of Endangered Species, Foreign Listed Species Index Web site: www.fws.gov/r9endspp/fornspp.html.
World Wildlife Fund Web site: www.panda.org.

ore. *See* mineral.

organic, Any substance that is composed of or derived from living ORGANISMS; also any chemical that contains CARBON (in various combinations with HYDROGEN, OXYGEN, NITROGEN, and/or other elements). PETROCHEMICALS are examples of organic chemicals. Today, the word organic is used as a catchword for anything that does not contain, or has not been produced with, synthetic (human-made) chemicals. Organic fruits and vegetables, for instance, are grown using FERTILIZERS derived from organisms or their remains, such as compost or animal manure, and without the use of synthetic PESTICIDES. Similarly, many people choose to eat organic beef, which means that the cattle have not been fed growth hormones, antibiotics, or synthetic chemicals of any kind. *See also* AGRICULTURE; BIOLOGICAL CONTROL; COMPOSTING; GREEN MANURE; INTEGRATED PEST MANAGEMENT; MULCH.

organic farming. *See* agriculture.

organism, Any living thing. Currently, most scientists recognize five large groups, or kingdoms, of organisms—the Monera (or BACTERIA), the Protista (or PROTISTS), the FUNGI, the Plantae (or PLANTS), and the Animalia (or animals). To be considered an organism, the following characteristics, which identify something as living, must be present: It must have one or more CELLS, the ability to respond to changes in the ENVIRONMENT, the ability to use ENERGY, the ability to reproduce, and mechanisms for growth and development. *See also* ADAPTATION; BIODIVERSITY; BIOLOGICAL COMMUNITY; CYANOBACTERIA; EVOLUTION; GENETIC DIVERSITY; POPULATION; SPECIES.

Organization of Petroleum Exporting Countries, A voluntary intergovernmental organization founded in 1960, which coordinates and unifies the PETROLEUM policies of its member countries. Founder members were Iran, Iraq, Kuwait, Saudi Arabia, and Venezuela. Member countries in 1998 were Algeria, Indonesia, Iran, Iraq, Kuwait, Libya, Nigeria, Qatar, Saudi Arabia, Abu Dhabi, United Arab Emirates, and Venezuela. The present member countries supply more than 40% of the world's oil and 14% of its NATURAL GAS. They possess about 78% of the world's total confirmed crude oil reserves.

The member countries promote stability and prosperity in the petroleum market. Their organizational structure includes the Secretariat, directed by the Board of Governors and the secretary general, and various bodies including the Economic Commission and the Ministerial Monitoring Committee.

The Organization of Petroleum Exporting Countries (OPEC) conference meets twice each year. The Secretariat functions as the headquarters of OPEC. Member countries respond to market needs and forecast developments by coordinating their petroleum policies. *See also* ALTERNATIVE ENERGY RESOURCES; ALTERNATIVE FUELS; FOSSIL FUELS; PETROCHEMICALS.

EnviroSource

Organization of Petroleum Exporting Countries Website: www.opec.org.

organochlorine. *See* chlorinated hydrocarbons.

organophosphate, Any ORGANIC chemical containing CARBON, HYDROGEN, and PHOSPHORUS. Organophosphates are primarily used as PESTICIDES to control INSECT pests. Organophosphates are BIODEGRADABLE—they degrade quickly in the ENVIRONMENT and in living ORGANISMS. Organophosphates work by disrupting the nervous system function of insects. Organophosphate pesticides are used to treat sheep for ectoparasites (external parasites), such as lice and scab, and to treat sea lice in SALMON raised on fish farms. The organophosphate malathion has also been used in California orchards to control the Mediterranean fruit fly (commonly called the Medfly).

Organophosphates may be harmful to human health and aquatic ECOSYSTEMS. In humans, high levels of organophosphate pesticides can cause nervous system disorders such as headaches, dizziness, language impairment, convulsions, and even coma. Improper disposal of the spent chemical can contaminate SURFACE WATER and GROUNDWATER. *See also* AQUACULTURE; PARASITISM.

TechWatch

Concern over the effects of ORGANOPHOSPHATES on humans has led to the development of alternatives for some uses. For example, a new method involves injecting sheep with antiparasitic drugs to control sheep scab, one of the leading DISEASES of British sheep. So far there has not been any major health or environmental concerns over the injection method, but more testing and evaluations are being done.

EnviroTerm

ectoparasite: A parasitic ORGANISM that lives on a host organism from which it ingests nourishment. The parasite is detrimental to its host.

oryx, An ENDANGERED SPECIES (*Oryx leucoryx*) of antelope native to the Arabian Peninsula. In the mid-nineteenth century, POPULATIONS of oryx began to disappear because of overhunting for horns and hides. By 1914, few oryx existed outside Saudi Arabia. The decline accelerated after World War I along with the proliferation of modern guns and faster motor vehicles, which facilitated hunting. By the early 1960s, the SPECIES was confined to only two small areas: the Rub al-Khali DESERT, or Empty Quarter of southwestern Saudi Arabia and northeastern Oman. The last wild oryx were probably killed in 1972 in the Jaddat al-Harasis area of Oman.

Oryx have long straight horns, black markings on the legs and face, relatively long legs, broad hooves, and a tasseled tail. In the wild, the species inhabits arid gravel plains and semidesert regions. Feeding primarily on grasses, herbs, fruits, and shoots, the oryx is able to exist for weeks without water. (*See* herbivore.) Oryx are herd animals, living in groups of 8 to 20, composed of both sexes and all ages. Oryx live up to 20 years.

To save the species from EXTINCTION, Operation Oryx was launched in the early 1960s by the Fauna and Flora Preservation Society of London, with support from the WORLD WILDLIFE FUND (WWF) and the sultan of Oman. Three wild oryx were captured in Oman in 1962 and transported to the Phoenix

Zoo in Arizona. Six captive animals from the London Zoo, Kuwait, and Saudi Arabia also were donated to form the nucleus of a "World Herd." These nine oryx bred successfully, and by 1984, more than 200 individuals existed in ZOOS around the world. In 1974, the White Oryx Project was launched by the sultan of Oman, with the aim of reestablishing a wild population. In 1980, the first oryx were returned to Oman for acclimatization and eventual reintroduction at Yalooni in the Jiddat-al-Harasis area. In 1982, the first herd of 10 individuals was released into the wild. Additional releases were made in 1984, 1988, and 1989. Numbers increased steadily, and by 1990, there were 109 free-ranging oryx, of which 80% were born in the wild. Oryx living in the wild now number more than 1,500. They are protected from POACHING within the Arabian Oryx Sanctuary, and the animals are covered by strict legislation whose provisions are enforced by rangers. *See also* ADAMSON, JOY GESSNER; CALIFORNIA CONDORS; CAPTIVE PROPAGATION; ENDANGERED SPECIES LIST; NATIVE SPECIES; RED LIST OF ENDANGERED SPECIES; WILDLIFE REFUGE; WOLVES.

EnviroSources

Saudi Arabia Geography, The Arabian Oryx Web site: www.arab.net/saudi/geography/sa-oryx.html.
U.S. Fish and Wildlife Service Web site: www.fws.gov.

osmosis. *See* reverse osmosis.

overdraft, The removal of water from an AQUIFER at a faster rate than the water is replaced by nature. People in many regions rely on aquifer water (GROUNDWATER) to meet their water needs. When such water is used faster than it is replaced by nature, people and other ORGANISMS may not have enough fresh water to meet their needs. In addition, environmental problems such as SUBSIDENCE and SALT-WATER INTRUSION may result. Subsidence occurs when rock and soil overlying an aquifer sink to fill the space once occupied by water. A large hole, called a SINKHOLE, may result. The formation of such holes can cause great property damage; it also potentially threatens life as structures lying above an aquifer collapse. Salt-water intrusion can occur from overdraft of aquifers located in coastal regions. This problem results when OCEAN water moves into an aquifer to fill the void created by the removal of fresh water. The intrusion of salt water into an aquifer makes its water unfit for use by humans and other ORGANISMS. *See also* OGALLALA AQUIFER; POTABLE WATER; WELL; ZONE OF AERATION; ZONE OF DISCHARGE; ZONE OF SATURATION.

overfishing, The process of continuing to fish in an area until all of the fish are harvested and the breeding grounds are almost depleted or destroyed. Overexploitation of fishing can push a fish POPULATION into a J-SHAPED CURVE toward EXTINCTION. According to the United Nations FOOD AND AGRICULTURAL ORGANIZATION (FAO), 69% of the world's fisheries are currently overexploited. Since the early 1950s, as the demand and supply of fish for human consumption has increased, many fishing companies have invested in trawler fleets and OCEAN-going, factory-type processing vessels that can travel to fishing grounds all over the ocean. These vessels, equipped with sophisticated, on-board electronic tools and satellite data receivers can track large schools of fish day and night. Harvesting fish has also improved with the use of long DRIFT NETS, long liners, and purse seines that can catch thousands of kilograms of fish in a brief period of time. The size of the world's fishing fleet doubled between 1970 and 1992, putting more pressure on fishing stocks. Some scientists are concerned that the global fishing fleet is at least 30% larger than it should be to sustainably harvest fish.

Overfishing is more severe in some regions than in others. One of the major areas in which fish stocks have been declining is the North Atlantic Ocean. Thousands of fishers in the United States and Canada have lost their jobs because of severe declines of cod, flounder, and haddock populations off the coasts of New England and Canada. In the northwestern section of the Pacific Ocean, coastal Asian fish stocks have been depleted. There have also been severe declines of SALMON stock in the Pacific Ocean and declines of grouper in the Gulf of Mexico. Other overexploited SPECIES include Gulf of Mexico shrimp, Maine lobster, California anchovies, and Texas red snapper. Bluefin TUNA have decreased by 75% over the past 30 years according to the American Fisheries Society. More than 30% of the 280 species of fish in U.S. waters are in trouble. To add to the problem, many marine FOOD WEBS will be destroyed if the rate of declining fish stocks continue. Many lakes, including the NORTH SEA, ADRIATIC SEA, BLACK SEA, and BALTIC SEA, have also suffered from overfishing.

Some measures may be effective in curbing overfishing and bringing fish stocks back to acceptable levels. An FAO study, "The State of World Fisheries and Aquaculture," suggested that it might be possible to return to historical peak productivity levels by implementing measures such as prohibiting the capture of juvenile fish, increasing mesh sizes, and temporarily or permanently closing those areas where young fish are concentrated.

Although most nations agree that overfishing is one of the main threats to Earth's oceans, there is little international agreement on how to solve the problem. The United Nations Fish Stocks Agreement, adopted in 1995, promoted precautionary measures to protect the world's imperiled marine fisheries and also set new standards for managing these limited resources in a sustainable manner; however, as of 1997, only 4 out of the top 20 fishing nations had ratified the agreement, and 30 nations must sign and ratify the agreement for it to enter into force. The nations that have signed and ratified it include the United States, Russia, Norway, and Iceland. As of 1997, the fishing nations that had signed but had yet to ratify it included China, Japan, Indonesia, South Korea, the Philippines, Denmark, Spain, Canada, and Bangladesh. The fishing nations that have not signed include Peru, Chile, India, Thailand, North Korea, Mexico, and Vietnam. Until there is international cooperation for high seas fishing, marine fish will continue to be depleted, one species after another. *See also* SUSTAINABLE DEVELOPMENT; SUSTAINED YIELD.

EnviroSources

Berrill, Michael, and David Suzuki. *The Plundered Seas: Can the World's Fish Be Saved?* San Francisco, CA: Sierra Club Books, 1997.

Iudicello, Suzanne, et al., *Fish, Markets, and Fishermen: The Economies of Overfishing.* Washington, DC: Island Press, 1999.

National Aeronautics and Space Administration, Ocean Planet Web site: www.seawifs.gsfc.nasa.gov.

National Marine Fisheries Service Web site: www.nmfs.gov.

National Oceanic and Atmospheric Administration Web site: www.noaa.gov.

United Nations Food and Agricultural Organization Web site: www.fao.org.

United Nations System Web site: www.unsystem.org.

owls, northern spotted, Nocturnal birds of prey (*Strix occidentalis caurina*) that live primarily in OLD-GROWTH FORESTS in the Pacific Northwest and northern California. Northern spotted owls were listed as a THREATENED SPECIES in June 1990 because of past and projected losses of suitable HABITAT resulting primarily from timber harvesting and inadequate state and federal regulatory mechanisms, exacerbated by periodic catastrophic natural events such as FOREST FIRES, wind storms, and volcanic eruptions. (*See* volcano.) The northern spotted owl's situation sparked a complicated debate over logging in

the Pacific Northwest. Under the ENDANGERED SPECIES ACT, logging of many old-growth forests has been suspended to protect the bird and its remaining habitat. The federal government sharply restricts logging within a 810-hectare (2,000-acre) zone surrounding known spotted owl nests, requires that at least 202 hectares (500 acres) of the largest trees within that zone be left uncut, and prohibits logging of any kind within 28 hectares (70 acres) of a nest.

The northern spotted owl depends on old-growth forests for its nesting, roosting, hunting, and mating territory. The age of a FOREST is not as important in determining habitat suitability for the northern spotted owl as forest structure and composition. Northern interior forests typically require 150–200 years to attain the attributes of suitable spotted owl nesting and roosting habitat. The tall trees, some more than 500 years old, form a high, protective canopy and provide habitat for many rare animals, such as the red tree vole, a valuable prey SPECIES for the owl.

Because the spotted owl can live only in old-growth forests, it is considered an INDICATOR SPECIES for the health of that ECOSYSTEM. When an old-growth forest has been clear cut and converted into a less biologically diverse, second-growth forest, the owl's habitat is destroyed, and its POPULATION dwindles.

Debate over reserving old-growth forests as habitat for the northern spotted owl has intensified the general controversy over the practice of CLEAR-CUTTING, which is widespread in the national forests of the Pacific Northwest; however, fragmentation of spotted owl habitat caused by even-aged FOREST MANAGEMENT and the continuing decrease in old-growth acreage also are concerns. Primarily because of historic timber harvest patterns, approximately 75% of the known population of spotted owls inhabits federal lands. Owl site centers on nonfederal lands usually are found in remnant stands of older forest or in younger forests that have had time to regenerate following harvest. On April 2, 1993, President Bill Clinton held a Forest Conference in Portland, Oregon, to address federal forest management and protection of species associated with old-growth forests in the Pacific Northwest and northern California. Following the conference, the president established a Forest Ecosystem Management Assessment Team to develop management options for federal forest ecosystems to provide habitat for stable populations of species associated with late-successional forests. The president chose a management option based on a system of late-successional reserves, riparian reserves, adaptive management areas, and a matrix of federal lands interspersed with nonfederal lands. This option, adopted on April 13, 1994, became the Forest Plan.

In light of the Forest Plan's adoption, the U.S. FISH AND WILDLIFE SERVICE assessed the CONSERVATION needs of the northern spotted owl on nonfederal lands in Washington and California. The agency concluded that since the Forest Plan's commitment to a comprehensive habitat-based strategy would accomplish or exceed the standards set for federal contributions to recovery of the owl and assurance of adequate habitat, it was no longer necessary to prohibit incidental take (the accidental killing) of the owl on all nonfederal lands within its range. *See also* BIODIVERSITY; DEFORESTATION; ECOLOGICAL SUCCESSION; ENDANGERED SPECIES ACT.

EnviroSource

U.S. Fish and Wildlife, Division of Endangered Species Web site: www.fws.gov/r9endspp/endspp.html.

oxidant. *See* oxidation.

oxidation, In general, a process that occurs when a substance combines chemically with OXYGEN. Oxidation is the loss of electrons by atoms or compounds. The rusting of an iron nail is an example of oxidation. The addition of oxygen can break down waste chemicals such as cyanides and PHENOLS and can aid in the breakdown of ORGANIC SULFUR compounds in SEWAGE by MICROORGANISMS and chemical means. An oxidant is usually a substance containing oxygen that reacts chemically with other materials to produce a new substance. Oxidants are the primary ingredients of photochemical SMOG. *See also* INCINERATION; OXIDIZING AGENT; WEATHERING.

oxidizing agent, A chemical containing OXYGEN that decomposes easily, providing oxygen that can react with something else. An oxidizing agent can be an atom, ion, or molecule that in a chemical reaction tends to take electrons from other atoms, ions, or molecules causing them to be oxidized. An oxidizing agent includes NITROGEN DIOXIDE and OZONE. *See also* OXIDATION.

oxygen, An odorless, colorless, and tasteless gas essential for most living ORGANISMS. Oxygen gas makes up about 21% of the ATMOSPHERE and is the most abundant element in Earth's crust. Oxygen gas contains two atoms (O_2) in the atmosphere and three atoms (O_3) in the OZONE layer. It may also be present in combination with other elements, such as oxides of SULFUR and NITROGEN.

EnviroTerm

oxide, A compound made up of oxygen. There are oxides of SULFUR and NITROGEN.

ozone, An OXYGEN molecule composed of three oxygen atoms, which forms when oxygen gas (O_2) in the ATMOSPHERE is exposed to ultraviolet (UV) RADIATION. About 90% of ozone (O_3) is located in a layer of the stratosphere located 10–50 kilometers (6–31 miles) above Earth's surface. This ozone forms when UV radiation splits O_2 molecules into individual oxygen atoms, which in turn collide with other oxygen molecules to form O_3. When an O_3 molecule absorbs UV radiation, it breaks apart to form O_2 and an atom of oxygen. The free oxygen atom can then collide with an O_3 molecule to form two O_2 molecules. These reactions occur repeatedly and simultaneously, and the total amount of ozone and oxygen in the stratosphere remains balanced as long as ozone forms and breaks apart at the same rate. Ozone in the stratosphere provides a protective layer that shields Earth from much of the UV radiation, especially UV-B radiation, given off by the sun. Such radiation is harmful to the CELLS and tissues of ORGANISMS and has been identified as a major source of CANCER.

About 10% of atmospheric ozone is located in the troposphere, the layer of the atmosphere nearest Earth. Much of the ozone in the troposphere forms when NITROGEN OXIDES (NO_x) emitted during the combustion of FOSSIL FUELS or PETROCHEMICAL products, react with sunlight. (*See* emissions.) This ground-level ozone is a major component of photochemical SMOG and is one of the CRITERIA POLLUTANTS for which the U.S. ENVIRONMENTAL PROTECTION AGENCY (EPA) sets standards under the CLEAN AIR ACT of 1970, as amended. Warm temperatures and stagnant high-pressure WEATHER systems with low wind speeds contribute to harmful ozone accumulation. Ground-level ozone can be harmful to the human respiratory system. Ozone pollution is not just an urban problem. Average rural summertime ozone concentrations in the southern and eastern United States are among the highest in North America. The EPA recommends that people limit the time they spend outdoors when measurements of ground-level ozone reach 120 parts per billion (ppb). Even ozone counts above 50 ppb can cause respiratory illnesses and asthma attacks.

Ozone is sometimes used commercially to kill MICROORGANISMS during the water purification process. It is also used as a bleaching agent. *See also* NATIONAL AMBIENT AIR QUALITY STANDARDS; OZONE DEPLETION; OZONE HOLE; PRIMARY, SECONDARY, AND TERTIARY TREATMENTS.

EnviroSources

National Aeronautics and Space Agency Web site: www.earth.nasa.gov/science/ozone.html.

National Oceanic and Atmospheric Administration (commonly asked questions about ozone) Web site: www.publicaffairs.noaa.gov/grounders/ozo1.html.

Ozone Action: 1621 Connecticut Avenue, NW, Suite 400, Washington, DC 20009; (202) 265-6738; e-mail cantando@essential.org.

U.S. Environmental Protection Agency (science of ozone depletion) Web site: www.epa.gov/ozone/science.

U.S. Environmental Protection Agency, Stratospheric Ozone Information: 501 3rd Street NW, Suite 260, Washington, DC 20460; (800) 296-1996.

ozone depletion, The thinning or reduction in the concentration of OZONE in portions of the ozone layer of Earth's stratosphere as a result of the destruction of ozone (O_3) molecules. Ozone layer depletion is believed to result largely from the release of a variety of chemicals into the ATMOSPHERE. The best known of the ozone-depleting substances are the CHLOROFLUORCARBONS (CFCs), which when acted upon by ultraviolet (UV) RADIATION, release CHLORINE that reacts with and breaks apart ozone molecules. Other major ozone-depleting substances include the HALONS, which are used in fire extinguishing equipment, CARBON TETRACHLORIDE (CCl_4), METHYL BROMIDE, and methyl chloroform.

Changes in ozone concentration also result from natural events including cycles of the sun, changes in winds, and seasonal changes. (*See* climate.) Volcanic eruptions may eject chemical substances such as chlorine and METHANE gases that react with and break apart stratospheric ozone in the atmosphere. Large decreases in ozone concentrations measured in 1992 and 1993 are believed to be related to the eruption of MOUNT PINATUBO in the Philippines. When the VOLCANO erupted, it released large amounts of SULFATE AEROSOLS, which may have accelerated ozone depletion in the stratosphere. *See also* GLOBAL WARMING; GLOBAL WARMING POTENTIAL; MONTREAL PROTOCOL; OZONE HOLE.

Ozone Depletion Process

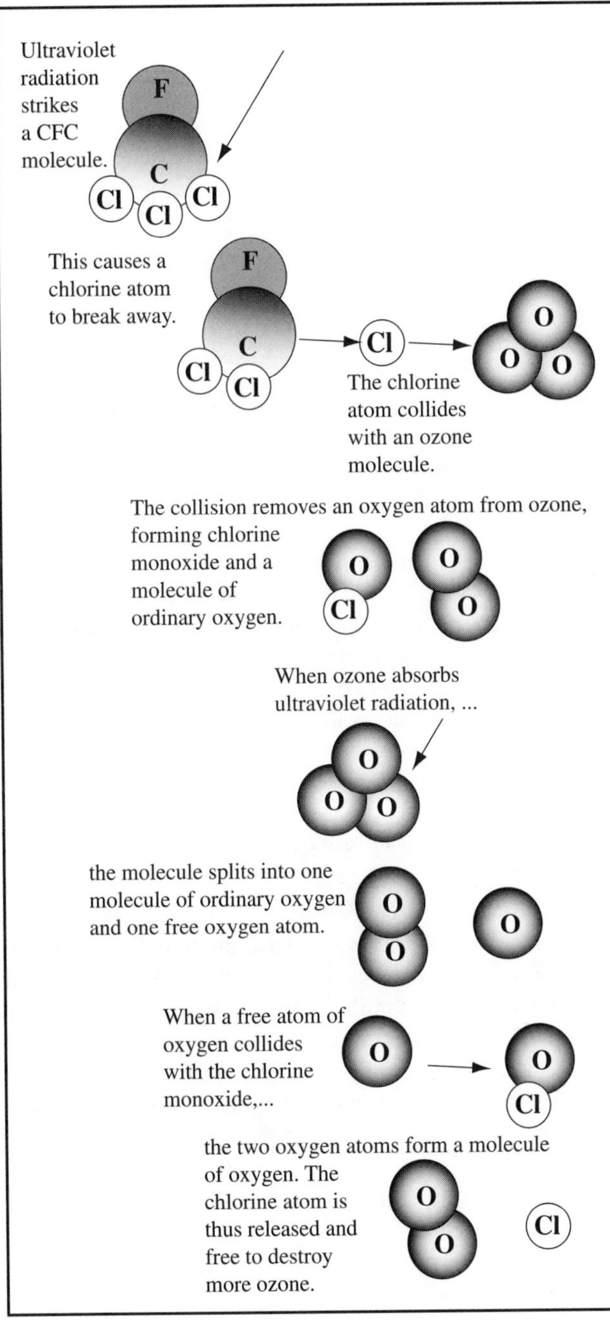

Source: NASA. 1993

EnviroSources

Mortensen, Lynn, ed. *Global Change Education Resource Guide.* Educational Programs Manager, Earth Sciences Directorate (Code 900) NASA Goddard Space Flight Center, Greenbelt, MD 20771.

National Oceanic and Atmospheric Administration, Network for the Detection of Stratospheric Change Web site: www.climon.wwb.noaa.gov.

National Oceanic and Atmospheric Administration (stratospheric ozone depletion) Web site: www.al.noaa.gov/WWWHD/pubdocs/StratO3.html.

Ozone Action: 1621 Connecticut Avenue, NW, Suite 400, Washington, DC 20009; (202) 265-6738; e-mail: antando@essential.org.

U.S. Environmental Protection Agency, Stratospheric Ozone Information Hotline: 501 3rd Street NW, Suite 260, Washington, DC 20460; Hotline: D.M. Saunders, (800) 296-1996.

ozone hole, A thinning or reduction in the concentration of OZONE in a large section of the ozone layer above ANTARCTICA, which many scientists believe will impact global CLIMATES, leading to conditions that may threaten the survival of some SPECIES. The term "hole" is widely used in popular media when reporting on ozone; however, the phenomenon is more correctly described as a low concentration of ozone. The British Antarctica Survey station located at Halley Bay, Antarctica, revealed decreases in the extent of the ozone layer in the 1980s. Measurements from ground-based and satellite instruments indicate that at certain times of the year, the levels of ozone in the stratosphere over Antarctica have decreased by more than 50%.

The "hole" in the ozone layer occurs yearly in late September, which is the beginning of spring in the Southern Hemisphere. The hole is roughly the size of the continental United States and has increased in size each year since being discovered. Many scientists believe that low stratospheric temperatures lead to the formation of icy clouds, which bring about chemical changes resulting in rapid OZONE DEPLETION during September and October. The Total Ozone Mapping Spectrometer (TOMS) on board an Earth satellite has mapped in detail the Antarctica ozone hole and has also mapped the distribution of ozone over the entire globe. Recently, ozone holes have also appeared during the spring in Russia, east of the Ural Mountains, and in an area comprising the Baltic states, Moscow, and St. Petersburg. One of these holes is twice the size of the state of Texas. *See also* CHLOROFLUOROCARBONS; MONTREAL PROTOCOL; SPECTROMETRY.

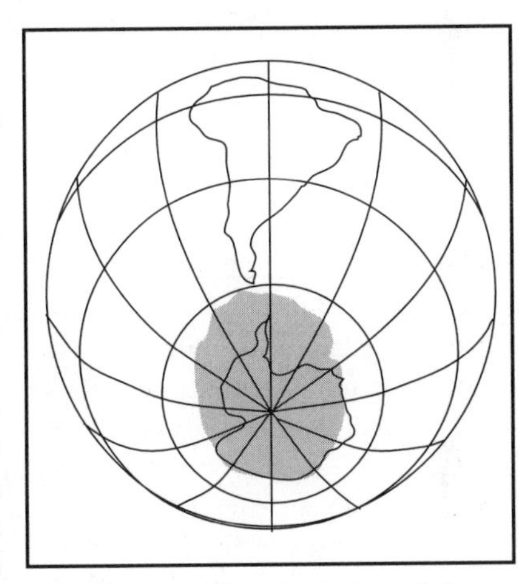

In the early 1970s, there was no ozone hole over the Antarctica. By 1991, the hole extended over the entire continent as shown.

EnviroSources

National Oceanic and Atmospheric Administration, Network for the Detection of Stratospheric Change Web site: www.climon.wwb.noaa.gov.

National Oceanic and Atmospheric Administration (stratospheric ozone depletion) Web site: www.al.noaa.gov/WWWHD/pubdocs/StratO3.html.

ozone pollution. *See* ozone.

P

pampas, A vast temperate GRASSLAND ENVIRONMENT located in southeastern South America. Temperate grasslands are terrestrial ECOSYSTEMS often found in the interiors of continents. Other temperate grasslands include the PRAIRIES of the midwestern United States and the STEPPES of Russia. All temperate grasslands are characterized by low rainfall, periodic DROUGHTS, and high temperatures. The major threat to the pampas and other grassland areas is farming and overgrazing by cattle, horses, and sheep. The grain crops, such as alfalfa and wheat, that have replaced the native grasses cannot hold soil in place very well because their root systems are very shallow. This results in soil EROSION, which reduces the overall fertility of soils. In addition, overgrazed grasses do not easily regenerate, making the soil very accessible to further erosion.

The word "pampa" is a Quechua Indian term meaning "flat surface." The vast grass-covered plains that make up the South American pampas extend westward across central Argentina from the Atlantic coastline to the foothills of the Andes mountains. The Argentine pampas cover an area of approximately 760,000 square kilometers (295,000 square miles) and can be divided into two separate sections. The western dry section is largely uninhabited by people, with sandy DESERTS and dry, scrubby vegetation. Rainfall is much higher in the eastern pampas. Vegetation is more abundant and it is the country's most populated area.

Grasses are the dominant form of life in the pampas, just as in all grassland ecosystems. Grasses are perfectly suited to the low precipitation and high temperatures. Their root systems form dense mats that can survive droughts and fire and can hold soil together. Few trees can survive in this kind of environment because of drought, fire, and high winds that often roar across the open plains. Characteristic animals of the pampas include hawks, a number of burrowing rodent SPECIES, and guanacos, large herbivorous members of the camel family. *See also* BIOME; CHAPARRAL; HERBIVORE.

parasite. *See* parasitism.

parasitism, A relationship between two SPECIES of ORGANISMS in which one (the parasite) lives in or on the other (the host), as a means of obtaining nutrition or some other benefit. Every kingdom of organisms—BACTERIA, PROTISTS, FUNGI, PLANTS, and animals—has some parasitic species, and every species of organism serves as host to some type of parasite. Thus, parasitism is present in all ECOSYSTEMS. Parasitism is a density-dependent LIMITING FACTOR because parasites thrive in dense POPULATIONS, where they can easily move among hosts.

The term "parasite" is derived from the Greek word "parasitos," which means "one who eats at the table of another." The effects of parasitism on a host may range from insignificant or minor to severe illness, or in rare cases, death. Typically, as a parasite derives nutrition, its actions weaken the host, without causing its death. Ectoparasites derive their benefit by living in the hair, feathers, scales, skin, or the blood of their host. Examples of such parasites include fleas, ticks, some flies, lice, and mosquitoes. Generally, these parasites must be present in large numbers to significantly harm their host by robbing it of nutrients; however, a single ectoparasite may harm its host indirectly, by transmitting DISEASE. Lyme disease, malaria, African sleeping sickness, and Rocky Mountain spotted fever are examples of diseases transmitted to humans by ectoparasites.

Endoparasites live inside their hosts. Often these parasites live in tubes or ducts within the digestive or respiratory systems, or embed in muscles, liver, or other body tissues. Some parasitic bacteria and protists invade the CELLS of their host. Many endoparasites coevolve with their hosts, developing ADAPTATIONS in response to changes that occur in their hosts, that permit them to survive in a hostile ENVIRONMENT. For example, many endoparasites have evolved cell walls or body walls that are highly resistant to digestive enzymes or antibodies produced by their hosts. Some have evolved specialized hooks or sucking mechanisms by which they anchor to their host's tissues, while accessing blood or nutrients. Many endoparasites have evolved extreme specializations, which include the loss of organs or complete organ systems. For example, the tapeworm lacks a digestive tract and instead feeds by surrounding itself with its host's digestive products, which it readily absorbs. *See also* BIOLOGICAL CONTROL; BLIGHT; COMMENSALISM; DUTCH ELM DISEASE; *ESCHERICHIA COLI;* FAMINE; INTEGRATED PEST MANAGEMENT; PATHOGEN; SYMBIOSIS; VIRUS.

EnviroSources

Abrahamson, Warren G. *Evolutionary Ecology across Three Trophic Levels: Goldenrods, Gallmakers, and Natural Enemies.* Princeton, NJ: Princeton University Press, 1997.
Berenbaum, May R. "Two Horns, Six Legs, and One Voracious Appetite." *Audubon* 102:1 (January–February 2000): 74–79.
Bower, Joe. "The Persistent Parasite." *Audubon* 100, no. 2 (March–April 1998): 16.
Gittleman, Ann Louise. *Guess What Came to Dinner: Parasites and Your Health.* New York: Avery, 1993.
Schmid-Hempel, Paul. *Parasites in Social Insects.* Princeton, NJ: Princeton University Press, 1998.

parrots, A group of about 350 SPECIES of tropical birds, of which many are either on the endangered or threatened list. Familiar parrots include macaws, cockatoos, and parakeets. The largest number of species is found in South America, but parrots are also found in Central America, Australia, New Guinea, New Zealand, and the Himalayas as well.

These multicolored birds have been valued as pets in the Western world for many centuries. Parrots have the ability to imitate human sounds. They are mainly PLANT eaters and range in size from about 8.5 centimeters (3.5 inches) in length to about 100 centimeters (40 inches).

Parrots have become one of the most threatened groups of birds in the world. Of the 350 species, more than 90 were under threat of EXTINCTION as of 1998 while other parrot species are declining. There are now fewer than 5,000 hyacinth macaws, the largest species of parrot, which lives in South America; in 1950, the total POPULATION was estimated at 100,000. The Lear's macaw, found in Brazil, is also endangered; there may be less than 130 of these birds in existence. Fewer than 75 yellow-eared parrots are known to exist, though they once populated forests throughout the northern portions of South America. The most serious threats to parrot species are the growing international legal pet trade, a business worth more than a $100 million a year, and HABITAT destruction in the RAINFORESTS. POACHING and illegal trade also add to the problem.

The yellow-naped parrot belongs to a group of about 350 species of tropical birds, of which many are either on the endangered or threatened list.

Some action has taken place to preserve these birds. The WILD BIRD CONSERVATION ACT passed by Congress in 1992, eliminated trade in South American parrots. It banned the import of 10 THREATENED SPECIES, including the red-headed and yellow-headed Amazon parrots. In October 1993, the act was expanded to include all birds listed by the CONVENTION ON INTERNATIONAL TRADE IN ENDANGERED SPECIES OF WILD FLORA AND FAUNA. Mexico has outlawed export of its parrots because of their dwindling populations. *See also* ENDANGERED SPECIES; ENDANGERED SPECIES LIST; RED LIST OF ENDANGERED SPECIES; WORLD WILDLIFE FUND.

EnviroSources

The Online Book of Parrots Web site: www.ub.tu-clausthal.dep/p_welcome.html.
U.S. Fish and Wildlife Service, Species List of Endangered and Threatened Wildlife Web site: www.fws.gov/r9endspp/lsppinfo.html.
World Parrot Trust Web site: www.worldparrottrust.org.
World Wildlife Fund Web site: www.panda.org.

particulate matter, Solid particles such as smoke, dust, ASH, and soot that are present in the AIR. Scientific evidence suggests that when inhaled, particulates can cause respiratory DISEASES and other health problems. Particulate matter in the air has been linked to between 10,000 and 100,000 premature deaths each year.

Most particulates in the air result from the combustion of FOSSIL FUELS in COAL-fired power plants, boilers, and other industrial sources, as well as AUTOMOBILES. Other sources of particulate matter include dust created when land is cleared, smoke and soot resulting from FOREST or agricultural fires, and smoke, dust, and ash that result from natural events such as volcanic eruptions, windstorms, TORNADOES, or other severe storms. (*See* volcano.)

To remain suspended in air, particulates must be small and light in mass. Particulate size is measured in micrometers, or microns, and may range between 0.005 to 100 microns. For comparison, the width of a human hair is about 100 microns. Particles smaller than 10 microns can travel deep into the respiratory system and become trapped on membranes within the lungs. The particles can cause excessive growth of fibrous lung tissue, leading to permanent injury. Because of the health problems associated with particulates, the size of the particles released into the air by power plants, industrial processes, and automobiles is regulated by law. The U.S. ENVIRONMENTAL PRO-

TECTION AGENCY (EPA) considers any particle less than 2.5 microns in diameter to be a health hazard for humans. The standard set by the EPA is 150 micrograms (mg) of dust per cubic meter (m^3) of air. Therefore, if a site has a reading of more than 150 mg/m^3 of particulate matter, it has exceeded the standard set to protect human health.

The EPA would like to strengthen the rules governing AIR POLLUTION by developing laws that would regulate EMISSIONS of some of the tiniest airborne dust particles. The rules would require utility companies and the PETROLEUM, automobile, and MINING industries to retrofit machinery or add new SCRUBBERS to smokestacks that would prevent the emission of such particles. *See also* CLEAN AIR ACT; CRITERIA POLLUTANTS; DUST BOWL; NATIONAL AMBIENT AIR QUALITY STANDARDS; SILTATION.

EnviroTerm

micron: A unit of length equal to one-millionth of one meter. The diameter of airborne particles are measured in microns.

parts per billion and parts per million, Units of measure commonly used to describe the concentration of a TOXIC CHEMICAL in another material, such as a PESTICIDE in a food product. The units can be used to establish the maximum permissible amount of a CONTAMINANT in water, soil, or AIR. The unit of measure is expressed in the number of parts contained within a billion or million of another substance. According to the U.S. ENVIRONMENTAL PROTECTION AGENCY (EPA), parts per billion (ppb) would be comparable to one kernel of corn in a filled silo that is 9 meters (45 feet) tall and 5 meters (16 feet) in diameter. Parts per million (ppm) would be comparable to one drop of GASOLINE in a tankful of gas in a full-sized AUTOMOBILE.

passenger pigeons, An extinct bird SPECIES that was once abundant throughout North America. As late as the mid-1800s, many scientists believe, as many as two billion of these birds lived throughout North America; however, only half a century later, in 1914, the last known passenger pigeon, a female named Martha, died in captivity at the Cincinnati Zoo. One of the main causes leading to the EXTINCTION of this bird was overhunting. HABITAT loss due to encroaching human POPULATIONS and DISEASE were other factors that contributed to its extinction. The rapid decline and extinction of the passenger pigeon is often cited as an example of how human activities can adversely affect natural populations and highlights the need to monitor population sizes of ORGANISMS to help protect Earth's BIODIVERSITY. *See also* BIRTH RATE; DEATH RATE; DODO BIRDS; ENDANGERED SPECIES; ENDANGERED SPECIES ACT; ENDANGERED SPECIES LIST; POPULATION DENSITY.

In 1914, the last known passenger pigeon, a female named Martha, died in captivity at the Cincinnati Zoo.

passive solar heating system. *See* solar heating.

pathogen, Any ORGANISM that is capable of causing DISEASE in humans or other organisms. Pathogenic organisms include disease-causing BACTERIA, PROTISTS, and FUNGI. Some scientists also consider disease-causing VIRUSES, prions, and viroids to be pathogens, even though there is great disagreement throughout the scientific community about whether these agents of disease are living things. Pathogens can move throughout an ENVIRONMENT to infect people or other organisms in a variety of ways and are thus of concern to scientists who study the environment. For example, many pathogens are spread to humans through contaminated water. The bacteria that cause cholera in many parts of Asia, Africa, and Latin America are spread in this way. Other bacteria that may be present in water and thus make the water unpotable include COLIFORM BACTERIA and some types of *ESCHERICHIA COLI*. In addition to water, pathogens also may be spread throughout a POPULATION in food. *Escherichia coli*, *SALMONELLA* bacteria, and *Listeria* bacteria often are spread throughout human populations in this way. Still other pathogens are transmitted from one organism to another through the bites of animals. For example, the virus that causes rabies is carried in the saliva of infected animals. The virus is then transmitted from one host to another through the bite of an infected animal. Similarly, the malaria, yellow fever, and African sleeping sickness, which are common in many of the less developed nations of the world, often are spread through the bites of INSECTS or other arthropods. *See also* BIOASSAY; BIODIVERSITY; BIOLOGICAL CONTROL; BIOREMEDIATION; BLIGHT; BOVINE SPONGIFORM ENCEPHALOPATHY; BRUCELLOSIS; CANCER; CARCINOGEN; CELL; CHLORINATION; CONTAMINANT; CRYPTOSPORIDIUM; CYANOBACTERIA; DEATH RATE; DECONTAMINATION; DISINFECTANT; DUTCH ELM DISEASE; ENVIRONMENTAL MEDICINE; FAMINE; GENETIC ENGINEERING; *GIARDIA LAMBLIA;* HAZARDOUS WASTE; INFECTIOUS WASTE; INTEGRATED PEST MANAGEMENT; MICROORGANISM; NEUROTOXIN; PARASITISM; PESTICIDE; POTABLE WATER; PRIMARY, SECONDARY, AND TERTIARY TREATMENTS; RED TIDE; SEWAGE; SEWAGE TREATMENT PLANT; STERILIZATION, PATHOGENIC; SYMBIOSIS; WORLD HEALTH ORGANIZATION.

PCBs. *See* polychlorinated biphenyl.

peat, A fibrous brown acidic mass of partially decayed PLANT remains found in BOGS, often used as a BIOMASS FUEL. Peat deposits form in water-logged ENVIRONMENTS where there is a rich variety of vegetation debris, including everything from plant parts (roots, bark, and stems), mosses, grasses, to decayed plants. Peat is cut into blocks and dried as a low-grade fuel. In some countries where peat is abundant, like Scotland, Finland, and Ireland, peat is used for heating homes and to generate ELECTRICITY. *Sphagnum* peat (formed from Sphagnum moss) is used for agricultural and horticultural purposes.

Peat is the first stage of COAL formation. In order for the peat to become coal, it must be buried by sediment. Over time, the peat is compacted and pressed beneath sediments and rock layers. During the process of metamorphosis, water is squeezed out of the peat and gases such as METHANE are expelled into the ATMOSPHERE. Over thousands of years, the continued burial and compression causes the peat to alter into higher grades of coal. The stages of this trend proceed from peat, lignite, subbituminous coal, bituminous coal to anthracite coal. *See also* ACID; ROCK CYCLE; WETLAND.

pelagic zone. *See* oceans.

perchlorine. *See* tetrachloroethylene.

percolation, The movement of water through soil layers or the slow SEEPAGE of any liquid through a filter. (*See* filtration.) Gravity generally causes water to move downward through subsurface soil layers. As it flows, the water may leach MINERALS or other substances from soil, or pick up particles, such as HEAVY METALS or other POLLUTANTS, transporting them to new locations. (*See* leaching.) When water reaches an impermeable material, such as clay or some types of rock, it may move radially (sideways) as well. Eventually, water and any dissolved substances it carries percolate down through soil and reach an area where it collects above and between soil layers. Water that collects beneath Earth's surface is called GROUNDWATER. In many areas, this water is used for drinking and other purposes. Thus, any CONTAMINANTS present in the water may be passed on to humans or other ORGANISMS.

In some cases, water may percolate upward through soil. This occurs naturally in hot springs or geysers. Because water moving up through soil moves against the force of gravity, such movement occurs only when the water is under pressure. *See also* ACID MINE DRAINAGE; AQUIFER; PERMEABILITY; SUBSIDENCE; WATER TABLE; WELL.

percolation pond, A pond (usually human made) designed to allow WASTEWATER to percolate slowly into the ground. The pond acts as a holding facility, while gravity draws the water down through the soil or other unconsolidated medium into the local WATER TABLE and lower AQUIFERS. *See also* PERCOLATION.

peregrine falcons, Birds of prey that are sometimes called duck hawks and were once found throughout the world except in ANTARCTICA and on some tropical islands. POPULATIONS of peregrine falcons declined during the mid-twentieth century to the point where the birds became considered rare throughout much of the world. Peregrine falcons (*Falco peregrinus*) are members of the Falconidae (falcon) family and are known for their strength and rapid flight. The principal prey of peregrines are other birds, specifically ducks and shorebirds. (*See* predation.) The birds generally grow to a length of 35–48 centimeters (14–19 inches) and have a blue-gray coloring on the back and a yellow-white underside with black stripes.

There are several subspecies of peregrine falcons throughout the world. The American peregrine falcon (*Falco peregrinus anatum*) once ranged from as far north as central Alaska, throughout Canada to as far south as central Mexico, with birds migrating to South America during winters; however, the SPECIES declined significantly in number during the mid-twentieth century and was considered extinct in the eastern United States and eastern Canada. (*See* extinction.) The chief HABITAT of the peregrine was open country near shorelines with high cliffs nearby where the birds could construct nests on the ledges. During breeding, two to four eggs are generally produced and the young are ready to leave the nest less than two months following hatching.

Because of the sharp declines in its population, the American peregrine falcon was first recognized as an ENDANGERED SPECIES by the U.S. FISH AND WILDLIFE SERVICE (FWS) in 1970. The bird also was listed as an endangered species on the RED LIST OF ENDANGERED SPECIES compiled by the WORLD CONSERVATION UNION. The major cause in the decline of the American peregrine falcon was the increase of levels of CHLORINATED HYDROCARBONS in the environment. These chemicals were once widely used in the PESTICIDE DICHLORODIPHENYLTRICHOROETHANE (DDT). The chlorinated hydrocarbons present in the DDT built up in the tissues and CELLS of the animals the falcons used as food. Through the process of BIOACCUMULATION, these substances became more concentrated in animals at higher levels of the FOOD CHAIN, culminating in a high concentration of the chemicals in the tissues of the peregrine falcon. In peregrine falcons, DDT did not directly poison the birds, but instead interfered with their successful reproduction: Birds that had high levels of DDT in their tissues laid eggs with thin shells that broke easily during incubation, resulting in a decrease in offspring. Loss of habitat also contributed to the bird's decline.

The principal prey of peregrine falcons are other birds, specifically ducks and other shore birds.

Major efforts were made to help restore populations of peregrine falcons throughout the United States. Toward this effort, DDT was banned in the United States in 1972. Other nations soon followed this effort and banned the use of the chemical. Despite these efforts, the chemical still remains in use in some parts of the world. In addition to the ban on DDT, several CAPTIVE PROPAGATION programs were established throughout the United States, leading to the births of numerous peregrine falcons in captivity. Mature birds resulting from these efforts were later released back into the wild in areas where the bird was considered extinct. Following the releases, the captive-bred birds successfully colonized new habitats and bred, thus increasing their numbers. These efforts have proved so successful in the case of the American peregrine falcon that the bird was finally removed from the endangered species list of the FWS in August 1999.

The American peregrine falcon is not the only peregrine falcon that has been identified as an endangered species. The Eurasian peregrine (*Falco peregrinus peregrinus*) also has been recognized as an endangered species by the FWS since 1976. This subspecies ranges throughout Europe and parts of Asia to locations as far south as the Middle East and Africa. Efforts to prevent extinction of this subspecies have been similar to those in the United States. *See also* AGRICULTURAL POLLUTION; AUDUBON, JOHN JAMES; BALD EAGLES; BIOCONCENTRATION; BIODIVERSITY; BIODIVERSITY TREATY; ENDANGERED SPECIES ACT; NATIONAL AUDUBON SOCIETY; WORLD CONSERVATION MONITORING CENTRE.

EnviroSources

Chadwick, Douglas H., and Joel Sartore. *The Company We Keep: America's Endangered Species.* New York: National Geographic Society, 1995.

U.S. Fish and Wildlife Service Web site: www.fws.gov.

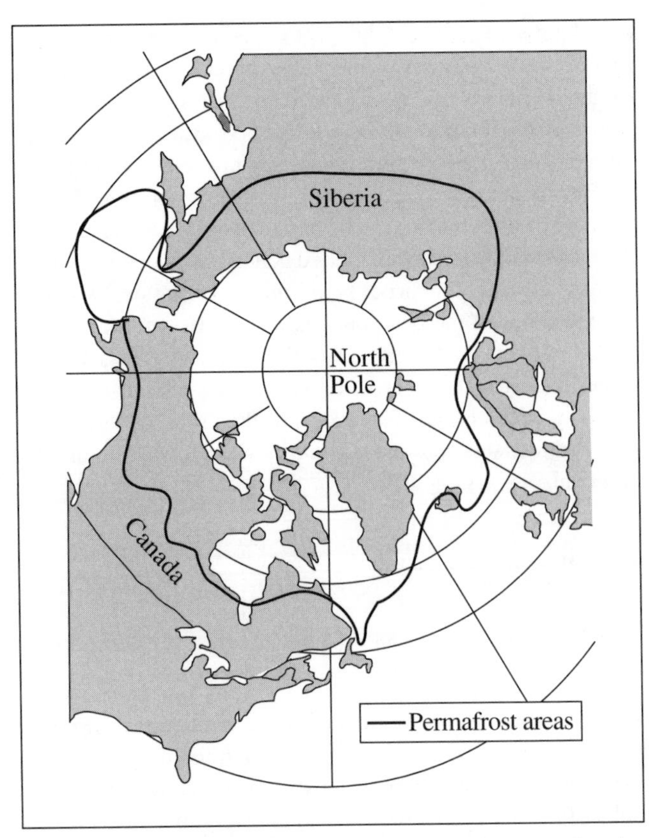

An estimated 20 to 26% of Earth's land area is covered by permafrost that likely formed during the last ice age

permafrost, The layers of soil or subsoil of ARCTIC and sub-Arctic TUNDRA regions that remain frozen throughout the year. An estimated 20–26% of Earth's land area is covered by permafrost, which likely formed during the last ICE AGE. Depending upon the temperature range in an area, permafrost may be discontinuous or continuous and can range in thickness from less than 1 meter (3.3 feet) to more than 400 meters (1,312 feet.) Virtually all of ANTARCTICA and Greenland rest atop continuous permafrost. Much of Canada, Alaska, SIBERIA, and the portions of northern Europe and Asia located above 45°–50° LATITUDE also have large regions of continuous permafrost. Portions of North American, European, and Asian countries located at (or just below) 45°–50° latitude generally have discontinuous permafrost—regions of permafrost intermingled with regions of thawed ground or BOGS.

In many places, the soil layer above the permafrost thaws during summer. The thickness of this "active layer" depends on the average temperature of the region, the slope of the land, and the soil's exposure to sunlight. It is in this thawed soil that PLANTS can grow, sometimes in large enough numbers to warm the soil and the permafrost below it. The presence or absence of plants, which, along with LICHENS and ice ALGAE, are tundra PRODUCERS, plays a vital role in determining which animal SPECIES can survive in a permafrost region.

The tundra is a fragile ECOSYSTEM. Studies indicate that temperatures in parts of the Alaskan, Canadian, and Siberian tundra are increasing by about 1°C (1.8°F) every decade, possibly as a result of GLOBAL WARMING. If this trend continues, the top 10–15 meters (33–50 feet) of the thicker permafrost regions could thaw and significantly change the tundra ecosystem. Thawing of permafrost also leads to SUBSIDENCE, which destroys roads, homes, and other structures. Moreover, as permafrost thaws, trapped METHANE gas is released into the ATMOSPHERE. Since methane is a GREENHOUSE GAS, its release could contribute to greater global warming.

Scientists worldwide are studying permafrost to determine what can be done to protect and maintain the stability of the tundra ecosystem and as a means of monitoring global warming trends. To help prevent accelerated thawing of large expanses of permafrost, scientists support initiatives that help reduce greenhouse gas EMISSIONS such as pollution control technologies and wider use of ALTERNATIVE ENERGY RESOURCES that reduce the use of FOSSIL FUELS. In addition, many scientists support abandonment of the search for additional PETROLEUM deposits in tundra regions and discourage construction of additional facilities and pipelines for collecting and transporting oil in these regions. *See also* ALASKA PIPELINE; ARCTIC NATIONAL WILDLIFE REFUGE.

EnviroSources

Dybas, Cheryl Lyn. "Houses Bow to Thawing Permafrost." Washington, DC: National Science Foundation, 1995. (National Science Foundation Web site: www.nsf.gov/stratare/egch/nws195.htm.)

Keller, Edward A. *Environmental Geology.* New York: MacMillan, 1995.

Thomas, Mary Powel. "Climate Change Hits Home." *Audubon* 99, no. 5 (September–October 1997): 19.

permeability, A term that describes the ease with which water or another liquid can flow through a filter, soil and rock layers, or another solid. Materials and substances through which liquids pass easily are described as being permeable or having high permeability; those through which liquids cannot pass easily are described as impermeable or having low permeability. Permeability greatly affects the ability of soil to support PLANT growth or permit LEACHING.

Permeable soil is composed of about equal amounts of large, medium, and small particles (sand, silt, and clay, respectively). Soils composed of large particles or of mostly large and medium-sized particles have the highest permeability. Such soils are generally good for plant growth, because they allow water and AIR to collect in the spaces between soil particles and provide spaces that allow plant roots to move easily through soil. In contrast, soils composed primarily of clay do not allow liquids to pass through them easily and are thus impermeable. Such soils are generally poor for plant growth, but are suitable for use in SANITARY LANDFILLS. In fact, clay often is used as a liner for landfills to prevent potentially harmful liquids from leaching through the landfill site and escaping into nearby soil or GROUNDWATER systems. *See also* FILTRATION; INFILTRATION; PERCOLATION; PRIMARY, SECONDARY, AND TERTIARY TREATMENTS; REVERSE OSMOSIS; ZONE OF AERATION; ZONE OF DISCHARGE; ZONE OF SATURATION.

persistent pesticide. *See* pesticide.

pesticide, A chemical used for pest control in AGRICULTURE and gardening. Pesticides are pollution hazards and are often highly toxic to humans and other ORGANISMS. There are three main groups of pesticides: Insecticides are used to control and kill INSECTS. Herbicides are used to control unwanted PLANTS such as weeds. Fungicides are applied to kill FUNGI and other parasites. (*See* parasitism.) Pesticides also include rodenticides.

The widespread use and disposal of pesticides by farmers, institutions, and the general public provide many possible sources of pesticides in the ENVIRONMENT. After being released into the environment, pesticides may have many different final destinations. Pesticides that are sprayed can move through the AIR and may eventually end up in other parts of the environment, such as in soil or water. Pesticides that are applied directly to the soil may be washed off the soil into nearby bodies of SURFACE WATER or may percolate through the soil to lower soil layers and GROUNDWATER. (*See* percolation.) Pesticides that are injected into the soil may also be subject to the latter two outcomes. The application of pesticides directly to bodies of water for weed control, or indirectly as a result of LEACHING from boat paint, RUNOFF from soil, or other routes, may lead not only to the build up of pesticides in water, but may also contribute to AIR POLLUTION through evaporation. These examples suggest that the movement of pesticides in the environment is very complex with transfers occurring continually among different ECOSYSTEMS. In some cases, these exchanges occur not only between areas that are close together (such as a local pond receiving some of the herbicide application on adjacent land) but may also involve transportation of pesticides over long distances.

The worldwide distribution of DICHLORODIPHENYLTRI-choroethane (DDT) and the presence of pesticides in bodies of water far from their primary use areas are good examples of the vast potential for such movement. While all of the above possibilities exist, this does not mean that all pesticides travel long distances or that all compounds are threats to groundwater. To understand which ones are of most concern, it is necessary to understand how pesticides move in the environment and what characteristics must be considered in evaluating contamination potential.

Two things may happen to pesticides once they are released into the environment. They may be broken down, or degraded, by the action of sunlight, water, other chemicals, or MICROORGANISMS, such as BACTERIA. This degradation process usually leads to the formation of less harmful breakdown products but in some instances can produce more toxic products. The second possibility is that the pesticide will be very resistant to degradation by any means and thus remain unchanged in the environment for a long period of time.

The ones that are most rapidly broken down have the shortest time to adversely affect the environment or people or other organisms. The ones that last the longest, the so-called "persistent pesticides," can move over long distances and can build up in the environment leading to greater potential for adverse effects.

In addition to resistance to degradation, there are a number of other properties of pesticides that determine their behavior and fate. One is how volatile they are—in other words, how easily they evaporate. The ones that are most volatile have the greatest potential to evaporate into the ATMOSPHERE and, if also persistent, to move long distances. Another important property is solubility in water—that is, how easily they dissolve in water. If a pesticide is very soluble in water, it is more easily carried with rainwater as runoff or through the soil as a potential groundwater CONTAMINANT (leaching). In addition, the water-soluble pesticide is more likely to stay mixed in the surface water where it can have adverse effects on fish and other organisms. If the pesticide is very insoluble in water, it usually tends to stick to soil and also to settle to the bottom of bodies of surface water, making it less available to organisms.

From a knowledge of these and other characteristics, it is possible to predict in a general sense how a pesticide will behave. Unfortunately, more precise prediction is not possible because the environment itself is very complex. There are, for example, huge numbers of soil types, which vary with respect to the percentage of sand, ORGANIC matter, metal content, acidity, and so on. All of these soil characteristics influence the behavior of a pesticide so that a pesticide that contaminates groundwater in one soil may not do so in another. Similarly, surface waters vary in their properties, such as acidity, depth, temperature, clarity (suspended soil particles

A tractor is used to spray pesticides on a cotton crop. *Credit:* United States Department of Agriculture. Photo by Bill Tarpenning

or biological organisms), flow rate, and general chemistry. These properties and others all can affect pesticide movement and fate.

How persistent are pesticides in soil? Persistence is measured as the time it takes for half of the initial amount of a pesticide to break down. Thus, if a pesticide's HALF-LIFE is 30 days, half will be left after 30 days, one-quarter after 60 days, one-eighth after 90 days, and so on. It might seem that a short half-life would mean a pesticide would not have a chance to move far in the environment; however, if the pesticide is soluble in water and the conditions are right, it can move rapidly through certain soils. As the residual pesticide moves away from the surface, it moves away from agents that can degrade it, such as sunlight and bacteria. As the pesticide moves deeper into the soil, it degrades more slowly and thus has a chance to enter groundwater.

The measures of soil persistence in the accompanying table describe pesticide behavior at or near the surface only. The downward movement (leaching) of nonpersistent pesticides may not occur. Several pesticides with short half-lives, such as Aldicarb, have spread in groundwater. In contrast, very persistent pesticides may have other properties that limit their potential for movement throughout the environment. Many of the CHLORINATED HYDROCARBON pesticides are very resistant to breakdown but are also very water insoluble, tending not to move down through the soil into groundwater. Less soluble pesticides can, however, become problems in other ways, since they remain on the surface for a long time where they may be subject to runoff and possible evaporation. Even if a pesticide is not very volatile, the tremendously long time that it can persist can lead, over time, to measurable concentrations moving through the atmosphere and accumulating in remote areas.

Living organisms may also play a significant role in pesticide distribution, especially for pesticides that can accumulate in living creatures. CHLORDANE, for example, which is very water-insoluble, can accumulate in a creature living in water. As this pesticide is continually absorbed and stored in the

Pesticide Persistence in Soil

Low (half-life <30 days)	Moderate (half-life 30–100 days)	High (half-life >100 days)
aldicarb	aldrin	bromacil
captan	atrazine	chlordane
calapon	carbaryl	lindane
dicamba	carbofuran	paraquat
malathion	diazinon	picloram
methyl-parathion	endrin	TCA
oxamyl	heptachlor	trifluralin

organism, its concentration increases. If this organism is eaten by another organism that also stores this pesticide, Chlordane levels can become much higher in the organism that feeds higher up the FOOD CHAIN than levels in the water in which it lives. Concentrations in fish, for example, can be ten times to hundreds of thousands of times greater than ambient water concentrations of the same pesticide. This type of accumulation is called BIOACCUMULATION. In this regard, it should be remembered that humans are at the top of the food chain and so may be exposed to these high levels when they eat food animals that have bioaccumulated pesticides and other ORGANIC chemicals. Domestic farm animals can also accumulate pesticides, and so care must be taken in the use of pesticides in agricultural situations.

Pesticides have global impacts because of their widespread use and chemical properties that make many pesticides long-lived. It has been estimated that 65% of all agricultural land in Europe contains high enough concentrations of pesticides to warrant health concern. Groundwater resources may be threatened by the presence of residual pesticides in soil.

Pesticides also contribute to air pollution, entering the air during application or through volatilization (evaporating into the air.) Studies have demonstrated that some types of pesticides travel long distances through the atmosphere, polluting air, fog, and rain many miles from the source area. Pesticides that become airborne as dust particles can settle out of the atmosphere in precipitation (i.e., rain and snow) over a long period of time. One such study estimated that 56 metric tons (62 tons) of Parathion, 127 metric tons (140 tons) of Atrazine, and 33 metric tons (36 tons) of Lindane fall into the North Sea from the atmosphere annually.

METHYL BROMIDE, a common fumigant, accelerates the depletion of Earth's protective OZONE layer. Chemicals such as methyl bromide may contribute to between one-tenth and one-twentieth of total ozone loss.

Farm workers who mix and apply pesticides are at particular risk from exposure through inhalation or skin contact. The WORLD HEALTH ORGANIZATION estimates that pesticides cause three million acute poisonings among agricultural workers worldwide each year. Experts estimate that the number of agricultural laborers poisoned by pesticides in the United States may range from 27,000 to 300,000 cases a year. Research has indicated that European farmers may be at increased risk of certain CANCERS including malignant lymphoma, leukemia, multiple myeloma, and cancers of the testicles, gastrointestinal tract, lung, and brain. Pesticide factory workers, and residents near pesticide factories, also may face greater risk of pesticide exposures, as the environmental disaster at Bhopal, India, demonstrated. (*See* Bhopal incident.)

During the 1980s, researchers became increasingly aware of the global phenomenon of "endocrine disruption"—the disruption of hormonal systems in WILDLIFE and humans, which appeared to occur in part as a result of man-made chemicals, including pesticides. The effects of endocrine disruption may have contributed to declines in the populations of many wildlife SPECIES since the 1950s. The symptoms of endocrine disruption include adverse effects on the development of offspring, reproductive abnormalities, and weakened immune systems. Chlorinated pesticides rank among the most widespread and persistent endocrine disrupters identified. *See also* HAZARDOUS WASTE; PESTICIDE REGULATION.

EnviroSources

Agency for Toxic Substances and Diseases, Registry Division of Toxicology: 1600 Clifton Road, NE Mailstop E-29 Atlanta, GA 30333.

Bourne, Joel. "Bugging Out." *Audubon* 101:2 (March–April 1999): 71–73.

———. "The Organic Revolution." *Audubon* 101:2 (March–April 1999): 64–70.

Mike Kamrin, Institute for Environmental Toxicology, C231 Holden Hall, MSU; (517) 353-6469.

Schnoor, J.L., ed. *Fate of Pesticides and Chemicals in the Environment.* Environmental Science and Technology. New York: John Wiley and Sons, 1991.

U.S. Environmental Protection Agency (toxics and pesticides) Web site: www.epa.gov/oppfead1/work_saf/. (pesticides in the atmosphere) Web site: www.p510dcascr.wr.usgs.gov/pnsp/atmos/atmos_4.html.

Wargo, John. *Our Children's Toxic Legacy: How Science and Law Fail to Protect Us from Pesticides.* New Haven, CT: Yale University Press, 1998.

pesticide regulation, The control of PESTICIDE use by various laws and agency rules. In the United States, no less than 14 different federal acts control some aspect of the manufacture, registration, distribution, use, consumption, and disposal of pesticides. Until recently, most of the pesticide regulation fell under the FEDERAL INSECTICIDE, FUNGICIDE, AND RODENTICIDE ACT (FIFRA). This legislation governs the registration, distribution, sale, and use of pesticides.

In 1996, with the passage of the Food Quality Protection Act (FQPA), new regulations governing pesticide levels in foods came into effect. The U.S. ENVIRONMENTAL PROTECTION AGENCY (EPA) is responsible for the administration of both of these acts and for establishing rules and regulations consistent with their intent. The three broad categories of pesticide regulation focus on the registration of new pesticides and the reregistration of existing pesticides, on establishing and monitoring pesticide levels in food products, and on monitoring pesticide levels in the ENVIRONMENT, especially in GROUNDWATER and SURFACE WATER.

The EPA is responsible for registering new pesticides to ensure that, when used according to label directions, they will not pose unreasonable risks to human health or the environment. The FIFRA and FQPA require the EPA to balance the risks of pesticide exposure to human health and the environment against the benefits of pesticide use to society and the economy.

Depending on the type of pesticide, the EPA can require up to 70 different tests. For a major pesticide to be used on food products, testing can cost the manufacturer many millions of dollars. Testing is needed to determine whether a pesticide has the potential to cause adverse effects in humans, WILDLIFE, fish, and PLANTS. Potential human risks, which are identified by laboratory tests, include acute toxic reactions, as well as possible long-term effects like CANCER, birth defects, and reproductive system disorders.

The food supply of the United States is among the safest in the world. Although many of the foods we consume may contain low levels of pesticide residues as a result of their use, numerous safeguards are built into the regulatory process to ensure that the public is protected from unreasonable risks posed by eating pesticide-treated crops and livestock. The EPA regulates the safety of the food supply by setting TOLERANCE LEVELS, or maximum legal limits, for pesticides on food commodities and in animal feed available for sale in the United States.

Pesticide tolerance levels are enforced by the FOOD AND DRUG ADMINISTRATION (FDA), which monitors all domestically produced and imported foods traveling in interstate commerce except meat, poultry, and some egg products. The FDA conducts a Total Diet Study, also known as a Market Basket Study, which measures the average American consumer's daily intake of pesticide residues from foods that are bought in typical supermarkets and grocery stores, and are prepared or cooked as they would be in a household setting. The findings of the ongoing Total Diet Study show that dietary levels of most pesticides are less than 1% of the safe level. Imported foods receive special attention in the FDA's monitoring program. If a single shipment from a given source is found to violate U.S. tolerance regulations, all shipments from the same source are subject to automatic detention. *See also* BIOACCUMULATION; BIOMAGNIFICATION; HAZARDOUS WASTE; TOXIC CHEMICAL.

EnviroSources

The following pamphlets are available from the Environmental Protection Agency. Write to U.S. EPA/NSCEP, P.O. Box 42419, Cincinnati, OH 45242 (1-800-490-9198); Web site: www.epa.gov/pesticides.

Citizen's Guide to Pest Control and Pesticide Safety: EPA 730-K-95-001.

Consumer Products Treated with Pesticides: EPA 735-F-98-004.

National Research Council, Committee on Scientific and Regulatory Issues Underlying Pesticide Use Patterns and Agricultural Innovation, Board on Agriculture, National Research Council. *Regulating Pesticides in Food: The Delaney Paradox.* Washington, DC: National Academy Press.

Pesticide Safety Tips: EPA 735-F-98-010.

10 Tips to Protect Children from Pesticide and Lead Poisonings: EPA-F-97-001.

Peterson, Roger Tory (1908–1996), An artist whose passion for nature made him one of the most well known NATURALISTS of the twentieth century. He was well known as an artist, editor, writer, scientist, lecturer, traveler, photographer, film-maker, and conservationist. In an interview, Peterson reflected on his professional life as a painter first, second as a writer, and third as a naturalist.

His *Field Guide to the Birds* was first published in 1934. The series, which includes more than 50 other guides, brought natural history to the general public for the first time. He will be best known for stimulating millions of people worldwide to observe birds.

The Roger Tory Peterson Institute was established in Jamestown, New York, his hometown, to honor his achievements in nature and CONSERVATION. *See also* AUDUBON, JOHN JAMES; LEOPOLD, ALDO; MUIR, JOHN; THOREAU, HENRY DAVID.

EnviroSources

Peterson, Roger Tory. *A Field Guide to the Birds: Giving Field Marks of All Species Found in Eastern North America.* New York: Houghton Mifflin, 1996.

Peterson, Roger Tory, and Virginia Marie Peterson. *A Field Guide to the Birds: A Completely New Guide to All the Birds of Eastern and Central North America.* Peterson Field Guides. New York: Houghton Mifflin, 1998.

Roger Tory Peterson Institute, 311 Curtis St., Jamestown, NY 14701; (716) 665-2473; e-mail: webmaster@rtpi.org.

petrochemical, An ORGANIC compound derived from PETROLEUM or NATURAL GAS—NONRENEWABLE RESOURCES that are often referred to as FOSSIL FUELS. Petrochemicals are obtained when crude oil or natural gas is refined, or separated, into GASOLINE, heating oil, asphalt, and other useful substances. Some petrochemicals are put to direct use as FUELS, SOLVENTS, PESTICIDES, drugs, and cosmetic preparations. Most petrochemicals, however, serve as raw materials, or intermediates, in the production of synthetic substances, particularly plastics.

Ethylene, a highly reactive gas, is perhaps the most widely used petrochemical. Ethylene used in the production of plastics, synthetic fibers, and antifreeze. Other important petrochemicals include BENZENE, which is used to make synthetic rubber and latex paints, and PHENOLS, important chemicals used in the manufacture of perfumes, artificial flavorings, and pesticides.

Petrochemical products are used in just about every industry today, from AGRICULTURE to medicine. Unfortunately, the production and use of petrochemicals causes a variety of environmental problems. When these substances are produced, for example, a number of POLLUTANTS, including SULFUR DIOXIDE and particulates, are released into the AIR. EMISSIONS of sulfur dioxide are a major cause of ACID RAIN. Certain petrochemicals themselves, such as benzene and toluene, are also highly toxic to humans and other ORGANISMS.

Many of the environmental problems associated with petrochemicals occur after they are purchased by consumers. For example, leaking AUTOMOBILE radiators and their improper disposal result in the release of large amounts of ethyl glycol contained in antifreeze. Ethyl glycol is directly toxic to dogs, cats, other animals, and children who ingest the spilled substance. In addition, the TOXIC CHEMICAL is often washed into nearby aquatic ECOSYSTEMS when it is carried into storm drains in RUNOFF. Similarly, runoff from lawns and land used for agriculture carries pesticides and FERTILIZERS containing petrochemicals into aquatic ecosystems or leaches the substances into deeper soil layers. (*See* leaching.)

In the United States, the ENVIRONMENTAL PROTECTION AGENCY tightly monitors the safety of petrochemical plants. Strict environmental laws, including the CLEAN AIR ACT, help control the amount of pollutants released by petrochemical plants; however, there will always be a tradeoff between the environmental dangers produced by petrochemicals and the need of society to use these products. *See also* AGRICULTURAL POLLUTION; AIR POLLUTION; HYDROCARBONS; OIL SPILL.

petroleum, A flammable, liquid FOSSIL FUEL that occurs naturally in deposits, usually underground, and which is also known as crude oil. Developed nations depend heavily on petroleum for power, FUEL, and synthetic products such as plastics. About 70% of the ENERGY consumed in the United States comes from petroleum and NATURAL GAS.

The composition of petroleum varies with locality, but it is mainly a mixture of HYDROCARBONS containing 5 to more than 60 CARBON atoms each, with SULFUR, NITROGEN, and OXYGEN as impurities. Petroleum is liquid at the earth's surface and varies in density. It is described as heavy, average, or light; light oils are the most valuable because they produce the most GASOLINE. Petroleum volume is usually measured in barrels, each representing 159 liters (about 42 gallons).

Petroleum was formed over millions of years by the chemical and physical alteration of PLANT and animal remains buried under thick rock layers. The primary requirements for the development of petroleum were extensive growth of ALGAE, standing water, and long burial by younger sediments.

Periods of GLOBAL WARMING may have accelerated petroleum formation. The late Jurassic, 150 million years ago, was one such period, responsible for the major petroleum deposits in the Middle East, the NORTH SEA, and parts of SIBERIA. In the mid-Cretaceous period, 90 million years ago, petroleum deposits were formed in northern South America. Much of the petroleum deposits in the United States were formed during the Permian era, 230 million years ago.

Geophysical methods, usually seismic, magnetic, or gravity surveys, are used to determine promising sites for oil WELLS, some of which must be dug several miles deep to reach a deposit. Most petroleum is found in sedimentary rock basins, of which there are about 700 worldwide. (*See* rock cycle.) Sandstone and limestone, generally folded and faulted, are common RESERVOIR rocks. Oil migrates through porous rock until it is trapped at the top of a fold, against a fault, or where a bed pinches out. Materials such as salt form seals that prevent oil and gas from leaking out of structural traps.

Petroleum became a major commercial source of fuel in the mid-1850s. The first oil wells were drilled where oil naturally seeped at the surface. Primitive rotary drilling rigs were introduced in the 1880s. In 1901, the first modern rotary rig was used at the Spindletop oil field, on a salt dome in Texas. During the next few years, new AUTOMOBILES were built with gasoline-fueled, internal-combustion engines, and gasoline quickly became the most important product of crude oil.

Some of the leading producers of petroleum include Russia, the United States, Saudi Arabia, and Iran. Worldwide, the largest reserves are in the Middle East, with the Persian Gulf region producing 27% of the world's oil. As of 1993, the U.S. GEOLOGICAL SURVEY estimated that worldwide, there were 1,103 billion barrels of oil remaining in discovered reserves and 471 billion barrels in undiscovered reserves, almost entirely in oil fields that have already been discovered. In the early 1990s, the annual world consumption of refined petroleum was more than 23 billion barrels. Because the United States consumes about twice as much crude oil as it produces, it must import supplies from other countries.

Petroleum can also be obtained from oil shales, which yield 95–114 liters (about 25 gallons) of oil per ton when heated. They are very expensive to mine and heat, however, and disposing of the spent shale poses environmental problems. Tar sands, another source of petroleum, contain thick, heavy oils that can be removed by injection of steam.

When large volumes of petroleum products are burned, the GREENHOUSE GASES and other byproducts that are released into the air cause serious environmental problems such as AIR POLLUTION, SMOG, ACID RAIN, and an enhanced GREENHOUSE EFFECT. Scientists generally believe that the combustion of fossil fuels and other human activities are the primary reason for the increased concentration of CARBON DIOXIDE in the ATMOSPHERE. The combustion of petroleum products to run vehicles, heat buildings, and power factories is responsible for about 80% of the world's carbon dioxide EMISSIONS, about 25% of U.S. METHANE emissions, and about 20% of global nitrous oxide emissions.

Processing petroleum and use of its products also create many other air POLLUTANTS, including airborne PARTICULATE MATTER. In addition, oil spilled from tankers and offshore wells has damaged OCEAN and coastline ENVIRONMENTS. (*See* oil spill.) The environmentally disruptive effects of oil wells have also sometimes led to strong opposition to new drilling, as in WILDERNESS areas of northern Alaska and in the ARCTIC NATIONAL WILDLIFE REFUGE. *See also* ALASKA PIPELINE; CARBON MONOXIDE; OIL POLLUTION; OIL TANKER; ORGANIZATION OF PETROLEUM EXPORTING COUNTRIES; SEDIMENTATION; VOLATILE ORGANIC COMPOUNDS.

EnviroSources

American Petroleum Institute Web site: www.api.org.
Oil & Gas Journal Online Web site: www.ogjonline.com.
Petroleum Information: Web site: www.petroleuminformation.com.
U.S. Department of Energy, Energy Information Administration Web site: www.eia.doe.gov.
U.S. Department of Energy, Office of Fossil Energy Web site: www.fe.doe.gov.
U.S. Geological Survey ("Changing Perceptions of World Oil and Gas Resources as Shown by Recent USGS Petroleum Assessments" fact sheet) Web site: www.greenwood.cr.usgs.gov/pub/fact-sheets/fs-0145-97/fs-0145-97.html.
U.S. Geological Survey, Energy Resources Program Web site: www.energy.usgs.gov/index.html.

pH, A logarithmic scale used to measure the intensity or strength of an ACID or alkaline solution. pH tests can be used to determine whether a chemical POLLUTION problem exists in SURFACE WATER or in soil.

The pH value of a solution in water is expressed on a logarithmic scale to the base of 10. The pH scale ranges from 0 to 14, zero is the most acid, 7 is neutral, and 14 is the most alkaline. A solution with a pH of 5 is 10 times more acidic than a

The pH Levels of Various Substances

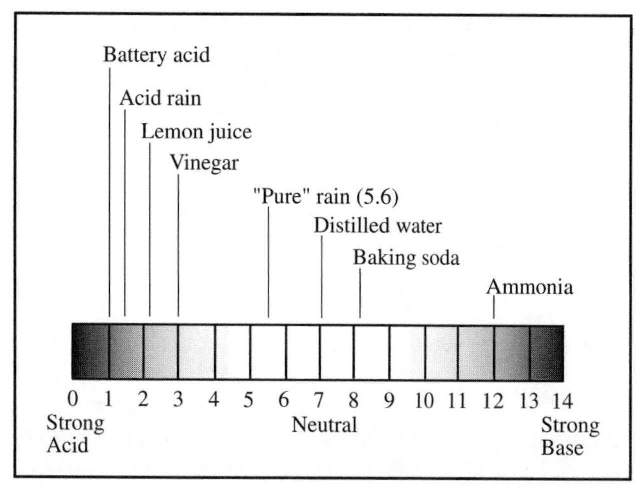

solution with a pH of 6. A difference of two units, for example, between pH 5 and pH 7, would indicate that the pH solution is 100 times more acidic than the pH 7 solution. Knowing the pH of water or soil is important. As an example, trout and SALMON cannot survive in water with a pH value of 6. Farm soils can be too acidic to grow certain crops. Acidic solutions require the use of limestone (calcium carbonate) products or other ALKALIS that can neutralize acidic soils and water. *See also* ACID MINE DRAINAGE; ACID RAIN.

phenol, Any ORGANIC compound that is a byproduct of PETROLEUM refining or tanning, textile, dye, and resin manufacturing. The chemical is also used for making plastics, wood preservatives, and pharmaceuticals. Low concentrations of phenols cause taste and odor problems in water. Higher concentrations are toxic to humans and have an adverse effect on aquatic life. *See also* PETROCHEMICAL; TOXIC CHEMICALS.

phosphates, Any chemical compound that contains the elements PHOSPHORUS and OXYGEN. Phosphates are important ingredients in many products, most notably in industrial DETERGENTS and commercial PLANT FERTILIZERS.

phosphorus, A natural chemical element that is essential for every form of life on Earth. Phosphorus forms the basis of a very large number of chemical compounds, the most important group of which are the PHOSPHATES. Phosphates are utilized by all living ORGANISMS, being involved in ENERGY transfer, metabolism, PHOTOSYNTHESIS, nerve function, muscle action, and in the formation of deoxyribonucleic acid (DNA) and other CELL components. Human bone is composed largely of phosphorus.

Almost three-fourths of the total phosphorus, in all of its forms, used in the United States goes into the production of FERTILIZERS. Other important uses for phosphorus include nutrient supplements for cattle and other animal feeds, water softeners, food additives, coating agents for metals, and some PESTICIDES.

Much of the phosphorus on Earth is tied up in sedimentary deposits and a number of phosphate-bearing rocks. Phosphorus cycles naturally in nature through the phosphorus cycle, continuously moving back and forth between the living (BIOTIC) and the nonliving (ABIOTIC) parts of the ENVIRONMENT. The phosphorus cycle begins when WEATHERING and EROSION break down phosphate-containing rocks. When phosphates dissolve in water, they form a solution that becomes available for ALGAE and terrestrial PLANTS. Phosphorus is then passed on through the FOOD CHAIN as animals eat plants or other animals. When organisms decompose or excrete wastes, phosphorus is once again released into the environment for recycling.

Disruption of the phosphorus cycle due to human activity has been a major concern of environmental scientists for decades. Scientists are concerned that several activities are adding too much phosphorus to the environment, altering its natural balance. Some of the excess phosphorus comes from SEWAGE TREATMENT PLANTS, and some is the result of RUNOFF from cattle feedlots. Most, however, comes from the commercial fertilizers that wash from farms into rivers and lakes. Excess phosphorus is known to result in poor water quality and the EUTROPHICATION of lakes and rivers. To reduce levels of phosphorus in the environment, scientists recommend the use of natural fertilizers such as manure as opposed to the chemical fertilizers that contain phosphorus. More effective WASTEWATER treatment methods also need to be developed in order to further reduce excessive levels of phosphorus. *See also* AGRICULTURE; BIOGEOCHEMICAL CYCLE; PRIMARY, SECONDARY, AND TERTIARY TREATMENTS.

photic. *See* ocean.

photic zone. *See* ocean.

photochemical smog. *See* smog.

photodegradable plastic, Plastic than can break down and decompose when exposed to light over a period of time. (*See* decomposition.) Many trash bags are made from photodegradable and BIODEGRADABLE plastics. Biodegradable products are used heavily in an effort to save the ENVIRONMENT, but many of these products do not break down as easily as was predicted.

There are two basic kinds of degradable plastics: photodegradable (breaks down via light) and biodegradable (breaks down via the action of BACTERIA.) Most of the plastic waste ends up buried in landfills where the light needed for photodegradation is absent. The biodegradable plastics are unlikely to break down in landfills within a reasonable amount of time because the physical conditions (AIR and water) needed for biodegradation are missing. *See also* DECOMPOSER; NONBIODEGRADABLE.

photosynthesis, A complex biochemical process that directly or indirectly provides most ORGANISMS on Earth with their nutrition, ENERGY, and OXYGEN needs. During photosynthesis, PLANTS, CYANOBACTERIA, ALGAE, and certain other PROTISTS use the energy in sunlight to drive a chemical reaction that converts CARBON DIOXIDE (CO_2) and water (H_2O) into the simple sugar glucose ($C_6H_{12}O_6$). (Energy from the sun is captured by a pigment called chlorophyll that is present in the CELLS of photosynthetic organisms.) This carbohydrate is used by most organisms as their energy source. Photosynthesis also releases oxygen (O_2) into the ENVIRONMENT as a byproduct. Because photosynthetic organisms use carbon dioxide for this process and release oxygen as a byproduct, they play a vital role in maintaining Earth's oxygen–carbon dioxide balance.

Organisms that synthesize their food via photosynthesis are vitally important to ECOSYSTEMS because they provide most other organisms with food and energy. Thus, photosynthetic organisms are the PRODUCERS at the base of most FOOD CHAINS. Glucose made by these organisms that is not used by them for food or energy is stored in the organism's cells or tissues as starch, a carbohydrate composed of simple sugars. Much of this starch is passed to a CONSUMER that eats the photosynthetic organism as food. The consumer may use some of the glucose contained in the starch for its life processes or store unused glucose in its cells, tissues, and organs. This stored glucose may then be passed on to another consumer that uses the first consumer as food, thus continuing the movement of food through subsequent TROPHIC LEVELS of the food chain.

The FOSSIL FUELS people use to meet many of their energy needs are composed of HYDROCARBONS formed through the photosynthetic activities of plants that lived on Earth millions of years ago. As these plants died, their remains, including the glucose and starch they contained, were buried in mud or between rock layers. Over time, additional sediments collected atop these remains burying them deeper in the Earth. Over millions of years, heat and pressure exerted by the layers of rock and soil acted upon these remains, changing them in form into the fossil fuels that people use today. *See also* AUTOTROPH; BIOMASS; CHEMOAUTOTROPH; CHEMOSYNTHESIS; COAL; ECOLOGICAL PYRAMID; FOOD WEB; FUEL WOOD; HETEROTROPH; NATURAL GAS; PETROLEUM.

EnviroTerm

carbohydrate: An ORGANIC compound composed of CARBON, oxygen, and HYDROGEN. Sugars and starches are carbohydrates.

EnviroSource

Kirk, John T.O. *Light and Photosynthesis in Aquatic Ecosystems.* Cambridge, UK: Cambridge University Press, 1994.

photovoltaic cell. *See* solar cell.

phytoplankton, Tiny aquatic ORGANISMS that derive their nutrition and ENERGY through PHOTOSYNTHESIS. Phytoplankton are PRODUCERS and are present in virtually all aquatic ECOSYSTEMS. They are of great environmental importance because they form the base of the FOOD CHAINS and FOOD WEBS in these ecosystems.

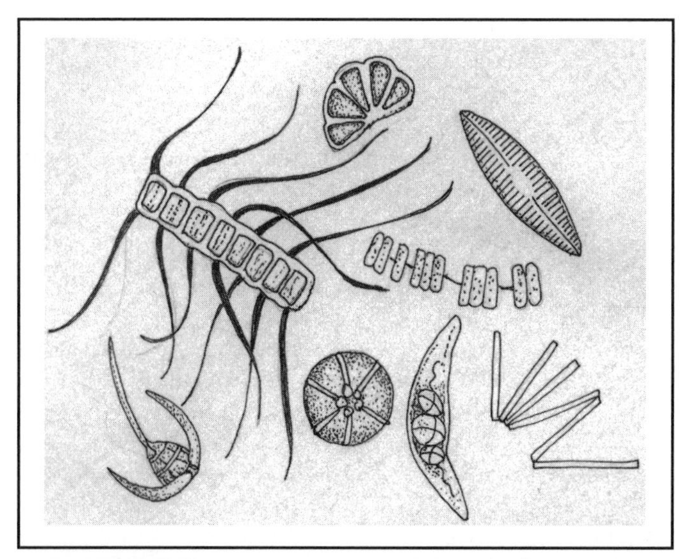

Phytoplankton are of great environmental importance because they form the base of the food chains and food webs.

Phytoplankton generally lack structures that enable independent movement and instead drift at or near the surface of water, where they are exposed to sunlight. Movement of these organisms, which include photosynthetic BACTERIA and ALGAE, generally occurs only as a result of movement of the water (due to winds, TIDES, or currents) in which they live. As producers, phytoplankton are the primary food source for most ZOOPLANKTON, tiny marine organisms incapable of carrying out photosynthesis, which, in turn, serve as a food source for larger aquatic animals. Phytoplankton also are the main food for many of Earth's largest aquatic organisms, including several SPECIES of WHALES. *See also* ALGAL BLOOM; CYANOBACTERIA; OCEAN; PLANKTON; ZOOXANTHELLAE.

phytoremediation, Natural processes carried out by PLANTS and trees in cleaning up and stabilizing contaminated soil and GROUNDWATER. Metals and ORGANIC chemicals from PETROCHEMICAL spills, mine TAILINGS, radioactive and nuclear wastes, ammunition wastes, and PESTICIDES have been treated by phytoremediation. Plants such as poplar trees, grasses, WETLAND SPECIES, and LEGUMES have been used in highly contaminated ENVIRONMENTS. Special enzymes called phytochelatins allow plants to absorb high levels of CONTAMINANTS with their roots, leaves, or stems and to store and biodegrade many of the POLLUTANTS. In studies, HYBRID poplar trees removed organic compounds such as trichloroethylene from soil and water and did not release any of the chemicals into the AIR. Phytoremediation is an effective way of cleaning up contaminated soil and water that otherwise would be transported to landfills or left to cause potential environmental harm. *See also* BIOREMEDIATION.

EnviroTerm

phytochelatins: Special enzymes that PLANTS produce to protect themselves against CELL damage from HEAVY METALS such as CADMIUM and LEAD.

Pinchot, Gifford (1865–1946), A public official of the United States, who served as the chief of the Division of Forestry (now the U.S. FOREST SERVICE) of the U.S. DEPARTMENT OF AGRICULTURE (USDA) from 1898 to 1910, was a professor of FORESTRY at Yale University from 1903 to 1936, and was the governor of Pennsylvania from 1923 to 1927 and again from 1931 to 1935. Gifford Pinchot was born in Simsbury, Connecticut. During his employment with the Division of Forestry, he authored the books *Primer of Forestry* (1899) and *The Fight for Conservation* (1909). During the latter part of his tenure with this office, he also began his teaching career as a professor of forestry. Through his writings, work with the Division of Forestry, and teaching, Pinchot became a leading expert on forestry and an early supporter of CONSERVATION. He stressed the need to conserve NATURAL RESOURCES and recommended that the government protect FORESTS and other natural resources through regulation of land use. Under President THEODORE ROOSEVELT, many of Pinchot's ideas became national policy. *See also* AGROFORESTRY; DEFORESTATION; FOREST MANAGEMENT; GREENBELT MOVEMENT; MUIR, JOHN.

EnviroSources

Chang, Chris. "Champions of Conservation: *Audubon* Recognizes 100 People Who Shaped the Environmental Movement and Made the Twentieth Century Particularly American." *Audubon* 100, no. 6 (November–December 1998): 131.

Pinchot, Gifford. *Breaking New Ground.* Washington, DC: Island Press, 1998.

pioneer species, The first group of ORGANISMS to colonize a barren, lifeless area, thus beginning the process of ECOLOGICAL SUCCESSION. During primary succession, pioneer species may colonize areas composed largely of bare rock, such as those left exposed by a retreating GLACIER or in a cooled lava field resulting from a volcanic eruption. In primary succession, primary species are those whose activities promote soil formation. In secondary succession, pioneer species are the first to colonize areas left barren and lifeless as a result of NATURAL DISASTERS such as FOREST FIRES, FLOODS, and EARTHQUAKES, or through the activities of humans such as agricultural abandonment (the cessation of crop growth in fields formerly used for this purpose) or DEFORESTATION.

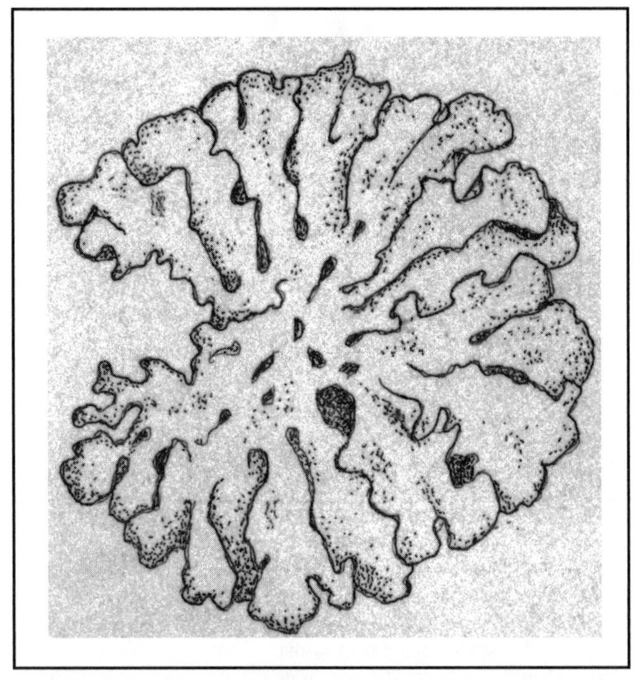

Pioneer species vary according to the type of succession an area is undergoing. In primary succession, such species generally include lichens (above), microorganisms, and a few small, hardy plant species.

Pioneer species vary according to the type of succession an area is undergoing. In primary succession, such SPECIES generally include LICHENS, MICROORGANISMS, some INSECTS, and a few small, hardy PLANT species such as mosses that are able to colonize bare rock. In secondary succession, pioneer species often include plants with shallow root systems such as grasses, particularly those hardy grasses that are generally described as weeds. *See also* CLEAR-CUTTING; DECOMPOSER; DECOMPOSITION; EVERGLADES NATIONAL PARK; SLASH AND BURN; VOLCANO; WETLAND.

plankton, Term used to describe tiny marine and fresh-water ORGANISMS that drift at or near the water's surface. Some plankton have structures for locomotion; however, most are dependent upon wind, TIDES, and water currents for their movement. As a group, plankton are extremely important to aquatic ECOSYSTEMS because of their roles in the complex FOOD CHAINS and FOOD WEBS of these ecosystems.

Plankton are divided into two principal groups: PHYTOPLANKTON and ZOOPLANKTON. Phytoplankton include microscopic organisms that derive their nutrition and ENERGY through PHOTOSYNTHESIS and include some SPECIES of BACTERIA, FUNGI, and ALGAE. Phytoplankton are PRODUCERS and thus form the base of aquatic food chains. Zooplankton are CONSUMERS. They consist primarily of protozoa (*see* protists), small CRUSTACEANS such as KRILL, worms, jellyfishes, and the eggs and larvae of many aquatic animals whose adult forms are recognized as nekton. Most zooplankton feed on phytoplankton; however, some may feed on other zooplankton. These small organisms, in turn, are used as a food source by many larger fishes and aquatic MAMMALS, such as WHALES.

Plankton are extremely numerous. In fact, a single liter of pond or lake water may contain more than 500 million planktonic organisms. Marine plankton may exist in such great numbers that they actually cause an apparent change in the color of the water. RED TIDES, for example, result when billions of algae known as dinoflagellates are present in OCEAN water. These tides can be dangerous because dinoflagellates produce a TOXIN that is poisonous to fishes, shellfishes, and humans. *See also* ALGAL BLOOM; NEUROTOXIN; ZOOXANTHELLAE.

EnviroTerm

nekton: Free-swimming aquatic organisms, such as fishes, WHALES, and the adult forms of cnidarians such as jellyfishes. Nekton are capable of swimming under their own power, in contrast to PLANKTON which generally drift or float according to the movement of the water in which they live.

plants, The kingdom of ORGANISMS comprised of SPECIES that are multicellular, eukaryotic, and able to synthesize their food through the process of PHOTOSYNTHESIS. Eukaryotic organisms are those whose internal CELL structures (e.g. a nucleus) are contained within membranes. The plant kingdom is comprised of approximately 250,000 known species. These species are widely distributed throughout Earth and are adapted to virtually every type of ENVIRONMENT. The widespread distribution of plants along with the diversity of NICHES they occupy account for their great environmental importance. Plants also are of great importance to humans because they provide people with a diversity of products, including food and medicines.

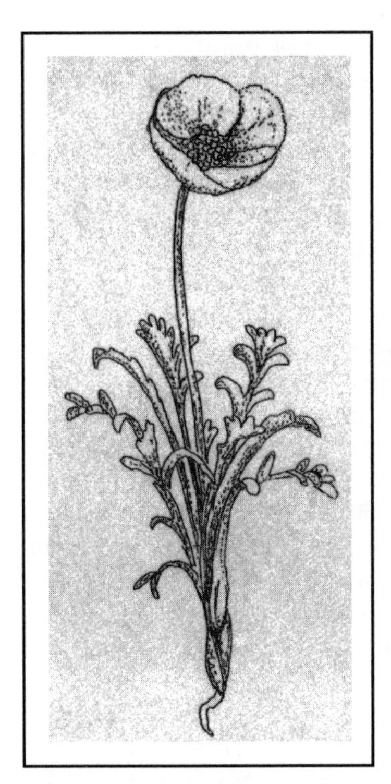

The Arctic poppy belongs to the plant kingdom which is composed of species that are multicellular, eukaryotic, and are able to synthesize their food through the process of photosynthesis.

Plant Classification. The plant kingdom is divided into two broad groups: the bryophytes, or nonvascular plants, and the tracheophytes, or vascular plants. Bryophytes include the mosses, the liverworts, and the hornworts. These plants do not have well-developed vascular systems for the transport of water and nutrients. As a result, these plants tend to be relatively small in size and live in moist environments, where they rely on diffusion from one cell to the next for transport of substances throughout the plant.

The tracheophytes, or vascular plants, have vascular systems comprised of two main types of tissue—xylem and phloem. These tissues are responsible for the movement of water and nutrients from the roots of a plant upward to its stem and leaves and for the movement of food (glucose) made in a plant's leaves throughout the rest of the plant. The development of the vascular system in tracheophytes enables these plants to live in environments that are less moist than those required by bryophytes. As a result, tracheophytes have been successful in colonizing virtually every type of terrestrial HABITAT on Earth. Most of the plants people are most familiar with are tracheophytes. This group includes the ferns, conifers, cycads, gingkoes, and the flowering plants, or angiosperms.

Role in the Food Chain. Because they manufacture their own food, plants are AUTOTROPHS. The food-making process of plants is called photosynthesis and involves using the ENERGY in sunlight to combine CARBON DIOXIDE from the AIR and water obtained from the soil to make glucose, the substance plants (and most organisms) use as food. During respiration, plants, like other organisms oxidize glucose to obtain the energy needed to carry out their life processes. (*See* oxidation.) Glucose that is not used by the plant is stored in its cells and tissues.

Glucose stored in plants is used either directly or indirectly by HETEROTROPHS, organisms that cannot make their own food. Thus, plants are producers and form the base of many FOOD CHAINS. Food stored in the cells and tissues of plants is taken in by organisms (HERBIVORES) that obtain their food by eating plants or plant products. These organisms, in turn, are eaten by other organisms that feed at higher TROPHIC LEVELS of the food chain.

Although plants are able to make their own food, a few species of plants have ADAPTATIONS that enable them to obtain nutrients from other organisms. For example, the Venus's flytrap and the pitcher plant are two types of carnivorous plants (*see* carnivore): Both of these plants have adaptations that allow them to capture INSECTS, which they can digest and use as a source of nutrition. A few species of plants have lost their chlorophyll and obtain their nutrition by feeding on the remains and wastes of other organisms. Other plants exist as parasites and obtain their nutritional needs by obtaining nutrients from a host organism, often another plant. An example of a parasitic plant is the rafflesia, a flowering plant that lives on the roots of vines that grow in the jungles of Borneo and Sumatra in Southeast Asia. This plant produces the largest flower of any known plant. The flower emits an odor that attracts flies, which help to pollinate the plant, thus aiding in reproduction. Another more common parasitic plant is mistletoe, which attaches itself to the branches of trees from which it steals nutrients.

Role in Biogeochemical Cycles. To carry out photosynthesis, plants take in carbon dioxide from the air, usually through their leaves. They also take in water from the soil in which they live through their roots. During photosynthesis, plants produce OXYGEN in addition to glucose. Much of this oxygen is released back into the environment through the leaves of the plant.

Like most other organisms, plants use oxygen from the air during their respiratory processes. During respiration, oxygen is used to obtain energy from the glucose the plant makes via photosynthesis; carbon dioxide and water vapor are produced as waste products of this process. These waste products are released back into the environment through small openings in the leaves of the plant. The exchange of oxygen, carbon dioxide, and water that take place between plants and the environment during photosynthesis and respiration are important components of the oxygen, carbon, and WATER CYCLES that occur on Earth.

In addition to their roles in these cycles, many plants also are an important component of the NITROGEN CYCLE. NITROGEN, present in the atmosphere as a gas, is an essential element for the survival of all living things; however, most organisms are unable to use nitrogen in its gaseous form. Instead, the nitrogen must be changed into nitrogen compounds, such as NITRATES and nitrites, which are usable to organisms.

Plants called LEGUMES are involved in the conversion of free nitrogen from the air into usable compounds of nitrogen. Legumes are plants that have small growths called nodules on their roots. These nodules provide a habitat for the BACTERIA that carry out NITROGEN FIXATION, the conversion of nitrogen gas into nitrogen compounds that can be used by plants and most other organisms. Nitrogen compounds that are absorbed by the plants are passed through the food chain to other organisms; compounds not absorbed by plants are released into the soil.

Agriculture. People throughout the world cultivate plants on farms both as a source of food for humans and domestic animals and to obtain plant parts that have a variety of uses to humans. Major food crops throughout the world include grains such as rice, wheat, oats, barley, and corn. These grains may be eaten as foods themselves or changed in form to produce breads, cereals, or other food products. Vegetables, fruits, seeds, and nuts also are major food products derived from plants.

In addition to raising plants for food, many farmers grow plants because they produce substances that are used to make a variety of products. Cotton, for example, is grown to obtain the fibers it produces. These fibers can then be woven into fabric for use in clothing and other products. Indigo is widely cultivated for the blue dye that can be obtained from the plant. This dye is used to color fabrics. Throughout the world, tobacco is grown for its leaves, which are harvested, dried, and used to make tobacco products that people smoke, inhale, or chew. Although use of many tobacco products has been determined to be a major cause of CANCER of the mouth and lungs, these products remain in use by people of many cultures throughout the world.

Trees are sources of a variety of food and nonfood products used by people. For example, maple syrup is obtained from the sugar maple. Other trees including MAHOGANY, ebony, various pines, oaks, birches, and cherry are major sources of lumber throughout the world. The lumber is used for the construction of houses and other buildings as well as in the manufacture of furniture. Wood from trees also is used extensively for the manufacture of paper and paper products.

Medicines. Substances derived from plants are frequently used to treat various illnesses and DISEASES. The cardiac drug known as digitalis, for example, is derived from the foxglove plant. The salicylate commonly known as aspirin, though made synthetically today, was originally derived from the meadowsweet and white willow. The world's first promising antimalarial drug, quinine, is derived from a tree called the yellow cinchona. This drug is widely used throughout tropical regions of the world where more than 400 million people each year become infected with malaria.

Uses of plants for medicinal purposes dates back to ancient times and occurs in virtually all cultures. Because disease is so common throughout the world and scientists are constantly seeking new ways to treat illness, the field of ethnobotany has grown substantially in recent years. Ethnobotany is a branch of science that combines knowledge of the cultures of people from around the world with the field of BOTANY, the science devoted to the identification and classification of new plants. Among the major goals of ethnobotanists is communication with the shamans, or healers, from various cultures to learn about the healing properties of plants that grow in remote areas. Using such knowledge, ethnobotanists hope to discover new drugs that can be used in the treatment of illnesses throughout the world.

Threatened and Endangered Plants. Many plants throughout the world are recognized as either THREATENED SPECIES or ENDANGERED SPECIES. Among the major threats to POPULATIONS of native plants is the loss of HABITAT, as land in FOREST areas and other areas is cleared for use by humans. Other threats include being outcompeted for resources by introduced or EXOTIC SPECIES, destruction by grazing animals, and overcollecting by humans. Scientists worldwide are very concerned about the potential EXTINCTION of many plant species, especially those living in tropical RAINFORESTS, many of which may not have been discovered yet. Although many scientists are concerned about preserving plant species because they hope to maintain Earth's BIODIVERSITY, others emphasize that because so much remains to be learned about many plants, loss of a plant species also may result in the loss of products such as foods and medicines that would greatly benefit humans in the future. *See also* AGROECOLOGY; AGROECOSYSTEM; AGROFORESTRY; ALGAE; ALLERGY; BIODIVERSITY TREATY; BIOGEOCHEMICAL CYCLE; BIOLOGICAL COMMUNITY; BIOMASS; BIOME; BLIGHT; CACTI; CHAPARRAL; CLEAR-CUTTING; COAL; COMPETITION; COMPOSTING; CONIFEROUS FOREST; CONSERVATION; COVER CROP; DEBT-FOR-NATURE SWAP; DECIDUOUS FOREST; DEFOLIANT; DEFORESTATION; DESERT; DETRITUS; DOUGLAS, MARJORY STONEMAN; DROUGHT; DUTCH ELM DISEASE; ECOLOGICAL PYRAMID; ECOLOGICAL SUCCESSION; ECOLOGY; ECOSYSTEM; ENDANGERED SPECIES ACT; ENDANGERED SPECIES LIST; EPIPHYTE; EVERGLADES NATIONAL PARK; FERTILIZER; FIRE ECOLOGY; FLORA; FOOD AND AGRICULTURAL ORGANIZATION OF THE UNITED NATIONS; FOOD WEB; FOREST FIRE; FOREST MANAGEMENT; FOREST SERVICE.; FORESTRY; FOSSIL FUEL; FUEL WOOD; GASOHOL; GERM PLASM BANK; GRASSLAND; IRRIGATION; MANGROVE; MARSH; MONOCULTURE; OLD-GROWTH FOREST; PARASITISM; PEAT; PLANKTON; POLYCULTURE; PRESCRIBED BURNING; RECLAMATION; RED LIST OF ENDANGERED SPECIES; REFORESTATION; RESTORATION ECOLOGY; SALT MARSH; SAVANNA; SEED-TREE CUTTING; SELECTION CUTTING; SHELTERWOOD HARVESTING; SOIL CONSERVATION; SOY INK; STRIP CROPPING; SUBSISTENCE AGRICULTURE; SUSTAINABLE AGRICULTURE; SWAMP; TEMPERATE DECIDUOUS FOREST; TEMPERATE RAINFOREST; TREE FARM; URBAN FOREST; WETLAND; WILDERNESS.

EnviroSources

Carolin, Roger, ed. *Incredible Plants.* New York: Time-Life Books, 1997.

Flora of North America Web site: www.fna.org.

Herb Research Foundation Web site: www.herbs.org/index.html.

Margulis, Lynn, and Karlene V. Schwartz. *Five Kingdoms: An Illustrated Guide to the Phyla of Life on Earth.* 2d ed. New York: W.H. Freeman, 1996.

A Modern Herbal Web site: www.botanical.com/botanical/mgmh/mgmh.html.

Rees, Robin, ed. *The Way Nature Works.* New York: MacMillan, 1998.

plate tectonics, A theory stating that the movements of the large plates comprise Earth's LITHOSHERE. The CONTINENTAL DRIFT is when these moving plates cause the continents to move. Some of the major tectonic plates include the South American Plate, North American Plate, African Plate, Eurasian Plate, Antarctic Plate, and the Indo-Australian Plate. Some of the smaller plates include Nazca Plate and Scotia Plate.

Each plate is about 100 kilometers (60 miles) thick and moves, on average, between 2.5 and 15 centimeters (about 1–6 inches) a year. The boundaries between the plates is where most volcanic, EARTHQUAKE, and mountain-building activity takes place.

The movement of the plates started about 200 million years ago, when Earth was just one supercontinent called Pangaea, that broke up and spread apart, forming the continents that are present today.

plume, A measurable or visible DISCHARGE of a CONTAMINANT into AIR, water, or soil from a known source of origin. (*See* point source.) A plume in water may be visible as it extends away from its source, as often occurs when PETROLEUM leaks from a tanker. In air, a plume may be visible as smoke emitted from a chimney or factory smokestack. A plume may also involve the release of unwanted ENERGY, such as hot water or RADIATION, into the ENVIRONMENT. Such plumes are generally detected through measurements. A thermal plume in water, for example, may be detected by measuring water temperature.

Many communities in which industries are located that could release potentially harmful materials into the ENVIRONMENT during an accident use a technique called plume mapping to plot the probable path of airborne plumes. The graphics are used by emergency officials to identify areas that would likely be affected by a toxic release. The information is used to set up emergency or evacuation plans to protect the health of citizens should an accident occur. *See also* AIR POLLUTION; OIL SPILL; TOXIC RELEASE INVENTORY.

EnviroSources

Citizens Environmental Coalition. "How to Create a Toxic Plume Map." 33 Central Avenue, Albany, NY 12210; (518) 462-5527.

U.S. Environmental Protection Agency. "Technical Guidance for Hazard Analysis." U.S. Environmental Protection Agency Hotline: (800) 535-0202.

plutonium-239, A highly toxic, radioactive element produced synthetically from the element URANIUM. Pu-239 is easily produced in BREEDER REACTORS by bombarding uranium-238 with neutrons. The fissionable isotope is used as a FUEL in some reactors or can be used to develop NUCLEAR WEAPONS.

Plutonium is one of the most toxic and most radioactive elements. The substance has a HALF-LIFE of 24,000 years. Plutonium is so dangerous to handle, it is processed only in small

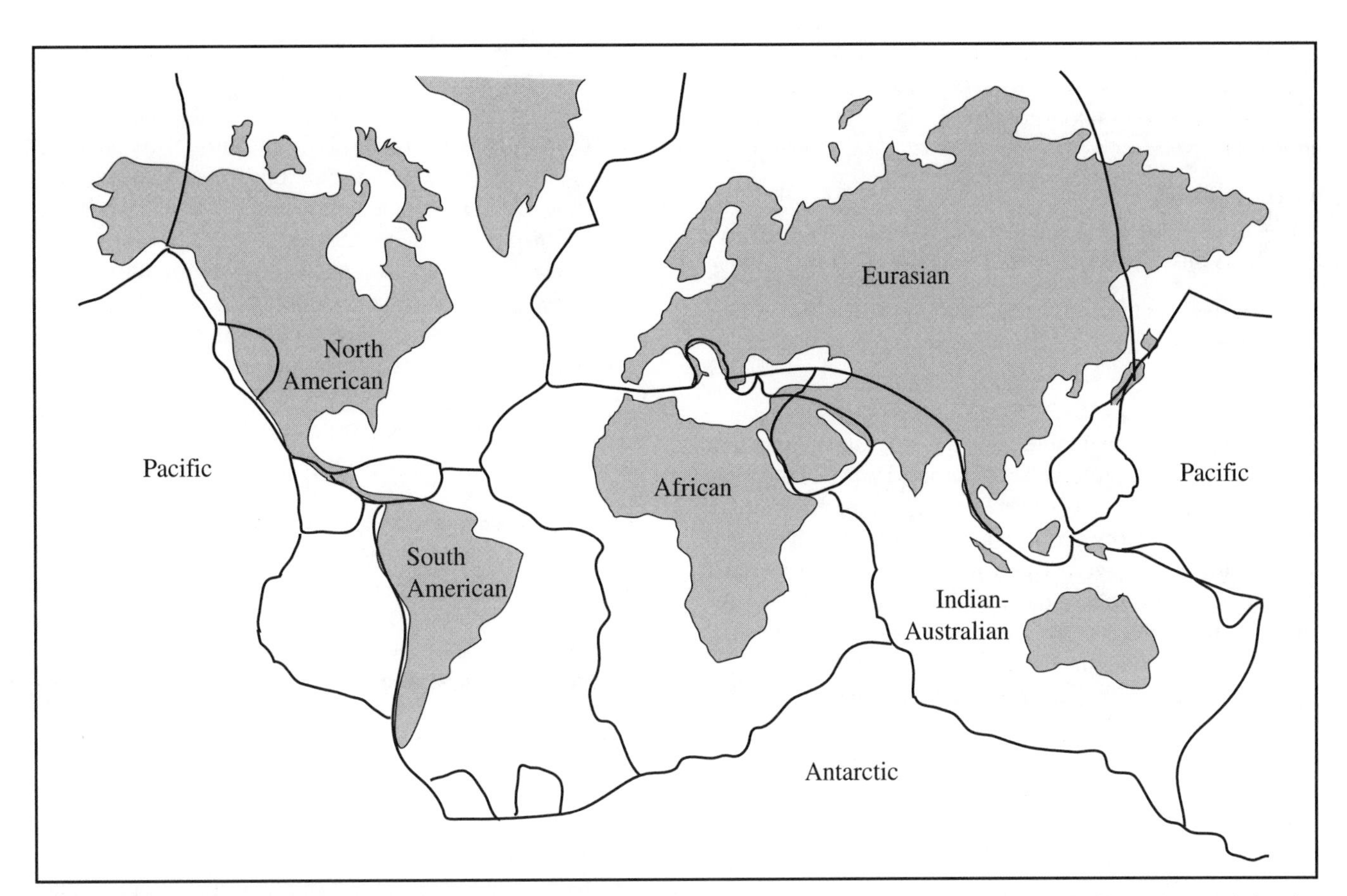

Major Tectonic Plates of the Lithosphere. Earth's lithosphere, the solid outer layer, is covered by a number of tectonic plates of different shapes and sizes that move relative to each other. Some of the major plates include North American, South American, Eurasian, Indian, African, Pacific, and Antarctic.

Plutonium Storage Sites

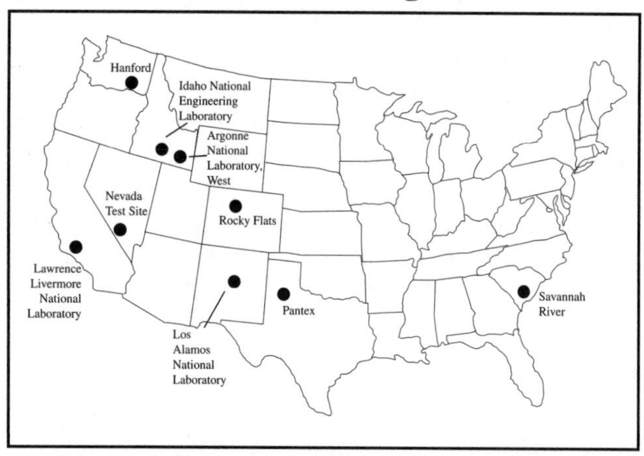

One of the largest laboratories ever built for plutonium production was the Hanford site in the state of Washington.

amounts. One of the largest laboratories ever built for plutonium production was the Hanford site (now the HANFORD NUCLEAR WASTE SITE) in the state of Washington. The plutonium made there was used to make fuel for nuclear weapons during and after World War II. Many countries want to see more controls on plutonium production because it can be used to make nuclear weapons. Only a few kilograms of this material are needed to make a nuclear bomb. Locating permanent disposal sites for this nuclear byproduct is a major international problem because plutonium is being produced faster than it is being used. In June 2000, the Russian and U.S. governments agreed to dispose of 34 tons of weapons-grade plutonium each. *See also* MANHATTAN PROJECT; RADIATION; RADIOACTIVE DECAY; RADIOISOTOPES OR RADIOACTIVE ISOTOPE; VITRIFICATION.

EnviroSources

Carter, Luther J. *Nuclear Imperatives and Public Trust: Dealing with Radioactive Waste.* Washington, DC: Resources for the Future, 1987.

U.S. Department of Energy, Civilian Radioactive Waste Information Center. P.O. Box 44375, Washington, DC 20026; (800) 225-6972.

poaching, The illegal hunting or trapping of WILDLIFE in a no-hunting area or preserve or at restricted times of the year. Poaching has caused many animals to be designated as threatened or ENDANGERED SPECIES, and often remains a significant threat to the species survival even after they are so designated. The greatest threat to both the Bengal and Siberian TIGERS, for instance, is poaching of the animals to obtain their pelts and body parts (including the tail, teeth, bones, whiskers, and reproductive organs), which have traditionally been used for a variety of purposes throughout Asia. For example, tiger bone has long been ground into a powder and used in medicines. Poaching also threatens GORILLAS in areas where the animals are hunted for food or for body parts used to make ornaments to sell and trade. *See also* CONVENTION ON INTERNATIONAL TRADE IN ENDANGERED SPECIES OF WILD FLORA AND FAUNA; ENDANGERED SPECIES ACT; ENDANGERED SPECIES LIST; FOSSEY, DIAN; RED LIST OF ENDANGERED SPECIES; THREATENED SPECIES; TRADE RECORDS ANALYSIS OF FLORA AND FAUNA IN COMMERCE; WILD BIRD CONSERVATION ACT.

point source, A stationary location where POLLUTANTS are DISCHARGED into the water or AIR from pipes, WELLS, ditches, tunnels, SEWERS, or stacks. Point sources are easily identifiable. Common point sources of pollution are factories, power plants, MINING operations, and municipal SEWAGE TREATMENT PLANTS. The term "point source" is legally and precisely defined in federal regulations. *See also* NONPOINT SOURCE.

pollutant, Any form of ENERGY or matter that is potentially harmful to ORGANISMS when released into the ENVIRONMENT. Forms of energy that act as pollutants include ionizing RADIATION, such as X-RAYS and gamma rays, the ultraviolet-B radiation given off by the sun, and excessive thermal, or heat, energy. Ionizing radiation and ultraviolet (UV) radiation harm organisms by altering the deoxyribonucleic acid (DNA) in their CELLS, causing the cells to function improperly. Excessive thermal energy can change the temperature in an environment, making the environment unfit for most organisms. Matter is a pollutant when it is irritating, harmful, or toxic to living things. Such matter may exist in solid, liquid, or gaseous form and may derive from natural sources or from the activities and products of humans. For example, pollen may be classified as a pollutant when it irritates the eyes or respiratory systems of organisms. The ASH, smoke, and soot released from a VOLCANO also act as pollutants. Human activities that release pollutants to the environment include the combustion of FUELS for energy; the use of PESTICIDES, DETERGENTS, and FERTILIZERS, MINING; and improper disposal of waste products. *See also* ACID; AEROSOL; AGRICULTURE; AIR POLLUTION; AIR POLLUTION CONTROL ACT; ALDRIN; ALKALI; ALLERGY; ARSENIC; ASBESTOS; BENZENE; BERYLLIUM; CADMIUM; CARBON MONOXIDE; CARBON TETRACHLORIDE; CHLORDANE; CHLORINATED HYDROCARBONS; CHLORINE; CHLOROFLUOROCARBON; CHROMIUM; CLEAN AIR ACT; CLEAN WATER ACT; COPPER; CRITERIA POLLUTANT; DICHLORODIPHENYLTRICHOROETHANE; DIOXIN; ETHANOL; FALLOUT; FLUE GAS; FLUORINE; FLUOROCARBON; GREENHOUSE GAS; HALON; HAZARDOUS SUBSTANCE; HEAVY METAL; HEPTACHLOR; HYDROCARBONS; HYDROCHLOROFLUOROCARBONS; HYDROGEN SULFIDE; INFECTIOUS WASTE; LEAD; MANGANESE; MERCURY; METHANE; METHANOL; METHYL BROMIDE; METHYLENE CHLORIDE; METHYL MERCURY; MULTIPLE CHEMICAL SENSITIVITY; NICOTINE; NITROGEN DIOXIDE; NITROGEN OXIDE; NOISE POLLUTION; OIL SPILL; ORGANOPHOSPHATE; OZONE; PARTICULATE MATTER; PETROCHEMICAL; PHENOL; PHOSPHATES; POLYCHLORINATED BIPHENYL; THERMAL POLLUTION; VOLATILE ORGANIC COMPOUNDS.

polonium, A natural radioactive element that occurs in pitchblende and other ores that contain URANIUM. Polonium, which has a HALF-LIFE of about three minutes, is a daughter product of RADON. The isotope can attach itself to dust, which can be inhaled into the lungs where damage can occur as the substance decays and releases RADIATION. Polonium has been linked to lung CANCER, particularly in those who work in uranium mines.

EnviroTerm

daughter, The material resulting from the radioactive decay of a radioactive element.

polychlorinated biphenyl, Any of a group of synthetic, chemical compounds that can be a health hazard to humans, other ORGANISMS, and the ENVIRONMENT. There are no known natural sources of polychlorinated biphenyls (PCBs). PCBs have been widely used in transformers, capacitors, and other electrical equipment. They were also once used in the manufacture of paints and inks.

In 1979, the sale of PCBs was banned by law in the United States because of evidence that PCBs built up in the environment and concerns about health effects. Consumer products that may contain PCBs include old fluorescent bulbs, old electrical devices, and appliances containing PCB capacitors that were made before PCB use was stopped. Prior to 1977, PCBs entered the AIR, water, and soil during the manufacture and use of these devices. Wastes containing PCBs were often placed in dump sites. PCBs also entered the environment from accidental spills and leaks during transport of the chemicals, or from leaks or fires in transformers, capacitors, or other PCB-containing products. Today, PCBs are released into the environment from poorly maintained HAZARDOUS WASTE sites that contain PCBs, through illegal or improper dumping of PCB wastes such as transformer fluids, from leaks or releases from electrical transformers containing PCBs, and from disposal of consumer products that contain PCBs into municipal or other landfills that are not designed to handle hazardous waste.

Despite the ban on PCB use, people can still be exposed to PCBs in some older transformers (which may have a useful life of more than 30 years), old fluorescent lighting fixtures, old electrical devices, and old appliances such as television sets and refrigerators. PCBs enter the bodies of fishes when they ingest contaminated water, sediment, particulates, or prey. PCBs can also be present in meat and milk and their byproducts. Contact with PCBs at hazardous waste sites can happen when workers breathe air or contact soil containing PCBs. Based on CANCER studies, the U.S. Department of Health and Human Services has determined that PCBs may reasonably be anticipated to be CARCINOGENS. Likewise, the International Agency for Research on Cancer and the U.S. ENVIRONMENTAL PROTECTION AGENCY have determined that PCBs are probably carcinogenic to humans. *See also* BIOACCUMULATION; CHRONIC TOXICITY.

EnviroSource

Inform: 120 Wall Street, New York, NY 10005; e-mail: inform@igc.apc.org.

polyculture, An agricultural method that makes use of crop diversity—the cultivation of more than one PLANT SPECIES in an area. Polyculture farming has several benefits over MONOCULTURE farming—cultivation of a single plant species. First, polyculture may reduce the need for PESTICIDES because increased crop diversity reduces the chances of pest ORGANISMS finding a host. (*See* parasitism.) Second, when combined with crop rotation, the growth of different crops on the same parcel of land from year to year, polyculture may reduce the need for FERTILIZERS by preventing the depletion of certain soil nutrients. One form of polyculture called intercropping involves the growth of plant varieties that mature at different times of the year. Use of intercropping benefits farmers by providing income-generating sources over a greater part of the year. *See also* AGRICULTURE; AGROFORESTRY; BIOLOGICAL CONTROL; GREENBELT MOVEMENT; INTEGRATED PEST MANAGEMENT; LIMITING FACTOR; PREDATION; TREE FARM.

population, All of the ORGANISMS of the same SPECIES that live in the same place at the same time. ECOSYSTEMS generally are comprised of many different species, and the members of each make up a distinct population. For example, all of the blue spruce trees in a FOREST ecosystem comprise one population, while the sphagnum mosses of the same forest comprise another population.

Population studies are extremely important to ecologists because such studies provide an indication of the overall health and stability of ecosystems. As a rule, ecosystems comprised of many diverse populations are more environmentally stable than those comprised of only a few species. This stability derives from the many complex interactions (feeding relationships, etc.) that exist among the various populations as well as the interactions between each population and the ABIOTIC factors of the environment. For example, if an ecosystem was comprised of only four or five populations, it is likely that the removal of any one species would significantly effect all the others, since the removal of this species would likely eliminate the main food source for at least one of the other populations. This population would quickly decrease in size as its members became unable to meet their food needs. The decline in this species, in turn, would lead to a decline in the population that is next in the FOOD CHAIN and so on, until all the populations were forced to either completely move out of the area or die out. By contrast, in an ecosystem comprised of hundreds of species, the removal of one species may have a less dramatic effect on the remaining populations, since members of these populations might be able to find an alternative food source. (*See* adaptation.)

The study of the changes that occur in population sizes and the BIOTIC and abiotic factors that contribute to such changes is called population dynamics. In addition to providing an indication of the overall health and stability of an ecosystem, such studies also can be used to predict how the availability of resources in an area is likely to change in response to changes in population sizes. *See also* AGE STRUCTURE; BIRTH RATE; CARRYING CAPACITY; DEATH RATE; EMIGRATION; IMMIGRATION; POPULATION DENSITY.

population density, The number of individuals of a given POPULATION that are present in an area of a specific size (square meter, square mile, etc.) at a given time. For example, if a total of 10 lions live in a part of a SAVANNA that measures 1 square hectare, the population density of lions in the area would be indicated as 10 lions per square hectare. Studies of population density provide environmental scientists with important data about the overall health of ECOSYSTEMS.

Every ecosystem has a CARRYING CAPACITY, a maximum number of individuals in each population that it can support. The carrying capacity is determined primarily by the amounts of essential resources available to that population in that ecosystem. Studies of population density provide scientists with an indication of what the carrying capacity of a given ecosystem is for each SPECIES.

Population density studies also can alert scientists about when some species might be at increased risk of PREDATION, PARASITISM, or DISEASE. For example, because a parasite has a greater number of host ORGANISMS in an area that is densely populated by its host species, the parasite population is more likely to thrive in such an environment, while the host population is at greater risk of being harmed by the parasites. For the same reasons, communicable diseases, those capable of being transmitted from one organism to another, also are more likely to spread throughout dense populations than sparse populations. Such increased rates of transmission of parasites and disease are one of the main reasons that scientists recommend POLYCULTURE, the growth of a diversity of crops, as a more sustainable agricultural practice compared to MONOCULTURE, the growth of only a single crop. *See also* BIRTH RATE; COMPETITION; DEATH RATE; EMIGRATION; IMMIGRATION; LIMITING FACTOR; SUSTAINABLE AGRICULTURE.

porosity, The ability of a substance or material to absorb liquids and gases. Stated another way, porosity is the ratio of the total volume of AIR space in a rock to the total volume of the rock. Knowing the volume of pore space is important in estimating the volume of GROUNDWATER to the total volume of an AQUIFER, a major source of drinking water.

In soil, clay particles have a porosity of about 45%, while the porosity of gravel is about 25%. Although both are porous, the clay would hold more water than gravel. In rocks, igneous rocks such as granite would have a porosity of 1% while the porosity of sandstone, a sedimentary rock, would be higher—about 15%.

Porosity is different from PERMEABILITY. Permeability is the ability of a rock or rock layer to allow water to flow through the rock or rock layer. Although clay is porous it has a low permeability. Clays soak up and hold water. Water cannot flow easily through clays. That is one reason why clays are used as liners in SANITARY LANDFILLS. *See also* LEACHING; ZONE OF AERATION; ZONE OF SATURATION.

porpoises. *See* dolphins and porpoises.

potable water, Water that is safe for use by people as drinking water. Potable water must be fresh (have a salinity lower than 0.35 grams per liter) and free of contamination by POLLUTANTS or PATHOGENS. Potable water also is generally free of MINERALS that give the water a bad taste, make it too acidic or too alkaline, or produce an objectionable odor. (*See* acid; alkali; pH.)

The demand for potable water often exceeds the supply in remote areas and in developing nations. In such places, people often rely on SURFACE WATER for their water supplies. This water may be shared with WILDLIFE, which use the water as HABITAT, for drinking, or for cooling, and may be contaminated with pathogens, suspended particulates, or ALGAE. Despite these problems, people may be forced to use such water for drinking, cooking, and bathing, because it is the only available water source.

Use of unclean water is responsible for the transmission of many DISEASES. For example, giardiasis is an intestinal disease caused by ingesting water containing *GIARDIA LAMBLIA*. Other serious diseases such as cholera and typhoid also are transmitted in contaminated water. For example, much of the surface water in Bangladesh and Somalia is contaminated with BACTERIA that cause cholera. Despite widespread knowledge of the problem, inhabitants continue to use this water, resulting in thousands of deaths each year. Infection from such pathogens often can be avoided by boiling the water, which kills them, or through FILTRATION.

In industrialized nations, water supplies often are provided to homes and businesses through a community or city water company. In such cases, water potability is usually monitored at water treatment plants; however, accidents or NATURAL DISASTERS that allow harmful substances to enter the water supply can still occur. For example, flooding following severe storms can damage parts of water transport systems, allowing pathogens from SEWAGE or other sources to enter the water supply. (*See* floods.) When such problems arise, people may be instructed to boil water prior to use or to purchase bottled water until the system can be flushed, cleaned, and returned to proper working order. Chemical or ENERGY pollutants also can enter water supplies. The community's ability to cleanse the water of such pollutants and make it potable using current water treatment technologies depends upon the nature of the pollutant. *See also* ARSENIC; PRIMARY, SECONDARY, AND TERTIARY TREATMENTS.

Powell, John Wesley (1834–1902), Early U.S. conservationist, geologist, and explorer. He was the leader of the first group to explore the Colorado River by boat. During this famous expedition in 1869, he and his party journeyed 1,450 kilometers (900 miles) with four boats, traveling from the Green River in Wyoming down through the Grand Canyon. He repeated his travels again in 1871 and in 1872 to make a more thorough study of the Green and Colorado Rivers. Powell was the first to report that valleys can be formed by river EROSION. Between 1880 and 1894, Powell was the director of the

U.S. Geological Survey, which became one of the largest scientific organizations in the world at that time.

While Powell is most widely known as the first explorer of the Colorado River, he also made significant contributions as an administrator and as an advocate for conservation and careful planning in the use of western lands. He promoted the concept that because of their arid climate, western public lands should be classified as to their potential use for irrigation, grazing, timber, and mineral extraction. The land use of arid climate lands should be regulated differently than those lands in a nonarid area.

John Wesley Powell was born in Mount Morris, New York, and was educated at Wheaton and Oberlin Colleges. He began geological work with a series of field trips, including a trip down the Mississippi River in a rowboat. In 1861 he enlisted in the Union Army and at the battle of Shiloh he lost his right arm at the elbow. In 1867 he commenced a series of expeditions to the Rocky Mountains and the canyons of the Green and Colorado Rivers. He founded and was named the first director of the Smithsonian Institution's Bureau of Ethnology (1879–1902). John Wesley Powell was the author of *Exploration of the Colorado River of the West and Its Tributaries* (1875), which was revised and enlarged as *Canyons of the Colorado* in 1895. *See also* Leopold, Aldo; Muir, John; Sierra Club.

EnviroSources

Bruns, Roger A. *John Wesley Powell: Explorer of Grand Canyon, Historical American Biographies.* Berkeley Heights, NJ: Enslow, 1997.

Powell, John Wesley. *The Exploration of the Colorado River and Its Canyons.* Introduction by Wallace Earle Stegner. New York: Penguin, 1997.

Powell, John Wesley, Wallace Earle Stegner, and Bernard Devoto. *Beyond the Hundredth Meridian and the Second Opening of the West.* New York: Penguin, 1992.

prairie, A North American biome characterized by grasses and forbs. Elsewhere in the world, equivalent grasslands are known as steppes (Asia), pampas (South America), and veldt (South Africa). Prairie once covered vast areas of North America, in a mid-continent band from south-central Canada to Texas, and from the Rocky Mountains to Ohio. Prairies, like other grasslands, are suffering from extensive agricultural activities, overgrazing, and deforestation.

Prairies exist in tall-grass, short-grass, and intermediate forms. Fertile tall-grass prairies occur in the easternmost section of the prairie's range, extending into relatively humid regions. Its name derives from the prevalence of upright bluestem grasses, which reach heights of over 2 meters (6 feet). Today, tall-grass prairie has been almost completely converted to cropland for agriculture.

Short-grass prairie is the least fertile form, and is located in the drier, westernmost extent of the prairie's range. Vegetation is mainly short, bunch grasses ("sod grasses"). Today, these grasslands are cattle rangeland; many portions have experienced severe overgrazing and have been subsequently invaded by prickly pear cactus and other thorny plants.

Undisturbed prairie provides habitat for grazing species such as deer, wild horses, and bison (once numbering in the millions, but now limited to private or re-introduced herds). Predators include the coyote and the black-footed ferret (an endangered species), which feed on prairie dogs, jackrabbits, and squirrels.

Frequent burning by fire has played a key role in the creation and maintenance of prairies; in fact, some grasses require periodic fires in order to grow new vegetation and exclude exotic species. Native American plains tribes also traditionally set fires to promote vegetative growth or to drive wildlife for hunting. Fire suppression efforts since the early 1900s have changed the species composition of many fire-adapted grassland communities and have reduced prairie biodiversity.

By 1900, conversion of tall-grass prairie to agricultural lands had left only an estimated 1% undisturbed by settlers. Interest in conservation and restoration of prairie as a heritage ecosystem began a few decades later. The Tall-grass Prairie National Preserve, the nation's only national park devoted to the preservation of the tall-grass prairie, was signed into existence in 1996 as part of the Omnibus Parks and Public Lands Act. Once private land, the preserve's 4,437 hectares (11,500 acres) were purchased by the nonprofit National Park Trust, which will manage the land in partnership with the National Park Service. Other ongoing prairie restoration projects are the Konza Prairie (Kansas) and Walnut Creek (Iowa.) *See also* biological community; fire ecology; predation.

EnviroSources

Madson, John. *Tall Grass Prairie.* Selangor Darul Ehsan, Malaysia: Falcon, 1993.

Manning, Richard. *Grassland: The History, Biology, Politics and Promise of the American Prairie.* New York: Penguin, 1997.

U.S. Geological Survey (Postcards from the Prairie) Web site: www.nrwrc.usgs.gov/postcards/postcards.html.

precipitate, An insoluble solid particle that separates from a liquid solution and usually settles to the bottom of the solution. Precipitation is a process of removing hazardous solids from liquid wastes to permit safe disposal in wastewater treatment plants. *See also* primary, secondary, and tertiary treatments.

precipitation. *See* water cycle.

predation, An ecological interaction between organisms of different species in which one organism hunts, kills, and eats the other as a means of obtaining nutrition. In this relationship, the organism that does the capturing and eating is the predator, while the organism that is eaten is called the prey. Because they feed on other animals, predators are carnivores.

Examples of predation exist in all ECOSYSTEMS and are a part of all feeding levels above the first TROPHIC LEVEL. For example, a spider that uses a web to capture INSECTS that it consumes is a predator, while the captured insects are its prey. The spider, in turn may become the prey of another animal, such as a bird, that captures it for use as food.

Predation helps to maintain balance within ecosystems. For example, predators are most likely to capture and feed on old, weak, or sick members of a POPULATION. In this way, predation helps eliminate the least well adapted members of a population, while allowing the most fit members to survive and pass their ADAPTATIONS for survival on to successive generations. Similarly, several species of predator may be in COMPETITION with each other for the same prey. When this occurs, the most fit individuals in either population are more likely to be successful at obtaining food and surviving to produce offspring. The size of a prey population is often a LIMITING FACTOR in regulating the size of the predator population. If a prey population increases in size, it can support a greater number of predators; however, a significant decrease in the size of prey populations can result in similar decreases in the sizes of predator populations as individuals who are unable to find adequate food are forced to leave the ecosystem or die of starvation. *See also* ECOLOGICAL PYRAMID; FOOD CHAIN; FOOD WEB; POPULATION DENSITY; SYMBIOSIS.

predator. *See* predation.

prescribed burning, A FOREST MANAGEMENT practice in which fires are set under controlled conditions in wooded areas to remove FOREST litter, such as fallen branches and leaves. Prescribed burning is used to prevent uncontrolled FOREST FIRES by removing matter that can act as a FUEL if a blaze does occur. Removal of these materials also improves growing conditions for small PLANTS by removing debris that prevents sunlight from reaching the ground, and by maintaining open spaces for WILDLIFE HABITAT. Worldwide, forest fires destroy billions of acres of forest annually; however, forest fires also generally improve plant growth conditions by releasing nutrients, such as calcium, magnesium, potassium, and PHOSPHORUS, into the soil. In some cases, fires directly aid plant growth. For example, the seeds of the lodgepole pine require exposure to fire to germinate. Fires also assist root growth in aspens and many shrubs. *See also* DETRITUS; FIRE ECOLOGY; OLD-GROWTH FOREST; YELLOWSTONE NATIONAL PARK.

prey. *See* predation.

primary and secondary air pollutants, Classifications of AIR POLLUTANTS considered to be harmful to the ENVIRONMENT and public health. The CLEAN AIR ACT defined two categories of air pollutants and established standards for them in the NATIONAL AMBIENT AIR QUALITY STANDARDS. Primary pollutants are materials released before combining with other components. Some of these materials would include CARBON MONOXIDE and CARBON DIOXIDE. The standards for primary air pollutants were established to protect public health including the health of "sensitive" POPULATIONS such as children, the elderly, and asthmatics. Secondary pollutants are materials formed from chemical reactions in the environment involving primary pollutants. OZONE is a secondary air pollutant. Secondary air pollutant standards are established to protect public welfare, including protection against decreased visibility, damage to crops, animals, vegetation, and buildings. *See also* ACID RAIN; NITROGEN OXIDES; PARTICULATE MATTER; SMOG.

primary consumer. *See* food chain.

primary, secondary, and tertiary treatments, The three stages used in the processing of SEWAGE or WASTEWATER to separate wastes from the water and make the water safe for release back into the ENVIRONMENT, usually through DISCHARGE into a SURFACE WATER system such as a lake, stream, or OCEAN. One or more of the individual processes involved in these three stages of treatment are carried out at wastewater or SEWAGE TREATMENT PLANTS. In the United States, the methods involved in primary and secondary treatments are required by federal law before cleaned water may be discharged back into the environment. Tertiary treatment, though not required by law, is practiced in regions where primary and secondary treatments may not sufficiently address the water quality concerns of a community.

Primary Treatment. Primary treatment occurs after wastewater or sewage that has been collected by SEWER systems is carried to a treatment plant through a network of pipes. Primary treatment begins when the wastewater is passed through a series of screens that are designed to remove larger debris (leaves, twigs, or other solid matter) from the water. Water and substances that pass through the screens is then carried to a large container known as a grit chamber.

In a grit chamber, sand, grit, and other relatively large PARTICULATE MATTER present in water are removed from the water. These substances are then collected for disposal. The remaining water is then piped from the grit chamber into another containment area called a SEDIMENTATION tank. Wastewater that enters a sedimentation tank still contains some sediment. In the sedimentation tank, particles are permitted to settle out of the water. In some cases, this process occurs naturally as water is allowed to stand undisturbed and gravity causes heavier particles in the water to settle out. In other cases, chemicals such as alum are added to the water to cause particles in the water to clump together and settle out of the water. The matter that settles to the bottom of the sedimentation tank is called SLUDGE. After sedimentation is complete, the water above the sludge is pumped out of the sedimentation chamber and the sludge is left behind for later removal.

Primary treatment of wastewater ends in the sedimentation chamber. In most cases, 30 to 40% of the wastes in the water are removed through the primary treatment phase. The remaining water is then generally piped to an AERATION tank, where secondary treatment begins.

Secondary Treatment. In the most common form of wastewater treatment, wastewater from the sedimentation tank is piped into a container known as an aeration tank. As its name suggests, AIR is added to the water in the aeration tank. In addition, sludge containing BACTERIA that decompose ORGANIC matter also are added to the water in the aeration tank. Once the processes of aeration and DECOMPOSITION are considered complete, water from the aeration tank is pumped out of the tank and into a second sedimentation tank. Here, sludge is permitted to again settle out of the water. Once the sludge has settled out, water from the sedimentation tank is piped to an area where it will undergo treatment with DISINFECTANTS. The sludge that collected at the bottom of the sedimentation tank is removed, dried, and disposed of.

In most cases, disinfection of wastewater involves treating the water with CHLORINE, a process known as CHLORINATION. In some cases, disinfection may involve the use of OZONE. Use of either substance is intended to kill any MICROORGANISMS that may remain in the water. After disinfection takes place, secondary treatment is complete.

The entire process of primary and secondary treatment removes about 90% of the POLLUTANTS that were present in the water that first entered the treatment plant. Following this process, the water left behind may follow one of two courses. If its treatment is considered complete, the water, now considered purified and safe for use by ORGANISMS, may be discharged into a lake, stream, or the ocean. In other cases, this water may undergo tertiary, or advanced treatment, to remove additional matter from the water.

Tertiary Treatment. Some wastewater, particularly that from industry or RUNOFF from agricultural lands, may require additional treatment to remove nutrients, HEAVY METALS, and some organic chemicals that are dissolved in the water. Such water is likely to undergo tertiary treatment before being released back into the environment. Tertiary treatment makes use of methods designed to remove specific types of pollutants from water. For example, filters made from sand or CARBON may be used to capture organic compounds that remain in water. The water may then be treated with additional chemicals that encourage coagulation and may undergo additional sedimentation of substances such as heavy metals or organic compounds.

Tertiary treatment methods remove an additional 5% of dissolved matter from wastewater. Once these substances have been removed, the water is considered safe for return to the environment. Treatment of wastewater is one of the ways that communities ensure that their drinking water supplies remain safe for use by people and other organisms. In regions where wastewater treatment does not occur or where people do not have access to water that has been cleaned and purified, the population often is forced to obtain their drinking water from surface water supplies that are contaminated with pathogenic organisms such as cholera or TOXIC CHEMICALS such as ARSENIC. Drinking water from such sources claims millions of lives throughout the world each year, especially in the developing nations of Asia, Latin America, and Africa. *See also* ACID MINE DRAINAGE; ACID RAIN; ACUTE TOXICITY; ADSORPTION; AEROBIC; ALGAL BLOOM; ANAEROBIC; ANAEROBIC DECOMPOSITION; AQUIFER; ARTESIAN WELL; BIOCHEMICAL OXYGEN DEMAND; CHEMICAL OXYGEN DEMAND; CHRONIC TOXICITY; CLEAN WATER ACT; COLIFORM BACTERIA; CONTAMINANT; CRYPTOSPORIDIUM; DAM; DECONTAMINATION; DISSOLVED OXYGEN; DRAINAGE BASIN; EFFLUENT; ESCHERICHIA COLI; EUTROPHICATION; FILTRATION; FLOCCULATION; *GIARDIA LAMBLIA;* HYDROLOGY; INFECTIOUS WASTE; INFILTRATION; LANGELIER INDEX; LEACHING; LEACHING FIELD; NATIONAL PRIMARY DRINKING WATER REGULATIONS; ORGANOPHOSPHATE; PERCOLATION; PERCOLATION POND; PERMEABILITY; POTABLE WATER; PRECIPITATE; RESERVOIR; SAFE DRINKING WATER ACT; SEPTIC TANK; WATER POLLUTION CONTROL ACT; WATER TABLE; WATERSHED.

EnviroSources

Botkin, Daniel B., and Edward A. Keller. *Environmental Science: Earth as a Living Planet.* 2d ed. New York: John Wiley and Sons, 1998.

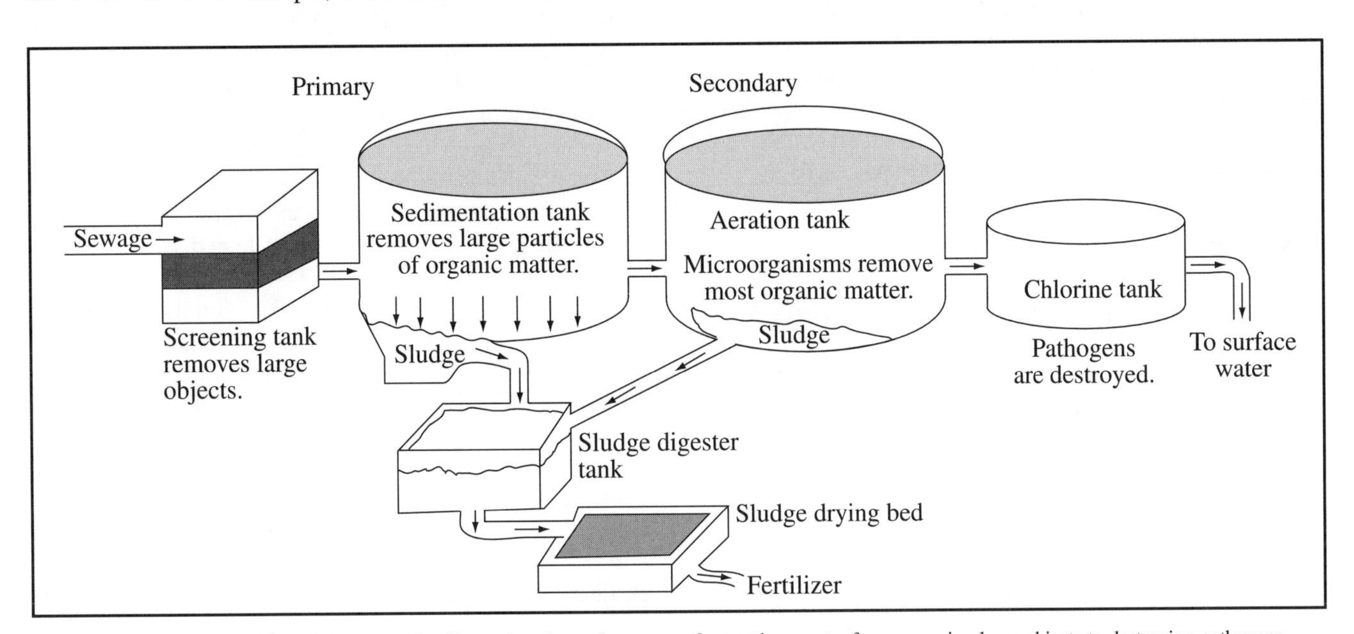

Primary and Secondary Waste Treatment. The illustration shows the process for treating wastes from removing large objects to destroying pathogens.

Chiras, Daniel. *Environmental Science: A Framework for Decision Making.* Menlo Park, CA: Addison-Wesley, 1989.

primary succession. *See* ecological succession; pioneer species.

producer, Any ORGANISM capable of making its own food using compounds obtained from the ENVIRONMENT. Producers are a vital component of all ECOSYSTEMS, because all other organisms derive their ENERGY (via food) either directly or indirectly from producers. Thus, producers form the base of all FOOD CHAINS and FOOD WEBS.

Most producers, which are also called AUTOTROPHS, synthesize their food through PHOTOSYNTHESIS. Organisms that carry out photosynthesis contain a green pigment called chlorophyll in their CELLS. This pigment captures the energy in sunlight, which is used to drive a chemical reaction that combines CARBON DIOXIDE (CO_2) and water (H_2O) to make glucose ($C_6H_{12}O_6$), a sugar that is used as food by the producer; OXYGEN (O_2) formed by this reaction is released into the environment as a waste. Producers that manufacture their food via photosynthesis include all PLANTS, ALGAE and some other PROTISTS, and the BACTERIA known as CYANOBACTERIA. In addition to forming the base of the food chain for most ecosystems, these organisms also serve a vital role in maintaining Earth's oxygen—carbon dioxide balance by helping to cycle these substances through the environment.

Producers that do not carry out photosynthesis derive their food through CHEMOSYNTHESIS. Such organisms obtain the energy that drives their food-making process by oxidizing chemicals, specifically compounds such as HYDROGEN SULFIDE (H_2S), METHANE (CH_4), and ammonia (NH_3). (*See* oxidation.) Organisms that obtain their nutrition in this way are known as chemotrophs or CHEMOAUTOTROPHS. This group is comprised of a small number of bacteria. Producers that derive their food through chemosynthesis often serve as the principal producers in regions that could not support photosynthetic organisms. For example, chemosynthetic organisms have been found living near deep OCEAN vents (areas sunlight cannot reach), where they form the base of the food chain for a variety of unusual organisms. *See also* CONSUMER; ECOLOGICAL PYRAMID; HYDROTHERMAL OCEAN VENTS; LICHENS; TROPHIC LEVEL.

protected area, A land or sea area that is protected by the WORLD CONSERVATION UNION (IUCN) because of its BIODIVERSITY, NATURAL RESOURCES, or cultural resources. Individual nations also may provide protected status to land or water areas using criteria similar to the IUCN. Worldwide, the IUCN has designated more than 50,000 locations as protected areas. These areas are deemed worthy of protection and maintenance through legal or other means. The IUCN has identified six categories of protected areas. These categories include strict nature reserves and WILDERNESS areas, national parks, natural monuments, HABITAT or SPECIES management areas, protected landscapes or seascapes, and managed resource protected areas. The status of areas deemed as protected is monitored regularly by the IUCN, and a list of such areas and their status is published every three years by the United Nations. *See also* BIOREGION; BIOSPHERE RESERVE; CARARA BIOLOGICAL RESERVE; ENVIRONMENTALLY SENSITIVE AREA; EVERGLADES NATIONAL PARK; LAND TRUST; MAN AND THE BIOSPHERE PROGRAMME; SERENGETI NATIONAL PARK; TSAVO NATIONAL PARK; YELLOWSTONE NATIONAL PARK; YOSEMITE NATIONAL PARK.

EnviroSources

Conservation Union Web site: www.ontarioparks.com/iuc.html.
World Conservation Monitoring Centre Web site: www.wcmc.org.

protist, An ORGANISM classified in the kingdom Protista, the members of which fill a variety of environmental NICHES. At one time, only unicellular organisms—distinguished from BACTERIA (monerans) by the presence of a membrane-bound nucleus and membrane-bound organelles—were classified as protists; more recently, the Protista has also come to include some multicellular organisms. Protists usually are classified into five groups on the basis of how they obtain nutrition and their mechanisms for locomotion. These groups are the protozoans (protozoa), the ALGAE, the euglenoids, the sporozoans, and the slime molds. Members of each group perform a variety of roles in the ENVIRONMENT.

Protozoans. The protozoans, or protozoa, are animal-like because they are CONSUMERS and usually have structures enabling independent movement. Most protozoa live in moist HABITATS, including soil, fresh-water lakes and ponds, and the OCEAN. Common examples of protozoa are amoebas, paramecia, and foraminiferans.

As unicellular consumers, protozoa generally feed at the second TROPHIC LEVEL of FOOD CHAINS. They, in turn, serve as a food source for other organisms. A few protozoan SPECIES are important because they cause DISEASE in humans or other organisms. Examples include the amoebas and ciliated protozoans that cause the intestinal disorder known as dysentery. Dysentery results from drinking water contaminated with protozoans and is most common in developing nations that lack water treatment plants. Other protozoa (the sarcodines) are important for their role in building Earth's surface. These protozoa are encased within a hard covering that is deposited at the bottoms of aquatic ECOSYSTEMS as the organisms die. Over many years, DEPOSITION of these "shells" may create structures of great heights, as evidenced by the famed White Cliffs of Dover in England.

Algae. Unicellular and multicellular protists that obtain their nutrients through PHOTOSYNTHESIS and have CELLS enclosed within a cell wall are classified as algae. Algae are plantlike protists that usually live in aquatic or moist terrestrial environments and are classified as PLANTS by some scientists. Common algae include the unicellular dinoflagellates and the multicellular kelp (sometimes called seaweed). As a group, algae are the main PRODUCERS of aquatic ecosystems. They also are important for their role in releasing OXYGEN—a waste product of photosynthesis—into the environment. It has been estimated that as much as 80% of Earth's atmospheric oxygen is generated through photosynthesis carried out by algae.

Euglenoids. Unicellular aquatic protists that have traits that are both plantlike and animal-like are classified as euglenoids. Many euglenoids obtain their nutrients through photosynthesis; however, unlike algae and plants, these organisms lack a cell wall. When sunlight is not available, euglenoids obtain their nutrients by eating other organisms such as PHYTOPLANKTON and ZOOPLANKTON. Thus, euglenoids are both producers and consumers. In addition, euglenoids usually have structures that make independent movement possible. The *Euglena* is the most common euglenoid.

Sporozoans. Unicellular protists that lack structures for independent movement and obtain their nutrients through PARASITISM are classified as sporozoans. As parasites, sporozoans often cause disease in their hosts. Examples of such sporozoans are the *Trypanosoma* that causes African sleeping sickness and the *Plasmodium* that causes malaria.

Slime molds. The slime molds are fungus-like protists. Like FUNGI, many slime molds obtain their nutrients through DECOMPOSITION, the breaking down of the remains of other organisms. Thus, slime molds help to cleanse the environment of waste matter and also help to recycle nutrients through the environment. Some slime molds are disease-causing parasites. For example, the potato BLIGHT which resulted in the Irish potato FAMINE of the 1840s was caused by a water mold that infected the potato crop. The three types of slime molds are the plasmodial slime molds, the cellular slime molds, and the water molds. *See also* ALGAL BLOOM; LICHENS; MICROORGANISMS; PATHOGEN; PLANKTON; RED TIDE.

EnviroSources

Margulis, Lynn, and Karlene V. Schwartz. *The Five Kingdoms: An Illustrated Guide to the Phyla of Life on Earth.* 2d ed. New York: W.H. Freeman and Co., 1996.

protozoa. *See* protist.

public land, Any land or interest in land owned by the United States and administered by the secretary of the interior through the BUREAU OF LAND MANAGEMENT. This includes public domain and acquired lands, but not lands held for the benefit of Native Americans, including Aleuts and Eskimos. Many public domain lands have never left federal ownership. The states that contain the most public domain land areas include Alabama, Alaska, Arizona, Arkansas, California, Colorado, Florida, Idaho, Illinois, Indiana, Iowa, Kansas, Louisiana, Michigan, Minnesota, Mississippi, Missouri, Montana, Nebraska, Nevada, New Mexico, North Dakota, Ohio, Oklahoma, Oregon, South Dakota, Utah, Washington, Wisconsin, and Wyoming. *See also* ARCTIC NATIONAL WILDLIFE REFUGE; YELLOWSTONE NATIONAL PARK; YOSEMITE NATIONAL PARK.

pyramid of biomass. *See* ecological pyramid.

pyramid of energy. *See* ecological pyramid.

pyramid of number. *See* ecological pyramid.

pyrite, A MINERAL composed of iron sulfide that is present in COAL seams and in the rock layers overlying the coal. When pyrite-bearing material in coal, ore, and TAILINGS is exposed to OXYGEN and water, the pyrite is oxidized to form compounds such as SULFURIC ACID. The DISCHARGE of the acidic water, called ACID MINE DRAINAGE (AMD), is highly toxic to aquatic life and ECOSYSTEMS in and around the MINING area. Stream bottoms can also be covered with iron hydroxide displaying a yellow-orange color.

Much of the acid mine drainage occurs during SURFACE MINING when the overlying rocks are broken up and removed to get at the coal. AMD can also occur in deep mines, which allow the entry of oxygen in the AIR to contact pyrite-bearing coal seams.

To determine whether or not a mine will cause acidic drainage, coal companies must analyze how much pyrite and neutralizers (limestone) are in the rocks that will be disturbed by mining. One idea to control this problem is to develop materials that could react with pyrite to form PRECIPITATES that would coat the mineral grains, preventing exposure to oxygen or water. *See also* ACID; ALKALI; OXIDATION.

pyrolysis, DECOMPOSITION of solid ORGANIC domestic wastes using extreme high temperatures in a reduced- or no-OXYGEN ENVIRONMENT. Products such as British termal unit (BTU) gas, oil, and other FUELS can be produced using the process of pyrolysis in municipal and commercial waste treatment facilities. The remaining solid material is equivalent to a low-grade COAL. Pyrolysis is also referred to as "destructive distillation." Temperatures during pyrolysis can reach more than 1,000°C. Pyrolysis has been successful in the INCINERATION of rubber tires and plastic wastes in the United Kingdom. Some of the concerns associated with pyrolysis include the EMISSIONS of DIOXINS, HEAVY METALS, and other airborne POLLUTANTS.

R

rad (radiation absorbed dose.) *See* radiation sickness.

radiation, ENERGY that exists as electromagnetic waves and particles that are emitted from various natural sources. Such sources include distant astronomical phenomena (e.g., stars) and radioactive elements on Earth. Electromagnetic radiation is categorized by its characteristic wavelength, which can range from less than 10–12 (0.000000000001) to 1,000 meters (1 kilometer, or 3,000 feet) in length. From shortest to longest, these waves include cosmic ray photons, gamma rays, X-RAYS, ultraviolet radiation (UV radiation), visible light, infrared radiation, microwaves, radio waves, and electric currents.

Some naturally occurring elements are sources of radiation that have existed in Earth since its formation. Radioactive elements have the property of radioactivity: They emit radiation. Radioactivity is the result of an unstable atomic nucleus (center of an atom) that disintegrates, forming a new element, ISOTOPE, or RADIOISOTOPE in the process and giving off radiation. Many of the large, heavy elements have a tendency to be radioactive. Common radioactive elements include RADIUM, RADON, and URANIUM. Many artificial radioactive elements, created by researchers, also exist.

The nucleus of an atom is composed of protons (positively charged particles) and neutrons (neutral particles). An isotope is one of at least two varieties of an element, each of which differs in the number of neutrons in its atomic nucleus but contains the same number of protons. Radioisotopes are radioactive isotopes. The sum of the number of protons and neutrons in the nucleus approximately equals the atomic mass of the isotope. For instance, the element potassium, with an atomic mass of about 39 amu, has 19 protons and 20 neutrons in its nucleus. One radioisotope, potassium-40, has 19 protons and 21 neutrons in its nucleus, making a total atomic mass of 40 because of the extra neutron.

The disintegration of radioactive elements, termed "RADIOACTIVE DECAY," occurs naturally at various rates characteristic of each element. As naturally occurring radioactive elements decay, they create a series of "daughter" products (elements or isotopes) in a specific sequence. When radium decays, it emits radiation and becomes a daughter product that is a different element, radon. Uranium undergoes a decay series that eventually, after several transformations, ends with the element LEAD.

Radium occurs in oil field brine, wood pulp waste, PHOSPHATE fertilizer, and radium paint wastes. Radium paints, which glow in the dark, have been used since the nineteenth century for painting the dials of aeronautical instruments and watches. Uranium is contained in some rocks (e.g., granite) and is used as a FUEL for NUCLEAR REACTORS. The uranium decay series includes the formation of radium, which in turn yields radon gas.

The term "HALF-LIFE" is used to indicate the length of time required for half of a quantity of a radioactive element or isotope to decay into the next daughter product in its decay series. This means that, in each successive half-life, half of the remaining amount of radioactive material will become the daughter product. Half-lives of radioactive isotopes can range from less than a day to billions of years. For instance, radium, with a half-life of 1,600 years, decays into radon gas, which in turn decays in a half-life of 3.8 days. So, it would take 1,600 years for 8 grams of radium to be reduced to 4 grams and another 1,600 years of decay for the radium to become 2 grams, and so on.

The three main types of radiation emitted from naturally occurring isotopes are alpha particles, beta particles, or gamma rays, depending on the type of radioactive element that decays. The higher energy (shorter wavelength) types of radiation are gamma rays, which are similar to x-rays, and alpha particles, which are similar to helium atoms. Beta particles, related to electrons, have very little mass and less energy than alpha particles. Gamma rays can penetrate through living tissue very easily and can cause CELL damage. Alpha particles, being heavy, can penetrate only the very surface of skin, posing little external threat, but can be harmful if swallowed or inhaled. Beta particles, which are capable of traveling approximately 6 millimeters (.25 inches) into the skin, can cause burns in sufficient doses.

Three Principal Radiation Types

Name	*Characteristics*	*Ability to Penetrate Tissue*
Alpha particle	large, similar to helium atom	Does not penetrate skin
Beta particle	similar to electron	Penetrates slightly
Gamma ray	high energy, similar to X-ray	Passes through body

Radiation measurements are expressed in several ways. One measure of radiation is called the "curie," which represents approximately the number of nuclei that decay per second in 1 gram of radium, an amount equal to 37 billion disintegrations per second. A smaller radiation unit, the bequerel, is equivalent to one decay per second. The unit called a "rem" (roentgen equivalent man) is a measure of radiation dosage, used to indicate how much radiation a person may receive from an x-ray or gamma ray source.

The modern world contains many natural, artificial, or human-induced sources of radiation that have potentially adverse effects on the ENVIRONMENT or health, if not carefully managed. Human beings are constantly exposed to naturally occurring radiation from Earth and the cosmos. Additional radiation exposure may occur when workers or others receive a dose of radiation from a source, such as a radioactive material used in industry, or devices that produce radiation. Large enough doses of radiation can cause burns, RADIATION SICKNESS, and CANCER.

The increase in UV radiation reaching Earth's surface observed during the 1990s has caused concern over the dangers of UV radiation exposure on human health. Excess UV exposure can lead to skin cancer. Earth's OZONE layer controls the amount of solar UV radiation that reaches the surface.

DISCHARGES of radioactive substances from the SELLAFIELD nuclear materials reprocessing plant in England into the Irish Sea are believed to have caused contamination of marine life in southwest Scotland. At Port William, for instance, radiation monitoring in 1996 detected 2,600 bequerels per kilogram of radiation in seaweed. Concentrations of the radioactive isotope technitium-99 in Solway Firth lobsters were found to have increased during the 1990s. OCEAN CURRENTS caused radioactive materials discharged from Sellafield to migrate toward Scotland, Ireland, and the NORTH SEA during a period of several months.

EnviroSource

U.S. Nuclear Regulatory Commission Web site: www.nrc.gov.

radiation sickness, Overexposure to RADIATION EMISSIONS from such sources as X-RAYS, gamma rays, alpha particles, beta particles, NUCLEAR WEAPONS, and FALLOUT that can damage cells or disrupt CELL division preventing the normal replacement or repair of blood cells, skin, and other tissues. Loss of hair, vomiting, and gastrointestinal illnesses are symptoms of radiation sickness. Radiation sickness can lead to death within a few days. Radiation exposures are measured in SIEVERTS (SV); 0.05 sv is the maximum radiation dose the human body should absorb in one year. Radiation sickness can occur after the body absorbs 1,000–2,000 millisieverts (mSv). Other units of measurement, the rad and the gray, which indicate the amount of radiation absorbed by material, are used to express doses of radiation also.

URANIUM MINING is an occupation that places workers at immediate risk of radiation poisoning from natural environmental sources. Miners in Jadugoda, in the state of Bihar, India, have suffered radiation sickness as a result of prolonged occupational exposures in uranium mines that supply FUEL for India's nuclear power industry.

The atomic bomb that fell on the Japanese city of Hiroshima in 1945, during World War II, killed 45,000 people instantly; however, other lingering effects of the radiation remained long afterward. Within a 3.4-kilometer (2-mile) radius of the blast center, all recorded pregnancies resulted in miscarriage or stillbirth. The death count attributable to radiation sickness and CANCERS that occurred in the ensuing decades reached an estimated total of 200,000 people.

The well-known CHERNOBYL nuclear power plant disaster of 1986 in Ukraine contaminated vast areas of land and resulted in extensive public health effects among nearby residents and cleanup workers. Thousands of Ukrainian's perished or suffered radiation sickness from the Chernobyl incident, demonstrating the dangers of nuclear accidents in populated areas.

The Kyshtym explosion, considered the worst nuclear accident before the Chernobyl incident, occurred in Chelyabinsk in September 1957. This accident involving the explosion of tanks containing liquid HIGH-LEVEL RADIOACTIVE WASTE, released more than 20 million curies of radioactivity and contaminated an area at least 15,000 kilometers (9,300 miles) across. The explosion released STRONTIUM-90, which reached a concentration of 0.1 curies per square kilometer, a concentration twice as great as that caused by nuclear fallout.

A Russian radioactive materials reprocessing plant at Chelyabinsk disposed of high- and intermediate-level radioactive wastes in the Techa River and Lake Karachay, severely contaminating these local water bodies. The Institute for Biophysics of the Russian Federation estimates that over 8,000 people have died from radiation exposures in the Techa River area.

radioactive. *See* radiation.

radioactive decay, A natural change that occurs in the nucleus of an unstable radioactive element as nuclear forces cause the nucleus to emit particles and ENERGY. During radioactive decay, an atomic nucleus or subatomic particle (proton or neutron) breaks up into fragments of matter and energy. These fragments are usually radioactive alpha particles or beta particles that may be accompanied by gamma radiation. As these radiation particles are emitted, the original element changes into a different element, ISOTOPE or RADIOISOTOPE, because the number of protons and neutrons contained within its nucleus changes. Often this new element ("the daughter product") is also radioactive and continues to change through the process of radioactive decay. These changes in atomic makeup repeat, creating new elements, isotopes or radioisotopes, in a predictable pattern called a RADIOACTIVE SERIES, until finally a stable, nonradioactive atom forms.

Radioactive decay occurs continually, at a constant rate, and cannot be controlled or prevented. Because of this fact, the environmental effects of RADIATION can be managed only through containment or isolation of radioactive materials, using methods that limit the migration of and reduce exposures to such substances.

Depending on the original element, radioactive decay can take from a fraction of a second to billions of years to form a stable element. URANIUM-238 (U-238) nuclei undergo radioactive decay over approximately 4.5 billion years before forming a stable isotope of LEAD. In this process, a U-238 atom first changes into a radioactive thorium atom and an alpha particle. The thorium, in turn, undergoes radioactive decay to become a stable isotope of lead.

A HALF-LIFE is the amount of time required for half of a quantity of a particular radioactive material to decay and become the next product in its radioactive series. The long half-lives and severe health effects of some radioactive materials make them difficult to store, manage, and dispose.

The naturally occurring radioactive decay of uranium presents a particular danger because this process yields RADIUM and RADON gas, a daughter product of radium. When radon gas is inhaled it can cause lung CANCER; if radium is present in food or water, people can develop other types of cancer. The effects of radioactive uranium decay are widespread. *See also*; HIGH-LEVEL RADIOACTIVE WASTE; LOW-LEVEL RADIOACTIVE WASTE; PLUTONIUM-239; RADIOACTIVE WASTE (NUCLEAR WASTE).

radioactive series, A series of radioactive elements in which each succeeding member is produced by the RADIOACTIVE DECAY of the preceding element. Each series ends in a nonradioactive element that is stable, such as LEAD. As an example, one radioactive series starts with URANIUM-238 and ends as lead-206 after several intermediate steps that take 4.5 billion years to complete. *See also* HIGH-LEVEL RADIOACTIVE WASTE; LOW-LEVEL RADIOACTIVE WASTE; PLUTONIUM-239; RADIATION; RADIOACTIVE WASTE (NUCLEAR WASTE); RADIOISOTOPES OR RADIOACTIVE ISOTOPE.

radioactive waste (nuclear waste), Low-level or high-level radioactive material that comes from NUCLEAR REACTORS, NUCLEAR WEAPONS, the MINING of URANIUM ore, and from industrial and research waste. Radioactive wastes are categorized as high-level wastes, transuranic wastes, uranium mill TAILINGS, and low-level wastes.

HIGH-LEVEL RADIOACTIVE WASTES (HLWs) would include spent nuclear FUEL from nuclear power reactors and waste from reprocessing the SPENT FUEL. Spent nuclear fuel is a solid material that is composed of irradiated uranium oxide pellets encased in metal tubes called FUEL RODS. All of the currently operating nuclear power reactors in the United States are temporarily storing spent fuel in water pools at the reactor site. Spent fuel is highly radioactive and produces considerable heat and therefore must be cooled and shielded. Ten U.S. nuclear power plants are also storing spent fuel in what is called dry storage. The spent fuel is placed in special casks made of metal or concrete. The casks are then stored on a concrete pad above ground. Approximately 30,000 metric tons (33,000 tons) of spent nuclear fuel are stored at commercial nuclear power reactors as of 1995, and the amount grows each year. By 2005, this amount is expected to increase to 52,000 metric tons (57,200 tons).

Transuranic wastes are produced during nuclear weapons research and fabrication and reactor fuel assembly. Transuranic wastes include contaminated equipment, tools, protective clothing, glassware, soils, SLUDGE, and other materials used in laboratories and research centers.

The term "transuranic" is derived from "trans," meaning beyond, and "uranic," which refers to uranium. Transuranic elements are beyond or heavier than uranium on the periodic table of the elements. These elements include PLUTONIUM, neptunium, americium, curium, and californium. Transuranic wastes contain transuranic elements or are contaminated with industrial RADIOISOTOPES heavier than uranium. These wastes decay slowly and need long-term waste storage.

Most of the transuranic waste can be packaged and stored in metal drums or in metal boxes. They can be handled under controlled conditions without any shielding beyond the container itself. The waste emits primarily alpha particles that are easily shielded. About 3% of transuranic waste, however, must be both handled and transported in shielded casks. This waste emits gamma radiation, which is very penetrating and requires concrete, LEAD, or steel to block the RADIATION.

Other radioactive wastes include uranium mill tailings from the mining and processing of uranium ore. The tailings consist of rock and soil containing small amounts of RADIUM and other radioactive materials. Uranium mill tailings become a radioactive waste disposal problem because RADON, a radioactive gas, is produced when radium decays. The Department of Energy Grand Junction Project Office in Colorado monitors uranium mill tailings disposal sites.

LOW-LEVEL RADIOACTIVE WASTE (LLW) comes from contaminated industrial or research waste. This waste does not include high-level wastes, transuranic waste, or uranium mill tailings. Such waste is generated by uranium enrichment processes, contaminated lab equipment, ISOTOPE production, and research and development activities. LLWs include materials such as tools, clothing, rags, papers, filters, equipment, soil, and construction rubble that is contaminated with low-levels of radioactivity. They require little or no shielding and no cooling during handling and transporting. LLWs are less hazardous than HLWs because the radioactivity of LLWs diminishes to harmless levels, through RADIOACTIVE DECAY, in only a few years.

The NUCLEAR REGULATORY COMMISSION (NRC) has developed a classification system for LLWs based on its potential hazards and has specified disposal requirements for each of the three general classes of waste—A, B, and C. Class A waste contains lower concentrations of radioactive material than Class C waste. Most LLWs are put into drums and buried at commercial disposal sites at the bottom of special trenches. When full, the trenches are covered with clay and topsoil. IN SITU VITRIFICATION is also used to treat and safely dispose of LLWs using on-site burial places at the waste processing facility. Approximately 19,000 cubic meters (about 690,000 cubic feet) of LLWs were disposed of in 1995.

LLW disposal facilities must be licensed by either the NRC or states in accordance with health and safety requirements. The facilities are to be designed, constructed, and operated to meet safety standards, with consideration given to the performance of the facility for thousands of years into the future.

Safely disposing of radioactive wastes that may remain radioactive for thousands of years is a major concern and top priority for many countries who use NUCLEAR ENERGY. According to the INTERNATIONAL ATOMIC ENERGY AGENCY, about 10,000 cubic meters (350,000 cubic feet) of HLWs accumulate each year. Unfortunately, not one country has any long-term plan or program to store radioactive wastes safely. The United States has investigated the possibility of constructing a HLW depository in the Yucca Mountains in Nevada; however, the plan is open to controversy. Environmental groups and the state believe that there are too many environmental and health issues associated with constructing a facility in that location.

EnviroTerm

decommissioning: The safe removal of a facility from service and reduction of residual radioactivity to a level that permits release of the property for unrestricted use and termination of the license to operate a nuclear power reactor plant. The NRC is currently overseeing the decommissioning of 15 nuclear power reactors. In 1999, the largest commercial reactor in Orgeon was taken off line, dismantled, and delivered to a burial site in eastern Washington.

EnviroSources

International Atomic Energy Agency ("Managing Radioactive Waste" Fact Sheet) Web site: www.iaea.org/worldatom/inforesource/factsheets/manradwa.html.

National Research Council, Board on Radioactive Waste Management Web site: www4.nas.edu/brwm/brwm-res.nsf.

U.S. Department of Energy, Office of Civilian Radioactive Waste Management Web site: www.rw.doe.gov.

U.S. Environmental Protection Agency (mixed-waste) Web site: www.epa.gov/radiation/mixed-waste.

U.S. Nuclear Regulatory Commission (radioactive waste) Web site: www.nrc.gov/NRC/radwaste.

radioactivity. *See* radiation.

radioisotope or radioactive isotope, A radioactive form of an ISOTOPE that is unstable and emits RADIATION that at high enough levels may cause RADIATION SICKNESS, tumors, CANCERS, leukemia, and birth defects. The U.S. ENVIRONMENTAL PROTECTION AGENCY (EPA) has set maximum CONTAMINANT levels of radioisotopes in public water systems.

A radioisotope emits radiation as it disintegrates or decays. The three major forms of radiation include gamma radiation, alpha particles, and beta particles. Gamma rays are the most penetrating form of radiation and the most dangerous of the three. Gamma rays can penetrate through most objects except for LEAD and thick concrete. An alpha particle has mass and is the same as a helium nucleus but is a very slow form of radiation that cannot penetrate through paper. A beta particle has some mass but is much smaller than an alpha particle. It can penetrate through paper but not through aluminum foil.

Radioisotopes of the same element have the same atomic number but different masses because they have the same number of protons but different numbers of neutrons. As an example, uranium-238 and uranium-235 are isotopes of URANIUM. Each has 92 protons, but uranium-238 has 3 more neutrons than uranium-235. A few radioisotopes occur naturally but most are produced artificially by bombarding stable isotopes with neutrons during nuclear FISSION. A radioisotope will emit radiation until it becomes a stable isotope; as an example, radioisotopes of uranium will decay and eventually be transformed to lead, a stable element. *See also* RADIOACTIVE DECAY; RADIOACTIVE SERIES.

Radioisotopes of Carbon

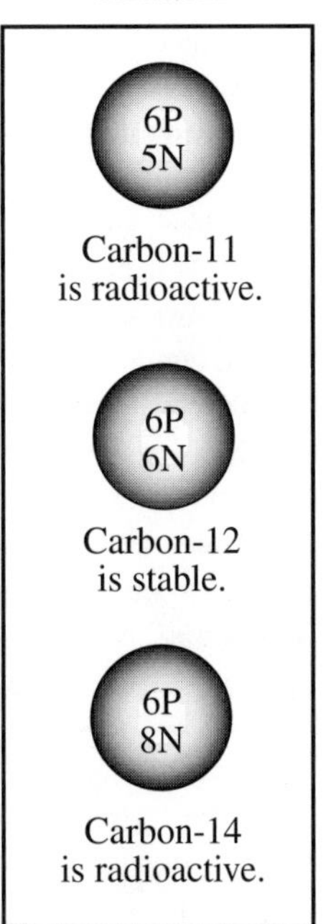

Radioisotopes of the same element have the same atomic number but different masses because they have the same number of protons but different numbers of neutrons.

radium, A rare, natural metallic element that is highly radioactive. Radium has a HALF-LIFE of about 1,620 years. As radium decays or disintegrates, it forms RADON gas, which can enter homes, buildings, and WELLS and if not ventilated may cause RADIATION SICKNESS. In turn, the radon will decay, transforming into "daughters"—other elements, ISOTOPES, or RADIOISOTOPES. At one time, radium was used in special paints applied to the hands of wristwatches or clocks to make them glow in the dark; however, many of the painters ingested the paint and over a period of time some of them developed bone CANCER. *See also* POLONIUM; RADIOACTIVE DECAY; URANIUM.

radon, A radioactive element that exists as a colorless, odorless gas. The element forms naturally when RADIUM disintegrates. (Radium itself is formed by the decay of URANIUM.) The "daughter products" of the decay of radon, such as POLONIUM-218, can attach themselves to airborne dust and other particles and, if inhaled, can damage the lining of the lungs. In the outdoor ENVIRONMENT, radon is not generally present in concentrations that pose a health risk; however, radon can accumulate inside buildings that have poor ventilation, creating a health hazard. In fact, in the United States, radon may be one of the most dangerous of all indoor POLLUTANTS. The U.S. ENVIRONMENTAL PROTECTION AGENCY (EPA) estimates that radon gas is responsible for about 15,000 lung CANCER deaths each year in the United States. After cigarette smoking, radon gas is estimated to be the second leading cause of lung cancer.

In 1988, the EPA established a radon division and reported a wide distribution of radon gas in the basements of homes. The hazard results when radon gas seeps up through soil layers and enters buildings through foundation cracks and openings in foundations. If a building has poor ventilation, the gas may build up to harmful levels. The gas can also be present in GROUNDWATER.

Homes and other buildings built on rocks and soil that contain granite and dark shales may have particularly high concentrations of radon gas. Such rocks and soils are common in the New England states and in New York, Pennsylvania, New Jersey, and the Midwest. The EPA advises homeowners in these regions to regularly have their homes tested for radon gas using a measurement device over a period of two to seven days. *See also* CARCINOGEN; RADIATION; SICK BUILDING SYNDROME.

Radon at a Glance

Sources of Radon
Earth and rock beneath home; well water; building materials.

Health Effects from Exposure to Radon
No immediate symptoms. Estimated to contribute to 7,000–30,000 lung cancer deaths each year. Smokers are at higher risk of developing radon-induced lung cancer.

Radon Levels in Homes
Based on a national residential radon survey completed in 1991, the average indoor radon level is 1.3 picocuries per liter (pCi/L) in the United States. The average outdoor level is about 0.4 pCi/L. The EPA places the maximum level at no more than 4 pCi/L.

Steps to Reduce Exposure to Radon
Testing your home for radon is easy and inexpensive. Screening measuring devices can test homes for radon. Costs can run between $50 to $100. Fix your home if your radon level is 4 pCi/L or higher. Radon levels of less than 4 pCi/L still pose a risk, and in many cases, may be reduced by sealing of suspected radon entry routes in the building's foundation and improving the ventilation system in the building as well. Opening windows, doors, and vents by bringing outdoor air into the house can help reduce radon levels.

EnviroSources

National Service Center for Environmental Publications: P.O. Box 42419, Cincinnati, OH 45242-2419; (800) 490-9198; Fax (513) 489-8695.

U.S. Environmental Protection Agency. *A Citizen's Guide to Radon: What It Is and What to Do about It.* Booklet.

U.S. Environmental Protection Agency, National Radon Hotline, c/o National Safety Council: 1019 19th Street, NW, Washington, DC 20036; (800) 767-7236; Homeowners (800) 557-2366.

U.S. Environmental Protection Agency, Office of Ground Water and Drinking Water Web site: www.epa.gov/safewater/radon.html.

U.S. Environmental Protection Agency. *Reducing Radon Risks: There Are Two Ways To Protect Your Family From Radon: First, The Hard Way; Holding Your Breath.* Pamphlet.

U.S. Environmental Protection Agency. *To Protect Your Family, Here Are Some Simple Things You Shouldn't Be Without, And Here's One More: Radon Detection Kit.* Pamphlet.

U.S. Geological Survey (radon in Earth, air, and water) Web site: sedwww.cr.usgs.gov:8080/radon/radonhome.html.

EnviroTerm

daughter: The material resulting from the radioactive decay of a radioactive element.

rainforest, A dense, heavily wooded FOREST that grows in areas of high precipitation and is threatened by DEFORESTATION. Rainforests typically contain a large variety of PLANT and animal life. The best-known and most ecologically diverse rainforests are the tropical rainforests of South America, Africa, and Asia. The term "rainforest" has also been used to describe some temperate forests, such as the northwest coastal CONIFEROUS FOREST of North America, where rainfall and humidity are high and winters mild. These temperate forests, however, are dominated by one or two SPECIES of large trees and are thus quite different from the highly diverse tropical rainforests.

Tropical rainforests are warm, moist, densely wooded areas with high BIODIVERSITY. They are restricted to the equatorial regions of the world where precipitation is high and temperatures are warm year round. The largest rainforests are in Brazil (South America), Zaire (Africa), and Indonesia (Southeast Asia). Other tropical rainforests are in Hawaii and on the islands of the Pacific and Caribbean Oceans. The largest continuous rainforest lies in South America, where about 6.9 million square kilometers (2.7 million square miles) of forest cloak the Amazon Basin. Tropical rainforests once covered more than 1.6 billion hectares (4 billion acres) of the Earth. Today, nearly half the tropical rainforests are gone.

Viewed from the air, a tropical rainforest looks like a rumpled blanket of foliage. About 70% of all plant species in these forests are trees. Biologists estimate that two-thirds of the world's plant species grow in tropical rainforests today. Tropical rainforests are especially fertile ECOSYSTEMS because the combination of steady heat and moisture creates an ENVIRONMENT that encourages growth of many kinds of plants. On average, tropical rainforests receive from 150–400 centimeters (55–160 inches) of rain every year. The temperature ranges from 25° to 35°C (77° to 95°F). With few significant seasonal changes—no wide fluctuations in temperature, no long dry spells—tropical rainforests essentially have an endless growing season.

A tropical rainforest has three layers: the forest floor, the understory, and the canopy. The canopy is made up of treetops, which may rise 15–45 meters (50–150 feet) above the ground. Most of the animals of the rainforest such as monkeys, birds, tree frogs, and even snakes, live in the canopy. The canopy also contains an entangled maze of woody vines

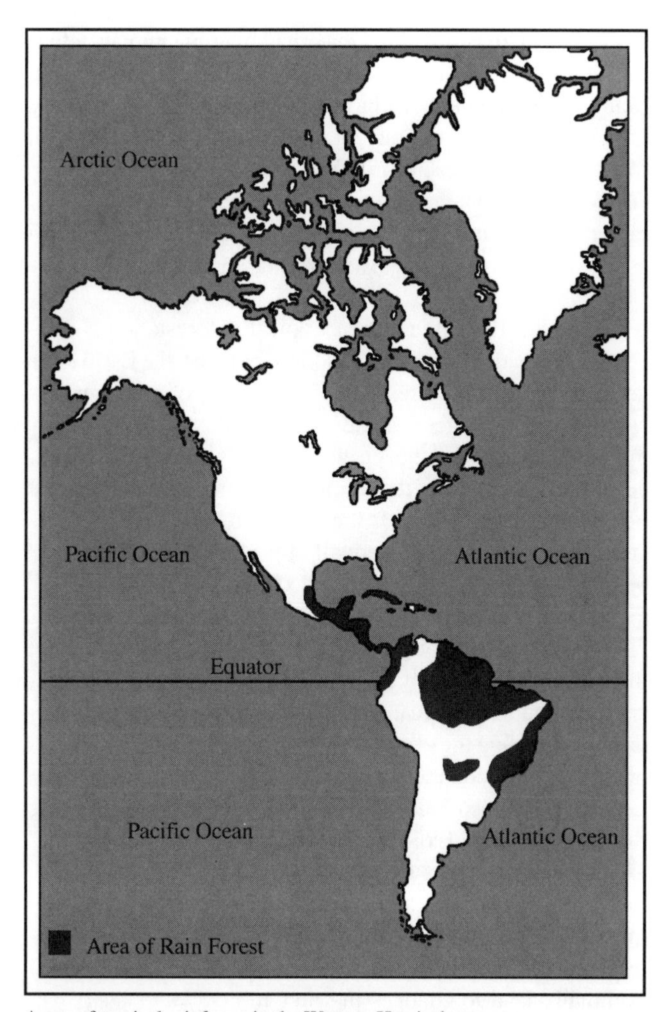

Areas of tropical rainforest in the Western Hemisphere.

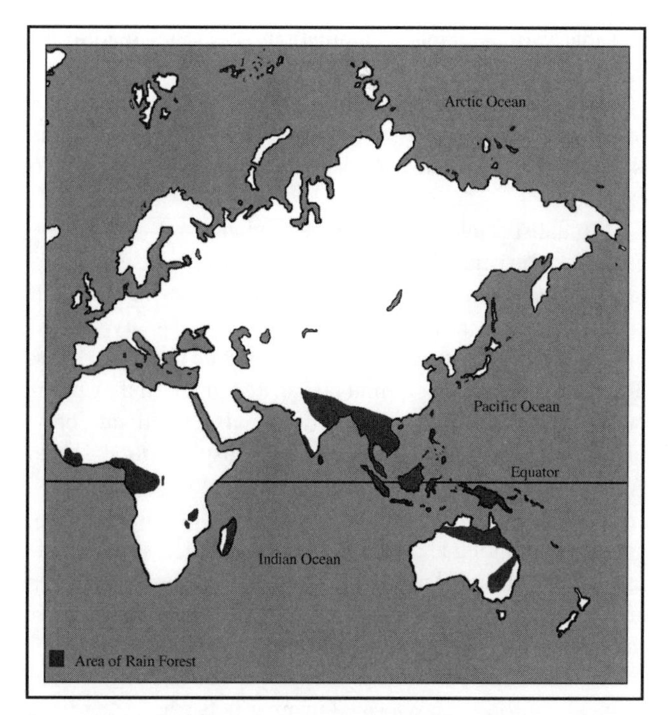

Areas of tropical rainforest in the Eastern Hemisphere.

called lianas and a rich EPIPHYTE community, made up of orchids and bromeliads that grow on trunks and branches of trees—locations that allow them access to more sunlight. Less than 2% of the sunlight filters down through the canopy, thus the understory consists of young trees, ferns, and shrubs that never grow to adult size. The forest floor is the bottom layer of the rainforest. Except for rotting vegetation, which nourishes the thin tropical soil, and large MAMMALS, such as jaguars, tapirs, apes, and monkeys, the forest floor is almost bare. This is because in the warm, damp CLIMATE of the tropical rainforest, DECOMPOSITION occurs rapidly.

Tropical rainforests are the most diverse ecosystems on Earth. The abundant vegetation provides food and shelter for millions of animal species. Most of this animal diversity is made up of INSECTS, but many other animal groups are also represented. It is estimated that 1,000 hectares (2,500 acres) of rainforest in South America contains thousands of insect species, including 150 different butterflies. The same patch of forest also contains dozens of species of poisonous snakes and frogs, and hundreds of varieties of brightly colored birds, such as toucans, PARROTS, and hummingbirds. Mammals include the tree-dwelling sloths, monkeys, and fruit BATS, as well as jaguars, tapirs, ocelots, and GORILLAS. The great diversity of many animal groups can in large part be explained by the

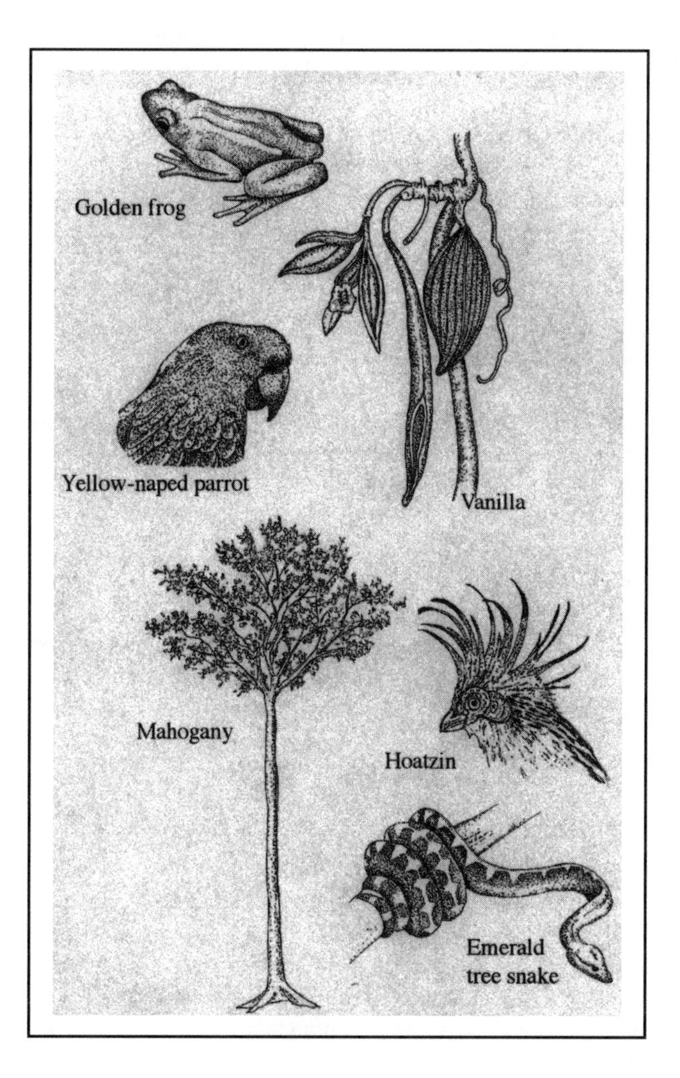

fact that various unique combinations of species tend to inhabit different layers of rainforest.

Rainforest animals also tend to specialize on particular foods in order to avoid COMPETITION between species. In this way, rainforests contain many unique NICHES for animals to exploit.

Mutualistic interactions between plants and animals are also characteristic of tropical rainforests. MUTUALISM is an interaction between two or more different species in which all species benefit in some way. For example, many animal groups, especially insects and birds, pollinate rainforest trees. The insects receive food from nectar, and in return they pollinate the next flowers they visit. After fruit is formed, rainforest plants rely on birds and mammals to disperse their seeds. Animal groups can also provide protection to a plant species, while the host plants provide the animals with a home. Many species of stinging and biting ants, for instance, use the hollow stems of tropical plants and trees as their homes. The ants supply nutrients to the trees and in many cases also protect the trees from leaf and seed predators. (*See* predation.)

TEMPERATE RAINFORESTS, which are fewer in number than tropical rainforests, are located in areas between 32° and 60° north or south latitude that receive at least 200 centimeters (80 inches) of rainfall annually. Nearly all of the world's original temperate rainforests have been cut down for timber or destroyed by pollution and only a small fraction remains in North and South America, New Zealand, Tasmania, portions of Japan, northwest Europe, and Turkey. Temperate rainforests have a more seasonal CLIMATE, with less constant temperatures and less rain, than do tropical rainforests near the equator.

The Pacific Northwest is home to North America's only temperate rainforest, the Olympic rainforest in Washington State. The Olympic rainforest is one of the world's largest and most impressive coniferous temperate rainforests. Although plant and animal life are abundant, the species are not as diverse as they are in warmer, tropical rainforests. The forest is dominated by towering evergreen trees such as Douglas fir and Sitka spruce, some of which are more than 1,000 years old. Mosses, LICHENS, and epiphytes drape the tree branches and cover massive tree trunks, and the damp, dark, forest floor is literally carpeted with ferns, fallen and decaying logs, mushrooms, and new seedlings. Insects, including mosquitoes and flies, are common inhabitants of temperate rainforests. Few cold-blooded animals, such as frogs, salamanders, turtles, and snakes, live in these forests due to the low temperatures; however, temperate rainforests typically contain many types of mammals and birds, including woodpeckers, hawks, owls, squirrels, shrews, moose, deer, and WOLVES.

Although the Olympic rainforest is located between 50° and 60° north latitude, the forest never freezes because the Pacific Ocean helps regulate temperatures. In this area, winds travel from west to east, picking up a lot of moisture from the Pacific. As the AIR moves up the western side of the Olympic Mountains, its moisture is "wrung out" over the coastal forest. Up to 415 centimeters (167 inches) of rain falls annually in the Olympic forest, more rainfall than anywhere else in the continental United States.

Today, there is growing concern about how human activities and development are damaging rainforests worldwide. Many natural phenomena have had an impact on rainforests. CYCLONES, FOREST FIRES, DISEASE, and landslides are all natural forces that have impacted rainforest ecosystems; however, the environmental damage caused by these phenomena are of rather minimal influence compared to human activities such as logging, road-building, MINING, and large-scale DEFORESTATION.

Since the 1970s the rate of tropical deforestation has accelerated greatly. According to some estimates, the Earth loses about 31 million hectares (78 million acres) of rainforest every year, an area larger than the country of Poland. Some highly unique rainforests have been almost totally destroyed. For example, less than 2% of the original Atlantic coastal rainforest of Brazil remains today. If the current rate of rainforest destruction continues unabated, all of the remaining forest will be gone by the early part of the next century.

Scientists are concerned because the rate of rainforest loss far exceeds the rate of growth. The large-scale deforestation that takes place in nearly all rainforest areas today is so extensive that hundreds of years would probably be needed for anything resembling the original vegetation to return. When acres of trees are lost, countless animals are driven from their homes. Loss of trees also leads directly to DROUGHT, flooding (*see* flood), and soil EROSION.

The consequences of deforestation reach far beyond the rainforest. Rainforests contribute significantly to global climate by playing a major role in Earth's WATER CYCLE. Scientists fear that continued destruction of the rainforests may eventually lead to changes in wind and OCEAN CURRENT patterns, as well as changes in global rainfall distribution. In addition, because rainforests contain the Earth's greatest diversity of plants and animals, they also represent giant gene banks that can provide new drugs, foods, and other products for people. Medicinal substances already discovered in rainforests include diosgenin, an active agent in contraceptive pills; reserpine, used to treat cardiac problems; and curare, used in heart and lung surgery. Scientists fear that rainforests will disappear before they can learn what other medicines can be found.

Last, but certainly not least, the lives and livelihoods of many indigenous peoples are being destroyed by an endless demand for goods found in the rainforest. In the year 1500, some 6–9 million indigenous people inhabited the Brazilian rainforest. Today, less than 200,000 remain. In the 1980s, CHICO MENDES, a Brazilian environmental activist spoke out on the tropical deforestation of the Brazilian rainforest. Chico Mendes organized the Seringeiros, a community of rubber tappers, who lived in the state of Acre. Mendes advocated that the Seringeiros become more independent of the landowners and to help stop the unnecessary destruction of the rainforests by cattle ranchers. After Mendes's death, the Brazilian government established the Reserva Extractavisitas (extractive reserves) a program that set aside several forest preserves in northwestern Brazil. These reserves were established to protect the livelihood of the Seringieros and to save and preserve the trees from any further deforestation. One of the reserves was named the Chico Mendes Extractive Reserve, an area that covers 1 million hectares (2.5 million acres).

Governments, scientific organizations, conservationists, and other citizens are deeply concerned about the loss of rainforests. The future of Earth's rainforests may depend on management plans that preserve some areas of rainforest while allowing people to cut trees selectively in other areas. Better land use practices, education, and wiser planning may slow deforestation, but experts worry that the rainforest will be virtually gone by the time these changes can be widely implemented. *See also* BIOME; CARBON CYCLE; EXTINCTION; GLOBAL WARMING; SLASH AND BURN.

EnviroSources

Greenpeace International (forests) Web site: www.greenpeace.org/~forests.
Hollowell, Christopher. "Rainforest Pharmacist." *Audubon* 101:1 (January–February 1999): 28–30.
Lewington, Anna. *Atlas of Rainforests.* (Atlases Series.) Chatham, NJ: Raintree/Steck Vaughn, 1997.
Rainforest Action Network (RAN): 450 Sansome, Suite 700, San Francisco, CA 94111.
Rainforest Alliance (RA), 65 Bleeker Street, New York 10012.
Rajala, Richard A. *Clearcutting the Pacific Rainforest: Production, Science, and Regulation.* University of British Columbia, 1999.
Revkin, Andrew. *The Burning Season: The Murder of Chico Mendes and the Fight for the Amazon Rainforest.* New York: Plume, 1994.
Smith, Nigel J.H. *The Amazon River Forest: A Natural History of Plants, Animals, and People.* Oxford, UK: Oxford University Press, 1999.
Struhsaker, Thomas T. *Ecology of an African Rainforest: Logging in Kibale and the Conflict between Conservation and Exploitation.* Gainesville: University Press of Florida, 1999.
Terborgh, John. *Diversity and the Tropical Rainforest.* Scientific American Library, no. 38. New York: W.H. Freeman and Co., 1992.
U.S. Forest Service Web site: www.fs.fed.us.
Whitmore, Timothy. *An Introduction to Tropical Rainforests.* Oxford, UK: Oxford University Press, 1998.
World Wildlife Fund (forests for life campaign) Web site: www.panda.org/forests4life.

Ramsar Convention, A treaty signed in 1971 by representatives of 18 nations who convened in Ramsar, Iran, to halt the worldwide loss of WETLANDS and to conserve those that remain. The Convention on Wetlands of International Importance, commonly referred to as the Ramsar Convention, aims to conserve one of the most threatened HABITATS—wetlands. These are shallow, open waters, such as lakes, rivers, and coastal fringes, and any land that is regularly saturated by water, such as MARSHES, SWAMPS, and FLOODPLAINS. Presently, over 80 countries have signed the Ramsar Convention representing 75% of the world's lands.

Members are encouraged to nominate specific wetlands that are unique and are valued for their BIODIVERSITY and their potential as a waterbird habitat and to promote the wise use of all wetlands within their territory. This includes promoting the training of wetland managers, consultation and cooperation with each other (particularly in the case of a shared wetland, water system, or resource such as migratory waterbirds), and the creation and management of wetland reserves. *See also* BOG; ESTUARY; MANGROVE; NATIONWIDE 26; NORTH AMERICAN WETLANDS CONSERVATION ACT; SWAMPBUSTER PROVISION.

raw sewage. *See* sewage.

Ray, Dixy Lee (1914–1994), A scientist, politician, university professor, and writer who gained attention during the 1990s for the views she expressed in her books *Trashing the Planet* (1990) and *Environmental Overkill* (1993), written with co-author Lou Guzzo. In her writings, Ray suggested that much of the information disseminated by the media, environmentalists, and government officials about environmental problems such as ACID RAIN, OZONE DEPLETION, GLOBAL WARMING, use of PESTICIDES, and the potential dangers associated with the use of NUCLEAR ENERGY, was exaggerated, intended to scare the public, and not supported by scientific evidence. Ray encouraged her readers to view these problems and their potential impact on humans with great skepticism. She also recommended that people demand accurate and supported scientific evidence of the existence and damage caused by these factors before allowing their tax dollars to be spent on costly and restrictive legislation intended to solve any purported problems. To support her views, Ray presented brief synopses of various environmental problems and cited scientific principles and research results to support her view that these problems should be of little concern.

Both of Ray's books were popular with the general public; however, scientists, conservationists, and environmentalists were quick to refute Ray's claims. Many of these people accused Ray of citing only scientific reports and information that supported her views. To bolster this claim, scientists reviewing Ray's books often pointed out that the materials Ray selected to support her premise were outdated or had been proven inaccurate and then abandoned by the scientific community.

Dixy Lee Ray was born in Tacoma, Washington. From 1963–1972, she served as the director of the Pacific Science Center of Seattle, Washington, a position she left to assume the post as chairperson of the Atomic Energy Commission (*see* Nuclear Regulatory Commission), in Washington, D.C. In 1975, Ray left the commission to serve as the assistant secretary of state for the Bureau of Oceans, International, and Scientific Affairs. She later returned to Washington, where she served as governor from 1977 until 1981. Following her service as governor, Ray divided her time between teaching at the University of Washington and writing books. *See also* ABBEY, EDWARD; BROWER, DAVID ROSS; FOREMAN, DAVID.

EnviroSources

Anderson, Daniel W. "Effects of DDT on Birds: Does Dixy Know Something the Experts Do Not?" *Environmental Review Newsletter* 1, no. 7 (July 1994.)
Ray, Dixy Lee. *Environmental Overkill: Whatever Happened to Common Sense?* Washington, DC: Regnery Gateway, 1993.

Ray, Dixy Lee with Lou Guzzo. *Trashing the Planet: How Science Can Help Us Deal with Acid Rain, Depletion of the Ozone, and Nuclear Waste (Among Other Things.)* Washington, DC: Regnery Gateway, 1990.

reclamation, Restoring land to its natural setting after MINING operations have been completed; also the process of recovering materials from discarded REFUSE for reuse. Reclaimed materials are often reprocessed and used again for their original purpose, as occurs in the RECYCLING of ALUMINUM beverage cans to make new cans. Materials may also be reclaimed for a purpose other than their original use, as occurs when plastic from beverage bottles is shredded and used as insulating material in sleeping bags or clothing, or when aluminum from cans is used to make lawn furniture. *See also* CONSERVATION; NATURAL RESOURCE; NATURAL RESOURCES CONSERVATION SERVICE; OFFICE OF SURFACE MINING, RECLAMATION, AND ENFORCEMENT; RESOURCE CONSERVATION AND RECOVERY ACT; SURFACE MINING.

Reclamation Act. *See* Surface Mining Control and Reclamation Act.

recombinant DNA. *See* genetic engineering.

recycling, A method for conserving NATURAL RESOURCES that involves collecting discarded materials and reprocessing these materials to make new products. A common example of recycling is the collection of discarded ALUMINUM beverage cans for reprocessing into new aluminum products. Recycling natural resources provides several benefits to the ENVIRONMENT. First, the recycling of materials reduces the need to use raw materials, extending the useful life of reserves of such resources. This benefit is particularly important to the CONSERVATION of NONRENEWABLE RESOURCES, which exist in fairly fixed amounts on Earth and are used by humans at a rate that exceeds the time needed for their formation through natural processes.

Reusing materials such as MINERALS reduces the need to mine for new sources of these materials. MINING can damage the environment by disturbing or destroying the natural HABITATS of ORGANISMS. In addition, mining practices often release POLLUTANTS into the environment. Recycling also helps to reduce problems associated with waste disposal by reducing the volume of materials that must be deposited in landfills or disposed of in some other way. Like nonrenewable resources, land available for landfills or other means of waste disposal is in limited supply. Thus, finding ways to reduce the overall amount of materials that need to be discarded through such practices can help to conserve land space, while also reducing the potential for discarded materials to release toxic or otherwise harmful substances into the environment as they break down. (*See* decomposition.)

Another major benefit of recycling is that it often helps to conserve ENERGY. For example, it has been estimated that reprocessing beverage cans to reclaim the aluminum they contain uses as much as 95% less energy than obtaining aluminum from ore. This energy savings helps to conserve the vital FOSSIL FUELS (COAL, PETROLEUM, and NATURAL GAS), which are burned to generate the heat needed to process ores. The burning of such fuels release AIR pollutants, including CARBON DIOXIDE, SULFUR DIOXIDE, smoke, and ASH. Burning less of these FUELS helps to reduce the amounts of these EMISSIONS, while also helping to conserve vital reserves of these fuels.

In the last quarter of the twentieth century, many nations initiated recycling campaigns to help conserve natural resources. Among the most common recycled materials are paper, aluminum, glass, plastic, steel, and motor oil. More recently, methods for recycling rubber, LEAD, zinc, COPPER, and brass also have been implemented. Once recycled, these materials are used to make a variety of products, including many of the component parts used in the manufacture of new AUTOMOBILES, resurfacing materials for roadways, and reprocessed versions of the original products (glass containers, cans, and newspapers) from which the recycled materials were obtained. *See also* ACID MINE DRAINAGE; ACID RAIN; BIOGEOCHEMICAL CYCLE; BOTTLE BILL; BUREAU OF RECLAMATION; COMPOSTING; DEEP-WELL INJECTION; DETRITUS; GARBAGE; HAZARDOUS WASTE; HAZARDOUS WASTE TREATMENT; INCINERATION; LITTER; NATURAL RESOURCES CONSERVATION SERVICE; NONBIODEGRADABLE; OCEAN DUMPING; PHOTODEGRADABLE PLASTIC; RECLAMATION; RESOURCE CONSERVATION AND RECOVERY ACT; SANITARY LANDFILL; SURFACE MINING CONTROL AND RECLAMATION ACT.

EnviroSources

Saign, Geoffrey C. *Green Essentials: What You Need to Know about the Environment.* San Francisco, CA: Mercury House, 1994.
Seymour, John, and Herbert Girardet. *Blueprint for a Green Planet: Your Practical Guide to Restoring the World's Environment.* New York: Prentice Hall, 1987.

Red List of Endangered Species, A global survey maintained by the WORLD CONSERVATION UNION (IUCN) of SPECIES at risk of EXTINCTION. The Red List is prepared by the Species Survival Commission (SSC) and is similar to, but broader in scope than, the ENDANGERED SPECIES LIST compiled and maintained by the U.S. FISH AND WILDLIFE SERVICE (FWS). The primary purpose for preparing such lists is to increase global awareness of species that are at risk of becoming extinct so actions may be taken to prevent such occurrences.

The IUCN maintains separate lists for PLANTS and animals and updates these lists every few years. In 1994, the IUCN created six classification categories for the ORGANISMS on its lists: extinct, extinct in the wild, critically endangered, endangered, vulnerable, and lower risk. Extinct species are those for which there is no doubt that all individuals of that species have ceased to exist. Extinct in the wild refers to organisms that are believed to now exist only in captivity (ZOOS, WILDLIFE REFUGES, captive breeding facilities), when cultivated (grown in nurseries or on cropland), or as a naturalized POPULATION—one that survives outside its original range. The term "critically endangered" refers to organisms that are at high risk of extinction in the immediate future. This listing is comparable to the listing of ENDANGERED SPECIES used by the FWS. A species is deemed vulnerable when it is not critically endangered, but there is a risk of its extinction in the intermediate future. The vulnerable species category is similar to the THREATENED SPECIES designation used by the FWS. Species are at lower risk when their population sizes indicate that they are not vulnerable, endangered, or critically endangered.

Through its lists, the IUCN and SSC have helped increase awareness of the numbers of species currently facing extinction and how human activities such as the introduction of EXOTIC SPECIES into an ECOSYSTEM, pollution, and fragmentation or loss of HABITAT affect wild populations. For example, in its 1996 publication of the Red List of Threatened Animals, the SSC indicated that almost one-fourth of Earth's MAMMALS (about 1,000 species) are facing extinction. Other animal and plant populations also are showing an increase in the number of species at risk of extinction.

See also ADAMSON, JOY GESSNER; ADDO NATIONAL ELEPHANT PARK; AFRICAN WILDLIFE FOUNDATION; ALLIGATORS; AMERICAN ZOO AND AQUARIUM ASSOCIATION; ARCTIC NATIONAL WILDLIFE REFUGE; BALD EAGLES; BATS; BIODIVERSITY; BIODIVERSITY TREATY; BIOSPHERE RESERVE; BISON; CACTI; CALIFORNIA CONDORS; CAPTIVE PROPAGATION; CARARA BIOLOGICAL RESERVE; CHEETAHS; CONVENTION ON INTERNATIONAL TRADE IN ENDANGERED SPECIES OF WILD FLORA AND FAUNA; DODO BIRDS; DOLPHINS AND PORPOISES; ELEPHANTS; ENDANGERED SPECIES ACT; FAUNA; FLORA; FLORIDA PANTHERS; FOSSEY, DIAN; GALÁPAGOS ISLANDS; GALDIKAS, BIRUTÉ; GOODALL, JANE; GORILLAS; INTERNATIONAL BIOLOGICAL PROGRAM; INTERNATIONAL CONVENTION FOR THE REGULATION OF WHALING; INTERNATIONAL WHALING COMMISSION; KEYSTONE SPECIES; LACEY ACT; MADAGASCAR; MAHOGANY; MANATEES; MARINE MAMMAL PROTECTION ACT; MIGRATORY BIRD TREATY ACT; NATIONAL AUDUBON SOCIETY; NATIONAL MARINE FISHERIES SERVICE; NATIONAL WILDLIFE REFUGE SYSTEM; ORANGUTANS; ORYX; OVERFISHING; OWLS, NORTHERN SPOTTED; PARROTS; PEREGRINE FALCONS; POACHING; RHINOCEROSES; SALMON; SEA TURTLES; SEALS AND SEA LIONS; SERENGETI NATIONAL PARK; TAXONOMY; TIGERS; TRADE RECORDS ANALYSIS OF FLORA AND FAUNA IN COMMERCE; TSAVO NATIONAL PARK; TUNA; WHALES; WILD BIRD CONSERVATION ACT; WILDERNESS; WILDLIFE REFUGE; WOLVES; WORLD WILDLIFE FUND; ZEBRA.

EnviroSource

World Conservation Monitoring Center: 219 Huntington Road, Cambridge CB3 0DL, United Kingdom; Web site: www.unep-wcmc.org

red tide, A POPULATION explosion of toxic ALGAE (dinoflagellates) that impacts the OCEAN FOOD CHAIN. Red tides occur when water temperature, salinity, and nutrients reach levels that support a massive increase in the dinoflagellate population. This type of ALGAL BLOOM can lead to high mortalities among marine ORGANISMS as toxins produced by the algae are released into the water. The large numbers of algae can also turn the water red, olive green, or yellow.

The optimum salinity for the formation of a red tide is 30–40 parts per 1,000 (ppt). A concentration of 250,000 red tide organisms per liter of salt water is lethal to fishes. As the algae die off, there is an increase in the BACTERIA of decay, which facilitates DECOMPOSITION. This population explosion depletes the OXYGEN in the water, causing death to huge numbers of fishes, shellfishes, and CRUSTACEANS. The red tide was linked to the deaths of 39 MANATEES in Florida in the late 1970s. Both during and following a red tide, the beaches and shoreline may be covered with dead and dying fishes, closing beaches for days or weeks. Swimming in red tide water is not seriously harmful to humans, but can cause throat and nose irritation and burning eyes. The poison released by the algae can also accumulate in the CELLS of shellfishes such as clams and mussels, making them toxic to humans. Red tides are common on the northeast coast of the United States and in the BALTIC SEA and ADRIATIC SEA.

In 1999 scientists reported that pollution may not be the cause of the red tides. Their research reveals that it is the distribution of the MICROORGANISM, *Alexandrium tamarensis*, that causes the red tides. The organism, which are usually in a dormant stage as a cyst in the ocean, becomes active as a swimming dinoflagellate under the right conditions such as temperature changes in the water. The dinoflagellates can wash into beds of mussels, oysters, and clams where the shellfish ingest the microorganisms making the shellfish poisonous to eat. *See also* PROTISTS.

reef. *See* coral reef.

reforestation, The deliberate replanting of an area with FOREST cover, such as trees. Reforestation is often used by the timber and logging industries to replace trees that are removed as a means of aiding the sustainability of forest resources and to help prevent EROSION in forest areas from which large numbers of trees had been removed. In 1933, the U.S. government established the Civilian Conservation Corps (CCC) as part of the New Deal program of President Franklin Delano Roosevelt. The agency was developed for the CONSERVATION of NATURAL RESOURCES and provided employment to about three million young men during the Great Depression. Reforestation was a major project of the CCC until the agency was abolished in 1942.

Reforestation projects have not been limited to the United States. Reforestation is now used in many countries to replace forests that were clear-cut to make land available for AGRICULTURE. In recent years, Israel has devoted human resources to reforestation projects intended to slow the process of DESERTIFICATION. Almost 5% of Israel's land is now occupied by natural woodlands and reforested areas. In Kenya in the 1970s, environmentalist Wangari Mutta Maathai began the GREENBELT MOVEMENT, which is committed to restoring the forests in that country. Like the CCC, Maathai's efforts also were designed to create jobs, in this case mostly for women. Since its inception, the greenbelt movement has been greatly successful and has expanded to several other countries throughout the African continent and beyond. *See also* CLEAR-CUTTING; DEFORESTATION; HABITAT; RAINFOREST; ROOSEVELT, THEODORE; SLASH AND BURN.

refuse, A term used to describe items or materials (SOLID WASTES) that are deemed worthless and thus are discarded or thrown away. Refuse is generally distinguished from GARBAGE, which contains food wastes. When improperly disposed of on land, refuse often is called LITTER and is considered a form of pollution. Proper disposal of refuse includes INCINERATION, burial in SANITARY LANDFILLS, or RECYCLING. "TRASH" and "rubbish" are other terms used for refuse. *See also* CONSERVATION; RECLAMATION.

relative biological effectiveness, A term used to describe the effectiveness of a particular type of RADIATION is in producing a response in living tissue. Relative biological effectiveness (RBE) is important to scientists who want to compare the effects of different types of radiation.

rem. *See* radiation.

remedial investigation, An in-depth environmental study designed to gather data needed to determine the nature and extent of contamination at a SUPERFUND site. Remedial investigations typically include drilling test bores, collection of soil samples, the construction of GROUNDWATER-monitoring WELLS, groundwater testing, and laboratory analyses of environmental samples. AIR-quality monitoring or sediment sampling and analysis may also occur.

The results of a remedial investigation are used to establish cleanup criteria, identify preliminary alternatives for remedial action, and support technical and cost analyses of alternatives. The remedial investigation is usually done in conjunction with a feasibility study. Together, the two studies are usually referred to as the "RI/FS." *See also* COMPREHENSIVE ENVIRONMENTAL RESPONSE, COMPENSATION, AND LIABILITY ACT; RESOURCE CONSERVATION AND RECOVERY ACT; TOXIC RELEASE INVENTORY; TOXIC WASTE.

renewable resource, A NATURAL RESOURCE that is in abundant supply or is regularly formed and replaced through natural processes and thus has the potential to be available for use by living things for many years in the future or even indefinitely. Examples of renewable resources include ORGANISMS, AIR, nutrients, water, and soil. Some ENERGY sources such as wind, sunlight, and GEOTHERMAL ENERGY also are renewable natural resources.

Although renewable resources are regularly replaced through natural processes, it is necessary to practice CONSERVATION, preservation of the quality and quantity of resources in the use of these resources because many of them are being used faster than they are naturally replaced, degraded, and made unfit for use by organisms. For example, the gases that make up air are continuously recycled through the OXYGEN, CARBON, and NITROGEN CYCLES; however, the release of many chemical POLLUTANTS into the ATMOSPHERE degrades air quality in many parts of the world, making the air potentially harmful to humans, other organisms, and many ABIOTIC components of the ENVIRONMENT. For example, a reaction between NITROGEN OXIDES—which are released into the air as a result of the combustion of FOSSIL FUELS and other activities—with sunlight often leads to the formation of a chemical haze that is commonly called SMOG. This reaction is most commonly observed near large cities, but the substances that form smog can be carried by wind and air currents to other areas. When severe, smog can make breathing difficult for people suffering from lung disorders; smog also poses a health threat to people with compromised immune systems. In addition, the chemicals that contribute to the formation of smog also contribute to the formation of ACID RAIN. Acid rain that falls into lakes or FOREST ECOSYSTEMS often causes significant harm to the organisms that make their homes in these environments. In addition, the ACIDS in the rainfall also cause many physical structures made from rock or metals to deteriorate and break apart.

Organisms are considered renewable resources because they replenish themselves through their reproductive processes; however, many SPECIES of organisms are at risk of EXTINCTION because their POPULATIONS are decreasing faster than they are being replaced. Although extinction is a natural process, the rate of extinction has increased substantially over the last century, largely as a result of the activities of humans. Currently, the greatest threat to organisms is HABITAT loss—changing the environment in ways that make it unsuitable for the continued existence of the populations that once lived there. Some habitat loss occurs from natural events such as volcanic eruptions, EARTHQUAKES, FLOODS, and FOREST FIRES sparked by lightning; however, most habitat loss results from human activities such as the clearing of land or the filling of WETLANDS for agricultural development, road and housing construction, or other activities. Other human activities that threaten the survival of many organisms include overhunting, the collecting of rare species, the introduction of EXOTIC SPECIES that outcompete NATIVE SPECIES, and the release of pollutants into the environment.

To prevent the misuse or overuse of renewable resources, the conservation practices of reducing, reusing, and RECYCLING are encouraged for renewable resources. These are the same practices currently encouraged for use of NONRENEWABLE RESOURCES. Reducing the use of resources involves using less of them when possible. For example, although water is constantly recycled through the environment, supplies of fresh water are sometimes jeopardized in different regions of the world as a result of conditions such as DROUGHT, floods, and contamination. Thus, scientists recommend that people try to practice water conservation methods such as taking shorter showers and fixing any leaks in pipes. In addition, environmentalists believe that people should reuse and recycle renewable resources when possible as a means of preventing such resources from being overused as the global human population continues to grow. *See also* AGRICULTURAL POLLUTION; AGROECOLOGY; AIR POLLUTION; AIR POLLUTION CONTROL ACT; BIOGEOCHEMICAL CYCLE; BUREAU OF RECLAMATION; CALCIUM CYCLE; CARBON CYCLE; CARRYING CAPACITY; COMPOSTING; COMPREHENSIVE ENVIRONMENTAL RESPONSE, COMPENSATION, AND LIABILITY ACT; DEFORESTA-

TION; DESERTIFICATION; ENVIRONMENTAL EDUCATION ACT; EROSION; FORESTRY; GLOBAL WARMING; HAZARDOUS WASTE; NATIONAL ENVIRONMENTAL POLICY ACT; NATIONAL POLLUTANT DISCHARGE ELIMINATION SYSTEM; OCEAN DUMPING; RED LIST OF ENDANGERED SPECIES; SOIL CONSERVATION; SURFACE MINING CONTROL AND RECLAMATION ACT; SUSTAINABLE AGRICULTURE; SUSTAINABLE DEVELOPMENT; SUSTAINED YIELD.

EnviroSources

Bernstein, Leonard, Alan Winkler, and Linda Zierdt-Warshaw. *Environmental Science: Ecology and Human Impact.* Menlo Park, CA: Addison-Wesley, 1996.
Saign, Geoffrey C. *Green Essentials: What You Need to Know about the Environment.* San Francisco, CA: Mercury House, 1994.
Seymour, John, and Herbert Girardet. *Blueprint for a Green Planet: Your Practical Guide to Restoring the World's Environment.* New York: Prentice Hall, 1987.

reptiles, A class of VERTEBRATES comprised of approximately 5,000 SPECIES that include snakes, lizards, ALLIGATORS, CROCODILES, turtles, tortoises, and the rare tuatara. Of these, approximately 112 species are identified as ENDANGERED SPECIES or THREATENED SPECIES according to the ENDANGERED SPECIES LIST maintained by the U.S. FISH AND WILDLIFE SERVICE. Reptiles are characterized as being ectothermic, primarily egg laying,

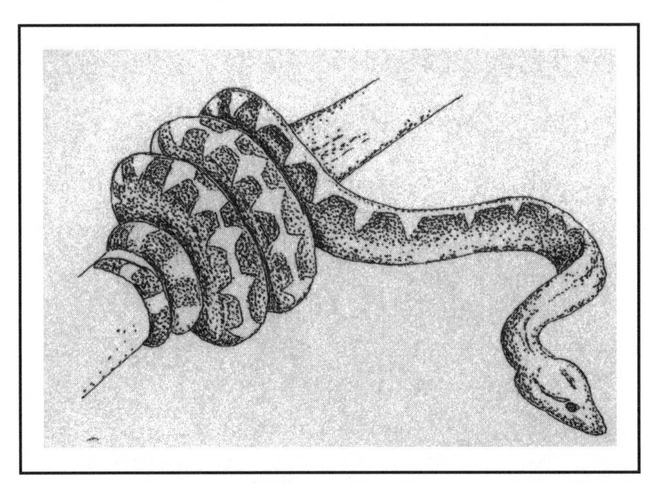

The Emerald Tree Snake belongs to a reptile class of vertebrates composed of approximately 5,000 species that include other snakes, lizards, alligators, crocodiles, turtles, tortoises, and the rare tuatara.

and having a body covering of scales or horny plates. Most reptiles live on land and breathe using lungs, although a few species of snakes and turtles are adapted to life in fresh-water or marine ENVIRONMENTS. (*See* ocean.) As a group, reptiles are widely distributed throughout the world and live in virtually all environments, except ARCTIC or sub-Arctic CLIMATES. *See also* DESERT; EXTINCTION; GALÁPAGOS ISLANDS; RAINFOREST; SEA TURTLES.

EnviroTerm

ectothermic: Animals having a dependence upon external environmental factors for the regulation of their body temperature: cold-blooded.

EnviroSource

Williams, Ted. "The Terrible Turtle Trade." *Audubon* 101, no. 2 (March–April 1999): 45–51.

reservoir, A pond, lake, tank, or basin where water is collected, stored, and regulated for use. A reservoir can have a natural origin or can be created by humans; for example, the construction of a DAM creates a reservoir. Large bodies of GROUNDWATER are called groundwater reservoirs.

The term "reservoir," or "SINK," can also refer to a storage place or zone for chemicals; for instance, the OCEAN is a reservoir for CARBON DIOXIDE. *See also* AQUIFER; WELL.

EnviroTerm

reservoir evaporation: The amount of water lost to the ATMOSPHERE from a reservoir through direct evaporation and sublimation losses during below-freezing temperatures.

resource. *See* alternative energy resource; natural resources; nonrenewable resource; renewable resource.

Resource Conservation and Recovery Act, A 1976 amendment to the Solid Waste Disposal Act (1965) of the United States that established guidelines for disposing SOLID WASTES. When it was created, the Resource Conservation and Recovery Act (RCRA) had four main goals: to protect human health and the ENVIRONMENT from possible hazards involving waste disposal, to reduce the total amount of generated waste material, to conserve ENERGY and NATURAL RESOURCES, and to ensure environmentally sound waste management practices. The primary authority for achieving these goals rests with the U.S. ENVIRONMENTAL PROTECTION AGENCY (EPA).

In 1984, Congress built upon the RCRA by passing the Hazardous and Solid Waste Amendments. These amendments require the EPA to manage disposal of all solid wastes, including nonhazardous household wastes, and to manage HAZARDOUS WASTES from CRADLE TO GRAVE. Thus, the EPA regulates the generation, handling, storage, transport, and disposal of hazardous wastes. The RCRA also regulates the design and operation of underground storage tanks that contain PETROLEUM or other chemical substances. In 1988, the RCRA was further amended to include the management of INFECTIOUS WASTES. Such wastes include animal waste, human blood, and blood products.

The hazardous waste disposal regulations under the RCRA involve only current and future wastes and those designated for RECYCLING. Responsibilities for the cleanup of such sites is covered by the COMPREHENSIVE ENVIRONMENTAL RESPONSE, COMPENSATION, AND LIABILITY ACT, or SUPERFUND. *See also* CONSERVATION; DEEP-WELL INJECTION; HAZARDOUS MATERIALS TRANSPORTATION ACT; NATIONAL PRIORITIES LIST.

EnviroSources

RCRA, Superfund, Emergency Planning and Community Right-to-Know Act Hotline: (800) 424-9346; from the Washington, DC, area (703) 412-9810 Web site: www.epa.gov/epaoswer/hotline.

U.S. Environmental Protection Agency. *RCRA: Reducing Risk from Waste.* Washington, DC, September 1997.

restoration ecology, The transformation of an ECOSYSTEM that has been changed for use by humans back to its predisturbed state. Restoring ecosystems was largely initiated by environmentalist ALDO LEOPOLD in the early twentieth century. The practice may be used to reclaim lands that were altered by MINING, AGRICULTURE, or other human activities.

To restore a damaged ecosystem, scientists working in a diversity of fields must assess the BIOTIC and ABIOTIC components of the region that will undergo restoration to determine the characteristics of the original POPULATIONS for that area and how the conditions of the present ENVIRONMENT must be changed to support those populations. Often, restoration then begins with the use of natural or synthetic FERTILIZERS to replace nutrients lacking in the soil. In addition, the soil may be mixed with sand or other materials to improve its ability to retain or drain water. These activities are conducted to ensure that the soil can support the growth of PLANTS.

Once the soil in an area undergoing restoration is prepared, plants that would naturally be present in that environment are reintroduced to the region and allowed to grow. The area is then permitted to undergo ECOLOGICAL SUCCESSION, which allows a series of new BIOLOGICAL COMMUNITIES to colonize the area. In some cases, animal SPECIES representative of the original populations of the ecosystem may be introduced to the area to help control the type of ecosystem that develops. In other cases, the animal populations are determined according to the natural process of succession the area undergoes.

Ecological restoration is a slow process. Like natural communities, it may take hundreds of years for a final climax community to be established in an area; however, scientists hope that by restoring some regions to their predeveloped conditions, EXTINCTIONS resulting from loss of HABITAT will be slowed, while an understanding of the interactions among ORGANISMS and their environments will improve. *See also* BIODIVERSITY; BUREAU OF RECLAMATION; CONSERVATION; EVERGLADES NATIONAL PARK; RECLAMATION; SURFACE MINING CONTROL AND RECLAMATION ACT.

reverse osmosis, A process used in some DESALINATION plants to produce fresh water from BRACKISH water, which is less salty than ocean water. Reverse osmosis is the opposite of the natural process of osmosis, which involves the flow of purer water towards a more concentrated solution.

Reverse osmosis forces the more concentrated solution through a semipermeable membrane, a type of screen that holds back dissolved substances. The liquid flows through the membrane into a body of fresh water, which has a lower concentration of particles. Applied pressure is needed in order to force the salt water through the membrane to the freshwater side. The process increases the volume of fresh water and decreases the volume of salt water, which also becomes more concentrated. Reverse osmosis is also used in WASTEWATER treatment systems to remove impurities.

EnviroTerm

osmosis: The diffusion of a liquid through a semipermeable membrane from an area containing fewer dissolved or suspended particles to the other side containing a greater concentration of particles. In this process, the fresh water will flow through the membrane to the salty side in an effort to equalize the concentrations on both sides of the membrane. This process occurs naturally in the root CELLS of PLANTS.

Rhine River, A major European waterway that was once threatened by severe water pollution problems and is now recovering. The Rhine River begins in Switzerland and flows 1,320 kilometers (820 miles) through several European countries, including Germany, France, and the Netherlands, and then empties into the NORTH SEA. The river is used by barges and other boats to transport goods and supplies such as PETROLEUM, COAL, finished products, and AUTOMOBILES to and from major cities on the Rhine, such as Strasbourg, Bonn, and Rotterdam.

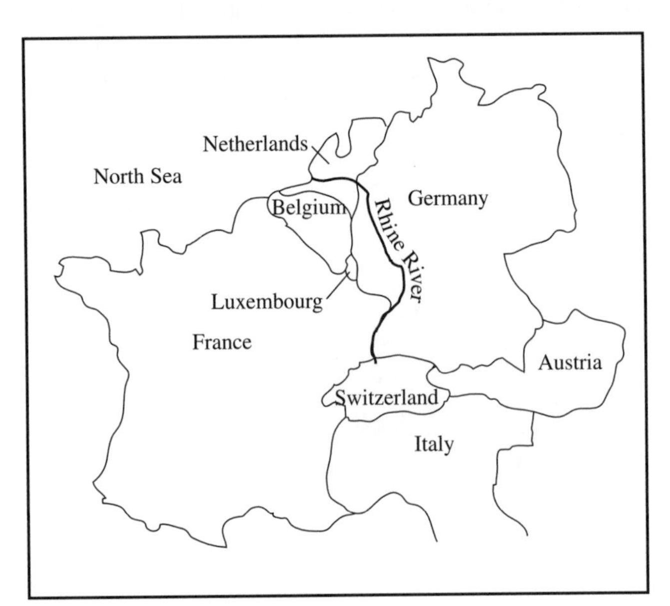

The Rhine River watershed covers an area of 190,000 square kilometers and includes a variety of habitats for many plant and animal species.

The Rhine River WATERSHED covers an area of 190,000 square kilometers (73,000 square miles) and includes a wide variety of HABITATS for many PLANT and animal SPECIES. The river is also an important drinking water supply for eight million people.

In the 1960s and 1970s, sections of the Rhine became polluted by wastes from industrial and agricultural sources located along the river's banks. Industrial DISCHARGES included HEAVY METALS such as CADMIUM, MERCURY, and COPPER. The RUNOFF of FERTILIZERS, PESTICIDES, and animal wastes from farmlands also caused pollution problems. As a result, many fish SPECIES and other aquatic life became extinct, endangered, or threatened. The river water became unsafe for swimming and drinking.

The countries bordering the Rhine set up a number of programs to clean up and protect the river. One of the programs, called the Rhine Action Program, established in the 1980s, set out to protect drinking water supplies, eliminate the TOXINS in the sediments of the river, and reintroduce POPULATIONS of fish. The program also called for actions to protect the North Sea. As a result of this program and other measures, discharges into the river have been cut in half and some fish have returned to the Rhine in large numbers. *See also* AGRICULTURAL POLLUTION; ENDANGERED SPECIES; SANDOZ CHEMICAL SPILL; THREATENED SPECIES.

rhinoceroses, Large, three-toed, single- or double-horned herbivorous MAMMALS from Africa and southern Asia that are on the endangered list. Rhinos first roamed Earth during the Miocene era 50 million years ago and whose ancestors once roamed North America and Europe. Modern rhinos grow to a shoulder height of 1.5 meters (5 feet) and weigh more than 1,800 kilograms (4,000 pounds). Rhinos are HERBIVORES whose HABITAT ranges from GRASSLANDS and open SAVANNAS to dense, woody areas. They live 35 to 40 years and have no natural predators in the wild. (*See* predation.) The rhino's appearance is distinguished by one or two "horns" that grow atop the animal's snout; not true horns, these structures actually are composed of thickly matted hair that grows from the skull without skeletal support.

In recent decades, rhinos have been hunted nearly to EXTINCTION by poachers who sell the animals' "horns" for medicinal and ornamental purposes in Asia (particularly China) and the Middle East (particularly north Yemen). Since 1970, the world rhino POPULATION has declined by 90%, with only 5 SPECIES remaining today—the black (*Diceros bicornis*), white (*Ceratotherium simum*), Indian (*Rhinoceros unicornis*), Javan (*Rhinoceros sondaicus*), and Sumatran (*Dicerorhinus sumatrensis*)—all of which are ENDANGERED SPECIES.

In the short time between 1970 and 1994, 95% of Africa's black rhinos were killed; by 1993, only 2,550 black rhinos remained in the wild. Zimbabwe is the only country where there are any large numbers. The other African species, the white rhino, totals about 4,000 in number. The number for the

Since 1970, the world rhino population, such as the white rhino above, has declined by 90%, with five species remaining today—black/white (above), Indian, Javan, and Sumatran—all of which are endangered species.

Asian rhinos totals are Indian (2,000), Javan (50–100), and Sumatran (200–500). Without drastic action, four species of rhino could become extinct in the wild within the next 10 years. Only 13,565 of these creatures survive in the wild, with another 1,000 in captivity. Of these totals, more than half are the white rhino. There are fewer than 5,000 of the other four species combined. To prevent their extinction, many rhinos have been relocated to fenced sanctuaries since the early 1990s. This effort appears to be succeeding: In 1994, for the first time in 20 years, rhino numbers did not decline.

Because of the urgent actions often needed in the fight to save rhinos, the WORLD WILDLIFE FUND (WWF) has set up an emergency fund for African rhino CONSERVATION. This fund allows WWF personnel to respond quickly and assist government agencies, park officials, and law enforcement officials to meet critical rhino conservation needs. *See also* ADDO NATIONAL ELEPHANT PARK; AFRICAN WILDLIFE FOUNDATION; CONVENTION ON INTERNATIONAL TRADE IN ENDANGERED SPECIES OF WILD FLORA AND FAUNA; POACHING; RED LIST OF ENDANGERED SPECIES; ZOO.

EnviroTerm

ungulate: Any animal that has hoofs. Ungulates are classified in the orders Perissodactyla and Artiodactyla and include MAMMALS such as cattle, horses, deer, swine, ELEPHANTS, and rhinos.

EnviroSources

African Wildlife Foundation Web site: www.awf.org.
International Rhino Foundation Web site: www.rhinos-irf.org.
U.S. Fish and Wildlife Service, Division of Endangered Species, Foreign Listed Species Index Web site: www.fws.gov/r9endspp/fornspp.html.
World Wildlife Fund (Worldwide Fund for Nature) Web site: www.panda.org.

ribonucleic acid, A complex ORGANIC compound present in the CELLS of ORGANISMS that is involved in protein synthesis and thus plays a key role in determining the traits of an organism. In some VIRUSES, RNA replaces deoxyribonucleic acid (DNA) as the genetic material. Most RNA is made in a cell's nucleus. The RNA molecule consists of a single strand of nucleotides in which ribose is the sugar and the bases are adenine, cytosine, guanine, and uracil.

There are three types of RNA, which are identified by their functions. Messenger RNA (mRNA) carries the genetic code—determined by a cell's DNA—to the ribosomes of the cell. In the ribosomes, ribosomal RNA (rRNA), translates the code carried by the mRNA and "instructs" transfer RNA (tRNA), to assemble amino acids to make proteins. Each tRNA molecule is specific for a particular protein. Once the code carried by the mRNA has been translated by the rRNA, tRNA can begin its work and assemble the amino acids needed to create the protein. *See also* ADAPTATION; GENETIC DIVERSITY; MUTATION.

riparian habitat, A WILDLIFE HABITAT adjacent to a body of water. River and stream banks and lake shores are typical riparian areas. Riparian habitats and WETLANDS are an important source of BIODIVERSITY.

riparian rights, A doctrine or common-law concept that has existed since early Colonial times that states that a landowner located on or adjacent to the banks of a stream, riverbed, or lake has the rights to withdraw water as long as there is no disruption in the flow or degradation of the quality of the water. The doctrine was established to help protect the water rights of other landowners who were located adjacent to the same water and thus also had access to it. Riparian rights are applied to many of the states east of the Mississippi River. Modern interpretations of the doctrine have produced disagreements between landowners. As an example, if a state wants to build a DAM or RESERVOIR, the riparian rights of the local landowners upstream or downstream maybe be usurped. *See also* RIPARIAN HABITAT.

Rivers and Harbors Appropriation Act of 1899, Legislation commonly known as the Rivers and Harbors Act of 1899 that prohibits the construction of any bridge, DAM, dike, or causeway over or in navigable waterways of the United States without congressional approval. This includes the construction of any wharfs, piers, jetties, and other structures. Structures authorized by state legislatures may be built if the affected navigable waters are totally within one state, provided that the plan is approved by the chief of engineers and the secretary of the Army.

RNA. *See* ribonucleic acid.

rock cycle, A series of changes which over many years produce rocks and soil. In the rock cycle, rocks change by the processes of heat, pressure, WEATHERING, EROSION, melting, cementation, and compaction. Rocks are classified as igneous, metamorphic, and sedimentary. Igneous rocks are formed by the cooling and hardening of molten material from a VOLCANO or deep inside Earth's crust. Examples of igneous rocks include basalt, granite, and gabbro. About 75% of the rocks on Earth's surface are sedimentary. Sedimentary rocks consist of sediments from rock materials and PLANT remains that have hardened under pressure and chemical action. Examples of sedimentary rocks include sandstone, shale, and limestone. Metamorphic rocks are formed from existing igneous and sedimentary rocks that undergo additional heat, pressure, and chemical action. Examples include slate, marble, and gneiss.

Materials in Earth's crust may pass through the igneous, sedimentary, and metamorphic stages repeatedly and in several possible sequences. As an example, molten material that cools and hardens at Earth's surface becomes igneous rocks. Through chemical and mechanical weathering the igneous rocks at the surface can be worn down into sediments and carried away by water, wind, or moving ice. (*See* glacier.) In time, some of these sediments will accumulate in layers and be subject to pressure and chemical action, changing the sedi-

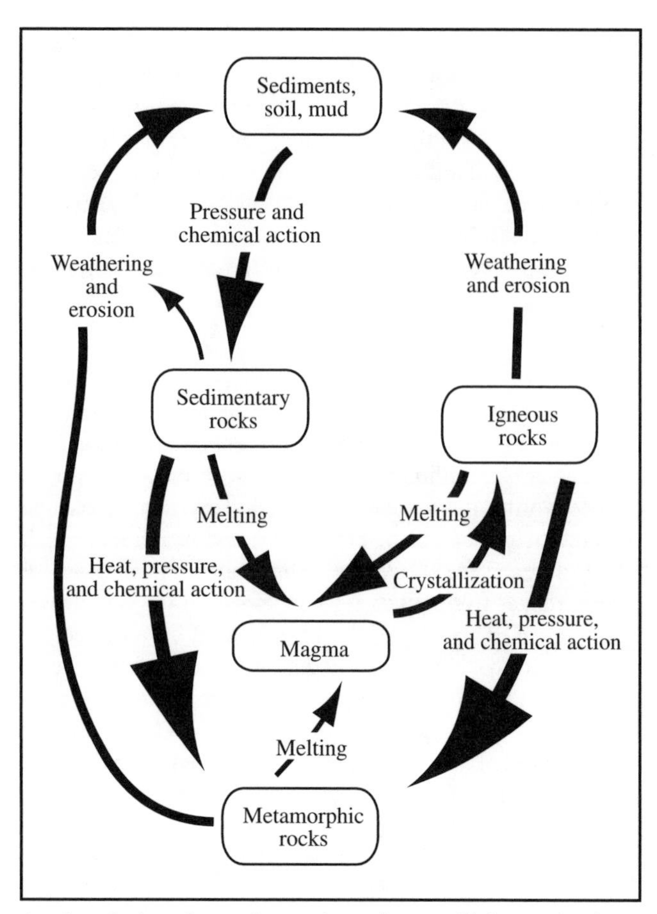

A rock cycle shows how rocks can change from one kind to another due to factors such as heat, pressure, melting, weathering, chemical action, erosion, and climate.

ments into sedimentary rocks. If the sedimentary rocks undergo further changes from heat and pressure, they will become metamorphic rocks. Sometimes internal heat and pressure inside the crust can melt the metamorphic rocks to molten rocks to start the cycle again. *See also* GEOLOGY.

rodenticide. *See* pesticide.

Roosevelt, Theodore (1858–1919), The twenty-sixth president of the United States, who during his two terms in office (1901–1909) established the CONSERVATION of NATURAL RESOURCES as a national goal. Roosevelt's dedication to conservation was part of his Square Deal program. This program had four major concerns: addressing social problems of the nation, increasing regulation of big business, increasing control of the railroads, and conserving natural resources.

To achieve his conservation goals, Roosevelt focused his attention on FORESTS and WILDLIFE. During his first term in office, he used the provisions of the Forest Reserve Act of 1891 to withdraw 51 million hectares (125 million acres) of western timberland and an additional 34 million hectares (85 million acres) of forest in Alaska from sale. Roosevelt set aside this land as national forest land, placing it under the administration of the U.S. FOREST SERVICE, which was headed by GIFFORD PINCHOT. In 1902, Roosevelt signed the National Reclamation Act into law. This act authorized money obtained from the sale of land in 16 semiarid states to be used to develop IRRIGATION systems. The law resulted in the construction of DAMS that allowed land in such states to be reclaimed for use by people.

During his second term in office, Roosevelt built on his earlier efforts by holding a conservation conference at the White House in 1908. The aim of the conference was to consider what policies should be adopted to preserve national resources for future generations. The conference was attended by scientists, governors, university presidents, and representatives of business, and resulted in the creation of a National Conservation Commission. Similar commissions also were established in 41 states.

Roosevelt also addressed growing concerns about food quality, which had developed, in part, as a result of *The Jungle*, a fact-based novel written by Upton Sinclair about the meatpacking industry in 1906. To address these concerns, Roosevelt passed the Pure Food and Drug Act in 1906. The act established the U.S. FOOD AND DRUG ADMINISTRATION (FDA).

After leaving office, Roosevelt's interest in the ENVIRONMENT continued. Between October 1913 and May 1914, he led an expedition into the Amazon region of Brazil, during which he observed regional wildlife and explored the River of Doubt. During the trip, Roosevelt broke his leg and contracted malaria. His health steadily declined following his return to the United States until his death in 1919. The River of Doubt was later renamed the Rio Roosevelt in his honor. In 1978, Roosevelt was honored again when the NATIONAL PARK SERVICE established the Theodore Roosevelt National Park in North Dakota. *See also* WILDLIFE REFUGE.

EnviroSources

Seidman, David. "Preaching from the Bully Pulpit: Theodore Roosevelt." *Audubon* 100, no. 6 (November–December 1998): 89.

Stefoff, Rebecca. *Theodore Roosevelt: Twenty-Sixth President of the United States.* Deerfield Beach, FL: Garrett, 1988.

runoff, Water from rain, ice, or snow that is not absorbed into the ground but runs off into SURFACE WATER bodies. The National Water Quality Inventory reports that runoff from urban areas is the leading source of POLLUTANTS in ESTUARIES and the third-largest source of pollutants in lakes. In addition, POPULATION and development trends indicate that by 2010 more than half of the people in the United States will live in coastal towns and cities, some of which will have tripled in population. Runoff from these areas will continue to degrade coastal waters. In 1999, the NATIONAL OCEANIC AND ATMOSPHERIC ADMINISTRATION and the U.S. ENVIRONMENTAL PROTECTION AGENCY created a joint program to help more than 25 coastal states and territories develop procedures to reduce runoff from agricultural fields and city streets. The two agencies encouraged states to manage the application of nutrients and FERTILIZERS applied to farm areas and city lawns, to control stormwater runoff from new developments, and to reduce pollution from recreational boats and marinas.

In soil, runoff moves toward receiving waters gradually. In contrast, nonporous urban landscapes such as roads, bridges, parking lots, and buildings do not absorb runoff. Water remains at the surface, accumulates, and runs off in large amounts. When leaving the drainage system and emptying into a stream, it erodes streambanks, damages streamside vegetation, and widens stream channels. This will result in lower water depths during nonstorm periods, higher-than-normal water levels during wet WEATHER periods, increased sediment loads, and higher water temperatures.

Native fish and other aquatic life cannot survive in urban streams that have been severely impacted by urban runoff. Urbanization also increases the variety and amount of pollutants transported to receiving waters; the list of pollutants may include sediment from development and new construction; oil, grease, and TOXIC CHEMICALS from vehicles; nutrients and PESTICIDES from turf management and gardening; VIRUSES and BACTERIA from failing SEPTIC TANK systems; road salts; and HEAVY METALS. Sediments and solids constitute the largest volume of pollutant loads to receiving waters in urban areas.

Laws that help control urban runoff focus either on urban POINT SOURCES or on urban NONPOINT SOURCES. Point sources are addressed by the NATIONAL POLLUTANT DISCHARGE ELIMINATION SYSTEM permit program of the CLEAN WATER ACT, which regulates storm-water DISCHARGES. Urban nonpoint sources are covered by nonpoint source management programs developed by states, territories, and tribes under the Clean Water Act. *See also* AGRICULTURAL POLLUTION; CHESAPEAKE BAY; EFFLUENT; FLOOD; SEWER; WASTEWATER.

S

Safe Drinking Water Act, A 1974 federal law enacted in response to outbreaks of waterborne DISEASE and increasing chemical contamination of public drinking water in the United States. This law focuses on the protection of all waters that are actually or potentially suitable for public drinking use, from both above-ground and underground sources. The act protects drinking water supplies by establishing water quality standards for drinking water, monitoring public water systems, and guarding against GROUNDWATER contamination from injection wells. (*See* deep-well injection.) Before 1974, each state ran its own drinking water program and set required standards at the local level; as a result, drinking water protection standards differed from state to state.

The act authorized the U.S. ENVIRONMENTAL PROTECTION AGENCY (EPA) to establish safe standards of water purity and required all operators of public water systems to comply with primary (health-related) standards. A "public water system" is defined as providing piped water for human consumption and having at least 15 service connections or regularly serving at least 25 persons. In virtually all states, the EPA has relinquished enforcement of the Safe Drinking Water Act to state regulators and now supervises the state programs approved to take its place; however, 1986 amendments to the act gave the EPA increased enforcement authority if a state took no action within 30 days of receiving EPA notification that water quality standards had been violated.

The 1986 amendments directed the EPA to set standards for 83 specified CONTAMINANTS, to ban the use of LEAD pipes and solder in new drinking water distribution systems, and to establish well-head protection programs for all public water supply WELLS. To date, the EPA has established enforceable standards for 80 contaminants, including ORGANIC compounds, such as BENZENE; certain BACTERIA, VIRUSES, and PROTISTS such as CRYPTOSPORIDIUM; and inorganics, such as lead and COPPER. MAXIMUM CONTAMINANT LEVELS (MCLs) have been established to minimize the levels of dangerous chemicals and waterborne PATHOGENS in the public's drinking water. HCLs also are frequently incorporated into criteria for site cleanup programs, such as those managed by SUPERFUND.

The Safe Drinking Water Act Amendments of 1996 established training and certification programs for water treatment plant operators and bolstered communities' abilities to upgrade obsolete water FILTRATION and purification systems. Under the new law, communities can apply for state-administered loans and grants from the multibillion-dollar Drinking Water State Revolving Fund to construct new facilities or upgrade old ones. This law includes provisions to protect the well-being of children and at-risk POPULATIONS, ensures community right-to-know policies, and encourages the collection of data to improve the quality and efficiency of community drinking water systems. The EPA also now requires public water utilities to notify their customers of the source and health effects of tap-water contaminants and requires public water systems serving more than 10,000 customers to notify those customers annually of the levels of federally regulated contaminants in their water. This could help educate Americans about drinking water quality and become a powerful tool for mobilizing citizen support for river and WATERSHED CONSERVATION.

Today, the EPA is committed to working with the states on drinking water issues. One of the formal means by which the EPA solicits the assistance of states is through the National Drinking Water Advisory Council (NDWAC). The council, comprised of citizens, state and local agencies, and private groups concerned with safe drinking water, advises the EPA administrator on the agency's actions relating to drinking water. All NDWAC working-group meetings and all full NDWAC meetings are open to the public. The Safe Drinking Water Act is due for reauthorization in the year 2003. *See also* AQUIFER; EMERGENCY PLANNING AND COMMUNITY RIGHT-TO-KNOW ACT; ENVIRONMENTAL RACISM; NATIONAL PRIMARY DRINKING WATER REGULATIONS; POTABLE WATER; PRIMARY, SECONDARY, AND TERTIARY TREATMENTS; WATER TABLE.

EnviroSources

Cronin, John, and Robert F. Kennedy, Jr. *The Riverkeepers.* New York: Scribner, 1997.

National Drinking Water Clearinghouse Web site: www.estd.wvu.edu/ndwc/ndwc_homepage.html.

U.S. Environmental Protection Agency, Office of Groundwater and Drinking Water Web site: www.epa.gov/OGWDW.
U.S. Environmental Protection Agency, Safe Drinking Water Hotline: (800) 426-4791.

saline. *See* salinization.

salinization, The process by which soluble salts, such as sodium chloride, calcium chloride, and magnesium chloride, accumulate in soils to a level that harms PLANTS or prevents their growth. Salinization has occurred in farming areas worldwide; between 20 million and 30 million hectares (50–75 million acres), or 8–12% of irrigated areas have been seriously salinized, and each year, an additional 1–1.5 million hectares (2.5–3.5 million acres) are affected. About 30% of the irrigated farmlands in southern Asia, South America, Mexico, and northern Africa have suffered from salinization leading to DESERTIFICATION. In the United States, salinization has been a problem in the agricultural areas of the San Joaquin Valley of California, the Colorado River basin, and farmland along the Rio Grande. Twenty-eight to thirty percent of irrigated croplands in the United States suffer from salinization.

The growth of salt-tolerant EXOTIC SPECIES interspersed among native plants or crops is an early sign of high salinity. Saline or salty soils also tend to have a white, crusty surface due to the precipitation of salts. Factors that control soil salinity include soil parent materials (the type of rock from which the soil formed), GROUNDWATER HYDROLOGY, precipitation levels, and agricultural practices.

In saline soils, plants are forced to work against osmosis—the process by which fresh water moves toward salt water solutions, which are more concentrated. This decreases the plants' ability to extract water from the soil. Symptoms of salinization include burned leaf edges and wilting due to poor water absorption. Salinization also degrades the quality of shallow groundwater and lakes and ponds, where salt may be concentrated to levels that pose a risk to the health of livestock and WILDLIFE.

Salinization frequently occurs in irrigated soils where salts brought in with IRRIGATION water or in water-soluble FERTILIZERS remain behind as the water evaporates or is absorbed by plants. Saline soils also occur naturally in DESERTS and semiarid regions, where evaporation or evapotranspiration rates exceed precipitation rates. (*See* transpiration.) In order for saline soils to form in these ENVIRONMENTS, the WATER TABLE must generally be within 2 meters (about 6 feet) of the soil surface to allow capillary action (the rise of groundwater into unsaturated soil) to sufficiently raise salt-laden groundwater. The lack of rainfall allows such salts to build up in the soil and crystallize at the surface.

Proper irrigation management can slow down or stop salinization in cultivated areas. The amount of salinization that crops can endure and still produce adequate yields can be measured; therefore, irrigation on a sustainable basis must maintain the salt balance at a level that is under critical salinity levels.

The most effective and expensive remedial technique to abate salinization is to combine the use of LEACHING and artificial drainage of irrigation water. Drained water must be disposed of carefully, however: Improper drainage of irrigation water can cause local or regional salinity problems by increasing salt content in water tables and downstream AQUIFERS.

Increasing irrigation efficiency by minimizing evaporation is the most cost-effective solution to the problem of salinization. Evaporation can be decreased significantly by placing MULCH or plastic over irrigated fields. *See also* AGRICULTURE; REVERSE OSMOSIS; SURFACE WATER; WATER CYCLE.

EnviroTerms

saline: Water containing a high levels of salts (e.g. sodium chloride, calcium chloride, magnesium chloride.)
salinity: The quantity of salts dissolved in water. The salinity of sea water is approximately 35,000 parts per million.

salmon, A type of fish found in North America, Asia, and Europe, of which some species are threatened by OVERFISHING and loss of HABITAT. In 1999, the NATIONAL MARINE FISHERIES SERVICE of the United States announced that eight wild salmon species in the Pacific Northwest were placed on the list of THREATENED SPECIES of fish and one was placed on the ENDANGERED SPECIES LIST. Wild salmon are located in most of the WATERSHED areas in Washington State and parts of the Willamette Valley in Oregon. The fish on the threatened list include the Puget Sound chinook, the Lower Columbia River chinook, Lake Ozette sockeye, Hood Canal summer chum, Lower Columbia chum, and the mid-Columbia steelhead. The Upper Columbia spring chinook is on the endangered list. Under U.S. environmental law it is a felony to harm a listed species or its habitat.

Wild salmon have been threatened by a variety of conditions. Excessive NITROGEN from animal waste and FERTILIZERS has reduced the DISSOLVED OXYGEN in streams and rivers. Some streams and rivers have become too warm as a result of the cooling operations of much industry, which dissipates waste heat into waterways. Streams and rivers have become toxic from NONPOINT SOURCES, PESTICIDES, and inadequate SEWAGE treatment. Overfishing has also depleted salmon stocks. Plans to protect the salmon will include measures to stop the un-

Wild salmon have been threatened by a variety of conditions. Excessive nitrogen from animal waste and fertilizers has reduced the dissolved oxygen in streams and rivers.

necessary RUNOFF of soil as a result of EROSION and fertilizer from farms and timber harvesting operations, to improve SEWAGE TREATMENT PLANTS, to set a limit on fishing, and to enforce better monitoring of pesticides and water use by homeowners.

The life cycle of salmon begins when the eggs hatch and larvae are released into fresh-water streams and rivers. Within a few months, the larvae grow into young fish that reach a length of 2–3 centimeters (5–7 inches). The young fish remain in the fresh water for about two years, then they begin their migration to the OCEAN. In the salt water, the salmon grow to adults, which weigh 2–50 kilograms (4–110 pounds). After about three to four years, the adults return to the fresh-water streams and rivers where they spawn. Most of the adults will die after spawning, but a few will return to the ocean. Then the cycle begins again. Most of the salmon eaten in the United States is flown in from Alaska or raised in domestic pens in the United States, Canada, and South America. *See also* COMMERCIAL FISHING; TUNA.

EnviroTerm

fish ladder: A series of step-like pools built by humans to allow fish to migrate up and over such obstacles as dams.

EnviroSources

National Oceanic and Atmospheric Administration (fisheries) Web site: www.nmfs.gov.

National Oceanic and Atmospheric Administration, National Marine Fisheries Service (salmon) Web site: www.nwr.noaa.gov/1salmon/salmesa/index.htm.

Salmonella, A genus of rod-shaped BACTERIA (bacilli) that includes more than 1,400 SPECIES, many of which are pathogenic to humans. (*See* pathogen.) *Salmonella* bacteria are often present in polluted water, on kitchen surfaces, in soil, within the bodies of INSECTS or other animals, and on surfaces in food processing or food packaging facilities. Infection with *Salmonella* bacteria is called salmonellosis. Such infection occurs in three forms: as enteric (typhoid) fever, as septicemia (blood poisoning), and as acute gastroenteritis (stomach and intestinal distress). Most salmonellosis is caused by the ingestion of contaminated food and results in acute gastroenteritis. Acute symptoms of *Salmonella* food poisoning are nausea, vomiting, abdominal cramps, diarrhea, fever, and headache. These symptoms generally last for only 1–2 days; however, infection by some species of *Salmonella* results in a fatal form of food poisoning.

Salmonella can be transmitted by a variety of foods. Some of the most common vectors of *Salmonella* are raw or undercooked meats, raw or undercooked poultry, eggs, milk and dairy products, fish, shrimp, yeast, coconut, and egg- or dairy-based sauces and salad dressings. It is estimated that 2–4 million cases of salmonellosis infection occur in the United States annually. The incidence of salmonellosis appears to be rising both in the United States and in other developed nations.

To help prevent outbreaks of *Salmonella* food poisoning, the U.S. FOOD AND DRUG ADMINISTRATION (FDA) encourages people to thoroughly cook all foods that could be infected with the bacteria. Guidelines recommend that meat and poultry products be cooked until their juices run clear. In addition, the FDA cautions people to avoid eating foods made with raw eggs, such as salad dressings and uncooked cake and cookie batters. *See also* BOVINE SPONGIFORM ENCEPHALOPATHY; *ESCHERICHIA COLI*; PARASITISM.

salt marsh, A marine ECOSYSTEM characterized by fluctuating water levels due to TIDES and the growth of salt-tolerant PLANTS. Salt marshes are a type of WETLAND ENVIRONMENT. A wetland is an area of land where water is the dominant ABIOTIC factor in the environment. In a wetland, water covers or saturates the soil all year or for varying periods of time during the year. Like all wetlands, salt marshes are transitional areas between land and water. In fresh-water wetlands that occur inland, rainfall and overflow from lakes and rivers are the primary sources of water. Salt marshes, on the other hand, occur along coasts, in bays, lagoons, and other protected coastlines and are therefore influenced by the tides. Salinity and the frequency and extent of flooding determine the types of plants and animals that live in the salt marsh. Salt marshes are valuable but unappreciated resources. Although they provide people with food and help improve the quality of the environment (as plants help filter out some POLLUTANTS from water), throughout history salt marshes have been drained, filled, polluted, and generally regarded as mosquito-infested wastelands of little value.

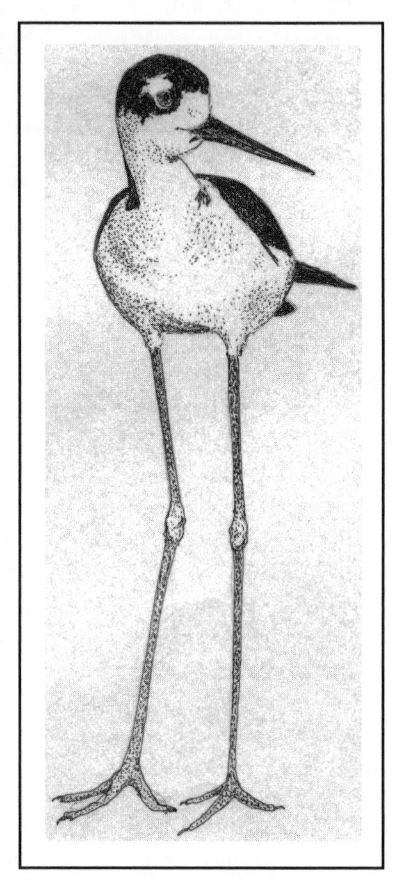

The Black-necked stilt lives in salt marshes. Salt marshes are among the most productive ecosystems in the world.

Salt marshes are places of extreme conditions that change twice daily with the tides. Many animals, such as snails and crabs, adapt by burrowing themselves in the sand or by climbing tall plant stems during high tide. Salt marsh plants possess ADAPTATIONS that help them survive the very salty conditions of the marsh. Spartina, for instance, is a common marsh grass with stiff, pointed leaves and specialized glands that help it excrete excess salt absorbed by the roots.

Salt marshes are among the most productive ecosystems in the world. More plants sprout, grow, and die in salt marshes than in almost any other kind of environment. HABITATS with high productivity can support many animals because there are lots of plants for them to eat. But in a salt marsh, cordgrass and other plants are not a direct source of food for the majority of salt marsh animals. Rather, most decaying plants are consumed by BACTERIA and other MICROORGANISMS, through the process of DECOMPOSITION. SCAVENGERS, including worms, fishes, shrimps, and crabs, may also feed on the decaying plant material, or DETRITUS, and then excrete the undigested plant remains in feces that can also be colonized by microorganisms. As the microorganisms reduce the detritus into smaller and smaller pieces, it becomes FERTILIZER for more growing plants. Daily movement of the tides also enhances the decomposition process. As a result, the entire salt marsh is constantly bathed in a type of fertilizing soup as the nutrients are recirculated.

Salt marshes perform many functions that are valuable to human beings. They serve as nursery grounds for numerous commercially and recreationally important fish and shellfish SPECIES. They act as buffers for the mainland by slowing waves, thereby reducing EROSION of the coastline. Like all wetlands, they act as filters and help remove sediments and TOXINS from the water. Unfortunately, a number of threats to salt marsh habitats exist. One of the most important threats to salt marshes, as well as most other marine ecosystems, is RUNOFF from bridges and roads (which contains PETROLEUM products from cars) and from farms and lawns (which contains PESTICIDES and fertilizers). Laws, such as the CLEAN WATER ACT, regulate POINT SOURCE pollution from single sources such as an industrial plant or boat, but NONPOINT SOURCE pollution, such as runoff, is more difficult to monitor and control due to its multiple sources. Pollution disrupts the FOOD WEB in the salt marsh by killing entire populations of some species and causing others to greatly increase in number. *See also* ESTUARY; MANGROVE; MARSH; NORTH AMERICAN WETLANDS CONSERVATION ACT; SWAMP.

EnviroSources

Coultas, Charles L., et al., eds. *Ecology and Management of Tidal Marshes: A Model from the Gulf of Mexico.* Boca Raton, FL: Saint Lucie Press, 1996.

Hay, John, Christopher Merrill, ed. *The Way to the Salt Marsh:* A John Hay Reader. Hanover, NH: University Press of New England, 1998.

Packham, John R. *Ecology of Dunes, Salt Marsh and Shingle.* Boca Raton, FL: Chapman and Hall, 1997.

Stoddart, D.R., ed. *Salt Marshes and Coastal Wetlands.* Institute of British Geographers Special Publications. Oxon, UK: Blackwell, 1999.

Teal, John. *Life and Death of the Salt Marsh.* New York: Ballantine Books, 1991.

U.S. Geological Survey (coastal and marine geology) Web site: marine.usgs.gov.

U.S. Geological Survey, National Wetlands Research Center Web site: www.nwrc.usgs.gov/educ_out.html.

salt-water intrusion, A phenomenon that occurs in coastal areas when salt water moves into a fresh-water AQUIFER. Fresh water is less dense than salt water (an amount of fresh water weighs less than the same amount of salt water), and as a result, fresh water will float on top of salt water before mixing. In an aquifer by the coast, fresh GROUNDWATER is found beneath the ground surface to a certain depth; however, deeper beneath the fresh groundwater, salt water is sometimes found intruding from the OCEAN. The fresh water floating above the salt water displaces (pushes down on) the intruding salt water. Near the coast, the depth at which the fresh water meets the salt water depends on the amount of fresh water and the stage of the TIDE.

Salt-water intrusion sometimes becomes a problem when salt water is drawn into water supply WELLS that pump groundwater from coastal aquifers. If too much fresh water is pumped out, the salt water rises in the aquifer and intrudes farther inland, contaminating drinking water. When a large river that opens into the ocean is dammed, further salt-water intrusion can sometimes occur because the river then delivers less fresh water to the coast to displace the salt water. Some places where salt-water intrusion occurs include Cape Cod, Massachusetts, and Long Island, New York, in the northeastern United States. *See also* POTABLE WATER; SINKHOLE; SUBSIDENCE.

Sandoz chemical spill, The accidental release of tons of TOXIC chemicals from a warehouse owned by the Sandoz pharmaceutical company into the RHINE RIVER near Basel, Switzerland, in November 1986. Release of the chemicals occurred as water being used to extinguish a fire at the company's warehouse mixed with and carried the chemicals into the nearby Rhine River. The release involved an estimated 27 metric tons (30 tons) of water mixed with a variety of highly TOXIC CHEMICALS, including PESTICIDES containing MERCURY. Within days, the environmental effects of the spill became apparent as thousands of dead fishes and other WILDLIFE began appearing at the river's surface and along its banks.

The Sandoz chemical spill provides an excellent example of how an environmental problem created in one area can adversely affect areas far away from the source. The Rhine River winds its way through several European countries, including Switzerland, France, Germany, and Luxembourg. Prior to the spill, water from the Rhine was used by people living in the many towns and cities located along the river; however, within days of the spill, cities in countries located downriver of the spill had to stop using water from the river. In response to this international crisis, environmental groups and authorities from several nations, with support from the United Nations, increased their efforts at monitoring pollution along the Rhine. Through these efforts, it was discovered that many other companies along the river were deliberately dumping or accidentally releasing harmful POLLUTANTS into the river. These companies have been required to correct these problems. At the same time, efforts are being made to clean and restore the Rhine to its prepolluted state. *See also* DISEASE; EFFLUENT; MINAMATA DISEASE; OCEAN DUMPING; OIL SPILL; POINT SOURCE; POTABLE WATER.

sanitary landfill, A disposal site for deposits of nonhazardous REFUSE such as SOLID WASTE, TRASH, rubbish, and GARBAGE. Sanitary landfills are built to reduce the nuisance of solid waste and the threat of unsanitary conditions to human health. The refuse brought to the landfill is spread out over an area and then flattened down by heavy machinery, making the material tightly compressed. The compacting is done to reduce the volume of the refuse. At the end of each day, the crushed refuse is covered by soil or other material. Doing this on a daily basis helps reduce odors, wind-blown debris, and other nuisances. Some large landfills can handle more than 3,500 metric tons (3,850 tons) of garbage a day. The buried material is graded into a mound that slopes to about a 30° angle. Over time, landfills grow in size, some developing into great mounds of debris. The Fresh Kills landfill, probably the world's largest landfill, located on Staten Island in New York, has been operating since 1948 and has reached the height of a 17-story building spread over 1,200 hectares (3,000 acres). It is scheduled to close in the year 2001. The standards and regulations for landfills are monitored and supervised by state governments. Landfills may be operated by private businesses or by counties, cities, or towns. The U.S. ENVIRONMENTAL PROTECTION AGENCY (EPA) regulates only those landfills that are used by more than 100,000 people.

The ORGANIC material in a landfill will gradually degrade naturally, producing various byproducts. Liquid produced by the degrading waste combined with water filtering through the landfill is called "leachate." (*See* leaching.) Leachate is highly polluted. The POLLUTANTS in leachate can include organic (CARBON-based) and inorganic (noncarbon) substances, and can contaminate GROUNDWATER or SURFACE WATER. In addition, biological CONTAMINANTS, such as BACTERIA, can be present in leachate. Various gases may be produced in landfills also, such as CARBON DIOXIDE, METHANE, and HYDROGEN SULFIDE. Several of these landfill gases can be hazardous if they accumulate. Methane may migrate from a landfill to nearby homes or underground structures, such as SEWERS, where explosions can occur. Hydrogen sulfide is a foul-smelling and poisonous gas that can accumulate underground and pose a hazard in SEWERS or basements.

Sanitary landfills are lined on the bottom with dense clay (termed a "clay liner") and layers of plastic that act as a barrier to prevent leachate from entering groundwater. The daily spreading of soil over the landfill also prevents rain from leaching the dumped wastes. Heavy rainfall can cause serious EROSION of landfills and polluted RUNOFF. Often, a leachate collection system is built around a sanitary landfill to capture leachate before it escapes to the ENVIRONMENT. Any leachate that collects at the bottom of the landfill is pumped out and then treated to remove contaminants. The left-over sludgelike material can be burned, used as FERTILIZER, or if it contains toxic materials can be transported to a HAZARDOUS WASTE site.

Landfills are a large source of AIR POLLUTION emitted by the decomposing garbage. The major gas emitted is methane, which contributes to the formation of SMOG and is also highly explosive. To collect the methane in sanitary landfills, a collection system of pipes are built into the landfill site to allow gases to be vented into the ATMOSPHERE or collected and used as a FUEL. This feature prevents the migration of hazardous gases. In 1996, the EPA required the largest landfill operations, about 280 sites, to reduce their air pollutants by 90%. To achieve this, operators will need to drill WELLS into the sites to collect the gases before they are released into the atmosphere.

"Landfill caps," specially designed covers, are constructed to seal the tops of landfills. The cap reduces leachate production by preventing water from entering the landfill, reduces erosion, and helps control landfill gases. Sometimes the finished landfill is purchased by local communities and used as a recreational site or golf course. *See also* LOVE CANAL.

EnviroSources

Environmental Industry Associations. *Landfill Capacity in North America* Report: 4301 Connecticut Ave., NW, Suite 300, Washington, DC 20008.

saturated zone. *See* zone of saturation.

savanna, A tropical or subtropical HABITAT that consists mainly of grasses, bushes, and scattered small trees. The word "savanna" is from the Spanish *zavana* (now spelled *sabana),* meaning a "treeless plain." Savannas are located between the edges of FORESTS and dry regions. The year-round temperatures are high, ranging from 38°C to 45°C (70°F to 85°F) and the precipitation, most of which falls only once or twice a year, averages about 12 centimeters (30 inches) in a year. Savannas are located primarily in Africa, and they cover large areas of that continent. Some of the most popular WILDLIFE REFUGES in the world include Tanzania's SERENGETI NATIONAL PARK and Kenya's TSAVO NATIONAL PARK, both well-known African savannas. Other savannas are in South America, Australia, and parts of Asia.

Grasses and shrubs play an important role in the ECOSYSTEM of the savanna. The short and tall grasses, underground stems and roots, and other vegetation support large groups of grazing HERBIVORES including giraffes, antelopes, ELEPHANTS, wildebeest, gazelles, and ZEBRAS. During the year, many of these animals migrate in herds numbering in the thousands in search of food and water. In turn, the herbivores provide food for predators that include CHEETAHS, lions, leopards, and wild dogs, as well as, SCAVENGERS such as vultures and hyenas. The savanna also supports a variety of birds, lizards, and INSECTS. The Australian and African savanna are home to the ostrich, the largest bird, which can weigh more than 160 kilograms (350 pounds) and reach a height of 2 meters (6 feet).

Savannas and the animals that live there are in danger because of overgrazing of domestic herds, POACHING of elephants and other animals, excessive harvesting of FUEL WOOD that can lead to DEFORESTATION, and lack of precipitation. The increase in human POPULATION in the savanna has led to a loss of wildlife habitat forcing more animals to live in smaller and smaller areas. *See also* ECOTOURISM; FIRE ECOLOGY; GRASSLAND; PAMPAS; PRAIRIE; PREDATION; WATER CYCLE.

Savannah River Site, A U.S. government facility (300 square miles) that was constructed during the early 1950s to produce the basic materials, primarily tritium and PLUTONIUM-239, used in the production of NUCLEAR WEAPONS. The Savannah River Site (SRS) is located in western South Carolina bordering the Savannah River. Five NUCLEAR REACTORS were built on the site to produce plutonium. Also built were support facilities including two chemical separations plants, a heavy water extraction plant, and WASTE MANAGEMENT FACILITIES. Today, the five nuclear reactors are permanently shut down. The DEPARTMENT OF ENERGY (DOE) and private contractors are focusing on the environmental cleanup and waste management of nuclear waste materials on this site.

During the 1950s, the SRS produced unusable byproducts such as RADIOACTIVE WASTE. The HIGH-LEVEL RADIOACTIVE WASTE, about 130 million liters (35 million gallons), is stored in waste tanks on site. The Defense Waste Processing Facility (DWPF) will bond the radioactive elements in borosilicate glass, to stabilize and store radioactive material. (*See* vitrification.)

In addition to the high-level wastes handled by the DWPF, other wastes at the site are low-level solid and liquid radioactive wastes; transuranic wastes, which are contaminated with ISOTOPES that emit alpha particles and have decay rates and concentrations exceeding specified levels; HAZARDOUS WASTES, which are any toxic, corrosive, reactive, or ignitable material that could affect human health or the ENVIRONMENT; mixed wastes, which contain both hazardous and radioactive components; and sanitary wastes, which are neither radioactive nor hazardous.

There are now over 400 waste and GROUNDWATER units included in the site's environmental restoration program. Waste sites range in size from a few meters to tens of hectares and include basins, pits, piles, burial grounds, landfills, tanks, and groundwater contaminations. The SRS is regulated under the RESOURCE CONSERVATION AND RECOVERY ACT (RCRA) and the COMPREHENSIVE ENVIRONMENTAL RESPONSE, COMPENSATION, AND LIABILITY ACT (CERCLA). To date, more than 30 hectares (80 acres) of land have been treated and closed. Also, almost 7.4 billion liters (2 billion gallons) of groundwater have been treated, with more than 135,000 kilograms (300,000 pounds) of ORGANIC compounds removed. Even though there has been some success in cleaning up the SRS, there is a tremendous amount of environmental restoration work that still needs to be done. This cleanup process is expected to take decades.

The site also houses the Savannah River Ecology Laboratory (SREL), an environmental research center operated for the DOE by the University of Georgia. The SREL, founded in 1951, conducts research programs in basic and applied ECOLOGY as well as on the effects of various CONTAMINANTS such as HEAVY METALS on groundwater and SURFACE WATERS and the effects of DROUGHT and flooding on BIOLOGICAL COMMUNITIES. Other research at the SREL focuses on fresh-water and terrestrial ECOSYSTEMS.

EnviroSources

Savannah River Ecology Laboratory: Drawer E, Aiken, SC 29802.
Savannah River Site Web site: www.srs.gov.

scavenger, An animal that obtains nutrients and ENERGY by feeding on animal wastes, such as dung, or the remains of PLANTS and animals. Unlike predators, scavengers do not kill the animals upon which they feed but instead use carrion for food. Scavengers exist in both aquatic and terrestrial ECOSYSTEMS and serve the vital function of cleansing the ENVIRONMENT of decaying ORGANIC matter. As they feed, scavengers break down complex organic substances and return simpler substances to the environment. Common aquatic scavengers include snails and the fishes commonly called "catfishes." Terrestrial scavengers include foxes, the hyenas and vultures of the African plains, and the larvae, or maggots, of various SPECIES of flies. The matter scavengers use as food is called DETRITUS. Thus, scavengers, along with the MICROORGANISMS known as DECOMPOSERS, are sometimes called detritus feeders. *See also* DECOMPOSITION; DETRIVORE; FOOD CHAIN; FOOD WEB; PREDATION; SAVANNA.

Schumacher, Ernst Friedrich (1911–1977), A German-born English economist and philosopher who proposed small-scale technologies for economic development. His international best-selling book, *Small Is Beautiful: A Study of Economics as Though People Mattered* (1973), highlighted Schumacher's ideas that developing countries can do quite well using small-scale and low-cost production methods with local materials and labor. Schumacher believed that developed countries, or the Western nations, placed too much emphasis on economies that require high production, high consumption, labor-saving efforts, and finite resources. Schumacher's other books include *Economic Development and Poverty* (1966), *A Guide for the Perplexed* (1978), and *Good Work* (1979).

Schumacher was a popular lecturer and the recipient of many awards and honorary degrees from several universities. Schumacher helped establish the Intermediate Technology Development Group in London which staffs experts in the field of farming and ENERGY throughout the world.

EnviroSources

Beaud, Michel. *Economic Thought Since Keynes: A History and Dictionary of Major Economists.* New York: Routledge, 1997.
Schumacher, E.F. *Small Is Beautiful*: *Economics As if People Mattered.* Point Roberts, WA: Hartley & Marks Publishers, Inc., 1999.

scrubber, A device that removes SULFUR DIOXIDE and other particulates from FLUE GASES before they are emitted from tall stacks. Scrubbers are installed in COAL-burning power plants, asphalt industries, concrete factories, and a variety of other facilities that emit sulfur dioxides, HYDROGEN SULFIDES, and other gases. In the scrubber process, particulates, vapors, and gases are circulated and passed through a liquid solution containing lime. The lime reacts with the sulfur dioxide and forms SULFATE compounds and other residue that is collected and disposed of in landfills. *See also* ACID RAIN; CATALYTIC CONVERTER; ELECTROSTATIC PRECIPITATOR.

Scrubber

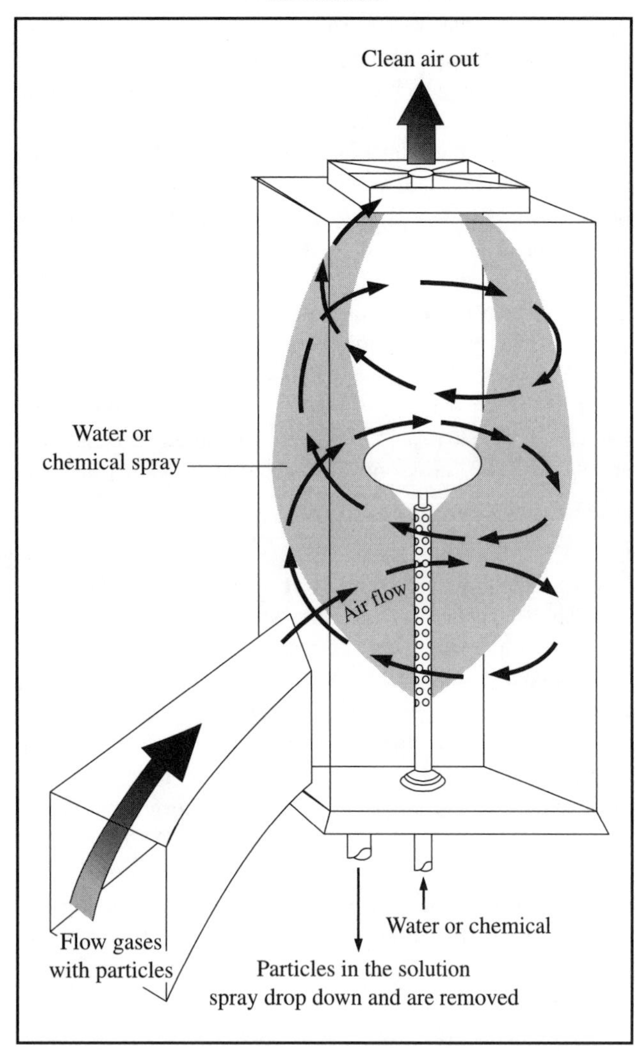

This illustration demonstrates how a typical scrubber is used to remove particulates from flue gases before they are emitted into the atmosphere.

sea level rise, An increase of OCEAN surface levels that is related to CLIMATE change. Climate studies have documented that the surface of the ocean has risen 10–25 centimeters (4–10 inches) in the last 100 years. Presently, ocean levels are rising about 1–2 millimeters (a fraction of 1 inch) each year. Scientists predict they will continue to rise in the twenty-first century.

According to the Information Unit on Climate Change, forecasts of rising ocean levels are based on climate model studies that show that Earth's average surface temperature may increase between 1.5°C and 4.5°C (1°F–7°F) over the next 100 years. Higher temperatures will increase the amount of water in the oceans due to the melting and shrinking of ice sheets and GLACIERS and the thermal expansion of ocean waters. Based on the climate models, some researchers project that ocean levels will rise 15–95 centimeters (6–37 inches) by the end of the twenty-first century.

Major changes in sea level rise will cause disruptions in human activity by displacing millions of people who live in densely populated coastal areas around the world. The first effects will be substantial EROSION of many coastlines and beaches, destroying coastal water HABITATS, and forcing SPECIES to migrate inland or to higher locations or be faced with elimination. SALT-WATER INTRUSION will increase in coastal areas when salt water moves into fresh water AQUIFERS. ECOSYSTEMS such as tidal WETLANDS, MANGROVE FORESTS, ESTUARIES, and fresh- and salt-water MARSHES will suffer serious losses as a result of erosion and flooding. Low-lying countries such as the Netherlands, Bangladesh, Indonesia, and Thailand could be inundated with major FLOODS. Productive farmlands along major rivers such as the Ganges, Indus, Mississippi, Niger, Nile, Yangtze, and Mekong would be affected and could suffer serious losses.

Most scientists agree that sea level rise is caused by GLOBAL WARMING. Others believe the evidence is still not conclusive. To find answers and ways to forecast sea level rise, the U.S. ENVIRONMENTAL PROTECTION AGENCY has been conducting computer model projections of sea level rise since 1980. The National Science Foundation and the DEPARTMENT OF ENERGY is sponsoring research that will increase the knowledge of ice movements and changes. The NATIONAL OCEANIC AND ATMOSPHERIC ADMINISTRATION (NOAA) is using satellites to measure tides. The NOAA is also planning to measure sea level rise through satellite-based remote sensing. Other public and private organizations are also conducting their own research. More time, research, and field work, however will be needed to make accurate sea level forecasts so that countries can prepare to deal with this potential environmental problem. *See also* LANDSAT.

EnviroSources

Greenpeace International (climate) Web site: www.greenpeace.org/~climate.

United Nations Intergovernmental Panel on Climate Change Web site: www.ipcc.ch.

U.S. Environmental Protection Agency (global warming) Web site: www.epa.gov/globalwarming.

sea turtles, Marine REPTILES of the order Chelonia that generally live in tropical and subtropical waters of the Atlantic, Pacific, and Indian Oceans. There are seven SPECIES of sea turtles: the green, the leatherback, the loggerhead, the flatback, the hawksbill, the olive ridley, and the Kemp's ridley. Of these, the green turtle, the hawksbill, the Kemp's ridley, and the olive ridley are identified as ENDANGERED SPECIES by the U.S. FISH AND WILDLIFE SERVICE (FWS) and thus are protected by law.

Sea turtle POPULATIONS are in decline for a variety of reasons. PREDATION is one of the major threats to the animals. Like most other reptiles, sea turtles lay their eggs on land. Often female turtles migrate hundreds of miles through the ocean to lay and bury their eggs on sandy beaches. The eggs are then left behind as the female turtle returns to the OCEAN. Many unattended eggs fall victim to predators such as foxes and ocean birds. Young turtles that emerge from the remaining eggs, immediately begin a journey to the ocean. These young turtles risk being eaten by land predators before reaching the water and also face the risk of predation by sharks and other fishes once they reach the water.

The Kemp's ridley sea turtle, pictured here, is identified as an endangered species by the U.S. Fish and Wildlife Service and thus is protected by law.

Humans also have contributed to the decline in sea turtle populations. In heavily populated areas, many sea turtle hatchlings become victims of AUTOMOBILES as they are drawn toward roadways by the bright lights of streets or automobiles. In some parts of the world, sea turtles and their eggs are hunted by people. The eggs are a valuable food in some cultures; adult turtles are hunted for their meat, shells, and skins, which are used to make leather.

Another danger to sea turtles are fishing nets. Many sea turtles are killed each year as they become entangled in the nets of COMMERCIAL FISHING vessels. GARBAGE discarded into oceans, such as plastic bags and the plastic rings that hold beverage cans together, are often eaten by sea turtles that mistake these objects for the jellyfish upon which they feed. Once ingested, these objects become lodged in the digestive systems of the animals, preventing them from getting the food they need. As a result, the animals die. Chemical POLLUTANTS and infections caused by VIRUSES also are responsible for the deaths of many turtles.

One of the major threats to sea turtles is destruction of their nesting HABITAT. As people construct houses or other structures along beaches, many of these areas become unsuitable as breeding grounds or nurseries. Many organizations and governments worldwide have become involved in protecting beach areas commonly used by sea turtles in an effort to increase turtle populations. For example, two beaches in Florida have been established as national WILDLIFE reserves, primarily for the purpose of protecting sea turtles. One of these, the Archie Carr National Wildlife Reserve, is a 12-hectare (29-acre) area set aside for protection of the green sea turtle and the loggerhead sea turtle. The same species also receive protection at the 397-hectare (980-acre) Hobe Sound National Wildlife Reserve. Costa Rica has established the world's largest green turtle refuge with its protected beach in the Tortuguero National Park. In the Virgin Islands, 132 hectares (327 acres) of beach have been set aside as the Sandy Point National Wildlife Reserve for the protection of leatherback turtles. *See also* CONVENTION ON INTERNATIONAL TRADE IN ENDANGERED SPECIES OF WILD FLORA AND FAUNA; ENDANGERED SPECIES ACT; ENDANGERED SPECIES LIST; RED LIST OF ENDANGERED SPECIES; WORLD CONSERVATION MONITORING CENTRE; WORLD CONSERVATION UNION.

EnviroSources

Elsa Wild Animal Appeal: (818) 761-8387.

Sea Turtle Restoration Project of the Earth Island Institute: (415) 788-3666.

seals and sea lions, Pinnipeds (fin-footed MAMMALS) of the families Otariidae (sea lions and fur seals, or eared seals) and Phocidae (true seals). Seals and sea lions, along with walruses, are marine mammals. As such, all SPECIES of seals and sea lions, regardless of their POPULATION size, receive protection in U.S. waters under the MARINE MAMMAL PROTECTION ACT of the United States.

Thirty-three different species of seals and sea lions exist around the world, in the ARCTIC, ANTARCTICA, temperate zones, and the tropics. Their global population sizes vary widely by species, from the Caribbean monk seal (500 individuals) to the southern elephant seal (600,000), to the ringed seal (almost 7 million).

Seals and sea lions all have four limbs that are modified to flippers. Some species have unusual diving abilities. The Weddell's seal of Antarctica, for example, has been known to remain at depths of nearly 610 meters (2,000 feet) for as long as 25–35 minutes. The seal's pulse rate slows when it submerges, reducing its need for OXYGEN. Although most of their lives are spent in the water, pinnipeds, unlike WHALES and DOLPHINS, also are dependent on land. Pinnipeds come ashore periodically to rest and bask in the sun, and at least once each

The sea lion is protected under the Marine Mammal Protection Act of the United States.

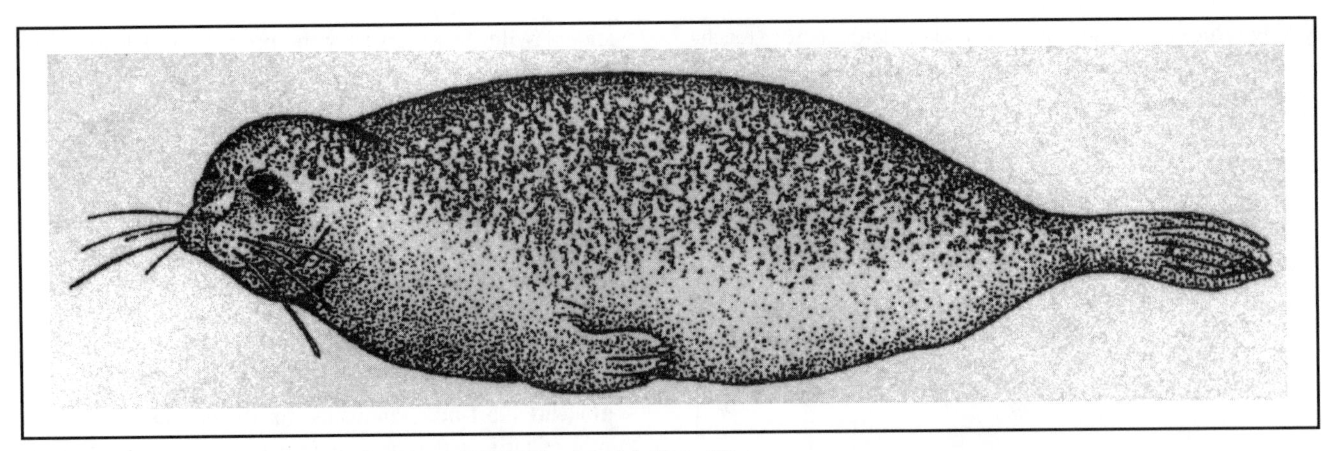

The harbor seal is protected under the Marine Mammal Protection Act of the United States.

year most species congregate on beaches or sea ice to breed and give birth. A pinniped's body is insulated with a thick layer of blubber located just under the skin. In addition to reducing heat loss to cold sea water, blubber contributes to a pinniped's streamlined shape and stores ENERGY.

Pinnipeds evolved from land-dwelling mammals that eventually returned to the sea. The earliest pinniped fossils date from the early Miocene epoch, about 20 million years ago. Although much of pinniped evolution remains unknown, there is general agreement among scientists that seals, sea lions, and fur seals evolved from a bear-like CARNIVORE.

Sea lions and fur seals have small external ear flaps and long front flippers that are usually hairless. Their hind flippers are large, paddle-like, webbed, and hairless, with nails on the center three digits. Sea lions and fur seals can turn their hind flippers forward for movement on land. By contrast, true seals have tiny ear holes with no external ear flaps. Their front flippers are short, blunt, and covered with hair, and the digits have claw-like nails. Their hind flippers are short, paddle-like, webbed, and covered with hair, with nails on all digits. True seals cannot turn their hind flippers forward. On land, they move by belly hopping without the support of their limbs. The smallest true seals are about 1.2 meters (4 feet) long and weigh about 90 kilograms (200 pounds). The largest seal, and the largest pinniped, is the elephant seal; adult males may reach 6 meters (20 feet) in length and weigh 3,700 kilograms (8,160 pounds).

All pinnipeds are carnivorous, feeding on fishes, cephalopods such as squids and cuttlefishes, and small CRUSTACEANS. Some species also feed on birds and mammals, including other seals. True seals usually leave the water daily and do not travel far from land. Eared seals, especially fur seals, migrate extensively across the OCEAN, remaining at sea for long periods of time. Most pinnipeds are gregarious; they travel in small groups or large herds and often rest together on land. Large numbers congregate at traditional breeding grounds, forming rookeries that sometimes contain thousands of individuals.

The meat, hides, and blubber of adult seals and the fur of certain pups once supported large-scale commercial sealing operations, particularly in Europe and North America, and as a result certain species were brought close to EXTINCTION. In 1972, the United States prohibited the hunting of seals in its waters. In Canada and Norway, controversial commercial hunts of harp seal pups—which were clubbed to death for their white fur—continued to take place each year. Protesters, associated with GREENPEACE and other organizations, mobilized to stop these hunts, and in 1987, Canadian legislators banned the killing of seal pups, although they continued to allow the killing of adult seals by using rifles. Elsewhere during the twentieth century, the commercial sealing industry has declined, though subsistence hunting by local peoples continues. In addition to hunting, many seals also are killed as a result of contact with toxic POLLUTANTS such as those associated with OIL SPILLS, in boating accidents, or from ingesting or becoming entangled in plastics that are discarded into the water by careless boaters. Ingestion of such materials disrupts the digestive processes of seals, ultimately leading to their deaths. Today, the Steller sea lion, Caribbean monk seal, Guadalupe fur seal, and Hawaiian monk seal all are recognized as threatened or ENDANGERED SPECIES. *See also* NATIONAL MARINE FISHERIES SERVICE; THREATENED SPECIES.

EnviroSources

Katona, Steven K. *A Field Guide to Whales, Porposies, and Seals from Cape Cod to Newfoundland.* Washington, DC: Smithsonian Institution Press, 1993.

Laws, R.M. *Antarctic Seals: Research Methods and Techniques.* Cambridge, UK: Cambridge University Press, 1993.

Momatiuk, Yva, and John Eastcott. "Keepers of the Seals." *Audubon* 100, no. 2 (March–April 1998): 46–53.

National Oceanic and Atsmospheric Administration, National Marine Fisheries Service, Office of Protected Resources Web site: www.nmfs.gov/prot_res/pinniped/pinniped.html.

National Oceanic and Atmospheric Administration, National Marine Mammal Laboratory Web site: nmml01.afsc.noaa.gov.

Riedman, Mariane. *The Pinnipeds: Seals, Sea Lions, and Walruses.* Los Angeles: University of California Press, 1991.

U.S. Fish and Wildlife Service, Office of Endangered Species Web site: www.fws.gov/r9endspp/endspp.html.

second law of thermodynamics. *See* laws of thermodynamics.

secondary consumer. *See* consumer.

secondary standards, National Ambient Air Quality Standards (NAAQS) that are designed to prevent environmental degradation, such as damage to soils, crops, water, wildlife, and property and transportation hazards. The NAAQS established by the Environmental Protection Agency contains emissions standards for six air pollutants known as criteria pollutants. They include ozone (O_3), carbon monoxide (CO), sulfur dioxide (SO_2), lead, nitrogen oxide (NO), and suspended particulates. *See also* air pollution; particulate matter.

secondary succession. *See* ecological succession.

sediment. *See* sedimentation.

sedimentary rock. *See* rock cycle.

sedimentation, The process of depositing sediment (organic or inorganic solids) in new places. Air, water, and ice that is in motion can carry with it large amounts of solid matter which may then be deposited in a location far from its source. Gravity also is responsible for the deposition of rock, soil, sediment, and other solid matter in new locations. These new deposits help to build up Earth's surface; however, sedimentation can create problems in the environment by changing the physical features of an area, thus forcing organisms that live in the area to adapt to these changes or possibly risk extinction. Sedimentation can be particularly destructive when it occurs quickly and involves large amounts of solid matter, as occurs during a mudslide or avalanche. These mass movements of matter, resulting from the erosive action of gravity, can bury everything in its path, greatly altering habitat and killing organisms.

Sedimentation also is a problem in aquatic ecosystems. Sediments that accumulate in aquatic ecosystems may fill in these ecosystems, making the water more shallow. As the water becomes more shallow, plants that could not previously grow in the ecosystem may find a habitat that supports their growth. In time, if sedimentation continues and nonnative plants continue colonizing the area, the ecosystem may be completely changed in form from an aquatic ecosystem to a terrestrial ecosystem, through the process of ecological succession. Even if ecological succession does not occur, an increase in sediment in water may suspend the ability of sunlight to penetrate the water, which can affect the ability of the producers in the water to carry out photosynthesis and can also affect water temperature. In addition, tiny particles that enter the water may be ingested by organisms or enter their respiratory systems and interfere with the organism's ability to carry out these vital life processes. *See also* algal bloom; biochemical oxygen demand; decomposition; erosion; food chain; rock cycle; siltation; weathering.

seed banks. *See* germ plasm banks.

seed-tree cutting, A type of selection harvesting or selection cutting in which mature, economically desirable trees are left standing in an area that has been cleared of other trees. The mature trees, called seed trees, are left uncut as a source of seeds to revegetate the area. Seed-tree cutting is a sustainable forestry practice designed to replenish tree populations in forest areas where trees are harvested as a source of timber. *See also* clear-cutting; conservation; deforestation; germ plasm bank; shelterwood harvesting; silviculture; slash and burn; sustainable development; sustained yield; tree farm.

seepage, Water slowly emerging from the ground along a line or surface. Seepage from groundwater flows into streams, lakes, and rivers; however, if the water table is lowered due to excessive groundwater removal, the seepage will dry up as well as the streams or rivers it is feeding. Seepage is different from a spring where the water emerges from a localized spot. *See also* artesian well; subsidence; zone of discharge.

selection cutting, A sustainable forestry practice, sometimes called selection harvesting, in which only trees meeting certain criteria are harvested, so as to promote the growth of other trees. Unlike clear-cutting, which involves the removal of most or all of the trees from an area, the smaller number of trees removed through selection cutting poses less damage to the environment by minimizing habitat loss and placing soil at less risk of erosion. Selection cutting practices usually involve harvesting trees of a certain age or a certain species, or those that are diseased. The removal of such trees helps promote the growth of younger and healthier trees by reducing competition. At the same time, it encourages the growth of large numbers of trees of differing ages and species, helping to achieve sustainable development. *See also* blight; Dutch elm disease; seed-tree cutting; shelterwood harvesting; slash and burn; sustained yield; tree farm.

selenium, A naturally occurring element that is released into the environment in wastewater derived from irrigation runoff and in wastes from petroleum refineries. Selenium from irrigation runoff can harm aquatic ecosystems. Depending upon its concentration and the organisms present in water, selenium may kill fishes and aquatic wildlife directly or cause deformities in their offspring.

Scientists are working to develop crops that can remove selenium from runoff. The use of plants to cleanse toxic wastes is called phytoremediation. Some crop plants, such as broccoli, cabbage, and rice, and some marsh plants have been discovered to remove selenium from wastewater and store it in their leaves and seeds or release it as a gas into the atmosphere. The selenium-containing plants can be harvested for use as cattle feed because selenium is needed in trace amounts in the animals' diet. Some species of FUNGI also have been useful in removing excess selenium from soil. *See also* AGRICULTURAL POLLUTION; SEWAGE.

Sellafield, A NUCLEAR REACTOR situated on the northwest coast of England that reprocesses spent nuclear FUEL from other nuclear plants. The Sellafield Nuclear Plant, operated by the British government, generates approximately one-quarter of the ELECTRICITY used in the United Kingdom, and produces nuclear material used to manufacture NUCLEAR WEAPONS. The Sellafield facility presents a controversial environmental dilemma: It serves important economic and technological functions for the United Kingdom's ENERGY programs and at the same time poses great potential environmental and human health risks.

Evidence exists that RADIOACTIVE WASTE from Sellafield has been disposed of in the Irish Sea since the 1950s. Older nuclear power plants in the United Kingdom have delivered SPENT FUEL rods to the Sellafield facility, where the FUEL RODS are dissolved in an ACID solution. Radioactive wastes generated by this process are stored at the Sellafield facility. A nuclear waste management program has been operating at Sellafield since the early 1990s.

The future of the Sellafield facility is the subject of much debate in the United Kingdom and in nearby Ireland. The public has shown great concern about the safety of the Sellafield power plant in the event of a nuclear accident: It has a history of incidents involving nuclear materials since 1957, when a fire occurred in the reactor core of the plant releasing POLONIUM and iodine-131 into the ATMOSPHERE. As many as 30 people may have perished from the accident. Records also indicate that SOLVENT mixtures containing radioactive wastes were discharged into the Irish Sea in 1983.

The Irish Sea is contaminated by radioactive waste such as CESIUM and PLUTONIUM, which contain relatively high levels of RADIATION. Through analysis of biological samples, scientists have confirmed that fish and marine PLANTS from the Irish Sea contain higher concentrations of radioactive elements than would naturally occur. The Republic of Ireland has expressed concern about the potential health effects of radioactive releases into the OCEAN. Of further concern to the Irish is the fact that Ireland receives no direct energy benefit from the Sellafield plant, while suffering from adverse effects attributable to the facility. *See also* CONTAMINANT.

septic tank, A specially designed underground tank used to collect and hold domestic SEWAGE and that is not connected to a SEWER line. Household WASTEWATER is piped to the underground tanks where ANAEROBIC ORGANISMS decompose the material. The liquid portion of the wastes flow out of the tank into the ground through perforated pipes arranged in a LEACHING FIELD. The SLUDGE in the tank is pumped out periodically. Defective septic tanks and inadequate leaching fields may cause GROUNDWATER or SURFACE WATER pollution. *See also* DECOMPOSER; DECOMPOSITION.

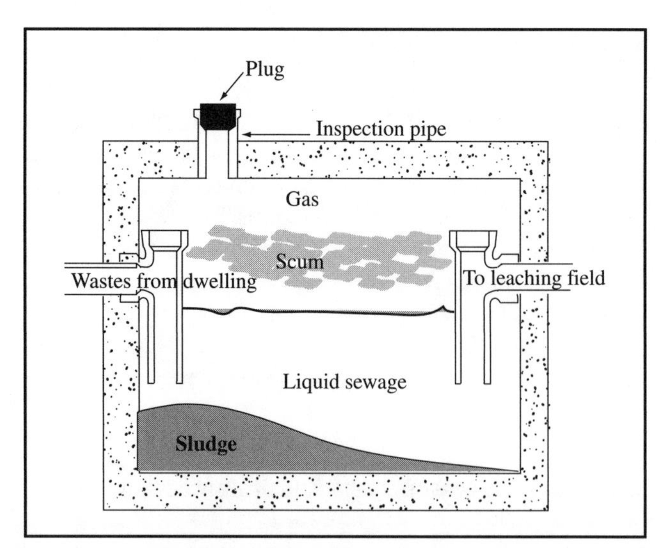

The septic tank is designed as an underground container used to collect and hold domestic sewage. The excessive water flows out of the outlet into a leaching field. The tank is not connected to a sewer line.

TechWatch

One new type of septic tank consists of three parts: a septic tank, a recirculation tank, and a sand filter. The new system cleans WASTEWATER before it is allowed to flow into the LEACHING FIELD. The wastewater from the septic tank is pumped into an underground recirculation tank. At the outlet of the recirculation tank, another pump sprays the wastewater into the sand filter. Then the water seeps through the sand into a pit. The water from the pit is channeled through the leaching field in a purer form. Wastewater experts believe the three-part septic system is less expensive, more economical to operate, and less harmful to the ENVIRONMENT than conventional septic systems.

EnviroTerm

septage: Wastes removed from cesspools and septic tanks.

Serengeti National Park, The largest and oldest of Tanzania's 12 national parks. It is an area where POACHING can be a problem, particularly for ELEPHANTS and the black RHINOCEROS. However, every effort is being made by the Tanzanian government to protect these animals and conserve this unique park.

The Serengeti National Park is located in the vast subtropical GRASSLAND or SAVANNA in east Africa and is one of the most famous WILDLIFE parks in Africa. Its name is derived from the Maasai word, *siringet*, meaning "wide open space," or "extended place." Established in 1951, the park is about 14,000 square kilometers (5,400 square miles) or about the size of the state of Connecticut. It is located near Lake Victoria, one of the largest lakes in the world. The Serengeti's CLIMATE is usually warm and dry. The main rainy season is from March to May, with short rainy seasons from October to November. The yearly rainfall can average between 500 millimeters (20 inches) to 1,200 millimeters (48 inches).

The Serengeti Research Institute, founded in 1962 has provided valuable data for the management and conservation of this park. Probably more is now known about dynamics of the Serengeti than any other ECOSYSTEM in the world.

Lions are the major predators in Tanzania's Serengeti National Park. *Credit:* Jane E. Mongillo

The Serengeti park has a unique combination of diverse HABITATS which support more than 30 SPECIES of large HERBIVORES and nearly 500 species of birds. The Serengeti savanna is home for more than two million wildebeest, half a million Thomson's gazelle, and a quarter of a million ZEBRA. The park has the greatest concentration of plains game in Africa. Hundreds of thousands of wildebeest and zebra provide a unique spectacular for tourists as the animals migrate from the southeastern plateaus of the Serengeti northward in search of food and water. Apart from the rhinos, which have been decimated by poachers, the wildebeest and buffalo populations have multiplied, benefiting the main predators—lion, CHEETAH, and hyena. But the ecosystem is delicate and volatile, easily affected by DROUGHT, DISEASE, or overgrazing. Other animals include elephants, giraffes, warthogs, jackals, Thomson's gazelles, ostriches, lizards, ORYX, eagle leopards, hippopotamuses, and vultures. *See also* ADDO ELEPHANT PARK; EVERGLADES NATIONAL PARK; TSAVO PARK; YELLOWSTONE NATIONAL PARK; YOSEMITE NATIONAL PARK.

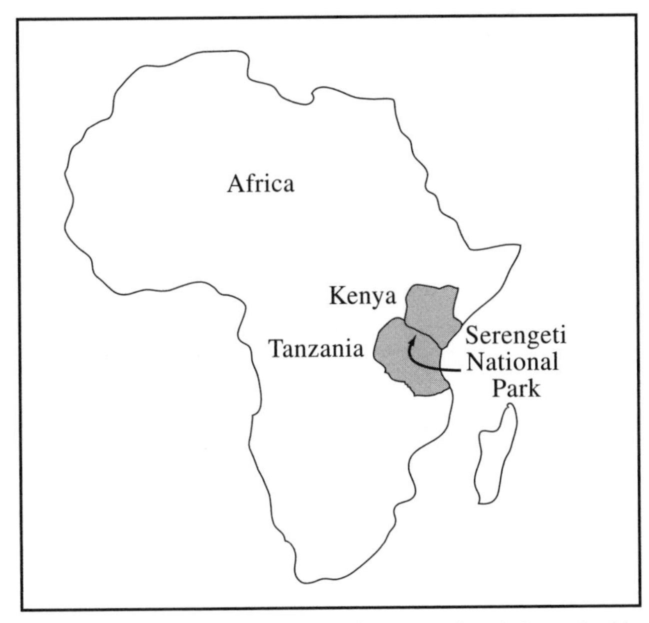

The Serengeti National Park is located in a vast subtropical grassland in Tanzania near the Kenya border.

sewage, WASTEWATER, including solid, and semisolid wastes it carries, derived from homes, businesses, and SURFACE WATER. Sewage is generally classified into three categories depending upon its source: Domestic sewage is wasterwater derived from homes and apartments; industrial sewage is wastewater derived from facilities engaged in manufacturing and chemical processes; and storm sewage is the RUNOFF from roads and land resulting from rainwater.

If released into the ENVIRONMENT, raw sewage can be harmful to the health of humans and other ORGANISMS. For this reason, domestic and industrial sewage generally are collected in public SEWER systems that transport the waste to SEWAGE TREATMENT PLANTS. At such facilities, much of the water contained in the sewage is reclaimed and returned to ponds or streams in the environment. Sewage from homes may also be collected in a SEPTIC TANK, where wastes are broken down by the activities of MICROORGANISMS. Generally, domestic sewage is composed of between 95 and 99% pure water by weight. The remaining weight is due to dissolved and suspended impurities, including microorganisms, ORGANIC matter, and various inorganic elements and compounds. The microorganisms in domestic sewage are primarily COLIFORM BACTERIA derived from the human intestinal tract; however, sewage also may contain disease-causing microorganisms. (*See* pathogens.) Sewage also may contain compounds of NITROGEN and PHOSPHORUS, substances that are essential PLANT nutrients. In addition to water, industrial wastewater may contain a variety of chemical compounds. The types of compounds are dependent upon the nature of the industrial processes from which the wastes originated. *See also* ADSORPTION; ALGAL BLOOM; ANAEROBIC DECOMPOSITION; CRYPTOSPORIDIUM; *ESCHERICHIA COLI*; PRIMARY, SECONDARY, AND TERTIARY TREATMENTS; SOLID WASTE.

sewage treatment plant, A facility designed for the collection, treatment, and sanitary disposal of SEWAGE (WASTEWATER) from municipal and/or industrial sources. A major role of a sewage treatment plant is to separate solids, semisolids, and other dissolved substances carried in wastewater from the water, and then clean and purify the water to allow its safe release back into the ENVIRONMENT, where it may be reused by people and other ORGANISMS.

Most municipalities in industrialized nations have some type of wastewater disposal procedures that make use of sewage treatment plants. Methods used for reclaiming water from sewage include the removal of solids from wastewater by screening, filtering, SEDIMENTATION, flotation, chemical coagulation, and FLOCCULATION, processes collectively known as primary treatment. Next, the sewage is acted on by BACTERIA and other MICROORGANISMS that convert ORGANIC matter contained in the sewage into such compounds as water, CARBON DIOXIDE, NITRATES, PHOSPHATES, and other organic and inorganic matter, a process known as secondary treatment. Following secondary treatment, the remaining water (which may contain microbes) may be chemically treated to remove any PHOSPHORUS and dissolved solids remaining in the water. In addition, the water also may be treated with CHLORINE or OZONE to kill any potential PATHOGENS present in the water. This phase of treatment, known as tertiary treatment, purifies the water, rendering it essentially microbe free, and is carried out primarily when the reclaimed water is intended to be reused by humans.

Most sewage treatment plants make use of only some of these methods in their treatment of wastewater. Once wastewater has undergone treatment and is deemed free of pathogens or other potentially harmful substances, the water generally is carried away from the plant through a series of pipes for DISCHARGE into a receiving lake or stream. In some areas, treated wastewater is discharged into the recharge area of an AQUIFER, from which it may be drawn for use in crop IRRIGATION or industrial processes. *See also* AERATION; CHLORINATION; COLIFORM BACTERIA; CRYPTOSPORIDIUM; FILTRATION; GROUNDWATER; OVERDRAFT; PRIMARY, SECONDARY, AND TERTIARY TREATMENTS; SEPTIC TANK; SEWER; SOLID WASTE; SUBSIDENCE.

sewer, A conduit that transports WASTEWATER to a disposal site, such as a SEWAGE TREATMENT PLANT, where it can be properly treated to prevent the release of potentially harmful substances into the ENVIRONMENT. Prior to the development of sewer systems, wastes, which can contain ORGANIC or inorganic POLLUTANTS, were released directly into the environment, such as a stream, where they contributed to the spread of DISEASE and were unpleasant sights that released foul odors. There are three types of sewers: Sanitary sewers carry household, commercial, and industrial wastes to a sewage treatment plant; storm sewers carry RUNOFF, along with dissolved substances, to either a receiving stream or a sewage treatment plant; and combined sewers transport wastes typically carried by both sanitary sewers and storm sewers. SEWAGE and wastewater can also be disposed of in a SEPTIC TANK. *See also* COLIFORM BACTERIA; CRYPTOSPORIDIUM; PRIMARY, SECONDARY, AND TERTIARY TREATMENTS; SEWAGE.

Most surface runoff from streets and paved parking lots is carried away by storm sewers. *Credit:* Linda Zierdt-Warshaw.

shale. *See* rock cycle.

shelterwood harvesting, A SUSTAINABLE DEVELOPMENT practice of the logging industry that involves the systematic removal of all mature and fully grown trees from a FOREST, generally, over a period of several decades. As the taller, more mature trees, called shelterwood, are removed, the shade once created by their canopies also is eliminated. Thus, younger, shorter trees receive increased exposure to sunlight, helping to promote their growth. The removal of shelterwood is an example of SELECTION CUTTING based on tree age or size. *See also* AGROFORESTRY; CLEAR-CUTTING; COMPETITION; CONSERVATION; DEFORESTATION; FORESTRY; OLD-GROWTH FOREST; SEED-TREE CUTTING; SILVICULTURE; SLASH AND BURN; TREE FARM.

Siberia, A part of Russia located in the north-central part of Asia, which covers 10 million square kilometers (4 million square miles), the combined size of the United States and India. Much of the northern region of Siberia lies on top of PERMAFROST. Siberia has a POPULATION of around 25 million and has a diverse ECOSYSTEM containing many NATURAL RESOURCES. The land mass contains thousands of rivers including some of the longest in the world. The topography includes plateaus, mountains, TUNDRA, and geysers. There are one million lakes, including LAKE BAIKAL, which holds 20% of the world's fresh-water reserves. The FORESTS cover an area of 2.1 million square kilometers (800,000 square miles). Foresters believe that 25% of the forests are original. Other natural resources include PETROLEUM, NATURAL GAS, gold, diamonds, COAL, and ferrous and nonferrous metals.

Major industrial development has taken place in Siberia, notably the new large oil and natural gas fields in western Siberia. Other industrial sites include nuclear power plants, paper mills, SMELTING furnaces, and lumber and MINING operations. Many pollution problems have emerged with the growth. RADIOACTIVE WASTES from the use of PLUTONIUM in NUCLEAR REACTORS have polluted waterways. Industrial complexes in Siberia are a major source of AIR POLLUTION. Large smelting operations emit SULFUR CONTAMINANTS and other HEAVY METALS into the air. Other environmental concerns include the massive CLEAR-CUTTING of forests and the POACHING of animals, particularly the Siberian TIGER. The tiger is declining in population and is on both the ENDANGERED SPECIES list maintained by the U.S. FISH AND WILDLIFE SERVICE and the RED LIST OF ENDANGERED SPECIES maintained by the WORLD CONSERVATION UNION. Fish harvests in Lake Baikal have declined, and changes in the water level of the lake have led to shore EROSION and lost HABITAT in nearby WETLANDS.

What concerns most Russian environmentalists is that their country, in need of boosting its economy, may be unable to provide strong measures in safeguarding Siberia's vast natural resources from unnecessary exploitation. Therefore, finding the right balance between economic growth and preserving the land and its ecosystems will be a continuing problem in the new millennium.

EnviroSource

Center for New Information Technologies Web site: www.cnit.nsk.su/univer/english/siberia.html.

Siberia

The landmass of Siberia contains thousands of rivers, including some of the longest in the world. The topography also includes plateaus, mountains, tundra, geysers, ice wastes, as well as one million lakes including LAKE BAIKAL.

sick building syndrome, A condition in which building occupants experience acute health symptoms that do not fit the pattern of any particular DISEASE, appear to be linked to time spent in a particular building, cannot be traced to any specific source, and continue for more than two weeks. (By contrast, "building-related illness" refers to a diagnosable illness that can be attributed directly to a defined indoor AIR POLLUTION source.) Comparative risk studies performed by the U.S. ENVIRONMENTAL PROTECTION AGENCY Science Advisory Board have ranked indoor AIR pollution among the top five environmental risks to public health. The National Institute for Occupational Safety and Health estimates that the problem costs tens of billions of dollars each year in health care, absenteeism, reduced worker productivity, building investigations, and building improvements.

Some WORLD HEALTH ORGANIZATION experts believe that up to 30% of new or remodeled commercial buildings have unusually high rates of occupant health and comfort complaints that may potentially be related to indoor air quality. Pollutants that contribute to poor indoor air quality include tobacco smoke, chemicals such as formaldehyde released from carpeting and foam insulation, ammonia from cleaning agents, and HYDROCARBONS from SOLVENTS and adhesives. The U.S. OCCUPATIONAL SAFETY AND HEALTH ADMINISTRATION estimates that of the 70 million employees who work indoors in the United States, 21 million are exposed to poor-quality indoor air (air that is continually recycled through an enclosed building without an adequate influx of fresh air) and millions more to environmental tobacco smoke.

Victims of sick building syndrome may complain of one or more of the following symptoms: dry or burning nose, eyes,

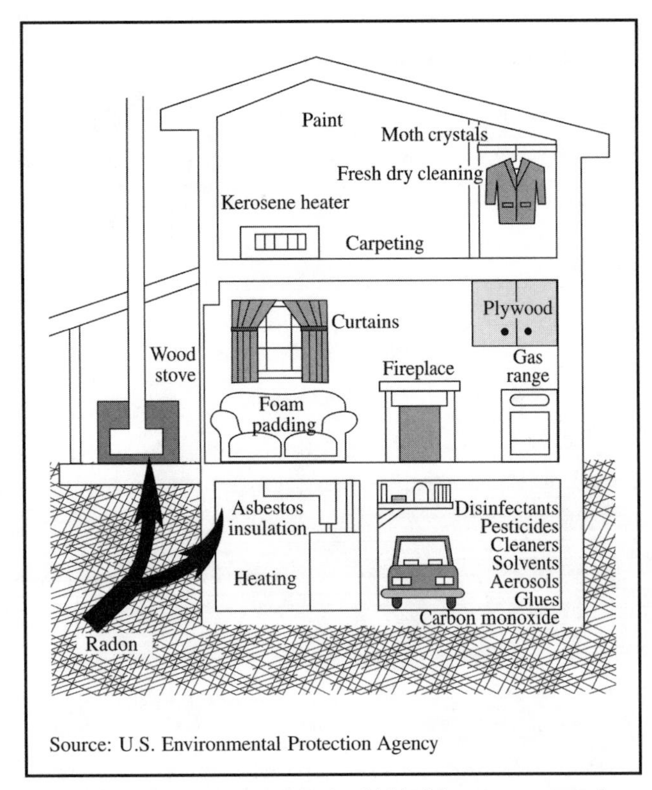

Sick building syndrome is a condition in which building occupants experience acute health symptoms. Environmental illnesses may occur from poor lighting, noise, vibrations, inadequate ventilation, and pollutants such as volatile organic compounds (VOCs) and radon gas.

and throat; sneezing; stuffy or runny nose; fatigue or lethargy; headache; dizziness; nausea; irritability; and forgetfulness. Poor lighting, noise, vibrations, inadequate ventilation, POLLUTANTS such as VOLATILE ORGANIC COMPOUNDS, and psychological stress also may cause or contribute to these symptoms.

There is no single manner in which these health problems appear. In some cases, problems begin when workers enter their offices and diminish when workers leave; in other cases, symptoms continue until the illness is treated. If large numbers of people suffer from the same health problems, which is attributed to a definite cause, the building may undergo renovation to remove the problem. Sometimes there are outbreaks of illness among many workers in a single building; in other cases, health symptoms show up only in a single worker.

Often, sick building syndrome is temporary, but some buildings have long-term problems that result when a building is operated or maintained in a manner that is inconsistent with its original design or prescribed operating procedures. Sometimes indoor air problems are a result of occupant activities or poor building design.

In attempting to "cure" sick buildings, investigators examine four basic building factors that influence indoor air quality: the occupants; the heating, ventilation, and air conditioning systems; possible pollutant pathways; and possible sources of contamination. The solution to the condition often lies in changing characteristics of a combination of these factors. *See also* ALLERGY; CONTAMINANT; MULTIPLE CHEMICAL SENSITIVITY.

EnviroSources

National Institute for Occupational Health Web site: www.cdc.gov/niosh/homepage.html.

U.S. Environmental Protection Agency, Indoor Air Quality Information Clearinghouse: P.O. Box 37133, Washington DC 20013-7133; (703) 356-4020 or (800) 438-4318.

U.S. Environmental Protection Agency (indoor air quality) Web site: www.epa.gov/iaq.

Sierra Club, A nonprofit, member-supported, public-interest organization founded in 1892 by JOHN MUIR. The Sierra Club promotes CONSERVATION of the natural ENVIRONMENT by influencing public policy decisions—legislative, administrative, legal, and electoral. The Sierra Club's mission is to explore, enjoy, and protect the wild places of Earth; to practice and promote the responsible use of Earth's ECOSYSTEMS and resources; to educate and enlist humanity to protect and restore the quality of the natural and human environment; and to use all lawful means to carry out these objectives. The Sierra Club campaigns against suburban sprawl and confined animal feeding operations and for the control of RADIOACTIVE WASTE.

Sixty chapters with 400 local groups have formed in the United States and Canada with a membership of 550,000. The National Outings program offers the chance for members to experience WILDERNESS on more than 300 trips across the United States and around the world. *See also* FRIENDS OF EARTH; NATIONAL AUDUBON SOCIETY.

EnviroSource

Sierra Club: 85 Second Street, San Francisco CA, 94105-3441; (415) 977-5500; fax 415-977-5799.

sievert, An international unit of measure for small DOSES of RADIATION absorbed in the body. The sievert (Sv) has been recommended by the International Commission on Radiation Units and Measurements to replace the rem. One sievert is equivalent to 100 rems. Radiation doses are frequently measured in millisievert (mSv), or one-thousandth (0.001) of a sievert. Each form of radiation—X-RAYS, gamma rays, neutrons—has a slightly different effect on living tissue. The average person in the United States receives about 1.5 mSv a year of natural radiation. The maximum possible dose of radiation in the workplace is 50 mSv. When doses reach 500 mSv in humans, damage can occur. Doses of 1,000–2,000 mSv can cause RADIATION SICKNESS, vomiting, and other health problems. A dosage of 5,000 mSv or more can be fatal.

EnviroTerms

rem: A unit of measure for small RADIATION doses received by humans. A millirem, or one-thousandth of a rem, is the unit most often used to measure doses. In the United States, Americans receive about 360 millirems of radiation a year mostly from natural radiation. They receive about 15% of that amount from radiation exposure, 3% from consumer materials, and about 1% from activities associated with nuclear industry.

roentgen: A unit that is used to measure large doses of X-RAYS and gamma rays.

Silkwood, Karen (1946–1974), A chemical technician at the Kerr-McGee nuclear FUELS production plant in Crescent, Oklahoma, and a member of the Oil, Chemical, and Atomic Workers' Union. Karen Silkwood also was an activist who was critical of allegedly inadequate safety provisions at the Kerr-McGee facility at which she worked. During the week prior to her death, Silkwood reportedly had been gathering supporting evidence for the union's claims that Kerr-McGee was negligent in maintaining plant safety. During the same period, she was the victim of several unexplained exposures to PLUTONIUM-239, approximately 300 micrograms of which was ultimately found in her apartment.

Karen Silkwood died in a suspicious, one-car accident in November 1974, while on her way to meet a *New York Times* investigative reporter. The circumstances of her death have been the subject of great speculation, and her story achieved worldwide fame as the subject of many books, articles, and the motion picture, *Silkwood*. Following the accident, an autopsy was performed on Silkwood at University Hospital in Oklahoma City, Oklahoma. Organs from her body also were examined by the Los Alamos Tissue Analysis Program at the request of the Atomic Energy Commission (AEC) (*see* Nuclear Regulatory Commission) and the Oklahoma City Medical Examiner. The highest concentrations of radioactive material in her body was found to be in the contents of her gastrointestinal tract, demonstrating that she had ingested plutonium prior to her death. Silkwood's case was important to the AEC's program because it was one of very few involving exposure to plutonium within 30 days before death. It also confirmed the accuracy of contemporary techniques for the measurement of plutonium in the body.

The Kerr-McGee nuclear fuel plants closed in 1975; however, Karen Silkwood's estate filed a civil suit against Kerr-McGee for the allegedly inadequate plant health and safety program that led to Silkwood's exposure to RADIATION. The first trial ended in 1979, with the jury awarding the estate $10.5 million in personal injury and punitive damages. This finding was later reversed by the Federal Court of Appeals in Denver, Colorado, which instead awarded a mere $5,000 for the personal property Silkwood lost during the DECONTAMINATION of her apartment. In 1986, 12 years after Silkwood's death, the suit was again headed for retrial before being finally settled out of court for $1.3 million. *See also* BERYLLIUM; NUCLEAR ENERGY; OCCUPATIONAL SAFETY AND HEALTH ADMINISTRATION; RADIATION SICKNESS; RADIOISOTOPE OR RADIOACTIVE ISOTOPE.

EnviroSource

Rashke, Richard. *The Killing of Karen Silkwood: The Story Behind the Kerr-McGee Plutonium Case.* Boston, MA: Houghton Mifflin, 1981.

silt. *See* siltation.

siltation, The gradual DEPOSITION of tiny particles of soil and rock (silt) in aquatic ECOSYSTEMS. Siltation is a natural process that generally occurs slowly; however, it can greatly change the characteristics of an ecosystem. For example, as siltation occurs and soil builds up on the floor of a pond, lake, or WETLAND, the water depth is changed. As the water gets more shallow, PLANTS may begin to take root and grow in the soil. If the process continues over a long period of time, the aquatic ecosystem can completely fill with new soil, allowing a terrestrial ecosystem to replace it through a process known as ECOLOGICAL SUCCESSION.

Siltation is accelerated when land surrounding aquatic ecosystems is cleared of vegetation for construction of roads, houses, or other structures. This practice loosens soil making it more vulnerable to EROSION. As the soil is carried into the water, it can adversely affect the ORGANISMS living there. Soil particles suspended in water can prevent sunlight from penetrating the water. Without sunlight, photosynthetic organisms living in the water (ALGAE, PHYTOPLANKTON, and plants) may not be able to make enough food for their survival. If POPULATIONS of these organisms diminish, populations of other organisms that use them for food also will decrease in size. In coastal marine ecosystems, disruption of the photosynthetic processes of ZOOXANTHELLAE (organisms that live in SYMBIOSIS with coral polyps) has killed large portions of some CORAL REEFS. Since these reefs provide HABITAT to a great many organisms and are among the most diverse ecosystems on Earth, scientists are very concerned about this process. In addition, the small size of silt particles allows them to easily be taken into the digestive and respiratory systems of water organisms. This can disrupt the organism's ability to carry out these vital life functions.

Siltation also frequently occurs behind DAMS as water carrying particles downstream is slowed to a point where it can no longer carry particles. To prevent the soil buildup from becoming a problem, it is often necessary to remove soil that builds up behind a dam through DREDGING. *See also* BIODIVERSITY; FOOD CHAIN; FOOD WEB; PHOTOSYNTHESIS.

silviculture, The management of a FOREST that includes the growing and cultivation of trees as a crop. Silviculture includes the planning stages, the planting of seeds or seedlings, fertilizing the land, and harvesting mature trees. *See also* TREE FARM.

sink, A part of the BIOSPHERE where chemical substances, including POLLUTANTS, are stored. A sink is a place in the ATMOSPHERE, LITHOSPHERE, or HYDROSPHERE where elements and compounds naturally collect and are taken out of the BIOGEOCHEMICAL CYCLE for a period of time. A sink, such as a FOREST or OCEAN or other HABITAT, can trap nutrients, TOXIC CHEMICALS, and other pollutants for a period of time. As an example, CARBON DIOXIDE dissolved in OCEANS can be stored 2–10 years, decreasing the amount of carbon dioxide released into the atmosphere. The ocean is a "heat sink." It may help delay GLOBAL WARMING.

sinkhole, A depression in the ground that forms when acidic GROUNDWATER dissolves away underlying layers of carbonate rocks, such as marble or limestone, causing the ground above to sink into the newly created space. A sinkhole may also form when the roof of a shallow underground cavern collapses. Sinkholes are common in areas that have a topography called Karst topography, which is characterized by abundant limestone caves, springs, and underground streams. Sinkholes also may result from human activities such as the removal of water from an AQUIFER at a rate faster than it can be replenished.

Rainwater is normally slightly acidic, containing small amounts of NITRIC ACID, CARBONIC ACID, and SULFURIC ACID. Groundwater also becomes acidic as it percolates through the soil and picks up CARBON DIOXIDE produced by ORGANISMS living in the soil. (*See* percolation.) The carbon dioxide dissolves in the water to form carbonic acid. The carbonic acid then attacks and eats away the marble or limestone. As the process continues, underground caves and passageways form. If one of these underground cavities collapses, a sinkhole forms.

In the United States, Florida and Texas are areas where sinkholes are a common geologic hazard. In fact several sinkholes form nearly every year in central Florida. When sinkholes occur, they may damage houses and other buildings, SEWAGE lines, highways, and farmers' fields. Some sinkholes are large and can severely disrupt a field's productivity. One of the largest sinkholes in the United States was formed in 1972 near Montevallo, Alabama. It measured 120 meters (370 feet) wide and 45 meters (140 feet) deep. In the past, many sinkholes were used as dumping sites for TRASH, vehicles, and other GARBAGE. This practice contributed to pollution of groundwater and is no longer allowed. Today, when sinkholes are found they are filled with soil or cement to help protect the groundwater supply. *See also* ACID MINE DRAINAGE; ACID RAIN; GEOLOGY; ROCK CYCLE; SUBSIDENCE.

slash and burn, A practice that involves cutting down large numbers of trees from a FOREST to clear the land for agricultural uses. Once cut, trees are permitted to dry before being burned, which releases nutrients contained in the wood to the soil. Slash-and-burn practices are most common in tropical RAIN FORESTS, such as in Southeast Asia, South America, and some regions of Africa. Typically, the nutrient content of soils in such areas is low because nutrients that are released to soil through processes such as DECOMPOSITION are immediately absorbed by the numerous large trees that inhabit the area. In regions where slash-and-burn methods are used, the practice often is accompanied by shifting cultivation, an agricultural process in which plots of land that have been cultivated for only a few years are abandoned, while new plots are created. Through the process of ECOLOGICAL SUCCESSION, the abandoned area is permitted to be taken over by weeds and later bushes, and may again be reclaimed many years later through slash-and-burn methods. *See also* AGROFORESTRY; CLEAR-CUTTING; DEFORESTATION; GREENBELT MOVEMENT; HABITAT; SELECTION CUTTING; SHELTERWOOD HARVESTING; SILVICULTURE; SUSTAINABLE AGRICULTURE.

sludge, A thick, slushy, mudlike solid produced as a byproduct of WASTEWATER in municipal SEWAGE TREATMENT PLANTS and SEPTIC TANKS. Sludge is a mixture of SOLID WASTES and BACTERIA removed from the wastewater at various stages of the treatment process. It can be categorized as "primary sludge" and "secondary sludge." Primary sludge is about 4% solids and 96% water. It consists of the material that settles out of wastewater in the primary SEDIMENTATION tanks, before bacterial digestion takes place. Secondary, or activated, sludge is much more liquid—about 1% solid and 99% water. Secondary sludge consists of bacteria and ORGANIC materials on which the bacteria feed. About 30% of the secondary sludge produced is returned to the AERATION tanks to assist with the biological process of SEWAGE treatment. The remaining 70% is disposed of by INCINERATION or COMPOSTING, or delivered to SANITARY LANDFILLS. Sludge can also be treated and marketed as a FERTILIZER, although environmentalists are concerned that sludge can contain TOXIC CHEMICALS and should be tested before it is used. *See also* PRIMARY, SECONDARY, AND TERTIARY TREATMENT.

slurry, A pastelike mixture of liquid and some solids that is produced in WASTEWATER treatment plants and COAL plants or is collected from farm animals. Slurry is stored in tanks or artificial lagoons or is burned in incinerators. It may also be used as a FERTILIZER on farmland. *See also* INCINERATION.

smelting, The process of using heat to extract MINERALS or metals from ore. Smelting releases SULFUR DIOXIDE EMISSIONS into the ATMOSPHERE. COPPER and nickel ores contain a high percentage of SULFUR. As a result, smelters of these ores also emit nickel and copper PARTICULATE MATTER and other TOXIC CHEMICALS into the AIR. These air POLLUTANTS can contaminate the soil and damage FORESTS as well. *See also* AIR POLLUTION; CONTAMINANT.

smog, A term originally derived from the words "smoke" and "fog" that was used to describe the cloudy haze formed from an accumulation of visible AIR POLLUTION in the ATMOSPHERE overlying many large cities. When the term "smog" was first used in England in the early 1900s, it referred primarily to the visible POLLUTANTS that were released to the AIR from the smokestacks of factories and industry. The potential health problems associated with smog became apparent in 1911 when Des Voeux, a scientist, issued a report to the Manchester Conference of the Smoke Abatement League of Great Britain Sources in which he attributed the numerous deaths that had occurred in 1909 among people of Edinburgh, Scotland, and Glasgow, England, to the smoke and fog that had plagued both cities. The recognition that the presence of smog could adversely affect the health of people in conjunction with its unpleasant appearance led to the recognition of smog as an environmental concern.

Scientists recognize two forms of smog: sulfurous smog and photochemical smog. Sulfurous smog is characterized by a high concentration of SULFUR oxides in the air. The presence of this substance is most closely associated with the combustion of FOSSIL FUELS, especially COAL, which often contains sulfur. Sulfurous smog was the first type of smog to be recognized and is most likely to occur in regions where the air overlying a city also has a high concentration of suspended PARTICULATE MATTER and moisture. Such areas include regions that often experience fog such as London, England, and San Francisco, California.

Unlike sulfurous smog, photochemical smog does not require moist air to form. This type of smog occurs with the greatest frequency in the air above urban centers that have large numbers of AUTOMOBILES on their roads and many industrial centers. Photochemical smog originates when NITROGEN OXIDES and HYDROCARBONS are emitted to the air by automobiles and other sources that burn fossil fuels. Once in the lower atmosphere, these substances undergo a photochemical reaction as they are acted upon by sunlight to produce NITROGEN DIOXIDE and OZONE, a highly toxic gas. This chemical reaction causes an atmospheric haze with a light brown color overlying the area. The substances produced during the formation of photochemical smog can lower visibility in the area and can result in health problems for humans and other ORGANISMS. For example, ozone is extremely toxic to PLANTS, which play a vital role in the cycling of OXYGEN and CARBON DIOXIDE throughout the ENVIRONMENT. In addition, photochemical smog is an irritant to the eyes and nose, and can make breathing very difficult for individuals suffering from respiratory disorders such as emphysema, lung CANCER, and asthma. *See also* ACID RAIN; AEROSOL; AFTERBURNER; AIR POLLUTION CONTROL ACT; ALTERNATIVE ENERGY RESOURCE; ALTERNATIVE FUEL; ATTAINMENT AREA; CATALYTIC CONVERTER; CHLOROFLUOROCARBON; CLEAN AIR ACT; CLIMATE; ELECTROSTATIC PRECIPITATOR; EMISSIONS STANDARDS; FLUOROCARBON; GASOHOL; GREENHOUSE EFFECT; GREENHOUSE GAS; HYDROCHLOROFLUOROCARBON; NATIONAL AMBIENT AIR QUALITY STANDARDS; NATIONAL EMISSIONS STANDARDS FOR HAZARDOUS AIR POLLUTANTS; NITROGEN CYCLE; NONATTAINMENT AREA; OZONE DEPLETION; OZONE HOLE; PETROCHEMICAL; PETROLEUM; PRIMARY AND SECONDARY AIR POLLUTANTS.

soil conservation, A variety of techniques and practices that are intended to protect soil from degradation as a result of such conditions as EROSION, DESERTIFICATION, and nutrient deficiencies. Many soil conservation methods are employed by farmers who depend upon quality soil to support the growth of crops; however, soil conservation methods also are used by state and local governments and by individuals when planning construction of new structures and roads or when landscaping properties or maintaining open space.

Soil Erosion. Erosion is the transportation of soil or other surface materials by such agents as moving water, wind, gravity, and GLACIERS. Most erosion of soil involves the movement of topsoil (the upper layer of soil) from an area by wind or running water. Some estimates indicate that in the United States alone, as much as one-third of the topsoil used for agricultural purposes is lost to erosion each year. Erosion is even more drastic in other countries, including China, Thailand, Guatemala, and Ethiopia. The loss of so much topsoil from farmland makes meeting the food needs of a growing global human POPULATION difficult. In many parts of the world, the problem is compounded by severe WEATHER conditions such as DROUGHT or FLOODS.

To combat soil erosion on agricultural lands, many farmers and governments have developed methods designed to help reduce the loss of topsoil caused by wind and moving water. These methods include the use of CONTOUR FARMING, STRIP CROPPING, TERRACING, no-till AGRICULTURE, and the use of shelter belts. In contour farming, land is plowed across a slope rather than along the slope. As water flows down the slopes, it is slowed by and collected in the furrows between the rows of crops, making the water available to the PLANTS. This method of soil conservation often is practiced in conjunction with strip cropping, the alternation of bands of crops with land covered by vegetation, the roots of which help to hold soil in place. In addition, the roots also absorb water that flows across the soil, preventing the soil from becoming water-logged.

Terracing is practiced in regions where, because of the topography of the land, farming is practiced on hillsides. To prevent water that flows down the hillside from rushing down the hill and carrying away soil, platforms, called terraces, are built into the hillside in a steplike fashion. When rainwater does flow down the hillside, it is slowed by the plant-covered terraces, allowing the water to soak into the soil, where it can be used by the plants.

Shelter belts are intended to prevent topsoil loss resulting from the action of wind. A shelter belt is a row of trees that is planted along the boundaries of a field. The foliage of the trees provides a barrier to the wind, helping to prevent it from carrying away topsoil. Solid fences also can serve this same function. No-till or low-till agriculture also helps to reduce the erosion of topsoil by wind. This process prevents erosion by eliminating the plowing and tilling of soil that loosens soil prior to planting. Loose soil is more vulnerable to the effects of wind and running water than compacted soil. Instead of tilling an entire field, no-till agricultural methods instead use machinery to create only a small opening, or hole, in the soil into which the seeds of the crops to be grown are placed. At the same time, FERTILIZER, water, and other substances that may help the plant grow may also be placed into the hole with the seed, while the soil surrounding the planted seed is left intact.

Desertification. Poor soil management practices such as overgrazing and improper IRRIGATION in semiarid regions can lead to desertification. Desertification occurs when a semiarid region is transformed into a DESERT as a result of human activities. Desertification usually begins when the plant cover in an area is removed through practices such as overgrazing, deforestation, or extensive plowing in preparation for the cultivation of crops. When this occurs, the soil loses the natural plant cover that helped hold it in place while also covering the soil in a manner that helped to retain moisture. Without this plant cover, the soil can be quickly eroded by wind or running water. Soil beneath the topsoil becomes hard and dry. In addition, the hard, dry soil begins to reflect more of the sun's heat. This in turn, drives away clouds, changing the local WEATHER patterns to create a microclimate that is hot and dry. Without moisture, the area is slowly transformed into a desert, and becomes unable to support the growth of most plants. This process of desertification occurs slowly (over a period of years) and has been observed in the United States, China, central Asia, the western portion of South America, Australia, and throughout much of the African continent.

Irrigation, the process by which water is provided to plants through means other than rainfall, can contribute to desertification. In areas where heavy irrigation occurs, much of the water that is dispersed over the land is evaporated by the sun, thus irrigation can use huge amounts of water. As the water evaporates, salts and other MINERALS that were present in the water remain in the soil. Over time, these salts and minerals build up, changing the characteristics of the soil, which will not support the growth of plants. Some estimates indicate that worldwide as much as 6 million hectares (14.8 million acres) of land each year are transformed into desert. To help slow this development, farmers in many areas have been encouraged to reduce their dependence on irrigation, graze fewer animals per acre of land, and to practice AGROFORESTRY, rather than employ methods such as CLEAR-CUTTING or SLASH AND BURN to make land available for crop development.

Nutrient Deficiencies. Each type of plant has slightly different nutrient requirements. When the same type of plant is grown on the same parcel of land year after year, the same nutrients are repeatedly removed from the soil. In time, the soil may be completely depleted of these nutrients, unless they are replaced through the use of chemical fertilizers. The use of such fertilizers helps provide crop plants with their needs but may have adverse effects on nearby aquatic ECOSYSTEMS, when fertilizers are carried off the land in RUNOFF. To prevent this problem, many farmers now alternate the types of crops planted on a particular parcel of land from year to year, making use of plants that have different nutrient requirements. In addition, plants called LEGUMES may be planted periodically to help return valuable nutrients to the soil. Legumes are plants that have nodules on their roots that provide a HABITAT for nitrogen-fixing BACTERIA. (*See* nitrogen fixation.) These bacteria are able to change NITROGEN from the AIR into nitrogen compounds that can be used by ORGANISMS.

In addition to alternating crop growth from year to year, many farmers are now reducing their use of synthetic chemical fertilizers and increasing their use of natural materials, such as manure, ORGANIC MULCHES, and compost, that provide the same benefits to soil as fertilizers. Each of these materials returns valuable nutrients to the soil as it decomposes. In addition to eliminating synthetic fertilizers, many farmers also are eliminating their use of synthetic PESTICIDES in response to growing consumer concerns about the potential long-term dangers to the ENVIRONMENT and the health of humans who consume crops grown using such substances. To reduce crop damage caused by pests, many of these farmers are making use of BIOLOGICAL CONTROLS. The growth of crops using only natural organic materials to stimulate growth is known as organic farming. This technique is widely used in many parts of Germany and is becoming more common in many parts of the United States. *See also* AGRICULTURAL POLLUTION; AGROCLIMATOLOGY; AGROECOLOGY; ALGAL BLOOM; ARAL SEA; BENNET, HUGH HAMMOND; CLIMATE; COMPOSTING; DECOMPOSITION; DEPARTMENT OF AGRICULTURE; DETRITUS; FOOD AND AGRICULTURAL ORGANIZATION OF THE UNITED NATIONS; GLOBAL WARMING; GREEN MANURE; GREENBELT MOVEMENT; HERBIVORE; HYDROPONICS; INTEGRATED PEST MANAGEMENT; LICHENS; MYCORRHIZAL FUNGI; NATURAL DISASTER; PIONEER SPECIES; ROCK CYCLE.

EnviroSources

Abromovitz, Janet, et al. *State of the World, 1998.* New York: Norton, 1998.

Bernstein, Leonard, Alan Winkler, and Linda Zierdt-Warshaw. *Environmental Science: Ecology and Human Impact.* Menlo Park, CA: Addison-Welsey, 1995.

Bourne, Joel. "The Organic Revolution." *Audubon* 101, no. 2 (March–April 1999): 64–70.

Brown, Lester, et al. *Vital Signs, 1997.* New York: Norton, 1997.

Soil Conservation and Domestic Allotment Act, A 1935 federal law that authorized the U.S. DEPARTMENT OF AGRICULTURE (USDA) to conduct soil EROSION surveys and implement erosion prevention measures. The act resulted in the establishment of the Soil Erosion Service (the precursor to today's NATURAL RESOURCES CONSERVATION SERVICE) to conduct these activities. Emphasis was given to engineering operations, cultivation methods, growth of vegetation, and other land uses as preventive measures to protect topsoil. This act also authorized the secretary of agriculture to provide financial assistance to agricultural producers for implementing SOIL CONSERVATION, water conservation, restoration, erosion prevention, and environmental enhancement measures. It also prohibited the secretary from entering into certain WETLAND drainage agreements if drainage would adversely affect WILDLIFE.

During the DUST BOWL catastrophe of the 1930s, the U.S. Congress recognized that the deterioration of soil and water resources occurring on farmland, pastures, and FOREST lands as a result of soil erosion was a threat to the national welfare. Congress intended to permanently provide for the control and prevention of soil erosion, thereby preserving NATURAL RESOURCES, controlling FLOODS, preventing SILTATION of RESERVOIRS, maintaining the navigability of rivers and harbors, protecting public health and PUBLIC LANDS, and relieving unemployment. Early amendments to the act were passed in 1936 and 1937. The act has also been amended by the Food and Agriculture Act of 1962, the Rural Development Act of 1972, the Food Security Act of 1985, and the Farm Bill of 1996, among other statutes. *See also* AGRICULTURE; BENNETT, HUGH HAMMOND; DEPARTMENT OF THE INTERIOR; DROUGHT; IRRIGATION.

EnviroSource

Natural Resources Conservation Service Web site: www.nrcs.gov.

Soil Erosion Service. *See* Natural Resources Conservation Service.

soil erosion. *See* erosion.

solar cell, A device that converts SOLAR ENERGY into ELECTRICITY in a manner that does not release any POLLUTANTS into the ENVIRONMENT. Solar cells, which are also called photovoltaic (PV) cells, have made the ENERGY of sunlight a valuable ALTERNATIVE ENERGY RESOURCE that has many applications.

In 1839, the French physicist Edmond Becquerel observed that when light was absorbed by certain materials, the materials generated electricity. Becquerel also recorded that the amount of electricity varied with the intensity of the light. Despite these early findings by Becquerel, PV research did not begin in earnest until the late 1950s. PV systems were first used by National Aeronautics and Space Administration in 1958 to power the radio of the U.S. *Vanguard I* space satellite with less than 1 watt of electricity. Since then, PV systems have become increasingly more common. For example, very small PV systems are currently used to power about 5% of all calculators and small watches. Larger systems provide at least some of the electrical needs of about 40% of homes worldwide. In Japan alone, about 10% of all new homes make use of photovoltaics to provide their electric needs. Many water pumping stations also use photovoltaics for their energy.

Most solar cells are made from two layers of silicon that have been chemically treated using a process called doping, which gives one silicon layer a negative charge and the other a positive charge. In simple terms, a photovoltaic cell is simply a wafer of semiconductor in which there is a junction between negative and positive materials. Because the layers have different charges, they create an electric field at the junction, or place where the two layers meet, that allows direct current (DC) to flow. Part of the solar cell also has metal terminals, similar to those of a BATTERY, which conduct the flow of electricity from the negative layer to the positive layer.

All PV systems require light, which contains energy in the form of photons, or particles of light. When photons strike a PV cell, they cause electrons to be ejected from the silicon atoms located near the junction. The stream of electrons move freely from the negative layer to the positive layer through the metal terminals. A 10-centimeter (4-inch) solar cell can produce about 1 watt of DC electricity when exposed to sunlight. To generate more electricity, many solar cells are wired together in a panel called a solar array that is encased in a water-tight container. The panels, in turn, can be wired together to generate an even greater amount of electricity. *See also* ALTERNATIVE FUEL; RENEWABLE RESOURCE.

EnviroSources

American Solar Energy Society: 2400 Central Avenue, Suite G-1, Boulder, CO 80301

Solar Energy Industries Association: 122 C Street, NW, 4th Floor, Washington, DC 20001; Web site: www.seia.org/main.html.

University of Southhampton, England (solar energy) Web site: www.soton.ac.uk/~solar/.

solar energy, Energy derived from the radiation of the sun. Radiant ENERGY from the sun can be converted into other forms of energy to provide SOLAR HEATING and ELECTRICITY. The two kinds of solar heating systems are passive and active. Passive solar heating systems rely largely on the GREENHOUSE EFFECT, trapping heat inside a building much as a closed AUTOMOBILE traps heat when parked in an unshaded area on a sunny day. Active systems require the use of some new materials and technologies and also makes use of devices such as fans to circulate heat. Sunlight can also generate electricity with the use of SOLAR CELLS in a manner that does not release any POLLUTANTS to the ENVIRONMENT. *See also* ALTERNATIVE ENERGY RESOURCE; ALTERNATIVE FUEL; RENEWABLE RESOURCE; SOLAR THERMAL POWER SYSTEM.

EnviroTerm

solar collector: A device used in an active solar heating system to collect the sun's rays. The collector is usually placed on the roof and positioned to optimize its exposure to the sun.

EnviroSources

Schaeffer, John, et al., eds. *The Real Goods Solar Living Sourcebook: The Complete Guide to Renewable Energy Technologies and Sustainable Living.* White River Junction, VT: Real Goods, 1999.

Solar Energy Industries Association Web site: www.seia.org/main.html.

Strong, Steven J. *The Solar Electric House: Energy for the Environmentally-Responsive, Energy-Independent Home.* White River Junction, VT: Chelsea Green, 1994.

solar heating, The use of ENERGY from the sun to provide heat and hot water for homes and buildings. SOLAR ENERGY is an ALTERNATIVE ENERGY RESOURCE that can be used to provide both heat and ELECTRICITY, thus reducing or eliminating the use of FOSSIL FUELS for these purposes. The main advantages of solar energy over fossil fuels are that solar energy is both a RENEWABLE RESOURCE and is nonpolluting.

The use of solar energy to provide heat and hot water for buildings increased in the United States in the 1970s, following an energy crisis that resulted from a decreased supply of PETROLEUM from the Middle East. (*See* Organization of Petroleum Exporting Countries.) Use of the sun's energy for heat can employ either a passive system or an active system.

Passive Solar Heating Systems, Passive solar heating systems rely largely on the GREENHOUSE EFFECT, trapping heat inside a building much as a closed AUTOMOBILE traps heat when parked in an unshaded area on a sunny day. The heating effect occurs when sunlight is absorbed by materials within the enclosed space, such as the seats in the automobile, and the absorbed light energy is converted into heat energy, which is then radiated back into the ENVIRONMENT.

The simplest passive solar heating system makes use of the position of a building (the direction it faces) and the locations of its windows. Buildings designed to make use of passive solar heating generally are constructed so that all or most of their windows are on the south-facing side of the structure. South-facing windows are exposed to the greatest number of hours of daily sunlight throughout the year. As sunlight enters the building through the windows, light energy is absorbed by the walls, floors, and furnishings of the room, which are generally composed of dark materials with rough surfaces, which enable the greatest amount of light energy to be converted into heat energy. The resulting heat energy is then slowly released back into the rooms of the structure. At night, when no sunlight is available, window shades are used to cover the windows to prevent the heat energy from escaping back into the outside environment.

Some passive heating designs feature a thermal storage wall or water-filled containers that store heat as it is generated. Vents and registers placed near these structures gather some of the heat and direct it into other rooms. Because buildings are not generally exposed to consistent amounts of sunlight throughout the year, passive solar heating systems do not completely eliminate the need for other types of heating systems, which may be fueled either directly or indirectly by fossil fuels; however, studies have indicated that, on average, the use of passive solar heating strategies in homes can reduce utility costs by as much as 20 to 30%.

Active Solar Heating Systems, An active solar system is more complicated than a passive system. Active systems require the use of new materials and technologies and also make use of devices such as fans that can circulate heat. A typical simple active solar system might include three subsystems. The first subsystem is a device that collects sunlight and changes it to heat. This device is usually placed in a south-facing position on the roof of a structure. The collector is composed of one or two panels of glass or plastic, a blackened plate, and rows of metal tubes filled with AIR or liquid. The metal tubes are fastened to the black plate. Sunlight entering the glass or plastic panes is absorbed and converted to heat energy by the plate. The metal tubes absorb this heat and transfer it to the gas or liquid they contain. The heated gas or liquid is then transported to the next subsystem—a storage system.

The storage system consists of containers filled with rocks or water. Of the two materials, water works best. Water can hold heat much longer than most other materials, making it ideal for heat storage. The storage tanks are insulated to keep the heat contained inside the tank from escaping to the environment.

The last part of the subsystem is the transport system, which is composed of a network of pipes and possibly pumps or fans. The transport system circulates heated water or air from the storage system to rooms located throughout the building through a network of pipes. Pumps or fans may be used to help move the heated water or air through the pipes. As the materials move through the pipes, they give up some of the heat they carry to the surroundings. Once the heat carried by the liquid or air is given off, the cool air or water returns to the storage tank to be reheated.

Like a passive solar heating system, an active system may also be accompanied by a backup heating system. The backup system is used during cold periods of the year or during cloudy weather. An active solar energy system can distribute heat more effectively throughout a dwelling because of its pumps and fans; however, the pumps and fans employed by the system require electricity to operate, thus increasing the amount of energy needed to operate the system. In some areas, this electricity is provided by SOLAR CELLS, devices that convert the energy of sunlight into electricity. These cells are often arranged in groups called solar arrays.

EnviroSources

American Solar Energy Society: 2400 Central Avenue, Suite G-1, Boulder, CO 80301.

Solar Energy Industries Association: 122 C Street, NW, 4th Floor, Washington, DC 20001.

U.S. Union of Concerned Scientists: 2 Brattle Square, Cambridge, MA 02238.

World Resources Institute, 1709 New York Avenue, NW, Suite 700, Washington, DC 20006; e-mail: info@wri.org.

solar pond, A body of water that collects and stores sunlight for the purpose of providing renewable ENERGY. The sunlight is "captured" and stored in the bottom layer of a body of water where large quantities of salt have been dissolved. The stored salt water can attain temperatures high enough to heat buildings. The hot water can also be used in a Rankine-cycle engine generator to produce ELECTRICITY that could be used by nearby companies. Solar pond technology does not damage the ENVIRONMENT or produce waste materials.

Generally, the solar pond has three main layers. The top layer is cold and has relatively little salt content. The bottom layer is hot—up to 100°C (212°F)—and is very salty. Separating these two layers is the important gradient zone. The gradient is between 1 and 2 meters thick. Under normal conditions, water will rise when heated. The increased salinity stops this process. When large quantities of salt are dissolved in the hot, bottom layer of water, it becomes too dense to rise to the surface and cool. The stable gradient zone suppresses convection and acts as a transparent insulator, permitting sunlight to be trapped in the hot bottom layer from which useful heat may be withdrawn or stored for later use. Even covered with ice, the El Paso Solar Pond's (see below) lower zones produce temperatures of 68°C (154°F)—hot enough to generate electricity.

Solar ponds can provide a supplemental energy source for electrical production and heat for greenhouses and livestock buildings. Solar ponds can process heat for production of chemicals, foods, textiles, and other industrial products and for the separation of crude oil from brine in oil recovery operations. Solar ponds may also be used to protect fish from "cold kill" in AQUACULTURE applications or as receptacles for brine, a waste product of crude oil production and power plant COOLING TOWER blowdown systems.

The El Paso Solar Pond is a research, development, and demonstration project operated by the University of Texas at El Paso and funded by the U.S. BUREAU OF RECLAMATION and the state of Texas. The project, which is located on the property of Bruce Foods, Inc., a food canning company, was initiated in 1983 in cooperation with the Bureau of Reclamation. Since 1985, the El Paso Solar Pond had been continuously operated for seven years. The El Paso Solar Pond is the first solar pond in the world to deliver heat to a commercial manufacturer in 1985, the first such electric power facility in the United States in 1986, and the nation's first experimental water desalting facility in 1987.

A solar pond at the University of Texas, El Paso, is used to produce heat and electricity for nearby buildings. *Credit:* Courtesy of the Solar Pond Project at the University of Texas at El Paso.

According to experts, solar pond technology has potential especially for areas in which there is a unique combination of solar energy, salt, and BRACKISH water to make the technology a viable energy source. One such area is the southwestern United States. Experts believe the potential for solar pond technology is immense in this location because of the abundance of underground salt resources, brackish water, and natural salt lakes, which represent a potentially significant, untapped resource. Australia, Canada, India, Israel, and Saudi Arabia are just a few of the countries conducting solar pond research. *See also* ALTERNATIVE ENERGY RESOURCE.

EnviroSources

David Ben-Gurion National Solar Energy Center Web site: www.ramat-negev.org.i/solar.html.

Department of Mechanical and Industrial Engineering, University of Texas at El Paso, El Paso TX 79968; (915) 747-5450; fax: 915-747-5019; e-mail: aswift@cs.utep.edu.

National Renewable Energy Laboratory: 1617 Cole Blvd., Golden, CO 80401.

University of Texas at El Paso Web site: www.cerm.utep.edu/solarpond/epsp.html.

solar radiation. *See* solar energy.

solar thermal power system, An experimental technology that uses concentrated SOLAR ENERGY to drive generators that provide ELECTRICITY and hot water at the same time. One kind of system, called a central receiver system uses mirrors called heliostats that redirect the sun's solar energy to a receiver mounted on top of a building or tower. The receiver normally consists of a large number of metal tubes that contain a fluid (water, molten salts, etc.) or AIR, which absorbs the heat. The outer surfaces of the tubes are black, which optimizes light absorption and its conversion to heat. The sun heats the fluid in the receiver to very high temperatures, which produces steam. The steam is used directly to power a conventional steam TURBINE/generator to produce electricity. One such system, the National Solar Thermal Test Facility was operated by Sandia National Laboratories for the Department of Energy, near Barstow, California. France, Spain, Italy, Switzerland, Israel, Germany, Japan, and Russia are also working on central receiver systems.

solid waste, All nonhazardous discarded GARBAGE, REFUSE, SEWAGE, and SLUDGE—products of a solid nature that result from homes, AGRICULTURE, FORESTRY, and MINING. Worldwide, individuals and businesses generate billions of tons of solid wastes each year. Finding adequate means of disposal for these wastes in a manner that protects the health of people, ORGANISMS, and the ENVIRONMENT is an international problem that becomes increasingly difficult as the global human POPULATION continues to increase in size.

Types of Solid Waste. The types of materials that typically comprise solid waste vary throughout the world as do the amounts of wastes generated. In the United States, the amount of solid waste generated averages about 2 kilograms (4.4 pounds) per person per day; Canada generates almost 3.5 kilograms (7 pounds) of waste per person per day, and Japan about 1 kilogram (2.2 pounds) per person per day. In the United States, the largest amount of solid waste collected in most municipalities is paper and paper products. In fact, as much as half of the solid waste produced by a typical community in the United States may consist of these materials. In addition to paper, the United States produces large amounts of plastic (10%), metal (6%), glass (1%), and ORGANIC materials (13%) as part of their solid wastes. The remaining 20% of the waste materials are comprised of a variety of materials.

Means of Disposal. How to adequately dispose of solid wastes is a major problem. The earliest organized means of waste disposal for garbage, refuse, and sludge involved open dumps, areas of land or water in which people piled materials for which they no longer had a use. As communities increased in size and open dumps followed suit, the drawbacks became apparent: Odors were emitted by the decaying garbage; there was an unsightly buildup of matter that had not decomposed; and numerous, often DISEASE-carrying pest organisms, such as rats, cockroaches, and flies, were drawn to these areas by their unlimited food sources.

To resolve concerns about open dumps, many communities began to burn much of the solid wastes generated in their communities in large incinerators. INCINERATION reduced the volume of wastes disposed of in open dumps, but did not address odor and pest problems associated with already existing open dumps and also created a new pollution problem, as smoke, soot, and other substances were emitted into the AIR as byproducts of combustion. In addition, unburned materials and ASH remained a disposal problem and often again cluttered the land in open dumps. Because of the pollution problems associated with incineration, many municipalities enacted legislation making the incineration of solid wastes illegal.

An alternative to incineration and open dumps was the development of the landfill. In a landfill, solid wastes are buried beneath a layer of soil. The development of the landfill was considered an improvement over open dumps because it eliminated the unsightly appearance of wastes in the environment associated with open dumps; however, refuse in landfills continued to produce noxious odors and attract a variety of undesirable INSECTS and other animals prior to burial. In addition, rainwater that fell onto open landfills picked up POLLUTANTS as it moved through the wastes and carried these pollutants into the soil, the GROUNDWATER supply, and in some cases into nearby waterways. (*See* leaching.) Although landfills were considered a major improvement over many earlier waste disposal methods, concern about the types of pollutants that might enter groundwater supplies as well as unpleasant odors and unwanted pests led many areas to pass laws making this form of garbage disposal illegal.

Today's alternative to the landfill is the SANITARY LANDFILL. Unlike earlier landfills, sanitary landfills are lined with clay and or plastic liners that are intended to prevent water that percolates through the landfill from escaping back into the environment. Solid waste placed into such landfills is deposited in layers and compacted by bulldozers. The layer of disposed items is then covered by a layer of soil, which is again compacted. Pipes often are positioned among the waste layers to capture METHANE gas that is produced during the DECOMPOSITION process. In some communities this gas is reclaimed and sold for use as FUEL. This process of alternative layers of waste with soil is repeated until the landfill reaches a predetermined height. The landfill is then closed and the area atop the landfill may be planted with vegetation and landscaped for use as a park or other recreational area.

Sanitary landfills address some of the problems associated with solid waste disposal, but not all of them. For example, while a sanitary landfill is in operation, people living in communities immediately surrounding the landfill must deal with the unsightly appearance of huge amounts of wastes as well as the odors and pest animals associated with these wastes. In addition, the location where such landfills might be constructed are limited to stable areas to ensure that the liner is not torn by shifts in the ground below. In general, people do not want such facilities located in their communities because of the odors and unsightly appearances associated with these facilities.

Reducing Waste Volume. To address some of the problems associated with disposal of solid wastes, people in many communities are now encouraged to reduce, reuse, and recycle as many of their solid wastes as possible. For example, methods of RECYCLING newsprint and paper, which make up the largest amount of solid waste in the United States, have been in use for a number of years and have greatly reduced the amounts of such materials that must be disposed of in landfills. Much of the glass, ALUMINUM, plastic, steel, and other metals disposed of by people also can be recycled to make new products. In addition, such methods extend the life of reserves of NATURAL RESOURCES and reduce the amount of pollution that is generated when FOSSIL FUELS must be burned to generate the ENERGY needed to process new raw materials.

Other methods that help to reduce problems associated with solid waste disposal include making use of septic systems or community SEWAGE TREATMENT PLANTS for the processing of SEWAGE. These methods help to reduce the numbers of disease-causing organisms in the environment by preventing fecal wastes and other household wastes from being released to terrestrial or aquatic environments. Sludge produced as a result of sewage treatment processes may be reclaimed for use as FERTILIZER or disposed of in sanitary landfills.

To reduce the amounts of garbage and yard wastes that require disposal in landfills, many people now practice

COMPOSTING, a process in which organic wastes are mixed together and permitted to decompose through natural processes. Matter that is composted helps to build new soil, while also returning valuable nutrients to soil. Compost materials also are useful as fertilizers. In addition to composting, many yard wastes can be used as MULCHES to help control moisture loss and weed growth around PLANTS. Grass, leaves, and wood that has been processed through a chipper are especially useful for this purpose. Use of such materials also benefits the environment by returning valuable nutrients to the soil as these wastes decompose and helps to build new soil. *See also* ANAEROBIC DECOMPOSITION; BACTERIA; BIOCONVERSION; BIODEGRADABLE; BIOFUEL; BIOGEOCHEMICAL CYCLE; BIOMASS; BOTTLE BILL; COGENERATION; COLIFORM BACTERIA; CONSERVATION; CONTAMINANT; DECOMPOSER; DECONTAMINANT; DETRIVORE; EMISSIONS; FILTRATION; FLUIDIZED BED COMBUSTION; FUNGI; HAZARDOUS WASTE; HAZARDOUS WASTE TREATMENT; INFILTRATION; LITTER; NONBIODEGRADABLE; OCEAN DUMPING; PATHOGEN; PHOTODEGRADABLE PLASTIC; RESTORATION ECOLOGY.

EnviroSources

Abromovitz, Janet, et al. *State of the World: 1998.* New York: Norton, 1998.

Bernstein, Leonard, Alan Winkler, and Linda Zierdt-Warshaw. *Environmental Science: Ecology and Human Impact.* Menlo Park, CA: Addison-Wesley, 1995.

solvent, A substance that dissolves another substance to form a solution. Water is a common solvent. Many ORGANIC solvents are liquids that are distilled from PETROLEUM. Solvents include alcohols, lacquers, resins, shellac paint strippers, glues, BENZENE, GASOLINE, and CHLORINATED HYDROCARBONS used in paint strippers and dry cleaning fluids. METHYLENE CHLORIDE, another organic solvent, is used in aerospray containers and in highway paint. At high levels, the vapors from some solvents can have adverse effects on human health including respiratory problems and damage to the nervous system.

source separation, Separating wastes into recyclable and nonrecyclable materials. After separation, the materials are recycled or disposed of, respectively. Source separation takes place at the source where the waste is produced. The process removes all designated recyclable materials from the waste stream.

Source separation, and the subsequent increase in RECYCLING, can foster competition among recycling companies, thereby keeping costs low and quality of services high. Source separation encourages each individual to become more responsible for SOLID WASTE disposal. This can foster further source reduction and recycling activities at work, home, and school.

soy ink, A nontoxic ink extracted from soybean oil that is used for printing magazines, newspapers, and other products. Soy ink is naturally low in VOLATILE ORGANIC COMPOUNDS. Soy ink was developed by the American Newspaper Publishers Association.

speciation, The development of new SPECIES. Speciation is an evolutionary process through which genetic differences arise that prevent successful breeding between POPULATIONS that were once able to interbreed. As a result, two distinct species emerge. Often, speciation results when an interbreeding population of ORGANISMS becomes divided by a physical barrier. Through natural selection, the populations in different HABITATS develop traits different enough from each other that they can no longer successfully breed. *See also* ADAPTIVE RADIATION; BIODIVERSITY; CONVERGENT EVOLUTION; DARWIN, CHARLES ROBERT; EVOLUTION; GENE POOL; GENETIC DIVERSITY; MUTATION.

species, The smallest taxonomic division of ORGANISMS, which consists of a single kind of organism. A species is a naturally occurring POPULATION or group of populations that can interbreed only with others of their species. To date, scientists have identified more than 1.5 million species of organisms; however, many believe the actual number of species that exist or have existed on Earth may be in the trillions. *See also* BACTERIA; BIODIVERSITY; BOTANY; ENDANGERED SPECIES; EXTINCTION; FUNGI; GENETIC DIVERSITY; PLANTS; PROTISTS; SPECIATION; SPECIESISM; TAXONOMY; THREATENED SPECIES.

species diversity, The number of different SPECIES living in a BIOLOGICAL COMMUNITY or ECOSYSTEM. Most often, measures of species diversity are expressed in conjunction with the relative abundance of species present. For example, a biological community comprised of 50 different species is considered more diverse than a community with only 15 different species. When data for such measures are available for a region over a period of time, they may be useful in evaluating the overall health of the ENVIRONMENT in a region. *See also* ADAPTIVE RADIATION; BIODIVERSITY; GENE POOL; GENETIC DIVERSITY; POPULATION; WILSON, EDWARD OSBORNE.

Species Survival Commission. *See* World Conservation Union.

speciesism, The belief, which some claim that humans hold, that humans are inherently superior to or more important than ORGANISMS of other SPECIES. In this belief system, humans, as ANTHROPOCENTRIC beings, tend to view other species only in terms of their usefulness or desirability. The idea of speciesism was popularized by Australian ANIMAL RIGHTS activist Peter Singer in his book, *Animal Liberation* (1975). Supporters of animal rights and practitioners of DEEP ECOLOGY work in opposition to speciesism by encouraging people to recognize that all living beings have their own intrinsic value and right to exist. *See also* BIOCENTRIC.

spectrometry, A process that uses a spectrometer to identify and monitor concentrations of chemicals in the ENVIRONMENT. The technology is being used to monitor concentrations of atmospheric OZONE as well as to identify the locations of harmful substances such as HEAVY METAL wastes that have been released into the environment as a result of MINING activities, particularly those involving zinc, silver, gold, and LEAD. Heavy metals are metallic elements with fairly high molecular masses. Most of these elements are toxic to ORGANISMS when ingested. In addition, chronic exposure to the ionic forms of many of these elements, including CADMIUM, ARSENIC, MERCURY, COPPER, and CHROMIUM have been shown to have harmful effects on the health of many types of ORGANISMS. (*See* chronic toxicity.)

A spectrometer is an optical instrument that analyzes the spectrum emitted by a substance. Analysis of the spectrum can be used to identify the composition of the substance. To search for heavy metal contamination in the environment, a spectrometer is placed inside a camera that is mounted on a high-altitude airplane. The camera photographs areas where mining operations have taken place. Scientists then study the photographs to identify what substances are present in the environment and use this data to identify areas in need of cleanup.

Spectrometry was used to map the Gulch SUPERFUND site in Leadville, Colorado. Spectrometry is also being used to map the global distribution of Earth's atmospheric ozone. For example, the Total Ozone Mapping Spectrometer program uses a spectrometer on board a satellite to provide high-resolution maps of global ozone amounts on a daily basis. The satellite has helped scientists detect small, but long-term, damage to the ozone layer over several parts of the globe, including some of the heavily populated areas in the Northern Hemisphere. *See also* COMPREHENSIVE ENVIRONMENTAL RESPONSE, COMPENSATION, AND LIABILITY ACT; GLOBAL WARMING; HAZARDOUS WASTE.

EnviroSource

National Aeronautics and Space Administration, Goddard Space Flight Center, Greenbelt, MD 20771.

spent fuel, FUEL RODS used in a NUCLEAR REACTOR that are no longer able to produce nuclear FISSION but are still hazardous. Spent nuclear fuel consists of metal tubes called fuel rods containing irradiated URANIUM oxide pellets. Spent fuel is highly radioactive and produces considerable heat. For these reasons spent fuel must be cooled and shielded. With time, cooling and some shielding requirements decrease as a result of the natural RADIOACTIVE DECAY. The solid spent fuel pellets will not explode spontaneously, catch fire, or burn.

For years, nuclear power plants have temporarily stored used FUEL in water pools at the reactor site, but storage in pools was becoming a problem. In 1982, the U.S. Congress passed the NUCLEAR WASTE POLICY ACT, which addressed the spent fuel storage problem. The act directed the NUCLEAR REGULATORY COMMISSION (NRC) to approve means of interim dry spent storage. The NRC now authorizes nuclear power plant licensees to store spent fuel at reactor sites in NRC-approved dry storage casks. Casks can be made of metal or concrete and are either placed horizontally or stand vertically on a concrete pad above ground. The casks used in the dry storage systems are designed to resist FLOODS, EARTHQUAKES, TORNADOES, and temperature extremes. They hold spent fuel already cooled in the spent fuel pool for at least five years. Ten nuclear power plants in the United States are currently storing spent fuel under the dry storage option; however, before using the dry storage casks, a licensee must be approved by the NRC.

Federal regulations permit the transportation of spent nuclear fuel only in very strong robust metal containers, called "Type B" transportation casks. Type B casks are designed and constructed to safely contain their radioactive contents under normal and severe accident conditions. Cask designs are reviewed and certified by the NRC. Tests have demonstrated that this type of cask will survive the forces that it would likely experience in an earthquake, train collision and derailment, highway accident, or fire. Since 1965, there have been more than 2,500 shipments of spent nuclear fuel in the United States without injury or environmental consequences.

Spent Fuel Stored at U.S. Nuclear Plants

State	1995
Alabama	1,439 metric tons
Arizona	465
Arkansas	581
California	1,391
Colorado	15
Connecticut	1,254
Florida	1,440
Georgia	1,019
Illinois	4,292
Iowa	235

Ten nuclear power plants in the United States are currently storing spent fuel under the dry storage option. *Source:* Nuclear Regulatory Commission.

EnviroSource

U.S. Department of Energy Web site: www.ntp.doe.gov; www.rw.doe.gov/pages/resource/facts/transfct.html.

S-shaped curve, A line formed on a graph showing a pattern of a successful POPULATION. The graph illustrates the size of a population over a period of time. The S-shaped curve shows a rapid growth rate followed by slower GROWTH RATE displaying a stable population. If the population rises above the CARRYING CAPACITY of the ENVIRONMENT, the

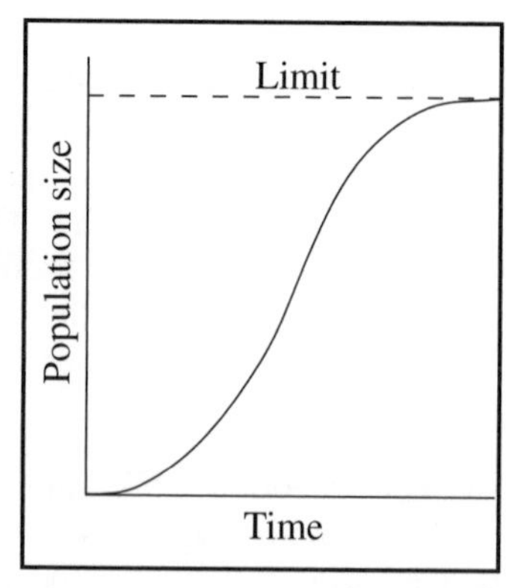

A line formed on a graph showing a pattern of a successful population.

population will decline until the population stabilizes near the carrying capacity.

steppe, A type of GRASSLAND with few if any trees that is located across central Eurasia. Overgrazing and extensive farming operations in the steppe cause degradation of this BIOME. The vast steppe lands stretch from southern Russia, through Kazakhstan, and southern SIBERIA, west to Mongolia, and south to include Western China. The Mongolian Red Data Book lists 23 MAMMALS, 19 birds, 4 REPTILES, 2 AMPHIBIANS, and 2 fish SPECIES that live in the steppe as endangered, vulnerable, or rare. Included is the snow LEOPARD. The steppe CLIMATE is characterized by light rainfall in the spring and early summer—the growing season—and dry, sunny late summers—when crops ripen and are harvested.

EnviroSource

Colorado State University Web site: www.lternet.edu/network/sites/05_cpr.html.

sterilization, pathogenic, The removal or destruction of all MICROORGANISMS present on or in an object, ORGANISM, or part of the ENVIRONMENT, usually to prevent the spread of DISEASE. Sterilization is used primarily to destroy PATHOGENS that pose a threat to human health, but it may also be used to protect PLANTS and animals from infectious agents. Heat, RADIATION, treatment with chemicals, and FILTRATION are common means of sterilization. *See also* BACTERIA; BLIGHT; CHLORINATION; CONTAMINANT; DECONTAMINATION; DETERGENTS; DISINFECTANT; *ESCHERICHIA COLI*; FUNGI; IRRADIATION; VIRUS.

sterilization, reproductive, The purposeful removal of an organism's ability to produce offspring, often as a means of controlling POPULATION growth. Sterilization techniques vary according to SPECIES and purpose. For example, INSECTS that carry DISEASE or damage crops may be exposed to RADIATION or chemicals that destroy their ability to successfully reproduce. Similarly, rodents and other wild animals may be provided with food baited with contraceptive chemicals to prevent reproduction.

In larger animals, including humans, sterilization often is accomplished through surgical procedures. Dog and cat owners often have their pets surgically spayed (females) or neutered (males) to prevent large, roaming packs of feral animals. Sterilization may also be performed on an animal to reduce the risk that it might develop a potentially harmful medical condition. For example, spaying a female dog reduces its risk of developing breast CANCER. Neutering male cats reduces their risk of developing certain types of kidney stones.

Purposeful sterilization in humans typically requires surgical procedures. Most commonly, women undergo a tubal ligation (sealing the Fallopian tubes to prevent fertilization) or hysterectomy (the removal of the uterus, Fallopian tubes, and ovaries). In men, sterilization usually involves a vasectomy—a surgical procedure in which the vas deferens is sealed to prevent sperm transmission. Surgical or chemical castration, which eliminates sperm production and/or the reproductive impulse, also is a means of sterilization in males.

In most countries, decisions regarding human sterilization are made by the affected individuals; however, in some countries where population growth places a strain on resource availability, governments encourage sterilization by enacting laws or through incentive programs. In Singapore, for example, government employees who undergo sterilization to limit their family size to two or fewer children receive better housing than do couples who do not undergo sterilization procedures. In 1975, the government of India passed legislation mandating sterilization of men who had sired three children. Men who did not undergo the procedure could be imprisoned. This law was later abolished as a result of public pressure; however, India continues to provide incentives to people who undergo sterilization procedures. *See also* AGE STRUCTURE; BIRTH RATE; CARRYING CAPACITY; DEATH RATE; POPULATION DENSITY; ZERO POPULATION GROWTH.

EnviroSources

Ehrlich, Paul. *The Population Explosion.* New York: Touchstone Books, 1991.
Saw, Swee-Hock. *Population Control for Zero Growth in Singapore.* Cambridge, UK: Oxford University Press, 1980.
Zero Population Growth: 1400 16th St., NW, Suite 320, Washington, DC 20036; (202) 332-2200; e-mail zpg@igc.apc.org.

stratosphere. *See* atmosphere.

strip cropping, An agricultural technique that reduces soil EROSION resulting from wind or running water. In strip cropping, land is plowed in a direction that is perpendicular to its greatest wind exposure or to its slope in areas vulnerable to erosion by running water. Two or more types of crops are then planted in alternating bands that run in the same direction in which the land was plowed. The crops chosen for growth and their placement is determined by the time of year that the PLANTS are harvested to ensure that, year round, por-

A combine is harvesting corn that was planted using strip cropping. *Credit:* United States Department of Agriculture. Photo by Lynn Betts.

tions of the land remain covered by plant growth. The coverage of ground by plants helps to reduce erosion by slowing wind and running water, thus reducing the ability of those forces to carry soil. In addition, the roots of growing plants help reduce erosion by holding soil in place. *See also* AGRICULTURE; AGROFORESTRY; DUST BOWL; POLYCULTURE; SOIL CONSERVATION.

strip mining, A MINING method used to obtain rock or MINERALS located very near Earth's surface. Strip mining is a type of SURFACE MINING that is most often used to obtain COAL. The method is destructive to the ENVIRONMENT because it involves removing all vegetation, soil, and rock laying above a mineral deposit to expose the deposit. In addition, coal deposits typically are rich in SULFUR which can be transformed into SULFURIC ACID—a strong ACID that is harmful to PLANTS, animals, and aquatic ORGANISMS—by rainwater that passes through mining wastes. (*See* tailings.)

In the United States, strip mining is used mostly to obtain bituminous and anthracite coal; in Europe, particularly England and Germany, the technique most often is used to expose deposits of lignite or subbituminous coal (called brown coal). To correct or reverse problems associated with strip mining, the U.S. government and several states have legislated that lands on which such mining is practiced be restored to their premining condition, when mining is completed. *See also* ACID MINE DRAINAGE; LEACHING; RECLAMATION; SURFACE MINING CONTROL AND RECLAMATION ACT.

strontium-90, A RADIOISOTOPE of strontium that is released in large quantities in FALLOUT from a NUCLEAR WEAPON or is produced as a FISSION product in nuclear reactions. It has a HALF-LIFE of about 28 years. The ISOTOPE contaminates water and land and is absorbed by PLANTS; as a result the isotope is passed along the FOOD CHAIN. Because it can chemically replace calcium, long-term exposure to strontium-90 results in the substitution of calcium in bones. At high levels this can result in the destruction of the blood-cell-forming bone marrow in the human body, which in turn may lead to leukemia. In addition, some studies have shown that in populated areas where above-ground nuclear weapons have been tested, there have been increases in thyroid CANCERS following the radioactive fallout.

subsidence, The sinking of land as a result of natural processes or human activities. The formation of a SINKHOLE through Karst activity, such as the collapse of a cavern, is an example of natural subsidence. Karst activity occurs when acidic subsurface water wears away limestone to form a hole beneath Earth's surface. (*See* acid; groundwater.) Over time, subsurface rock may become too weak to support the soil and rock above it. The surface layers collapse to form a large crater called a sinkhole. Subsidence also occurs along coastlines as waves crashing along the shoreline undercut sand and soil. Eventually, this erosive action causes the land along the coast to sink relative to sea level. (*See* erosion.)

MINING, pumping oil from WELLS, and removing excessive amounts of water from AQUIFERS are human activities that may lead to subsidence. In the case of mining, digging beneath the surface and the removal of subsurface sediments, ores, and other material weakens rock layers that support rock and soil located nearer the surface. If the rock layers are weakened too much, the land above the mined area may collapse into the mine.

The removal of excessive amounts of water from an aquifer is called OVERDRAFT. When overdraft occurs, soil and rock layers located above the WATER TABLE may sink to fill the available space. The removal of too much oil from a well also may lead to subsidence as soil and rock layers sink to fill the space once occupied by oil. *See also* ACID MINE DRAINAGE; PERMAFROST; ROCK CYCLE.

subsistence agriculture, The practice of AGRICULTURE—raising crops or livestock—on a small scale for the purpose of meeting the immediate food needs of a family or small community. Because subsistence agriculture need only provide enough crops or livestock to feed a small POPULATION, the practice generally is not as degrading to the ENVIRONMENT as most commercial agriculture. For example, subsistence farmers generally grow a diversity of crops, which does not deplete the soil of nutrients as readily as does the growth of a single crop. (*See* monoculture.) This also reduces the reliance on chemical FERTILIZERS. When fertilizers are used, they generally are ORGANIC fertilizers derived from plowing under PLANT wastes left after harvest or manure derived from livestock. Growing a diversity of crops also reduces dependence on chemical PESTICIDES by limiting the number of host ORGANISMS available to pests. (*See* polyculture.)

Areas that have suffered from the practice of subsistence agriculture are generally those in which FOREST land has been cleared through CLEAR-CUTTING or SLASH AND BURN methods to make land available for crop growth or livestock. These practices are most harmful to tropical RAINFORESTS because they result in HABITAT loss for native WILDLIFE and because land that supports the growth of rainforest trees generally does not support the growth of crops over time. Soils in such areas become quickly depleted of nutrients, thus preventing successful crop growth. In addition, without tree cover, such soil becomes dry as moisture is evaporated by the hot sun. This leaves the soil vulnerable to EROSION by wind and running water. Over a period of many years, these regions may be transformed into "dust bowls." Such soils also are at risk of DESERTIFICATION as salts left behind by evaporating water accumulate in the soil, leading the soil to become dry. As such soils become unable to support crop growth, farmers often move their crops to a new location by clearing another parcel of forest land, destroying more habitat and eliminating more forest.

Prior to the Industrial Revolution, much of the agriculture practiced throughout the world occurred at the subsistence level. Today, the practice generally occurs in sparsely populated, rural, and underdeveloped areas, where people live well below the poverty level. Today, subsistence-level agriculture continues to be practiced throughout rural areas of Indonesia, Mali, Kenya, Nigeria, Rwanda, Zaire, and in parts of North America inhabited by Alaskan Indians and Eskimos. *See also* AGROFORESTRY; DUST BOWL; GREENBELT MOVEMENT; GREEN MANURE.

succession. *See* ecological succession.

sulfate, Salts of SULFURIC ACID that are made of SULFUR and OXYGEN. Common sulfates include sodium sulfate (Na_2SO_4), copper sulfate ($CuSO_4$), iron sulfate ($FeSO_4$), zinc sulfate ($ZnSO_4$), and calcium sulfate ($CaSO_4$). Sulfates are released when COAL is burned. They contribute to ACID RAIN and if inhaled can cause respiratory illnesses.

sulfur, A yellow, odorless, nonmetallic element that occurs in elemental form or combined with metals such as COPPER, zinc, and MERCURY. It is found in all FOSSIL FUELS. Sulfur burns with a blue flame and is used in the manufacturing of SULFURIC ACID, gunpowder, and FERTILIZERS. Sulfur oxides are formed when one sulfur atom combines with different numbers of OXYGEN atoms. Sulfur oxides are a major source of AIR POLLUTION. SULFUR DIOXIDE is a sulfur compound that contains one sulfur atom and two oxygen atoms. At high levels, sulfur dioxide can cause respiratory and cardiovascular diseases.

sulfur dioxide, A colorless gas with a characteristic acrid odor at high concentrations that is a common POLLUTANT emitted when FOSSIL FUELS, containing SULFUR, are burned. Major EMISSIONS of sulfur dioxide (SO_2) in the United States derive from power plants east of the Mississippi River, particularly those in the Ohio Valley. When released into the ATMOSPHERE, sulfur dioxide reacts with water vapor to form SULFURIC ACID—a major component of ACID RAIN. SULFATE particles can be deposited on Earth as a dry CONTAMINANT that reacts with moisture in soil to form sulfuric acid.

Sulfur dioxide is used to treat wool, silk, and linen for textiles and to bleach wood pulp for paper. Some sulfur dioxide used for these purposes may be released to the ENVIRONMENT. Sulfur dioxide emissions can be reduced by using low-sulfur COAL instead of high-sulfur coal, or other more refined fuels, when economically feasible. Washing coal to remove its sulfur is also effective but expensive. One option that eliminates emissions of sulfur includes coal gasification, a process used by manufacturing plants to "scrub" the gas and remove sulfur compounds. Other options include FLUIDIZED BED COMBUSTION, scrubbing, and FLUE GAS desulfurization.

Sources of Sulfur Dioxide Emmissions in the United States

Source	Percent
Burning of high-sulfur coal	82%
Sulfuric acid industries	14%
Transportation	3%
Miscellaneous	1%

ACIDS formed from sulfur dioxide can be damaging to PLANTS, aquatic ECOSYSTEMS, and structures made from rock and metal. Exposure to high levels of sulfur dioxide gas cause significant constriction of air passages in the lungs leading to respiratory DISEASES such as bronchitis, particularly among asthmatics, children, and elderly people. Symptoms of exposure include wheezing, shortness of breath, and coughing. *See also* AIR POLLUTION; CLEAN AIR ACT; CRITERIA POLLUTANTS; NATIONAL AMBIENT AIR QUALITY STANDARDS; SCRUBBER.

sulfuric acid, A solution of HYDROGEN SULFATE (H_2SO_4) in water, which is formed in the ATMOSPHERE from SULFUR DIOXIDE EMISSIONS. Sulfuric acid is a major component of ACID RAIN. Sulfuric acid is also widely used in chemical industries.

Superfund, The fund established by the COMPREHENSIVE ENVIRONMENTAL RESPONSE, COMPENSATION, AND LIABILITY ACT (CERCLA) of 1980, a U.S. law mandating cleanup of the worst hazardous and TOXIC WASTE sites on land and water, which constitute threats to human health and the ENVIRONMENT. The Superfund provides funding for cleanups in which no responsible parties are found. The U.S. ENVIRONMENTAL PROTECTION AGENCY (EPA) administers the Superfund program through its Office of Solid Waste and Emergency Response in cooperation with individual states and tribal governments.

HAZARDOUS WASTE sites are discovered by federal, state, and local agencies, businesses, the EPA, the U.S. Coast Guard, and ordinary citizens. The Superfund program is the most aggressive hazardous waste cleanup program in the world. Where possible, old hazardous waste sites are being restored to productive use, and millions of people have been protected by Superfund cleanup actions. Under CERCLA, responsible individuals and companies, or potentially responsible parties (PRPs), pay for cleanup work at Superfund sites. Current landowners who had no part in disposing of waste or polluting the land or water can also be held liable.

Years ago, people were less aware of how dumping chemical wastes might affect public health and the environment. Wastes were left in the open, where they seeped into the ground, flowed into rivers and streams, and contaminated soil and GROUNDWATER. On thousands of properties where these practices had been intensive or continuous, such as abandoned warehouses, manufacturing facilities, processing plants, and landfills, the result was uncontrolled or abandoned hazardous waste sites. Citizen concern led to the passage of CERCLA, which created a tax on the chemical and PETROLEUM industries and provided broad federal authority to respond directly to releases or threatened releases of HAZARDOUS SUBSTANCES. Over a five-year period, $1.6 billion was collected, and the tax went to a trust fund—the Superfund—to be used to clean up abandoned or uncontrolled hazardous waste sites. The Superfund trust continues to increase each year. In 1990, 10 years after the project was started, the fund had increased to $15.2 billion. It consumes 25% of the EPA's yearly budget, making it the EPA's single largest program. Unfortunately, as of 1994, more than 20% of Superfund expenditures were being spent on litigation related to assignment of liability.

Hazardous waste site cleanups are very complex and require the efforts of many experts in various disciplines. CERCLA's six-stage cleanup process starts with preliminary assessment/site inspection, continues through remedial action, and ends with deletion of the site from the National Priorities List (NPL). In July 1999, there were 1,226 uncontrolled hazardous waste sites listed on the NPL. According to the Congressional Budget Office, it requires an average of 12 years and $30 million for a typical NPL site to reach the "construction complete" stage.

In 1986, the Superfund Amendments and Reauthorization Act (SARA) amended CERCLA. SARA stressed the importance of permanent remedies and innovative technologies, required that Superfund actions consider the standards of other environmental laws, provided new enforcement authorities and settlement tools, increased state involvement, increased the focus on human health problems, encouraged greater citizen participation in decision-making, and increased the size of the trust fund.

In early 1997, a Superfund reform proposal was introduced to the U.S. Senate. Two bills in the House of Representatives have been considered; however, efforts to reach consensus on Superfund reform legislation in 1997 and 1998 were largely unsuccessful because of differences with respect to provisions to be contained in the final bill. The various bills under consideration address such issues as liability caps, NATURAL RESOURCE damage restoration, and protection of uncontaminated GROUNDWATER. *See also* EMERGENCY PLANNING AND COMMUNITY RIGHT-TO-KNOW ACT; HAZARDOUS WASTE TREATMENT.

EnviroSources

National Response Center Hotline: (800) 424-8802.

RCRA, Superfund, Emergency Planning and Community Right-to-Know Act Hotline: (800) 424-9346; Washington, DC, area (703) 412-9810; Web site: www.epa.gov/epaoswer/hotline.

Superfund Information Line: (800) 424-9346.

U.S. EPA (superfund program) Web site: www.epa.gov/superfund/index.html.

surface mining, The removal of MINERALS from regions located at or near Earth's surface. Large equipment such as bulldozers are generally used to remove soil and rock and to expose minerals, such as COAL that are located near the surface. Surface mining may involve removing 10–25 meters (33–82 feet) of topsoil and rock to reach the underground minerals. Compared to subsurface mining, surface mining generally costs less, is safer for miners, and usually results in the removal of a greater percentage of the minerals underground; however, it also results in extensive disruption of the land.

Surface mining can cause serious environmental problems. Common problems include the destruction of ECOSYSTEMS and HABITATS. In addition, the removal of vegetation involved in surface mining makes an area more prone to soil EROSION and landslides. Water pollution can also occur in nearby streams. As water percolates through TAILINGS left over from MINING operations, it picks up minerals and carries them into GROUNDWATER reserves or to lakes and streams. (*See* leaching.) ACID MINE DRAINAGE is also a pollution problem that results from many types of surface mining. In many countries, including the United States, government agencies enforce RECLAMATION operations to restore mined land to the environmental conditions that existed before the mining operations. *See also* BUREAU OF LAND MANAGEMENT; OFFICE OF SURFACE MINING, RECLAMATION, AND ENFORCEMENT; SINKHOLE; SOIL CONSERVATION; SURFACE MINING CONTROL AND RECLAMATION ACT.

Coal is loaded into a truck from surface mining. *Credit:* Tom Repine, West Virginia Geological and Economic Survey

Surface Mining Control and Reclamation Act, U.S. legislation enacted in 1977 that requires companies involved in SURFACE MINING operations to restore mined lands back to their natural conditions after the MINING operations cease. The law established a coordinated effort between the states and the federal government to prevent the abuses that had characterized surface and underground COAL mining in the past. It also required a balance between the nation's call for environmental quality and its need for ENERGY resources. Two major programs were created by the act: an environmental protection program to establish standards and procedures for approving permits and inspecting active coal mining and RECLAMATION operations both on the surface and underground, and a program funded by fees that operators pay on each ton of coal mined, to reclaim abandoned land and water resources adversely affected by pre-1977 coal mining.

In 1990, Congress expanded the act to include reclamation of mines abandoned after passage in 1977. Provisions in the law also protect people and the ENVIRONMENT before any coal mining operation is conducted. As an example, a coal mining operator must post a bond sufficient to cover the cost of reclaiming the site. Upon approval of the bond, the operator is granted a permit. If there is a violation of the law, an inspector issues a notice to the operator to correct the violation within a specific period of time. If the violation is not corrected, a cessation order is issued to stop any active mining. The law also prohibits mining within the boundaries of national parks, refuges, trails, wild and scenic rivers, and WILDERNESS and recreation areas. *See also* BUREAU OF LAND MANAGEMENT; MINERAL; OFFICE OF SURFACE MINING, RECLAMATION, AND ENFORCEMENT.

surface water, Any body of water—salt or fresh—that is naturally exposed to the ATMOSPHERE. Fresh-water bodies of surface water include rivers, lakes, ponds, streams, and RESERVOIRS. These bodies of water are of great environmental importance because they serve as the HABITAT for many living things and also provide drinking water to most of Earth's ORGANISMS, including humans. RUNOFF, rainwater, or meltwater that moves across the land also is a form of surface water. Seas, OCEANS, and ESTUARIES are bodies of salt water that are considered surface water. WELLS, springs, and aquifers that lie beneath Earth's surface and are directly influenced by surface water, such as runoff, also are classified as surface water. Such bodies of water are often the source of water used by humans. *See also* ACID MINE DRAINAGE; AGRICULTURAL POLLUTION; ALGAL BLOOM; AQUIFER; CLEAN WATER ACT; GROUNDWATER; POTABLE WATER; SAFE DRINKING WATER ACT; SUBSIDENCE.

sustainable agriculture, Any type of farming practice that helps protect the ENVIRONMENT by conserving water and ENERGY, and by limiting the use of synthetic PESTICIDES and FERTILIZERS. Sustainable agriculture practices stress the long-term CONSERVATION of resources and are designed to balance the human need for food with concerns for the environment. Today, the U.S. DEPARTMENT OF AGRICULTURE offers funding to farmers and researchers toward the development of new sustainable agriculture methods.

Sustainable agriculture is more widespread these days because of the concerns farmers have about pesticides, soil EROSION, and the effects of chemical fertilizers on aquatic ECOSYSTEMS. For instance, damage to the environment often occurs when fertilizers enter aquatic HABITATS as a result of RUNOFF from farms. Fertilizers are PLANT nutrients. When these chemicals are introduced into aquatic ecosystems they sometimes cause tremendous POPULATION explosions of ALGAE, BACTERIA, and other MICROORGANISMS. The rapidly growing populations use up a lot of OXYGEN in the water, which can lead to massive kills of fishes and other aquatic ORGANISMS. Pesticides can also harm organisms when they enter the FOOD CHAIN. When eaten, pesticides build up inside organisms. The dangerous chemicals can then be passed on through the ecosystem as organisms eat one another.

Sustainable agriculture is similar to organic farming (the growth of crops without the use of synthetic chemicals) but on a larger scale. Like organic farming, sustainable agriculture tries to establish a long-term relationship with the environment by limiting the harmful effects of more conventional farming methods. Examples of sustainable agriculture practices include crop rotation, INTEGRATED PEST MANAGEMENT, and no-till farming.

In crop rotation, different crops are alternately planted on a plot of land in order to maintain soil fertility. Corn, tobacco, and cotton, for instance, deplete nutrients, especially NITROGEN, from the soil. On the other hand, LEGUMES, such as alfalfa, barley, and beans, add nutrients to the soil. By alternating these crops, soil fertility is maintained. In addition to eliminating the need for fertilizers, crop rotation also limits the need for pesticides because INSECT pest populations have less time to increase before a different crop is planted.

INTEGRATED PEST MANAGEMENT (IPM) is another important part of sustainable agriculture. IPM is a system of pest management that uses a combination of methods to help reduce pest problems. One common IPM method is the use of a BIOLOGICAL CONTROL, which eliminates the need for synthetic pesticides. A biological control is a natural enemy of insect pests. For example, ladybugs often are used as biological controls against aphids. Other techniques of IPM that help eliminate the use of synthetic pesticides include insect sterilization, which controls the growth of insect populations by rendering insects infertile through exposure to radiation or certain chemicals; genetically engineered crops that are resistant to insects; and the use of natural insecticides, such as pyrethrin, which is a liquid extract of chrysanthemum flowers.

In a no-till agricultural practice, the seeds of crop plants are planted along with the remains of the previous crop. Tillage refers to the mechanical manipulation of soil, such as plowing. In no-till agriculture, soils are not tilled at the end of the growing season to remove old plants. The main benefit of no-till farming is that soils are never left bare and exposed to wind and water erosion. *See also* AGRICULTURE; AGROECOLOGY; SOIL CONSERVATION.

sustainable development, The management of environmental resources in a manner that will ensure that such resources will remain available for use long into the future. Sustainable development policies have been embraced and adopted by many nations throughout the world. During the 1992 UNITED NATIONS EARTH SUMMIT, the development and adoption of such policies were initiated in the document that came to be known as AGENDA 21. This document was signed by more than 90 nations who agreed to not only begin establishing sustainable development policies for their countries, but to also help developing nations implement similar policies within their borders as they strive to become more industrialized.

The major goal of sustainable development is the CONSERVATION OF NATURAL RESOURCES, including living things. This not only centers on using such resources in a wise manner that will prevent their depletion, but also on using them in a manner that will not degrade their quality for future generations. For example, fresh water is a resource that exists in limited amounts on Earth's surface, but is needed by most of Earth's ORGANISMS for their survival. In many places on Earth, bodies of fresh water have become polluted through the activities of humans or have become unsafe for use because of contamination with PATHOGENS. Sustainable development practitioners seek to ensure that potable water is used wisely, so that it remains in adequate supply to meet the demands of a growing POPULATION. At the same time, it is essential that practices that threaten to diminish the quality of fresh water be identified and eliminated as a means of protecting this valuable resource, while other methods are employed to improve the quality of water that already has suffered in quality because of human activities.

Preserving BIODIVERSITY is another goal of sustainable development. To achieve this goal, scientists hope to increase awareness of the ways in which the activities of humans impact other organisms that share the planet. It is hoped that by increasing such awareness, activities that now threaten the survival of many SPECIES will diminish and help slow the rate of EXTINCTION. Other goals of sustainable development include slowing human population growth, developing new methods of AGRICULTURE that will ensure that soils can sustain the growth of crops in a manner that does not degrade the ENVIRONMENT, and developing and making use of ALTERNATIVE ENERGY RESOURCES that will decrease international dependence on FOSSIL FUELS and result in the release of fewer POLLUTANTS into the environment. *See also* AGROECOLOGY; AGROFORESTRY; ALTERNATIVE FUELS; BIODIVERSITY TREATY; BIOMONITORING; BIOREGION; BIOSPHERE RESERVE; CLEAN AIR ACT; CLEAN WATER ACT; COUNCIL ON ENVIRONMENTAL QUALITY; DEBT-FOR-NATURE SWAP; DEEP ECOLOGY; EARTH DAY; EARTHWATCH; ECOTOURISM; GREENBELT MOVEMENT; INTERNATIONAL BIOLOGICAL PROGRAM; MAN AND THE BIOSPHERE PROGRAMME; MONTREAL PROTOCOL; NATIONAL ENVIRONMENTAL POLICY ACT; NATIONAL WILDLIFE REFUGE SYSTEM; NATURAL RESOURCES CONSERVATION SERVICE; RED LIST OF ENDANGERED SPECIES; TRADE RECORDS ANALYSIS OF FLORA AND FAUNA IN COMMERCE; UNITED NATIONS DEVELOPMENT PROGRAM; UNITED NATIONS ENVIRONMENT PROGRAMME; WORLD BANK; WORLD CONSERVATION MONITORING CENTRE; WORLD CONSERVATION UNION; WORLD RESOURCES INSTITUTE; WORLD WILDLIFE FUND.

EnviroSources

Sitarz, Daniel. *AGENDA 21: The Earth Summit Strategy to Save Our Planet.* Boulder, CO: Earth Press, 1993.

United Nations Earth Summit Web site: www.un.org/esa/earthsummit/index.html; www.un.org/dpcsd/earthsummit.

sustained yield, A harvest level for NATURAL RESOURCES, such as timber or crops, during a given time period that can be continued over time without jeopardizing the ability of the ENVIRONMENT to recover. The level of harvest that permits the environment to fully recover should allow that environment to continue producing a predictable supply of resources over time. The term "sustained yield" most often is applied to FOREST MANAGEMENT, particularly as it relates to timber; however, the term also applies to other natural resources such as crops and WILDLIFE. If harvests that exceed the maximum sustained yield are taken from an area, the environment will not completely recover and yields in subsequent years will decrease. *See also* SEED-TREE CUTTING; SELECTION CUTTING; SUSTAINABLE AGRICULTURE; SUSTAINABLE DEVELOPMENT.

swamp, A WETLAND in which the dominant PLANT life is woody vegetation, such as trees or shrubs. As a wetland, a swamp remains at least partly covered by water throughout the year and provides HABITAT to a variety of ORGANISMS, including waterfowl, migratory birds, AMPHIBIANS, and often large INSECT POPULATIONS. Unlike BOGS, swamps generally lack PEAT deposits on their floors. Depending upon their water source, swamps may be fresh-water or marine, and marine swamps can be tidal or nontidal in nature. *See also* ECOLOGICAL SUCCESSION; ESTUARY; MANGROVE; MARSH; SALT MARSH; SURFACE WATER; SWAMPBUSTER PROVISION.

EnviroTerm

fen: A WETLAND that accumulates PEAT and supports marsh-like (herbaceous) vegetation around its edges. A fen receives some drainage from nearby soil that is rich in MINERALS.

swamp gas. *See* methane.

swampbuster provision, A section of the Food Security Act of 1985 designed to protect WETLAND areas of the United States from conversion into farmland as a means of preserving these unique ECOSYSTEMS and the SPECIES that live there. The Food Security Act, or Farm Bill, provides federal funding to help farmers maintain their farms during periods of increasing agricultural costs. The swampbuster provision reduces or eliminates funding to farmers who grow crops on wetlands that have been filled in with soil to create new croplands. *See also* BOG; DOUGLAS, MARJORY STONEMAN; ECOLOGICAL SUCCESSION; EVERGLADES NATIONAL PARK; HABITAT; SWAMP.

symbiosis, A close association between two ORGANISMS of different SPECIES that benefits one or both organisms. If both organisms benefit from the association, the association is considered MUTUALISM. If only one organism benefits, while the other is not significantly affected, the relationship is called COMMENSALISM. The organisms involved in a symbiotic relationship are called symbionts. Symbiotic relationships exist in all ECOSYSTEMS because organisms continuously interact with each other and with their ENVIRONMENTS. For example, COMPETITION is a symbiotic relationship that exists when organisms of different species try to make use of the same NATURAL RESOURCES, such as food, living space, and water. Feeding relationships such as PREDATION and PARASITISM are also symbiotic associations.

Many mutualistic symbiotic relationships are obligatory because the organisms involved cannot survive without each other. For example, PLANTS called LEGUMES have nodules on their roots in which *Rhizobium* BACTERIA live. The bacteria carry out NITROGEN FIXATION, in which NITROGEN gas from the AIR is changed into a form of nitrogen the plant can use as a nutrient. The bacteria benefit from the HABITAT provided by the nodules. The plant benefits from the nutrients provided by the activities of the bacteria.

In a commensal symbiotic relationship, only one species benefits from the association. The other species neither benefits nor is harmed. A commensal association occurs between some species of WHALES and barnacles. The barnacles (which are sessile) live attached to the whale's skin and benefit by being moved throughout the world's OCEANS, providing them with greater access to food and aiding in the distribution of their species. The whale does not appear to benefit or be harmed from the relationship. *See also* ALGAE; CYANOBACTERIA;

ECOSYSTEM; FOOD CHAIN; FOOD WEB; FUNGI; INTEGRATED PEST MANAGEMENT; LICHENS; MYCORRHIZAL FUNGI; NICHE.

synfuel. *See* synthetic fuel.

synthetic fuel, Artificial liquid or gaseous FUEL derived particularly from COAL. Coal gasification is the process of converting coal into a gas. Coal gasification has provided fuel for municipalities since the mid-nineteenth century—though, by the 1950s, other sources predominated. Known as synthetic NATURAL GAS, or SNG, this fuel burns more cleanly than coal and can be transported via pipeline. SNG is produced by burning powdered coal and combining the waste products with HYDROGEN gas to form METHANE, the chief ingredient of natural gas. Coal liquefaction involves converting coal into liquid HYDROCARBONS such as METHANOL or synthetic GASOLINE. Synthetic fuels, or synfuels, can also be made from oil shale and BIOMASS (animal and PLANT wastes), but most are produced from solid coal.

Synfuels have several advantages over using solid coal. Synthetic fuels are generally less costly to use, easier to transport, and much less damaging to the ENVIRONMENT; however, the production of synthetic fuels is a very expensive process. It is much less expensive to build a coal-burning power plant that makes use of state-of-the-art technology than it is to build a synthetic fuel plant. Another problem with synfuels is that they contain only about one-third the ENERGY content of solid coal. Therefore, any large-scale production of synthetic fuels would rapidly deplete the world's coal supplies. *See also* ALTERNATIVE ENERGY RESOURCES; FOSSIL FUEL; NONRENEWABLE RESOURCE.

T

taiga, A CONIFEROUS FOREST BIOME characterized by very cold winters, high soil moisture, densely packed trees, and abundant snow. The taiga has been extensively exploited for timber in the southern portions of its range; most American softwood lumber comes from this type of FOREST. Taiga forests take a long time to regenerate (50 years or more) because of the short growing season. Replanting after logging leads to single-SPECIES conifer MONOCULTURES, resulting in a loss of BIODIVERSITY.

Other disruptions in the taiga ECOSYSTEM have included high-intensity hunting and trapping, which have reduced the numbers of fur-bearing animals, such as mink and marten. MINERAL development activities have caused HABITAT loss and pollution.

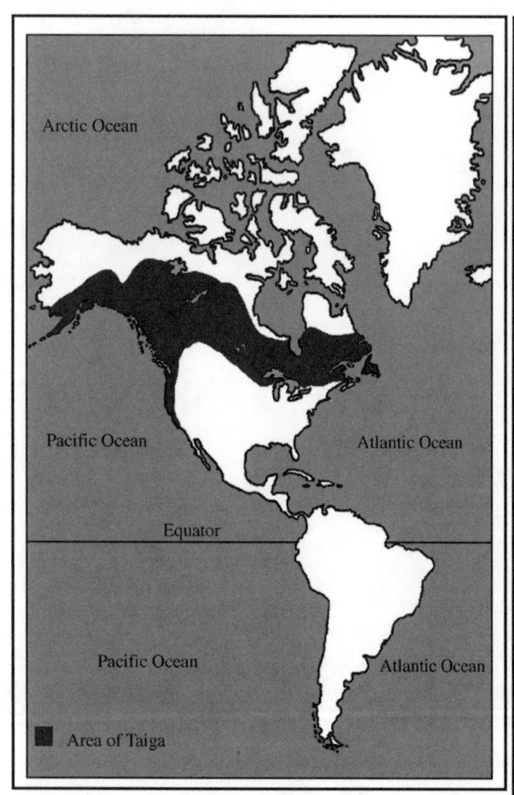

Areas of the taiga in the Western Hemisphere.

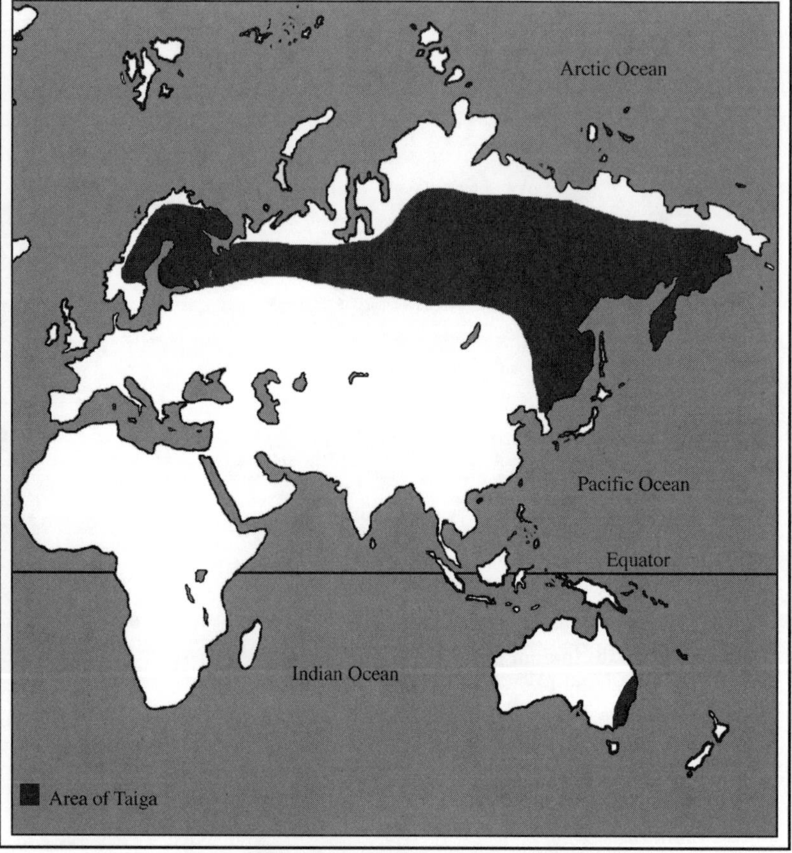

Areas of the taiga in the Eastern Hemisphere.

The taiga, also known as boreal forest and northern coniferous forest, is located in a nearly continuous belt in northern Eurasia and North America, between 45° and 57° north latitudes. The northern edge of the taiga is defined by a treeline south of the ARCTIC TUNDRA.

There is low tree species diversity in this ecosystem. In fact, most taiga tree species fall into only four genera: spruce, fir, pine, and tamarack or larch. (Tamarack is a deciduous conifer.) In general, plant species within these genera differ between North American and Eurasian taigas. If taiga conifers are cleared by fire or logging, broad-leaved species, such as aspen and poplar, invade.

There is a short growing season of around 130 days in the taiga, with a wide annual temperature range. Patchy PERMAFROST exists in about 65% of the taiga's range. Mineral soils are thin and poorly drained, while ORGANIC soils may range from several meters to 35 meters (105 feet) thick. Much of this biome is covered with lakes and ponds, which eventually fill with mineral and ORGANIC matter to form BOGS.

Typical MAMMALS include moose, deer, snowshoe hare, and bobcat. Taiga birds include migratory INSECT eaters, such as wood warblers, and year-round seed eaters, such as finches. *See also* PREDATION.

EnviroSource

U.S. Fish and Wildlife Service, Species List of Endangered and Threatened Wildlife Web site: www.fws.gov/r9endspp/lsppinfo.html.

tailings, Waste products from MINING activities, including small particles of COAL, zinc, COPPER, and LEAD. In some subsurface mining operations, about half the tailings are returned underground and mixed with cement to be used as backfill in the mine. The remaining tailings are stored in a special area on the surface. If sulfide-bearing tailings are left to weather on the surface, the small amounts of sulfide MINERALS (pyrite) that remain in the tailings could combine with AIR and water to produce ACID MINE DRAINAGE that would ultimately kill fish and PLANTS in surrounding waters. *See also* LEACHING; RECLAMATION.

Tailings are materials left over after mining. *Credit:* Tom Repine, West Virginia Geological and Economic Survey

taxonomy, The branch of biology that is primarily concerned with identifying and classifying the numerous ORGANISMS that contribute to Earth's BIODIVERSITY, on the basis of their evolutionary relationships. Taxonomists, scientists who study taxonomy, also are concerned with naming and describing organisms.

Organisms studied by taxonomists may be living or extinct. *(See* extinction.) To determine evolutionary relationships—how organisms are related to each other—taxonomists look at such features as CELL structure and function, anatomy, similarities and differences among the structures of proteins and nucleic acids, and the sizes, shapes, and numbers of chromosomes. Once all of these factors are studied, organisms are placed into groups called taxa, based on how their characteristics compare to those of other organisms. *See also* BOTANY; GENETIC DIVERSITY; WILSON, EDWARD OSBORNE; ZOOLOGY.

tectonic plates. *See* plate tectonics.

temperate deciduous forest, A Northern Hemisphere BIOME characterized by moderate CLIMATE, relatively high rainfall, well-defined seasons, and mostly deciduous trees. Worldwide, there is little virgin FOREST of this type left. The biome has been extensively affected by human activity. Much of it has been converted into agricultural fields, and sources of degradation include timber harvesting, fragmentation caused by highways and rights of way, PESTICIDE use, ACID RAIN, AIR POLLUTION, and suppression of natural fires.

The temperate deciduous forest occupies portions of the United States, Canada, central Europe, Japan, Korea, and China, south of the TAIGA's range. (Regions of similar climate occur in the Southern Hemisphere, but their FLORA and FAUNA differ from those of the northern biome. Instead, a mixed evergreen FOREST occurs.) In China, intensive agriculture has been practiced for at least 4,000 years; as a result, this forest is almost completely gone and is known primarily from the fossil record.

Annual leaf fall before the cold season is a defining feature of deciduous ECOSYSTEMS. Leaves change color in fall because shortening days cause trees to withdraw chlorophyll from their leaves and halt PHOTOSYNTHESIS in preparation for winter dormancy.

Dominant PLANTS in this forest are broad-leaved, thick-barked, deciduous SPECIES, including maple, beech, oak, hickory, and elm. Some conifers, mostly pines, are usually mixed in with hardwoods. Because the biome covers large geographical areas, it is generally subdivided into regions based on the dominant tree species or association of species, such as beech-maple or oak-hickory. Associations result from the varying tolerances of different tree species for specific water and temperature conditions.

In a mature deciduous forest, there is obvious stratification. An understory of saplings, shrubs, and herbs is typically well developed, and canopy trees are 20–35 meters (60–100 feet) high. Some understory species, particularly small species, flower in the spring before the tree canopy leafs out. Fire is a major factor in maintaining forest health and the diversity of herbs and shrubs on the forest floor. (*See* prescribed burning.)

Temperate deciduous forest fauna include black bears, brown bears, deer, moose, raccoons, skunks, and resident and migratory birds. Large CARNIVORES have been almost completely eliminated through deliberate human hunting, but untouched temperate deciduous forests would include timber WOLVES, cougars, and bobcats. The coyote has dispersed eastward and taken over the timber wolf's NICHE in many such forests.

Many of the FOOD CHAINS in this biome are DETRITUS-based; numerous ORGANISMS depend on the nutrients and ENERGY released from the leaves that drop to the forest floor each fall. Soils are rich; by comparison, a temperate area with poor soil and little water will support a temperate CONIFEROUS FOREST.

In temperate deciduous forests, precipitation ranges from 75 centimeters (30 inches) per year in northern portions to 150 centimeters (60 inches) in southern portions. Frost occurs throughout the biome. The growing season in temperate deciduous forests ranges from 140–300 days per year, and the average temperature is 10°C (about 50°F).

Almost all the temperate deciduous forests of eastern North America are second-growth, but they preserve the world's greatest BIODIVERSITY of temperate deciduous forest flora and fauna. For example, the Great Smoky Mountains National Park in Tennessee and North Carolina has been designated a world BIOSPHERE RESERVE to protect its rich assortment of species. *See also* FIRE ECOLOGY; WATER CYCLE.

EnviroSources

Society of American Foresters Web site: www.safnet.org.
U.S. Forest Service Web site: www.fs.fed.us.
World Resources Institute Forest Frontiers Initiative Web site: www.wri.org/ffi.

temperate rainforest, A primarily coastal CONIFEROUS FOREST BIOME characterized by thick stands of trees, high rainfall, and high humidity. This ECOSYSTEM, also known as moist coniferous forest, occurs in wet, cool CLIMATES where marine AIR meets coastal mountains. Commercial timber activities have dramatically reduced the extent of this FOREST type, but the technique of CLEAR-CUTTING—which caused much of the destruction—is generally no longer practiced. In recent years, fire suppression efforts have increased tree losses by enabling the build-up of FUEL, resulting in intense, uncontrollable FOREST FIRES.

Temperate rainforests sustain the highest standing BIOMASS of all terrestrial ecosystems and are among the most productive softwood forests in the United States. Most American OLD-GROWTH FORESTS are located in this biome. The temperate rainforest is considered an ancient relictual forest, featuring "living fossils," PLANTS that were once much more widely distributed around the world. Primary tree SPECIES include Douglas fir, California redwood, and western red cedar.

Temperate rainforests have always had very limited worldwide distribution. Today, undisturbed forests of this type exist in Chile and in the Pacific Northwest region of North America, where they extend in a narrow band from the southern part of Alaska to central California. Because of prevailing moisture-laden winds, there is also some temperate rain forest on the western slopes of the Rocky Mountains in Idaho and British Columbia.

Abundant precipitation—over 200 centimeters (80 inches) per year in places—is characteristic, in the form of both rain and snow. Fog is also an important component in the southern portion of the range. Cool temperatures are usual, with only a moderate range from summer to winter. In most places, the forest is thick, with a dark floor; understory species are limited to mosses and ferns because of the lack of light.

The Goshawk lives in a temperate deciduous forest which is a Northern Hemisphere biome characterized by moderate climate, relatively high rainfall, well-defined seasons, and mostly deciduous trees.

Soils in the temperate rainforest are deep and rich, but MINERAL-poor, supporting diverse conifers and hardwoods, many of which are very tall and old. Redwoods, for example, have thick, INSECT- and PATHOGEN-resistant bark, which helps some individual trees live to be 1,000 to 2,000 years old. There are many NATIVE SPECIES of shrubs and wildflowers, and the forest provides HABITAT for numerous resident birds and MAMMALS. Fur-bearing mammals, such as otters, were once common before being overhunted. *See also* WATER CYCLE.

EnviroSources

Greenpeace International (forests) Web site: www.greenpeace.org/~forests.
Society of American Foresters Web site: www.safnet.org.
U.S. Forest Service Web site: www.fs.fed.us.
World Conservation Monitoring Centre Web site: www.wcmc.org.uk.
World Resources Institute, Forest Frontiers Initiative Web site: www.wri.org/ffi.
World Wildlife Fund (Worldwide Fund for Nature), Forests for Life Campaign Web site: www.panda.org/forests4life.

temperature inversion, A condition that occurs in the ATMOSPHERE when AIR temperature increases with an increase in altitude. As an example, a temperature inversion can occur when a warm layer of air moves over a cooler layer of SMOG, which contains high levels of POLLUTANTS. The warm air traps the cooler air so the smog cannot disperse. As a result, the smog continues to build up, causing harmful and dangerous air conditions at ground level.

Tennessee Valley Authority, The nation's largest electric power producer, a regional economic development agency, and a national center for environmental research. The Tennessee Valley Authority (TVA) is an independent agency of the executive department of the U.S. government that was founded by Congress in 1933. Major functions of the corporation include management of the Tennessee River system, generation of ELECTRICITY, sale and transmission of electricity, investment in economic development activities, stewardship of TVA assets, and research and technology development.

The TVA serves as a testing ground for agricultural, environmental, education, economic, and industrial programs. The agency has been involved in developing 75% of the FERTILIZER used in the United States. The TVA has developed new clean COAL technologies that reduce ACID RAIN and other AIR TOXINS related to burning coal. The TVA operates one of the nation's largest facilities for researching, developing, and demonstrating how WETLANDS can alleviate water pollution problems. TVA foresters are working to monitor the health of 14 million hectares (33 million acres) of the nation's FORESTS.

EnviroSource

Tennessee Valley Authority (TVA) Web site: www.tva.gov.

terracing, A method of reshaping hilly terrain in order to grow crops and inhibit soil EROSION. Terracing is commonly practiced in the mountainous regions of South America, Africa, China, Japan, the Philippines, and other places that experience heavy rainfall. In these areas where water can flow rapidly downhill, soil erosion is a constant problem. Terracing reduces soil erosion by slowing waters to nonerosive speeds.

In the early practice of terracing, the land was typically shaped into a series of broad, step-like platforms. This method, called bench terracing, which is still commonly used today on steeper slopes, is very effective at reducing soil degradation because water gradually flows from one nearly level terrace to the next. Bench terracing is also a good technique for growing moisture-loving crops, such as rice. When water runs down a hill, it gets trapped on each terrace. Any water that does run off simply cascades to the next terrace. In this way, large pools of water can be created—an ENVIRONMENT perfect for growing rice and other crops.

Most modern terracing practices involve building wide terraces with shallow channels and levees that help carry water at slow speeds. Such broad terraces are typically used on gently sloping land. Because these terraces are so wide, crops can be grown and worked with modern-day machinery. *See also* AGRICULTURE; SOIL CONSERVATION.

tetrachloroethylene, A colorless, nonflammable liquid at room temperature that evaporates easily. It is also known as perchloroethylene (PCE) or "perc" and is a member of a group of ORGANIC chemicals, most of which are SOLVENTS, known as chlorinated organic compounds. Exposure to tetrachloroethylene happens mostly from breathing contaminated AIR and drinking contaminated water. Short-term exposure to high levels of this chemical can cause damage to the central nervous system and may lead to death.

Tetrachloroethylene is used for dry cleaning fabrics, processing and finishing textiles, as a metal parts degreaser, in the manufacture of other chemicals (primarily FLUOROCARBONS), and as a heat-exchange fluid. Perc is also found in small amounts in many household products such as rug and upholstery cleaners.

The main source of perc in the ENVIRONMENT is from evaporation during dry cleaning and degreasing uses. Other sources include evaporation and LEACHING from waste disposal sites, EMISSIONS from sites where it is produced or used, evaporation from textile manufacturing processes, and the many household products that include perc as an ingredient.

The OCCUPATIONAL SAFETY AND HEALTH ADMINISTRATION limits the amount of perc in workplace air to 100 parts per million (ppm) averaged 8 over a 40-hour workweek. The National Institute for Occupational Safety and Health recommends that perc be handled as a potential CARCINOGEN and states that workplace air levels should be as low as possible. *See also* PARTS PER BILLION AND PARTS PER MILLION.

EnviroSource

U.S. EPA (chemical summary) Web site: www.epa.gov/opptintr/chemfact/s_perchl.

thermal desorption, A method used in SUPERFUND cleanups to remove HAZARDOUS WASTES from contaminated soils by heating the soil to 100–540°C (200–1,000°F). This allows CONTAMINANTS with low boiling points to vaporize (turn into a gas) and separate from the soil. Any contaminants not vaporized must be treated by other means. The vaporized contaminants are then treated before release into the ATMOSPHERE. Thermal desorption is most effective at removing ORGANICS from contaminated soils. It can separate SOLVENTS, PESTICIDES, POLYCHLORINATED BIPHENYLS (PCBs), DIOXINS, and FUEL oils from contaminated soil. Thermal desorption is not effective on most metals, and its effectiveness may be reduced depending on the composition of the soil being processed.

thermal pollution, A DISCHARGE of heated water into lakes, rivers, or other natural waterways that can cause harm to aquatic ORGANISMS and destroy ECOSYSTEMS. Most discharges result from industrial processes, particularly from FOSSIL FUEL and nuclear power plants. The discharged water increases the water temperature and decreases the amount of DISSOLVED OXYGEN in the lake or river, which can disrupt FOOD WEBS and prevent fish from spawning. The warm water can also encourage undesirable organisms to reproduce and live. In some cases, the warmer water may attract some SPECIES during the cold season. Heated water can be treated by the use of COOLING TOWERS and COOLING PONDS. These systems allow the discharged water to cool off by evaporation.

thermocline, A layer of water present in OCEANS and lakes and ponds that divides a relatively warm layer of water above, called the epilimnion, and a bottom layer of cold water called the hypolimnion. The thermocline is one of the three major water layers in aquatic ECOSYSTEMS. The thermocline may occur at varying depths. Its position, or depth, within the ecosystem is linked to WEATHER, CLIMATE changes, and nutrient availability, and plays a role in determining the distribution of ORGANISMS within oceans, lakes, and ponds.

In the thermocline, water temperature rapidly decreases as depth increases. Because of the temperature changes that occur within the thermocline, it is often called the discontinuity layer. It is also sometimes called the metalimnion.

The thermocline of the Pacific Ocean in areas near the equator is usually located near the surface; however, during EL NIÑO years, the thermocline sinks deeper than normal into the ocean because of the downward pressure exerted by waves moving from west to east. The warm water above the thermocline prevents the deeper layer of nutrient-rich cold water from UPWELLING to the surface. As a result of this absence of nutrients, POPULATIONS of marine ORGANISMS that depend on the cold-water nutrients die off because of a lack of food or move to different regions of the ocean. This mass movement of ocean organisms, in turn, greatly affects the fishing industry, worldwide, as some regions begin experiencing higher-than-normal harvests, while other areas, nearer the equator have smaller-than-normal catches.

thermodynamics. *See* laws of thermodynamics.

thermosphere. *See* atmosphere.

third law of thermodynamics. *See* laws of thermodynamics.

Thoreau, Henry David (1817–1862), An American writer, teacher, NATURALIST, businessman, engineer, surveyor, and lecturer who is considered by many to be the father of the American preservation and CONSERVATION movement. A key figure in the transcendentalist movement of nineteenth-century American literature, Thoreau was a contemporary of writers Ralph Waldo Emerson and William Ellery Channing. He is best known for his retreat to the woods around Walden Pond in Concord, Massachusetts, where he wrote extensively in a series of journals, during a period of two years, about his experience living alone. These journal entries later became his book, *Walden* (1854). Thoreau's observations of nature enrich all of his work.

Thoreau strongly believed in an individual's obligation to determine right from wrong. He wrote about this in his widely read essay, "Civil Disobedience" (1849), in which his doctrine of passive resistance inspired later intellectuals and political figures, including Mahatma Gandhi of India.

Thoreau considered the ownership of material possessions beyond the basic necessities of life to be an obstacle. Nature was Thoreau's favorite writing theme. The question of how we should live was his second favorite. All of his writing except his poetry is expository (nonfiction).

One of his most widely quoted phrases is from *Walden*: "If a man does not keep pace with his companions, perhaps it is because he hears a different drummer. Let him step to the music which he hears, however measured or far away."

EnviroSources

Paul, Sherman. *The Shores of America: Thoreau's Inward Exploration.* Urbana, IL: University of Illinois Press, 1959.
Richardson, Robert D. *Henry Thoreau: A Life of the Mind.* Berkeley, CA: University of California Press, 1986.
Roger Tory Peterson Institute of National History Web site: www.rtpi.org.
Thoreau, Henry David. *Wild Fruits: Thoreau's Rediscovered Last Manuscript.* New York: W.W. Norton, 1999.

threatened species, A term used to identify SPECIES whose POPULATIONS are larger and more stable than those recognized as ENDANGERED SPECIES, but have the potential of becoming an endangered species. The U.S. ENDANGERED SPECIES ACT of 1973 identifies threatened PLANTS and animals as those that are "likely to become endangered in the foreseeable future throughout all or a significant portion of its range." In the United States, species that have been identified as threatened species through their inclusion on the ENDANGERED SPECIES LIST of the U.S. FISH AND WILDLIFE SERVICE (FWS) are offered protection under the Endangered Species Act. Such protections include bans on hunting of these species as well as restrictions on disturbing or destroying the critical HABITATS of such species. Critical habitats are those that are deemed to be essential to the survival of the species.

The endangered species list of the FWS includes species not native to the United States; however, a more comprehensive list of such species is maintained by the WORLD CONSERVATION UNION (IUCN) in its Red List of Threatened Animals and Red List of Threatened Plants. These lists, which are updated about every three years, indicate that the total number of plant and animal species that are considered threatened exceeds 5,200. It identifies the greatest threat to these species as habitat destruction, especially that related to logging and CLEAR-CUTTING in the world's tropical RAINFORESTS. To help combat this problem, the IUCN is working with many nations and CONSERVATION groups to establish WILDLIFE REFUGES and nature reserves throughout the world to help protect the critical habitats of species. Other major threats to species include the use of PESTICIDES, the release of POLLUTANTS into the ENVIRONMENT, the collection of rare plant species, overhunting, and POACHING. *See also* ADDO NATIONAL ELEPHANT PARK; AFRICAN WILDLIFE FOUNDATION; AMERICAN ZOO AND AQUARIUM ASSOCIATION; ARCTIC NATIONAL WILDLIFE REFUGE; BIODIVERSITY; BIODIVERSITY TREATY; BIOLOGICAL COMMUNITY; BIOME; BIOREGION; BIOSPHERE; BIOSPHERE RESERVE; CARARA BIOLOGICAL RESERVE; EARTHWATCH; ENVIRONMENTAL IMPACT STATEMENT; EVERGLADES NATIONAL PARK; FOSSEY, DIAN; GALÁPAGOS ISLANDS; GALDIKAS, BIRUTÉ; GOODALL, JANE; GREENPEACE; INDICATOR SPECIES; INTERNATIONAL BIOLOGICAL PROGRAM; KEYSTONE SPECIES; MADAGASCAR; MAN AND THE BIOSPHERE PROGRAMME; MARINE MAMMAL PROTECTION ACT; NATIONAL AUDUBON SOCIETY; NATIONAL WILDLIFE REFUGE SYSTEM; UNITED NATIONS EARTH SUMMIT; WORLD CONSERVATION MONITORING CENTRE.

Three Gorges Dam Project, The current construction of a great DAM on China's Yangtze (Chang Jiang) River that could provide HYDROELECTRIC POWER, navigation, and FLOOD control. It is an enormous engineering project that will profoundly alter the character of the area. It may serve as one of the most striking examples of human modification of the ENVIRONMENT by controlling a body of SURFACE WATER. The project is also a major concern to many environmental organizations in China and elsewhere.

The planning stage of the Three Gorges Dam began in the mid-twentieth century with the evaluation of possible construction sites and effects. The Chinese government, recognizing the great economic potential and power of the Yangtze River, supported the idea of a single, colossal dam. An alternative view, which was not accepted, envisioned a series of smaller dams on tributary rivers of the Yangtze, which, if constructed as preliminary projects, might provide the ENERGY and flood control without the risks presented by a single dam.

The Yangtze River, the third longest of the world's great rivers, stretches almost 6,400 kilometers (4,000 miles). The river has periodically flooded throughout human history, caus-

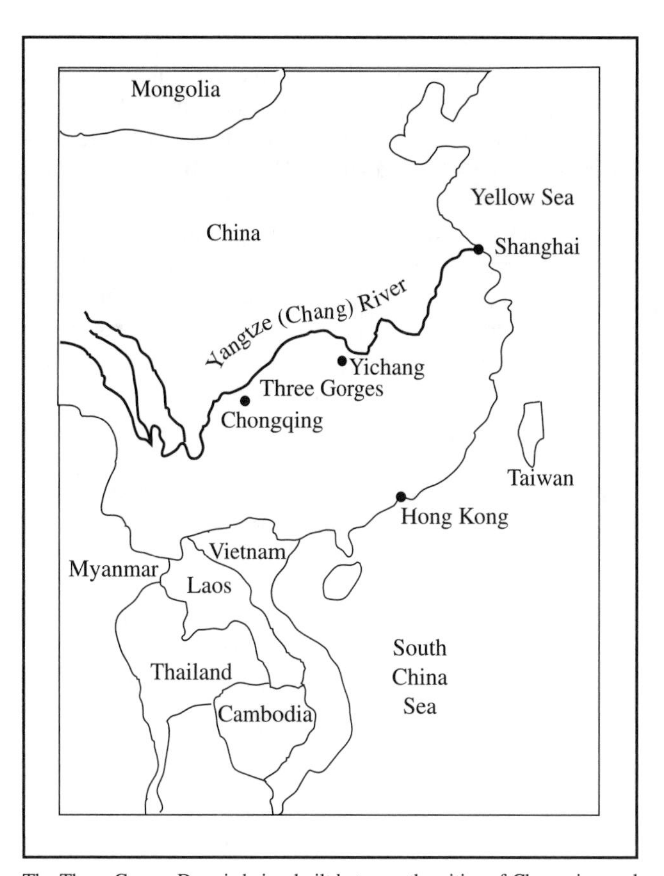

The Three Gorges Dam is being built between the cities of Chongqing and Yichang. The great dam on China's Yangtze (Chang Jiang) River is being constructed to provide hydroelectric power, better navigation, and flood control.

ing great death and destruction. Dams can provide flood control by releasing excess water as needed to buffer the effects of heavy precipitation or RUNOFF. The proposed Three Gorges Dam will create a lake approximately 600 kilometers (375 miles) long and up to 180 meters (600 feet) deep. This great water body would permit deep-water vessels to travel 2,400 kilometers (1,500 miles) into the interior of China. The unquestionable destructive power of the Yangtze is also the source of its enormous potential for hydroelectric power. The Three Gorges Dam could produce an estimated 18,200 megawatts of ELECTRICITY.

In the opinion of the dam's critics, the construction of the Three Gorges Dam likely will result in a variety of social and environmental problems. The economic costs of the construction, the displacement of communities, and the loss of historical sites may cause a great deal of discontent. The environmental aftermath of dam construction could include geological and biological impacts. Large accumulations of sediment behind the dam could lead to structural failure—Chinese dams have a higher rate of such failures than do comparable dams elsewhere in the world. The lake formed by the project could suffer from severe water pollution resulting from flooded landfills and SEWAGE DISCHARGES. Such pollution could contaminate drinking water supplies, kill aquatic life, and encourage outbreaks of disease. The Three Gorges Dam is scheduled to be completed by 2009.

Three Mile Island, A NUCLEAR REACTOR in Middleton, Pennsylvania, whose Unit 2 (TMI-2) was the site of the worst U.S. nuclear power plant accident. On the morning of March 28, 1979, equipment failure, design deficiencies, and worker error led to severe damage to the TMI-2 reactor core and small releases of RADIATION that were transported into neighboring communities in the air. The event began around 4 A.M. when a shutdown occurred in the secondary (nonnuclear) section of the plant. Mechanical or electrical failure caused the main feed-water pumps to stop running, thus preventing the steam generators from dispersing heat. First the TURBINE and then the reactor automatically shut down. Immediately, the pressure in the primary system (the nuclear portion of the plant) began to increase, prompting the pressurizer relief valve to open. The valve should have closed when the pressure decreased by a certain amount, but did not, and signals available to the operator failed to indicate this. The stuck-open valve allowed the pressure in the system to continue decreasing.

Meanwhile, another problem occurred elsewhere in the plant. As part of a routine test, a valve in the emergency feed-water system had been closed and through either administrative or human error did not reopen. This prevented the emergency feed-water system from functioning. The valve was discovered closed about eight minutes into the accident. Once it was reopened, the emergency feed-water system began working correctly again, allowing cooling water to flow into the steam generators.

As pressure in the primary system continued decreasing, water redistributed and filled the pressurizer. The level indicator that informs the operator of the amount of coolant capable of heat removal falsely indicated that the system was full, prompting the operator to stop adding water. Because adequate cooling was unavailable, the nuclear fuel overheated to the point where some of the zirconium cladding (the long metal jackets that hold nuclear FUEL pellets) reacted with the water, generating HYDROGEN, which was released into the reactor containment building. By March 30, two days after the chain of events began, some hydrogen still remained within the containment vessel, forming a bubble above the reactor core. The concern was that if reactor pressure decreased, the hydrogen bubble would expand and interfere with the flow of cooling water through the core, or possibly cause an explosion. Over the next few days, the bubble was reduced in volume by adjusting AIR and water pressure within the pressurizer.

The primary reactor damage occurred two to three hours into the accident as a result of the inadequate coolant level. Although fuel did not "melt down" through the floor of the containment building or through the steel reactor vessel, a significant amount of fuel did, in fact, melt. Radioactivity in the reactor coolant increased dramatically, and small leaks occurred in the reactor coolant system. This resulted in high radiation levels in other parts of the plant, triggering small releases, primarily of xenon gas, into the ENVIRONMENT.

Some water contaminated with fuel debris and nuclear FISSION products also escaped from the reactor coolant system and flowed into the reactor building basement. The radioactive water in the basement was heated by residual heat from the reactor vessel, evaporated, condensed on the building's interior walls, and drained back down into the basement. Radionuclides then permeated the porous concrete and layers of iron, corroding them. This area of the plant became a major focus of the subsequent cleanup and DECONTAMINATION.

The U.S. NUCLEAR REGULATORY COMMISSION (NRC) was notified at 7:45 A.M. on March 28. The regional office promptly dispatched the first team of inspectors to the site, and other agencies, such as the U.S. DEPARTMENT OF ENERGY (DOE) and the U.S. ENVIRONMENTAL PROTECTION AGENCY (EPA), mobilized their response teams. At 9:15 A.M., the White House was notified, and at 11:00 A.M., all nonessential personnel were ordered off the plant's premises.

On Friday, March 30, Governor Thornburgh of Pennsylvania ordered a precautionary evacuation of preschool children and pregnant women from within a 8-kilometer (5-mile) radius of the plant. Most evacuees returned to their homes by April 4, by which time the situation at the reactor had been brought under control.

Detailed studies of the radiological consequences of the accident were conducted by the NRC, the EPA, the Department of Health, Education, and Welfare (now the Department of Health and Human Services), the DOE, and the state of Pennsylvania. Several independent studies also were conducted. The average dose of radiation received by the two million people in the area was estimated about 1 millirem. Compared to the area's natural radioactive background dose of about 100–125 millirems per person per year, the collective dose to the community from the accident was very small:

The maximum dose to a person standing at the site boundary would have been less than 100 millirems.

In the months following the accident, questions arose about possible adverse effects from radiation on human, animal, and PLANT life in the TMI area. Thousands of samples of air, water, milk, vegetation, soil, and food were collected by groups monitoring the area; only very low levels of radionuclides could be attributed to releases from the accident. Several comprehensive investigations and assessments have concluded that in spite of serious damage to the reactor, most radiation was contained and the actual release had negligible effects on the health of individuals or the environment.

Today, the TMI-2 reactor is permanently shut down and defueled; the reactor coolant system has been decontaminated, radioactive liquids treated, and most components shipped to a licensed, LOW-LEVEL RADIOACTIVE WASTE (LLW) disposal site. The site is still being monitored. The plant's owner, General Public Utilities Nuclear Corporation, says it will keep the facility in long-term "storage" until the operating license for the TMI-1 plant expires in 2014, at which time both plants will be decommissioned.

Although the accident did not result in death or injury among plant workers or community members, it nevertheless brought about sweeping changes in emergency response planning, reactor operator training, human-factor engineering, radiation protection, and other areas of nuclear power plant operations related to safety and environmental protection. It also caused the NRC to tighten its regulatory oversight. For example, at each nuclear plant operating in the United States today, two certified inspectors must live nearby, work exclusively in that plant, and perform daily surveillance for adherence to NRC regulations. Plant design and equipment requirements also have been upgraded significantly. *See also* CHERNOBYL; CONTAMINANT; DOSE/DOSAGE; HAZARDOUS WASTE; MELTDOWN; NUCLEAR ENERGY; RADIATION SICKNESS; RADIOACTIVE WASTE.

EnviroSources

Nuclear Regulatory Commission: "Answers to Frequently Asked Questions about Cleanup Activities at Three Mile Island, Unit 2." NUREG-0732.

Nuclear Regulatory Commission. "Investigation into the March 28, 1979, Three Mile Island Accident by the Office of Inspection and Enforcement." NUREG-0600.

Nuclear Regulatory Commission, Public Document Room: 2120 L Street, NW, Washington, DC, 20037; (202) 634-3273; (800) 397-4209.

Nuclear Regulatory Commission. *Report of the President's Commission on the Accident at Three Mile Island.* October, 1979.

Nuclear Regulatory Commission Web site: www.nrc.gov/OPA/gmo/tip/tmi.html.

Public Broadcasting System. "The American Experience: Meltdown at Three Mile Island." Web site: www.pbs.org/wgbh/pages/amex/three.

Rogovin, Mitchell, and George T. Frampton. *Three Mile Island: A Report to the Commissioners and to the Public.* Vols. 1–2, 1980.

Stephens, Mark. *Three Mile Island.* New York: Random House, 1980.

tidal power, The production of ELECTRICITY by harnessing the mechanical ENERGY in TIDES. The first tidal power plant in North America was built in 1984 on the bay at Annapolis Royal, Nova Scotia. The power plant uses tidal energy to generate about 20 megawatts of electricity. Twice each day, along most coastlines, the sea level slowly rises and falls a few meters due to the gravitational forces exerted on Earth by the sun and the moon. High tide is when sea level is high and the tide flows in to shore. Low tide occurs when the sea level is low and water flows out toward the sea. Tidal power is a form of HYDROELECTRIC POWER. In all hydroelectric power plants, electrical energy is generated when water flows over a TURBINE, a type of fan blade. As water pushes on the blades, the turbine spins. This spinning action is transferred to coils of wire inside a generator, which produces electricity.

On most coastlines, tides move in and out very slowly. As a result, the energy in these tides is insufficient to spin a turbine generator. To solve this problem, most tidal power plants use DAMS to capture and later release water with more force. The rising and the falling of the water level is used to spin a turbine generator.

Tidal power, like SOLAR ENERGY and WIND POWER, is a relatively "clean" energy source. It is nonpolluting and does little if any other damage to the ENVIRONMENT. Tidal power is also advantageous because it is a RENEWABLE RESOURCE. In other words, it will never run out, unlike nonrenewable sources of energy, such as PETROLEUM and NATURAL GAS. The main drawback to this tidal energy, however, is that it is very expensive. The cost of building a tidal power plant is far greater than the cost of building a conventional power plant with the same capacity to generate electricity. *See also* ALTERNATIVE ENERGY RESOURCE; NONRENEWABLE RESOURCE.

EnviroSource

Friends of the Earth Web site: www.foe.co.uk/CAT/publicat/waterpower/tidal.html.

tide, The periodic rise and fall of waters in the OCEAN as well as in some large lakes and inland seas due to the gravitational pull of the sun and moon on Earth. Low tide and high tide alternate in a continuous and predictable cycle. Along most coastlines throughout the world, two high tides, or flow tides, and two low tides, or ebb tides, occur each day. The enormous amount of mechanical ENERGY associated with tidal movement is now being utilized by TIDAL POWER plants throughout the world to generate ELECTRICITY. *See also* ALTERNATIVE ENERGY RESOURCE; BAY OF FUNDY; HYDROELECTRIC POWER.

tigers, Endangered, predatory CARNIVORES; the largest members of the cat family. Tigers are recognized by their distinctive coats, which are usually yellow-orange with dark vertical stripes. All tigers belong to the SPECIES *Panthera tigris* of which there are two major subspecies—the Siberian tiger (*Panthera tigris altaica*) and the Bengal tiger (*Panthera tigris tigris*). Both subspecies are at risk of EXTINCTION; however, the Siberian tiger is in more imminent danger.

The Siberian tiger is the largest living member of the cat family. It makes its home in temperate FORESTS and lives in the far eastern (Asian) regions of Russia, with some members of the subspecies ranging as far north as the Arctic Circle. Current estimates indicate that only about 200 Siberian tigers remain in the wild; many others are raised in captivity at ZOOS and theme parks throughout the world.

The Bengal tiger once roamed widely throughout Southeast Asia and India, but most of these tigers today live in the tropical RAINFORESTS of India. Estimates indicate that about 4,000 Bengal tigers still exist in the wild. One activity threatening these tigers is DEFORESTATION; however, the greatest threat to both the Bengal and the Siberian tiger is POACHING to obtain their pelts and body parts (including the tail, teeth, bones, whiskers, and reproductive organs), which have traditionally been used for a variety of purposes throughout the Asian continent. For example, tiger bone long has been ground into a powder and used in medicines. Whiskers are used as protective charms. The tail of the tiger is ground and mixed with soaps used to treat skin DISEASES. Reproductive organs are used as aphrodisiacs.

Throughout the world, various legislation has been enacted that make it illegal to capture, kill, and sell tigers and their parts. Since 1972, tigers have been included on the ENDANGERED SPECIES LIST of the United States. Tigers also are included on the RED LIST OF ENDANGERED SPECIES maintained by the WORLD CONSERVATION UNION (formerly the International Union for Conservation of Nature). Additional international protection is offered to tigers through the CONVENTION ON INTERNATIONAL TRADE IN ENDANGERED SPECIES OF WILD FLORA AND FAUNA. All of these legislative actions contain provisions for heavy fines for buying, selling, or trading in tiger products as well as for activities that involve the capture, injury, or killing of the animals. In addition, several nonprofit environmental groups, such as the TRADE RECORDS ANALYSIS OF FLORA AND FAUNA IN COMMERCE established and funded by the WORLD WILDLIFE FUND, work to make the public knowledgeable about the plight of the tiger and also employ people to help identify poachers and traffickers of ENDANGERED SPECIES. *See also* ADAMSON, JOY GESSNER; FOSSEY, DIAN; GALDIKAS, BIRUTÉ; GOODALL, JANE; GORILLAS; ORANGUTANS; WORLD CONSERVATION MONITORING CENTRE.

Endangered Tigers At a Glance	
Bengal tiger (*Panthera tigris tigris*)	
Length	3 m (10 ft)
Weight	290 kg (650 lbs)
Estimated Population	4,000
Status	Endangered
Habitat	Mainland of Southeast Asia; India
Siberian tiger (*Panthera tigris altaica*)	
Length	4 m (13 ft)
Weight	300 kg (700 lbs)
Estimated Population	200
Status	Endangered
Habitat	Asia, as far north as Arctic Circle

EnviroSources

Elsa Wild Animal Appeal: P.O. Box 4572, North Hollywood, CA 91617-0572; (818) 761-8397.

World Wildlife Fund: 1250 Twenty-fourth St., NW, Washington, DC 20037-1175.

The tiger is an endangered, predatory carnivore that comprises the largest member of the cat family.

Times Beach Superfund Site, A Missouri HAZARDOUS WASTE site that was targeted for cleanup as required by the federal program known as SUPERFUND. The 1980 COMPREHENSIVE ENVIRONMENTAL RESPONSE, COMPENSATION, AND LIABILITY ACT (CERCLA) established the Superfund to facilitate the cleanup of hazardous waste sites such as Times Beach, under the auspices of the U.S. ENVIRONMENTAL PROTECTION AGENCY (EPA).

Investigations of Times Beach in the 1980s revealed that oil contaminated with DIOXIN, a highly toxic substance, had been used to treat the town's streets. Dioxin waste from a chemical manufacturing operation had been mixed with waste oil, making the oil a hazardous waste. A contractor then sprayed the waste oil on town roads to keep the dust down. In 1983, the residents of Times Beach were ordered to evacuate the community.

In 1990, the EPA and Missouri environmental officials agreed to a cleanup plan for Times Beach that entailed construction of an incinerator to burn soil excavated from the contaminated portions of the town and from 27 other waste sites in eastern Missouri. From 1996 to 1997, a specially designed incinerator was used to destroy approximately 241,000 metric tons (265,000 tons) of the dioxin-containing waste materials. By the fall of 1997, all cleanup work at the site was completed by the EPA and Syntex Agribusiness, the company that assumed responsibility for the site's cleanup in 1990. *See also* INCINERATION.

EnviroSources

Lixley, Bruce (producer.) "Times Beach, Missouri." (videotape.) 57 mins. Media Process Educational Films, n.d.

U.S. Environmental Protection Agency (Returning Superfund Sites to Productive Use: Times Beach, Missouri) Web site: www.epa.gov/oerrpage/superfund/programs/recycle/1-pagers/timesbch.html.

Wildavsky, A.B. *But Is It True? A Citizen's Guide to Environmental Health and Safety Issues.* Cambridge, MA: Harvard University Press, 1995.

tolerance level, The approved residue levels for PESTICIDES in processed foods and raw agricultural products. The U.S. ENVIRONMENTAL PROTECTION AGENCY establishes a tolerance level whenever a pesticide is registered for use on food or feed crops. *See also* TOXIC CHEMICALS.

toluene, A colorless, liquid chemical that occurs naturally in PETROLEUM oil and in the tolu tree. Toluene belongs to a group of ORGANIC chemicals called "aromatic HYDROCARBONS." Exposure to toluene happens most often through breathing AIR in a contaminated workplace and from AUTOMOBILE exhaust. Breathing high levels of toluene can have serious effects on the brain and can cause headaches, confusion, dizziness, sleepiness, and memory loss.

Toluene is used to manufacture paints, paint thinners, nail polish, laquers, certain chemicals, adhesives, rubber, and in some printing and leather tanning processes. Because it is a common SOLVENT found in many consumer products and is volatile, anyone can be exposed to it at home or outdoors.

The OCCUPATIONAL SAFETY AND HEALTH ADMINISTRATION has set a limit of 200 parts per million toluene for air in the workplace averaged over a 40-hour workweek. *See also* PARTS PER BILLION AND PARTS PER MILLION.

EnviroSources

Agency for Toxic Substances and Diseases Registry, Division of Toxicology: 1600 Clifton Road NE, Mailstop E-29, Atlanta, GA 30333; (404) 639-6000

U.S. Environmental Protection Agency (chemical summary) Web site: www.epa.gov/opptintr/chemfact/s_toluen.txt.

tornado, A severely violent and damaging thunderstorm that is generally characterized by a funnel-shaped cloud of rapidly whirling wind. The funnel clouds generally form when warm AIR is forced rapidly upward, causing a sudden drop in air pressure. These drastic changes cause a convection current of winds that whirl around the low-pressure area to form a funnel. The strong winds nearest the core of a funnel can move at speeds greater than 500 kilometers (310 miles) per hour, and the funnel-shaped cloud can extend from a thundercloud down to Earth's surface. Tornadoes that reach Earth's surface move along the ground in a narrow path, destroying virtually everything along their course. In the United States, there are on the average about 750 tornadoes a year. The most destructive tornadoes are categorized as F5 with wind speeds of 345 to 510 kilometers (216 to 319 miles) per hour and F4 with wind speeds of 330 to 415 kilometers (207 to 260 miles) per hour. *See also* CLIMATE; CYCLONE; HURRICANE; NATURAL DISASTER; TYPHOON; WEATHER.

toxic chemical, A substance that is harmful to the health of living things. TOXICOLOGY is the study of the toxic effects of chemicals. Toxicologists have developed methods to rate the TOXICITY of substances. Generally, toxicologists estimate the toxicity of a substance by conducting experiments to measure the effects of increasing DOSES of the substance on a POPULATION of laboratory animals.

One measure of toxicity is the LETHAL DOSE–50% (LD_{50}) of a chemical. Basically, the LD_{50} of a substance is the amount that, when given to a group of laboratory animals, kills half (50%) of the animals used in the experiment within 14 days. The more toxic a chemical, the smaller the LD_{50}. Laboratory animals may be exposed to the dose in AIR, water, or food, or the substance may be directly placed into the animals. Commonly, doses are measured in units of milligrams of a toxic chemical per kilogram of body weight of the ORGANISM.

Many toxic effects will occur in populations of laboratory animals exposed to chemicals at doses that are less than the LD_{50}. Toxicity experiments can determine the harmful effects that occur when such doses are given to a population. Effects that occur after a short exposure period are called "acute" effects. Effects that develop gradually over time are called "chronic" effects. Acute effects may include temporary or permanent damage to various organs of the body. Chronic effects may include organ damage and DISEASES, such as CANCERS.

Different chemicals may affect various systems and organs in the body, depending on the toxic properties of each chemical and the ways in which it comes into contact with the body. In toxicity studies, toxicologists consider how an animal is exposed to a chemical: the "exposure route." Exposure routes include direct contact with the skin, inhalation (breathing), and ingestion (eating). Toxic effects may be seen in the respiratory system, blood, or central nervous system. Toxicologists identify "target organs" for toxic substances, meaning the parts of the body that are adversely affected by the substances. Target organs include the liver, kidneys, lungs, and skin. Exposure to some toxic chemicals may harm the reproductive system and unborn offspring.

There are three important types of chronic toxicological effects that chemicals may have on humans or other organisms. Carcinogenic substances (CARCINOGENS) are substances that have either been found to cause cancer or are likely to cause cancer in animals or humans. For example, some forms of the metal CHROMIUM are carcinogenic. Teratogenic substances (teratogens) are those that have been observed to cause defects in the unborn offspring of animals when the parent has been exposed to the substance. Chemicals containing MERCURY may be teratogenic. Mutagenic substances (MUTAGENS) cause MUTATIONS (genetic defects) in the CELLS or cell materials of animals. Radioactive materials, such as URANIUM, are known to be mutagenic.

Among the substances of highest concern identified by the U.S. ENVIRONMENTAL PROTECTION AGENCY and the U.S. Department of Health and Human Services are ARSENIC, LEAD, mercury, BENZENE, POLYCHLORINATED BIPHENYLS (PCBs), CADMIUM, chromium, TRICHLOROETHANE (TCE), and the PESTICIDES dieldrin and CHLORDANE. (*See* aldrin.)

Based on the known or suspected toxicities of chemicals, toxicologists and industrial hygienists (experts in worker protection) develop recommended safe exposure limits to protect workers from health effects. *See also* CHRONIC TOXICITY; HAZARDOUS SUBSTANCES.

EnviroSources

Agency for Toxic Substances and Disease Registry, Division of Toxicology Web site: www.atsdrl.atsdr.cdc.gov.

Cohen, Gary, and John O'Connor. *Fighting Toxics: A Manual for Protecting Your Family, Community, and Workplace*. Washington, DC: Island Press, 1990.

U.S. Environmental Protection Agency, Integrated Risk Information System Web site: www.epa.gov/iris/subst/index.html.

U.S. Occupational Safety and Health Administration Web site: www.osha-slc.gov/sltc/hazardoustoxicsubstances/index.html.

Toxic Release Inventory, A database compiled by the U.S. ENVIRONMENTAL PROTECTION AGENCY (EPA) that records the annual releases of toxic substances into the ENVIRONMENT by various industries. The Toxic Release Inventory (TRI) contains information about releases of more than 650 chemicals. The TRI categorizes data by state, manufacturer, year and chemical, and medium of release (AIR, water, underground injection, land disposal or transfer to off-site facilities [DEEP-WELL INJECTION].)

Manufacturers that make or use chemicals must report annually to the EPA and their states the amounts of chemicals in various categories their plants release. The data contained in the TRI are available to the public as mandated in the EMERGENCY PLANNING AND COMMUNITY RIGHT-TO-KNOW ACT (EPCRA). The TRI provides citizens with accurate information about potentially hazardous chemicals and their use so that communities have more power to hold companies accountable and make informed decisions about how chemicals are to be managed. This helps communities prepare to respond to chemical spills and similar emergencies. The goal is to reduce risk for communities as a whole.

EnviroSources

EPCRA Hotline: (800) 532-0202; fax (703) 412-3333.

U.S. Environmental Protection Agency (TRI information) Web site: www.epa.gov.

Toxic Substance Control Act, Legislation enacted by the U.S. Congress in 1976 requiring that all chemicals produced in or imported into the United States be tested to determine potential adverse effects on the health of humans, other ORGANISMS, and the ENVIRONMENT. Thousands of new chemicals are developed each year, of which many have unknown toxic or dangerous characteristics. To prevent tragic consequences, the Toxic Substances Control Act (TSCA) requires that any chemical that will reach the consumer marketplace be tested for possible toxic effects prior to commercial manufacture and distribution. The provisions of the TSCA supplement other federal statutes, including the CLEAN AIR ACT and the EMERGENCY PLANNING AND COMMUNITY RIGHT-TO-KNOW ACT.

EnviroSources

U.S Environmental Protection Agency, Toxic Substances Control Act, Assistance Information Service: 401 M Street, SW, Mail Code 7408, Washington, DC 20460.

toxic waste, Liquid, solid, or gaseous waste produced during chemical and industrial processes that is poisonous to human health when ingested or inhaled. Toxic wastes include substances that are carcinogenic (causing CANCER), teratogenic (causing defects of a fetus) and mutagenic (causing genetic defects in the CELLS or cell materials of animals). Among the toxic wastes of highest concern identified by the U.S. ENVIRONMENTAL PROTECTION AGENCY (EPA) and the U.S. Department of Health and Human Services are ARSENIC, LEAD, MERCURY, BENZENE, ASBESTOS, POLYCHLORINATED BIPHENYLS (PCBs), CADMIUM, CHROMIUM, TRICHLOROETHANE (TCE), and the PESTICIDES dieldrin and CHLORDANE. (*See* aldrin.) The illegal dumping of toxic wastes is a major source of stream, river, and lake pollution.

Some of the methods for managing toxic wastes include landfills, BIOREMEDIATION, INCINERATION, and cleanup action through SUPERFUND legislation. Some countries even export their toxic wastes to other countries. According to GREENPEACE, Australia, Canada, Germany, the United States, and the United Kingdom shipped more than 5.4 million metric tons (5.9 million tons) of toxic wastes to countries in Asia between 1990 and 1993. In 1995, however, a global treaty barring wealthy countries from dumping toxic waste in the Third World was signed, reinforcing an existing voluntary ban. In 1998 the ban included materials exported for RECYCLING. The treaty was approved by members of the 1989 Basel Convention, which controls trade in HAZARDOUS WASTE among members of the Organization for Economic Cooperation and Development. *See also* LOVE CANAL; TOXIC CHEMICAL; TOXIC RELEASE INVENTORY; TOXIC SUBSTANCE CONTROL ACT.

EnviroSource

Institute for Global Communications Web site: www.igc.org/igc/issues/tw.

toxicity, The quality of being toxic (poisonous). The adverse effects of toxicity can range from slight symptoms such as headaches or nausea to severe symptoms such as coma, convulsions, and death. The science of TOXICOLOGY is largely concerned with the study of toxicities of various substances by evaluating the toxic effects of substances on ORGANISMS. In a similar fashion, the field of industrial hygiene deals in part with toxicities of chemicals used in the workplace. If a substance is said to have "liver toxicity," this means the substance has an adverse or toxic effect on the liver.

Toxicity is normally divided into various types, based on the number of exposures to a poison and the time required for toxic symptoms to develop. The two types most often referred to are acute and chronic. ACUTE TOXICITY is due to short-term exposure and occurs within a relatively short period of time. CHRONIC TOXICITY is due to long-term exposure and occurs over a longer period.

The study of toxicities makes up an extremely important part of modern ENVIRONMENTAL SCIENCE, with respect to synthetic (human-made) chemicals in the ENVIRONMENT. When scientists assess the impacts of chemical pollution, they must evaluate the toxicity of various chemicals and how they affect people. People could suffer toxic effects from various POLLUTANTS that are present in AIR, water, soil, or sediment, depending on the toxicity of the pollutant, the type of exposure, and the DOSE of the pollutant. Understanding the toxicity of each pollutant enables scientists to determine the best ways to protect people from toxic effects by reducing exposures or cleaning up pollution.

Most toxic effects are reversible and do not cause permanent damage, but in some cases, complete recovery may take a long time. Some poisons, however, cause permanent damage. Poisons can affect just one particular organ system, or they may produce generalized toxicity by affecting a number of physiological systems. Poisons can change the speed of different body functions, such as increasing the heart rate or sweating or decreasing the rate of breathing. For example, people poisoned by Parathion, a PESTICIDE, may experience increased sweating as a result of a series of changes in the body.

toxicology, The science and study of the adverse effects of poisons on living ORGANISMS. Toxicologists study how poisons enter the bodies of organisms, what happens to those poisons once they are in the body, and what effects they have on the body. One of the goals of toxicology is to predict a chemical's toxic effects on humans. Before a chemical is allowed to be sold or used, it must undergo TOXICITY tests to determine possible adverse effects. These effects fall into three basic categories—ACUTE TOXICITY, subacute toxicity, and CHRONIC TOXICITY. *See also* BIOASSAY; NEUROTOXIN; TOXIN.

toxin, A poison produced by certain PLANTS, animals, or MICROORGANISMS that is harmful or dangerous to humans and other ORGANISMS. Most plants produce toxins to defend themselves against animals, INSECTS, and FUNGI. The production of toxins is an ADAPTATION that helps protect plants from predators. (*See* predation.) When a toxin is injected into a human, the body may react by producing counteracting antibodies called antitoxins. *See also* NEUROTOXIN; RED TIDE.

trace metal, Any metals present in food found naturally in AIR, water, and soil. Some of these metals are toxic, while others are essential (in small amounts) for good health in humans and other ORGANISMS. Zinc, iron, COPPER, MANGANESE, and CHROMIUM are trace metals if consumed in small amounts in a healthy diet but could be considered HEAVY METALS if used in large quantities. *See also* ACUTE TOXICITY; CHRONIC TOXICITY; MINERAL.

EnviroTerm

trace elements: Elements including metals and nonmetals from food or found naturally that are essential for ORGANISMS but only in small amounts.

Trade Records Analysis of Flora and Fauna in Commerce, A network of offices whose purpose is to ensure that WILDLIFE trade does not exceed sustainable levels and is in accordance with domestic and international laws and agreements. In essence, it monitors illegal trade. Trade Records Analysis of Flora and Fauna in Commerce (TRAFFIC) is a joint CONSERVATION program of the WORLD CONSERVATION UNION and encourages implementation of the CONVENTION ON INTERNATIONAL TRADE IN ENDANGERED SPECIES OF WILD FLORA AND FAUNA (CITES).

TRAFFIC monitors all wildlife trade—including trade in animals and PLANTS and any products produced by them. During 1997–98, TRAFFIC published studies about the trade in sharks and shark products in North America. It assisted in the implementation of new international trade controls for all sturgeons and their products, including caviar, under CITES. The trade controls require that all sturgeon SPECIES and their products in international trade now be regulated under a system of permits.

TRAFFIC closely monitored and assisted in efforts to ensure a sustainable trade in big-leafed MAHOGANY. This tree, also known as American mahogany, is declining throughout its range, particularly in Central America. TRAFFIC played a lead role in auditing Africa's IVORY stocks and in establishing comprehensive monitoring systems to track international trade in ELEPHANT products and the illegal killing of elephants in both Africa and Asia. TRAFFIC also undertook a variety of activities to help develop and implement wildlife trade laws and projects to help build enforcement capacity. *See also* ENDANGERED SPECIES; RED LIST OF ENDANGERED SPECIES; THREATENED SPECIES.

EnviroSource

Trade Records Analysis of Flora and Fauna in Commerce Web site: www.traffic.org/about.

Tragedy of the Commons, An article written by Garrett Hardin in 1968, the title of which became an operative metaphor for the negative impact of individual users on common resources. It led to the strong belief among many citizens that regulatory control was needed to protect environmental resources. Harden proposed the name "the tragedy of the commons" from an original idea laid out in 1832 by William Forster Lloyd, a professor of political economy at Oxford in England.

Hardin created the following model of the tragedy of commons: "On the pasture open to all, each herdsman seeks to maximize his gain, so each of them thinks about his cost and benefit of adding one more animal to his herd. Each of them concludes that he will get the full profit from selling the additional animal, but he will share only part of the cost of overgrazing. So, thinking rationally, each herdsman will continuously increase the number of animals until tragedy occurs." Hardin postulated an agrarian community where all citizens graze their livestock on a commonly owned field. The field can only support a limited number of animals before it is denuded and ruined.

Hardin pointed out that the cost-benefit analysis performed by an individual townsperson using the commons will always lead to the conclusion that the immediate benefit of adding another animal far outweighs the remoter, less visible harm of degradation of the commons. The benefit from the additional animal belongs to the townsperson alone, while the harm to the commons is shared proportionally across all its users. Hardin's essay emphasized the importance of protecting the ENVIRONMENT for the common good rather than for self-interest. *See also* NATURAL RESOURCES; NONRENEWABLE RESOURCES; RENEWABLE RESOURCES.

EnviroSource

Harding, Garrett. *Filters Against Folly.* New York: Penguin Books. 1985

trait. *See* adaptation.

transpiration, The process by which ORGANISMS release water vapor to the AIR. Transpiration is a biological process that moves water between Earth and its ATMOSPHERE, and thus plays a part in the WATER CYCLE. Organisms that carry out AEROBIC respiration produce water vapor as a waste during of this process. The structures used by such organisms for transpiration vary. For example, MAMMALS, REPTILES, birds, and AMPHIBIANS have lungs from which they release water vapor to the air when they exhale. Amphibians also lose water vapor to the air through their skin, as do many INVERTEBRATE animals, such as worms. Single-celled organisms and many simple, multicellular organisms release water vapor directly to the air from their CELLS by diffusion. Most PLANTS have special cells, called guard cells, on the undersides of their leaves through which they release water vapor to the atmosphere.

In addition to respiration, some animals, such as humans, excrete water in the form of sweat or perspiration. This water, along with some dissolved salts, is released by special glands in the skin called sweat glands. Water contained in sweat is removed from the surface of the skin through evaporation and thus also plays a role in Earth's water cycle. *See also* BIOGEOCHEMICAL CYCLES.

transuranic element. *See* radioactive waste (nuclear waste).

transuranic waste. *See* radioactive waste (nuclear waste).

trash, Solid household waste that includes paper products, wood products, and cardboard, glass, plastic, tin, and ALUMINUM containers. Trash does not include food waste. Trash is usually picked up by vehicles and brought to local SANITARY LANDFILLS. Much of it can be recycled, such as paper and aluminum and glass containers. In the United States, some states export their trash to other states. Pennsylvania is one of the leading states in importing trash. In 1998, the state imported about 6 million metric tons (6.6 million tons) of trash. Other importers of trash include Virginia, Indiana, Michigan, and Illinois. The U.S. Congress wants to limit the interstate transport of trash. *See also* DETRITUS; EFFLUENT; GARBAGE; HAZARDOUS WASTE; INCINERATION; RADIOACTIVE WASTE; RECYCLING; REFUSE; SOLID WASTE; TOXIC WASTE.

tree farm, An area of land on which trees are grown and managed for commercial uses, such as for use as pulp wood or Christmas trees, or for resale for residential landscaping. Most tree farms have stands of trees of the same age. Often, a tree farm consists of only one or a few SPECIES of trees. In some areas, tree farming is conducted to maintain a SUSTAINABLE YIELD of wood or wood products; for example, after most of the white pine FORESTS in the state of Maine disappeared as a result of overharvesting, several large paper companies began tree farming to provide the pulpwood needed for their products. One of the earliest tree farming operations was started by British scientist Henry Wickham in the late 1800s. Wickham collected more than 70,000 seeds from the rubber plant *Hevea brasiliensis* in the Amazon jungle for the purpose of growing trees in Kew Gardens (London, England). Later, the saplings were transported to Ceylon, India, and

Malasia where they were used to establish the rubber industry. *See also* MONOCULTURE; POLYCULTURE.

trichloroethane (or 1,1,1-trichloroethane), A colorless, liquid synthetic chemical that does not occur naturally in the ENVIRONMENT and evaporates readily. Under provisions of the 1990 amendments to the CLEAN AIR ACT and the MONTREAL PROTOCOL, the production of 1,1,1-trichloroethane in the United States and Canada was stopped in 1996 because production and use of the chemical depletes Earth's OZONE layer. It is still found in AIR samples taken from all over the world. Adverse health effects from high levels of exposure to 1,1,1-trichloroethane include dizziness and decreased blood pressure.

1,1,1-trichloroethane is also known as methyl chloroform, methyl trichloromethane, trichloromethylmethane, and trichloromethane. It was used in a large variety of products such as adhesives, PESTICIDES, AEROSOLS, lubricants, drain cleaners, shoe polishes, spot cleaners, printing inks, and electrical equipment.

EnviroSources

Agency for Toxic Substances and Disease Registry, Division of Toxicology: 1600 Clifton Road NE, E-29, Atlanta, GA 30333; (404) 639-6000.

Mike Kamrin, Institute for Environmental Toxicology, C231 Holden Hall, MSU; (517) 353-6469.

trophic level, A term used to describe the layer within a FOOD CHAIN at which the different ORGANISMS in an ECOSYSTEM feed. In most ecosystems, there are between three and five trophic levels. PRODUCERS (organisms that synthesize their own food) form the base of all food chains and FOOD WEBS, and thus occupy the first trophic level in all ecosystems. The second trophic level is comprised of primary CONSUMERS, organisms that feed directly on producers. Organisms at this feeding level may be HERBIVORES (organisms that feed only on PLANTS or plant products) or OMNIVORES (organisms that feed on both plants and animals). Secondary consumers, organisms that use primary consumers as food, comprise the third trophic level. Organisms at this feeding level may be CARNIVORES (meat eaters), omnivores, or SCAVENGERS (organisms that feed on the remains of other organisms). The fourth trophic level is comprised of tertiary consumers, animals that feed on secondary consumers. As occurred at the third trophic level, organisms at this level may be carnivores, omnivores, or scavengers. Because most organisms rely on other organisms for their food and ENERGY needs, the organisms comprising each trophic level, except the first, within an ecosystem are completely dependent upon the organisms at the previous level. *See also* AUTOTROPH; DECOMPOSER; DETRIVORE; ECOLOGICAL PYRAMID; HETEROTROPH; INSECTIVORE.

tropical rainforest. *See* rainforest.

troposphere. *See* atmosphere.

Tsavo National Park, A park in Kenya, consisting of the Tsavo Park East and Tsavo Park West National Parks, that is a refuge for many endangered ELEPHANTS and other THREATENED SPECIES. The park, which was opened in 1948, covers about 20,000 square kilometers (7,720 square miles). The parks are known as one of the world's leading BIODIVERSITY strongholds, because they include a wide variety of vegetation and terrain. Over 600 SPECIES of birds have been recorded. Other animals that live in the parks include LEOPARDS, CHEETAHS, buffalos, RHINOCEROSES, elephants, CROCODILES, mongooses, giraffes, ZEBRAS, and lions. Elevations in the parks range from 160–2,000 meters (500–6,000 feet) above sea level. Included in Tsavo Park East is Yatta Plateau, the world's largest lava flow. The parks are popular sites for safaris because of their easy accessibility. The parks became famous as a result of the notorious "man eaters of Tsavo" incident at the turn of the century, when lions were preying on the workers building the great Uganda Railway. The incident was depicted in the movie *Ghost and the Darkness.* Other interesting sites in the parks include the Mudanda Rock, Luggard's Falls on the Galana River, Chaimu volcanic crater, the Shetani Laval Flow, the Nguli Rhino Sanctuary, and views of Mount Kilimanjaro.

The wildebeest live in Tsavo National Park in Kenya, a refuge for many endangered elephants and other threatened species.

tsunami, Powerful, fast-moving, long-length OCEAN waves associated with EARTHQUAKE or volcanic activity originating from below or near the ocean floor and which can cause extensive destruction to coastal homes and other structures. The word "tsunami" is derived from the Japanese word meaning harbor wave. A tsunami is also referred to as a tidal wave, but it has no connection to TIDES. Tsunamis can be caused by underwater volcanic eruptions and landslides that are triggered by earthquakes. Most tsunamis occur in the area that encircles the Pacific Ocean.

In the deep ocean away from the coastline, a tsunami is not easy to detect. People aboard large vessels and ships cannot feel the action of the waves. And from the AIR the waves cannot be seen. The movement of the waves in the deep water is fairly smooth. The distance between each wave crest of a tsunami may be 100 kilometers (62 miles) or more, and the height of each wave may be less than 1 meter (3 feet); however, the tsunami waves travel at speeds exceeding 800 kilometers (500 miles) per hour. As a tsunami reaches shallower coastal waters, wave heights can increase rapidly, reaching as high as 15 meters (45 feet) as they approach shore. This is what causes the destruction of homes and the injury and death to people on shore. Before a tsunami strikes shore, coastal waters are drawn out into the ocean. When this occurs, more shoreline may be exposed than even at the lowest tide.

Papua New Guinea was struck by a 10-meter (30-foot) high tidal wave in July 1998. The wave wiped out communities along a 45-kilometer (27-mile) coastline. About 2,000 people died, 9,500 people were homeless, and many villages were destroyed. The tsunami was triggered by an earthquake with a magnitude of 5.8 near the north coast of New Guinea.

Another large tsunami that caused widespread death and destruction throughout the Pacific was generated by an earthquake located off the coast of Chile in 1960. It caused loss of life and property damage not only along the Chilean coast but in Hawaii and as far away as Japan. In the last 180 years, about 40 tsunamis have struck the Hawaiian Islands. The Great Alaskan Earthquake of 1964 produced deadly tsunami waves in Alaska, Oregon, and California. The Mindoro Tsunami occurred on November 15, 1994, and was triggered by a powerful earthquake with a magnitude of about 7.1 near Verde Island, Philippines. This tsunami totally destroyed 1,530 houses and killed 41 people.

To keep tabs on tsunamis, the United States has two tsunami warning centers, one in Hawaii and the other in Alaska. The centers monitor sea-floor earthquakes and water levels and issues a tsunami watch or warning if conditions seem likely to cause a tsunami. *See also* NATURAL DISASTER; VOLCANO.

EnviroSources

National Oceanic and Atmospheric Administration, Pacific Marine Environmental Laboratory: 7600 Sand Point Way NE, Seattle, WA 98115; (206) 526-6239; fax (206) 526-6815.

National Oceanic and Atmospheric Administration, Pacific Marine Environmental Laboratory, National Tsunami Hazard Mitigation Program Web site: www.pmel.noaa.gov/tsunami-hazard; e-mail webmaster@pmel.noaa.gov.

West Coast and Alaska Tsunami Warning Center: 910 South Felton Street, Palmer, AK 99645; (907) 745-4212; fax (907) 745-6071; e-mail: atwc@alaska.net.

tuna, The common name for several large, marine game and food fishes of the mackerel family. Tuna live throughout much of the world's OCEANS and are economically important to many countries. The United States, Nova Scotia, Japan, Thailand, the Philippines, Brazil, Columbia, Australia, Portugal, and several other Mediterranean countries have large tuna industries. The most commercially important tuna belong to the genus *Thunnus,* which includes the albacore (a white-meat tuna), the yellowfin, and the bluefin. The skipjack (or striped) tuna, which belongs to the genus *Katsuwonus*, also has commercial importance.

Wide use of tuna as a food fish has had several distinct environmental impacts. One has been a reduction in worldwide tuna POPULATIONS as a result of OVERFISHING. As early as 1926, the California tuna-packing industry collapsed when albacore disappeared from the region. In the mid-1970s, yellowfin and bluefin populations along both coasts of the United States and Mexico showed significant declines as well. A decline was observed in the bluefin tuna population of Australia in the 1980s. Many countries now have quotas limiting annual tuna captures; however, populations continue to decline.

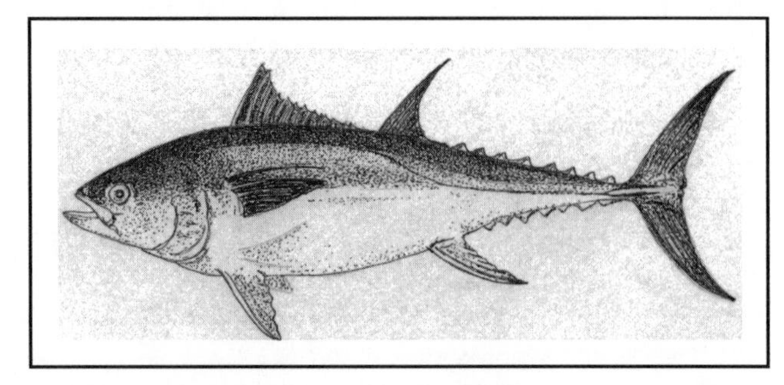

Tuna live throughout much of the world's oceans and are economically important to many countries.

Some scientists estimate that the Atlantic tuna population alone has decreased by as much as 90% over the last 30 years. If this trend continues, some tuna SPECIES may soon be in danger of EXTINCTION.

Another problem results from the fishing methods used to catch tuna. Traditionally, tuna, which travel in large schools, have been captured using DRIFTNETS. This results not only in the capture of great numbers of tuna, but also in the capture and deaths of other marine ORGANISMS, including dolphins, porpoises, and SEA TURTLES. In the 1970s, concerns over the capture of dolphins in tuna fishing nets brought about public protests and boycotting of tuna products. In response to the boycotts, many tuna fishers now use nets that capture tuna, while allowing dolphins and other nontuna species to escape. Companies that package tuna captured using the new nets often market their products in cans carrying "dolphin friendly" labeling. (*See* ecolabeling.)

Another problem affecting tuna populations is pollution of ocean waters with chemicals. Some of these chemicals are directly toxic to tuna and other marine organisms; others, such as MERCURY, collect in the tissues of organisms and increase in concentration as they move through the FOOD CHAIN, in a process called BIOACCUMULATION. Often, mercury ingested by tuna does not cause the death of the tuna; however, tuna captured in some areas have had mercury concentrations high enough to be potentially toxic to humans or other organisms that eat the tuna. In 1970, concerns about mercury in tuna led the U.S. FOOD AND DRUG ADMINISTRATION (FDA) to order a national recall of canned tuna because tests on some fish showed mercury levels above the acceptable standard of 0.5 parts per million. Testing of the recalled product showed that as much as 3% of the canned tuna exceeded the acceptable mercury level. Similar problems with mercury and other chemicals in water pollution have been observed in other fishes and aquatic species, including fresh-water species living in streams and rivers. *See also* CHRONIC TOXICITY; DOLPHINS AND PORPOISES; MINAMATA DISEASE; PARTS PER BILLION AND PARTS PER MILLION; RHINE RIVER; SANDOZ CHEMICAL SPILL.

EnviroSource

Center for Marine Conservation: (202) 429-5609.

tundra, A vast, treeless BIOME characterized by a very cold and dry CLIMATE, and such vegetation as LICHENS, mosses, herbs, small shrubs, and some dwarf trees. Tundra is found worldwide within the ARCTIC Circle (Arctic tundra) or at high ALTITUDES (alpine tundra). The tundra can be found in the northern half of Alaska, parts of China, Canada, SIBERIA, Scandinavia, and even the northern regions of Scotland and Ireland.

Although PLANTS and animals of the tundra adapt to very harsh conditions, the tundra is one of the most fragile biomes on Earth. FOOD CHAINS in the tundra are relatively simple and can be easily disrupted. And because environmental conditions are so extreme, the land is easily damaged and slow to recover. Until recently, tundra ENVIRONMENTS worldwide were left mostly undisturbed by humans; however, with the discovery of oil in some tundra locations, such as Prudoe Bay, Alaska, and the increasing popularity of recreational fishing and hunting, some regions have been experiencing environmental damage. Once pristine rivers are becoming overfished and polluted with waste from MINING activity. Nearby airports and airplanes are disturbing the nesting areas of migratory birds. Minor OIL SPILLS are reducing water quality, and off-road vehicles are leaving permanent tracks in the land that are eroding into deep gullies.

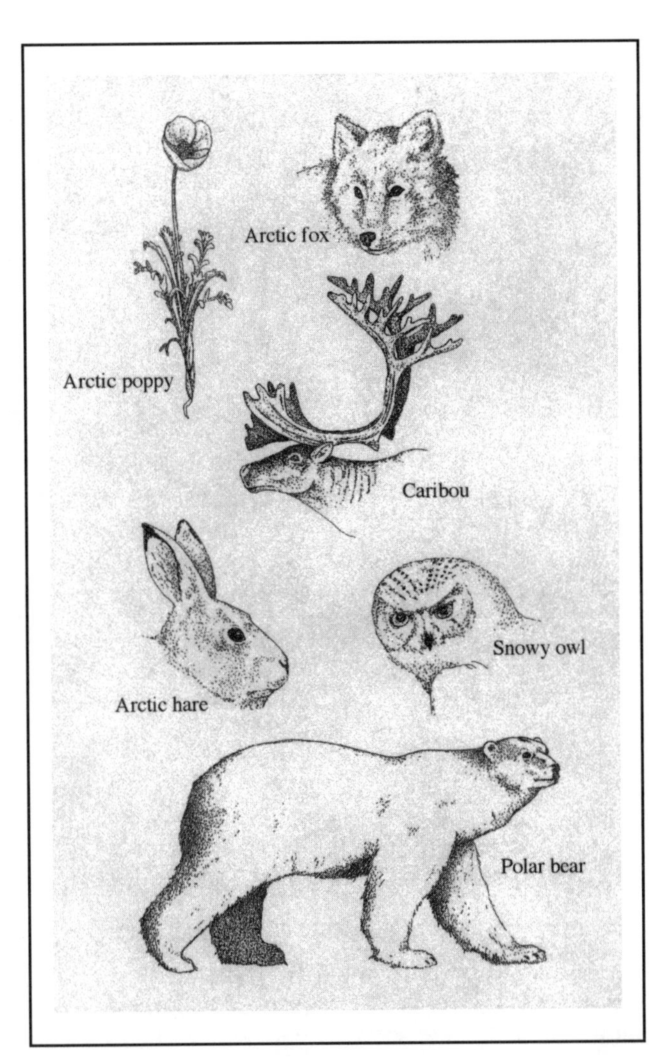

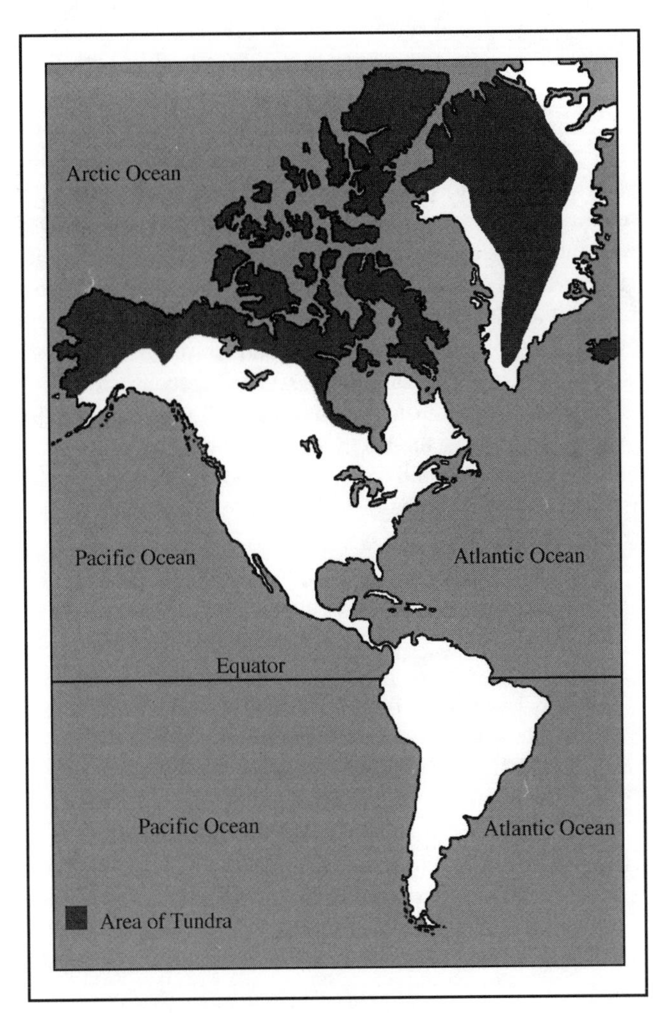

Areas of the tundra in the Western Hemisphere.

An extensive long pipeline carries oil from Prudoe Bay, home of the ARCTIC NATIONAL WILDLIFE REFUGE, across the entire state of Alaska, to the Gulf of Alaska. Oil company geologists believe that more oil deposits exist in Prudoe Bay and that oil drilling should continue if U.S. oil consumption needs are to be met. Advocates of oil exploration say that only a small amount of the area in the tundra will be disturbed. Conservationists and ecologists are concerned about the impact that further oil exploration will have on animals that make their homes in the refuge.

CLIMATE is the main determining factor of any biome. Temperature and precipitation, such as rainfall or snow, are the two primary factors that make up a region's climate. Climate influences what kinds of plants can grow, which, in turn, determines the types of animals that can live there. Climate in the tundra is characterized by low temperatures, –25°C (–13°F) in the winter to about 6°C (43°F) in the summer, and very

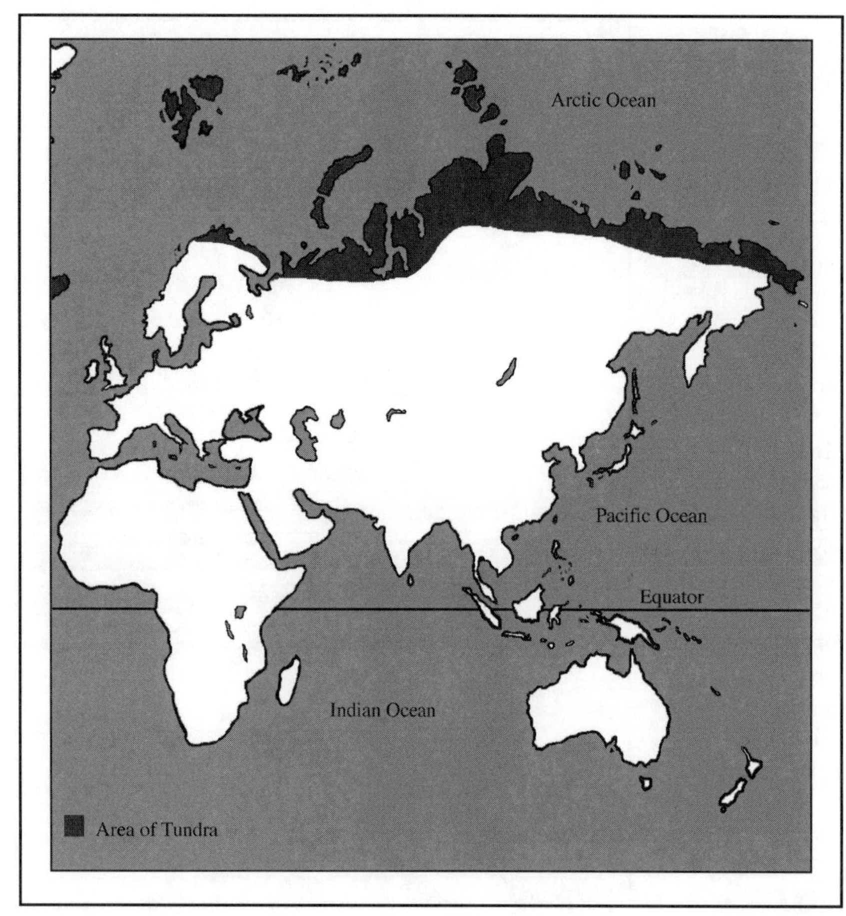

Areas of the tundra in the Eastern Hemisphere.

little precipitation, about 12 centimeters (4.7 inches) per year, mostly in the form of snow. In fact, other than DESERTS, the tundra is the driest place on Earth. Few plants and animals can live in such inhospitable conditions. Those that do, possess structural and behavioral ADAPTATIONS that help them survive the hostile climate.

The temperature of a land area is determined by the amount of sunlight that reaches it. Sunlight is important to ORGANISMS not only because it provides the light ENERGY plants need for PHOTOSYNTHESIS, but also because it heats Earth. Not all parts of the world, however, receive the same amount of sunlight. The amount of sunlight is primarily related to the latitude of a region. Because of Earth's curvature, sunlight that hits the Earth at higher latitudes (e.g., tundra) is spread over a wider area than is sunlight that hits near the equator. In other words, regions near the equator receive more light energy per unit area than areas north and south of the equator. The higher the latitude—that is, the farther away from the equator and the closer to the North and South poles—the colder is the climate. This is why tundra environments, located in the northernmost latitudes of Earth, have the lowest average annual temperatures of all biomes.

Lakes and ponds are plentiful in the tundra, but much of this environment resembles a treeless plain with short grasses and low-lying shrubs. Mostly this has to do with the fact that winters are very long in the tundra and there are many months of twilight and darkness, when the sun is below the horizon the entire day. As a result, the growing season is very short in the tundra. During the brief and cool summer season, plants sprout and flower quickly. Another reason for the relative lack of vegetation in the tundra is the poor soil quality. Below the surface of the ground is a layer of permanently frozen soil called PERMAFROST, which can reach depths of 90–456 meters (300–1,500 feet). The top layer thaws a little during the short summer, but the lower layers always remain frozen. Permafrost and the low precipitation level prevent the rooting and growth of FORESTS in the tundra. The plants that do grow in the tundra—lichens, mosses, shrubs, sedges, dwarf trees—adapt by growing very close to the ground, protected from the battering winds that whip across the open territory.

Tundra animals adapt to the difficult conditions as well. Many of the year-round animal residents, including reindeer, caribou, WOLVES, polar bears, and musk oxen, have thick skin and fur to protect them from the cold. Others, such as snowshoe hares, foxes, and snowy owls, are adorned with white fur or feathers to help them blend in with their surroundings and avoid predators. (*See* predation.) Still other animals, like mice, voles, lemmings, and other small rodents, build burrows under the snow. By far the most abundant animals in the tundra are INSECTS, especially in the southernmost regions where boggy soils, streams, rivers, and lakes are common. Huge swarms of mosquitoes and black flies make life almost unbearable for other animals. The tundra is also home to a great variety of birds. Some, such as the snowy owl, live there all year. Others, including many SPECIES of ducks and geese, nest in the tundra during the summer and migrate south by the millions during the harsh winter months when temperatures plummet and food sources are used up. Human POPULATIONS include the Inuit peoples. *See also* ALASKA PIPELINE; BOG; ECOSYSTEM; NATIONAL WILDLIFE REFUGE SYSTEM.

EnviroSources

Pipes, Rose. *Tundra and Cold Deserts*. World Habitats. Austin, TX: Raintree/Steck Vaughn, Springer Verlag, 1998.

Sayre, April Pulley. *Tundra (Exploring Earth's Biomes.)* New York: Twenty First Century Books, 1995.

Zwinger, Ann H., and Beatrice E. Willard. *Land above the Trees: A Guide to American Alpine Tundra.* Boulder, CO: Johnson Books, 1996.

turbine, A machine, usually with vanes, blades, or buckets that rotate about an axis driven by the pressure of a liquid or gas. The mechanical ENERGY produced can be used directly, or it can be converted to electrical power by linking the turbine's torque to an electrical generator. *See also* AEROGENERATOR; ELECTRICITY; HYDROELECTRIC POWER.

typhoon, A severe tropical CYCLONE that occurs in the western Pacific Ocean region in areas between 5° and 30° latitude. Typhoons can cause great damage to physical structures and property, be devastating to ORGANISMS, and bring about both short-term and long-term changes in the ENVIRONMENT. In the western Atlantic and Caribbean region, tropical cyclones are called HURRICANES. In western Australia, they are known as willy-willies.

All tropical cyclones are characterized by a calm, central area of low atmospheric pressure known as the eye, which averages 24 kilometers (15 miles) in diameter. Surrounding the eye is a wide circular whirl of rain, clouds, and very high winds, ranging anywhere from 80–800 kilometers (50–500 miles) in diameter. To be classified as a tropical cyclone, wind speeds must exceed 117 kilometers (74 miles) per hour; however, it is not uncommon for such storms to have sustained winds of 160 kilometers (100 miles) per hour. Winds twice this strong have been recorded for some extremely violent storms.

All tropical storms develop over warm tropical waters where there is an abundant supply of water vapor in the AIR. When the warm, moist air rises, it condenses into tiny droplets that form clouds. The thermal energy released during condensation warms the air and sends it rushing skyward in the storm's center. As the storm grows, Earth's rotation causes the storm to begin spinning. As it continually draws ENERGY from the condensation of water vapor, a mature cyclone forms after several days of intensification. *See also* CLIMATE; NATURAL DISASTER; TYPHOON; WEATHER.

understory. *See* forest.

United Nations Convention on the Law of the Sea, Legislation adopted by the United Nations in 1982 to address issues concerning use of the world's OCEANS. Specific issues addressed by the convention include navigation rights on and above the sea, environmental protection, fishing, scientific research, and development of sea-bed MINERALS. Although not yet enforced, the convention grants coastal nations the right to exercise sovereignty over a territorial sea extending up to 22 kilometers (14 miles) from their shores. The convention also provides these nations jurisdiction over an economic zone (a region the nation can utilize for economic benefits) extending as much as 370 kilometers (230 miles), and holds them responsible for environmental protection, management of resources, and scientific research in those areas.

The final provision of the convention addresses mineral development of the sea bed. This legislation was proposed, in part, in response to MINING methods developed in the 1970s for collecting MANGANESE nodules from the sea floor. In addition to manganese, these nodules are rich in cobalt, COPPER, and iron oxide. Mining the nodules involves connecting drag lines or suction hoses to barges that move along the ocean's surface. Provisions of the convention state that coastal nations must share revenue derived from such mining that occurs beyond their territorial zone with the international community. Objections to this policy prevented the United States from signing the accord; however, it has endorsed the other provisions of the convention. *See also* MINING; NATURAL RESOURCE; OCEAN DRILLING PROGRAM.

EnviroSource

United Nations Web site: www.un.org.

United Nations Development Program, An agency established within the United Nations in 1965 to aid developing nations in achieving sustainable human development. The United Nations Development Program (UNDP) has headquarters in New York, but maintains 136 offices worldwide from which it serves 175 nations. To achieve its goals, the UNDP tries to increase literacy, create jobs, reduce poverty, and improve technical cooperation between industrialized and nonindustrialized nations. At the same time, the agency also strives to help developing nations preserve and protect their ENVIRONMENT by helping communities to prevent and eliminate water pollution, to preserve BIODIVERSITY, and to reduce AIR POLLUTION.

The UNDP has an annual budget of approximately $1 billion that is used to administer the following six programs:

- The United Nations Sudano-Sahelian Office, the primary objective of which is to promote SUSTAINABLE DEVELOPMENT through SOIL CONSERVATION techniques (and the reduction of DESERTIFICATION) in the sub-Saharan region of Africa.
- The United Nations Revolving Fund for Natural Resources Exploration, which provides funding to several countries in Latin America, Asia, and Africa to help locate MINERAL deposits in those areas.
- The United Nations Capital Development Fund, which provides the world's most undeveloped nations with assistance in generating capital.
- The United Nations Fund for Science and Technology for Development, which promotes international cooperation among research organizations and helps countries apply the latest advances in science and technology to their needs.
- The United Nations Volunteers, which provides the funding needed to make field workers available to countries carrying out specific development projects.
- The United Nations Development Fund for Women, which supports projects that promote the concerns of women throughout the world.

Programs administered by the UNDP are intended to help developing nations improve their social and economic standing in the world in a manner that protects and regenerates the environment. To achieve this goal, the UNDP encourages involvement of a nation's people at both the national and local level. *See also* AGENDA 21; BIODIVERSITY TREATY; BIOSPHERE RESERVE; COALITION FOR ENVIRONMENTALLY RESPONSIBLE ECONOMIES; CONVENTION ON INTERNATIONAL TRADE IN ENDANGERED SPECIES OF WILD FLORA AND FAUNA; DEBT-FOR-NATURE SWAP; DEFORESTATION; ENDANGERED SPECIES; FOREST MANAGEMENT; GLOBAL ENVIRONMENT MONITORING SYSTEM; GREENBELT MOVEMENT; INTERNATIONAL BIOLOGICAL PROGRAM; INTERNATIONAL COUNCIL FOR LOCAL ENVIRONMENTAL INITIATIVES; MAN AND THE BIOSPHERE PROGRAMME; MONTREAL PROTOCOL; RED LIST OF ENDANGERED SPECIES; UNITED NATIONS EARTH SUMMIT; UNITED NATIONS ENVIRONMENT PROGRAMME; WORLD CONSERVATION MONITORING CENTRE; WORLD CONSERVATION UNION.

EnviroSources

United Nations Development Program Web site: www.undp.org.

United Nations Earth Summit, An international conference dealing with environmental issues that was held in Rio de Janeiro, Brazil, in 1992. The United Nations Earth Summit was attended by the heads of state of 108 nations, who worked together for the development and adoption of several major agreements designed to benefit the global ENVIRONMENT. Among the major agreements of the summit was a document known as AGENDA 21, which provided a plan of action for achieving SUSTAINABLE DEVELOPMENT. In addition to Agenda 21, agreements known as the Statement of Forest Principles and the Rio Declaration on Environment and Development also resulted from the 1992 Earth Summit. The Statement of Forest Principles provided a guide to more sustainable use of the FORESTS of the world; the Rio Declaration outlined the rights and obligations of states toward achievement of environmental development.

In addition to its major agreements, two conventions also emerged from the Earth Summit: The Convention on Climate Change included an agreement on the part of summit participants to reduce the EMISSION of gases that contribute to GLOBAL WARMING. The other convention resulted in the BIODIVERSITY TREATY, which requires signatory nations to develop plans to identify and protect the ENDANGERED SPECIES that reside within their borders. As part of the plan, nations agreed to try to protect the HABITATS of SPECIES identified as threatened or endangered as a means of lowering the risk of EXTINCTION for such species.

In 1997, a five-year review of the policies and initiatives resulting from the Rio Earth Summit was held during a special session of the United Nations General Assembly in New York. This session often is referred to as the "Earth Summit Plus Five." Results of this meeting indicated that while nations have made some progress in implementing the principles recommended at the earlier meeting, much still needs to be done in all areas. For example, some gains were made in slowing global POPULATION growth, but at the same time, Earth's fresh-water resources became more scarce and the amount of arable agricultural land decreased globally. In addition, while industrialized nations have made progress in decreasing their production of wastes and have increased RECYCLING, such nations continue to use more than their share of NATURAL RESOURCES. Some estimates indicate that 80% of the world's resources are used by only 20% of its people.

The Earth Summit Plus Five also indicated that although 166 nations agreed to decrease their production of GREENHOUSE GASES to 1990 levels by the year 2000, emissions of CARBON DIOXIDE and other greenhouse gases continues to rise globally. This is partly due to increased use of FOSSIL FUELS in developing nations, while use of such FUELS in industrialized nations has decreased slightly. Other increases in carbon dioxide emissions have been attributed to SLASH AND BURN agricultural practices and increases in DEFORESTATION. In response to these problems, the United Nations proposed increased involvement in governments toward issues related to sustainable development. *See also* ANTARCTIC TREATY; BIOSPHERE RESERVE; CARARA BIOLOGICAL RESERVE; CARTAGENA CONVENTION; CHLOROFLUOROCARBON; CONVENTION ON INTERNATIONAL TRADE IN ENDANGERED SPECIES OF WILD FLORA AND FAUNA; EARTH DAY; EARTHWATCH; FOOD AND AGRICULTURAL ORGANIZATION OF THE UNITED NATIONS; GLOBAL OCEAN ECOSYSTEMS DYNAMICS; GREENHOUSE EFFECT; INTERNATIONAL BIOLOGICAL PROGRAM; INTERNATIONAL CONVENTION FOR THE REGULATION OF WHALING; INTERNATIONAL COUNCIL FOR LOCAL ENVIRONMENTAL INITIATIVES; INTERNATIONAL DECADE FOR NATURAL DISASTER REDUCTION; INTERNATIONAL REGISTER OF POTENTIALLY TOXIC CHEMICALS; MAN AND THE BIOSPHERE PROGRAMME; MONTREAL PROTOCOL; RAMSAR CONVENTION; RED LIST OF ENDANGERED SPECIES; TRADE RECORDS ANALYSIS OF FLORA AND FAUNA IN COMMERCE; UNITED NATIONS CONVENTION ON THE LAW OF THE SEA; UNITED NATIONS DEVELOPMENT PROGRAM; UNITED NATIONS ENVIRONMENT PROGRAMME; WORLD BANK; WORLD CONSERVATION MONITORING CENTRE; WORLD CONSERVATION UNION.

EnviroSource

United Nations (Earth Summit) Web site: www.un.org/esa/earthsummit/index.html.

United Nations Environment Programme, An agency established within the United Nations in 1973 to guide and coordinate environmental activities of nations and nongovernmental organizations throughout the world. The United Nations Environment Programme (UNEP) is overseen by a 58-member governing council that is elected by the UN General Assembly. The council is charged with making policy recommendations that promote international environmental cooperation to address such concerns as GLOBAL WARMING, DEFORESTATION, preserving BIODIVERSITY, and developing future ENERGY resources.

To meet its goals, the UNEP relies on research and information gathered through the international program EARTHWATCH. Earthwatch is composed of the INTERNATIONAL REGISTER OF POTENTIALLY TOXIC CHEMICALS, the GLOBAL ENVIRONMENT MONITORING SYSTEM, and INFOTERRA, a global information network of which the U.S. ENVIRONMENTAL PROTECTION AGENCY is a part. These three groups monitor environmental problems and provide their data to the UNEP, who may then propose solutions to the General Assembly or plan conferences or conventions for the purpose of developing international legislation or treaties. *See also* AGENDA 21; ANTARCTIC TREATY; BIODIVERSITY TREATY; BIOSPHERE RESERVE; CARTEGENA CONVENTION; CONVENTION ON INTERNATIONAL TRADE IN ENDANGERED SPECIES OF WILD FLORA AND FAUNA; MAN AND THE BIOSPHERE PROGRAMME; RED LIST OF ENDANGERED SPECIES; UNITED NATIONS DEVELOPMENT PROGRAM; UNITED NATIONS EARTH SUMMIT; WORLD CONSERVATION MONITORING CENTRE; WORLD CONSERVATION UNION; WORLD HEALTH ORGANIZATION.

EnviroSource

United Nations Environment Programme Web site: www.unep.org.

unsaturated zone. *See* zone of aeration.

upwelling, The rise of deep OCEAN water carrying large amounts of dissolved nutrients to the ocean's surface. Upwelling regions are located off the west coasts of North and South America, the northwest and southwest coasts of Africa, and the Somalian and Arabian coasts. One of the best-known zones of upwelling is off the coast of Peru. Areas where upwelling occurs contain many of the world's most valuable commercial fisheries in the world. (*See* commercial fishing.)

Upwelling occurs when warm SURFACE WATER is drawn away by strong winds off the coastline. The warm surface water is replaced by colder water rising from the bottom of the ocean. The rising colder waters contain MICROORGANISMS, such as PHYTOPLANKTON and ZOOPLANKTON, which multiply in large numbers and provide abundant food for many kinds of fishes, which in turn provide food for larger animals such as MAMMALS and birds. Upwelling plays an important part of the marine FOOD WEB in those areas. If the coastal winds decline due to a change in WEATHER or climatic conditions the warm surface water remains and the upwelling does not occur. This phenomenon occurs during a period called the EL NIÑO and affects fish POPULATIONS and other marine life. The declines of commercially important fish stocks can be so severe that it affects the welfare of coastal communities and regions who depend on the fish for their livelihood. *See also* CLIMATE; OCEAN CURRENT; PLANKTON.

uranium, A metallic, radioactive element that is used as a FISSION FUEL in NUCLEAR REACTORS. Uranium and its byproducts are naturally present in AIR, soil, water, and in trace amounts in the food eaten by people. Uranium has natural ISOTOPES with atomic masses ranging from 234 to 239 atomic mass units. Uranium-235 (U-235) is one of the fissionable isotopes of uranium. The most common form of uranium is uranium-238 (U-238), which has a HALF-LIFE of 4.5 billion years.

U-238 cannot be used as a fuel; however, it can be changed into a usable fuel in a BREEDER REACTOR. The nuclear reactor emits fast-moving neutrons to induce fission in U-238, which produces PLUTONIUM-239. Plutonium-239 can be used as fuel in nuclear reactors and NUCLEAR WEAPONS.

Uranium minerals are widely distributed in Earth's crust. They are present in sandstones, in veins within rock fractures, and in deposits of ore materials in river deltas and streams. (*See* erosion; deposition.) Pitchblende is a major source of uranium. Most uranium mined in the United States derives from sandstone deposits. Worldwide, the richest deposits of uranium are present in France, the Russian Federation, Ukraine, Australia, Canada, and southern Africa. Some experts believe that there is about a 40-year supply of uranium reserves. *See also* NUCLEAR ENERGY; NUCLEAR WEAPON; RADIATION; RADIOACTIVE WASTE; URANIUM MILL TAILINGS RADIATION CONTROL ACT.

TechWatch

In the 1990s, the U.S. DEPARTMENT OF ENERGY (DOE) announced a program that would include building two special facilities that would be used to convert about 700,000 metric tons (770,000 tons) of depleted URANIUM into new products or wastes that can be stored safely. The depleted uranium, mostly U-238, is what remains after U-235 is extracted for use in the production of NUCLEAR WEAPONS and nuclear power. Most of the depleted uranium, which emits a relatively low amount of radiation, is stored in special canisters; however, the canisters have the potential to leak and emit toxic gases. Some of the depleted uranium has been stored since the time of the MANHATTAN PROJECT, which was in operation in the 1940s. The DOE plan would encourage private companies to build facilities that would take the depleted uranium and process the material into other products, such as pellets to reinforce concrete.

EnviroSource

Covalt, Ann, ed. *Affordable Cleanup? Opportunities for Cost Reduction in the Decontamination and Decommissioning of the Nation's Uranium Enrichment Facilities.* Washington, DC: National Academy Press, 1996.

uranium mill tailings. *See* Uranium Mill Tailings Radiation Control Act.

Uranium Mill Tailings Radiation Control Act, Federal legislation enacted in 1978 that allows the U.S. DEPARTMENT OF ENERGY (DOE) to control URANIUM mill TAILINGS from inactive sites as a means of minimizing health hazards. Uranium mill tailings are byproducts of uranium MINING and the processing of the ores. The tailings consist of rock and soil containing small amounts of RADIUM and other radioactive materials. Uranium mill tailings become a RADIOACTIVE WASTE disposal problem because RADON, a radioactive gas, is produced when radium decays. At one time, the tailings were used as fill in various locations; however, this practice was stopped when it was recognized that the tailings posed a health hazard.

The Uranium Mill Tailings Radiation Control Act authorizes the DOE to undertake remedial cleanup actions at designated inactive sites. One estimate indicates that there are more than 18 million metric tons (19.8 million tons) of uranium mill tailings located in inactive sites across the western United States. The tailings are disposed of in DOE repositories. *See also* HIGH-LEVEL RADIOACTIVE WASTES; NUCLEAR WASTE POLICY ACT; RADIATION; RADIOACTIVE DECAY.

EnviroSources

International Atomic Energy Agency ("Managing Radioactive Waste" fact sheet) Web site: www.iaea.org/worldatom/inforesource/factsheets/manradwa.html.
National Research Council, Board on Radioactive Waste Management Web site: www4.nas.edu/brwm/brwm-res.nsf.
U.S. Department of Energy, Office of Civilian Radioactive Waste Management Web site: www.rw.doe.gov.
U.S. Environmental Protection Agency (mixed waste) Web site: www.epa.gov/radiation/mixed-waste.
U.S. Nuclear Regulatory Commission (radioactive waste) Web site: www.nrc.gov/NRC/radwaste.

urban forest, The nation's forested lands in and adjacent to urban areas. Urban and community FORESTS are made up of trees and associated woody vegetation within the environs of populated places, including natural woodlands within the zone of influence of urbanization. The urban forest encompasses trees along streets and in greenbelts, city parks, municipal WATERSHEDS, and similar areas. There are 29 million hectares (72 million acres) of such vegetation in the United States; 80% of Americans live and work among them daily. They are small pockets of green in a gray landscape, which can often mitigate the environmental impacts of development.

Urban and community forests provide multiple benefits, including reduction of ENERGY costs through summer shading and winter wind protection. Urban areas have distinctly different temperature regimes from those of large parks and rural areas; they are usually warmer, a condition known as the urban HEAT ISLAND. Summertime studies have shown a 0.5°–1°C (0.9–1.8°F) decrease in temperature for every 10% increase in vegetation cover. Houses shaded by trees require 4–25% less energy for cooling than houses in the open. Winter energy savings of as much as 10,300 British thermal units for an individual home have been attributed to increases in surrounding urban vegetation. Urban trees will be increasingly important in energy CONSERVATION as FOSSIL FUELS for cooling and heating become scarcer.

Urban forests also control storm-water RUNOFF and EROSION; filter airborne PARTICULATE MATTER and POLLUTANTS, such as OZONE; consume CARBON DIOXIDE; produce OXYGEN through PHOTOSYNTHESIS; and cool the surrounding area through evapotranspiration. The cooling effects also reduce the need for utilities to generate power for cooling, reducing carbon dioxide production. In the United States, 400–900 million metric tons (440–990 million tons) of CARBON are stored as BIOMASS in urban forests. Planting urban trees is a good way to address the GLOBAL WARMING problem and is a more cost-effective energy conservation strategy than many other fuel-saving measures.

Urban forests also directly increase property values by making communities more attractive, improving recreational opportunities, and providing urban WILDLIFE HABITAT. There are also benefits to human physical and psychological health, community stability, and crime reduction efforts. These forested areas provide city dwellers the opportunity to experience, understand, and appreciate forests.

Establishing and maintaining urban and community forests is costly. Larger saplings and full-grown trees are expensive to purchase. In addition, there are high maintenance costs related to utility line clearance, storm damage repair, debris removal, and DISEASE protection. Some costs and potential problems can be reduced or avoided through proper selection and location of trees.

Tree SPECIES DIVERSITY is also an important consideration in terms of the ability of the urban forest to withstand stress. A diverse POPULATION may slow or prevent the spread of INSECTS or diseases, and PATHOGENS will have a lesser impact on mixed trees than strands of a single species. One well-known urban forest lesson involves the American elm, which once dominated urban forests throughout North American cities. DUTCH ELM DISEASE was able to devastate the urban population of American elm because the disease could spread from tree to tree with few intervening SPECIES to slow its progress.

The unique demands on the urban forest, its location within heavily populated and developed areas, and its potential as a medium to educate and engage the public in NATURAL RESOURCE issues mean that urban forests require unique management approaches. Management of these ENVIRONMENTS must be coordinated with both managers of "built" environments within cities and with forest managers in the rural-urban interface.

Over the past several years, shrinking municipal budgets have produced a crisis for the nation's urban forests. Traditionally, city trees were managed by city governments; however, downsizing has led to drastic cuts in spending on urban FOREST MANAGEMENT. More and more, the responsibility has shifted toward nonprofit organizations, which have also been instrumental in raising public awareness of and involvement in the urban forest. *See also* BLIGHT; MONOCULTURE; PARASITISM; POLYCULTURE.

EnviroSources

American Forests: P.O. Box 2000, Washington, DC 20013: Web site: www.amfor.org.
TreeLink Web site: www.treelink.org.

vadose zone. *See* zone of aeration.

vapor recovery system, A system attached to GASOLINE dispenser pumps and tanks to contain volatile gases that would otherwise be emitted into the ATMOSPHERE. Such systems are designed to collect vapors during refilling of underground tanks by tanker trucks and during the refueling of vehicles. *See also* AUTOMOBILE; CATALYTIC CONVERTER.

vertebrate, Any animal that has a backbone and an internal skeleton made of bone or cartilage. All vertebrates have a backbone made of bony or cartilaginous segments called vertebrae. It is from these structures that the group receives its name. Biologists have identified approximately 40,000 SPECIES of vertebrates on Earth. Scientists classify, or organize, these species into different groups according to similarities in structure and evolutionary histories. Biologists recognize five distinct types (classes) of vertebrates: fishes, AMPHIBIANS, REPTILES, birds, and MAMMALS. *See also* ADAPTATION; ANIMAL RIGHTS; FAUNA; INSECTIVORE; INVERTEBRATES; TAXONOMY; WILDLIFE; ZOOLOGY.

EnviroSources

U.S. Fish and Wildlife Service, Species List of Endangered and Threatened Wildlife Web site: www.fws.gov/r9endspp/lsppinfo.html.

vicuna, A member of the camel family that lives in South America and is closely related to llamas, alpacas, and guanacos. In 1970, the vicuna was listed as an ENDANGERED SPECIES as a result of overhunting for their skins. Most vicunas live in small, wild herds in the semiarid GRASSLANDS of the central Andes Mountains at ALTITUDES of 3,600–4,800 meters (12,000–16,000 feet). They are swift, graceful animals with long, slender legs, small heads, and large, pointed ears. They are extremely gentle and feed upon grass and small PLANTS. Vicunas are primarily valued for their fur, which is long, fine, and soft, and varies in color from light brown to white. Vicuna fur is very strong and resilient and is used to make some of the finest and most expensive fibers in the world. The government of Peru tried to encourage vicuna domestication in the late nineteenth century and early part of the twentieth century; however, these attempts were unsuccessful. However, international laws are protecting the animals. The Vicuna POPULATION is recovering, and they number about 100,000, which is an increase from 10,000–15,000 in 1970. *See also* HERBIVORE; MAMMALS; ORYX.

EnviroSource

U.S. Fish and Wildlife Service, Species List of Endangered and Threatened Wildlife Web site: www.fws.gov/r9endspp/lsppinfo.html.

The vicuna is a member of the camel family that lives in South America and is closely related to llamas, alpacas, and guanacos.

vinyl chloride, A gaseous PETROCHEMICAL consisting of CARBON, CHLORINE, and HYDROGEN. Vinyl chloride is used to make polyvinyl chloride (PVC), a component of a variety of plastic products, including pipes, wire and cable coatings, packaging materials, furniture, AUTOMOBILE upholstery, wall coverings, housewares, and automotive parts. Until the mid-1970s, vinyl chloride was also used as a coolant, a propellant in AEROSOL spray cans, and in some cosmetics. It is no longer used for these purposes because the U.S. Department of Health and Human Services, the International Agency for Research on Cancer, and the U.S. ENVIRONMENTAL PROTECTION AGENCY (EPA) have all determined that vinyl chloride is carcinogenic to humans. (*See* carcinogen.)

Worldwide, the greatest source of vinyl chloride in the ENVIRONMENT is the manufacture of plastics, which releases vinyl chloride into AIR or in WASTEWATER. The EPA limits the amount of vinyl chloride that industries may release into the ATMOSPHERE under the NATIONAL EMISSIONS STANDARDS FOR HAZARDOUS AIR POLLUTANTS provision of the CLEAN AIR ACT. Vinyl chloride is also emitted into the air from HAZARDOUS WASTE sites and landfills. Studies have shown that workers who are exposed to vinyl chloride over many years have a higher-than-normal risk of developing liver CANCER. Brain cancer, lung cancer, and some blood cancers are also suspected to result from the inhalation of vinyl chloride over long periods of time. To protect workers, the OCCUPATIONAL SAFETY AND HEALTH ADMINISTRATION supervises and monitors workplaces where vinyl chloride is manufactured. *See also* CHRONIC TOXICITY.

virus, A subatomic particle consisting of a nucleic acid (deoxyribonucleic acid—DNA or RIBONUCLEIC ACID—RNA) surrounded by a protein coat. Viruses generally are not considered ORGANISMS because they lack virtually all characteristics of living things; however, viruses depend on organisms for their existence because they are able to replicate (make more of themselves) only when inside living CELLS. Viruses are of environmental importance because they often serve as agents of DISEASE in PLANTS, animals, and BACTERIA and are easily spread throughout dense POPULATIONS of organisms. Some human diseases caused by viruses include cold, influenza (flu), measles, acquired immune deficiency syndrome (AIDS), herpes, tuberculosis, some forms of pneumonia, and even some forms of CANCER.

Thus far, scientists have identified more than 4,000 different viruses, which have been classified into more than 70 families according to their means of replication, structure, type of nucleic acid, and the types of organisms they infect. In addition, scientists have identified two types of PATHOGENS that may be related to viruses: viroids and prions. Viroids are disease-causing agents composed only of a circular RNA molecule that infect plants. Prions are pathogens that infect humans and animals and are composed of only protein. The infectious agent that causes Creutzfeldt-Jakob disease in humans, scrapie in sheep, and BOVINE SPONGIFORM ENCEPHALOPATHY in cattle is thought to be a prion that attacks brain tissue in each of these organisms. *See also* CARCINOGEN; FUNGI; MICROORGANISMS; POPULATION DENSITY; PROTISTS.

vitrification, A process in which RADIOACTIVE WASTE materials are mixed with molten glass to form solid glass logs or blocks that can be safely disposed of in a repository. Vitrification is a term that refers to the conversion of solids into glass.

Vitrification does not reduce radioactivity but changes the form of the wastes from a leachable liquid to an immobile solid. (*See* leaching.) The final product captures RADIOISOTOPES, preventing these potentially harmful wastes from contaminating soil, GROUNDWATER, or SURFACE WATER.

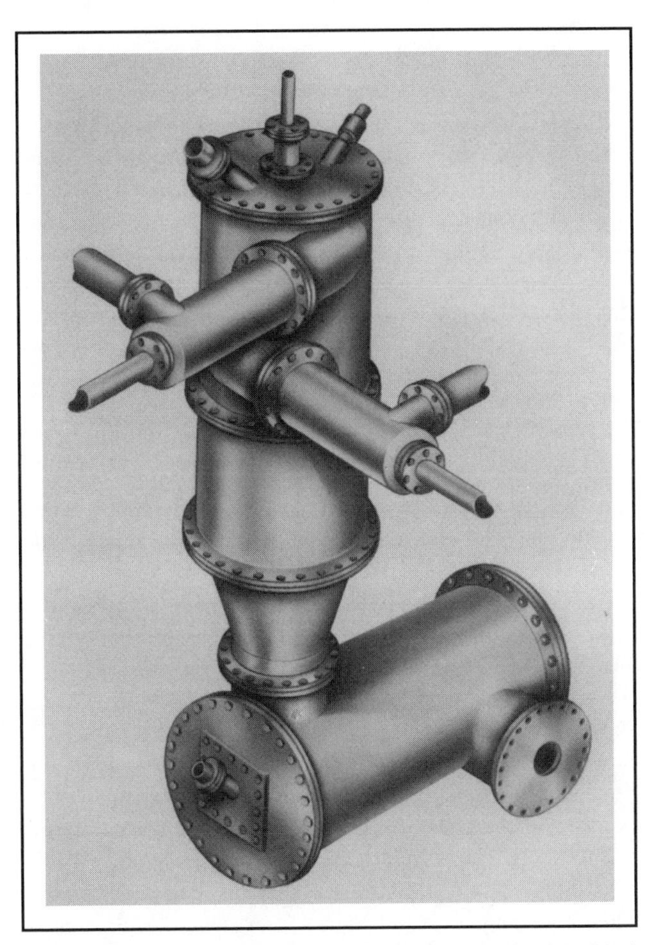

This is an illustration of a vitrification unit that vitrifies hazardous and residual wastes. The unit recycles incinerator wastes into products such as ceramic tile and glass fibers using glass melting technology. *Credit:* Vortec Corporation, Collegeville, PA

Although vitrification provides a means for containing RADIATION, simply vitrifying a material does not necessarily produce an environmentally stable product. To ensure that a waste material is stable for disposal, the chemistry of the material needs to be assessed and, possibly, modified. Not all radioactive wastes currently produced can undergo vitrification, and additional testing of this process is continuing.

One vitrification technology that is already being used immobilizes HIGH-LEVEL RADIOACTIVE WASTES (HLWs) in glass made from borosilicate sand, a material that is impermeable, durable, and not damaged by high temperatures. (*See* permeability.) The borosilicate glass is expected to remain stable for at least 1 million years. Borosilicate glass vitrification is currently in use at the Defense Waste Processing Facility of the SAVANNAH RIVER SITE in South Carolina. The technology at this plant involves melting sand mixed with HLWs to approximately 1,200°C (2,192°F) and then pouring the resulting molten glass mixture into a stainless steel canister that is 3 meters (9.8 feet) tall and 1 meter (3 feet) in diameter to cool and harden into a stable glass. Over the next 25 years, the Savannah River Site will fill about 6,000 canisters. The radioactive glass canisters will be stored on site until they can be shipped to a geologically stable repository for underground disposal.

Currently, vitrification is used only in the United States; however, the process is being explored for use in Europe. A Swiss company, Seiler Pollution Control, is planning to build the first European vitrification plant in Freiberg, Germany. This plant will be used to treat a variety of HAZARDOUS WASTES. Other European countries are interested in making use of the Swiss-based system to treat incinerator fly ASH, a major air POLLUTANT. *See also* AIR POLLUTION; DEEP-WELL INJECTION; HANFORD NUCLEAR WASTE SITE; INCINERATION; *IN SITU* VITRIFICATION.

EnviroSource

U.S. Department of Energy Web site: www.em.doe.gov/fs/fs3m.html.

volatile organic compounds, Any ORGANIC compound that quickly evaporates into the ATMOSPHERE and is potentially toxic to humans and other ORGANISMS. Many volatile organic compounds (VOCs) are derived from PETROCHEMICALS. They include alcohols, acetone, TRICHLOROETHYLENE, perchloroethylene, dichloroethylene, BENZENE, VINYL CHLORIDE, TOLUENE, and METHYLENE CHLORIDE. More than 250 VOCs are widely used in industry as SOLVENTS, degreasers, paint thinners, and FUELS.

When VOCs evaporate into the AIR, they are easily dispersed by wind, increasing the potential that humans will be exposed to the substances. Excessive exposure to some forms of VOCs can cause liver and kidney problems and CANCER risks. Some VOCs can move through soil into GROUNDWATER where they may be absorbed by soil organisms or organisms that drink or otherwise contact the groundwater. VOCs also contribute to the formation of photochemical SMOG. *See also* AIR POLLUTION; VAPOR RECOVERY SYSTEM.

volcano, An opening in Earth's crust through which lava, gases, and ASHES escape, sometimes explosively. Volcanoes are classified as active, dormant, or extinct. Volcanic activity is the result of enormous pressure at the boundary of crustal plates. (*See* plate tectonics.) Most volcanoes occur around the edge of the Pacific Ocean, known as the Ring of Fire, where there are several plate boundaries. The gases from volcanoes contain SULFUR DIOXIDE, CARBON DIOXIDE, and other elements such as OXYGEN, iron, calcium, sodium, potassium, and CARBON. The sulfur dioxide can combine with water vapor to produce SULFURIC ACID, a pollutant in ACID RAIN.

To learn more about volcanoes, volcanologists associated with the International Association of Volcanology and Chemistry of the Earth's Interior, in cooperation with the United Nations' INTERNATIONAL DECADE FOR NATURAL DISASTER REDUCTION (IDNDR), are studying the activity of 15 major volcanoes throughout the world, including Etna and Vesuvius in Italy, Colima in Mexico, and Unzen and Sakura-Jima in Japan. One of their goals is to establish better early warning systems.

Abnormal global WEATHER conditions can be caused by extensive volcanic activity. In 1883, KRAKATOA erupted in Indonesia. The explosion destroyed 60% of the island and caused TSUNAMIS. More than 30,000 people were killed. Large quantities of carbon dioxide, SULFUR, dust, and ash were ejected into the stratosphere from the eruption. When the material reached high into the ATMOSPHERE, winds circulated the dust around the globe. As a result, the volcanic debris or AEROSOLS blocked RADIATION from the sun and reduced global temperatures slightly for a period of three years. Similar CLIMATE changes happened again after the eruptions of El Cichon in 1982 and Mount Pinatubo in 1991. Not every volcanic eruption produces significant global climate conditions, however, some volcanoes produce just lava while others emit large quantities of dust and other EMISSIONS. Most scientists agree that the ability of an eruption to alter global climate conditions depends on the latitude of the volcano, the season in which it erupts, the height of the volcano, and the kinds of gases it produces.

Scientists are now using heat-sensitive equipment aboard Earth-orbiting weather satellites to predict when volcanoes will erupt in North and South America. The infrared sensors relay data to scientists at ground stations when vents open up or when there are new lava flows in or around the dome of an active volcano. The information can warn the scientists when a volcano is about to erupt. In 1999 scientists were able to forecast the eruption of the Pacaya volcano in Guatemala about a week before the volcano spewed ashes over the area. *See also* PLATE TECTONICS; NATURAL DISASTER; ROCK CYCLE.

EnviroSources

Decker, Robert, and Barbara Decker. *Volcanoes.* New York: W. H. Freeman & Co., 1997.

Fisher, Richard V. *Volcanoes: Crucibles of Change.* Princeton, NJ: Princeton University Press, 1998.

Sigurdsson, Haraldur. *Melting the Earth: The History of Ideas on Volcanic Eruptions.* Oxford, UK: Oxford University Press, 1999.

U.S. Geological Survey (about volcanoes) Web site: www.usgs.gov/education/learnweb/volcano/index.html.

U.S. Geological Survey (building safer structures) Web site: www.quake.wr.usgs.gov/QUAKES/FactSheets/SaferStructures.

U.S. Geological Survey (earthquake information) Web site: www.quake.wr.usgs.gov/QUAKES/CURRENT/current.html.

U.S. Geological Survey (volcano hazards) Web site: www.volcanoes.usgs.gov.

Major Volcanoes in the World

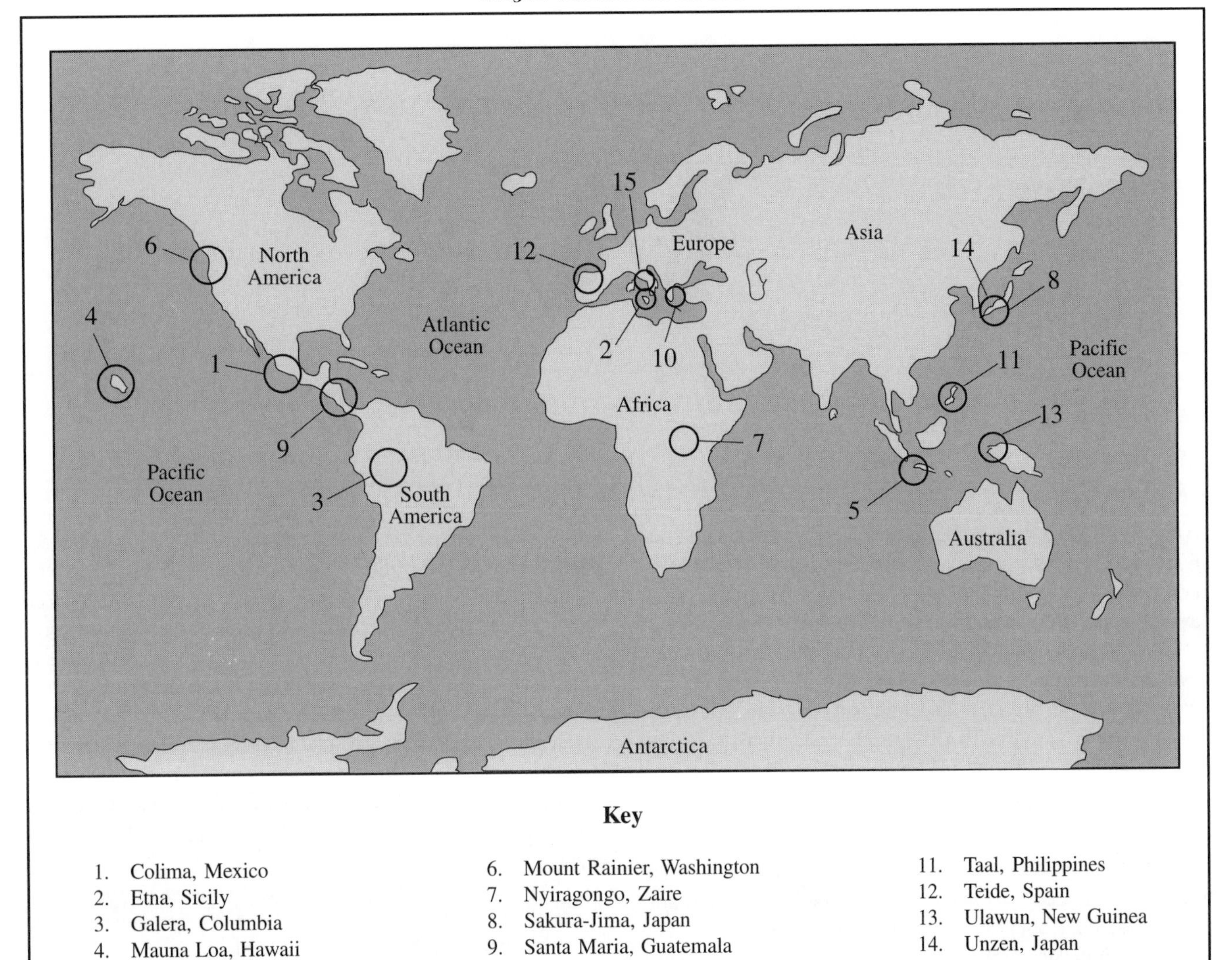

The International Association of Volcanology and Chemistry of the Earth's Interior, in cooperation with the United Nations' International Decade for Natural Disaster Reduction, are studying these major and potentially dangerous volcanoes. All of them are near heavy populated areas. *Source:* United Nations' International Decade for Natural Disaster Reduction.

Waldsterben, A German term meaning "FOREST death" due to AIR POLLUTANTS such as ACID RAIN, OZONE, photochemical oxidants, NITROGEN OXIDES, and SULFUR DIOXIDES. The term, popularized in the1970s was first used to describe the decline of trees in certain sections of the Black Forest of Germany, where conifers such as pine, fir, and spruce as well as some deciduous trees such as beech, maple, and oak became leafless, deformed, or covered with LICHENS. Forests in Britain and Switzerland are suffering also from Waldsterben. *See also* CONIFEROUS FOREST; DECIDUOUS FOREST.

warning label, A precautionary label used on PESTICIDES that indicates its TOXICITY as mandated by the FEDERAL INSECTICIDE, FUNGICIDE, AND RODENTICIDE ACT. Toxicity is rated from one to four: One indicates high toxicity; four indicates low toxicity. However, only three terms are used to indicate toxicity: The strongest label is DANGER; the next is WARNING; and the mildest is CAUTION. *See also* ECOLABELING; FOOD AND DRUG ADMINISTRATION.

waste management facility, A facility in the United States that is specially designed, constructed, and operated to manage various types of wastes. Special federal and state laws and regulations govern HAZARDOUS WASTE management facilities. SOLID WASTE management facilities are subject to specific requirements also.

Hazardous waste facilities, known as treatment, storage, and disposal facilities (TSDFs), are operated according to regulations established by the RESOURCE CONSERVATION AND RECOVERY ACT (RCRA) for management of hazardous wastes. Each TSDF must have an RCRA permit from the U.S. ENVIRONMENTAL PROTECTION AGENCY to accept specific wastes for storage, treatment, or disposal. Before construction of a hazardous waste facility, the RCRA and other regulations require a thorough evaluation of the proposed facility location to determine whether the facility might present a threat to human health or the ENVIRONMENT.

Typically, hazardous waste management facilities are designed to manage specific types of waste materials; each type of material handled is referred to as a "waste stream." A waste stream must have physical characteristics and a chemical composition that vary only within a defined range. Before hazardous waste is shipped from its source to a facility, the source must determine what type of waste stream the material is and must properly label the waste to identify it during shipment to the selected facility. When a facility receives a waste shipment, it may conduct tests to ensure that the waste received fits one of the waste streams the facility is permitted to manage.

TSDFs conduct various waste treatment operations or collect wastes, storing them temporarily before shipping the waste to another facility. Some TSDFs blend certain wastes to produce mixtures that may be recycled or may be burned to produce heat or ENERGY. (*See* incineration.) Other TSDF operations may include neutralization, oil/water separation, solidification, or solids separation.

Not all waste management facilities are hazardous waste facilities. Examples of SOLID WASTE management facilities include SANITARY LANDFILLS and transfer stations. RECYCLING facilities manage various recyclable materials, some of which might otherwise be handled as solid waste or even hazardous waste if not processed appropriately. Since the 1980s, a major trend in waste management has emphasized recycling and waste minimization, encouraging reductions in the amount of hazardous wastes produced and managed. *See also* HAZARD RANKING SYSTEM; HAZARDOUS WASTE TREATMENT; HAZARDOUS MATERIALS TRANSPORTATION ACT.

wastewater, Discharged water from homes, industry, agricultural areas, and commercial businesses that contains dissolved ORGANIC and inorganic materials, floating visible particles, suspended solids, as well as debris and coarse sand and gravel from storm SEWERS. *See also* DISCHARGE.

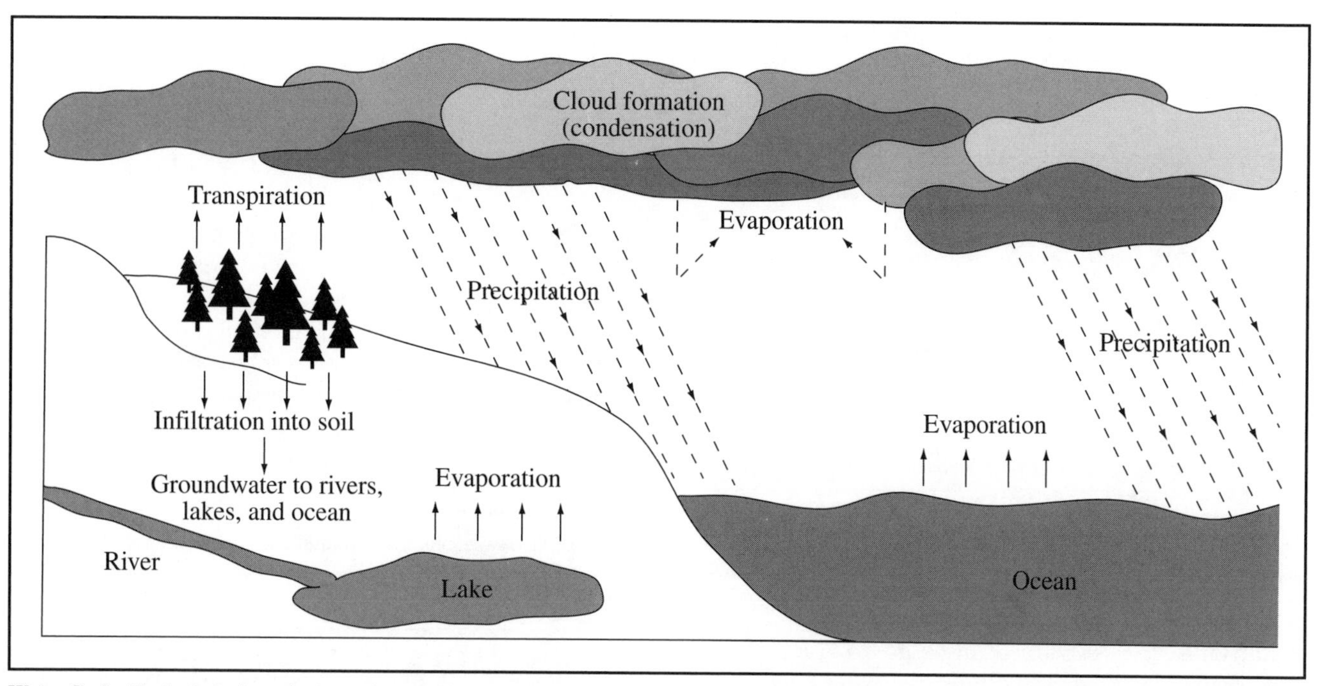

Water Cycle. The hydrologic cycle shows the natural pathway through which water moves between Earth's surface and atmosphere.

wastewater treatment. *See* primary, secondary, and tertiary treatment.

water cycle or hydrological cycle, A natural circulation of water in the form of a solid, liquid, or gas in the BIOSPHERE. The sun's RADIATION provides the ENERGY for the cycle. The radiation causes the water in OCEANS, lakes, streams, and ponds to change from a liquid to a gas (water vapor) through the process of evaporation. More than 85% of the water that evaporates from Earth's surface is from the ocean. Water vapor also enters the ATMOSPHERE by TRANSPIRATION, a process by which plants give off water vapor into the atmosphere.

The water vapor rises into the atmosphere and when it cools it condenses and changes into tiny water droplets forming clouds. If the droplets are heavy enough they will fall to Earth's surface as precipiation in the form of rain, snow, hail, and fog. Most of the precipitation, about 80% enters the oceans. The continuous process of evapotranspiration, condensation, and precipitation makes up the water cycle. Human activity can impact the water cycle. As an example, large CLEAR-CUTTING activities and overgrazing can reduce evapotranspiration while building reservoirs can increase evaporation.

Water Pollution Control Act, The U.S. Water Pollution Control Act was authorized in 1948 by the surgeon general of the Public Health Service, in cooperation with other federal, state, and local governments. The act was designed to prepare comprehensive programs for eliminating or reducing the pol-

Water Pollutants

Point sources	Bacteria	Nutrients	Total dissolved Ammonia	Solids	Acids	Toxics
Municipal sewage treatment plants	•	•	•			•
Industrial facilities				•		•
Combined sewer overflows	•	•	•			•
Nonpoint sources						
Agricultural runoff	•	•		•		•
Urban runoff	•	•		•		•
Construction runoff		•				•
Mining runoff				•	•	•
septic systems	•	•				•
Landfills spills						•
Forestry runoff		•				•

Source: U.S. EPA

lution of interstate waters and tributaries and improving the sanitary condition of surface and underground waters. *(See* groundwater.) The original statute was amended in 1970, and in 1972, the act was amended and renamed the CLEAN WATER ACT. More amendments followed in 1977, 1981, and 1987. About 60% of U.S. waters met the act's designated-use goals in 1992, compared with 36% in 1972, according to the U.S. ENVIRONMENTAL PROTECTION AGENCY. In the same period, the quality of 98% of river miles and 96% of lake acreage remained the same or improved, according to the Association of State and Interstate Water Pollution Control Administrators. *See also* DISCHARGE; SURFACE WATER; ZERO DISCHARGE.

EnviroSource

Early History of the Clean Water Act Web site: www.csi.cc.id.us/ip/ag/water/Clean%20Water%20Act/sld002.html.

water rights. *See* riparian rights.

water table, The location beneath the ground's surface where GROUNDWATER has completely filled the spaces between the soil or rock. In general, if a WELL were constructed in the ground, the level of the water in the well would represent the water table (except in the case of a confined AQUIFER). The area beneath the water table is known as the saturated zone, because the area is completely saturated with groundwater. Depending on local conditions, water tables may be located at the ground's surface or greater than 30 meters (100 feet) below the surface. *See also* ZONE OF SATURATION.

watershed, The area of land surrounding a river or lake that provides all the water that enters the river or lake. For example, when precipitation falls on a watershed, the water is delivered to the lake or river in smaller surface streams or through underground AQUIFERS. Watersheds are important because they help control the amount of water in a river or lake and also act as natural water purifiers.

The size, shape, and vegetation of a watershed influences how it functions. For instance, areas with few trees and other PLANTS often face terrible flooding problems as well as decreased water quality. This is most dramatically observed in areas where FORESTS have been cleared for AGRICULTURE or in areas where excessive cattle grazing has occurred. When there is little vegetation to break the force of falling rain, soil becomes very saturated and muddy. This prevents water from heavy rains or melting snow from fully soaking into the ground. Rivers near such disturbed watersheds often FLOOD because they cannot contain the excess amounts of water that flow into them as RUNOFF. When there are few plants, streams move quite rapidly, picking up soil as they flow, which gives the stream a brownish color.

In undisturbed watersheds containing a large number of trees and other plants, the waters of rivers, lakes, and streams are usually very clear. This clarity of water occurs because the soil in healthy ECOSYSTEMS can absorb a great amount of water. When water does flow along the surface in streams, such as after periods of heavy rain or during a spring thaw, it flows slowly and picks up fewer sediments due to the abundance of vegetation. *See also* DEPOSITION; DRAINAGE BASIN; EROSION; GROUNDWATER; SEDIMENTATION; WATER CYCLE; WATER TABLE; ZONE OF SATURATION.

weather, The conditions of the ATMOSPHERE at a specific time and place in terms of AIR pressure, temperature, water vapor or humidity, precipitation, wind speed and direction, cloud coverage, and pollution conditions. Weather occurs in the layer of the atmosphere called the troposphere and varies in different parts of the world. CLIMATE refers to weather patterns recorded over a long period of time in a particular region. *See also* WATER CYCLE.

EnviroSources

National Oceanographic and Atmospheric Administration Web site: www.noaa.gov.

National Weather Service: Web site: www.nws.noaa.gov.

weathering, A process in which rock and other surface materials of Earth are broken down as a result of contact with physical, chemical, or biological agents in the ENVIRONMENT. There are two types of weathering: physical weathering and chemical weathering. Physical weathering breaks matter, such as rock, into smaller pieces without changing its composition. Chemical weathering, breaks down matter by altering its chemical composition. The breaking down of a piece or iron or iron-containing MINERALS in rock through OXIDATION, or rusting, is an example of chemical weathering.

Weathering is both beneficial and harmful to the environment. For example, the formation of new rock and new soil is a beneficial result of weathering that results as rock is broken down into smaller and smaller pieces. (*See* rock cycle.) Rock often is broken down through physical processes such as extreme changes in temperature. Such temperature changes cause the rock to continuously expand and contract, eventually weakening the rock and allowing it to break apart. Extreme temperature changes also can allow water that collects in the cracks of rocks to repeatedly freeze and thaw. In this process, the water exerts pressure on the rock as it expands and contracts, helping to weaken and break the rock apart. Interactions with ORGANISMS also cause rocks and other surface matter to weather. Pressure exerted by the growing roots of PLANTS and burrowing animals are two activities of organisms that can cause physical weathering. The release of chemical substances by organisms into the environment can lead to the chemical weathering of rock and other surface materials. For example, LICHENS (an association of FUNGI and ALGAE) often make their HABITAT on bare rock. As they carry out their life processes, these organisms give off ACIDS that can slowly break down the rock, leading to the formation of new soil. This action is vitally important to the process of primary ECOLOGICAL SUCCESSION.

DECOMPOSITION is a chemical process in which organisms such as BACTERIA and fungi break down the wastes or remains of organisms in the environment to meet their nutritional and energy needs. These organisms break down their food source before it is ingested. Substances not used by the organisms as food are returned to the environment, where they may be used by other organisms. (*See* biogeochemical cycles.)

Not all of the weathering that occurs in the environment results from natural phenomena. For example, POLLUTANTS released to the environment through activities carried out by humans can speed or promote weathering. NITROGEN OXIDES (No_x) and SULFUR DIOXIDE (SO_2) are pollutants released to the environment as a result of the combustion of FOSSIL FUELS. Once in the ATMOSPHERE, these compounds can combine with water vapor to form acids that are carried back to Earth's surface in precipitation such as rain, snow, and fog. Acid precipitation that falls into lakes or onto FOREST soils can be extremely harmful to the organisms living in these ECOSYSTEMS. ACID RAIN that falls onto structures on Earth's surface (buildings, monuments, metal structures) can break down these substances, causing them to fall apart. Such action can be very destructive to property and cost a great deal to repair. *See also* ACID MINE DRAINAGE; ANAEROBIC DECOMPOSITION; CARBON CYCLE; DECOMPOSER; DETRITUS; DETRIVORE; EROSION; PIONEER SPECIES.

weed. *See* pesticides.

well, A human-made structure that allows people to obtain or remove water that is stored beneath Earth's surface as GROUNDWATER or oil or gas from underground RESERVOIRS. The two principal types of water wells are dug wells and drilled wells. Dug wells, the simpler type, consist of a hole dug into the ground that is often lined with rocks or other materials that keep the walls of the hole from collapsing. The many types of drilled wells are constructed with special drilling equipment and techniques selected for the particular location and uses of the well. Wells may be drilled into soil or solid rock. Typically, water is removed from a well by a pump located either above ground or inside the well.

The major uses of water wells include to provide drinking water for homes or communities, to provide water for agricultural IRRIGATION, and to supply industry with its fresh water needs. Other special well uses exist, such as groundwater testing, "dewatering" (lowering the WATER TABLE for construction), groundwater or soil treatment, heat exchange systems, and underground injection. Oil and gas wells serve very important functions in obtaining various types of PETROLEUM from underground oil and gas RESERVOIRS throughout the world. *See also* AQUIFER; NATURAL GAS; POTABLE WATER; ZONE OF DISCHARGE; ZONE OF SATURATION.

wetland, A HABITAT that is saturated with water for all or part of the year. A few wetlands are dry most of the year. Wetlands include BOGS, coastal salt-water MARSHES, fresh-water marshes, SWAMPS, vernal pools, and fens. Wetlands supply water directly to people, to AQUIFERS, and to other wetlands; they monitor water flow and help in FLOOD control; they protect against coastal EROSION and flooding from HURRICANES and other NATURAL DISASTERS; and they have the ability to remove TOXINS from EFFLUENTS and polluted water.

Each wetland is characterized by a variety of animal and PLANT communities and varies in types of soil, topography, CLIMATE, and water content. Wetlands are found on most continents, even in the TUNDRA. In the United States, wetlands are found in every state, from the coastal marshes of Alaska to the MANGROVE FORESTS of Florida.

Bogs. Bogs, also known as peatlands, are simply wetlands that have ORGANIC soils consisting of PEAT—the partially decomposed remains of plants and animals. Bogs are found in colder regions of the world where the temperatures and limited OXYGEN supply in the water discourage the breakdown of organic material. Bogs usually occupy shallow depressions associated with areas of water DISCHARGE, such as springs and brooks. Bogs serve important ecological functions in preventing downstream flooding by absorbing direct precipitation and runoff, protecting water quality by intercepting and filtering runoff, and providing critical habitat for unique plant and animal communities.

Fens are similar to bogs but are less acidic because the water in fens contains calcium and magnesium. Peat also accumulates in a fen.

Marshes. Coastal marshes are open, shallow wetlands that connect major water bodies with smaller lakes and streams. Coastal wetlands support a rich and diverse aquatic ECOSYSTEM, providing spawning and nursery grounds for a vast portion of U.S. commercial fish and shellfish harvest. Much of the COMMERCIAL FISHING industry depends directly or indirectly on coastal wetlands. In 1991, the value of fish marketed in the United States was $3.3 billion, which served as the basis of a $26.8 billion fish processing and sales industry, which in turn employs hundreds of thousands of people. Wetlands also provide plant food for commercial and recreational fish and shellfish industries. Coastal wetlands are found primarily along the Atlantic, Gulf, and Alaskan Coasts, and are thinly dispersed on the U.S. West Coast.

Fresh-water marshes, familiar to most Americans, make up nearly 90% of our nation's wetlands. Marshes are open areas, usually with few trees and shrubs, that provide food supplies for migratory birds and other WILDLIFE. The water in a marsh fluctuates, rising during the rainy season and disappearing during dry periods. Remaining fresh-water wetlands in suburban and urban areas are valuable for purifying storm water and reducing flooding.

Swamps. Swamps are shrubby or forested wetlands, located in poorly drained areas on the edges of lakes and streams. Forest swamps primarily exist in river FLOODPLAINS connected to major river systems. In particular, bottomland hardwood forests filter and purify the waters adjacent to rivers in the southeast region of the United States.

Other Wetlands. PRAIRIE potholes are saucer-shaped depressions formed by retreating GLACIERS in the ICE AGE. Although inundated with water for only a short period of time each spring, they play a vital role in aquatic and wildlife habitat. Prairie potholes are located in the U.S. upper plains states and are often called the "duck factories" of America because

Percentage of Wetland Acreage Lost Between 1780 to 1980

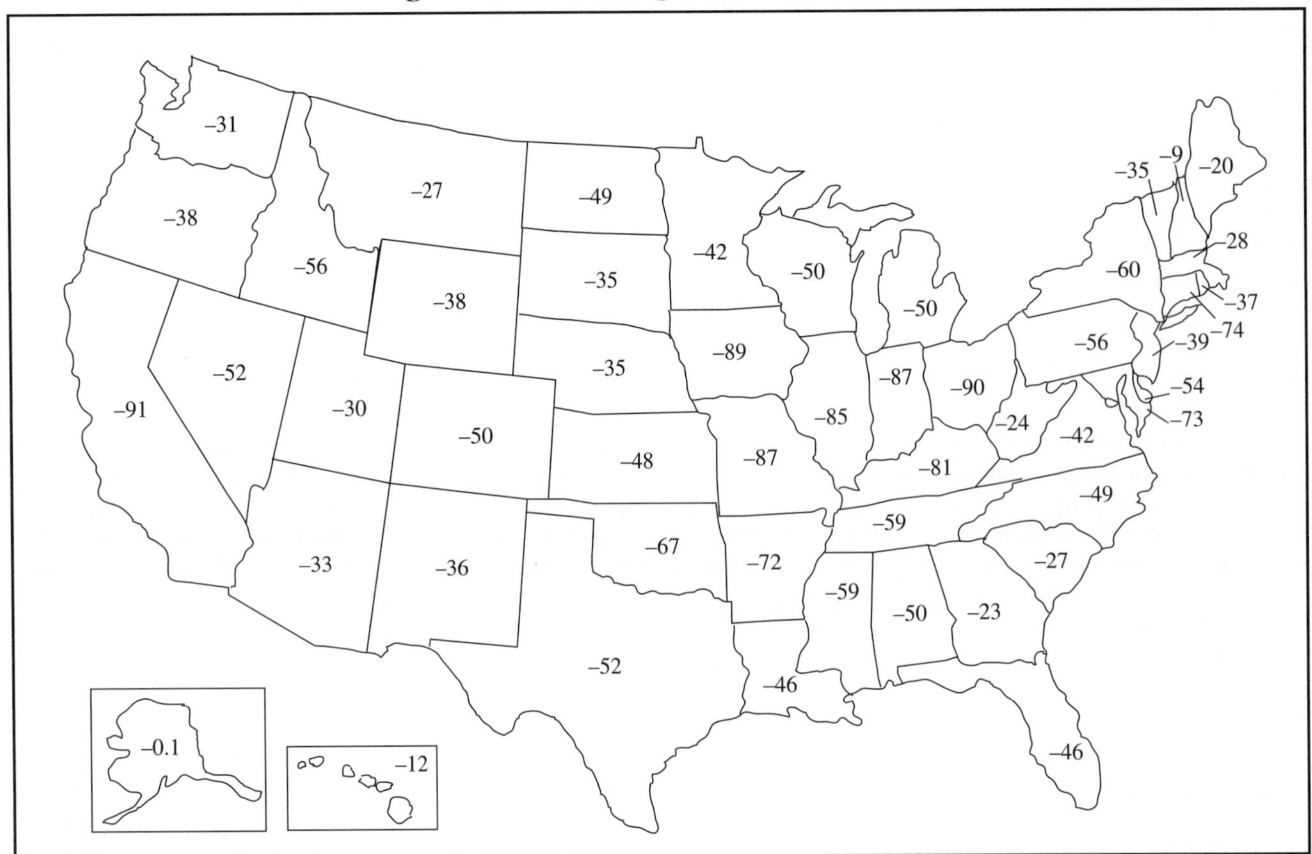

All of the states have lost wetlands between 1780 and 1980. Twenty-two states lost at least 50% of their original wetlands. *Source:* United States Environmental Protection Agency. 1989

of their importance to the livelihood of ducks and other migratory birds. In addition to supporting waterfowl hunting and birding, prairie potholes also absorb rain, snow melt, and flood waters and release their waters slowly throughout the watershed, thereby reducing the risk and severity of downstream flooding.

Vernal pools are small, isolated wetlands that retain water on a seasonal basis. The pools are vital to the survival of AMPHIBIANS; nearly 50% of the amphibians in the United States breed primarily in vernal pools because the pools are too shallow to support fish, the major predator of amphibian larvae. (*See* predation.) Vernal pools are also home to many endangered and rare plant SPECIES. (*See* endangered species; threatened species.) Many of these plants and animals spend the dry seasons as seeds, eggs, or cysts, and then grow and reproduce when the ponds are flooded again. In addition, birds such as egrets, ducks, and hawks use vernal pools as a seasonal source of food and water.

Changes to Wetlands. Over the past 200 years, human activities have directly caused the destruction of valuable wetlands. Historically, hundreds of thousands of hectares of wetlands were destroyed because of agricultural practices. Unconfirmed research indicates that Earth may have lost as much as 50% of its wetlands since 1900 due to agricultural production in many countries. By 1985, 55–65% of the wetlands in Europe and North America were drained. In the United States, the U.S. ENVIRONMENTAL PROTECTION AGENCY (EPA) reported that about 50% of the wetlands in the lower 48 states were lost between the late 1700s and the mid-1980s. States with the highest percentages of wetland losses included California, Illinois, Indiana, Ohio, and Missouri. The estimates for wetland losses in subtropical regions of Asia is about 27%, 6% in South American countries, and a very small percentage in Africa. These numbers are expected to climb as pressures to drain land for AGRICULTURE, AQUACULTURE, and the construction of buildings on wetlands intensify in the twenty-first century. Today, immediate threats to wetlands are more diverse than ever before.

Urbanization results in the filling and DREDGING of wetlands as well as the degradation of remaining wetlands. As more buildings, homes, and roads are constructed, ground surfaces become impervious to rainfall, and consequently, polluted RUNOFF and flooding increases. This makes the remaining urban wetland increasingly important to protect, even as their functions decline.

Urban pollution originates from many widespread sources, such as SEWAGE and storm-water overflow, parking lots, lawns, and the dumping of TOXIC WASTE. Highways and roads are often built through or adjacent to wetlands and can change the water flow in a WATERSHED, disrupting wetland functions and degrading the water quality of the water in wetlands. Car EMISSIONS and EROSION produce excess amounts of CONTAMINANTS and impact nearby wetlands and water bodies. TOXIC CHEMICALS from MINING operations can seap into the water system. STRIP MINING produces ACID MINE DRAINAGE and physically destroys streams and wetlands. Hard rock mining in arid areas leaves toxic waste and can lower WATER TABLES, drying up GROUNDWATER-fed wetlands. Sand and gravel extraction destroys habitats along the bottom of rivers, lakes, and wetlands.

Channeling streams—the physical modification of the natural course of streams and rivers—typically results in increased downstream SEDIMENTATION and increasingly severe flooding. It can also completely wipe out riparian wetlands. Channeling estuarine ENVIRONMENTS causes OCEAN water to move rapidly through SALT MARSHES into fresh-water systems, killing vegetation and removing valuable wetland soil in the process.

Timber harvesting in wetlands is the leading cause of wetland destruction in the southeast United States. Building forest roads threatens vital wetlands and RIPARIAN HABITATS in Alaska, the Pacific Northwest, and the Rocky Mountain states.

The worldwide loss and degradation of wetlands was the basis for the establishment of the RAMSAR CONVENTION on wetlands founded in 1971. The Ramsar Convention emphasized the importance of preserving wetlands. The convention provided an opportunity for governments, organizations, and citizens to raise public awareness of wetland values. There are presently 116 contracting parties to the convention, with 1,005 wetland sites, totaling 71.3 million hectares (177 million acres), designated for inclusion in the Ramsar List of Wetlands of International Importance. *See also* DOUGLAS, MARJORY STONEMAN; NATIONWIDE 26; NORTH AMERICAN WETLANDS CONSERVATION ACT.

EnviroSources

Batzer, Darol, et al., eds. *Invertebrates in Freshwater Wetlands of North America: Ecology and Management.* New York: John Wiley & Sons, 1999.

Groombridge, B., ed. *Global Biodiversity: Status of the Earth's Living Resources.* Norwell, MA: Kluwer Academic Publishers, 1992.

Kent, Donald M. *Applied Wetlands Science and Technology.* Boca Raton, FL: Lewis,1994

Niering, William A. *Wetlands.* Audubon Society Nature Guides. New York: Knopf, 1985.

Quammen, David. "Backwater Boodoggle." *Audubon* 100, no. 1. (January –February 1998): 52–63, 100–104.

National Service Center for Environmental Publications: P.O. Box 42419, Cincinnati, OH 45242-2419; (800) 490-9198; fax (513) 489-8695.

Ramsar List of Wetlands of International Importance Web site: www.ramsar.org/key_sitelist.html.

U.S. Environmental Protection Agency. *Citizen's Guide to Wetland Restoration: Approaches to Restoring Vegetation Communities and Wildlife Habitat Structure in Freshwater Wetland Systems.* Pamphlet. To order this pamphlet contact Nonpoint Pointers: Managing Wetlands to Control Nonpoint Source Pollution, Pointer No. 11

U.S. Environmental Protection Agency, Office of Wetlands, Oceans, Watersheds Web site: www.epa.gov/owow/wetlands.

U.S. Environmental Protection Agency, Public Information Center: EXA124, 1200 6th Ave, Seattle, WA 98101; (206) 553-1200; fax (206) 553-0149.

U.S. Environmental Protection Agency, Wetlands Hotline: (800) 832-7828; e-mail: wetlands-hotline@epamail.epa.gov.

U.S. Fish and Wildlife Service, North American Wetlands Conservation Act Web site: www.fws.gov/r9nawwo/nawcahp.html.

U.S. Fish and Wildlife Service, North American Wetlands Conservation Council Web site: www.fws.gov/r9nawwo/nawcc.html.

U.S. Geological Survey, National Wetlands Research Center Web site: www.nwrc.usgs.gov/educ_out.html.

Williams, Ted, "Who Can Save the Wetland?" *Audubon* 101:5 (September–October): 60–69.

World Conservation Union, Ramsar Convention on Wetlands (international) Web site: www2.iucn.org/themes/ramsar.

whales, A group of marine MAMMALS of the order *Cetacea* that have flippers and horizontal tails called flukes. Human beings have hunted whales for millennia to obtain food, bone, oil, and other products, but excessive commercial WHALING in the nineteenth and twentieth centuries brought several POPULATIONS of whales to the brink of EXTINCTION. Many types of whales are now THREATENED SPECIES or ENDANGERED SPECIES.

Preindustrial societies, such as Inuit and other Native American groups, have hunted whales as a source of food. Since the seventeenth century, Europeans, Americans, and other groups hunted whales to obtain oil (derived from the blubber of whales) and bone, especially the flexible, plastic-like substance called baleen, obtained from the mouths of certain whale SPECIES.

During the nineteenth century, whaling operations expanded their geographic range throughout the world's OCEANS, including the Atlantic, Pacific, and Arctic. The sperm whale (*Physeter catodon)* and right whale (*Balaena glacialis*) were among the most desirable species hunted for their oil. Important commercial species of baleen whales included the right whale, the bowhead whale (*Balaena mysticetus*), and the humpback whale (*Megaptera novaeangliae*). Modern improvements in whaling technology, such as the advent of factory ships, made the whaling process much more efficient and the loss of whales more rapid. By the end of the twentieth century, marine biologists estimated that 25% of the major whale POPULATIONS had been reduced to less than 5% of their initial population.

The Atlantic gray whale (*Eschrichtius glaucus*), has already been hunted to extinction. The humpback whale is currently the most endangered of whale species. Other endangered whale species include the blue whale (*Sibbaldus musculus*), eastern Arctic bowhead whale (*Balaena mysticetus)*, finback whale (*Balaenoptera physalus*), and sei whale (*Balaenoptera borealis*). The sperm whale, hunted intensively during the nineteenth century as a source of oil, has recovered remarkably well during the twentieth century and has reached a worldwide population of approximately two million.

In 1946, the INTERNATIONAL WHALING COMMISSION (IWC) was formed, enlisting 37 countries that shared a concern for whale CONSERVATION. The IWC established whaling quotas—limits on the number of whales that could be killed. In spite of conservation measures, however, whale populations declined still further until eight whale species became endangered by 1970. The IWC declared a moratorium on whaling

The humpback (top), right whale (middle) and bowhead (bottom) are part of a group of marine mammals of the order *Cetacea* that have flippers and tails with horizontal flukes.

in 1982, but some nations refused to participate. GREENPEACE, an environmental advocacy group, has advanced the cause of whale protection through action and protest. *See also* DOLPHINS AND PORPOISES; INTERNATIONAL CONVENTION FOR THE REGULATION OF WHALING; MARINE MAMMAL PROTECTION ACT; TUNA.

EnviroSources

Carwardine, Mark. *Eyewitness Handbooks: Whales Dolphins and Porpoises.* New York: Dorling Kindersley, 1995.
Connor, Richard C., and Dawn Micklewaite Peterson. *The Lives of Whales and Dolphins: From the American Museum of Natural History.* New York: Henry Holt, 1996.
Hoyt, Erich. *Meeting the Whales.* Ontario, Canada: Camden House, 1991.
Institute of Cetacean Research: Web site: www.whalesci.org.
Payne, Roger. *Among Whales.* New York: Dell Publishing, 1996
Reader's Digest Association. *Whales, Dolphins and Porpoises.* New York: Reader's Digest Association, 1997.
U.S. Fish and Wildlife Service, Species List of Endangered and Threatened Wildlife Web site: www.fws.gov/r9endspp/lsppinfo.html.
Wurtz, M., and N. Repetto. *Whales and Dolphins: A Guide to the Biology and Behavior of Cetaceans.* Thunder Bay, Canada: Thunder Bay Press, 1999.

whaling, Hunting WHALES for their meat, bone, and blubber. This industry is dominated by American, Japanese, Norwegian, and Russian whalers.

The INTERNATIONAL WHALING COMMISSION (IWC) was formed in 1946 and joined by 37 countries committed to the CONSERVATION of whale stocks and the orderly development of the whaling industry; however, until the 1960s, the IWC was largely ineffective. The committee set quotas (the amount of whales that were allowed to be killed) that were too high to enable endangered whale SPECIES to recover. By 1970 eight whale species were endangered. In 1982 a moratorium was declared by the IWC that took effect in 1986. Some countries such as Japan, Norway, and Iceland defied the commission or resigned. The United Kingdom had stopped whaling in 1963 and the United States in 1972.

The most endangered whale is the humpback, known for its songs and acrobatics. Also endangered are the blue, bowhead, finback, right, sei, and sperm whales. Vigorous efforts to stop whaling have been coordinated by the environmental group, GREENPEACE. *See also* ENDANGERED SPECIES; THREATENED SPECIES.

EnviroSources

Greenpeace Web site: www.greenpeace.org.
High North Alliance Web site: www.highnorth.no/iceland/th-in-to.html.
Hoyt, Erich. *Meeting the Whales.* Ontario, Canada: Camden House, 1991.

White, Gilbert (1720–1793), An eighteenth-century NATURALIST who is considered England's first ecologist. Because of his book, *The Natural History of Selbourne*, the village of Selbourne and its countryside became famous. In his book, he recorded observations on the PLANTS, birds and animals of this part of Hampshire. His home in Selbourne is now a 120-hectare (275-acre) woodland and common area, called the National Trust Meadow, that is open all year. *See also* AUDUBON, JOHN JAMES; LEOPOLD, ALDO; MUIR, JOHN; THOREAU, HENRY DAVID.

EnviroSources

Burton, Ian, et al., eds. *Geography, Resources, and Environment (Themes from the Work of Gilbert White).* Urbana, IL: University of Chicago Press, 1986.
Gilbert White House: Selborne, Alton, Hampshire, GU 34 3JH; (01420) 511275; Web site: www.hants.gov.uk/leisure/house/gibwhit/index.html.

Wild and Scenic Rivers Act, U.S. legislation enacted in 1968 that was designed to protect certain rivers from being dammed or damaged in other ways. (*See* dam.) As part of the Wild and Scenic Rivers Act, federal land management agencies, including the U.S. FOREST SERVICE and the NATIONAL PARK SERVICE, were directed to identify rivers that could be included in a National Wild and Scenic Rivers System. Rivers and river segments protected under this law fall into one of three categories: wild, scenic, or recreational. Wild rivers are undisturbed rivers that are accessible only by trails. Scenic rivers are mostly untouched, but are accessible by road. Recreational rivers are used for recreational activities such as swimming, fishing, and boating.

Today, there are 152 designated rivers and river segments within the National Wild and Scenic Rivers System, including the American and Klamath Rivers in California; the Delaware River in New York, New Jersey, and Pennsylvania; and the Rio Grande in Texas. The Wild and Scenic Rivers Act strictly prohibits the construction of dams, HYDROELECTRIC POWER plants, and other structures on and around these rivers. The act also limits the extraction of MINERALS from any river within the national system.

The act has done much to protect the ENVIRONMENT and beautify the land. Environmentalists, however, fear that the Wild and Scenic Rivers Act is much too flexible to be effective. Critics often point out that even when a river is included in the national system, any preexisting land use along that river, such as MINING, logging, and farming, is permitted to continue. *See also* BUREAU OF LAND MANAGEMENT; DEPARTMENT OF THE INTERIOR; NATIONAL PARK SERVICE; WILDERNESS; WILDERNESS ACT OF 1964.

EnviroSources

Doppelt, Bob, Mary Scurlock, Chris Frissell, and James Karr. *Entering the Watershed: A New Approach to Save America's River Ecosystems*. Washington, DC: Island Press. 1993.

Palmer, Tim. *The Wild and Scenic Rivers of America.* Washington, DC: Island Press, 1993.

Wild Bird Conservation Act (WBCA), Legislation passed by the U.S. Congress to protect wild POPULATIONS of exotic birds from the growing international pet trade. The act was supported by CONSERVATION and humane organizations, ornithologists, bird breeders, and the pet industry and prohibits the importation of SPECIES that are listed in the CONVENTION ON INTERNATIONAL TRADE IN ENDANGERED SPECIES OF WILD FLORA AND FAUNA (CITES).

After the passage of the WBCA, the volume of birds imported to the United States dropped. However, the United States is the largest importer of wild birds not on the CITES list. Some experts believe that as many as 80% of birds in trade die before reaching their final destination, due to insufficient space, food, water, or ventilation. Excessive trade and HABITAT destruction now threaten more than 1,000 species with EXTINCTION. In fact, for several endangered PARROT species, more birds exist in captivity in the United States than remain in the wild. Australia, Guyana, the Philippines, and Zimbabwe are countries designing laws to ban the exportation of their native birds. *See also* LACEY ACT; MIGRATORY BIRD TREATY ACT; POACHING; RED LIST OF ENDANGERED SPECIES.

The Nene bird of Hawaii is an endangered species. *Credit:* Linda Zierdt-Warshaw

EnviroSource

Defenders of Wildlife (Wild Bird Conservation Act) Web site: www.defenders.org.wbca.fact.html.

wilderness, An undisturbed roadless area free of human influence, intrusion, and development. Wilderness areas can include forests, deserts, wetlands, and mountains. Today, nearly 500 wilderness areas, covering almost 38.5 million hectares (95 million acres), exist throughout the United States and are protected by the WILDERNESS ACT.

Before the Wilderness Act was passed, people argued about whether these delicate ECOSYSTEMS should be preserved. Today, the debate is often about how much land should be protected. Many people fear that "locking up" too much land as wilderness will be harmful to businesses and the economy. In addition, such restrictions will limit availability of land needed for the construction of new roads, bridges, and homes as the human POPULATION increases. Wilderness areas can provide the NATURAL RESOURCES, such as MINERALS and timber, needed to complete these projects. Others believe that more wilderness areas are needed to protect the country's many endangered and THREATENED SPECIES. Even if it is illegal to trap, harm, or kill these species, they are still threatened indirectly from ACID RAIN, water pollution, and other problems caused by human activities. Still others stress that, compared to other public lands, relatively few places are even classified as wilderness areas. They argue that these remaining lands should be preserved for human enjoyment before they are gone forever. *See also* FOREST SERVICE; SUSTAINABLE DEVELOPMENT; WILDERNESS SOCIETY.

Wilderness Act of 1964, Legislation passed in 1964 by the U.S. Congress to protect and preserve more than 38.5 million hectares (95 million acres) of federal land. (*See* public land.) The lands, designated as WILDERNESS areas, are used entirely to sustain and protect BIODIVERSITY and for educational and research purposes. Under this law, the land shall remain as a preserve—unimpaired for future use. The law prohibits permanent structures, roads, motor vehicles, timber harvesting, and MINING. Some recreational activities such as hiking and similar activities are allowed. The last major wilderness bill was in 1994 when the U.S. Congress passed the California Desert Protection Act, which set aside 3.4 million hectares (8.5 million acres) in the Mojave Desert in the southern part of the state. However, there are groups who believe in amending the Wilderness Act to provide less restrictions for off-highway vehicles (OHV), logging, and mining. *See also* LEOPOLD, ALDO; MARSHALL, ROBERT; ROOSEVELT, THEODORE; WILDERNESS SOCIETY.

Wilderness Society, An organization cofounded in 1935 by ALDO LEOPOLD, ROBERT MARSHALL, and Benton Mackaye to expand and protect the nation's WILDERNESS areas. The society was intended to form the cornerstone of the movement needed to save America's vanishing wilderness. The Wilderness Society works to protect these areas and to develop a nationwide network of wild lands through public education, scientific analysis, and advocacy. Its goal is to ensure that future generations can enjoy clean AIR and water, WILDLIFE, and the beauty and opportunities for recreation and renewal that pristine FORESTS, rivers, DESERTS, and mountains provide.

The society has helped pass many federal laws, including the WILDERNESS ACT OF 1964 and the California Desert Protection Act of 1994, which protects some 3.5 million hectares (about 8 million acres) of fragile desert lands. The society has contributed several million hectares to the National Wilderness Preservation System.

The society publishes a number of fact sheets on such topics as the ARCTIC NATIONAL WILDLIFE REFUGE, EARTH DAY, forest road-building, hunting and fishing in wilderness, the Land and Water Conservation Fund, national forests, and the Sierra Nevada Mountains. *See also* BIOREGION; GREENPEACE; NATIONAL AUDUBON SOCIETY; NATIONAL PARK SERVICE; SIERRA CLUB; WILDERNESS ACT OF 1964; WILSON, EDWARD OSBORNE.

EnviroSource

Wilderness Society: 900 17th St, NW, Washington, DC 20006; (800) 843-9453; Web site: www.wilderness.org/newsroom/factsheets.html.

wildlife, According to federal regulations, any living or dead nondomesticated animals, even if bred, hatched, or born in captivity, including any part, product, egg, or offspring of such animals. The term applies to wild MAMMALS, birds, fishes, REPTILES, AMPHIBIANS, and other terrestrial or aquatic life. Some farmed or ranched animals such as BISON also are considered to be wildlife and are thus regulated by applicable federal laws, including the CONVENTION ON INTERNATIONAL TRADE IN ENDANGERED SPECIES OF WILD FLORA AND FAUNA (CITES). *See also* CAPTIVE PROPAGATION; ENDANGERED SPECIES; FISH AND WILDLIFE SERVICE; WILDLIFE REFUGE.

wildlife refuge, An area of land and water that provides food, water, shelter, and space for WILDLIFE and is maintained, usually by a government or nonprofit organization, for the preservation and protection of one or more wildlife SPECIES. Refuges also protect WETLANDS and historical and archaeological sites.

As of 1992, the U.S. NATIONAL WILDLIFE REFUGE SYSTEM, administered by the U.S. FISH AND WILDLIFE SERVICE, comprised some 503 areas, covering 36.5 million hectares (90 million acres) among the 50 states. U.S. refuges range in size from Minnesota's Mille Lacs, which is less than 1 hectare, to Alaska's Yukon Delta, which is about 9 million hectares (20 million acres).

One feature that sets refuges apart from many other federal lands is the fact that they are actively managed: the num-

Wildlife refugees, such as this one, were first established by President Theodore Roosevelt in 1903, to protect birds and other animals. *Credit.* Linda Zierdt-Warshaw.

ber and type of PLANTS and animals living in a refuge can be increased using water, plantings, and other techniques. Among the main goals of the National Wildlife Refuge System is the preservation, restoration, and enhancement of all species of endangered or threatened plants and animals in their natural ECOSYSTEMS (when practicable). All units of the refuge system are required to consider ENDANGERED SPECIES in developing and implementing management activities.

Established by President THEODORE ROOSEVELT in 1903, the oldest U.S. national wildlife refuge is Florida's Pelican Island, which was created as a protected area for egrets, herons, and other birds. Refuges have also been established for big game (e.g., elk), small resident game, waterfowl, and nongame birds (e.g., pelicans). By far the most numerous are the waterfowl refuges, which variously supply breeding areas, wintering areas, and resting and feeding areas along major flyways during migration. Although the main purpose of the refuge system is to ensure the survival of wildlife by providing suitable HABITAT and protection from humans, many refuges permit hunting and fishing in season, as well as other recreational activities, such as hiking and swimming. Some refuges have been designated as WILDERNESS areas.

Refuges have also been established by private individuals and societies, such as the Nature Conservancy and the NATIONAL AUDUBON SOCIETY, and by all levels of government. Other countries throughout the world also maintain parks, refuges, and game preserves. One of the oldest is Kruger National Park in South Africa, which was established in 1898 for the preservation of big game animals. *See also* ADDO NATIONAL ELEPHANT PARK; BATS; BIOSPHERE RESERVE; CARARA BIOLOGICAL RESERVE; MIGRATORY BIRD TREATY ACT; SERENGETI NATIONAL PARK; TSAVO NATIONAL PARK.

EnviroSources

Conservation International Web site: www.conservation.org.
National Audubon Society Web site: www.audubon.org.
The Nature Conservancy Web site: www.tnc.org.
U.S. Fish and Wildlife Service, National Wildlife Refuge System; (800) 344-WILD Web site: www.refuges.fws.gov.
World Conservation Union Web site: www.iucn.org.

Wilson, Edward Osborne (1924–), A Pulitzer Prize winning author and biologist who is an authority on social INSECTS, particularly ants, and the study of the distribution of SPECIES. Some have called him the champion of BIODIVERSITY.

Wilson was a professor and curator of comparative zoology at Harvard University and is the only person to receive two Pulitzer Prizes in literature and the highest award in science in the United States, the National Medal of Science. In 1990 he shared Sweden's Crafoord Prize with the American biologist PAUL EHRLICH for his work in ECOLOGY. The Crafoord Prize encompasses areas (general biology, OCEANOGRAPHY, mathematics, astronomy) not covered by the Nobel Prizes. Wilson is also noted for his work in sociobiology—a discipline in which human and animal behavior is studied in conjunction with DARWIN's theory of EVOLUTION.

Wilson is the author of several books, all of which reveal his unceasing efforts to research animals and their habits. In 1979 he won his first Pulitzer Prize for *On Human Nature* (1978), in which he described the ethics of human society. In 1990, Wilson and co-author Bert Holldobler wrote *The Ants*, an in-depth study of these insects for which they were awarded a Pulitzer Prize. In *The Diversity of Life* (1992), Wilson described how the world's living species became diverse and explained the cause of the massive EXTINCTION of PLANTS and animals in the twentieth century due to the increase in human POPULATION and activity. Wilson predicted that animal and plant species will decline by 20% by the year 2020. He published his autobiography, *Naturalist*, in 1994. In 1995, Wilson received the Audubon Medal for contributions to CONSERVATION and environmental protection. Wilson is a native of Alabama. *See also* MARSH, GEORGE PERKINS; MARSHALL, ROBERT.

EnviroSources

Wilson, Edward Osborne. *Biodiversity*. Washington, DC: National Academy Press, 1988.

———. *Biophilia.* Cambridge, MA: Harvard University Press, 1984.

———. *The Diversity of Life*. Questions of Science series. Cambridge, MA: Harvard University Press, 1992.

———. *Naturalist*. Washington, DC: Island Press, 1994.

wind power, An ALTERNATIVE ENERGY RESOURCE that uses the renewable ENERGY in moving AIR to generate ELECTRICITY. Although wind power currently produces less than 2% of the world's electricity, it is the fastest growing energy resource. The American Wind Energy Association estimates that by the year 2025 wind power will produce more than 10% of the electricity in the United States. Other countries working to increase their use of wind power include Germany, India, Denmark, England, Spain, and China. Presently, Germany and the United States lead the world in total wind power capacity.

Wind TURBINES, or AEROGENERATORS, are used to generate electricity from wind. Most often, aerogenerators are used in large numbers at wind power plants or wind farms. Sites suitable for use as wind farms must be located in areas that regularly receive sustained winds of at least 22.5 kilometers per hour (14 miles per hour) and must not be blocked by obstacles such as high mountains.

The American Wind Energy Association (AWEA) estimates that by the year 2025 wind power will produce more than 10 percent of the electricity in the United States. *Credit:* Linda Zierdt-Warshaw.

California has the greatest number of wind farms in the United States. The largest of them, Altamont Pass, is located east of San Francisco among a series of low hills that separate the San Francisco Bay area from the hot interior of the San Joaquin Valley. Altamont Pass contains the world's largest concentration of wind turbines; however, the turbines of the Tehachapi Pass, located in the Mojave Desert north of Los Angeles, generate more electricity. The more than 5,000 aerogenerators of the Tehachapi Pass wind resource area generate enough electricity to meet the residential needs of more than 500,000 southern Californians, making the site the world's largest producer of wind-generated electricity.

In recent years, research into additional U.S. sites suitable for wind farms has increased. A site currently under development is located in northwestern Iowa. When complete, the Iowa project's aerogenerators are expected to generate enough electricity annually to circumvent the EMISSION of 0.86 million kilograms (1.9 million pounds) of air POLLUTANTS that would be generated from powers plants that burn FOSSIL FUELS. The project is expected to provide sufficient clean, pollution-free electricity to power about 50,000 average homes.

The development of wind power technology is not unique to the United States. The resource is being seriously developed by many countries. Most economists predict that the largest growth markets for wind turbines are in Germany, India, Spain, Great Britain, and China. Many believe that in the next decade, China will become the dominant market for wind turbines.

The use of wind power is growing most rapidly in Germany. In 1994, the recently reunified country surpassed Denmark as the world's second-largest wind energy powerhouse. It is expected that Germany will exceed the United States as a leader in wind energy by the year 2000. Wind farms in Germany are concentrated along the coastal areas in the northwest portion of the country.

The Scandinavian country of Denmark is the world's largest supplier of aerogenerators and ranks third in total wind generating capacity. The Danes were the first people to regularly produce electricity using wind power and by World War I had a network of wind turbines that generated about 100,000 kilowatts of electricity. Today, Denmark contains the largest number of wind turbines outside the United States and Germany. In some provinces, wind turbines now provide as much as 7% of the region's electricity; 75% of these turbines are installed as single units or in small clusters, rather than on large wind farms. Most of these wind turbines are owned cooperatively by the people living in the region.

The WEATHER patterns and topography of the United Kingdom, particularly Great Britain, provide this region with excellent wind resources. The development of wind power as a source of electricity in Great Britain began in the early 1990s, when 10 wind turbines were installed on a farm in Cornwall. Today, most of the United Kingdom's wind plants are located in England and Wales. Great Britain's approach to wind development has made the region a model for uncluttered wind farm development. Unlike northern Europe, where most wind turbines are installed singly or in small clusters, aerogenerators in England and Wales tend to be placed on small wind power plants with 10 to 100 turbines. These wind plants are far smaller than those in California, which often contain at least 300 turbines.

India is the second-fastest-growing market for wind energy. By the mid-1990s India had installed enough aerogenerators to provide the country with more generating capacity than North America, Denmark, Great Britain, and the Netherlands.

The South American country of Brazil is also developing wind power technology as a major source of electricity. Brazil intends to install wind turbines that will produce 1,000 megawatts of electricity by the year 2005.

EnviroSources

American Wind Energy Association: 122 C Street, NW, 4th Floor, Washington, DC 20001; Web site: www.igc.apc.org/awea/news/html; (202) 383-2500; e-mail: awea@mcimail.com.

Center for Renewable Energy and Sustainable Technology, Solar Energy Research and Education Foundation: 777 North Capitol St., NE, Suite 805, Washington, DC 20002; Web site: www.solstice.crest.org.

Pacific Northwest Laboratories. *Wind Energy Resource Atlas.* American Wind Energy Association. Washington, D.C. 1987. Reprint.

Time Line of Wind Power

644 A.D.	The first windmill is used in Prussia to power a machine used for grinding corn.
1890	The first use of a windmill to generate electricity occurs in Denmark.
1931	The Darrieus aerogenerator, the first wind turbine to make use of blades resembling airplane propellers mounted on a tall pole, is built in France.

Wolman, Abel (1892–1989), U.S. sanitary engineer who worked to develop safe standards for drinking water and to promote the safe disposal of WASTEWATER. In the early 1900s, public drinking water was unsafe in many cities in the United States and elsewhere. Untreated water carried waterborne DISEASES such as cholera and typhoid fever. Thousands of people who ingested such water became ill and died. Wolman, who was a research scientist with the Maryland State Department of Health, tested and studied the application of CHLORINATION in drinking water to control PATHOGENS. Wolman's research proved that when CHLORINE was added to the drinking water there was a decrease in incidents of diseases.

Wolman was a worldwide advocate for developing programs to safeguard water supplies and for sanitation planning. As a member of the U.S. delegation to the first WORLD HEALTH ORGANIZATION, Wolman recommended that wastewater sanitation planning and programs to improve water supply be a part of the new organization. For many years, he was a professor of environmental engineering at Johns Hopkins University.

Throughout his career, he participated in many organizations and served as an editor of several professional journals. He was chairman of the Water Resources Committee of the Natural Resources Planning Board, which studied major river basins. He worked with the World and Pan-American Health Organizations clean water and sanitation programs. *See also* POTABLE WATER; PRIMARY, SECONDARY, AND TERTIARY TREATMENTS; SEWAGE TREATMENT PLANT.

wolves, The largest members of the dog family, which are endangered in the lower United States, except in Minnesota, where the animals are listed as threatened. There are two SPECIES: the gray, or timber, wolf and the red wolf. The gray wolf was among the first species to be officially considered an ENDANGERED SPECIES outside of Alaska under the first federal Endangered Species Law in 1967. By the late 1920s, gray wolves had been eliminated in the Rocky Mountains and in the East except for several hundred animals in Minnesota. The elimination of the animals was due to government programs that paid bounties for killing wolves because of their threat to livestock.

Like the gray wolf, the red wolf was hunted and destroyed by farmers and ranchers who believed it was a threat to their grazing animals. The last remaining red wolves were removed from the wild for CAPTIVE PROPAGATION in the 1970s. A few red wolves survive in the wild today only as a result of reintroduction programs established by the U.S. FISH AND WILDLIFE SERVICE.

Gray wolves range from 1.5 to 1.7 meters (4.5–5 feet) in length and weigh from 50–65 kilograms (110–145 pounds). Wolves are social animals that hunt during the day or at night in packs. Wolves mainly feed on wild game, including deer, elk, BISON, pronghorn antelope, and smaller animals such as beaver, rabbits, and mice.

The red wolf is smaller than the timber wolf, weighing 18–36 kilograms (40–80 pounds), and ranges throughout the southeastern part of the United States. The red wolf does not hunt in large packs, but rather in pairs; its prey includes rabbits, raccoons, rodents, and young deer.

In the 1990s, the U.S. Fish and Wildlife Service directed a program to reintroduce gray wolves to YELLOWSTONE NATIONAL PARK and central Idaho in an attempt to reestablish the spe-

In the 1990s, the U.S. Fish and Wildlife Service directed a program to reintroduce gray wolves to Yellowstone National Park and central Idaho.

cies after an absence of more than 60 years. Records indicate that the wolf POPULATION was eliminated in Yellowstone in the 1920s. Most of the wolves that have been reintroduced there were captured and transported from Alberta and British Columbia, Canada. Canada, which has nearly 50,000 gray wolves, supports the U.S. plan for reintroduction. Alaska also has nearly 7,000 gray wolves in the wild. The reintroduction program is a cooperative effort with the NATIONAL PARK SERVICE and U.S. FOREST SERVICE.

Mexican gray wolves are native to Arizona, New Mexico, and northern Mexico but were shot, poisoned, and trapped to near-EXTINCTION by the 1970s. Five packs with more than two dozen wolves are now in eastern Arizona's wilderness as part of the reintroduction program; however, in 1999, a pack of Mexican gray wolves that repeatedly harassed and attacked cattle in eastern Arizona were relocated. Officials interceded before the pups became used to livestock, instead of WILDLIFE as a food source. *See also* ALLIGATORS; BALD EAGLES; BATS; BISON; CACTI; CALIFORNIA CONDOR; CAPTIVE PROPAGATION; CHEETAH; CROCODILES; DODO BIRD; ELEPHANTS; FLORIDA PANTHER; GORILLAS; LEOPARD; MANATEES; ORANGUTANS; ORYX; PARROTS; SEALS AND SEA LIONS; TIGER.

EnviroSources

Putten, Mark Van. "Of Wolver, Wetlands, and Wisdom." National Wildlife 36, no. 4. (June/July 1998): 7.

U.S. Fish and Wildlife Service Web site: www.fws.gov.

U.S. Fish and Wildlife Service, Species List of Endangered and Threatened Wildlife Web site: www.fws.gov/r9endspp/lsppinfo.html.

World Wildlife Fund: 1250 24th St, NW, Washington, DC 20037; (800) 225-5993; Web site: www.worldwildlife.org.

World Bank, An agency of the United Nations, more appropriately known as the International Bank for Reconstruction and Development, that works to promote international trade and development. The World Bank was established in 1944 at the Bretton Woods Conference primarily for the purpose of providing aid to countries in Europe that had suffered damage as a result of World War II. Today, the World Bank uses much of its monetary resources to improve conditions in the poorest developing nations by helping their peoples gain access to health care, waste disposal facilities, clean drinking water, and adequate food and housing.

The chief objective of the World Bank is to aid in the development and reconstruction of member nations by helping these nations acquire capital investment from other nations. To aid in this process, a member nation may be granted a loan for a specific project, but only if a private concern from within the nation agrees to guarantee the loan. At the time of its creation, most loans granted by the World Bank went to European countries for the purpose of reestablishing industries that had been devastated by the activities of World War II. When such reconstruction was complete, most of the World Bank's loans were granted to developing nations throughout South America, Africa, and Asia. Most recently, loans have been granted to specific territories in these same regions to help increase their productivity in a global marketplace and also to provide peoples living in these regions with improved education, health care, and family planning facilities, as well as improved access to clean water, adequate waste disposal facilities, and better nutrition through improved agricultural methods. *See also* AGENDA 21; DEBT-FOR-NATURE SWAP; POTABLE WATER; UNITED NATIONS ENVIRONMENT PROGRAMME; WORLD HEALTH ORGANIZATION.

World Conservation Monitoring Centre, An organization with headquarters in Cambridge, United Kingdom, that serves as an international clearinghouse for data on CONSERVATION OF NATURAL RESOURCES, particularly endangered and threatened WILDLIFE. Working with the WORLD CONSERVATION UNION, the WORLD WILDLIFE FUND, and the GLOBAL ENVIRONMENT MONITORING SYSTEM of the United Nations, the World Conservation Monitoring Centre's mission is to "provide information services on conservation and the sustainable use of the world's living resources." In addition, the organization works with other agencies to help them develop similar information systems for their own use. *See also* BIOSPHERE RESERVE; ENDANGERED SPECIES; ENDANGERED SPECIES ACT; ENDANGERED SPECIES LIST; ENVIRONMENTALLY SENSITIVE AREA; FISH AND WILDLIFE SERVICE; PROTECTED AREA; RED LIST OF ENDANGERED SPECIES; THREATENED SPECIES; UNITED NATIONS DEVELOPMENT PROGRAM; UNITED NATIONS ENVIRONMENT PROGRAMME.

EnviroSource

World Conservation Monitoring Centre, Information Officer: 219 Huntington Road, Cambridge CB3 0DL, United Kingdom; Web site: www.wcmc.org.uk; e-mail: info@wcmc.org.uk.

World Conservation Union, An international organization established in 1948 by the United Nations (UN) with the goal of encouraging and promoting the CONSERVATION of WILDLIFE, wildlife HABITATS, and NATURAL RESOURCES as part of the national policies of UN member nations and states. At the time of its inception, the World Conservation Union was named the International Union for the Conservation of Nature and Natural Resources (IUCN). Although its name has changed, the goals of the organization and its acronym have remained the same. Currently, the IUCN is composed of 74 member nations and more than 700 nongovernmental organizations.

Working through its members, the IUCN helps to set up international meetings at which laws and policies affecting global conservation are established. Two major international meetings of which the IUCN played a significant role were the Convention on Wetlands of International Importance and the CONVENTION ON INTERNATIONAL TRADE IN ENDANGERED SPECIES OF WILD FLORA AND FAUNA. The IUCN also develops guidelines for its members aimed at conserving the virtue and diversity of nature and sponsors programs to help guarantee that all uses of natural resources are equitable and ecologically sustainable. (*See* biodiversity.) In addition, the organization also has set up research programs and advisory bodies, such as the Species Survival Commission (SSC), which gather data about the status of the many SPECIES that contribute to Earth's biodiversity. Such information is made available to UN members and to the public through publication of the IUCN RED LIST OF ENDANGERED SPECIES—a global listing of threatened and ENDANGERED SPECIES that is similar to, but broader in scope than, the ENDANGERED SPECIES LIST maintained by the U.S. FISH AND WILDLIFE SERVICE. In addition, the IUCN in 1980 implemented its World Conservation Strategy. Key objectives of this strategy include maintaining essential ecological processes and life-support systems for wildlife, preserving GENETIC DIVERSITY, and ensuring the sustainable use of species and ECOSYSTEMS. The strategy was adopted by at least 30 nations. *See also* BIOSPHERE RESERVE; MAN AND THE BIOSPHERE PROGRAMME; WETLAND; WORLD CONSERVATION MONITORING CENTRE.

EnviroSources

World Conservation Union Web site: www.iucn.org.

World Health Organization, A specialized agency of the United Nations (UN) founded in 1948, which now has 191 member states. The World Health Organization (WHO) promotes technical cooperation for health among nations. Its four main functions are to give worldwide guidance in the field of health, to set global standards for health, to cooperate with governments in strengthening national health programs, and to develop and transfer appropriate health technology, information, and standards.

One of the WHO's major achievements is the global eradication of smallpox. Similarly, polio and guinea-worm DISEASE are on the threshold of eradication largely due to the efforts of the WHO. The WHO promotes primary health care, delivers essential drugs, makes cities healthier, builds partnerships for health, and promotes healthy lifestyles and ENVIRONMENTS. WHO experts work to fight infectious diseases such as human immunodeficiency virus (HIV) infections and acquired immune deficiency syndrome (AIDS), viral hemorrhagic fevers such as Ebola, sexually transmitted diseases, as well as noncommunicable diseases such as cardiovascular disease, CANCERS, and respiratory diseases. The WHO gathers current data on conditions and needs. It also sponsors health promotion projects such as Healthy Cities and Villages; Health Islands and Health-Promoting Schools; Hospitals; and Work Sites. WHO works closely with other organizations in the UN system including UNICEF, and the WORLD BANK. *See also* FOOD AND DRUG ADMINISTRATION; UNITED NATIONS DEVELOPMENT PROGRAM; UNITED NATIONS ENVIRONMENT PROGRAMME.

EnviroSource

World Health Organization Web site: www.who.int.

World Resources Institute, An independent center for policy research on such topics as technology, FORESTS, CLIMATE, and economics that are related to global environmental issues. The mission of the World Resources Institute (WRI) based in Washington, DC, is to encourage human society to live in ways that protect Earth's ENVIRONMENT and Earth's capacity to provide for the needs and aspirations of current and future generations.

The WRI, founded in 1982, provides objective information and practical proposals for policy and institutional change that will foster environmentally sound, socially equitable development. The institute focuses on U.S. policies (since the United States is a major world producer, consumer, and polluter), and developing countries (where NATURAL RESOURCE deterioration is dimming development prospects and swelling the ranks of the poor and hungry).

The WRI focuses on issues supporting the CONSERVATION, sustainable use, and equitable distribution of benefits of BIODIVERSITY—the totality of genes, SPECIES, and ECOSYSTEMS—throughout the world. It publishes articles on ENERGY, climate change, PESTICIDES, forests, and environmental pollution. Its 113-member interdisciplinary staff is strong in the social and natural sciences and augmented by a network of advisors, collaborators, international fellows, and partner institutions in more than 50 countries. The board of directors has 38 members from 11 countries. Financial support comes from private foundations, government and intergovernmental institutions, private corporations, and interested individuals.

EnviroSource

World Resources Institute: 10 G Street, Suite 800, Washington, DC 20002; (202) 729-7600: Web site: www.wri.org/wri/biodiv.

World Wildlife Fund, The largest privately supported international CONSERVATION organization, with 50 national chapters across five continents. The World Wildlife Fund (WWF) was founded in 1961 and is recognized globally by its panda logo. WWF is dedicated to protecting the world's WILDLIFE and their HABITATS. Its three global goals are to protect ENDANGERED spaces, to save endangered species, and to address global threats to habitat of wildlife that is in danger of EXTINCTION. The WWF has worked to save the giant panda, TIGERS, RHINOCEROSES, ELEPHANTS, WHALES, and other endangered species. WWF scientists have identified 200 terrestrial, freshwater, and marine habitats that need protection. These ecoregions have been named the Global 200. Global campaigns of the WWF include species at risk, TOXIC CHEMICALS, GLOBAL WARMING, FORESTS for life, and endangered seas. To address these issues, the WWF has sponsored more than 2,000 projects in 116 countries.

EnviroSource

World Wildlife Fund: 1250 24th St, NW, Washington, DC 20037; (800) 225-5993; Web site: www.wwf.org.

xeriscape, A type of garden composed of DROUGHT-tolerant PLANTS that can survive in very hot and dry year-round conditions. The term "xeriscape" is derived from the Greek word "xeros," meaning dry. These kinds of gardens conserve water and are less susceptible to pests. Xeriscape methods include creative landscaping for water and ENERGY efficiency and lower maintenance. The seven xeriscape principles are good planning and design, practical lawn areas, efficient IRRIGATION, soil improvement, use of MULCHES, low water demand plants, and good maintenance. One study reported that by using the principles of xeriscape, gardeners would save 50–60% of the water that would otherwise be used for growing other plants. Drought-tolerant plants include Mexican sage, yellow potentilla, California poppies, African daisies, aloe vera, and the century plant. *See also* AGRICULTURE; CACTI; EXOTIC SPECIES.

x-ray, A high ENERGY form of electromagnetic RADIATION produced by charged particles. X-rays have short wavelengths—shorter than ultraviolet light. Although excessive radiation exposure from x-rays can cause RADIATION SICKNESS, increased risk of CANCER, and damage to the development of the fetus, the x-ray is the most widely used method of taking images of the human body to detect bone damage, heart defects, and problems of the brain. X-rays are measured in units of roentgen, or in the international system of units, coulombs per kilogram. *See also* CARCINOGEN; MUTATION.

xylene, A colorless liquid that evaporates readily, also known as xylol or dimethylbenzene. Xylene is produced from PETROLEUM and COAL tar and is used as a cleaning agent and as a thinner for paints and varnishes. (*See* solvent.) It also used as raw material in the chemical, plastics, and synthetic fiber industries and as an ingredient in coatings applied to fabrics and papers.

Xylene evaporates rapidly and can last for some days in the AIR until it is broken down by sunlight. As a liquid it can leak into the soil, SURFACE WATER, or GROUNDWATER, where it may remain for months before it breaks down into other chemicals.

Short-term exposure to high levels of xylene in humans can result in a variety of harmful effects, including damage to the liver and kidneys. Long-term high-level exposure to mixtures of xylene and other solvents like BENZENE can lead to chronic bronchitis; in addition it may cause infertility in women.

The U.S. ENVIRONMENTAL PROTECTION AGENCY (EPA) has established a maximum level of 10 parts per million for xylene in public drinking water. The EPA and the U.S. FOOD AND DRUG ADMINISTRATION specify conditions under which xylene may be used as a part of herbicides, PESTICIDES, or articles used in contact with food. *See also* PARTS PER BILLION AND PARTS PER MILLION.

EnviroSources

Agency for Toxic Substances and Diseases, Registry, Division of Toxicology: 1600 Clifton Road, NE Mailstop E-29, Atlanta, GA 30333.

Mike Kamrin, Institute for Environmental Toxicology, C231 Holden Hall, MSU; (517) 353-6469.

xylophagous. *See* herbivore.

Y

Yangtze River. *See* Three Gorges Dam Project.

Yellowstone National Park, The oldest national park in the world, created by the U.S. Congress in 1872. Yellowstone National Park is the largest national park in the lower 48 states, encompassing 898,714 hectares (3,472 square miles) of land in Wyoming, Montana, and Idaho. Ninety-nine percent of the park's area remains undeveloped, providing a wide range of HABITATS that support one of North America's most diverse large-MAMMAL POPULATIONS. Yellowstone may be one of the last WILDERNESS ECOSYSTEMS remaining in the world's temperate zones.

The natural features that initially led to preservation of Yellowstone as a national park were its geothermal phenomena, the Grand Canyon of the Yellowstone River, and Yellowstone Lake. There are nearly 10,000 thermal features in Yellowstone, including 200–250 active geysers (75% of the world's total), the most famous of which is Old Faithful. The park's thermal features are located in the only undisturbed geyser basins left in the world. The park also boasts one of the world's largest volcanic craters, a caldera measuring 45 kilometers by 75 kilometers (28 miles by 47 miles).

A small tribe of Shoshone were the only people who traditionally lived in this area year round. The written history of Yellowstone dates back 200 years to a reference in explorer William Clark's journal, although the Lewis and Clark expedition did not travel through the land during their northwest journey. Fur trappers, prospectors, and mountain men frequented the area, and their tales prompted local rancher, Nathaniel P. Langford, to gather a group of local leaders and set out in 1870 to separate myth from reality. These men spearheaded the campaign to save Yellowstone from private ownership, exploitation, and vandalism. At their urging, Ferdinand Hayden, director of the U.S. GEOLOGICAL SURVEY, mounted an official expedition. His team's resultant photographs and land survey confirmed the incredible treasures of Yellowstone, persuading Congress to set aside the area as national parkland.

Canadian geese stop by Yellowstone National Park before continuing their migration. *Credit:* U.S. Department of Interior, National Park Services.

Yellowstone is home to more than 300 SPECIES of animals: 60 different mammals (including elk, bighorn sheep, grizzly bear, and bobcat), 18 species of fishes, and more than 225 bird species. BISON are the largest mammals in the park, and the park preserves the only place in the lower 48 states where a population of wild bison has persisted since prehistoric times. By 1902, fewer than 50 of these animals existed. Fearing EXTINCTION of the species, the park imported 21 privately owned bison to add stock to the wild herd; by 1996, Yellowstone's bison numbers had increased to about 3,500.

Three birds of the park—the PEREGRINE FALCON, the BALD EAGLE, and the whooping crane—were listed as threatened or ENDANGERED SPECIES in this region as of May 1998. Efforts to restore the peregrine falcon and bald eagle populations within Yellowstone have been relatively successful. Since whooping cranes are one of the most endangered birds in North America, it is hoped that reintroduction and relocation efforts will increase the birds' population within the protected ENVIRONMENT of the park.

Two Yellowstone mammals—the gray WOLF and the grizzly bear—are listed as endangered in the lower 48 states. As of April 1999, about 110 wolves, many of which are monitored via radio collar, inhabited the Yellowstone ecosystem, thanks to a successful (and controversial) reintroduction program begun several years ago. About 250 grizzly bears live in and around Yellowstone Park. As a result of cooperative efforts of many land management agencies, the population is now at the recovery target set for Yellowstone by the NATIONAL PARK SERVICE. The U.S. FISH AND WILDLIFE SERVICE (FWS) recently proposed listing the Canada lynx under the ENDANGERED SPECIES ACT; however, evidence is too scant to reliably state that a resident population of lynx exists in the park today, or if it even did historically.

Yellowstone's landscapes have long been shaped by fire, including large-scale burns across the park's volcanic plateaus and hot, wind-driven FOREST FIRES climbing up to tree crowns at several-hundred-year intervals. Many of Yellowstone's PLANT species are fire-adapted. For example, some lodgepole pines, which make up nearly 80% of the park's FORESTS, have cones that are sealed by resin until the intense heat of fire cracks the bonds and releases the seeds inside. Fires also may stimulate regeneration of sagebrush, aspen, and willows. Though above-ground portions of plants may be consumed by flames, root systems typically remain unharmed, and for a few years after fires, plant productivity typically increases.

By the 1970s, Yellowstone and other parks had instituted a natural fire management plan to allow lightning-caused fires to continue to influence ECOLOGICAL SUCCESSION. This approach has not always met with public approval, with the most recent outcries erupting during the great Yellowstone fires of 1988. In June of that year, the greater Yellowstone area experienced a severe DROUGHT. FUEL accumulations (DETRITUS) grew progressively drier, and early summer thunderstorms produced lightning but no rain. By mid-July, because of continued dry conditions, the decision was made to suppress all fires. Even so, within a week, park fires consumed nearly 40,000 hectares (99,000 acres); by the end of the month, dry fuels and high winds combined to make larger fires uncontrollable. On the worst single day, high winds drove fires across more than 60,000 hectares (150,000 acres). Fires of such scale were unprecedented in the 125-year history of the park.

Gradually, the National Park Service has moved toward minimal interference with the park's natural state and also has recognized that Yellowstone Park is dynamically related to the Greater Yellowstone ecosystem. For example, the underground structure that sustains the park's complex geothermal features extends beyond park borders. Animals also migrate through adjacent ranchlands and national forests in search of food. (*See* emigration; immigration.) The environmental integrity of Yellowstone Park is dependent on the careful management of these lands, which in most instances must remain in a relatively natural condition for the BIOLOGICAL COMMUNITY of Yellowstone itself to remain viable. Some human activities on surrounding national forest, state, and private lands pose severe threats to the WILDLIFE, water, AIR, and geothermal features of the park.

Legislative restrictions on geothermal development (the harnessing or use of the heat within the Earth or the hot water resulting from such activity) around Yellowstone, such as the Old Faithful Protection Act introduced in 1992, have failed to receive congressional approval. In 1994, the National Park Service and the state of Montana agreed to monitor and control the use of GROUNDWATER in areas north of the park to ensure the continued flow of heat and water to Yellowstone's geysers and hot springs. *See also* ARCTIC NATIONAL WILDLIFE REFUGE; CONIFEROUS FOREST; FIRE ECOLOGY; GEOTHERMAL ENERGY; THREATENED SPECIES; YOSEMITE NATIONAL PARK.

EnviroSources

National Park Service (Yellowstone National Park) Web site: www.nps.gov/yell.
Peacock, Doug. "The Newest Place on Earth." *Audubon* 101, no. 2 (March–April 1999): 26–31.
Robbins, Jim. "Yellowstone Reborn." *Audubon* 100, no. 4 (July–August 1998): 64–68.
Watkins, T.H. "National Parks, National Paradox." *Audubon* 99, no. 4 (July–August 1997): 40–45.
Yellowstone Ecosystem Studies Web site: www.yellowstone.org.
Yellowstone National Park: P.O. Box 168, Yellowstone National Park, WY 82190; (307) 344-7381.
Yellowstone Net Web site: www.yellowstone.net.

Yosemite National Park, A California WILDERNESS area encompassing 1,931 square kilometers (1,200 square miles) set aside in 1890 to preserve the central Sierra Nevada Mountains along California's eastern flank. So magnificent is this land that it was the first territory ever set aside by Congress "for public use, resort, and recreation." President THEODORE ROOSEVELT described Yosemite as "the most beautiful place on Earth."

The park ranges in elevation from 600 meters (1,969 feet) above sea level to more than 4,000 meters (13,125 feet) and hosts myriad natural attractions in three general ENVIRONMENTS: alpine wilderness, giant sequoia groves, and the glacially carved Yosemite Valley with its impressive waterfalls, cliffs, and unusual rock formations. (*See* alpine tundra.) Yosemite Valley is the focus of most park visitors, even though it covers only 1% of the park's area. The valley's Yosemite Falls, which drop 745 meters (2,444 feet), is North America's highest waterfall, and the world's second highest. Other famous features include El Capitan, the largest single granite rock on Earth at 1,230 meters (4,035 feet), and MONO LAKE, one of the oldest lakes in North America. The giant sequoias near Wawona are some of the oldest and largest PLANTS on Earth. The diameter of the oldest tree's base is 9 meters (30 feet), it has a girth of 29 meters (95 feet), and a height of 62 meters (204 feet)—the height of a 20-story building.

At one time, this area was composed of gentle rolling hills, crisscrossed with a maze of stream systems. Over a period of about five million years, the Sierra Nevada range rose. The land tilted westward and streams carved deep, V-shaped river canyons. An ICE AGE caused GLACIERS to form, which transformed the canyons into U-shaped valleys. Scientists believe that at least three major periods of glacial advance and recession affected Yosemite.

Humans first inhabited the region 7,000 to 10,000 years ago. Various Native American tribes have lived in the area, the most recent of whom were the Miwok, who were relocated to reservations in 1851. By 1855, the first party of tourists arrived; nine years later, encouraged by a group of influential Californians, President Abraham Lincoln signed the Yosemite Grant, which set aside Yosemite Valley and the Mariposa Grove of giant sequoias as a state-supervised public reserve. In 1890, Robert Underwood Johnson and JOHN MUIR became concerned that the high country and WATERSHED for Yosemite Valley were being destroyed by grazing and logging activities. The two men launched a successful campaign to persuade Congress to set aside the area as a national park.

In Yosemite National Park, more than 247 bird SPECIES, 80 MAMMAL species, 40 REPTILE species, 37 species of native trees, and hundreds of NATIVE SPECIES of wildflowers have been recorded, though not all of them are commonly seen by park visitors. Opportunities to observe and study WILDLIFE vary throughout the park and are dependent on the season. While many animals stay within a particular zone or community year round, some, such as mule deer and some bird species, migrate to lower elevations in winter. *(See* emigration.) Others migrate northward or southward depending on the season. Two once-ENDANGERED SPECIES live in Yosemite: the BALD EAGLE and the PEREGRINE FALCON. Efforts to restore populations of both species proved successful enough to allow their removal from the ENDANGERED SPECIES LIST in 1999.

Yosemite National Park also is home to a large black bear POPULATION. The animals' natural behavior, foraging habits, distribution, and numbers have been altered by a long history of widely available human foods in the park. Bears habituated to these unnatural foods often lose their instinctive fear of humans and eventually may cause damage or injury. When bears become destructive and aggressive, they may have to be killed. The major emphasis of the bear program of the NATIONAL PARK SERVICE is to break the link between bears and human food sources and to restore the bears' natural foraging habits. To accomplish this, the Park Service has made all outdoor REFUSE containers bear-resistant and requires all visitors to comply with special food storage regulations.

Transportation, both to and within the park, has recently been the focus of Park Service efforts to reduce Yosemite Valley's infamous summer traffic snarls. Besides visitor frustration, the excessive traffic causes AIR POLLUTION and increases noise in the valley. (*See* noise pollution.) A General Plan was developed in 1980 to deal with the problem, but a lack of funds slowed its implementation. Congress has since approved emergency funds, and in 1997 and 1998 the park's administration circulated a plan for revising transportation within Yosemite Valley and in the greater Yosemite region. *See also* ARCTIC NATIONAL WILDLIFE REFUGE; CONIFEROUS FOREST; OLD-GROWTH FOREST; YELLOWSTONE NATIONAL PARK.

EnviroSources

National Park Service (Yosemite National Park) Web site: www.nps.gov/yose.

Yosemite Association Web site: www.yosemite.org.

Yosemite National Park: P.O. Box 577, Yosemite, CA 95389; (209) 372-0200.

Z

zebra, A member of *Equidae* or horse family who are medium-sized HERBIVORES with long heads and necks and slender legs. Some SPECIES are threatened or endangered. The zebras of Africa fall into three distinct species: Grevy's zebra (*Equus grevyi*), the plains zebra (*Equus burchelli*), and the mountain zebra (*Equus zebra zebra*).

The Grevy's zebra lives in the sub-desert, plains, and GRASSLAND areas of Ethiopia, Kenya, and Somalia. The zebra was named for Jules Grevy, President of France, who was given a gift of these zebras by the King of Ethiopia. The Grevy's zebra is listed as endangered SPECIES and is protected under the CONVENTION ON INTERNATIONAL TRADE IN ENDANGERED SPECIES OF WILD FLORA AND FAUNA (CITES).The primary cause of POPULATION decline is POACHING and the loss of grazing HABITATS and water holes due to the impact of expanding farming areas and the use of IRRIGATION. The largest concentration of animals, about several thousand, live in northern Kenya. There are about 300 Grevy's zebras living in captivity.

The Grevy's zebra weighs about 300–370 kilograms (600-800 pounds) and has narrow, black and white stripes, long, narrow head and prominent, broad ears. Like all zebras, the Grevy's zebra grazes primarily on coarse grasses and sedges but will eat bark, leaves, buds, fruits, and roots.

The plains zebra, the most common of Africa's large MAMMALS, is the most abundant and widespread of all zebras occurring throughout Africa. These are the ones that are in seen in zoos and in circuses. Plains zebras inhabit the open grasslands and the drier SAVANNAS of east Africa and southern Africa. The animals graze and travel in herds of up to 10,000 during the different seasons. The plains zebra has differentiated into several subspecies. One of them, Grant's zebra is the most common of the plains zebra subspecies.

The Grant's zebra is the most studied of the plains zebras and much of what is known of the behavior of zebras comes from scientific research of these animals. As an example, research findings indicated that zebra stripes are like human fingerprints. No two zebras have the same stripe pattern. The different stripe patterns makes it easy for scientists to observe and identify individuals and to learn how they behave differently from one another. There are about 300,000 Grant's zebra left in the wild, of which about 150,000 of them live on the Serengeti Plains.

The zebra population throughout the world is decreasing because of loss of habitat, hunting, and political instability.

The third zebra species is the mountain zebra of which two subspecies are recognized, Hartmann's mountain zebra and Cape mountain zebra who live in the mountainous areas of southwest Africa. The Hartmann's mountain zebra is on the threatened list and lives in Angola and Namibia. The current wild population is about 7,300 animals and another 140 are living in captivity. The Cape Mountain zebra is on the endangered list and is almost extinct. Estimates put their numbers at approximately 400. The Mountain Zebra National Park preserve in South Africa is home to about 200 mountain zebras. *See also* ELEPHANTS; TIGERS

EnviroSource

U.S. Fish and Wildlife Service, The Species List of Endangered and Threatened Wildlife Web site: www.fws.gov/r9endspp/lsppinfo.html

zero discharge, A requirement set forth in the 1972 amendments to the U.S. CLEAN WATER ACT that no toxic DISCHARGE may be released into lakes, ponds, or streams, a prohibition instituted to keep these waters safe for use by people and other ORGANISMS. The zero discharge limitation requires that businesses, industries, and municipalities ensure that only WASTEWATER carrying human sanitary wastes be discharged into municipal SEWER systems. *See also* *ESCHERICHIA COLI*; POTABLE WATER; SEWAGE.

zero population growth, A condition that occurs when the POPULATION of a specific area does not increase or decrease over time. To achieve zero population growth (ZPG), the number of births in a population must equal the number of deaths during the same period. At the same time, population increases resulting from IMMIGRATION must balance decreases resulting from EMIGRATION.

Zero population growth rarely occurs in nature; however, many people believe the world is overpopulated with humans and think zero population growth should be a goal for human societies. Maintaining zero population growth might ensure that resource usage remains at a constant level. In many regions of the world, however, human population sizes already place too much stress on the availability and sustainability of NATURAL RESOURCES. Thus, zero population growth in these areas will not achieve a sustainable outcome. *See also* BIRTH RATE; CARRYING CAPACITY; DEATH RATE; EHRLICH, PAUL RALPH; GROWTH RATE; POPULATION DENSITY; STERILIZATION, REPRODUCTIVE.

EnviroSources

Zero Population Growth: 1400 16th St., NW, Suite 320, Washington, DC 20036; (202) 332-2200; e-mail: zpg@igc.apc.org.

zone of aeration, The area of Earth's crust above the WATER TABLE where spaces between rock and soil particles are saturated with AIR rather than water. The zone of aeration extends from the water table to the soil's surface and determines the depth to which WELLS must be dug to obtain water from the ground. (*See* groundwater.) While some water may be present in the zone of aeration, especially in the area nearest the water table, the presence of much water in this region is unusual. Other terms used for the zone of aeration are the unsaturated zone and the vadose zone. *See also* AQUIFER; ZONE OF DISCHARGE; ZONE OF SATURATION.

zone of discharge, The part of an AQUIFER, an underground river or RESERVOIR, where GROUNDWATER reaches and breaks through the ground's surface. Zones of discharge occur naturally where groundwater empties into lakes, ponds, or rivers. A zone of discharge also may exist as a spring when the WATER TABLE reaches the surface, as often occurs on hillsides. In some regions, heat from deep within Earth may raise water temperature and place groundwater under enough pressure to periodically force water to the surface as occurs with a hot spring or geyser. (*See* geothermal energy.) Such activity is responsible for the periodic eruptions of the Old Faithful Geyser of YELLOWSTONE NATIONAL PARK.

Groundwater often provides individuals or entire communities with their drinking water supply. In such areas, WELLS are dug to access groundwater. If the water is under pressure, it may naturally move up through the well; if not under pressure, pumps may be used to raise the water to the surface. Wells dug to access groundwater form an artificial zone of discharge. *See also* IRRIGATION; OGALLALA AQUIFER; POTABLE WATER; ZONE OF AERATION.

zone of saturation, The area beneath Earth's surface where the spaces between soil and rock particles are completely filled with water instead of AIR. The zone of saturation, or saturated zone, is the storage area for much of Earth's GROUNDWATER and lies beneath the ZONE OF AERATION, or unsaturated zone. Gravity usually causes water in the zone of saturation to be under greater pressure than water at Earth's surface. The WATER TABLE generally represents the upper boundary of the zone of saturation. *See also* AQUIFER; OGALLALA AQUIFER; PERMEABILITY; POROSITY; SUBSIDENCE; WELL; ZONE OF DISCHARGE.

zoo, An animal facility. Zoos once simply displayed live animals for public entertainment, but they have evolved into sophisticated scientific and educational institutions that contribute to the understanding and CONSERVATION of wild animal POPULATIONS. Among challenges facing modern zoos are the costs of upgrading old facilities, the struggle to obtain sufficient operating funds, and the need to attract more visitors in new and entertaining exhibits.

Many older zoos in American cities have undergone renovations during the last decades of the twentieth century. Among the recent trends in zoo improvement is the construction of new enclosures that resemble natural HABITATS. The replacement of traditional steel bars and sawdust-covered concrete floors with appropriately designed surroundings improves visitor appreciation of the animals and helps educate the public about the animals' natural habitats. Such renovations may reduce stress on animals and allow them to interact with one another more naturally.

Several major zoos conduct CAPTIVE PROPAGATION programs: the intentional breeding of selected zoo or wild animals to obtain offspring, usually for release into the wild or for transfer to other zoos. Captive breeding constitutes one method of combating SPECIES EXTINCTION. The captive breeding program at the San Diego Zoo, for example, produced CALIFORNIA CONDORS (an ENDANGERED SPECIES) that were later successfully released into the wild. Some zoos have extended their activities to conservation programs in other countries, assisting in the establishment of WILDLIFE preserves.

Zoos have expanded and improved public education programs also, with education departments that develop programs related to zoo exhibits. Public outreach activities include in-school programs, zoo tours, special events, and Internet sites. The Zoological Society of New York, for example, conducted

The rare snow leopard and polar bear are found in some zoos. These animals occupy the Roger Williams Zoo in Providence, Rhode Island. *Credit:* Roger Williams Park Zoo, Providence, Rhode Island.

a joint project with the government of Cameroon in west Africa, to monitor an ELEPHANT herd as it moved throughout its range.

Critics contend that many remaining older zoos provide poor living conditions for the animals they keep. Visitors to these zoos may see old, unhealthy, or dysfunctional wildlife specimens. Unfortunately, some zoos have acted irresponsibly in accepting threatened or endangered wildlife species, or animals collected illegally. (*See* poaching.) Opponents of such zoos believe that the zoos seek to profit from the exploitation of animals, with the effect of rewarding poachers. Often, however, zoos provide a haven for animals confiscated from illegal wildlife traffickers, circuses, or illicit traveling zoos that improperly handle live animals. (*See* canned hunt.)

The importance of zoos increases continually, as natural habitat dwindles. Through their efforts in support of conservation, education, and environmental advocacy, zoos will continue to play a critical role in wildlife preservation throughout the world. *See also* AMERICAN ZOO AND AQUARIUM ASSOCIATION; ANIMAL RIGHTS; WILDLIFE REFUGE.

EnviroSources

Bronx Zoo Web site: www.bronxzoo.com.
San Diego Zoo Web site: www.sandiegozoo.org.

zoology, The branch of biology that deals with the study of animals and animal life, including their roles in the ENVIRONMENT. Scientists who work in this field are zoologists. One of the major concerns of zoology is animal classification, or TAXONOMY. Animals are classified into taxonomic groups, or taxa, based on their traits and evolutionary histories. Other major concerns of zoology are anatomy and physiology and animal development. *See also* BIODIVERSITY; BOTANY; ECOLOGY; EVOLUTION.

zooplankton, Tiny animals and animal-like ORGANISMS that are part of the PLANKTON in aquatic ECOSYSTEMS. Many zooplankton are microscopic PROTISTS known as protozoans; other zooplankton are macroscopic, although these organisms are often so tiny they are almost undetectable to the unaided eye. Among the macroscopic zooplankton are tiny adult animals, their eggs, and their larvae. For example, the tiny CRUSTACEANS known as KRILL and the larvae of mollusks, known as trochophores, are a component of zooplankton. Zooplankton may possess some structures for locomotion; however, most zooplankton rely on water currents as their primary vehicle of locomotion.

All zooplankton are CONSUMERS that feed on PHYTOPLANKTON, the photosynthetic component of plankton, or on other zooplankton. Thus, zooplankton generally occupy either the first or second TROPHIC LEVEL in aquatic FOOD CHAINS. Zooplankton, in turn, are a food source for consumers that feed at higher trophic levels, such as fishes and filter feeding mollusks, and thus constitute a vital component in aquatic food chains and FOOD WEBS. *See also* MICROORGANISMS; PHOTOSYNTHESIS; ZOOXANTHELLAE.

zooxanthellae, Microscopic, photosynthetic ORGANISMS that live in mutualistic association with corals. (*See* coral reef.) The zooxanthellae are a type of PROTIST belonging to the dinoflagellate family. Because they are photosynthetic, zooxanthellae can live among reefs located in regions of shallow OCEAN water, where they are exposed to sunlight. Zooxanthellae are symbionts of living coral polyps, so they also live in reef areas that have active coral growth. In the zooxanthellae and coral association, the coral provide the zooxanthellae with support and a HABITAT that protects them from ocean waves and currents. (*See* ocean current.) The zooxanthellae, in turn, carry out PHOTOSYNTHESIS, providing themselves and the corals with the nutrients needed for growth and the energy-generating process of respiration. *See also* AUTOTROPH; MUTUALISM; PHOTOSYNTHESIS; PHYTOPLANKTON; PRODUCER; SYMBIOSIS; ZOOPLANKTON.

Environmental Timeline

1798

British author Thomas Malthus publishes the *Principle of Population,* which states that the human population size, like that of other species, is limited by the availability of environmental resources.

1845

HENRY DAVID THOREAU* moves to Walden Pond to observe the fauna and flora of Concord, Massachusetts.

1849

U.S. DEPARTMENT OF THE INTERIOR (DOI) is established.

1857

FREDERICK LAW OLMSTED develops the first city park—New York's Central Park.

1859

British naturalist CHARLES DARWIN publishes *The Origin of the Species by Means of Natural Selection.* In time, the ideas in his book become the most widely accepted theory of EVOLUTION.

1866

German biologist Ernst Haeckel introduces the term "ECOLOGY."

1869

JOHN MUIR moves to the Yosemite Valley.

JOHN WESLEY POWELL travels the Colorado River through the Grand Canyon.

1872

YELLOWSTONE NATIONAL PARK is established as the first national park of the United States in Yellowstone, Wyoming.

U.S. Legislation: Passage of Mining Law permits individuals to purchase rights to mine public lands.

1876

Appalachian Mountain Club is founded.

1879

U.S. GEOLOGICAL SURVEY (USGS) is formed.

1882

The first hydroelectric plant opens on the Fox River in Wisconsin.

1883

KRAKATOA, a small island of Indonesia, is virtually destroyed by a volcanic explosion.

1890

The first windmill for generating electricity is constructed in Denmark.

Sequoia National Park, YOSEMITE NATIONAL PARK, and General Grant National Park are established in California.

1891

U.S. Legislation: Passage of Forest Reserve Act provides the basis for a system of national forests.

1892

JOHN MUIR founds THE SIERRA CLUB.

1893

The National Trust is founded in the United Kingdom. The group purchases land deemed of having natural beauty or considered a cultural landmark.

1895

The American Scenic and Historic Preservation Society is founded.

*Use of small capital letters indicates that word or person has an entry in the encyclopedia.

1898

Cornell University establishes the first college program in forestry.

U.S. Legislation: Passage of Rivers and Harbors Act bans pollution in rivers and oceans.

1900

U.S. Legislation: Passage of Lacey Act makes it unlawful to transport illegally killed game animals across state boundaries.

1902

U.S. Legislation: Passage of Reclamation Act establishes the BUREAU OF RECLAMATION.

1903

First federal U.S. wildlife refuge established on Pelican Island in Florida.

1905

The NATIONAL AUDUBON SOCIETY, named for wildlife artist JOHN JAMES AUDUBON, is founded.

1906

Yosemite Valley is incorporated into YOSEMITE NATIONAL PARK.

1907

GIFFORD PINCHOT is appointed the first Chief of the U.S. FOREST SERVICE.

Inland Waterways Commission is established.

1908

Grand Canyon is set aside as a national monument.

CHLORINATION is first used at U.S. water treatment plants.

President THEODORE ROOSEVELT hosts first Governors' Conference on CONSERVATION.

1914

The last PASSENGER PIGEON, Martha, dies in the Cincinnati ZOO.

1916

NATIONAL PARK SERVICE (NPS) is established.

1918

Hunting of migratory bird species is restricted through passage of the MIGRATORY BIRD TREATY ACT. The Act supports treaties between the U.S. and surrounding nations.

Save-the-Redwoods League is created.

1920

U.S. Legislation: Passage of Mineral Leasing Act regulates mining on federal lands.

1922

Izaak Walton League is organized under direction of Will H. Dilg.

1924

Environmentalist ALDO LEOPOLD wins designation of Gila National Forest, New Mexico, as first extensive wilderness area.

1925

Geneva Protocol is signed by numerous countries as a means of stopping use of biological weapons.

1928

Boulder Canyon Project (Hoover Dam) is authorized to provide irrigation, electric power, and a flood control system for Arizona and Nevada communities.

1930

CHLOROFLUOROCARBONS (CFCs) are deemed safe for use in refrigerators and air conditioners.

1931

France builds and makes use of the first Darrieus AEROGENERATOR to produce electricity from wind energy.

ADDO NATIONAL ELEPHANT PARK is established in the Eastern Cape region of South Africa to provide a protected habitat for African ELEPHANTS.

1933

TENNESSEE VALLEY AUTHORITY (TVA) is formed.

Civilian Conservation Corps (CCC) employs more than 2 million Americans in forestry, FLOOD control, soil erosion, and beautification projects.

1934

Greatest DROUGHT in U.S. history is recorded.

U.S. Legislation: Passage of Taylor Grazing Act regulates livestock grazing on federal lands.

1935

Soil Conservation Service (SCS), now called NATIONAL RESOURCES CONSERVATION SERVICE, is established,.

WILDERNESS SOCIETY founded.

1936

National Wildlife Federation (NWF) is formed.

1940

U.S. Wildlife Service is established to protect fish and wildlife.

U.S. Legislation: President Franklin Roosevelt signs the BALD EAGLE Protection Act.

1945

The United Nations establishes the FOOD AND AGRICULTURE ORGANIZATION (FAO).

1946

The INTERNATIONAL WHALING COMMISSION (IWC) forms to research WHALE populations.

U.S. Bureau of Land Management (BLM) and Atomic Energy Commission, now Nuclear Regulatory Commission, are created.

1947

Marjory Stoneman Douglas publishes *The Everglades: River of Grass* and serves as a member of the committee that gets the Everglades designated a national park.

1948

The United Nations (UN) creates the International Union for the Conservation of Nature (IUCN), now called World Conservation Union, as a special environmental agency.

Air pollution incident in Donora, Pennsylvania, kills 20 people; 14,000 become ill.

U.S. Legislation: Federal Water Pollution Control Law is passed.

1949

Aldo Leopold's *A Sand County Almanac* is published posthumously.

1950

Oceanographer Jacques Cousteau purchases and transforms a former minesweeper, the *Calypso*, into a research vessel that he uses to increase awareness of the ocean environment.

1951

Tanzania begins its national park system with the establishment of the Serengetti National Park.

1952

Clean air legislation is enacted in Great Britain after an air-pollution-induced smog kills nearly 4,000 people.

David Brower becomes the first executive director of the Sierra Club.

1953

Radioactive iodine from atomic bomb testing is found in the thyroid glands of children living in Utah.

1955

U.S. Legislation: Air Pollution Control Act is passed, the first federal legislation designed to control air pollution.

1956

U.S. Legislation: Water Pollution Control Act authorizes development of water treatment plants.

1959

Antarctic Treaty is signed.

1961

African Wildlife Foundation (AWF) established as an international organization to protect African wildlife.

1962

Rachel Carson publishes *Silent Spring*.

1963

Nuclear Test Ban Treaty between U.S. and USSR stops atmospheric testing of nuclear weapons.

U.S. Legislation: The first Clean Air Act (CAA) authorizes money for air pollution control efforts.

1964

U.S. Legislation: The Wilderness Act creates National Wilderness Preservation System.

1965

U.S. Legislation: The Water Quality Act authorizes the federal government to set water standards in absence of state action.

1966

Eighty people in New York City die from air pollution-related causes.

1967

Torey Canyon runs aground spilling 175 tons of crude oil off Cornwall, England.

Dian Fossey establishes the Karisoke Research Center in the Virunga Mountains, within the Parc National des Volcans in Rwanda, to study endangered mountain gorillas.

Environmental Defense Fund (EDF) forms to lead effort to save the osprey from DDT.

1968

U.S. Legislation: The Wild and Scenic Rivers Act and National Trails System Act identify areas of great scenic beauty for preservation and recreation.

Paul Ehrlich publishes *The Population Bomb.*

1969

Austrian wildlife photographer Joy Adamson establishes the Elsa Wild Animal Appeal, an organization dedicated to the preservation and humane treatment of wild and captive animals.

Greenpeace is created.

Blowout of oil well in Santa Barbara, California, releases 2,700 tons of crude oil into the Pacific Ocean.

U.S. Legislation: The National Environmental Policy Act (NEPA) requires all federal agencies to complete an environmental impact statement for any dam, highway, or other large construction project undertaken, regulated, or funded by the federal government.

Friends of the Earth (FOE) is founded in the United States.

1970

The first Earth Day is celebrated on April 22.

Construction of the Aswan High Dam on the Nile River in Egypt is completed.

U.S. Legislation: Passage of an amended CLEAN AIR ACT (CAA) expands air pollution control.

U.S. ENVIRONMENTAL PROTECTION AGENCY (EPA) is established.

1971

Canadian primatologist BIRUTÉ GALDIKAS begins her studies of ORANGUTANS through the Orangutan Research and Conservation Project in Borneo.

The United Nations Educational, Scientific and Cultural Organization (UNESCO) establishes the MAN AND THE BIOSPHERE PROGRAMME, developing a global network of biosphere reserves.

1972

Biological and Toxin Weapons Convention is adopted by 140 nations to stop the use of biological weapons.

EPA phases out use of DDT in the United States to protect several species of predatory birds. The ban was inspired by information from RACHEL CARSON's 1962 book, *Silent Spring*.

U.S. Legislation: Coastal Zone Management Act (CZMA) and the Environmental Pesticide Control Act are passed.

Oregon passes the first bottle-recycling law.

1973

Norwegian philosopher Arne Naess coins the term "DEEP ECOLOGY" to describe his belief that humans need to recognize natural things for their intrinsic value, rather than just for their value to humans.

The CONVENTION ON INTERNATIONAL TRADE IN ENDANGERED SPECIES OF WILD FAUNA AND FLORA (CITES) is signed by over 80 nations. The ENDANGERED SPECIES ACT of the United States also is enacted.

Congress approves constuction of the 1,300-kilometer pipeline from Alaska's North Slope oil field to the port of Valdez.

Energy crisis in the United States arises from Arab oil embargo.

A collision and resulting explosion between the *Corinthos* oil tanker and the *Edgar M. Queeny* release 272,00 barrels of crude oil and other chemicals into the Delaware River near Marcus Hook, Pennsylvania.

1974

Scientists report their discovery of a hole in the OZONE layer above ANTARCTICA.

U.S. Legislation: THE SAFE DRINKING WATER ACT sets standards to protect nation's drinking water. EPA bans most uses for aldrin and dieldrin and disallows production and importation of these chemicals into the United States.

1975

Unleaded gas goes on sale. Cars are equipped with anti-pollution catalytic converters.

EPA bans use of asbestos insulation in new buildings.

EDWARD ABBEY publishes *The Monkey Wrench Gang*, a novel detailing acts of ECOTAGE as a means of protecting the environment.

1976

Argo Merchant runs aground releasing 25,000 tons of fuel into the Atlantic Ocean near Nantucket, Rhode Island.

National Academy of Sciences reports that CHLOROFLUOROCARBON gases from spray cans are damaging the OZONE layer

U.S. Legislation: The Resource Conservation and Recovery Act empowers EPA to regulate the disposal and treatment of municipal solid and hazardous wastes. The TOXIC SUBSTANCES CONTROL ACT and RESOURCE CONSERVATION AND RECOVERY ACT are enacted.

Fire aboard *Hawaiian Patriot* releases nearly 100,000 tons of crude oil into Pacific Ocean.

1977

GREENBELT MOVEMENT is begun by Kenyan conservationist Wangari Muta Maathai on World Environment Day.

Blowout of Ekofisk oil well releases 27,000 tons of crude oil into the NORTH SEA.

Construction of the ALASKA PIPELINE, the 1,300-kilometer pipeline that carries oil from Alaska's North Slope oil field to the port of Valdez, is completed at a cost of more than $8 billion.

U.S. Legislation: SURFACE MINING CONTROL AND RECLAMATION ACT is passed.

DEPARTMENT OF ENERGY (DOE) is created.

1978

AMOCO CADIZ tanker runs aground spilling 226,000 tons of oil into ocean near Portsall, Brittany.

People living in the LOVE CANAL community of New York are evacuated from the area to reduce their exposure to chemical wastes that have surfaced from a canal formerly used as a dump site.

Rainfall in Wheeling, West Virginia, is measured at a pH of 2, the most acidic rain yet recorded.

AEROSOLS with FLUOROCARBONS are banned in United States.

EPA bans the use of ASBESTOS in insulation, fireproofing, or decorative materials.

1979

British scientist James E. Lovelock publishes *Gaia: A New Look at Life on Earth*. Lovelock suggest that Earth could be thought of as a living organism.

Collision of *Atlantic Empress* and *Aegean Captain* releases 370,000 tons of oil into the CARRIBBEAN SEA.

Convention on Long-Range Transboundary Air Pollution (LRTAP) signed by several European nations to limit SULFUR DIOXIDE emissions that cause ACID RAIN problems in other countries.

THREE MILE ISLAND Nuclear Power Plant in Pennsylvania experiences near-MELTDOWN.

EPA begins a program to assist states in removing flaking ASBESTOS insulation from pipes and ceilings in school buildings throughout the United States.

EPA bans the marketing of Agent Orange in the United States.

1980

Debt-for-nature swap idea is proposed by Thomas E. Lovejoy, whereby nations could convert debt to cash that would then be used to purchase parcels of tropical rainforest to be managed by local conservation groups.

Global Report to the President addresses world trends in population growth, natural resource use, and the environment by the end of the century, and calls for international cooperation in solving problems.

U.S. Legislation: Comprehensive Environmental Response, Compensation, and Liability Act (Superfund) and the Low-Level Radioactive Waste Policy Act are passed.

1981

Earth First!, a radical environmental action group that resorts to ecotage to gain objectives, forms.

1982

U.S. Legislation: Nuclear Waste Policy Act is passed.

1984

Toxic gases released from the Union Carbide chemical manufacturing plant in Bhopal, India, kill an estimated 3,000 people and injure thousands of others. (*See* Bhopal incident.)

The Jane Goodall Institute (JGI) is founded for studying chimpazees.

More than 1.8 million gallons of oil are spilled into the Gulf of Mexico by the British tanker *Alvenus.*

Famine in the Sahel, Africa.

U.S. Legislation: Hazardous and Solid Waste Amendments are passed.

1985

Primatologist Dian Fossey is discovered murdered in her cabin at the Karosoke Research Center she founded. Her death is attributed to poachers.

The *Rainbow Warrior* (a boat owned by Greenpeace that was part of a protest against nuclear testing by the French in the Pacific) is sunk in a New Zealand harbor by agents of the French government.

Antarctic ozone hole discovered

U.S. Legislation: Food Security Act is passed.

1986

Tons of toxic chemicals stored in a warehouse owned by the Sandoz pharmaceutical company are released into the Rhine River near Basel, Switzerland. Effects of the spill are experienced in several European countries, including Switzerland, France, Germany, and Luxembourg. (*See* Sandoz chemical spill.)

An explosion destroys a nuclear power plant in Chernobyl, Ukraine, immediately killing more than 30 people and leading to the permanent evacuations of more than 100,000 others.

Bovine spongiform encephalopathy (BSE), a neurodegenerative illness of cattle, also known as mad cow disease, comes to the attention of the scientific community when it appears in cattle in the United Kingdom.

Levels of dioxin 100 times the emergency level are found in town of Times Beach, Missouri, leading to evacuation.

U.S. Legislation: Emergency Planning and Community Right-to-Know Act and Superfund Amendments and Reauthorization Act (SARA) are passed.

1987

The Montreal Protocol, an international treaty that proposes to cut in half the production and use of chlorofluorocarbons, is approved by more than 30 nations.

The world's fourth largest lake, the Aral Sea of Asia, is divided in two as a result of diversion of water from its feeder streams, the Syr Darya and Amu Darya Rivers.

The *Mobro*, a garbage barge from Long Island, New York, travels 9,600 kilometers in search of a place to offload the garbage it carries.

1988

Use of ruminant proteins in the preparation of cattle feed is banned in the U.K. to prevent outbreaks of BSE.

Global temperatures reach their highest levels in 130 years.

The Marine Protection, Research, and Sanctuaries Act (MPRSA) legislates international dumping of wastes in the ocean. It is also known as the Ocean Dumping Ban Act.

Epa studies report that indoor air can be 100 times as polluted as outdoor air. Radon is found to be widespread in U.S. homes.

Beaches on the east coast of the United States are closed because of contamination by medical waste washed on shore.

The United States experiences its worst drought in 50 years.

Plastic ring six-pack holders are required to be made degradable.

U.S. Legislation: Plastic Pollution Research and Control Act bans ocean dumping of plastic materials.

1989

The United Kingdom bans the use of cattle brains, spinal cords, tonsils, thymuses, spleens, and intestines in foods intended for human consumption as a means of preventing further outbreaks of Creutzfeldt-Jakob Disease (CJD), the human version of mad cow disease, in humans.

Fire aboard *Kharg 5* releases 75,000 tons of oil into sea surrounding Canary Islands.

The Montreal Protocol treaty is updated and amended.

New York Department of Environmental Conservation reports that 25 percent of the lakes and ponds in the Adirondacks are too acidic to support fish.

Exxon Valdez runs aground on Prince William Sound, Alaska, spilling 11 million gallons of oil in one of the world's most fragile ecosystems.

1990

U.N. report forecasts a world temperature increase of 2 degrees Fahrenheit within 35 years because of GREENHOUSE GAS EMISSIONS.

U.S. Legislation: NEW CLEAN AIR ACT amendments include requirements to control the emission of SULFUR DIOXIDE and NITROGEN OXIDES.

1991

The Persian Gulf War concludes with hundreds of oil wells in Kuwait being set afire by Iraqi troops, resulting in extensive air and water pollution problems.

The United States accepts an agreement on ANTARCTICA that prohibits activities relating to mining, protects native species of flora and fauna, and limits tourism and marine pollution.

Eight scientists begin a two-year stay in the Biosphere 2 in Arizona, a test center designed to provide a self-sustaining habitat modeling Earth's natural environments. The experiment, which is repeated in 1993, meets with much criticism and is deemed largely unsuccessful.

1992

UNITED NATIONS EARTH SUMMIT is held in Rio de Janeiro, Brazil. Major resolutions resulting from the summit include the Rio Declaration on Environment and Development, AGENDA 21, BIODIVERSITY TREATY, Statement of Forest Principles, and the Global Warming Convention, which is signed by more than 160 nations.

The MONTREAL PROTOCOL is again amended with signatories agreeing to phase out CFC use by the year 2000.

1993

Sugar producers and government agree on a restoration plan for the Florida EVERGLADES.

1994

Failure of a dike results in the release of 102,000 tons of oil into the Siberian TUNDRA near Usink in northern Russia.

The Russian government calls for preventive measures to control the destruction of LAKE BAIKAL.

BALD EAGLE is reclassified from an ENDANGERED SPECIES to a THREATENED SPECIES on the U.S. ENDANGERED SPECIES LIST.

An 8.5 million gallon spill is discovered in Unocal's Guadalupe oil field in California.

1995

Government reintroduces endangered WOLVES to YELLOWSTONE PARK.

1999

Scientists report that the human population of Earth now exceeds six billion people.

PEREGRINE FALCON is removed from the U.S. ENDANGERED SPECIES LIST.

New Carissa runs aground off coast of Oregon, leaking some oil into Coos Bay. The tanker is later towed into the open ocean and sunk.

Japan suffers its worst nuclear accident in the city of Takaimura. An accident occurs at a URANIUM processing plant. Japan replies on NUCLEAR ENERGY for about a third of its electricity.

2000

The CHERNOBYL plant is scheduled to close permanently in December 2000.

EARTHDAY 2000: 4,000 groups in 169 countries join the Earthday Network in April 2000.

UNITED STATES FISH AND WILDLIFE SERVICE release a proposal for reintroducing 25 grizzly bears into the wilderness areas of Idaho and Montana.

Mongolia suffers the worst drought in 31 years.

EPA bans two common PESTICIDES used in the home

MTBE additive in gasoline to be banned by 2004.

Bibliography

Abbey, Edward. *Confessions of a Barbarian: Selections from the Journals of Edward Abbey 1951–1989.* Boston: Little Brown and Co., 1996.

———. *Desert Solitaire.* New York: Ballantine Books, 1991.

———. *The Monkey Wrench Gang,* New York: Avon Books, 1997.

Abrahamson, Warren G. *Evolutionary Ecology Across Three Trophic Levels: Goldenrods, Gallmakers, and Natural Enemies.* Princeton: Princeton University Press, 1997.

Abromovitz, Janet, Lester R. Brown, et al. *State of the World: 1998, A Worldwatch Institute Report on Progress toward a Sustainable Society*. New York: Norton, 1998.

Adamson, Joy Gessner. *Born Free.* New York: Harcourt, Brace, and World, 1960.

———. *Forever Free.* New York: Harcourt, Brace, and World, 1963.

———. *Living Free.* New York: Harcourt, Brace, and World, 1961.

Air and Waste Management Association. *Pollution Prevention for Our Land, Water and Air*. Pittsburgh: Air and Waste Management Association, 1993.

Allaby, Michael, ed. *The Concise Oxford Dictionary of Ecology.* Oxford, UK: Oxford University Press, 1994.

Anderson, Daniel W. "Effects of DDT on Birds: Does Dixy Know Something the Experts Do Not?" *Environmental Review Newsletter* 1, no. 7 (July 1994).

Anderson, Roger N. *Marine Geology*. New York: John Wiley and Sons, 1989.

Arms, Karen. *Holt Environmental Science.* Austin, TX: Holt, Rhinehart, and Winston, 1996.

Art, Henry W., ed. *The Dictionary of Ecology and Environmental Science*. New York: Henry Holt, 1993.

Atwood, Nicolas. "Feeding Flipper." *Audubon* 99, no. 6 (November–December 1997): 12.

Baird-Middleton, Bruce, prod. and ed. "Edward O. Wilson: Reflections on a Life in Science." Videotape. Film Study Center, Harvard University, 1992.

Batman, Robert, Anthony W. Diamond, and Rudolf L. Schreiber. *Save the Birds*. Newfoundland: Breakwater Books, 1989.

Batzer, Darol, et al., eds. *Invertebrates in Freshwater Wetlands of North America: Ecology and Management.* New York: John Wiley and Sons, 1999.

Beaud, Michel. *Economic Thought since Keynes: A History and Dictionary of Major Economists*. New York: Routledge, 1997.

Benarde, Melvin A. *Asbestos: The Hazardous Fiber.* Cleveland: CRC Press, 1990.

Bennett Information Group. *The Green Pages: Your Everyday Shopping Guide to Environmentally Safe Products*. New York: Random House, 1990.

Berenbaum, May R. "Two Horns, Six Legs, and One Voracious Appetite. *Audubon* 102, no. 1 (January–February): 74–79.

Berkow, Robert, ed. *The Merck Manual,* 16th ed., 2 vols. Rahway, NJ: Merck Research Laboratories, 1992.

Bernstein, Leonard, Alan Winkler, and Linda Zierdt-Warshaw. *Environmental Science: Ecology and Human Impact,* 2d ed. Menlo Park, CA: Addison-Wesley, 1996.

Berrill, Michael and David Suzuki. *The Plundered Seas: Can the World's Fish Be Saved?* San Francisco: Sierra Club Books, 1997.

Beveridge, C.E. et al, eds. *Frederick Law Olmsted: Designing the American Landscape*. New York: Rizzoli International, 1995.

Bickel, L. *The Deadly Element: The Story of Uranium.* New York: Stein and Day, 1979.

Bishop, James. *Epitaph for a Desert Anarchist: The Life and Legacy of Edward Abbey.* New York: Touchstone Books, 1995.

Bolt, Bruce A. *Earthquakes.* New York: W.H. Freeman and Co. 1993.

Botkin, Daniel B., and Edward A. Keller. *Environmental Science: Earth as a Living Planet.* 2d ed. New York: John Wiley and Sons, 1995.

Bourne, Joel. "Bugging Out." *Audubon* 101, no.2 (March–April 1999): 71–73.

———. "Gorillas in Our Midst." *Audubon* 100, no. 5 (September-October 1998): 70-72.

———. "The Organic Revolution." *Audubon* 101, no.2 (March–April 1999): 64–70.

Bower, Joe. "The Persistent Parasite." *Audubon* 100, no. 2 (March–April 1998): 16.

Boyce, Barry. *The Traveler's Guide to the Galápagos Islands.* Hunter, 1998.

Brandt, Anthony. "Not in My Backyard." *Audubon* 99, no. 5 (September–October 1997): 58–62, 86–87, 102.

Breton, Mary Joy. *Women Pioneers for the Environment.* Boston: Northeastern University Press, 1998.

Breymeyer, A. I. , D. O. Hall, and J. M. Melillo (ed.), *Global Change: Effects on Coniferous Forests and Grasslands.* Scope, No. 56, New York: John Wiley and Sons, 1997.

Bridgman, R. *Dark Thoreau.* Lincoln: University of Nebraska Press, 1982

Brink, Wellington. *Big Hugh: The Father of Soil Conservation.* New York: Macmillan, 1951.

Brower, David. *For Earth's Sake: The Life and Times of David Brower.* Layton, UT: Gibbs Smith, 1990.

———. *Let the Mountains Talk, Let the Rivers Run: A Call to Those Who Would Save the Earth.* New York: Harper Collins, 1995.

———. *Only a Little Planet*. New York: McGraw-Hill, 1972.

Brown, Katrina, et al. *The Causes of Tropical Deforestation: The Economic and Statistical Analysis of Factors Giving Rise to the Loss of the Tropical Forests*. Seattle: University of Washington Press, 1995.

Brown, Lauren. *Grasslands.* Audubon Society Nature Guides. New York: Knopf, 1985.

Brown, Lester, et al. *Vital Signs, 1997: The Environmental Trends that Are Shaping Our Future*. New York: Norton, 1997.

Bruns, Roger A. *John Wesley Powell: Explorer of Grand Canyon*. Historical American Biographies. Springfield, NJ: Enslow, 1997.

Bullard, Robert D. *Confronting Environmental Racism*. Boston: South End Press, 1993.

———. *Unequal Protection*. San Francisco: Sierra Club Books, 1994.

———., et al. *We Speak for Ourselves: Social Justice, Race, and Environment.* London: Panos Institute, 1990.

Bulloch, David K. *The Wasted Ocean.* Lyons and Burford, 1989.

Burton, Ian, et al., eds. *Geography, Resources, and Environment (Themes from the Work of Gilbert White).* Chicago: University of Chicago Press, 1986.

Cain, A. J. *Animal Species and Their Evolution.* New Jersey: Princeton University Press, 1993. Cambridge University Press, 1996.

Capula, Massimo. *Simon and Schuster's Guide to Amphibians and Reptiles of the World*. New York: Simon and Schuster, 1989.

Caro, Timothy and George Schaller, eds. *Cheetahs of the Serengeti Plains.* Chicago: University of Chicago Press, 1994.

Carolin, Roger, ed. *Incredible Plants.* New York: Time-Life Books, 1997.

Carson, Rachel L. *The Edge of the Sea.* Mariner Books, 1998.

———. *Lost Woods: The Discovered Writing of Rachel Carson.* ed. Linda Lear. Boston: Beacon Press, 1998.

———. *The Sea Around Us*. Cambridge, MA: Oxford University Press, 1989.

———. *Silent Spring.* Boston: Houghton Mifflin, 1987.

———. *Under the Sea Wind*. Penguin, 1996.

Carter, Luther J. "Nuclear Imperatives and Public Trust: Dealing with Radioactive Waste." *Resources for the Future* (1987): 473.

Carwardine, Mark. *Eyewitness Handbooks: Whales, Dolphins, and Porpoises.* Covent Garde, UK: Dorling Kindersley, 1995.

Castleman, Barry I., and Stephen L. Berger. *Asbestos: Medical and Legal Aspects.* 4th ed., Gaithersburg, MD: Aspen, 1996.

Cavell, S. *The Senses of Walden.* Chicago: University of Chicago Press, 1992

Chadwick, Douglas H., and Joel Sartore. *The Company We Keep: America's Endangered Species.* Washington, DC: National Geographic Society, 1995.

Chang, Chris, et al. "Champions of Conservation: *Audubon* Recognizes 100 People Who Shaped the Environmental Movement and Made the Twentieth Century Particularly American." *Audubon* 100, no. 6 (November–December 1998).

Chernousenko, V.M.M. *Chernobyl: Insight from the Inside*. New York: Springer-Verlag, 1992.

Chiras, Daniel D. *Environmental Science: A Framework for Decision Making*. Menlo Park, CA: Addison-Wesley, 1989.

Christianson, Gale E. *Greenhouse: The 200-Year Story of Global Warming.* New York: Walker and Co., 1999.

Citizens Environmental Coalition. "How to Create a Toxic Plume Map." (33 Central Avenue, Albany, NY 12210).

Cogger, Harold G., Richard G. Zweifel, eds., *Encyclopedia of Reptiles and Amphibians.* San Diego: Academic Press, 1998.

Colinvaux, Paul. *Ecology 2.* New York: John Wiley and Sons, 1993.

Colten, Craig E., and Peter N. Skinner. *The Road to Love Canal: Managing Industrial Waste Before EPA.* Texas: University of Texas Press, 1996.

Commoner, Barry. *The Closing Circle*. New York: Knopf, 1971.

Conefrey, Mick, and Tim Jordan. *Icemen: Mick Conefrey and Tim Jordan*. Companion Volume to the Documentary Series. New York: Harper Collins, 1999.

Connor, Richard C., and Dawn Micklewaite Peterson. *The Lives of Whales and Dolphins: From the American Museum of Natural History.* New York: Henry Holt, 1996.

Coultas, Charles L., et al., eds. *Ecology and Management of Tidal Marshes: A Model from the Gulf of Mexico.* Boca Raton, FL: Saint Lucie Press, 1996

Cousteau, Jacques Yves. *Exploring the Wonders of the Deep.* Austin, TX: Raintree/Steck-Vaughn, 1997.

———. *The Ocean World.* Abradale Press, 1985.

Covalt, Ann, ed. *Affordable Cleanup? Opportunities for Cost Reduction in the* CRC Press, 1990.

Cronin, John, and Robert F. Kennedy, Jr. *The River Keepers: Two Activists Fight to Reclaim Our Environment as a Basic Human Right.* New York: Scribner, 1997.

Crossley, Louise. *Explore Antarctica*. New York: Cambridge University Press, 1995.

Curtis, Jane. *The World of George Perkins Marsh, America's First Conservationist and Environmentalist: An Illustrated Biography.* Billings Farm and Museum, 1982.

Darwin, Charles. *The Autobiography of Charles Darwin 1809–1882.* ed. Nora Barlow. New York: W.W. Norton and Co. 1993.

———. *The Origin of the Species by Means of Natural Selection.* Cambridge, MA: Oxford University Press, 1998.

———. *The Voyage of the Beagle: Charles Darwin's Journal of Researches.* New York: Penguin, 1989.

Davidson, Osha Gray. *The Enchanted Braid: Coming to Terms with Nature on the Coral Reef.* New York: John Wiley and Sons, 1998.

Deblieu, Jan. "Whirling Hurricanes." *Audubon* 101, no. 5 (September–October 1999): 38–43.

DeCicco, John, and Martin Thomas. *The 1999 Edition of "The Green Guide to Cars and Trucks."* Washington, DC: American Council for an Energy-Efficient Economy.

Decker, Robert, and Barbara Decker. *Volcanoes.* W.H. Freeman and Co., 1997.

Decontamination and Decommissioning of the Nation's Uranium Enrichment. New York: Dell, 1998.

De Roy, Tui. *Galápagos: Islands Born of Fire.* Toronto, Canada: Warwick, 1998.

De Seiguer, J. Edward. *The Age of Environmentalism.* New York: McGraw-Hill, 1996

DeStefano, Susan. *Chico Mendes: Fight for the Forest.* Frederick, MD: Twenty-First Century Books, 1992

Devlin, John C., and Grace Naismith. *The World of Roger Tory Peterson*. New York: New York Times Books, 1977.

DiChristina, Mariette. "Mired in Tires." *Popular Science* 245, no. 4 (October 1994): 62–64, 83.

Douglas, Marjory Stoneman. *The Everglades: River of Grass.* 50th anniversary ed. Sarasota, FL: Pineapple Press, 1947.

———, and John Rothchild. *Voice of the River.* Sarasota, FL: Pineapple Press, 1988.

Dubos, Rene J. *The Dreams of Reason: Science and Utopias*. New York: Columbia University Press, 1961.

———. *So Human an Animal: How We Are Shaped by Surroundings and Events*. New York: Scribners, 1968.

———, and Barbara Ward. *Only One Earth: The Care and Management of a Small Planet*. New York: W.W. Norton, 1972.

Duffield, W. A., J. H. Sass, and M.L. Sorey. "Tapping Earth's Natural Heat." U.S. Geological Survey Circular 1125. Washington, DC: USGS, 1994.

Earle, Sylvia. "Cousteau Remembered." *Popular Science* 251, no. 4 (October 1997): 81.

———. *Dive! My Adventures Undersea in the Deep Frontier.* Washington, DC: National Geographic Society, 1999.

———. *Exploring the Deep Frontier.* 1980.

———. *Sea Change: A Message of the Oceans.* Fawcett, 1995.

Earthworks Group. *50 More Things You Can Do to Save the Earth*. Kansas City: Andrews and McMeel, 1991.

Ehrlich, Paul. *Extinction: The Causes and Consequences of the Disappearance of Species.* New York: Random House, 1981.

———. *The Population Bomb.* 1968. Reprint, New York: Amereon, 1976.

———. *The Population Explosion.* New York: Touchstone Books, 1991.

Ehrlich, Paul, and Anne Ehrlich, *Betrayal of Science and Reason: How Anti-Environment Rhetoric Threatens Our Future.* Washington, DC: Island Press, 1998.

Elsneer, J. B. *Hurricanes of the North Atlantic: Climate and Society.* New York: Oxford University Press, 1999.

Elton, Charles Sutherland. *The Ecology of Invasions by Animals and Plants.* London: Methuen, 1958.

Environmental Protection Agency. *Lead in Your Home: A Parent's Reference Guide.* Booklet.

———. *RCRA: Reducing Risk from Waste.* Pamphlet.

———. *Solid Waste and Emergency Response.* Pamphlet. September 1997.

———. *Technical Guidance for Hazard Analysis*. Pamphlet.

Escheverria, John D., et al. *Rivers at Risk: The Concerned Citizen's Guide to Hydropower*. Washington, DC: Island Press, 1990.

"Eugene Odum: An Ecologist's Life." Videotape, 30 min. Athens: University of Georgia Center for Continuing Education.

Fagan, Brian M. *Floods, Famines, and Emperors: El Niño and the Fate of Civilizations*. Basic Books, 1999.

Feldman, Andrew J. *The Sierra Club Green Guide: Everybody's Desk Reference to Environmental Information*. San Francisco: Sierra Club Books, 1996.

Fisher, Arthur. "Attack on Ozone Science." *Popular Science* 251, no. 4 (October 1997): 72–78.

Fisher, Richard V. *Volcanoes: Crucibles of Change.* Princeton: Princeton University Press, 1998.

Fitton, Laura J., John Choe, and Richard Regan. *Environmental Justice: Annotated Bibliography*. Washington, DC: Center for Policy Alternatives, 1993.

Foreman, David. *Confessions of an Eco-Warrior.* Crown, 1993.

———. *Ecodefense: A Field Guide to Monkeywrenching.* Chico, CA: Abbzug Press,1993.

———. "Its Time to Return to Our Wilderness Roots." *Environmental Action* 15, no. 5 (December–January 1984).

Fornasari, Lorenzo, and Renato Massa. *The Temperate Forest.* The Deep Green Planet. Austin, TX: Raintree Steck Vaughn, 1997.

Forster, Bruce. *The Acid Rain Debate: Science and Special Interest in Policy Formation.* Natural Resources and Environmental Policy Series. Ames: Iowa State University Press, 1993.

Fossey, Dian. *Gorillas in the Mist.* Boston: Houghton Mifflin, 1983.

———. "The Imperiled Mountain Gorilla." *National Geographic Magazine* 159 (April 1981): 501–23.

———. "Making Friends with Mountain Gorillas." *National Geographic Magazine* (January 1970): 48–67.

———. "More Years with Mountain Gorillas." *National Geographic Magazine* (October 1971): 574–85

Foster, Lynne. *Adventuring in the California Desert.* San Francisco: Sierra Club Adventure Travel

Foulkes, E.C., ed. *Biological Effects of Heavy Metals: Metal Carcinogenesis*. Boca Raton, FL: CRC Press.

Fox, Stephen. *John Muir and His Legacy: The American Conservation Movement.* Boston: Little Brown and Co., 1981.

Franck, Irene, and David Brownstone. *The Green Encyclopedia*. New York: Prentice-Hall, 1992.

Fraser, Clarence M., et al. *The Merck Veterinary Manual,* 7th ed. Rahway, NJ: Merck and Co., 1991.

Galdikas, Biruté. *Reflections of Eden: My Years with the Orangutans of Borneo.* Collingdale, PA: Diane Publishing Company, 2000.

Garab, G. *Photosynthesis: Mechanisms and Effects.* Norwell, MA: Kluwer Academic, 1999.

Garfunkel, Zvi, ed. *Mantle Flow and Plate Theory*. Benchmark Papers in Geology Series. New York: Van Nostrand Reinhold, 1984.

Gash, J.H.C., et al., eds. *Amazonian Deforestation and Climate.* New York: Wiley and Sons, 1996.

Gibbs, Lois Marie. *Love Canal: The Story Continues.* New Society, 1998.

Gittleman, Ann Louise. *Guess What Came to Dinner: Parasites and Your Health.* Wayne, NJ: Avery, 1993.

Glantz, Michael H. *Currents of Change: El Niño's Impact on Climate and Society*. New York: Cambridge University Press, 1996.

Glassman, Michael. *Pollution of the Environment: Can We Survive?* New York: Globe, 1974.

Glover, J. M. *A Wilderness Original: The Life of Bob Marshall*. Seattle, WA: Mountaineers Books, 1986.

Goodall, Jane, *Through a Window: My Thirty Years with the Chimpanzees of Gombe*. Boston: Houghton Mifflin, 1991.

———, et al. *In the Shadow of Man*. Boston: Houghton Mifflin, 1988.

Gore, Al. *Earth in the Balance: Ecology and the Human Spirit.* Boston: Houghton Mifflin, 1992.

Gradwohl, J., and Greenberg, R. *Saving the Tropical Forests.* Washington, DC: Island Press, 1988

Graedel, Thomas E. *Atmosphere, Climate and Change.* Scientific American Library Paperback, no. 55. New York: W.H. Freeman and Co., 1997.

Graham, Frank, Jr. "The Day of the Condor." *Audubon* 102, no. 1 (January–February 2000): 46–53.

———. "Sounding the Alarm: Rachel Carson." *Audubon* 100, no. 6 (November–December 1998): 83.

Graham, Ian. *Geothermal and Bio-Energy.* Energy Forever. Austin, TX: Raintree/Steck Vaugh, 1999.

Gralla, Preston. *How the Environment Works.* Emeryville, CA: Ziff-Davis Press, 1994. Great West Books, 1988.

Gray, Robert G. et al., eds. *Coal Conversion and the Environment: Chemical, Biochemical, and Ecological Considerations.* Washington, DC: U.S. Department of Energy, 1981.

Hallowell, Christopher. "Rainforest Pharmacist." *Audubon* 101, no. 1 (January–February 1999): 28, 30.

Hardin, Garrett. "The Tragedy of the Commons." *Science* 162 (1968): 1243–48.

Harding, W., *The Days of Henry Thoreau: A Biography,* Princeton, NJ: Princeton University Press, 1992

———. *A Thoreau Handbook.* New York: New York University Press, 1959

Hart, S.I. *The Elephant in the Bedroom: Automobile Dependence and Denial; Impacts on the Economy and Environment*. Pasadena, CA: Hope, 1993.

Hawaii Audubon Society. *Hawaii's Birds.* Honolulu: Hawaii Audubon Society, 1993.

Hay, John, and Christopher Merrill, eds. *The Way to the Salt Marsh: A John Hay Reader*. Hanover, NH: University Press of New England.

Heacox, Kim. *Antarctica: The Last Continent.* Washington, DC: National Geographic Society.

Hecht, S., and, A. Cockburn., *The Fate of the Forest: Developers, Destroyers, and Defenders of the Amazon.* London and New York: Verso, 1989.

Hocking, Colin, Jan Coonrod, and Jacqueline Barber. *Acid Rain Guide.* Berkeley: University of California Berkeley, Lawrence Hall of Science (GEMS) 1990.

Hocking, Colin, Cary Sneider, and Lincoln Bergman, eds. *Global Warming and the Greenhouse Effect.* Berkeley: University of California Berkeley, Lawrence Hall of Science (GEMS), 1992.

Houghton, John Theodore. *Global Warming: The Complete Briefing.* New York: Cambridge University Press, 1997.

——— et al., eds. *Climate Change 1995: The Science of Climate Change.* Cambridge University Press, 1996.

Hoyt, Erich. *Meeting the Whales: The Equinox Guide to Giants of the Deep*. Ontario: Camden House, 1991.

Hunkin, Jorie. *Ecology for All Ages*. Connecticut: Globe Pequot Press, 1994.

Huttermann, Aloys, and Douglas Godbold, eds. *Effects of Acid Rain on Forest Processes*. Wiley Series in Ecological and Applied Microbiology. New York: John Wiley and Sons, 1994.

Iudicello, Suzanne, et al. *Fish, Markets, and Fishermen: The Economies of Overfishing.* Washington, DC: Island Press, 1999.

Jackson, D.C., ed. *Dams*. Brookfield, VT: Ashgate, 1988.

Javna, John. *Fifty Simple Things Kids Can Do to Save the Earth.* Kansas City: Andrews and McMeel, 1990.

Katona, Steven K. *A Field Guide to Whales, Porposies, and Seals from Cape Cod to Newfoundland.* Washington, DC: Smithsonian Institution Press, 1993.

Kaufman, Kenn. "An Educator at Heart: Roger Tory Peterson." *Audubon* 100, no. 6 (November–December 1998): 87.

Kennedy, D. and R. Bates, eds.. *Air Pollution, the Automobile, and Public Health*. National Academy Press. 1989.

Kirk, John T.O. *Light and Photosynthesis in Aquatic Ecosystems.* Cambridge: Cambridge University Press, 1994.

Knapp, Alan K., et al., eds. *Grassland Dynamics: Long-Term Ecological Research in Tallgrass Prairie.* Long-Term Ecological Research Network Series, 1), 1985. Reprint, Oxford University Press, 1998.

Kosova, Weston. "Alaska: The Oil Pressure Rises." *Audubon* 99, no. 6 (November–December, 1997): 66-74.

Krutch, Joseph. *The Desert Year.* Tucson: University of Arizona Press, 1985.

———. *Forgotten Peninsula: A Naturalist in Baja California.* Tucson: University of Arizona Press, 1986.

———. *More Lives than One*. Tucson: University of Arizona Press, 1962.

Langewiesche, William. *Sahara Unveiled: A Journey Across the Desert.* Vintage Departures. Vintage Books, 1997.

Lanier-Graham, Susan D. *The Nature Directory: A Guide to Environmental Organizations.* New York: Walker and Co., 1991.

La Pierre, Yvette. "On the Edge." *National Parks* 71, no. 11–12 (November–December 1997): 44.

Laws, R.M. *Antarctic Seals: Research Methods and Techniques.* New York: Cambridge University Press, 1993.

Lebeaux, R. *Thoreau's Seasons.* Amherst: University of Massachusetts Press, 1984.

Leopold, Aldo. *A Sand County Almanac*. New York: Oxford University Press, 1949.

Let the Mountains Talk, Let the Rivers Run: A Call to Those Who Would Save the Earth. New York: Harper Collins, 1995.

Levin, Ted. "Defending the 'Glades: Marjory Stoneman Douglas." *Audubon* 100, no. 6 (November–December 1998): 84.

Longhurst, Alan R. *Ecological Geography of the Sea.* Academic Press, 1998.

Lowenthall, David. *George Perkins Marsh: Versatile Vermonter.* New York: Columbia University Press. 1980.

Lucas, Mike. *Antarctica.* New York: Artabras, 1999.

Luoma, Jon R. "It's 10:00 P.M. We Know Where Your Turtles Are." *Audubon* 100, no. 5 (September–October 1998): 52–57.

———. "Spilling the Truth." *Audubon* 101, no. 2 (March–April 1999): 52–55.

———. "Vanishing Frogs." *Audubon* 99, no.3 (May–June 1997).

Mabey, Nick. *Arguments in the Greenhouse: The International Economics of Controlling Global Warming.* New York: Routledge, 1997.

MacEachern, Diane. *Save Our Planet: 750 Everyday Ways You Can Help Clean Up the Earth.* New York: Dell, 1995.

Manning, Richard. *Grassland: The History, Biology, Politics and Promise of the American Prairie.* New York: Penguin, 1997.

Margulis, Lynn, and Karlene V. Schwartz. *Five Kingdoms: An Illustrated Guide to the Phyla of Life on Earth,* 2d ed. New York: W.H. Freeman and Co., 1996.

Markels, Alex. "The Next Great Eco-Trips." *Audubon* 100, no. 5 (September–October 1998): 66–69.

Marples, D. R. *Chernobyl and Nuclear Power in the U.S.S.R.* New York: St. Martin's Press, 1986.

Massa, Renato. *Along the Coasts.* The Deep Blue Planet. Austin: Raintree/Steck-Vaughn, 1998.

———. *The Breathing Earth*. The Deep Green Planet. Austin, TX: Raintree/Steck-Vaughn, 1997.

———. *The Coral Reef.* The Deep Blue Planet. Austin, TX: Raintree/Steck-Vaughn, 1998.

———. *Ocean Environments.* The Deep Blue Planet. Austin, TX: Raintree/Steck-Vaughn, 1998.

———. *The Tropical Forest.* The Deep Green Planet. Austin, TX: Raintree/Steck-Vaughn, 1997.

Massa, Renato and Monica Carabella. *The Coniferous Forest.* The Deep Green Planet. Austin, TX: Raintree/Steck-Vaughn, 1997.

Massa, Renato, Monica Carabella, and Lorenzo Fornasari. *From the Water to the Land.* The Deep Green Planet. Austin, TX: Raintree/Steck-Vaughn, 1997.

Marzulla, Nancie G., and Roger J. Marzulla. *Property Rights, Understanding Government Takings and Environmental Regulation*. Rockville, MD: Government Institutes, 1997.

Mazur, Allan. *A Hazardous Inquiry: The Rashomon Effect at Love Canal.* Boston: Harvard University Press, 1998.

McBride, L.R. *About Hawaii's Volcanoes*. Hawaii: Peteroglyph Press, 1992.

McMahon, James. *Deserts.* Audubon Society Nature Guides. New York: Knopf, 1985.

McMurray, Emily M., ed. *Notable Twentieth Century Scientists.* 4 vols. Detroit: Gale Research, Inc., 1995.

Mendes, Chico. *Fight for the Forest*. New York: Monthly Review Press, 1992.

Middleton, Susan, David Liittschwager, and the California Academy of Sciences. "The Endangered 100." *Life.* (September 1994): 50–63.

Moeller, Dade W. *Environmental Health.* Boston: Harvard University Press, 1997.

Momatiuk, Yva, and John Eastcott. "Keepers of the Seals." *Audubon* 100, no. 2 (March–April 1998): 46–53.

Monteath, Colin. *Antarctica: Beyond the Southern Ocean*. Hauppauge, NY: Barrons Educational Series. 1997.

Montgomery, John H. *Agrochemicals Desk Reference: Environmental Data.* New York: Lewis, 1993.

Montgomery, Sy. *Walking with the Great Apes.* Boston: Houghton Mifflin, 1991.

Moores, Eldridge. M., ed. *Shaping the Earth: Tectonics of Continents and Oceans*. New York: W.H. Freeman, 1990.

Mortensen, Lynn, ed. *Global Change Education Resource Guide.* Greenbelt, MD: NASA Goddard Space Flight Center.

Moss, N. *The Politics of Uranium.* New York: Universe Books, 1982.

Mould, R.F. *Chernobyl: The Real Story.* New York: Elsevier Science, 1988.

Mowat, Farley. *Woman in the Mists: The Story of Dian Fossey and the Mountain Gorillas of Africa.* New York: Warner Books, 1987.

Muir, John. *John Muir: The Eight Wilderness Discovery Books.* Seattle, WA: Mountaineers Books, 1992.

Murck, Barbara W., Brian J. Skinner, and Stephen C. Porter. *Environmental Geology*. New York: John Wiley and Sons, 1996.

Myers, Norman, ed., *Rainforests.* Emmaus, PA: Rodale Press, 1993.

Myerson, Joel, ed. *The Cambridge Companion to Henry David Thoreau.* New York: Cambridge University Press, 1995.

Naess, Arne. "The Shallow and the Deep, Long-Range Ecology Movements: A Summary." *Inquiry* (Oslo) 16 (1973): 95–100.

National Mine Reclamation Center. *Acid Mine Drainage: Control and Treatment.* Evansdale, WV: National Mine Reclamation Center.

National Oceanic and Atmospheric Administration, Office of Public Affairs. *25 Things You Can Do to Save Coral Reefs.* Washington, DC.

National Parks Conservation Association. "Citizens Protecting America's Parks." Video.

National Research Council, Board on Agriculture. "Regulating Pesticides in Food: The Delaney Paradox. Committee on Scientific and Regulatory Issues Underlying Pesticide Use Patterns and Agricultural Innovation." Washington, DC: National Academy Press, 1987.

Necker, M. *Gold, Siver, and Uranium from Seas and Oceans: The Emerging Technology*. Los Angeles: Ardor, 1991.

Norse, Elliot A. *Ancient Forests of the Pacific Northwest*. New York: Island Press, 1990.

O'Brien, Karen L. *Sacrificing the Forest: Environmental and Social Struggles in Chiapas.* Boulder, CO: Westview Press, 1998.

Olson, Richard K., ed. *Integrating Sustainable Agriculture, Ecology, and Environmental Policy.* Binghamton, NY: Food Products Press, 1992.

Pacific Northwest Laboratories. *Wind Energy Resource Atlas.* Washington, DC: American Wind Energy Association, 1991.

Packham, John R. *Ecology of Dunes, Salt Marsh and Shingle.* New York: Chapman and Hall, 1997.

Palmer, Tim. *The Wild and Scenic Rivers of America.* New York: Island Press, 1993.

Paul, Sherman. *The Shores of America: Thoreau's Inward Exploration*. Urbana: University of Illinois Press, 1959.

Payne, Roger. *Among Whales*. New York: Delta, 1996

Peacock, Doug. "The Newest Place on Earth." *Audubon* 101, no. 2 (March–April 1999): 26–31.

———. "The Yellowstone Massacre." *Audubon* 99, no. 3 (May–June 1997): 40–49.

Perney, Linda, and Pamela Emanoil. "Where the Wild Things Are." *Audubon* 100: 5 (September-October 1998): 82–89.

Peterson, Roger Tory. *A Field Guide to the Birds: Giving Field Marks of All Species Found in Eastern North America*. Boston: Houghton Mifflin, 1996.

Peterson, Roger Tory, and Rudy Hoglund, eds., *Roger Tory Peterson: Art and Photography of the World's Foremost Birder*. New York: Rizzoli, 1994.

Peterson, Roger Tory, and Virginia Marie Peterson. *A Field Guide to the Birds: A Completely New Guide to All the Birds of Eastern and Central North America.* Peterson Field Guides. Boston: Houghton Mifflin, 1998.

Pielou, E.C. *A Naturalist's Guide to the Arctic.* Chicago: University of Chicago Press, 1994.

Pierce, F.J., ed., *Advances in Soil and Water Conservation.* Sleeping Bear Press, 1998.

Pinchot, Gifford. *Breaking New Ground.* New York: Island Press, 1998.

Pipes, Rose. *Grasslands.* World Habitats. Austin, TX: Raintree/Steck Vaughn, 1998.

———. *Tundra and Cold Deserts* World Habitats. Austin, TX: Raintree/Steck Vaughn, Springer Verlag, 1998.

Pope, Andrew M. *Environmental Medicine: Integrating a Missing Element into Medical Education.* Washington, DC: National Academy Press, 1995.

Porter, Henry F., and Eric Wybenga. *Forecast: Disaster; The Future of El Nino.* New York: Dell, 1998.

Powell, John Wesley. *Beyond the Hundredth Meridian and the Second Opening of the West.* New York: Penguin, 1992.

———. *The Exploration of the Colorado River and Its Canyons.* New York: Penguin, 1997.

Putten, Mark Van. "Of Wolves, Wetlands, and Wisdom." *National Wildlife* 36, no. 4 (June–July 1998): 7.

Pyne, Stephen J. *Vestal Fire: An Environmental History, Told through Fire, of Europe and Europe's Encounter with the World.* Washington: University of Washington Press, 1998.

Quammen, David. "Backwater Boondoggle." *Audubon* 100, no. 1 (January–February 1998): 52–63, 100–104.

Rashke, Richard. *The Killing of Karen Silkwood: The Story Behind the Kerr-McGee Plutonium Case.* Boston: Houghton Mifflin, 1981.

Ray, Dixy Lee, with Lou Guzzo. *Environmental Overkill: Whatever Happened to Common Sense?* Washington, DC: Regnery Gateway, 1993.

———. *Trashing the Planet: How Science Can Help Us Deal with Acid Rain, Depletion of the Ozone, and Nuclear Waste (Among Other Things).* New York: Harper Perennial, 1992.

Reader's Digest Association. *Whales, Dolphins and Porpoises.* Pleasantville, NY: Reader's Digest Association, 1997.

Rees, Robin, ed. *The Way Nature Works.* New York: MacMillan, 1992.

Revkin, Andrew. *The Burning Season: The Murder of Chico Mendes and the Fight for the Amazon Rain Forest.* New York: Plume, 1994.

Richardson, Robert D. *Henry Thoreau: A Life of the Mind.* Berkeley: University of California Press, 1986.

Riedman, Mariane. *The Pinnipeds: Seals, Sea Lions, and Walruses.* Berkeley: University of California Press, 1991.

Ringholz, R. *Uranium Frenzy: Boom and Bust on the Colorado Plateau.* Albuquerque: University of New Mexico Press, 1991

Robinson, Gordon. *The Forest and the Trees: A Guide to Excellent Forestry.* New York: Island Press, 1988.

Rogovin, Mitchell, and George T. Frampton. *Three Mile Island: A Report to the Commissioners and to the Public.* Vols. 1–2. Washington, DC: Nuclear Regulatory Commission, 1980

Roosevelt, Theodore. *The Autobiography of Theodore Roosevelt.* London: Octagon, 1975.

Rosen, J.D. "Much Ado about Alar." *Issues in Science and Techology* 7, no. 1 (1990): 85–90.

Safina, Carl. *Song for the Blue Ocean: Encounters along the World's Coasts and beneath the Seas.* New York: Henry Holt and Co., 1998.

Saign, Geoffrey C. *Green Essentials: What You Need to Know about the Environment.* San Francisco: Mercury House, 1994.

Sale, Peter F., ed. *The Ecology of Fishes on Coral Reefs.* San Diego: Academic Press, 1994.

Salomons, W., et al., eds. *Heavy Metals: Problems and Solutions.* New York: Springer-Verlag, 1995.

Saw, Swee-Hock. *Population Control for Zero Growth in Singapore.* Cambridge: Oxford University Press, 1980.

Sayre, April Pulley. *Temperate Deciduous Forest.* Exploring Earth's Biomes.New York: Twenty First Century Books, 1995.

———. *Tundra* Exploring Earth's Biomes. New York: Twenty First Century Books, 1995.

Schmid-Hempel, Paul. *Parasites in Social Insects.* Princeton: Princeton University Press, 1998.

Schneider, Paul. "Clear Progress: 25 Years of the Clean Water Act." *Audubon* 99, no. 5 (September–October 1997): 36–47, 106–07.

Schnoor, J.L. ed. *Fate of Pesticides and Chemicals in the Environment.* New York: John Wiley and Sons, 1991.

Schumacher, E.F. *Small Is Beautiful.* New York: Harper, 1973.

Seidman, David. "Preaching from the Bully Pulpit: Theodore Roosevelt." *Audubon* 100, no. 6 (November–December 1998): 89.

———. "Swimming with Trouble." *Audubon* 99, no. 5 (September–October 1997): 78–82.

Seymour, John, and Herbert Girardet. *Blueprint for a Green Planet.* New York: Prentice-Hall Press, 1987.

Sieh, Kerry E., and Simon LeVay. *The Earth in Turmoil: Earthquakes, Volcanoes, and Their Impact on Humankind.* New York: W.H. Freeman and Co., 1998.

Sitarz, Daniel. *AGENDA 21: The Earth Summit Strategy to Save Our Planet.* Boulder, CO: Earth Press, 1993.

Smith, William K., and Thomas M. Hinckley, eds. *Ecophysiology of Coniferous Forests.* Physiological Ecology Series. San Diego: Academic Press, 1994.

Solo-Gabriele, H., and S. Neumeister. "U.S. Outbreaks of Cryptosporidiosis." *Journal of the American Water Works Association* (September 1996).

Somerville, Richard C.J. *The Forgiving Air: Understanding Environmental Change.* Berkeley: University of California Press, 1998.

Soule, Judith D., and Jon K. Piper. *Farming in Nature's Image: An Ecological Approach to Agriculture.* Washington, DC: Island Press, 1992.

Steen, H. K., and Tucker, R. P., eds. *Changing Tropical Forests: Historical Perspectives on Today's Challenges in Central and South America,* Durham, NC: Forest History Society, 1992.

Steene, Roger. *Coral Seas.* Buffalo, NY: Firefly Books, 1998.

Stephens, Mark. *Three Mile Island.* New York: Random House, 1980.

Stevenson, Harold L., and Bruce Wyman, eds. *The Facts on File Dictionary of Environmental Science.* New York: Facts on File, 1991.

Stewart, John A. *Drifting Continents and Colliding Paradigms.* Bloomington, IN: Indiana University Press, 1990.

Stoddart, D.R. ed. *Salt Marshes and Coastal Wetlands.* Institute of British Geographers Special Publications. Oxford: Blackwell, 1999.

Strong, Steven J. *The Solar Electric House: Energy for the Environmentally-Responsive, Energy-Independent Home.* White River, VT: Chelsea Green Publishing Company, 1994.

Stuart, Chris, and Tilde Stuart. *Africa's Vanishing Wildlife.* Washington, DC: Smithsonian Institution Press, 1996

Talen, Maria. *Ocean Pollution.* Lucent Overview Series. San Diego: Greenhaven Press, 1991.

Teal, John. *Life and Death of the Salt Marsh.* Ballantine Books, 1991.

Terborgh, John. *Diversity and the Tropical Rain Forest.* Scientific American Library, no. 38. New York: W.H. Freeman and Co., 1992.

Thomas, Clayton L., ed. *Taber's Cyclopedic Medical Dictionary,* 15th ed. Philadelphia, PA: F.A. Davis, 1985.

Thomas, Mary Powel. "Climate Change Hits Home." *Audubon* 99, no. 5 (September–October 1997).

Thoreau, Henry David. *Wild Fruits: Thoreau's Rediscovered Lost Manuscript*. New York: W.W. Norton, 1999.

Thorpe, James.*Thoreau's Walden.* San Marino, CA: Huntington Library, 1977

Tisdale, Sallie. "Wild in the City." *Audubon* 99, no. 4 (July–August 1997): 38–39.

Turner, Frederick. *Rediscovering America: John Muir in His Time and Ours*. New York: Viking, 1985.

Tuttle, Merlin D. *America's Neighborhood Bats*. Austin, TX: University of Texas Press, 1988.

Tyson, Peter. *Acid Rain.* Broomal, PA: Chelsea House, 1992.

U.S. Department of Energy, Office of Environmental Restoration. *Questions and Answers about the U.S. Department of Energy's Environmental Restoration Activities.* Washington, DC, November 1994.

U.S. Environmental Protection Agency. Office of Water, Office of Pesticides and Toxic Substances. *National Survey of Pesticides in Drinking Water Wells.*), 1990.

U.S. Geological Survey. "Changing Perceptions of World Oil and Gas Resources as Shown by Recent USGS Petroleum Assessments." Fact Sheet: greenwood.cr.usgs.gov/pub/fact-sheets/fs-0145-97/fs-0145-97.html

Van Dyke, Jon M., et al. *Freedom for the Seas in the Twentyfirst Century: Ocean Governance and Environmental Harmony.* Washington, DC: Island Press, 1993.

Van Dyne, G. M. *Grasslands, Systems Analysis and Man.* New York: Cambridge University Press, 1980.

Vernet, J.P. ed. *Heavy Metals in the Environment*. New York: Elsevier Science Publishers, 1991.

Voglino, Alex and Renato Massa. *Oceans and Seas.* The Deep Blue Planet. Austin, TX: Raintree/Steck-Vaughn, 1998.

Vos, J.G. *Allergic Hypersensitivities Induced by Chemicals: Recommendations for Prevention.* Cleveland: CRC Press, 1996.

Voss, Gilbert L., and Marjory Stoneman Douglas. *Coral Reefs of Florida.* Sarasota, FL: Pineapple Press, 1988.

Wadsworth, Ginger. *John Muir: Wilderness Protector.* Minneapolis: Lerner, 1992.

Wagner, Travis P. *The Complete Guide to Hazardous Waste Regulations: RCRA, TSCA, HMTA, OSHA, and Superfund.* New York: John Wiley and Sons, 1999.

Wall, Derek. *Earth First! and the Anti-Roads Movement: Radical Environmentalism and Comparative Social Movements.* New York: Routledge, 1999.

Ward, Nathalie. *Stellwagen Bank: A Guide to the Whales, Sea Birds, and Marine Life of the Stellwagen Bank National Marine Sanctuary.* Camden, ME: Down East Books, 1995.

Wargo, John. *Our Children's Toxic Legacy: How Science and Law Fail to Protect Us from Pesticides.* New Haven: Yale University Press, 1998.

Waterlow, Julia. *Grasslands.* (Habitats). Austin, TX: Raintree/Steck-Vaughn, 1996.

Watkins, T.H. "National Parks, National Paradox." *Audubon* 99, no. 4 (July–August 1997): 40–45.

———. "The Spiritual Father: John Muir." *Audubon* 100: 6 (November–December 1998): 86.

———. "Voice of the Land Ethic: Aldo Leopold. *Audubon* 100, no. 6 (November–December 1998): 85.

Wegener, Alfred. *The Origin of Continents and Oceans*. Translated by John Biram. New York: Dover Publications, 1966.

Wenz, Peter S. *Environmental Justice*. Albany, NY: State University of New York Press, 1988.

Wheeler, Sara. *Terra Incognita: Travels in Antarctica.* New York: Modern Library, 1999.

Williams, Ted. "Back from the Brink." *Audubon* 100, no. 6 (November–December 1998): 70–76.

———. "Fatal Attraction," *Audubon* 99, no. 5 (September–October 1997): 24–31.

———. "The Terrible Turtle Trade." *Audubon* 101, no. 2 (March–April 1999): 44–51.

———. "Who Can Save the Wetland?" *Audubon* 101, no. 5 (September–October 1999): 60–69.

Wilson, Edmund. *The American Earthquake.* Cambridge: Da Capo Press, 1996.

Wilson, Edward. *Biodiversity*. Boston: National Academy Press, 1988.

———. *Biophilia*. Boston: Harvard University Press. 1984.

———. *The Diversity of Life*. Questions of Science Series. Boston: Harvard University Press. 1992.

———. *The Diversity of Life.* New York: Belknap Press, 1992.

———. *Naturalist*. Washington, DC: Island Press, 1994.

Wilson, Tuzo. *Continents Adrift*. New York: Scientific American, 1972.

Windley, Brian. F. *The Evoloving Continents,* 3rd ed. New York: John Wiley and Sons, 1995.

Winter, C. K. "Pesticide Tolerances and Their Relevance as Safety Standards." *Regulatory Toxicology and Pharmacology* 15 (1992): 137–50.

Wolfson, Richard. *Nuclear Choices: A Citizen's Guide to Nuclear Technology.* Cambridge, MA: MIT Press, 1993.

Worldwatch Institute. *State of the World 1997, Report on Progress Toward a Sustainable Society.* New York: W.W. Norton and Co., 1997.

Wurtz, M. and N. Repetto. *Whales and Dolphins: A Guide to the Biology and Behavior of Cetaceans*. Thunder Bay, Canada: Thunder Bay Press. 1999.

Yahner, Richard H. *Eastern Deciduous Forest: Ecology and Wildlife Conservation.* Wildlife Habitats, vol 4. University of Minnesota Press, 1996.

Zwinger, Ann H., and Beatrice E. Willard. *Land above the Trees: A Guide to American Alpine Tundra.* Johnson Books, 1996.

Web Sites Sorted by Subject

Acid Mine Drainage

Agency for Toxic Substances and Disease Registry: www.atsdr.cdc.gov/cxcx3.html
National Reclamation Center's West Virginia University, Evansdale office: www.nrcce.wvu.edu/news/nsamd.html

Acid Rain

Acid Rain Program from the Environmental Protection Agency: www.epa.gov/docs/acidrain/effects/enveffct.html
USGS Water Science/Acid Rain: www.ga.usgs.gov/edu/acidrain.html

Agriculture

United States Department of Agriculture Web site: www.usda.gov

Alternative Fuels

Crest's Guide to the Internet's Alternative Energy Resources: solstice.crest.org/online/aeguide/aehome.html
Department of Energy Alternative Fuels Data Center: www.afdc.nrel.gov
www.afdc.doe.gov/
or www.fleets.doe.gov
Department of Energy Web site: www.doe.gov

Amphibians

FrogWeb: www.frogweb.gov/

Antarctica

Antarctica Treaty Visit Web site: www.sedac.ciesin.org/pidb/register/reg-024.rrr.html
Greenpeace International Antarctic Homepage: www.greenpeace.org/~comms/98/antarctic
International Centre for Antarctic Information and Research Homepage (includes text of Antarctic Treaty): www.icair.iac.org.nz
Virtual Antarctica: www.terraquest.com/va

Arctic

Arctic Circle (University of Connecticut): arcticcircle.uconn.edu/arcticcircle
Arctic Council Home Page: www.nrc.ca/arctic/index.html
Arctic Monitoring and Assessment Programme (Norway): www.gsf.de/UNEP/amap1.html
Arctic National Wildlife Refuge: energy.usgs.gov/factsheets/ANWR/ANWR.html
Institute of Arctic and Alpine Research: instaar.colorado.edu
Institute of the North (Alaska Pacific University): www.institutenorth.org/
Inuit Circumpolar Conference: www.inac.gc.ca/decade/circum.html
Nunavut: www.acs.ucalgary.ca/~dgwhite/nu.html
Smithsonian Institution Arctic Studies Center: www.mnh.si.edu/arctic
World Conservation Monitoring Centre Arctic Programme: www.wcmc.org.uk/arctic
NOAA Fisheries Contact Web site: www.nmfs.gov/
U.S. Fish and Wildlife Service: www.fws.gov

Automobiles

Cars and Their Enviromental Impact: www.environment.volvocars.com/ch1-1.htm
National Center for Vehicle Emissions Control and Safety (NCVECS): www.colostate.edu/Depts/NCVECS/ncvecs1.html
U.S. Enviromental Protection Agency, Office of Mobile Sources: www.epa.gov/oms
U.S. Environmental Protection Agency Fact Sheet (#EPA 400-F-92-004, August 1994): "Air Toxics From Motor Vehicles:" www.epa.gov/oms/02-toxic.htm

Biological Weapons

Chemical and Biological Defense Information Analysis Center: www.cbiac.apgea.army.mil
Federation of American Scientist Biological Weapons Control: www.fas.org/bwc

Biomes

Committee for the National Institute for the Environment Web site: www.cnie.org/nle/biodv-6.html

Brownfields

EPA Office of Solid Waste and Emergency Response—Brownfields: www.epa.gov/brownfields

Cheetah

Cheetah Conservation Fund: www.cheetah.org
World Wildlife Fund: www.worldwildlife.org/

Chemical Weapons

Chemical Stockpile Disposal Project (CSDP): www-pmcd.apgea.army.mil/graphical/CSDP/index.html
Tooele Chemical Agent Disposal Site Facility: www.deq.state.ut.us/eqshw/cds/tocdfhp1.htm

Clean Water Act

Sierra Club "Happy 25th Birthday, Clean Water Act": sierraclub.org/wetlands/cwabday.html

Climate Change and Global Warming

Bioremediation Consortium: www.rtdf.org/public/biorem
EPA Global Warming Site: www.epa.gov/globalwarming
Greenpeace International, Climate: www.greenpeace.org/~climate
United States Geological Survey, Climate Change and History: geology.usgs.gov/index.shtml
United Nations Intergovernmental Panel on Climate Change: www.ipcc.ch

Coal

Coal Age Magazine: coalage.com
Department of Energy Office of Fossil Energy: www.doe.gov
U.S. Geological Survey National Coal Resources Data System: energy.er.usgs.gov/coalqual.htm

Composting

Cornell Composting: www.cfe.cornell.edu/compost/Composting_Homepage.html
EPA Office of Solid Waste and Emergency Response - Composting: http:www.epa.gov/epaoswer/non-hw/compost/index.htm

Coral Reefs

Coral Reef Alliance: www.coral.org
Coral Reef Network Directory, Greenpeace, 1995: www.greenpeace.org

El Niño

El Nino/El Nina NOAA theme page: www.pmel.noaa.gov/toga-tao/el-nino/nino-home-low.html
National Center for Atmospheric Research: www.ncar.ucar.edu/
National Hurricane Center/Tropical Prediction Center: www.nhc.noaa.gov/
National Oceanographic and Atmospheric Administration: www.noaa.gov/
NOAA La Nina Homepage: www.elnino.noaa.gov/lanina.html
Scripps Institute of Oceanography: sio.ucsd.edu/supp_groups/siocomm/elnino/elnino.html

Electric Vehicles

The Electric Vehicle Association of the Americas: www.evaa.org
Electric Vehicle Technology: www.avere.org/

Elephants

African Wildlife Foundation: www.awf.org
World Wildlife Fund: www.wwf.org

Ethanol

Department of Energy, Energy Efficiency and Renewable Energy Clearinghouse: www.doe.gov

Everglades

National Park Service, Everglades National Park: www.nps.gov/ever

Fishing, Commercial

National Oceanographic and Atmospheric Administration Fisheries: www.nmfs.gov/
United Nations Food and Agriculture Organization Fisheries: www.fao.org/waicent/faoinfo/fishery/fishery.htm

Forests

American Forests: www.amfor.org
Greenpeace International—Forests: www.greenpeace.org/~forests
Society of American Foresters: www.safnet.org
U.S. Forest Service: www.fs.fed.us
U.S. Forest Service Research: www.fs.fed.us/links/research.shtml
World Conservation Monitoring Centre: www.wcmc.org.uk
World Resources Institute Forest Frontiers Initiative: www.wri.org/ffi
World Wildlife Fund (Worldwide Fund for Nature) Forests For Life Campaign: www.panda.org/forests4life

Geology

U.S. Geological Survey: www.usgs.gov/mapping.usgs.gov/www/products/mappubs.html

Geothermal Energy

Energy & Geoscience Institute, University of Utah: www.egi.utah.edu
Geothermal Energy Information: geothermal.marin.org
Geothermal Database USA and Worldwide: www.geothermal.org
International Geothermal: www.demon.co.uk/geosci/igahome.html
Solstice: Internet information service of the Center for Renewable Energy and Sustainable Technology (CREST): solstice.crest.org/

Glaciers, Shrinking

GLOBEC Educational Web site: cbl.umces.edu/fogarty/usglobec/misc/education.html
United States Geological Survey: Climate Change and History: geology.usgs.gov/index.shtml

Grasslands and Prairies

Postcards from the Prairie: www.nrwrc.usgs.gov/postcards/postcards.htm

University of California, Berkeley, World Biomes: Grasslands: www.ucmp.berkeley.edu/glossary/gloss5/biome/grasslan.html
Worldwide Fund for Nature: Grasslands & Its Animals: www.panda.org/kids/wildlife/idxgrsmn.htm

Groundwater

EPA Website: www.epa.gov/swerosps/ej/
Groundwater Atlas of the United States: wwwcapp.er.usgs.gov/publicdocs/gwa/

Hazardous Substances

United States Environmental Protection Agency [Superfund] Program: www.epa.gov/oerrpage/superfund/programs/er/hazsubs/index.htm
United States Occupational Safety and Health Administration (OSHA): www.osha-slc.gov/sltc/hazardoustoxicsubstances/index.html
Environmental Defense Fund: www.scorecard.org

Hazardous Waste Treatment

Federal Remedial Technologies Roundtable: www.frtr.gov
Hazardous Waste Clean-Up Information ("CLU-IN"): www.clu-in.org

Heavy Metals

U.S. Environmental Protection Agency Office of Pollution Prevention and Toxics: www.epa.gov/opptintr

Hurricanes

National Hurricane Center: www.nhc.noaa.gov

Hydroelectric Power

The United States Geological Survey: wwwga.usgs.gov/edu/hybiggest.html
U.S. Bureau of Reclamation Hydropower Information: www.usbr.gov/power/edu/edu.htm

Hydrogen

Hydrogen InfoNet: /www.eren.doe.gov/hydrogen/infonet.html
National Renewable Energy Laboratory: www.nrel.gov/lab/pao/hydrogen.html

Landsat and Satellite Images

Earthshots: Satellite Images of Environmental Change: www.usgs.gov/Earthshots/
Landsat Gateway Web site: landsat.gsfc.nasa.gov/main.htm

Lead

EPA Web site: www.epa.gov/lead/

Leopards

U.S. Fish and Wildlife Service, The Species List of Endangered and Threathened Wildlife: www.fws.gov/r9endspp/lsppinfo.html

Litter

Keep America Beautiful: Web site: www.kab.org

Mammals

U.S. Fish and Wildlife Service (See Vertebrate Animals): www.fws.gov/r9endspp/lsppinfo.html

Manatees

Save the Manatee Web Site: www.savethemanatee.org
Sea World: Manatees: www.seaworld.org/manatee/sciclassman.html

Marshes

Environmental Protection Agency's Office of Wetlands, Oceans, Watersheds: www.epa.gov/owow/wetlands/wetland2.html
North American Waterfowl and Wetlands Office: www.wetlands.fws.gov.
North American Wetlands Conservation Act: www.northamerican.fws.gov/nawcahp.html
North American Wetlands Conservation Council: www.fws.gov/r9nawwo/nawcc.html

Mendes, Chico

Web site: www.edf.org/chico

National Disasters

Center for Integration of Natural Disaster Information: cindi.usgs.gov/events/
Earthquakes web sites: quake.wr.usgs.gov/geology.usgs.gov/quake.html
National Hurricane Center: www.nhc.noaa.gov
U.S. Geological Survey Website: geology.usgs.gov/whatsnew.html

National Marine Fisheries

History of National Marine Fisheries Service: www.wh.whoi.edu/125th/history/century.html
National Marine Fisheries: kingfish.ssp.nmfs.gov
NOAA Fisheries Contact Web site: www.nmfs.gov/

Natural Gas

American Gas Association: www.aga.org
Oil & Gas Journal Online: www.ogjonline.com
U.S. Department of Energy, Energy Information Administration: www.eia.doe.gov
U.S. Department of Energy, Office of Fossil Energy: www.fe.doe.gov
U.S. Geological Survey Energy Resources Program: energy.usgs.gov/index.html

Noise Pollution

Noise Pollution Clearinghouse Web site: www.nonoise.org

Nonpoint Source Pollution

Nonpoint Source Pollution Control Program: www.epa.gov/OWOW/NPS/whatudo.html
www.epa.gov/OWOW/NPS/

Nuclear Energy and Nuclear Reactors

American Nuclear Society: www.ans.org
Nuclear Energy Institute: www.nei.org
Nuclear Information & Resource Service: www.nirs.org

U.S. Department of Energy, Office of Nuclear Energy, Science & Technology: www.ne.doe.gov
U.S. Nuclear Regulatory Commission: www.nrc.gov

Nuclear Waste

Hazard Ranking System: www.epa.gov/superfund/programs/npl_hrs/hrsint.htm
National Research Council, Board on Radioactive Waste Management: www4.nas.edu/brwm/brwm-res.nsf
Superfund Visit websites: www.pin.org/superguide.htm and www.epa.gov/superfund
U.S. Department of Energy, Office of Civilian Radioactive Waste Management: www.rw.doe.gov
U.S. Environmental Protection Agency, Mixed-Waste Homepage: www.epa.gov/radiation/mixed-waste
U.S. Nuclear Regulatory Commission, Radioactive Waste Page: www.nrc.gov/NRC/radwaste

Nuclear Waste Policy Act

American Nuclear Society: www.ans.org
Nuclear Energy Institute: www.nei.org

Ocean Thermal Energy Conversion (OTEC)

Natural Energy Laboratory of Hawaii: bigisland.com/nelha/index.html
National Renewable Energy Laboratory: http:llnrelinfo.nrel.gov

Oceans

National Oceanographic and Atmospheric Administration: www.noaa.gov/
Safe Ocean Navigation Page: anchor.ncd.noaa.gov/psn/psn.htm

Old Growth Forests

Greenpeace International—Forests: www.greenpeace.org/~forests
World Resources Institute Forest Frontiers Initiative: www.wri.org/ffi

Olmstead, Frederick Law

Website: fredericklawolmsted.com

Overfishing

National Aeronautics and Space Administration, Ocean Planet: seawifs.gsfc.nasa.gov/OCEAN_PLANET/HTML/peril_overfishing.html
National Marine Fisheries Service: www.nmfs.gov
NOAA Web page: www.noaa.gov
The UN System: www.unsystem.org
United Nations Food and Agricultural Organization: www.fao.org
United Nations Food and Agriculture Organization Fisheries: www.fao.org/WAICENT/FAOINFO/FISHERY/HTM

Ozone-Related Issues

Environmental Protection Agency Web site on the Science of Ozone Depletion: www.epa.gov/ozone/science/
NOAA, Commonly Asked Questions about Ozone: www.publicaffairs.noaa.gov/grounders/ozo1.html
NOAA, Network for the Detection of Stratospheric Change: climon.wwb.noaa.gov

Parrots

The Online Book of Parrots: www.ub.tu-clausthal.dep/p_welcome.html
World Parrot Trust: www.worldparrottrust.org
World Wildlife Fund: www.panda.org

Pesticides

Pesticides in the Atmosphere: p510dcascr.wr.usgs.gov/pnsp/atmos/atmos_4.html
Toxics and Pesticides: www.epa.gov/oppfead1/work_saf/

Peterson, Roger Tory

Roger Tory Peterson Institute of Natural History: www.rtpi.org/info/rtp.htm

Petroleum

American Petroleum Institute: www.api.org
Oil & Gas Journal Online: www.ogjonline.com
Petroleum Information: www.petroleuminformation.com
U.S. Department of Energy, Energy Information Administration: www.eia.doe.gov
U.S. Department of Energy, Office of Fossil Energy: www.fe.doe.gov
U.S. Geological Survey Energy Resources Program: energy.usgs.gov/index.html
U.S. Geological Survey Fact Sheet FS-145-97, "Changing Perceptions of World Oil and Gas Resources as Shown by Recent USGS Petroleum Assessments.": greenwood.cr.usgs.gov/pub/fact-sheets/fs-0145-97/fs-0145-97.html

Plutonium

U.S. Nuclear Regulatory Commission, Radioactive Waste Page: www.nrc.gov/NRC/radwaste

Radiation

United States Nuclear Regulatory Commission: www.nrc.gov/

Radioactive Wastes

International Atomic Energy Agency; "Managing Radioactive Waste" Fact Sheet: www.iaea.org/worldatom/inforesource/factsheets/manradwa.html
National Research Council, Board on Radioactive Waste Management: www4.nas.edu/brwm/brwm-res.nsf
U.S. Department of Energy, Office of Civilian Radioactive Waste Management: www.rw.doe.gov
U.S. Environmental Protection Agency, Mixed-Waste Homepage: www.epa.gov/radiation/mixed-waste
U.S. Nuclear Regulatory Commission, Radioactive Waste Page: www.nrc.gov/NRC/radwaste

Radon

Radon in Earth, Air, and Water: sedwww.cr.usgs.gov:8080/radon/radonhome.html

Rainforest

Greenpeace International—Forests: www.greenpeace.org/~forests
Rainforest Action Network (RAN): www.ran.org/
Rainforest Alliance (RA): www.rainforest.alliance.org

U.S. Forest Service—www.fs.fed.us
World Wildlife Fund (Worldwide Fund for Nature) Forests For Life Campaign: www.panda.org/forests4life

Salmon

National Marine Fisheries Service Salmon Web site: www.nwr.noaa.gov/1salmon/salmesa/index.htm
NOAA Fisheries: www.nmfs.gov/

Salt Marshes

National Wetlands Research Center: www.nwrc.usgs.gov/educ_out.html
USGS Coastal and Marine Geology: marine.usgs.gov/

Sanitary Landfills

Landfills, An Issue Confronting Our Sustainable Use of the Land: www.lalc.k12.ca.us/uclasp/ISSUES/landfills/landfills.htm

Solar Energy

Solar Energy Industries Association: www.seia.org/main.htm
Solar Energy Homepage: www.soton.ac.uk/~solar/
U.S. Department of Energy, Photovoltaic Program: www.eren.doe.gov/pv/text_frameset.html

Spent Fuel

DOE Office of Civilian Radioactive Waste Management: www.rw.doe.gov/pages/resource/facts/transfct.htm
National Transportation Program: www.ntp.doe.gov

Superfund

EPA Superfund Web site: www.epa.gov/superfund,
Recycled Superfund sites: www.epa.gov/superfund/programs/recycle/index.htm
Superfund hotline: www.epa.gov/epaoswer/hotline
U.S. EPA Superfund Program Home Page: Web site: www.epa.gov/superfund/index.htm

Tennessee Valley Authority

Web site: www.tva.gov

Thoreau, Henry

Web site: www.rtpi.org/

Threatened and Endangered Species

U.S. fish and Wildlife Service, The Species List of Endangered and Threatened Wildlife: www.fws.gov/r9endspp/lsppinfo.html

Toxic Chemicals

Environmental Defense Fund: www.scorecard.org
United States Department of Health and Human Services Agency for Toxic Substances and Disease Registry ("ASTDR"): atsdr1.atsdr.cdc.gov
United States Environmental Protection Agency Integrated Risk Information System ("IRIS"): www.epa.gov/iris/subst/index.html
United States Occupation Health and Safety Administration: www.osha-slc.gov/sltc/hazardoustoxicsubstances/index.html

Toxic Release Inventory

Environmental Defense Fund: www.scorecard.org
EPA TRI Public Data Release information: www.epa.gov
Teach with Databases: Toxic Release Inventory: www.nsta.org/pubs/special/pb143x01.htm

Toxic Waste

Environmental Defense Fund: www.scorecard.org
Institute for Global Communications: www.igc.org/igc/issues/tw/

Urban Forests

American Forests: www.amfor.org
TreeLink: www.treelink.org

Vitrification

U.S. Department of Energy: www.em.doe.gov/fs/fs3m.html

Volcanoes

USGS, Volcanoes in the Learning Web: www.usgs.gov/education/learnweb/volcano/index.html
Volcano Hazards: volcanoes.usgs.gov/

Water Conservation, Pollution, and Other Water-Related Topics.

Early History of the Clean Water Act: www.csi.cc.id.us/ip/ag/water/Clean%20Water%20Act/sld002.htm
Environmental Protection Agency's Office of Wetlands, Oceans, Watersheds for Nonpoint Source information: www.epa.gov/owow/wetlands/wetland2.html www.epa.gov/swerosps/ej/
National Groundwater Association Home Page: www.h2o-ngwa.org
Water Resources Information: water.usgs.gov/
Water Use Data: water.usgs.gov/public/watuse/

Weather

National Weather Service: www.nws.noaa.gov

Wetlands

National Wetlands Research Center: www.nwrc.usgs.gov/educ_out.html
Ramsar Convention on Wetlands (International): www2.iucn.org/themes/ramsar/
The Ramsar List of Wetlands of International Importance: ramsar.org/key_sitelist.htm

Whales

Institute of Cetacean Research (ICR): www.whalesci.org
U.S. Fish and Wildlife Service, The Species List of Endangered and Threathened Wildlife: www.fws.gov/r9endspp/lsppinfo.html

Wilderness

U.S. Forest Service: www.fs.fed.us
The Wilderness Society: www.wilderness.org/newsroom/factsheets.htm

Wildlife Refuges

Conservation International: www.conservation.org

IUCN—The World Conservation Union/International Union for the Conservation of Nature: www.iucn.org
The Nature Conservancy: www.tnc.org
U.S. Fish and Wildlife Service, National Wildlife Refuge System: refuges.fws.gov

Wind Energy

The American Wind Energy Association: www.igc.apc.org/awea/news/html
Center for Renewable Energy and Sustainable Technology (CREST), Solar Energy Research and Education Foundation: solstice.crest.org/

Wolves

United States Fish and Wildlife Service: www.fws.gov/
U.S. Fish and Wildlife Service, The Species List of Endangered and Threathened Wildlife: www.fws.gov/r9endspp/lsppinfo.html
World Wildlife Fund: www.worldwildlife.org/

Zoos

Bronx Zoo: www.bronxzoo.com/
San Diego Zoo: www.sandiegozoo.org/

Endangered Species by State

According to the U.S. Fish and Wildlife Service, there were 962 endangered animal and plant species in the United States in the summer of 2000. The below list shows a selected grouping of species for each state. For the full list of endangered and threatened species, and other information about endangered species and the Endangered Species Act, see the Endangered Species Program Web site at <endangered.fws.gov/listing/index.html>

Alabama

Animals—

E - Bat, gray
E - Bat, Indiana
E - Cavefish, Alabama
E - Clubshell, black
E - Combshell, southern
E - Darter, boulder
E - Fanshell
E - Kidneyshell, triangular
E - Lampmussel, Alabama
E - Manatee, West Indian
E - Moccasinshell, Coosa
E - Mouse, Alabama beach
E - Mussel, ring pink
E - Pearlymussel, cracking
E - Pigtoe, dark
E - Plover, piping
E - Shrimp, Alabama cave
E - Snail, tulotoma (=Alabama live-bearing)
E - Stork, wood
E - Turtle, Alabama redbelly (=red-bellied)
E - Woodpecker, red-cockaded

Plants

E - Alabama canebrake pitcher-plant
E - Alabama leather-flower (*Clematis socialis*)
E - Gentian pinkroot (*Spigelia gentianoides*)
E - Green pitcher-plant (*Sarracenia oreophila*)
E - Leafy prairie-clover (*Dalea* (=*Petalostemum*) *foliosa*)
E - Morefield's leather-flower (*Clematis morefieldii*)
E - Pondberry (*Lindera melissifolia*)
E - Tennessee yellow-eyed grass (*Xyris tennesseensis*)

Alaska

Animals

E - Curlew, Eskimo (*Numenius borealis*)

Plants

E - Aleutian shield-fern (=Aleutian holly-fern) (*Polystichum aleuticum*)

Arizona

Animals

E - Ambersnail, Kanab (*Oxyloma haydeni kanabensis*)
E - Bat, lesser (=Sanborn's) long-nosed (*Leptonycteris curasoae yerbabuenae*)
E - Bobwhite, masked (quail) (*Colinus virginianus ridgwayi*)
E - Chub, bonytail
E - Chub, humpback
E - Chub, Virgin River (*Gila robusta semidnuda*)
E - Chub, Yaqui
E - Flycatcher, Southwestern willow
E - Jaguarundi, Sinaloan
E - Ocelot
E - Pronghorn, Sonoran
E - Pupfish, desert
E - Rail, Yuma clapper
E - Squirrel, Mount Graham red
E - Sucker, razorback
E - Topminnow, Gila (incl. Yaqui)
E - Trout, Gila
E - Vole, Hualapai Mexican
E - Woundfin

Plants

E - Arizona agave
E - Arizona cliffrose
E - Arizona hedgehog cactus
E - Brady pincushion cactus
E - Kearney's blue-star
E - Nichol's Turk's head cactus
E - Peebles Navajo cactus
E - Pima pineapple cactus
E - Sentry milk-vetch

Arkansas

Animals

E - Bat, gray
E - Bat, Indiana
E - Bat, Ozark big-eared
E - Beetle, American burying (=giant carrion)
E - Crayfish, cave
E - Pearlymussel, Curtis'
E - Pink mucket
E - Pocketbook, fat
E - Pocketbook, speckled
E - Rock-pocketbook, Ouachita (=Wheeler's pearly mussel)
E - Sturgeon, pallid
E - Tern, least
E - Woodpecker, red-cockaded

Plants

E - Harperella (*Ptilimnium nodosum* (=fluviatile))
E - Pondberry (*Lindera melissifolia*)
E - Running buffalo clover (*Trifolium stoloniferum*)

California

Animals

E - Butterfly, El Segundo blue (*Euphilotes battoides allyni*)
E - Butterfly, Lange's metalmark (*Apodemia mormo langei*)
E - Chub, Mohave tui (*Gila bicolor mohavensis*)
E - Condor, California (*Gymnogyps californianus*)
E - Crayfish, Shasta (=placid) (*Pacifastacus fortis*)
E - Fairy shrimp, Conservancy (*Branchinecta conservatio*)
E - Fly, Delhi Sands flower-loving (*Rhaphiomidas terminatus abdominalis*)
E - Flycatcher, Southwestern willow (*Empidonax traillii extimus*)
E - Fox, San Joaquin Entire kit (*Vulpes macrotis mutica*)
E - Goby, tidewater (*Eucyclogobius newberryi*)
E - Kangaroo rat, Fresno (*Dipodomys nitratoides exilis*)
E - Lizard, blunt-nosed leopard (*Gambelia silus*)
E - Mountain beaver, Point Arena (*Aplodontia rufa nigra*)
E - Mouse, Pacific pocket (*Perognathus longimembris pacificus*)
E - Pelican, brown (*Pelecanus occidentalis*)
E - Pupfish, Owens (*Cyprinodon radiosus*)
E - Rail, California clapper (*Rallus longirostris obsoletus*)
E - Salamander, Santa Cruz long-toed (*Ambystoma macrodactylum croceum*)
E - Shrike, San Clemente loggerhead (*Lanius ludovicianus mearnsi*)
E - Shrimp, California freshwater (*Syncaris pacifica*)
E - Snail, Morro shoulderband (=banded dune) (*Helminthoglypta walkeriana*)
E - Snake, San Francisco garter (*Thamnophis sirtalis tetrataenia*)
E - Stickleback, unarmored threespine (*Gasterosteus aculeatus williamsoni*)
E - Sucker, Lost River (*Deltistes luxatus*)
E - Tadpole shrimp, vernal pool (*Lepidurus packardi*)
E - Tern, California least (*Sterna antillarum browni*)
E - Toad, arroyo southwestern (*Bufo microscaphus californicus*)
E - Vireo, least Bell's (*Vireo bellii pusillus*)
E - Vole, Amargosa (*Microtus californicus scirpensis*)

Plants

E - Antioch Dunes evening-primrose (*Oenothera deltoides ssp. howellii*)
E - Bakersfield cactus (*Opuntia treleasei*)
E - Ben Lomond wallflower (*Erysimum teretifolium*)
E - Burke's goldfields (*Lasthenia burkei*)
E - California jewelflower (*Caulanthus californicus*)
E - California Orcutt grass (*Orcuttia californica*)
E - Clover lupine (*Lupinus tidestromii*)
E - Cushenbury buckwheat (*Eriogonum ovalifolium var. vineum*)
E - Fountain thistle (*Cirsium fontinale var. fontinale*)
E - Gambel's watercress (*Rorippa gambellii*)
E - Kern mallow (*Eremalche kernensis*)
E - Loch Lomond coyote-thistle (*Eryngium constancei*)
E - Robust spineflower (includes Scotts Valley spineflower) (*Chorizanthe robusta*)
E - San Clemente Island larkspur (*Delphinium variegatum ssp. kinkiense*)
E - San Diego button-celery (*Eryngium aristulatum var. parishii*)
E - San Mateo thornmint (*Acanthomintha obovata ssp. duttonii*)
E - Santa Ana River woolly-star (*Eriastrum densifolium ssp. sanctorum*)
E - Santa Barbara Island liveforever (*Dudleya traskiae*)
E - Santa Cruz cypress (*Cupressus abramsiana*)
E - Solano grass (*Tuctoria mucronata*)
E - Sonoma sunshine (=Baker's stickyseed) (*Blennosperma bakeri*)
E - Stebbins' morning-glory (*Calystegia stebbinsii*)
E - Truckee barberry (*Berberis sonnei*)
E - Western lily (*Lilium occidental*)

Colorado

Animals

E - Butterfly, Uncompahgre fritillary
E - Chub, bonytail
E - Chub, humpback
E - Crane, whooping
E - Ferret, black-footed
E - Flycatcher, Southwestern willow
E - Sucker, razorback
E - Tern, least
E - Wolf, gray

Plants

E - Clay-loving wild-buckwheat
E - Knowlton cactus
E - Mancos milk-vetch
E - North Park phacelia
E - Osterhout milk-vetch
E - Penland beardtongue

Connecticut

Animals

E - Tern, roseate
E - Turtle, hawksbill sea
E - Turtle, Kemp's (=Atlantic) ridley sea
E - Turtle, leatherback sea
E - Wedgemussel, dwarf

Plants

E - Sandplain gerardia

Delaware

Animals

E - Squirrel, Delmarva Peninsula fox
E - Turtle, hawksbill sea
E - Turtle, Kemp's (=Atlantic) ridley sea
E - Turtle, leatherback sea

Plants

E - Canby's dropwort

Florida

Animals

E - Bat, gray
E - Butterfly, Schaus swallowtail (*Heraclides* (=Papilio) *aristodemus ponceanus*)
E - Crocodile, American (*Crocodylus acutus*)
E - Darter, Okaloosa (*Etheostoma okaloosae*)
E - Deer, key (*Odocoileus virginianus clavium*)
E - Kite, Everglade snail (*Rostrhamus sociabilis plumbeus*)
E - Manatee, West Indian (=Florida) (*Trichechus manatus*)
E - Mouse, Anastasia Island beach (*Peromyscus polionotus phasma*)
E - Mouse, Choctawahatchee beach (*Peromyscus polionotus allophrys*)
E - Panther, Florida (*Felis concolor coryi*)
E - Rabbit, Lower Keys marsh (*Sylvilagus palustris hefneri*)
E - Rice rat (=silver rice rat) (lower FL Keys) (*Oryzomys palustris natator*)
E - Sparrow, Cape Sable seaside (*Ammodramus maritimus mirabilis*)
E - Stork, wood (*Mycteria americana*)
E - Turtle, hawksbill sea (*Eretmochelys imbricata*)
E - Turtle, Kemp's (=Atlantic) ridley sea (*Lepidochelys kempii*)
E - Turtle, leatherback sea (*Dermochelys coriacea*)
E - Vole, Florida salt marsh (*Microtus pennsylvanicus dukecampbelli*)
E - Woodpecker, red-cockaded (*Picoides borealis*)
E - Woodrat, Key Largo (*Neotoma floridana smalli*)

Plants

E - Apalachicola rosemary (*Conradina glabra*)
E - Beautiful pawpaw (*Deeringothamnus pulchellus*)
E - Brooksville (=Robins') bellflower (*Campanula robinsiae*)
E - Carter's mustard (*Warea carteri*)
E - Chapman rhododendron (*Rhododendron chapmanii*)
E - Cooley's water-willow (*Justicia cooleyi*)
E - Crenulate lead-plant (*Amorpha crenulata*)
E - Etonia rosemary (*Conradina etonia*)
E - Florida golden aster (*Chrysopsis floridana*)
E - Fragrant prickly-apple (*Cereus eriophorus var. fragrans*)
E - Garrett's mint (*Dicerandra christmanii*)
E - Key tree-cactus (*Pilosocereus robinii* (=Cereus r.))
E - Lakela's mint (*Dicerandra immaculata*)
E - Okeechobee gourd (*Cucurbita okeechobeensis ssp. okeechobeensis*)
E - Scrub blazingstar (*Liatris ohlingerae*)
E - Small's milkpea (*Galactia smallii*)
E - Snakeroot (*Eryngium cuneifolium*)
E - Wireweed (*Polygonella basiramia*)

Georgia

Animals

E - Acornshell, southern
E - Bat, gray
E - Bat, Indiana
E - Clubshell, ovate
E - Clubshell, southern
E - Combshell, upland
E - Darter, Etowah
E - Darter, amber
E - Kidneyshell, triangular
E - Logperch, Conasauga
E - Manatee, West Indian (=Florida)
E - Moccasinshell, Coosa
E - Pigtoe, southern
E - Stork, wood
E - Turtle, hawksbill sea
E - Turtle, Kemp's (=Atlantic) ridley sea
E - Turtle, leatherback sea
E - Woodpecker, red-cockaded

Plants

E - American chaffseed
E - Black-spored quillwort
E - Canby's dropwort
E - Florida torreya
E - Fringed campion
E - Green pitcher-plant
E - Hairy rattleweed
E - Harperella
E - Large-flowered skullcap
E - Mat-forming quillwort
E - Michaux's sumac
E - Persistent trillium
E - Pondberry
E - Relict trillium
E - Smooth coneflower
E - Tennessee yellow-eyed grass

Hawaii

Animals

E - 'Akepa, Hawaii (honeycreeper)
E - Bat, Hawaiian hoary
E - Coot, Hawaiian
E - Creeper, Hawaii
E - Crow, Hawaiian
E - Duck, Hawaiian
E - Duck, Laysan
E - Finch, Laysan (honeycreeper)
E - Finch, Nihoa (honeycreeper)
E - Goose, Hawaiian
E - Hawk, Hawaiian
E - Millerbird, Nihoa (old world warbler)
E - Nukupu'u (honeycreeper)
E - Palila (honeycreeper)
E - Parrotbill, Maui (honeycreeper)
E - Petrel, Hawaiian dark-rumped
E - Snails, Oahu tree
E - Stilt, Hawaiian
E - Turtle, hawksbill sea
E - Turtle, leatherback sea

Plants
E - Abutilon eremitopetalum
E - Bonamia menziesii
E - Carter's panicgrass
E - Diamond Head schiedea
E - Dwarf iliau
E - Fosberg's love grass
E - Hawaiian bluegrass
E - Hawaiian red-flowered geranium
E - Kaulu
E - Kiponapona
E - Mahoe
E - Mapele
E - Nanu
E - Nehe
E - Opuhe
E - Pamakani
E - Round-leaved chaff-flower
E - Viola helenae

Idaho

Animals
E - Caribou, woodland (*Rangifer tarandus caribou*)
E - Crane, whooping
E - Limpet, Banbury Springs
E - Snail, Snake River physa
E - Snail, Utah valvata
E - Springsnail, Bruneau Hot
E - Springsnail, Idaho
E - Sturgeon, white
E - Wolf, gray

Plants
No plants on the endangered list.

Illinois

Animals
E - Bat, gray
E - Bat, Indiana
E - Butterfly, Karner blue
E - Dragonfly, Hine's emerald
E - Fanshell
E - Pink mucket
E - Plover, piping (Great Lakes Watershed)
E - Pocketbook, fat
E - Snail, Iowa Pleistocene
E - Sturgeon, pallid
E - Tern, least

Plants
E - Leafy prairie-clover

Indiana

Animals
E - Bat, gray
E - Bat, Indiana
E - Blossom, tubercled
E - Butterfly, Karner blue
E - Butterfly, Mitchell's satyr
E - Catspaw, white
E - Clubshell
E - Fanshell
E - Mussel, Ring pink (=golf stick pearly)
E - Pearlymussel, cracking
E - Pearlymussel, white wartyback
E - Pigtoe, rough
E - Pimple back orange foot
E - Pink mucket
E - Plover, piping (Great Lakes Watershed)
E - Pocketbook, fat
E - Riffleshell, northern
E - Tern, least

Plants
E - Running buffalo clover

Iowa

Animals
E - Bat, Indiana
E - Higgins' eye
E - Snail, Iowa Pleistocene
E - Sturgeon, pallid
E - Tern, least

Plants
No plants on the endangered list.

Kansas

Animals
E - Bat, gray
E - Bat, Indiana
E - Crane, whooping
E - Curlew, Eskimo
E - Ferret, black-footed
E - Sturgeon, pallid
E - Tern, least
E - Vireo, black-capped

Plants
No plants on the endangered list.

Kentucky-

Animals
E - Bat, gray
E - Bat, Indiana
E - Bat, Virginia big-eared
E - Blossom, tubercled
E - Catspaw
E - Clubshell
E - Darter, relict
E - Fanshell
E - Mapleleaf, winged
E - Pearlymussel, cracking
E - Pearlymussel, dromedary
E - Pearlymussel, little-wing
E - Pimple back, orange-foot
E - Mucket, pink
E - Pearlymussel, white wartyback
E - Pigtoe, rough
E - Pocketbook, fat
E - Riffleshell, northern
E - Riffleshell, tan
E - Ring pink (=golf stick pearly)
E - Shiner, Palezone
E - Shrimp, Kentucky cave

E - Sturgeon, pallid
E - Tern, least
E - Woodpecker, red-cockaded

Plants
E - Cumberland sandwort
E - Rock cress, Braun's
E - Running buffalo clover
E - Short's goldenrod

Louisiana

Animals
E - Manatee, West Indian (=Florida)
E - Mucket, pink
E - Pelican, brown
E - Plover, piping
E - Sturgeon, pallid
E - Tern, least
E - Turtle, hawksbill sea
E - Turtle, Kemp's (=Atlantic) ridley sea
E - Turtle, leatherback sea
E - Vireo, black-capped
E - Woodpecker, red-cockaded

Plants
E - American chaffseed
E - Louisiana quillwort
E - Pondberry

Maine

Animals
E - Tern, roseate
E - Turtle, leatherback sea

Plants
E - Furbish lousewort (*Pedicularis furbishiae*)

Maryland

Animals
E - Bat, Indiana
E - Darter, Maryland
E - Squirrel, Delmarva Peninsula fox
E - Turtle, hawksbill sea
E - Turtle, Kemp's (=Atlantic) ridley sea
E - Turtle, leatherback sea
E - Wedgemussel, dwarf

Plants
E - Canby's dropwort
E - Harperella
E - Northeastern (=Barbed bristle) bulrush
E - Sandplain gerardia

Massachusetts

Animals
E - Beetle, American burying (=giant carrion)
E - Tern, roseate
E - Turtle, hawksbill sea
E - Turtle, Kemp's (=Atlantic) ridley sea
E - Turtle, Plymouth redbelly (=red-bellied)
E - Turtle, leatherback sea
E - Wedgemussel, dwarf

Plants
E - Bulrush, Northeastern (=Barbed bristle)
E - Sandplain gerardia

Michigan

Animals
E - Bat, Indiana
E - Beetle, American burying (=giant carrion)
E - Beetle, Hungerford's crawling water
E - Butterfly, Karner blue
E - Butterfly, Mitchell's satyr
E - Clubshell (*Pleurobema clava*)
E - Plover, piping (Great Lakes Watershed)
E - Riffleshell, northern
E - Warbler, Kirtland's
E - Wolf, gray

Plants
E - Michigan monkey-flower

Minnesota

Animals
E - Butterfly, Karner blue
E - Higgins' eye
E - Mapleleaf, winged
E - Plover, piping (Great Lakes Watershed)

Plants
E - Minnesota trout lily

Mississippi

Animals
E - Bat, Indiana
E - Clubshell, black (=Curtus' mussel)
E - Clubshell, ovate
E - Clubshell, southern
E - Combshell, southern (=penitent mussel)
E - Crane, Mississippi sandhill
E - Manatee, West Indian (=Florida)
E - Pelican, brown
E - Pigtoe, flat (=Marshall's mussel)
E - Pigtoe, heavy (=Judge Tait's mussel)
E - Pocketbook, fat
E - Stirrupshell
E - Sturgeon, pallid
E - Tern, least
E - Turtle, hawksbill sea
E - Turtle, Kemp's (=Atlantic) ridley sea
E - Turtle, leatherback sea
E - Woodpecker, red-cockaded

Plants
E - American chaffseed
E - Pondberry

Missouri

Animals
E - Bat, gray
E - Bat, Indiana
E - Bat, Ozark big-eared
E - Pearlymussel, Curtis'
E - Higgins' eye
E - Pink mucket

E - Pocketbook, fat
E - Sturgeon, pallid
E - Tern, least

Plants

E - Missouri bladderpod
E - Pondberry
E - Running buffalo clover

Montana

Animals

E - Crane, whooping
E - Curlew, Eskimo
E - Ferret, black-footed
E - Sturgeon, pallid
E - Sturgeon, white
E - Tern, least
E - Wolf, gray

Plants

No plants on the endangered list.

Nebraska

Animals

E - Beetle, American burying (=giant carrion)
E - Crane, whooping
E - Curlew, Eskimo
E - Ferret, black-footed
E - Sturgeon, pallid
E - Tern, least

Plants

E - Blowout penstemon

Nevada

Animals

E - Chub, bonytail
E - Chub, Pahranagat roundtail (=bonytail)
E - Chub, Virgin River
E - Cui-ui
E - Dace, Ash Meadows speckled
E - Dace, Clover Valley speckled
E - Dace, Independence Valley speckled
E - Dace, Moapa (Moapa coriacea)
E - Poolfish (=killifish), Pahrump
E - Pupfish, Ash Meadows Amargosa
E - Pupfish, Devils Hole
E - Pupfish, Warm Springs
E - Spinedace, White River
E - Springfish, Hiko White River
E - Springfish, White River
E - Sucker, razorback
E - Woundfin

Plants

E - Amargosa niterwort
E - Steamboat buckwheat

New Hampshire

Animals

E - Butterfly, Karner blue
E - Turtle, leatherback sea
E - Wedgemussel, dwarf

Plants

E - Jesup's milk-vetch
E - Northeastern (=Barbed bristle) bulrush
E - Robbins' cinquefoil

New Jersey

Animals

E - Bat, Indiana
E - Tern, roseate
E - Turtle, hawksbill sea
E - Turtle, Kemp's (=Atlantic) ridley sea
E - Turtle, leatherback sea

Plants

E - American chaffseed

New Mexico

Animals

E - Bat, lesser (=Sanborn's) long-nosed
E - Bat, Mexican long-nosed
E - Crane, whooping
E - Gambusia, Pecos
E - Isopod, Socorro
E - Minnow, Rio Grande silvery
E - Springsnail, Alamosa
E - Springsnail, Socorro
E - Sucker, razorback
E - Tern, least
E - Topminnow, Gila (incl. Yaqui)
E - Trout, Gila
E - Woundfin

Plants

E - Holy Ghost ipomopsis
E - Knowlton cactus
E - Kuenzler hedgehog cactus
E - Lloyd's Mariposa cactus
E - Mancos milk-vetch
E - Sacramento prickly-poppy
E - Sneed pincushion cactus
E - Todsen's pennyroyal

New York

Animals

E - Butterfly, Karner blue
E - Tern, roseate
E - Turtle, hawksbill sea
E - Turtle, leatherback sea
E - Turtle, Kemp's (=Atlantic) ridley sea
E - Wedgemussel, dwarf

Plants

E - Sandplain gerardia

North Carolina

Animals

E - Bat, Indiana (*Myotis sodalis*)
E - Bat, Virginia big-eared (Plecotus townsendii virginianus)
E - Butterfly, Saint Francis' satyr (*Neonympha mitchellii francisci*)
E - Elktoe, Appalachian (*Alasmidonta raveneliana*)
E - Heelsplitter, Carolina (*Lasmigona decorata*)
E - Manatee, West Indian (=Florida) (*Trichechus manatus*)
E - Pearlymussel, little-wing (*Pegias fabula*)
E - Shiner, Cape Fear (*Notropis mekistocholas*)
E - Spider, spruce-fir moss (*Microhexura montivaga*)
E - Spinymussel, Tar River (*Elliptio steinstansana*)

E - Squirrel, Carolina northern flying (*Glaucomys sabrinus coloratus*)
E - Tern, roseate (*Sterna dougallii dougallii*)
E - Turtle, hawksbill sea (*Eretmochelys imbricata*)
E - Turtle, Kemp's (=Atlantic) ridley sea (*Lepidochelys kempii*)
E - Turtle, leatherback sea (*Dermochelys coriacea*)
E - Wedgemussel, dwarf wedge (*Alasmidonta heterodon*)
E - Wolf, red (*Canis rufus*)
E - Woodpecker, red-cockaded (*Picoides borealis*)

Plants
E - American chaffseed (*Schwalbea americana*)
E - Bunched arrowhead (*Sagittaria fasciculata*)
E - Canby's dropwort (*Oxypolis canbyi*)
E - Cooley's meadowrue (*Thalictrum cooleyi*)
E - Green pitcher-plant (*Sarracenia oreophila*)
E - Harperella (*Ptilimnium nodosum* (=fluviatile))
E - Michaux's sumac (*Rhus michauxii*)
E - Mountain sweet pitcher-plant (*Sarracenia rubra ssp. jonesii*)
E - Pondberry (*Lindera melissifolia*)
E - Roan Mountain bluet (*Hedyotis purpurea var. montana*)
E - Rock gnome lichen (*Gymnoderma lineare*)
E - Rough-leaved loosestrife (*Lysimachia asperulaefolia*)
E - Schweinitz's sunflower (*Helianthus schweinitzii*)
E - Small-anthered bittercress (*Cardamine micranthera*)
E - Smooth coneflower (*Echinacea laevigata*)
E - Spreading avens (*Geum radiatum*)
E - White irisette (*Sisyrinchium dichotomum*)

North Dakota

Animals
E - Crane, whooping (*Grus americana*)
E - Curlew, Eskimo (*Numenius borealis*)
E - Ferret, black-footed (*Mustela nigripes*)
E - Sturgeon, pallid (*Scaphirhynchus albus*)
E - Tern, least (*Sterna antillarum*)
E - Wolf, gray (*Canis lupus*)

Plants
No plants on the endangered list.

Ohio

Animals
E - Bat, Indiana
E - Beetle, American burying (=giant carrion)
E - Butterfly, Karner blue
E - Butterfly, Mitchell's satyr
E - Catspaw, purple
E - Catspaw, white
E - Clubshell
E - Dragonfly, Hine's emerald
E - Fanshell
E - Madtom, Scioto
E - Pink mucket
E - Plover, piping (Great Lakes Watershed)
E - Riffleshell, northern

Plants
E - Running buffalo clover

Oklahoma

Animals
E - Bat, gray
E - Bat, Indiana
E - Bat, Ozark big-eared
E - Crane, whooping
E - Curlew, Eskimo
E - Rock-pocketbook, Ouachita
E - Tern, least
E - Vireo, black-capped
E - Woodpecker, red-cockaded

Plants
No plants on the endangered list.

Oregon

Animals
E - Chub, Borax Lake
E - Chub, Oregon
E - Deer, Columbian white-tailed
E - Pelican, brown
E - Sucker, Lost River
E - Sucker, shortnose
E - Turtle, leatherback sea

Plants
E - Applegate's milk-vetch
E - Malheur wire-lettuce
E - Marsh sandwort
E - Western lily

Pennsylvania

Animals
E - Bat, Indiana
E - Clubshell
E - Pearlymussel, cracking
E - Orange-foot pimple back
E - Mucket, pink
E - Pigtoe, rough
E - Plover, piping (Great Lakes Watershed)
E - Riffleshell, northern
E - Ring pink (=golf stick pearly)
E - Wedgemussel, dwarf

Plants
E - Northeastern (=Barbed bristle) bulrush

Rhode Island

Animals
E - Beetle, American burying
E - Tern, roseate
E - Turtle, hawksbill sea
E - Turtle, Kemp's Ripley Sea
E - Turtle, leatherback sea

Plants
E - Sandplain gerardia

South Carolina

Animals
E - Bat, Indiana
E - Heelsplitter, Carolina
E - Manatee, West Indian (=Florida)

E - Stork, wood
E - Turtle, hawksbill sea
E - Turtle, Kemp's (=Atlantic) ridley sea
E - Turtle, leatherback sea
E - Woodpecker, red-cockaded

Plants
E - American chaffseed
E - Black-spored quillwort
E - Bunched arrowhead
E - Canby's dropwort
E - Harperella
E - Michaux's sumac
E - Mountain sweet pitcher-plant
E - Persistent trillium
E - Pondberry
E - Relict trillium
E - Rough-leaved loosestrife
E - Schweinitz's sunflower
E - Smooth coneflower

South Dakota

Animals
E - Beetle, American burying (=giant carrion)
E - Crane, whooping
E - Curlew, Eskimo
E - Ferret, black-footed
E - Sturgeon, pallid
E - Tern, least
E - Wolf, gray

Tennessee

Animals
E - Appalachian monkeyface
E - Bat, gray
E - Bat, Indiana
E - Combshell, upland
E - Crayfish, Nashville
E - Cumberland bean
E - Darter, amber
E - Fanshell
E - Lampmussel, Alabama
E - Madtom, Smoky Entire
E - Marstonia (snail), royalobese)
E - Moccasinshell, Coosa
E - Ring pink (=golf stick pearly)
E - Riversnail, Anthony's
E - Spider, spruce-fir moss
E - Squirrel, Carolina northern flying
E - Sturgeon, pallid
E - Tern, least
E - Wolf, red
E - Woodpecker, red-cockaded

Plants
E - Cumberland sandwort
E - Green pitcher-plant
E - Large-flowered skullcap
E - Leafy prairie-clover
E - Roan Mountain bluet
E - Rock cress, Braun's
E - Rock gnome lichen
E - Ruth's golden aster
E - Spring Creek bladderpod
E - Tennessee purple coneflower
E - Tennessee yellow-eyed grass

Texas

Animals
E - Bat, Mexican long-nosed
E - Beetle, Coffin Cave mold
E - Crane, whooping
E - Curlew, Eskimo
E - Darter, fountain
E - Falcon, northern aplomado
E - Jaguarundi, Gulf Coast
E - Manatee, West Indian (=Florida)
E - Minnow, Rio Grande silvery
E - Ocelot
E - Pelican, brown
E - Prairie-chicken, Attwater's greater
E - Pupfish, Comanche Springs
E - Salamander, Texas blind
E - Spider, Tooth Cave
E - Tern, least
E - Toad, Houston
E - Turtle, hawksbill sea
E - Turtle, Kemp's (=Atlantic) ridley sea
E - Vireo, black-capped
E - Warbler, golden-cheeked
E - Woodpecker, red-cockaded

Plants
E - Ashy dogweed
E - Black lace cactus
T - Hinckley's oak
E - Large-fruited sand-verbena
E - Little Aguja Creek pondweed
E - Lloyd's Mariposa cactus
E - Nellie cory cactus
E - Sneed pincushion cactus
E - South Texas ambrosia
E - Star cactus
E - Terlingua Creek cats-eye
E - Texas poppy-mallow
E - Texas snowbells
E - Texas wild-rice
E - Tobusch fishhook cactus
E - Walker's manioc

Utah

Animals
E - Ambersnail, Kanab
E - Chub, bonytail
E - Chub, humpback
E - Chub, Virgin River
E - Crane, whooping
E - Ferret, black-footed
E - Flycatcher, Southwestern willow
E - Snail, Utah valvata
E - Sucker, June
E - Sucker, razorback
E - Woundfin

Plants
E - Autumn buttercup
E - Barneby reed-mustard
E - Barneby ridge-cress (=peppercress)

E - Clay phacelia
E - Dwarf bear-poppy
E - Kodachrome bladderpod
E - San Rafael cactus
E - Wright fishhook cactus

Vermont

Animals
E - Bat, Indiana
E - Wedgemussel, dwarf

Plants
E - Jesup's milk-vetch
E - Northeastern (=Barbed bristle) bulrush

Virginia

Animals
E - Bat, gray
E - Bat, Indiana
E - Bat, Virginia big-eared
E - Blossom, green
E - Darter, duskytail
E - Falcon, American peregrine
E - Fanshell
E - Isopod, Lee County cave
E - Logperch, Roanoke
E - Monkeyface, Appalachian
E - Monkeyface, Cumberland
E - Pearlymussel, birdwing
E - Pearlymussel, cracking
E - Pearlymussel, dromedary
E - Pearlymussel, little-wing
E - Pigtoe, fine rayed
E - Pigtoe, rough
E - Pigtoe, shiny
E - Pink mucket
E - Riffleshell, tan
E - Salamander, Shenandoah
E - Snail, Virginia fringed mountain
E - Spinymussel, James
E - Squirrel, Delmarva Peninsula fox
E - Squirrel, Virginia northern flying
E - Turtle, hawksbill sea
E - Turtle, Kemp's (=Atlantic) ridley sea
E - Turtle, leatherback sea
E - Wedgemussel, dwarf
E - Woodpecker, red-cockaded

Plants
E - Northeastern (=Barbed bristle) bulrush
E - Peter's Mountain mallow
E - Shale barren rock-cress
E - Smooth coneflower

Washington

Animals
E - Caribou, woodland
E - Deer, Columbian white-tailed
E - Pelican, brown
E - Turtle, leatherback sea
E - Wolf, gray

Plants
E - Lomatium, Bradshaw's
E - Marsh sandwort

West Virginia

Animals
E - Bat, Indiana
E - Bat, Virginia big-eared
E - Blossom, tubercled
E - Clubshell
E - Fanshell
E - Pink mucket
E - Riffleshell, northern
E - Spinymussel, James
E - Squirrel, Virginia northern flying

Plants
E - Harperella
E - Northeastern (=Barbed bristle) bulrush
E - Running buffalo clover
E - Shale barren rock-cress

Wisconsin

Animals
E - Butterfly, Karner blue
E - Dragonfly, Hine's emerald
E - Mapleleaf, winged
E - Pearlymussel, Higgins' eye
E - Plover, piping
E - Warbler, Kirtland's
E - Wolf, gray

Plants
No plants on the endangered list

Wyoming

Animals
E - Crane, whooping
E - Dace, Kendall Warm Springs
E - Ferret, black-footed
E - Sucker, razorback
E - Toad, Wyoming
E - Wolf, gray

Plants
E - Penstemon, blowout

Environmental Organizations

Action for Animals
P.O. Box 17702
Austin, TX 78760
Telephone: (512) 416-1617
Fax: (512) 445-3454
Web site: http://www.envirolink.org/orgs/afa

African Wildlife Foundation (AWF)
African Wildlife Foundation
1400 Sixteenth Street, N.W., Suite 120
Washington, D.C. 20036
Telephone: (202) 939-3333
Fax (202) 939-3332
Web site: http://www.awf.org/home.html

Agency for Toxic Substances and Diseases,
Registry Division of Toxicology (ATSDR)
1600 Clifton Road
NE Mailstop E-29
Atlanta, GA 30333
Telephone ATSDR Information Center: (888)42-ATSDR or (888)422-8737
E-mail: ATSDRIC@cdc.gov
Web site: http://www.atsdr.cdc.gov/contacts.html

Alaska Forum for Environmental Responsibility
P.O. Box 188
Valdez, AK 99686
Telephone: (907) 835-5460
Fax: (907) 835-5410
Web site: http://www.accessone.com/~afersea

American Conifer Society (ACS)
P.O. Box 360
Keswick, VA 22947-0360
Telephone: (804) 984-3660
Fax: (804) 984-3660
E-mail: ACSconifer@aol.com
Web site: http://www.pacificrim.net/~bydesign/acs.html

American Forests
P.O. Box 2000
Washington, DC 20013
Telephone: (202) 955-4500
Web site: http://www.americanforests.org

American Nuclear Society
555 North Kensington Ave.
La Grange Park, IL 60525
Telephone: (708) 352-6611
Fax: (708) 352-0499
E-mail: NUCLEUS@ans.org
Web site: http://www.ans.org

American Oceans Campaign
201 Massachusetts Ave., NE #C-3
Washington, DC 20002
Telephone: (202) 544-3526
Fax: (202) 544-5625
E-mail: aocdc@wizard.net
Web site: http://www.americanoceans.org

American Rivers
1025 Vermont Avenue, NW, Suite 720
Washington, DC 20005
Telephone: (202) 347-7500
Fax: (202) 347-9240
E-mail: amrivers@amrivers.org
Web site: http://www.amrivers.org

American Society for Horticultural Science (ASHS)
600 Cameron St.
Alexandria, VA 22314-2562
Telephone: (703) 836-4606
Fax: (703) 836-2024
E-mail: webmaster@ashs.org
Web site: http://www.ashs.org

American Society for the Prevention of Cruelty to Animals (ASPCA)
424 East 92nd Street
New York, NY 10128
Telephone: (212) 876-7700
Web site: http://www.aspca.org

American Solar Energy Society
2400 Central Avenue, Suite G-1

Boulder, CO 80301
Telephone: (303) 443-3130
Fax: (303) 443-3212
E-mail: ases@ases.org
Web site: http://www.ases.org
Publication: *Solar Today*

American Wind Energy Association
122 C Street, NW, 4th Floor
Washington, DC 20001
Telephone: (202) 383-2500.
E-mail: awea@mcimail.com.
Web site: http://www.awea.org

Animal Legal Defense Fund (ALDF)
127 Fourth Street
Petaluma, CA94952
Telephone: (707) 769-7771
Fax: (707) 769-0785
E-mail: info@aldf.org
Web site: http://www.aldf.org

Animal Rights Network
P.O. Box 25881
Baltimore, MD 21224
Telephone: (410) 675-4566
Fax: (410) 675-0066
Web site: http://www.envirolink.org/arrs/aa/index.html
Publication: *Animals' AGENDA,* a bi-monthly magazine

Baron's Haven Freehold
104 South Main Street
Cadiz, OH 43907
Telephone: (740) 942-8405
Web site: http://bhfi.1st.net

Biodiversity Support Program (BSP)
1250 N. 24th Street NW, Suite 600
Washington, DC 20037
Telephone: (202) 778-9681
Fax: (202) 861-8324
Web site: http://www.BSPonline.org

Biosfera
Pres. Vargas 435, Suites 1104 and 1105
Rio de Janeiro, RJ 20077-900
Brazil
Telephone: 55 21 221-0155
E-mail: biosferaAbsiofera.com.br
Web site: none available

Birds of Prey Foundation
2290 South 104th Street
Broomfield, CO 80020
Telephone: (303) 460-0674
Fax: (303) 666-1050
E-mail: raptor@birds-of-prey.org
Web site: www.birds-of-prey.org

Build the Earth
3818 Surfwood Road
Malibu, CA 90265
Telephone: (310) 454-0963

Center for Marine Conservation
1725 DeSales Street, SW, Suite 600
Washington, D.C. 20036
Telephone: (202) 429-5609
Fax: (202) 872-0619
E-mail: cmc@dccmc.org
Web site: http://www.cmc-ocean.org

Center for Research of Endangered Wildlife (CREW)
Cincinnati Zoo and Botanical Garden
E-mail: terri.roth@cincyzoo.org

Centers for Disease Control (CDC)
Atlanta, Georgia
Telephone: (800) 311-3435
Web site: http://www.cdc.gov

Cheetah Conservation Fund (CCF)
P.O. Box 1380
Ojai, CA 93024
Telephone: (805) 640-0390
Fax: (815) 640-0230
E-mail: info@cheetah.org
Web site: http://www.cheetah.org

Clean Air Council (CAC)
135 South 19th Street, Suite 300
Philadelphia, PA 19103
Telephone: (888) 567-7796
Web site: http://www.libertynet.org/~clean air/

Coalition for Economically Responsible Economies (CERES)
11 Arlington Street, 6th Floor
Boston, MA 02116-3411
Telephone: (617) 247-0700
Fax: (617) 267-5400
Web site: http://www.ceres.org

Convention on International Trade in Endangered Species
of Wild Fauna and Flora (CITES)
CITES Secretariat,
International Environment House, 15, chemin des Anémones, CH-1219
Châtelaine-Geneva, Switzerland.
Telephone: (4122) 917-8139/40
Fax: (4122) 797-3417
E-mail: cites@unep.ch
Web site: http://www.cites.org/index.shtml

Conservation International
1015 18th Street, N.W. Suite 1000
Washington, DC 20036
Telephone: (202) 429-5660
Web site: http://www.conservation.org/
Publication: *Orion Nature Quarterly*

Council for Responsible Genetics
5 Upland Road, Suite 3
Cambridge, MA 02140
Web site: http://www.gene-watch.org

The Cousteau Society
870 Greenbriar Circle, Suite 402
Chesapeake, VA 23320
Telephone: (804) 523-9335

E-mail: cousteau@infi.net
Web site: http://www.cousteausociety.org/
Publication: *Calypso Log*

Defenders of Wildlife
1101 Fourteenth Street, NW, Room 1400
Washington, DC 20005
Telephone: (800) 441-4395
Web site: http://www.Defenders.org
Publication: *Defenders,* quarterly magazine

The Dian Fossey Gorilla Fund International
800 Cherokee Ave., SE
Atlanta, GA 30315-1440
Telephone: (800) 851-0203
Fax: (404) 624-5999
E-mail: 2help@gorillafund.org
Web site: http://www.gorillafund.org/000_core_frmset.html

Earth Day Network
E-mail: earthday@earthday.net
Web site: http://www.earthday.net

Earth Island Institute (EII)
300 Broadway, Suite 28
San Francisco, CA 94133
Telephone: (415) 788-3666
Fax: (415) 788-7324
Web site: http://www.earthisland.org/abouteii/abouteii.html
Publication: *Earth Island Journal*, quarterly magazine

Earth, Pulp, and Paper
P.O. Box 64
Leggett, CA 95585
Telephone: (707) 925-6494
E-mail: tree@tree.org
Web site: http://www.tree.org/epp.htm

EarthFirst! (EF!)
P.O. Box 5176
Missoula, MT 59806
For the address of the EF! chapter nearer you, contact the following Web site: http://www.webdirectory.com/General_Environmental_Interest/Earth_First_/

Earthwatch Institute
IN USA AND CANADA:
3 Clocktower Place, Suite 100
Box 75
Maynard, MA 01754
Telephone: (800) 776-0188 or (617) 926-8200
Fax: (617) 926-8532
IN EUROPE:
57 Woodstock Road
Oxford OX2 6HJ
UK
Telephone: 44 (0) 1865-311-600
Fax: 44 (0) 1865-311-383
E-mail: info@uk.earthwatch.org
Web site: http://www.earthwatch.org

EcoCorps
1585 A Folsom Avenue
San Francisco, CA 94103
Telephone: (415) 522-1680
Fax: (415) 626-1510
E-mail: eathvoice@ecocorps.org
Web site: http://www.owplaza.com/eco

Ecotourism Society
P.O. Box 755
North Bennington, VT 05257
Telephone: (802) 447-2121
Fax: (802) 447-2122
E-Mail: ecomail@ecotourism.org
Web site: http://www.ecotoursim.org

E.F. Schumacher Society
140 Jug End Road
Great Barrington, MA 01230
Telephone: (413) 528-1737
E-mail: efssociety@aol.com
Web site: http://members.aol.com/efssociety/index.html

Electric Vehicle Association of the Americas
701 Pennsylvania Avenue, NW Fourth Floor
Washington, DC 20004
Telephone: (202) 508-5995
Fax: (202) 508-5924
Web site: http://www.evaa.org

Environmental Defense Fund (EDF)
257 Park Avenue South
New York, NY 10010
Telephone: (800) 684-3322
Fax: (212) 505-2375
E-mail (for general questions and information: Contact@environmentaldefense.org.)
Web site: http://www.edf.org
Publication: *Nature Journal*, monthly magazine

Exotic Cat Refuge and Wildlife Orphanage
Rt. 3, Box 96A
Kirbyville, TX 75956
Telephone: (409) 423-4847

Federal Emergency and Management Agency (FEMA)
500 C Street, SW
Washington, DC 20472
Web site: http://www.fema.gov

Friends of the Earth (FOE)
1025 Vermont Ave., NW, Suite 300
Washington, DC 20005-6303
Telephone: (202) 783-7400
Fax: (202) 783-0444
E-mail: foe@foe.org
Web site: http://www.foe.org

Green Seal
1001 Connecticut Ave., NW, Suite 827
Washington, DC 20036-5525
Telephone: (202) 872-6400
Fax: (202) 872-4324
Web site: http://www.greenseal.org

Greenpeace USA, Inc.
1436 U Street, N.W.
Washington, DC 20009
Telephone: (202) 462-1177

Web site: http://www.greenpeaceusa.org/
Publication: *Greenpeace Magazine*

Hawkwatch International
P.O. Box 660
Salt Lake City, UT 84110
Telephone: (801) 524-8511
E-mail: hawkwatch@charitiesusa.com
Web site: http://www.vpp.com/Hawk Watch

Humane Society of the United States (HSUS)
2100 L Street, NW
Washington, DC 20037
Web site: http://www.hsus.org
Publications: *All Animals,* quarterly; *Humane Activist* quarterly newsletter

International Atomic Energy Commission
P.O. Box 100
Wagramer Strasse 5
A-1400, Vienna, Austria
Telephone: (43-1) 2600-0
Fax: (43-1) 2600 7
E-mail: Official.Mail@iaea.org
Web site: http://www.iaea.org

International Centre for Antarctic Information and Research
Web site: http://www.icair.iac.org.nz

International Council for Local Environmental Initiatives (ICLEI)
World Secretariat
16th Floor, West Tower, City Hall
Toronto, M5H 2N2, Canada
Fax: 1-416/392-1478
E-mail: iclei@iclei.org
Web site: http://www.iclei.org

International Rhino Foundation (IRF)
14000 International Road
Cumberland Ohio 43732
E-mail: IrhinoF@aol.com
Web site: http://www.rhinos-irf.org

International Whaling Commission (IWC)
The Red House
135 Station Road
Impington, Cambridge UK CB4 9NP
Telephone: 44 (0) 1223 233971
Fax: 44 (0) 1223 232876
E-mail: iwc@iwcoffice.org
Web site: http://ourworld.compuserve.com/homepages/iwcoffice

International Wolf Center
1396 Highway 169
Ely, MN 55731-8129
Telephone: (218) 365-4695
Fax: (218) 365-3318
Web site: http://www.wolf.org

The Jane Goodall Institute (JGI)
P.O. Box 14890
Silver Spring, MD 20911-4890
Telephone: (301) 565-0086
Fax: (301) 565-3188
E-mail: JGIinformation@janegoodall.org

Keep America Beautiful
1010 Washington Boulevard
Stamford, CT 06901
Telephone: (203) 323-8987
Fax: (203) 325-9199
E-mail: info@kab.org

League of Conservation Voters
1707 L. Street, N.W., Suite 750
Washington, DC 20036
Telephone: (202) 785-8683
Fax: (202) 835-0491
E-mail: lcv@lcv.org
Web site: http://www.lcv.org

Mountain Lion Foundation (MLF)
P.O. Box 1896
Sacramento, CA 95812
Telephone: (916) 442-2666
E-mail: MLF@moutainlion.org
Web site: http://www.mountainlion.org

National Alliance of River, Sound, and Bay Keepers
P.O. Box 130
Garrison, NY 10524
Telephone: (800) 217-4837
E-mail: keepers@keeper.org
Web site: http://www.keeper.org

National Anti-Vivisection Society (NAVS)
53 W. Jackson Street, Suite 1552
Chicago, IL 60604
Telephone: (800) 888-NAVS
E-mail: navs@navs.org
Web site: http://www.navs.org

National Arbor Day Foundation
100 Arbor Avenue
Nebraska City, NE 68410
Telephone: (402) 474-5655
Web site: http://www.arborday.org
Publication: *Arbor Day*, bi-monthly magazine

National Audubon Society (NAS)
700 Broadway
New York, NY 10003
Telephone: (212) 979-3000
Web site: http://www.audubon.org
Publication: *Audubon,* bi-monthly magazine

National Center for Environmental Health
Mail Stop F-29
4770 Buford Highway, NE
Atlanta, GA 30341-3724
Telephone NCEH Health Line: (888) 232-6789
Web site: http://www.cdc.gov/nceh/ncehhome.htm

National Parks and Conservation Association (NPCA)
1015 31st Street, N.W.
Washington, DC 20007

Telephone: (202) 944-8530; (800) NAT-PARK
E-mail: npca@npca.org
Web site: http://www.npca.org
Publication: *National Parks,* bi-monthly magazine

National Wildlife Federation (NWF)
8925 Leesburg Pike
Vienna, VA 22184-0001
Telephone: (800) 822-9919
Web site: http://www.nwf.org
Publication: *National Wildlife,* bi-monthly magazine

Natural Resources Defense Council (NRDC)
40 West 20th Street
New York, NY 10011
Web site: http://www.nrdc.org
Publications: *Amiscus Journal,* quarterly; *Nature's Voice*

The Nature Conservancy (TNC)
1815 North Lynn Street
Arlington, VA 22209
Telephone: (703) 841-5300
Fax: (703) 841-1283
Web site: http://www.tnc.org
Publication: *Nature Conservancy* magazine

Noise Pollution Clearinghouse
P.O. Box 1137
Montpelier, VT 05601-1137
Telephone (888) 200-8332
Web site: http://www.nonoise.org

North Sea Commission
Business and Development Office
Skottenborg 26, DK-8800 Viborg, Denmark
Web site: http:\\www.northsea.org

People for Animal Rights
P.O. Box 8707
Kansas City, MO 64114
Telephone: (816) 767-1199
E-mail: parinfo@envirolink.org
Web site: http://www.parkc.org

People for the Ethical Treatment of Animals (PETA)
501 Front St.
Norfolk, VA 23510
Telephone: (757) 622-PETA
Fax: (757) 622-0457
Web site: http://www.peta-online.org/

Orangutan Foundation International
822 S. Wellesley Ave.
Los Angeles, CA 90049
Telephone: (800) ORANGUTAN
Fax: (310) 207-1556
E-mail: ofi@orangutan.org
Web site: http://www.ns.net/orangutan

Ozone Action
1700 Connecticut Ave. NW, Third Floor
Washington, DC 20009
Telephone: (202) 265-6738
E-mail: cantando@essential.org
Web site: www.ozone.org

Peregrine Fund
566 West Flying Hawk Lane
Boise, Idaho 83709
Telephone: (208) 362-3716
Fax: (208) 362-2376
E-mail: tpf@peregrinefund.org
Web site: http://www.peregrinefund.org

Rachel Carson Council, Inc.
8940 Jones Mill Road
Chevy Chase, MD 20815
Telephone: (301) 652-1877
E-mail: rccouncil@aol.com
Web site: http://members.aol.com/rccouncil/ourpage

Rainforest Action Network
221 Pine Street, Suite 500
San Francisco, CA 94104-2740
Telephone: (415) 398-4404
Fax: (415) 398-2732
E-mail: rainforest@ran.org
Web site: http://www.ran.org

Range Watch
45661 Poso Park Drive
Posey, CA 93260
Telephone: (805) 536-8668
E-mail: rangewatch@aol.com
Web site: http://www.rangewatch.org

Raptor Resource Project
2580 310th Street
Ridgeway, IA 52165
E-mail: rrp@salamander.com
Web site: http://www.salamander.com~rpp

Reef Relief
201 William Street
Key West, FL 33041
Telephone: (305) 294-3100
Fax: (305) 923-9515
E-mail: reef@bellsouth.net
Web site: http://www.reefrelief.org

ReefKeeper International
2809 Bird Avenue, Suite 162
Miami, FL 33133
Telephone: (305) 358-4600
Fax: (305) 358-3030
E-mail: reefkeeper@reefkeeper.org
Web site: http://www.reefkeeper.org

Renewable Energy Policy Project—Center for
Renewable Energy and Sustainable Technology (REPP-CREST)
National Headquarters
1612 K St., NW, Suite 202
Washington, DC 20006
Web site: http://www.solstice.crest.org

Resources for the Future (RFF)
1616 P Street NW
Washington, DC 20036
Telephone: (202) 328-5000
Fax: (202) 939-3460
E-mail: info@rff.org
Web site: http://www.rff.org

Roger Tory Peterson Institute
311 Curtis St.
Jamestown, NY 14701.
Telephone: (716) 665-2473
E-mail: webmaster@rtpi.org

Sierra Club
85 Second St., Second Floor
San Francisco, CA 94105
Telephone: (415) 977-5630
Fax: (415) 977-5799
E-mail (General information): information@sierraclub.org
Web site: http:// www. Sierraclub.org
Publications: *Sierra* magazine, published 6 times per year

Smithsonian Institution Conservation & Research Center (CRC)
For more information about the work of the CRC, contact their Web site at:
http://www.si.edu/crc/brochure/index.htm

Society of American Foresters
5400 Grosvenor Lane
Bethesda, MD 20814
Telephone: (301) 897-8720
Fax: (301) 897-3690
E-mail: safweb@safnet.org
Web site: http://www.safnet.org

Surfrider Foundation USA
122 S. El Camino Real #67
San Clemente, CA 92672
Telephone: (949) 492-8170
Fax: (949) 492-8142
Web site: http://www.surfrider.org

Union of Concerned Scientists
National Headquarters
2 Brattle Square
Cambridge, MA 02238
Telephone: (617) 547-5552
E-mail: ucs@ucsusa.org
Web site: http://www.ucsusa.org
Publications: *Nucleus,* quarterly magazine; *Earthwise,* quarterly newsletter

United Nations Environment Programme
United Nations, Rm. DC2-0803
New York, NY 10017
Telephone: (212) 963-8138
Web site: http://www.unep.org

United Nations Food and Agricultural Organization (FAO)
Web site: http://www.fao.org

United Nations Man and the Biosphere Programme (UNMAB)
U.S. MAB Secretariat, OES/ETC/MAB
Department of State
Washington, DC 20522-4401
Web site: http://www.mabnet.org

United States Department of Agriculture (USDA)
Web site: http://www.usda.gov

United States Department of Energy (DOE)
Web site: http://www.doe.gov

United States Environmental Protection Agency (EPA)
Web site: http://www.epa.gov

United States Fish and Wildlife Service (FWS)
1849 C Street, NW
Washington, DC 20240
Telephone: (202) 208-5634

United States Geological Service (USGS)
Web site: http://www.usgs.gov

United States National Park Service (NPS)
Web site: http://www.nps.gov

United States Nuclear Regulatory Commission (NRC)
Web site: http://www.nrc.gov

The Wilderness Society
900 Seventeenth Street, N.W.
Washington, DC 20006-2506
Telephone: (800) THE-WILD
Web site: www.wilderness.org

The Wildlands Project (TWP)
1955 W. Grant Road, #145
Tucson, AZ 85745
Telephone: (520) 884-0875
Fax: (520) 884-0962
E-mail: information@twp.org
Web site: http://www.twp.org

World Conservation Monitoring Centre (WCMC)
219 Huntington Road
Cambridge CB3 ODL, United Kingdom
Telephone: (+44) 1223-277314
Fax: (+44) 1223-277136
E-mail: info@wcmc.org.uk
Web site: http://www.wcmc.org.uk

World Conservation Union (IUCN)
formerly the International Union for the Conservation of Nature
USA: Multilateral Office
1630 Connecticut Avenue NW, 3rd Floor
Washington, DC 20009-1053
Telephone(202)387-4826
Fax:(202)387-4823
E-mail: postmaster@iucnus.org
Web site: http://www.iucn.org

World Health Organization (WHO)
Web site: http://www.eho.int

World Parrot Trust
USA:
P.O. Box 50733 St. Paul, MN 55150
Telephone: +1 651-994-2581
Fax: +1 651-994-2580
Email: usa@worldparrottrust.org
UNITED KINGDOM:

Karen Allmann, Administrator, Glanmor House,
Hayle, Cornwall TR27 4HY
Telephone: +44 (0) 1736 753365
Fax: +44 (0) 1736 756438
Email: uk@worldparrottrust.org
AUSTRALIA:
Mike Owen, 7 Monteray St., Mooloolaba,
Queensland 4557
Telephone: +61 7 54780454
Email: australia@worldparrottrust.org
Web site: http://www.world parrottrust.org

World Resources Institute
1709 New York Avenue, NW
Washington, DC 20006
Telephone: (202) 638-6300.
E-mail: info@wri.org
Web site: http://www.wri.org/wri/biodiv

World Society for the Protection of Animals (WSPA)
P.O. Box 190
Jamaica Plain, MA 02130
Web site: http://www.wspa.org
United Kingdom Division
Web site: http://www.wspa.org.uk/home.html

World Wildlife Fund, U.S. (WWF)
1250 Twenty-fourth St., NW
P.O. Box 97180
Washington, DC 20077-7180
Telephone: (800) CALL-WWF
Web site: http://www.worldwildlife.org

WorldWatch Institute
1776 Massacusetts Avenue, N.W.
Washington, DC 20036
Telephone: (202) 452-1999
Web site: http://www.worldwatch.org/
Publications: *WorldWatch, State of the World, Vital Signs* (annuals)

Zero Population Growth
1400 16th St., NW, Suite 320
Washington, DC 20036
Telephone: (202) 332-2200
Fax: (202) 332-2302
E-mail: zpg@igc.apc.org.
Web site: http://www.zpg.org

The Zoe Foundation
983 River Road
Johns Island, SC 29455
Telephone: (803) 559-4790
E-mail: savage@awod.com
Web site: http://www.2zoe.com

Index

John Mongillo currently writes school science textbooks. He has worked as an editor-in-chief for McGraw-Hill and as a science editor for Houghton Mifflin. He has both a bachelor of science and a master of science and has taught science for several years.

Linda Zierdt-Warshaw is a science editor and writer. She has worked as senior science editor for Globe Book Company and managing editor of *The Paper Magazine*. She is co-author of *Environmental Science: Ecology and Human Impact* (1995).